"Goswami's stated goal in writing this book is 'to prov
representative, review of some of the most interesting pas...
cognitive development and developmental cognitive neuroscience.' She attains this goal in impressive fashion. Professors as well as students will learn a lot from reading this book; I know that I did."

Robert S. Siegler, Schiff Foundations Professor of Psychology and Education, Teachers College, Columbia University, USA

"In this revision of a now-classic overview of cognitive development, Goswami continues to provide an engaging and synthetic review of the best behavioural research in the area, informed by interchange with cognitive science, neuroscience, computer modelling and studies of non-human animals. Working through the lens of neuroconstructivism, she nevertheless provides a lucid account of other classic and contemporary theories."

Nora S. Newcombe, Laura H. Carnell Professor of Psychology, Temple University, Philadelphia, USA

"In this book, Goswami masterfully blends research from traditional approaches to the study of cognitive development with the latest insights into developmental processes emerging from neuroscience and genetics. Spanning the developmental arch from prenatal development through to adolescence, this book provides an engaging overview of the developmental processes associated with attention, learning, memory, language, reasoning and the emergence of concepts. A must read not only for those interested in cognitive development but also for instructors seeking an up-to-date, interdisciplinary synthesis of cognitive development."

Daniel Ansari, Department of Psychology & Faculty of Education, Western University, Canada

"While some texts have an encyclopedic feel in terms of including a one-sentence description of every study in the literature, giving rise to a disjointed reading experience, Goswami's *Cognitive Neuroscience and Cognitive Development* has a more natural narrative flow. Coherence is preserved in the movement across topics because the book has an organizing theme in which ideas from cognitive neuroscience are used to connect perception and cognition. The notion that cognitive development may emerge from distributed spatial and temporal patterns of activity in brain response to perceptual experience, activity that is enriched as language is acquired, is especially provocative."

Paul C. Quinn, Francis Alison Professor and Trustees' Distinguished Professor of Psychological and Brain Sciences, University of Delaware, Newark DE, USA

Cognitive Development and Cognitive Neuroscience

Cognitive Development and Cognitive Neuroscience: The Learning Brain is a thoroughly revised edition of the bestselling *Cognitive Development*. The new edition of this full-colour textbook has been updated with the latest research in cognitive neuroscience, going beyond Piaget and traditional theories to demonstrate how emerging data from the brain sciences require a new theoretical framework for teaching cognitive development, based on learning.

Building on the framework for teaching cognitive development presented in the first edition, Goswami shows how different cognitive domains such as language, causal reasoning and theory of mind may emerge from automatic neural perceptual processes *Cognitive Development and Cognitive Neuroscience* integrates principles and data from cognitive science, neuroscience, computer modelling and studies of non-human animals into a model that transforms the study of cognitive development to produce both a key introductory text and a book that encourages the reader to move beyond the superficial and gain a deeper understanding of the subject matter.

Cognitive Development and Cognitive Neuroscience is essential for students of developmental and cognitive psychology, education, language and the learning sciences. It will also be of interest to anyone training to work with children.

Usha Goswami is one of the leading researchers in childhood and is respected worldwide. She is Professor of Cognitive Developmental Neuroscience at the University of Cambridge, a Fellow of St John's College, Cambridge, UK, and a Fellow of the British Academy.

Cognitive Development and Cognitive Neuroscience

The Learning Brain

Second Edition

Usha Goswami

LONDON AND NEW YORK

Second edition published 2020
by Routledge
2 Park Square, Milton Park, Abingdon, Oxon OX14 4RN

and by Routledge
52 Vanderbilt Avenue, New York, NY 10017

Routledge is an imprint of the Taylor & Francis Group, an informa business

British Library Cataloguing-in-Publication Data
A catalogue record for this book is available from the British Library

Library of Congress Cataloging-in-Publication Data
A catalog record has been requested for this book

ISBN: 978-1-138-92390-4 (hbk)
ISBN: 978-1-138-92391-1 (pbk)
ISBN: 978-1-315-68473-4 (ebk)

Typeset in Bembo
by Newgen Publishing UK

Visit the companion website: www.routledge.com/cw/goswami

For Sylvia Defior Citoler

Contents

Acknowledgements

I began revising this book while on sabbatical at the Research Centre for Mind, Brain and Behaviour (CIMCYC) at the University of Granada. I would like to thank my hosts at CIMCYC, particularly Professor Sylvia Defior, for the wonderful support and encouragement that I received there. This book would not have happened without her. I would also like to thank St John's College, Cambridge, for providing me with a haven for continuing with my writing. The tradition of support for scholarship offered by the Cambridge Collegiate University system is quite wonderful, and I feel privileged to be a Fellow of St John's College and a Professor at the University. Finally, I would like to thank my outstanding Centre Administrator Viviana Fascianella for her extensive help with the bibliography.

Cambridge, April 2019

Foreword

COGNITIVE DEVELOPMENT AND COGNITIVE NEUROSCIENCE

When does cognitive development begin? Traditionally, it has been assumed that cognitive development – the development of attention, learning, memory, language, reasoning and conceptual development – can only commence *outside* the womb. This certainly seems true of language learning, for example, a fundamental cognitive system. Yet more recent studies show that cognitive development in many respects begins even inside the womb. The foetus exhibits learning, memory and volitional motor behaviour, and the foetus even shows rudimentary learning about language. I describe some relevant studies below.

Some of the advances in understanding foetal learning were made possible by techniques developed in the field of cognitive neuroscience. Cognitive psychology and cognitive neuroscience differ in their measurement techniques and in their aims, despite being intimately related. Cognitive psychology explains cognition via concepts and ideas held in the mind – cognitive representations. The content and quality of these cognitive representations is typically explored via experimental paradigms employing measures like accuracy and response time. By contrast, cognitive neuroscience employs measures such as changes in blood flow or in electrical activity in the brain, and devises experimental paradigms to explore how these measures change during cognitive processing. Brain activity during a given cognitive task (e.g., a memory task) is typically compared to resting state activity (no task), or to experimentally driven comparison conditions (e.g., remembering material with or without an extra cognitive load, such as counting backwards). In general, cognitive neuroscience studies are only useful to understanding mechanisms of learning or of cognitive development if the cognitive tasks are carefully chosen. Developmental psychology excels at designing paradigms for measuring attention, learning, memory and conceptual development in pre-verbal children. State-of-the-art behavioural paradigms are required for good cognitive neuroscience. This offers our field a window of opportunity for studies in developmental cognitive neuroscience.

There are many well-tested paradigms in cognitive developmental psychology, for example regarding executive function tasks (Chapter 9), which can lend themselves to neuroimaging. Well-designed neuroimaging studies can reveal mechanisms of development, as we will see for language learning (Chapter 6). Nevertheless, most current studies using cognitive neuroscience techniques with children reveal brain-behaviour *correlations* rather than information about developmental mechanisms. To gather novel information about *mechanism*, developmental cognitive neuroscience needs to focus more on *how* the brain encodes and processes information than is currently the case in the field. For example, many studies show that executive function tasks requiring inhibitory control activate the frontal cortex. This demonstration in itself takes us no further in understanding *how* changes in the brain improve (for example) a child's inhibitory control. We also cannot tell the direction of causation – it could be changes in the child's behaviour that are driving changes in the brain. However, documenting brain-behaviour correlations can be a useful first step in understanding mechanisms. For example, these correlations may differ when development is atypical.

In my view, in order to go beyond brain-behaviour correlations, developmental cognitive neuroscience should begin with very basic explorations of sensory processing and related neurocomputational processing, and track these processes over developmental time using longitudinal studies (Goswami & Szücs, 2011). It is plausible that basic sensory neurocomputations constrain learning in important ways that shape conceptual knowledge. For example, there has been a lot of recent interest in the role of *statistical learning* in cognitive development. However, statistical learning by the brain can only be as good as the quality of the sensory input upon which it is based. Although considered by some to be a domain-general mechanism, statistical learning as studied to date is in essence a reflection of how the brain automatically processes visual and acoustic sensory information (as described in Chapter 1). A child with intact vision but atypical hearing is likely to show better statistical learning from visual input than from auditory input. Further, longitudinal studies are key. Studying neural encoding processes in the developing brain may offer mechanistic insights into why developmental trajectories unfold as they do. One such encoding process is the electrical signalling of large cell networks that oscillate at different frequencies (neuroelectric oscillations), which can be shown to be related to cognitive processes (for example, language development, see Chapter 6). Goswami and Szücs (2011) argued that studies focusing on *mechanisms* of neural information processing will be central to providing a solid foundation for understanding cognitive development. Such studies should begin with sensory processing (see also Goswami, 2016, for a more recent overview).

At the same time, researchers in developmental cognitive neuroscience must not neglect the fact that social learning processes (culture, context) will have top-down effects on sensory and therefore neural processing, and will necessarily affect how children's sensory representations develop over time into rich and sophisticated cognitive systems. In particular, learning language, and learning other symbolic systems that are culturally transmitted (such as literacy), will have profound effects on cognitive development. Cognitive neuroscience offers methods for studying such 'top-down' modulation as well, as will be described. Indeed, studies in adult cognitive neuroscience show that the brain rapidly learns 'expectations' (via encoding statistical patterns) concerning the sensory world. Accordingly, the brain soon processes novel sensory input on the basis of these abstracted statistical dependencies, rather than processing the exact sensory information available at a particular instant in time. The infant brain works the same way. There is evidence for processing based on prior learning of statistical dependencies in previously-experienced sensory inputs as early as three months of age (Kaufman, Mareschal & Johnson, 2003b; see Chapter 1). Accordingly, infant neural processing of the statistical structure of sensory spatio-temporal information can *in principle* yield the abstracted dependencies traditionally discussed in cognitive developmental psychology as 'prototypes' of concepts or as 'causal knowledge' about physical systems. I explain this idea in more detail throughout this book. Neuroscience also offers methods for integrating information about developmental mechanisms from related fields such as epigenetics (e.g., environmental effects on genetic expression) and psychopharmacology (e.g., the effects of hormones on the brain). Nevertheless, to address all factors that affect cognitive development within one research programme is highly challenging (although it is being attempted, see Rueda, Pozuelos & Cómbita, 2015, for a good example).

Accordingly, cognitive neuroscience offers cognitive developmental psychology methods for understanding mechanisms of development in fine-grained detail. For example, in order to build a cognitive system such as conceptual knowledge from information gained by looking at, acting on and listening to the sensory world, the infant brain can maximise information acquisition by focusing on dynamic spatio-temporal structure. Infants are most interested in agents, and agents move about and say and do things. The goal-directed actions of agents provide integrated dynamic multi-sensory information that is attended to and stored by the brain. Understanding *how* the storage and organisation of such information results in cognitive representations for concepts such as 'bird', 'tree' and 'car' in mechanistic detail is now possible using brain imaging techniques such as multi-voxel pattern analysis (see Chapter 5). To date, however, such cognitive neuroscience techniques have not been used in developmental studies.

NEURAL IMAGING METHODS AND DEVELOPMENTAL DATA

Four neural imaging techniques are currently in relatively widespread use with infants and children. One is *electroencephalography* (EEG). This technique involves placing sensitive electrodes on the scalp in order to record the brain electrical activation taking place below it. The electrodes measure the low-voltage changes caused by cells firing action potentials during cognitive activity. While EEG is very time-sensitive, able to record changes in brain activity at the millisecond level, a drawback of the technique is that the signals are difficult to localise. As the signals are measured on the surface of the skull, the neural systems generating the signals may not lie beneath the electrodes that are showing particular effects. A second measure is *functional magnetic resonance imaging* (fMRI), which depends on measuring changes in blood flow in the brain. An increase in blood flow to particular brain areas causes the distribution of water in the brain tissue to change. fMRI works by measuring the magnetic resonance signal generated by the protons of water molecules in neural cells as the brain is functioning, generating a BOLD (blood oxygenation level dependent) response. The BOLD response peaks over time, hence fMRI lacks the millisecond resolution of EEG, with images typically acquired over 0.5–several seconds. However, fMRI offers very good spatial resolution in terms of where in the brain neural activity is taking place.

A third measure is *magnetoencephalography* (MEG), a direct measure of the magnetic fields that are created by the electrical signalling of neurons. These magnetic fields are not distorted as they pass through the head, thereby enabling measurements with high spatial and temporal resolution. The magnetic fields generated by the electrical activity in the brain are also tiny, estimated to be one billion times smaller than the magnetic field generated by the electricity in a lightbulb. They are measured by highly sensitive equipment which is very costly to run, and which combines the temporal information with MRI scans to localise the activity. A fourth neuroimaging technique offers a different means of measuring changes in blood flow, *functional near-infrared spectroscopy* (fNIRS). Changes in oxygen availability (blood oxygenation level) are also marked by changes in the quantity of haemoglobin in brain tissue. Near-infrared light is absorbed differentially by brain tissue depending on the concentration of haemoglobin. Hence if optodes emitting near-infrared light are placed at the electrode positions used in EEG, changes in haemodynamic response to neuronal activation can be measured. Accordingly, fNIRS enables the collection of data with better spatial quality than EEG and better temporal quality than fMRI, without a child needing to lie inside a large and noisy cylindrical magnet (as in fMRI). It is thus particularly suited to testing infants. Finally, a novel measure that is being used extensively with infants is *eye tracking*. Although not usually considered a brain imaging tool, anatomically the eye is part of the brain. Eye tracking devices use sensors to determine where the pupil of the eye is focused in the available visual field. What we are looking at is usually related to what we are thinking about, and so with the right experimental paradigms, eye tracking measures can provide a window into infant cognition.

When researching this revised volume, it was striking how some areas of developmental enquiry now have numerous brain imaging studies, for example memory and executive function, while others have almost none (for example, conceptual development). Accordingly, the main focus of my book still lies with behavioural studies. Indeed, behavioural studies continue to be of critical importance for our field, even in areas of development with more cognitive neuroscience activity. Cognitive neuroscience that is not informed by careful studies of behaviour is not very useful (Goswami & Szücs, 2011). In my view, behavioural developmental psychology as a field is still much richer in its insights regarding *mechanisms* of children's development than developmental cognitive neuroscience. Nevertheless, I believe that neuroscience has the potential to transform our understanding of human learning and cognitive development, and I try to offer suggestions for relevant research in each chapter.

THE FOETAL AND NEONATE BRAIN

Foetuses have surprisingly active lives. Ultrasonic scanning studies reveal that, by the 15th gestational week, the foetus has at least 15 distinctly different movement patterns at its command, including a yawn-and-stretch pattern and a 'stepping' movement that enables it to totally change its position in the womb (via rotation) within two seconds (DeVries, Visser & Prechtl, 1985). Foetuses suck their thumbs by the 15th gestational week, and prenatal thumb-sucking predicts later handedness. Babies who suck their right thumbs at 15 weeks become right-handed, and most babies who suck their left thumbs at 15 weeks become left-handed (Hepper, Wells & Lynch, 2005).

More recently, studies using functional magnetic resonance imaging (fMRI) and magnetoencephalography (MEG) of the foetal brain have become possible. Such studies reveal that primitive neural networks are functioning between 24 and 38 weeks of gestational age, for example in motor, visual, thalamic and temporal lobe networks (Thomason et al., 2015). Further, the foetal brain is forming neural 'hubs', sets of highly interconnected regions that are important for efficient neuronal signalling and communication, even within the womb. These include hubs in association cortex; that is, areas of the brain involved in more than basic sensory processing, such as Wernicke's area (a language processing area; van den Heuvel et al., 2018). This suggests that in utero, the foetal brain is not simply being 'wired up' ready for perception and action, but that it is being prepared for the higher-order functioning, such as linguistic functioning, that will develop later in life.

Consistent with these findings, other studies show that foetuses already have a cognitive life. Memory for the mother's voice is developed while the baby is in the womb (see Chapter 3), and there is also evidence for foetal learning of particular pieces of music (such as the theme tune of the TV soap opera *Neighbours*; Hepper, 1988). These responses are not only mediated by the brainstem (Joseph, 2000), but are also cortical. For example, there are functional hemispheric asymmetries in auditory evoked activity in the foetal cortex (Schluessner et al., 2004), with apparent left hemisphere specialisation for speech stimuli by 38 weeks, and a preference for the mother's voice over other female voices by 34 weeks (Jardri et al., 2012). The foetus also shows deceleration of heart rate to certain sounds while in the womb (thought to index attention) and habituation of heart rate to vibro-acoustic stimuli (thought to show rudimentary learning; see Hepper, 1992; Kisilevsky & Low, 1998).

Regarding brain structure, we already know that most of the brain cells (neurons) that a child has form before birth, by the seventh month of gestation (see Johnson & De Haan, 2015; Joseph, 2000; for overviews). Accordingly, the environment within the womb can affect later cognition. For example, certain poisons (e.g., excessive alcohol) have irreversible effects on brain development, and may have quite specific effects depending on which brain structures are affected most. To give an example, it has been suggested that maternal alcoholism affects later mathematical cognition because of its effects on the development of the parietal cortex (Kopera-Frye, Dehaene & Neissguth, 1996). Prenatal alcohol exposure is known to lead to abnormalities in gyrification, cortical thickness and cortical volume, abnormalities that are still present during the teenage years (Hendrickson et al., 2017). Parietal cortex is still a major affected locus in a teenager. Naturally, even though most of the cells in the brain are present before birth, there are enormous changes in the connectivity of these cells during the first months and years of life. Indeed, the functional connectivity of the brain at six months versus 12 months of age is so different that by studying the activity of the brain at rest, researchers can

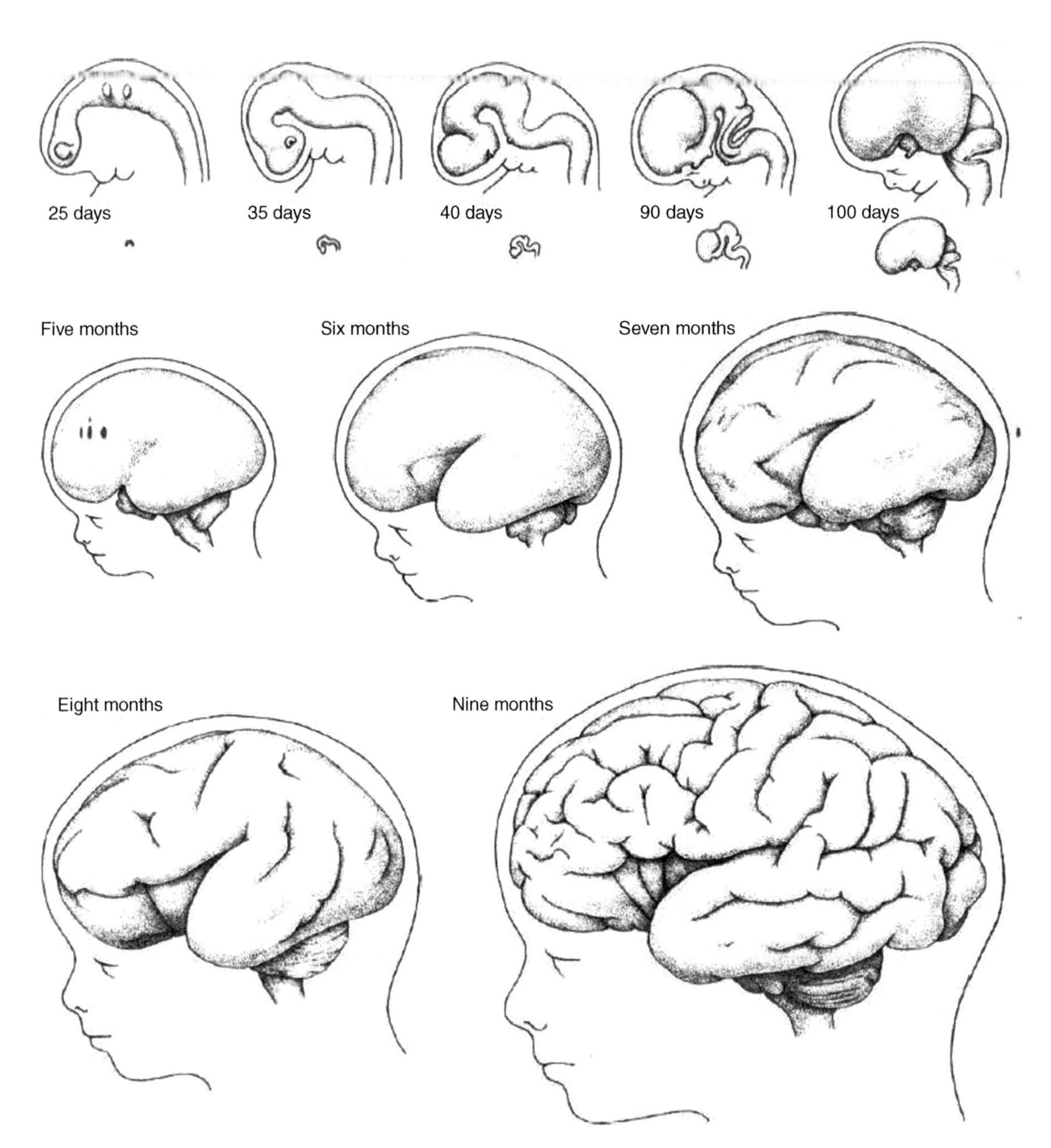

The developing brain in utero. The figure shows how the brain develops from the early neural tube in the first days of life to include cerebral cortices which overlay the midbrain, hindbrain and cerebellum.

identify via patterns of brain activity alone the scans of a six-month-old brain from a 12-month-old brain (Pruett et al., 2015). The 'resting state' network, the network of spontaneous intrinsic brain activity that can be observed when the brain is not engaged in any task, also changes dramatically in the first year of life (Gao et al., 2009). While even two-week-old infants exhibit a primitive resting state network, it is very incomplete (for example, the hippocampus, important for forming memories, is not yet included). By one year of age the resting state network is much more connected, and by two years of age the connectivity of the resting state network is already comparable to that of adults. Such information is important. This kind of structural and functional information about brain development constrains certain kinds of theorising in developmental psychology. An example is the notion of 'critical periods' for cognitive development, which are increasingly debated (Werker & Hensch, 2015). It is now recognised that the brain retains plasticity throughout the lifespan. Knowledge about when and how different neural regions develop may also offer new insights into longstanding and intriguing developmental problems (such as why infants appear to pass false belief tasks while young children do not, see Chapter 4).

After birth, therefore, brain development consists mainly of the growth of connections between neurons: *synaptogenesis*. Synaptogenesis leads the infant brain to double in size during the first year of life. Brain cells pass information to each other via low-voltage electrical signals, which travel from neuron to neuron via special junctions called synapses. As soon as the child is born, the brain is busy developing connections between neurons, proliferating some connections and pruning others. The main determinant of functional connectivity is the learning environment experienced by the child. It is environmental experiences that largely determine the development of specific neural pathways and networks as the infant and child learns about the world. These networks and pathways will be the basis of the child's conceptual system, and of perception, attention, language, learning and memory. Furthermore, environmental events that are experienced consistently will have a stronger effect on neural pruning and reorganisation than environmental events that are rare. The easiest and most effective way to support a child's development is by providing the best possible learning environments as consistently as possible in all aspects of their life – in the home and family, at nursery, at school, and in our wider culture and society.

In general, once the infant is born, it is the primary sensory systems that are established first (e.g., the visual and auditory systems, the motor system), while higher-order association areas mature later (Casey, Galvan & Hare, 2005). For example, the prefrontal cortex, a key structure for planning, reasoning and other 'executive functions', is one of the last brain regions to mature fully. However, the environment does not have to be especially rich to promote optimal development. Rather, the brain appears to be set to respond to *normative* visual and auditory experience. When many neurons in a network are 'firing' together, the patterns of neural activity are thought to correspond to particular mental states or mental 'representations'. Establishing which areas of the brain are specialised for different kinds of processing (e.g., linguistic versus visual processing) has been one of the first questions for the field of developmental cognitive neuroscience.

The neonate brain also appears to have certain inbuilt ways of processing information, described briefly below. At birth, infant brains are not so different. However, a brain that is born into an environment that offers early advantages can use these inbuilt ways of processing information to develop its neural architecture more efficiently. So a brain that is born into an optimal learning environment will do better over a person's lifetime than a brain that is faced with a less optimal early environment. The quality of early learning environments is hence critical for understanding individual differences in cognitive development. Further, there are always individual differences between brains. At birth, there may be individual differences in the efficiency of neural mechanisms for seeing or hearing, or in the lability (fussiness, or speed to respond to stimuli) of the infant. It is largely the environment that will determine whether these individual differences, which characterise all of us, have minor or more measurable consequences. Brains whose biology makes them less efficient in particular aspects of sensory processing seem to be at risk in specific areas of development. For example, when auditory processing is less efficient, this can carry a risk of later language impairment (see Chapter 6).

Nevertheless, individual differences are not *deterministic*. Reduced efficiency does not automatically result in a developmental impairment. Usually, as long as the environment is reasonably stimulating, sufficient learning experiences enable brains with less efficient sensory processing to reach similar developmental end points to more efficient brains. However, early awareness of impaired sensory efficiency can enable useful environmental interventions. An example is when an infant is born deaf. In some cases of deafness, a small microchip (called a cochlear implant) can be surgically inserted close to the ear to deliver information to the brain, and for some cochlear implant children, typically those implanted early, oral language development is equivalent to that of hearing children.

Even identical twins (monozygotic twins), who have developed from the same single egg, will have different brains. It is not yet clear why this is the case. One possibility is that individual differences between the brains of monozygotic twins may depend on the environment inside the womb. The intra-uterine environment will be experienced slightly differently by each baby. One twin is usually dominant, so perhaps may change their position in the womb first. The other twin must then accommodate to this change by changing their position. So the second twin has a different intra-uterine experience to the first twin. Nevertheless, identical twin research shows that biology is never destiny (Plomin, 2018). Although there are many cognitive similarities between identical twins, there are also many differences. For example, only one twin in a monozygotic twin pair may have developmental dyslexia.

TWO CORE DEVELOPMENTAL QUESTIONS

The study of children's cognitive development has traditionally focused around two major questions. The first is the apparently simple question of *what develops*. This question can be investigated by observing changes in children's cognitive abilities over time. For example, we can define certain principles of logical thought (such as the Piagetian principles of conservation and transitivity, see Chapter 11) and then track the development of these principles over time with experimental tests. At a very simple level, we can also investigate 'what develops' by using cognitive neuroscience techniques. We know, for example, that the sensori-motor cortices (vision, audition, action) mature earlier than the language and spatial areas (temporal and parietal cortices), with the prefrontal cortex (reasoning, problem-solving, monitoring one's behaviour) maturing last

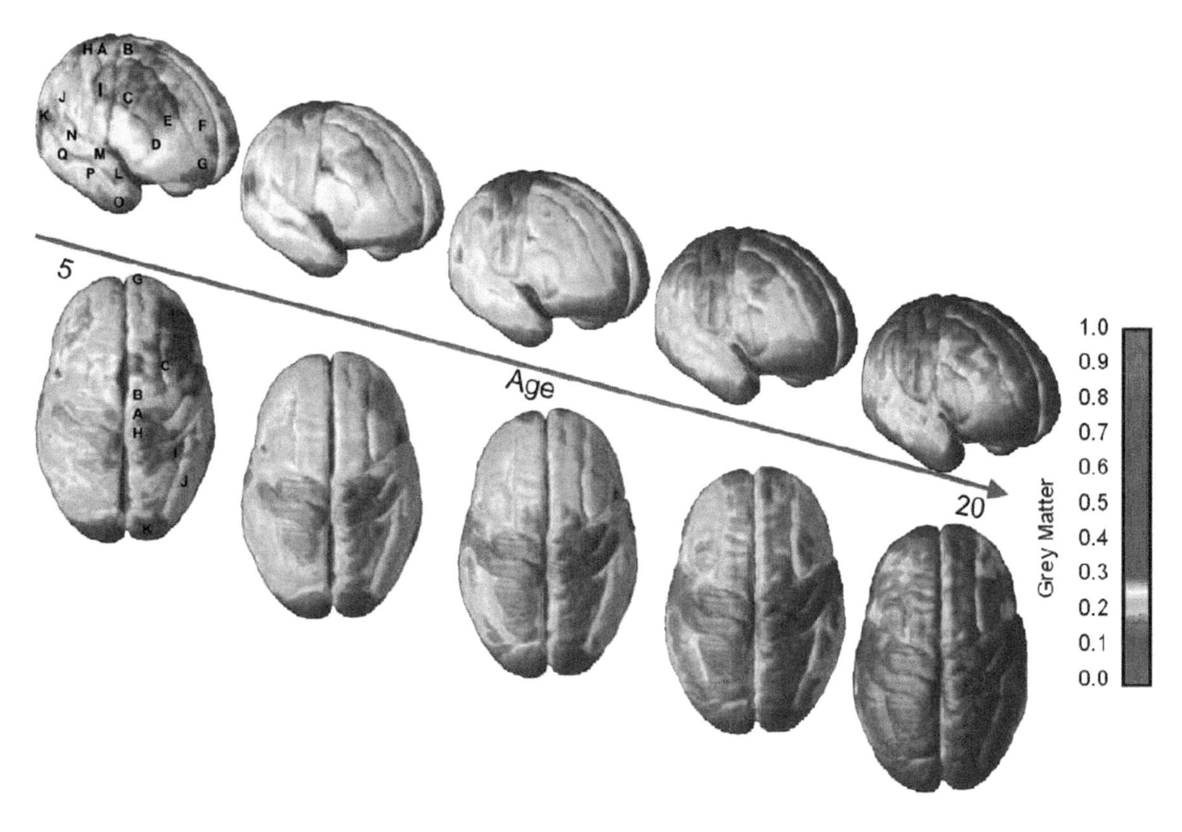

The developing brain: childhood and adolescence. The figure shows the maturation of grey matter during childhood and adolescence over the cortical surface, with time depicted from left to right. The colour bar shows units of grey matter volume.

of all, during adolescence and early adulthood. Such observations suggest that visual and auditory behaviour will approximate adult levels earlier than reasoning behaviour or self-monitoring behaviour. Remarkably, Piaget's theory of cognitive development began with a sensori-motor phase and ended with higher-order reasoning (see Chapter 11). In this sense, Piaget's theoretical framework parallels the course of brain development.

Information about *what develops* provides data for the second major question in cognitive developmental psychology, the less simple question of *why* development pursues its observed course. This question requires us to develop causal explanations for observed cognitive changes: we need *theories* of cognitive development. In developmental psychology, we try to understand why development pursues the course that it does via *experiments*. Most of this book will be concerned with such experiments. In the future, it may also be possible to develop causal explanations from neuroscience studies. Neuroconstructivism (Chapter 11) provides a current, albeit incomplete, example of such theorising.

DOMAIN-GENERAL VERSUS DOMAIN-SPECIFIC DEVELOPMENT

Two alternative (although not mutually exclusive) explanatory systems are typically developed to account for changes in children's cognition. The first type of theoretical account is based on the idea that core modes of learning or reasoning come online and are then applied across all cognitive domains. Whether a child is attempting to understand why another child is upset (the 'domain' of psychological causation), why animals usually have babies that look like them (the 'domain' of biology), or why objects fall when they are insufficiently supported (the 'domain' of physical reasoning), domain-general accounts postulate that certain types of learning, such as causal learning, or certain types of reasoning, such as the ability to make deductive inferences, are applied to the acquisition of all of these understandings. This is an example of a 'domain-general' explanation of cognitive development.

The second type of theoretical account postulates that the development of cognition is piecemeal, occurring at different time points in different domains. According to this view, cognitive development is 'domain-specific'. For example, deductive inferences may appear in the domain of physical causality long before they appear in the domain of psychological causality. The reason may be that a rich and principled understanding of the physical world is acquired before a rich and principled understanding of the psychological world (a 'theory of mind'). Domain-specific accounts of cognitive development acknowledge the importance of the *knowledge base* in children's cognition.

However, the knowledge that we have affects our cognition when we are adults as well as when we are children. While the ability to (for example) make deductive inferences per se may be a domain-general development, the use of deductive inferences may be domain-specific. Children may need sufficient *knowledge* in order to use their deductive abilities in different domains, just as adults do (most of us could not make coherent logical deductions in unfamiliar domains such as nuclear physics). This example illustrates why the two explanatory systems that have traditionally been developed as competing accounts of changes in children's cognition are not mutually exclusive. In the previous edition of this book, I argued that certain types of learning, such as associative learning, statistical learning and causal learning, were domain-general. I then argued that their use in different domains depended partly on the child's knowledge base. My view now is that these types of learning are inbuilt, because they are automatic mechanisms used by the brain to process information. In this edition, I will argue that associative learning, statistical learning and causal learning reflect the mechanisms of operation of neural sensory systems when encoding and processing information. However, and importantly, the *quality* of what is learned by these automatic neural mechanisms will depend on the quality of the sensory information upon which they operate. For example, the developmental product of acoustic statistical learning (such as the quality of some aspects of the child's language system) will vary depending on the quality of the processing of acoustic sensory information by that child. I expand on this view later in this book. Other factors will also affect the observed pattern of an individual's cognitive development. These include the child's existing knowledge, the richness of the child's environment, the time of maturation of certain neural structures such as the frontal cortex, and the quality of the support and teaching that a child receives at home and in school.

LEARNING AND CONSTRAINTS ON LEARNING

One of the most provocative questions raised by studies in developmental cognitive neuroscience is whether inbuilt mechanisms of neural information coding and inbuilt mechanisms for the neural transmission of information govern the development of processes that have traditionally been viewed as 'cognitive'. For example, a popular approach in child psychology has been to draw an analogy between infants and scientists. The analogy is based on the premise that babies and scientists have a similar approach to learning. Both make observations, carry out experiments, and draw conclusions. The description of 'baby as scientist' assumes cognitive processes. The core claim is that infants have naïve *theories* about how the world works, based on innate expectations.

In my view, studies in developmental cognitive neuroscience question such approaches. Instead, it is the *automatic* processing of information by the neural sensory systems of the infant and young child that comprise the building blocks of their cognitive systems. Accordingly, it is the mechanisms intrinsic to the brain for encoding and processing sensory information that constrain what is learned, not nascent cognitive *theories* held by the infant or young child. These mechanisms can be described metaphorically as theories, but they do not initially have cognitive content. In the first chapters of this book, I will illustrate how the notion of innate cognitive theories is questioned by certain findings in developmental cognitive neuroscience. Indeed, neural information processing may not prioritise the aspects of the environment that our cognitive analyses assume to be important for the development of particular cognitive systems. One example is neural processing of the speech signal, discussed more fully in Chapter 6. Here research in auditory neuroscience suggests that aspects of the speech signal assumed to be fundamental to linguistic processing by disciplines such as linguistics (for example, the phoneme) may not be fundamental to initial language learning. Rather, awareness of phonemes may be a cultural product of learning to be literate. As suggested in Chapter 6, consideration of the core features of infant-directed speech or *Parentese*, a cross-language modification of the speech signal used universally when speaking to infants, could provide important guidance regarding the key mechanisms of the early neural processing of speech information. Parentese is essentially a language teaching device, used (automatically) by adults and supporting infant learning. Accordingly, those aspects of the speech signal that are exaggerated by Parentese are probably fundamental to the neural processing of speech.

Developmental psychology has documented other such teaching devices, and another example is *motionese* or infant-directed action (Chapter 2). Behavioural work with infants has shown that adults modify their actions in important ways when they interact with infants. These modifications appear to facilitate infant learning, for example about goal-directed action. To my knowledge, there is as yet no cognitive neuroscience programme of research focused on motionese, yet motionese may hold important clues to the features of action that are fundamental to its neural encoding. These features may, for example, be important for the large field of 'mirror neuron' research (see Chapter 4)

According to an explanatory framework derived from data in cognitive neuroscience, the way that neural sensory systems acquire and process information underlies the 'constraints on learning' that have typically been considered to be cognitive in nature (Goswami, 2016). Examples of constraints that guide infant learning (or 'innate expectations' or 'principles') proposed in the cognitive behavioural field include expectations such as 'one object cannot be in two places at the same time'. These innate cognitive expectations are then thought to be elaborated via experience. For example, while one thing can never be in two places at the same time, two things can be in one place at the same time. This is possible if one object is inside another (the physical concept of *containment*).

It is interesting to consider whether inbuilt mechanisms of neural information coding and transmission rather than cognitive expectations are governing these 'constraints' on infant learning. Currently there is too little data from neuroscience to make a strong case for the 'inbuilt mechanisms' view, although I argue for it in this book. It is highly plausible that human neural sensory systems learn information in a way that enables cognitive *explanatory frameworks* to emerge and develop. Experience is still necessary to elaborate the skeletal frameworks produced by neural learning mechanisms, but by this account the explanatory frameworks in themselves are not innate. What is innate is the neural mechanisms of information processing that characterise the brain. Some of these neural mechanisms may be shared with the animal kingdom, potentially explaining some of the more surprising apparently cognitive abilities documented for other species (such as elephants recognising themselves in the mirror, Chapter 4, and fish doing approximate computations of magnitude, Chapter 10).

Infants are also *active* rather than passive learners. The learning brain appears to prioritise human faces and eyes, and is always interested in dynamic spatio-temporal information. Babies appear to be intrinsically interested in things that move (which may be biological kinds, like animals, or mechanical artefacts, like aeroplanes), and the learning brain is attracted to novelty and change in the environment. One reason may be that movement and change are very salient to the developing neural sensory systems. For example, perhaps an early focus on the movement of objects is driven in part by the number of motion-sensitive neurons (brain cells) already present at birth, and the kinds of inputs the visual system requires for further development. Simply looking at and listening to the world provides infants with an enormous amount of information. Further, by *choosing* what to attend to, such as prioritising people over objects, infants are active in prioritising their learning about certain kinds of information.

Accordingly, observation of the dynamic spatial and temporal behaviour of objects, people and animals may generate important evidence to support the development of explanatory systems for naïve physics and naïve psychology (as I describe in Chapters 1–4). In these chapters, I discuss how the infant brain is tracking all kinds of statistical co-dependencies, including co-dependencies *within* dynamic spatial and temporal events. This statistical database is a second source of knowledge for the development of explanatory systems, and statistical knowledge not only supplements event-based evidence but once it is established, may even *alter the processing* of event-based evidence. This is because the brain will make predictions about expected events on the basis of prior knowledge (learned statistical dependencies), and these predictions will begin to modulate primary sensory processing. Automatically acquired statistical knowledge about 'what usually happens' may thus support the development of what then appear to be abstract *cognitive* systems of knowledge, such as the 'object concept' or 'false belief'.

Regarding the infant's physical world, our growing understanding of how the adult brain processes dynamic spatio-temporal structure can be used to illustrate this point. Adult sensory systems actively use spatio-temporal structure to learn about the world and revise their current cognitive schemes. For example, using fMRI, Noesselt and his colleagues showed that the temporal patterning of inputs to different sensory modalities provided information about whether the inputs were related or not, and further showed that the resulting cross-modal processing of inputs that were experienced as corresponding then altered the *primary sensory processing* of the unimodal inputs (Noesselt et al., 2007). This is an important demonstration *in principle* for understanding the rapidity of infant learning about cross-modal information and infants' preference for congruence in sensory inputs (see further discussion in Chapter 1). Noesselt et al. (2007) used fMRI to explore how the adult brain might integrate related but not unrelated sensory information from individual sensory modalities. They devised a temporal correspondence paradigm, in which sequences of arrhythmic and unpredictable flashes of light either co-occurred perfectly in time with a sequence of arrhythmic and unpredictable tones, or did not. When the stream of auditory (A) tones was perfectly coincident with the visual (V) light flashes, the temporal patterning suggested a unique environmental source. This should support multi-sensory integration.

This mechanistic neuroscientific approach would accordingly predict increased neural activation in multi-sensory regions of the adult brain for the AV streams that had temporal correspondence, for example in superior temporal sulcus (STS). Noesselt and his colleagues also studied neural activation in the primary sensory cortices (A and V), both to the AV streams and to the tones and light flashes alone. Their data showed that a range of neural areas were active in the AV conditions, with STS uniquely responsive to temporal correspondence. Activation in STS increased significantly during coincident streams and decreased significantly during non-corresponding AV streams. Furthermore, in the temporal correspondence condition, *neural feedback* from STS modulated basic sensory processing in primary auditory and visual cortex. This is conceptually important. Noesselt et al.'s data demonstrate that basic sensory responses can be affected (modulated) by top-down information. Top-down modulation of basic sensory processing allows *prior experience* to modulate the processing of currently experienced events. Top-down modulation would enable (for example) the development of conceptual knowledge, where the brain is eventually responding to abstracted dependencies that are usually present in the environment when (for example) a bird flies into view, rather than only to the specific features of the currently experienced exemplar of a bird. Top-down neural modulation is also the source of various illusions.

This is shown by cognitive neuroscience studies showing that the adult brain will fill in gaps in sensory experience based on its prior knowledge. An example comes from studies of 'illusory' sounds, sounds that we believe that we hear even when they are not present in the input environment. An example would be that when a sound is interrupted by silence, we perceive it as discontinuous, but when a sound is interrupted by a noise burst which occurs during a silent gap in the sound, the sound is perceived as continuing through the noise. This is called the 'continuity illusion'. Riecke, van Opstal, Goebel and Formisano (2007) used fMRI to measure the neural response to tones that either evoked a continuity illusion or not, with different levels of masking noise. The brain data were analysed both in terms of the actual physical stimulus and then subsequently in terms of the perceived (yet not physically present) stimulus. It was found that stimulus-evoked brain activity did not correlate with the basic acoustic properties of the stimuli as experienced in the environment. Instead, the brain activity correlated with their *perceived* continuity. In other words, *identical* sensory stimuli evoked *different* neural responses depending on the listener's cognitive representation of their experience (i.e., whether the listener perceived the tones as being continuous or not). Riecke et al.'s data showed that the brain response was based on the statistical patterning (or *abstracted dependencies*) between the tones and noise, not on the physical tone/noise pairing that was presented during a specific trial. Hence the brain had developed abstract knowledge about prior perceptual experience, and it was this abstract knowledge (which we can think of as a cognitive representation) that was modulating basic sensory responding. There is no reason to think that the infant brain would behave differently, although this has yet to be tested.

Again, this adult fMRI study provides an important *in principle* demonstration regarding possible mechanisms of development. Even the brains of young infants show sustained activity based on abstracted dependencies (statistically learned patterns) in the absence of sensory input. An example is when a hidden object disappears unexpectedly (see Kaufman, Csibra & Johnson, 2003a; I discuss this study in Chapter 1). In my view, such *in principle* demonstrations from adult cognitive neuroscience are highly relevant to explaining cognitive development. If the neural processing mechanisms intrinsic to the brain operate *automatically* in ways that enable the development and storage of abstract knowledge about sensory experiences, then the brain's intrinsic information processing mechanisms enable and/or facilitate the development of the abstract knowledge structures that we refer to as 'cognitive'. The brain is in essence storing statistical dependencies in sensory input (spatio-temporal structure), extracted by instance-based learning, but using neural statistical processes that simultaneously create generalised representations of these spatio-temporal structures based on repeated experience. These abstracted dependencies then exert a role in basic sensory processing, so that 'top-down' information affects 'bottom-up' learning.

These generalised or abstracted dependencies are traditionally discussed in cognitive psychology as 'prototypes' of concepts, as 'causal knowledge' about physical systems and as 'naïve psychology' (e.g., Rosch, 1978; Shultz, 1982; Wellman & Gelman, 1998). Of course, these emergent knowledge systems are then enriched as the infant and child acquires further experience in these domains, and are transformed as the infant and child acquires language and other symbolic systems. The interesting possibility is that the *origins* of all these kinds of 'knowledge' is *not* innate *cognitive* constraints on learning. Instead, the brain has inbuilt mechanisms of neural information coding and inbuilt mechanisms for the neural transmission of information, and it is the fixed nature of these mechanisms that comprise 'constraints on learning'. I will attempt to illustrate and support this view periodically throughout this book.

INNATE VERSUS ACQUIRED ACCOUNTS OF COGNITION IN CHILDREN

A related theoretical issue to that of constraints on learning is that of nature versus nurture. Nativist accounts explain the underlying causes of development in terms of a rich genetic endowment of complex behavioural abilities, while acquired accounts explain development in terms of rich experience of the environment. The metaphor of the mind of the infant as a blank slate upon which experience writes has long been discredited. However, recent research demonstrating the relative sophistication of infant cognition has led to a renaissance of strong nativist views, as captured, for example, by the notion of constraints on learning. Certainly, research shows that the environment experienced by the infant and child will have a far bigger impact on cognitive development than genes per se. Yet even aspects of development that turn out to be totally under genetic control, such as tooth decay, can be dramatically altered by the environment. We can virtually eliminate tooth decay by looking after our teeth properly. We use environmental interventions, like brushing and flossing our teeth. The same principles apply to psychological development. Genes cannot 'code' psychological traits in any fixed or hard-wired fashion (see Turkewitz, 1995; Plomin, 2018). Although genes contribute to neural structures, these structures become active before they are fully mature, and this activity itself shapes development ('probabilistic epigenesis', see Gottleib, 2007). Knowing whether a particular ability is present at or near birth does not help us to understand its developmental origin. Instead, it is a starting point for the investigation of causes and consequences.

In my view, an important question for cognitive developmental psychology is how neural and genetic activity *interact* with the environment and with behaviour to produce development. We need to understand how the characteristics and limitations of infant motor, sensory, perceptual, and cognitive functioning produce modes of responding to the environment that help to shape the development of mature cognitive systems. In my view, cognitive neuroscience provides an additional set of rich tools to help us towards this goal. I will try and illustrate how in the rest of this book.

THE ORGANISATION OF THIS BOOK

This book still focuses on the question of *what develops* rather than on the question of *why*. The findings from a given experimental study (*what develops*) are generally fixed, but the interpretation of what particular findings mean (*why*) is fluid. This is one of the most exciting aspects of research. Some of the experiments that will be discussed have alternative interpretations, and every student interested in children's cognition is invited to develop their own ideas about what the different studies mean (preferably along with some ideas about how to find out whether the studies are right or not!). My aim is to provide a selective, but hopefully representative, review of some of the most interesting past and current work in cognitive development and developmental cognitive neuroscience. By considering research on perception and attention, learning and memory, conceptual and linguistic development and the development of logical, psychological and causal reasoning, the book will survey the different kinds of knowledge that children acquire, and how they acquire them. At the end of the book, we will assess the impact of recent findings in developmental psychology for the most famous theories of cognitive development, those of Jean Piaget and Lev Vygotsky. We will also consider newer theories such as neuroconstructivism.

A central theme will become apparent in our discussion of *what develops*, which will provide a partial answer to the question of *why*. As we will see, the human infant is born with certain kinds of learning mechanisms at its disposal. The infant brain can learn statistical patterns in the environment, enabling the extraction of an enormous amount of information. Infants are skilled at associative learning – for example, they readily learn that certain events co-occur. They are also skilled at learning conditional probabilities – that a certain event will reliably occur given that a specific prior event has occurred. The infant brain can also learn by imitation. Perception yields many examples of agents (e.g., parents, sisters and brothers) acting on the world, and infants can imitate what agents do, which appears to help them to represent and understand human action and its causes (social cognition). Infants can also learn by analogy. Imitation may involve an early form of analogy, as infants can recognise others as being 'like me' (see Chapter 3). Finally, infants have an apparently innate bias to learn about causal relations and to acquire causal explanations.

In my revised view, this 'innate causal bias' is again rooted in inbuilt mechanisms of neural information coding and inbuilt mechanisms for the neural transmission of information. It is not initially a *cognitive* bias. I will illustrate this when we consider Bayesian accounts of causal reasoning in Chapter 7. Nevertheless, whether its origins lie in mechanisms of neural information processing or not, the causal bias generates a tremendous amount of useful information for the young child. *Language* then becomes the key mechanism for generating causal understanding. Anyone with a child of their own or a young sibling is familiar with the constant tendency of young children to ask for causal information ("Why is the sky blue?" "How does the telephone call know which house to go to?" "How come the moon is big and orange now but other times it's little and white?", see for example Hood & Bloom, 1979; Callanan & Oakes, 1992). This relentless questioning is not just a device that children employ to keep a conversation going. Instead, causal questions such as these have an important developmental function, enabling a child to organise information across domains and deepen their understanding. I noted earlier that human sensory systems learn information in a way that enables *explanatory frameworks* to develop. For example, the explanatory system for naïve physics is organised around a core framework for describing the possible behaviour of cohesive, solid, three-dimensional objects. Partly by asking questions, young children also seek hidden features to help them to understand what makes objects and events similar. Children seek such features because they are actively learning 'causal explanatory frameworks' for interpreting the world around them. These kinds of perceptual and causal learning are supported and extended by the child's growing facility with language.

Over the developmental trajectory of infancy and childhood, the bias towards seeking causal information gives children the ability to explain, predict and eventually even to control events within their everyday worlds. As we will see in this book, this causal bias acts to organise early memory, it underlies conceptual development, it helps the child to understand the physical world, it helps to organise the social world of agents and their actions, and it acts as a pacesetter for logical thought. The kinds of objects and events that infants are prepared to link in a causal fashion appear to be constrained by the functioning of their sensory systems. For example, some kinds of movement appear more likely to be assigned a biological cause than others, even if the movements are of simple geometric shapes on a computer screen rather than of real entities in the 3D world (Chapter 3).

Inductive reasoning can be shown to be present at even younger ages. When we make inferences that are not necessarily deductively valid (when we 'go beyond the information given'), we are reasoning inductively. For example, we might make a generalisation on the basis of a known example, or use an analogy. Conceptual development and categorisation depend on inductive reasoning and analogy. For example, when children learn about the category *birds*, they may learn about one or two exemplars (e.g., the robins and sparrows in their back garden). However, they are happy to generalise properties like 'lives in a nest' to other birds, such as magpies, that they may not have seen before. These generalisations are made on the basis of inductive reasoning. When new exemplars (like magpies) appear typical of a category (like birds), then it seems natural to make generalisations about properties of a typical category member to other category members. Very young children do this all the time as they learn about the world around them, as we will see in Chapter 5.

The early chapters in this book will focus on cognitive development in the 'foundational' domains of human thought, the domains of physics, psychology and biology. These domains will provide a window for us to consider *what develops* in terms of learning, memory, problem-solving, reasoning, conceptual development, and causal reasoning. Given space limitations, I will focus largely on the years from birth to ten. Although we will return to the question of *why* children's cognition pursues its observed course in the final chapters, the explanatory systems or theories that have been used in different domains to explain what develops vary hugely, and many (in my view) are being supplanted as we gain more knowledge from cognitive neuroscience. This is not true of the theories of Piaget and Vygotsky, which will be covered in the final chapter. There I will also consider a new theoretical approach to understanding cognitive development that integrates cognitive neuroscience data, 'neuroconstructivism'.

CHAPTER 1

CONTENTS

Infancy: The physical world 1

1

What kinds of knowledge are central to human cognitive development? One proposal is that knowledge about the physical world of objects and events; knowledge about social cognition, self and agency; and knowledge about the kinds of things in the world, or conceptual knowledge, are the 'foundational' domains for cognitive development (Wellman & Gelman, 1998). These domains could be described as naïve physics, naïve psychology and naïve biology. Infants need to learn about objects and the physical laws governing their interactions, they need to learn and understand social cognition (interpreting and predicting people's behaviour on the basis of psychological causation), and they need to learn about the kinds of 'stuff' in the world (such as animate versus inanimate entities). Clearly, cognitive development in these foundational domains is also dependent on the development of perception, memory, attention, learning and reasoning. Most areas of cognition involve all of these processes at once.

It was once thought that young infants, who are immobile and whose perceptual abilities are still developing, have very limited cognition. For example, for many years it was thought that infants did not develop a full object concept until around 18 months of age (Piaget, 1954; a full object concept requires an understanding that objects are enduring entities that continue to exist when out of view). This seems to be far from the case. Recent work in infant perception demonstrates that a remarkable amount of information about the foundational domains is given simply by watching things happen in the world. As argued in the Foreword, this sensory information and the neural mechanisms used by the brain to process it are probably key sources of early cognitive development. The automatic processing and storage of spatio-temporal information about the world gives the infant emergent knowledge and experience-based *expectations* about how people and about objects behave, which we can call emergent cognitive frameworks.

These emergent cognitive frameworks are also supplemented by information gained through direct action. Much richer information becomes available when the infant becomes able to reach, grasp, sit and move. Usually, the ability to grasp objects unaided begins at around four months. Now infants can experience making a range of different actions on objects that they can hold, such as toys or their bottle of milk. Another important milestone is being able to sit up without support. Babies usually become 'self-sitters' between four and six months of age. Babies who can sit upright by themselves can expand their range of actions on the world of objects. For example, becoming a 'self-sitter' has been shown to be related to babies' understanding of objects as three-dimensional. Finally, the ability to crawl (or bottom shuffle) usually onsets at around nine months. Now babies can choose to move towards objects of interest and explore them further, and can experience

KEY TERMS

Naïve physics
An intuitive understanding we have about objects in the physical world; e.g., that objects that are dropped will fall, solid objects cannot pass through other solid objects, etc.

Full object concept
An understanding that objects are enduring entities that continue to exist when out of sight, or otherwise unavailable to our senses.

Associative learning
The ability to make connections between events that are reliably associated.

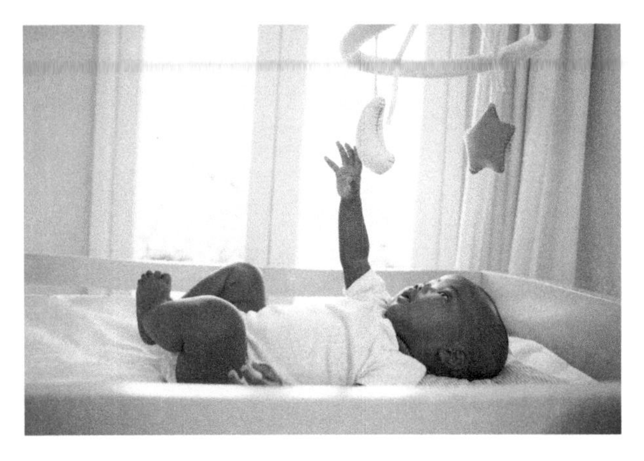

Once an infant is able to reach out and grasp objects, much richer information becomes available to him or her.

different viewpoints. This enables the development of 'allocentric' spatial frameworks or mental maps – an understanding of space based on salient landmarks in the environment rather than on the infant's own position in space.

There are a range of types of learning mechanism regarding the physical world that are functioning from very early in development and that support the emergence of early cognitive frameworks. One is associative learning. Babies appear to be able to make connections between associated events even while in the womb. Once outside the womb, they appear able to track statistical dependencies in the world, such as conditional probabilities between events. Statistical conditional learning turns out to be a very powerful mechanism, as we will see. Another type of learning that appears to be available early is learning by imitation. This may be particularly important for the development of social cognition. Learning by imitation is considered further in later chapters. Finally, infants appear able to connect causes and effects by using what machine learning theory calls 'explanation based' learning. This is a form of the 'causal bias' that was discussed in the Foreword. The causal inferences made apparently automatically by the brains of infants provide an extremely powerful mechanism for learning about the world. Infant's brains are not simply detecting causal regularities, but appear inherently to record spatio-temporal information about causal trajectories, perceptual information that includes causal structure. This eventually enables the construction of causal explanations for new phenomena on the basis of prior (statistically-based) knowledge. One cognitive mechanism that is used to help construct explanatory frameworks is learning by analogy. This type of learning is also considered further in later chapters.

MEMORY

Memory is a good place to begin to study infant perception and cognition. After all, without some form of memory, infants would live in a constant world of the 'here and now'. In order to remember, babies must learn what is familiar.

Memory for objects

Infant memory was originally investigated using rather mundane objects and events. For example, Bushnell, McCutcheon, Sinclair and Tweedie (1984) studied infants' memory for pictures of simple shapes such as red triangles and blue crosses, which were mounted on wooden paddles. The infants were aged three and seven weeks. Memory for a simple stimulus such as a yellow circle was first developed by asking the infants' mothers to present the stimulus daily for a two-week period. The mothers were encouraged to show their babies the stimulus 'actively' for two

15-minute sessions per day. The babies were then visited at home by an experimenter, who showed them the familiar or *habituating* stimulus, and also a random selection of other stimuli, varying either colour, shape, or colour *and* shape. The aim was to test the infants' memories for these different aspects of the stimuli. For example, to test colour memory, the baby might be shown a red circle rather than a yellow circle. To test memory for shape, the baby might be shown a yellow square instead of a yellow circle, and so on. Bushnell et al. found that the infants retained information about every aspect of the stimuli that they had been shown – shape, colour and size.

Cornell (1979) used pictures of groups of such stimuli to study recognition memory in infants aged from five to six months. In addition to pictures of patterns of geometric forms (see Figure 1.1), he also used photographs of human faces. The babies were first shown two identical pictures from Set 1 side by side, followed by two identical pictures from Set 2, followed by two identical pictures from Set 3, and were allowed to study each set for a period of up to 20 seconds. Two days later they were shown the pictures again, first in a brief 'reminder' phase in which each previously studied picture was presented on its own, and then for a recognition phase in which the familiar picture from each set was paired with an unfamiliar picture from the same set. Recognition memory was assumed if the infants devoted more looking time to the novel picture in each pair.

Cornell found a novelty preference across all the sets of stimuli that he used. Even though two days had passed since the infants saw the pictures, they remembered

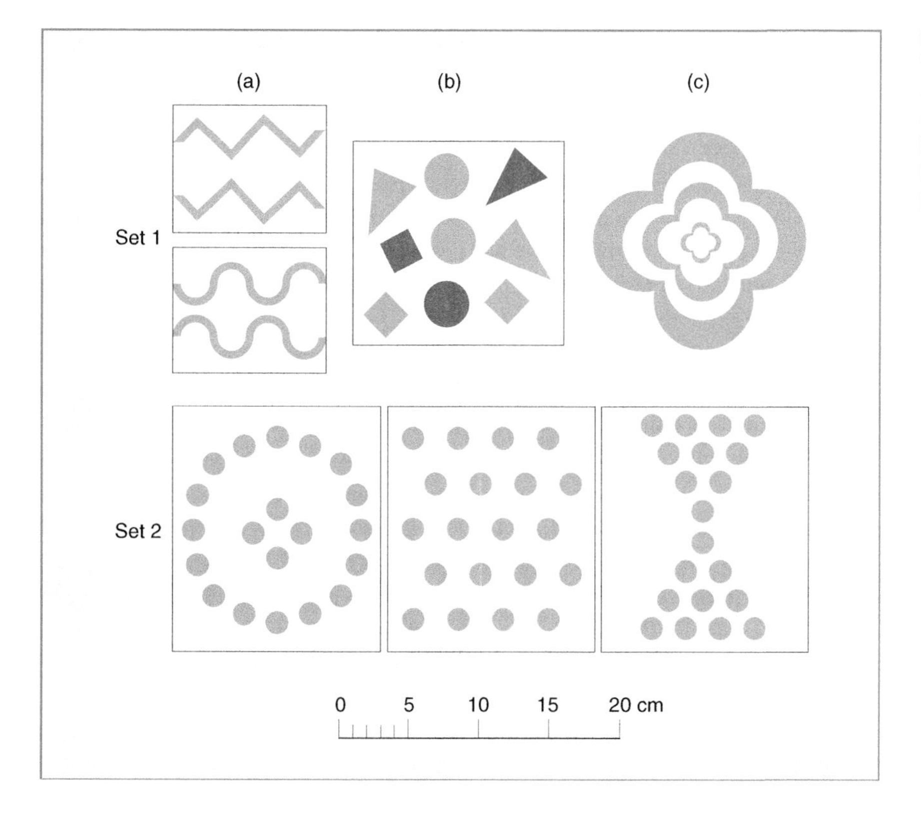

FIGURE 1.1
The stimulus sets used by Cornell (1979) to study recognition memory in infants. Copyright © 1979 Elsevier. Reproduced with permission.

those that were familiar and thus preferred to look at the novel pictures in the recognition phase of the experiment. Their recognition memory was not due to the brief reminder cue, as a control group who received the 'reminder' phase of the experiment without the initial study phase did not show a novelty response during the recognition test. Given that the stimuli were fairly abstract (except for the faces) and were presented for a relatively short period of time in the initial study phase, their retention over a two-day period is good evidence for well-developed recognition memory in young infants.

Working memory in infancy

The capacity to retain information over short periods of time is often called 'short-term memory' or 'working memory'. An influential model of memory in adult cognition is Baddeley and Hitch's (1974) model, which distinguishes a short-term from a long-term system. The short-term system, called *working memory*, is thought to enable the temporary maintenance of information while it is processed for further use (e.g., in reasoning or in learning). Working memory is thought to have both visuo-spatial and sound-based (phonological) subsystems, which maintain visual versus auditory information respectively.

KEY TERM

Working memory
The memory system that temporarily keeps in memory information just received that may be processed for further use.

Working memory abilities in babies were first studied by Rose and her colleagues (Rose, Feldman & Jankowski, 2001). Rose et al. (2001) measured how many items could be held in mind by infants as they developed, testing the same babies when they were aged five, seven and 12 months respectively. The infants were shown colourful toy-like stimuli, in sets of either one, two, three or four items. Once a particular set had been presented, recognition memory was tested by pairing each individual item with a novel item. Working memory capacity was measured by seeing how many objects the babies recognised as novel. For example, if a baby had been shown a set of four items, but only seemed to recognise two of them in the subsequent novelty preference pairings, memory span was assumed to be two items. *Primacy* and *recency* effects were also studied. In adult working memory experiments, participants find it easier to remember the first item of a set (primacy), and also to remember the last item (recency). The question was whether babies would show the same effects.

Rose et al. (2001) reported that memory span increased with age. When they were aged five and seven months, rather few babies could hold three or four items in working memory simultaneously (only around 25% of the sample achieved this span). By 12 months, almost half of the babies had a working memory span of three to four items. Recency effects were found at all ages tested – the babies showed better recall for the final item in the set. Primacy effects were not reported, but have been reported in seven-month-old infants by Cornell and Bergstrom (1983). Hence the working memory system of young infants appears to operate in a similar way to that of adults. Primacy and recency effects in adults are explained in terms of the extra cues to recall provided by being the first or the last item in the list. The development of working memory in infancy has also recently been studied using EEG. Many of these studies utilise delayed response paradigms (such as the A not B finding task) and examples are discussed in Chapter 2.

Memory for events

Some striking studies carried out by Clifton and her colleagues have shown that six-month-olds can also retain memories for events, and do so over very long time periods. For example, in one of Clifton's studies, 6.5-month-olds were able to retain a memory of a single event that had occurred once until they were 2.5 years of age (Perris, Myers & Clifton, 1990).

Perris et al. (1990) demonstrated this by bringing some infants who had taken part in an experiment in their laboratory as six-month-olds back to the laboratory at age 2.5 and retesting them. During the infancy experiment the babies had been required to reach both in the dark and in the light for a Big Bird finger puppet that made a rattle noise (the experiment was about the localisation of sounds). The reaching session had taken about 20 minutes. Two years later, the children were brought back to the same laboratory room and met the same female experimenter, who said that they would play some games. She showed them five plastic toys, including the Big Bird puppet, and asked which toy they thought would be part of the game. She then told them that Big Bird made a sound in the game, and asked them to guess which one it was out of a rattle noise, a bell and a clicker. Finally, the children played a game in the dark, which was to reach accurately to one of five possible locations for the sounding puppet. After five uninstructed dark trials, during which no instructions about what to do were given, the children were given five more trials in which they were told to "catch the noisy Big Bird in the dark". A group of control children who had not experienced the procedure as infants were also tested.

Perris et al. found that the experimental group showed little *explicit* recall of their experiences as infants. They were no more likely than the control group to select Big Bird as the toy who would be part of the game, or to choose the rattle noise over the bell and the clicker. However, they showed a clear degree of *implicit* recall, as measured by their behaviour during the game in the dark. They were more likely to reach out towards the sound than the children in the control group in the first five trials, and they also reached more accurately. If they were given a reminder of their early experience, by hearing the sound of the rattle for three seconds half an hour before the test in the dark, then they were especially likely to show the reaching behaviour. Again, this was not true of the control group. Finally, the children who had experienced the auditory localisation task as infants were much less likely to become distressed by the darkness during the testing than the children who had not experienced the auditory localisation task as infants. Nine of the latter children (out of 16) asked to leave before completing the uninstructed trials, compared to only two children in the experimental group. Children who had experienced reaching in the dark as infants thus showed evidence of remembering that event two years later in a number of different ways. Similar results were reported in a study by Myers, Clifton and Clarkson (1987), who showed that children who were almost three years old also retained memories of the laboratory and the auditory localisation testing procedures that they had encountered as infants. These children had had 15–19 exposures to the experimental procedures as infants, however, and so their memory is in some sense less surprising than that demonstrated in the experiment by Perris et al. (1990).

KEY TERM

Implicit recall
Recall of information that is not explicitly available.

Memory for causal events

Event memory can also be studied by teaching infants a causal *contingency* between an action and an outcome. Another experiment with rattles that made sounds, this time with eight-month-old babies, recorded an EEG response found in infants called the *mu rhythm* (Paulus, Hunnius, van Elk & Bekkering, 2012). As will be recalled, the EEG (electroencephalogram) is a measure of brain electrical activation obtained by sensitive electrodes that are held on the scalp via a headcap. The electrodes pick up electrical signalling by cell assemblies in the brain, although the exact generators of the signals are difficult to localise. The mu rhythm (also called an alpha rhythm in some infant literature) is an oscillatory brain response in the frequency band around 6–8 Hz. This is a relatively slow neural rhythm, with electrical activity peaking 6–8 times per second. Brain cells are thus oscillating or rhythmically alternating between activation and inhibition 6–8 times per second, hence the label mu *rhythm*. The mu rhythm is thought to be suppressed (desynchronised) during action-effect learning, an effect that is sometimes called *motor resonance* in this infant literature.

In Paulus et al.'s experiment, infants spent a week learning the sound made by a novel rattle when shaken. The sound was made by a bell for some infants, by tambourine disks for some infants, and by loose screws for other infants. Each infant played with their novel rattle for five minutes every day, and they were also exposed to one other sound for five minutes every day. So two sounds became familiar, the passively experienced sound, and the sound contingent on shaking the rattle.

In the test phase, the infants then listened to all three sounds (one from their particular rattle, one from the pre-exposure, and one heard for the first time, hence novel) while EEG was recorded. Paulus et al. found that the mu rhythm was desynchronised significantly more for the action-related sound than for the pre-exposed non-action-related sound or the novel control sound. Paulus et al. argued that they had produced neural evidence for acquisition of an action-effect association. Their neural data also show correlational evidence for the acquisition of a causal contingency.

A different technique for measuring learned causal relationships was devised by Rovee-Collier and her colleagues in some pioneering behavioural studies (e.g., Rovee-Collier, Sullivan, Enright, Lucas & Fagen, 1980). In their studies, the conditioned response was kicking, and the reward was the activation of an attractive mobile hanging over the infant's crib. The contingency was that kicking activated the mobile. Activation of the mobile occurred via a ribbon that was tied to the infant's ankle. As kicking comes naturally to young infants, the kicking response is present whether the mobile is there or not. The important point about Rovee-Collier's paradigm was that the infant must *learn* that kicking makes the mobile start to work. Memory for this cause–effect relation was then measured by returning the infants to the same crib after some time had passed, and seeing how much they kicked in the presence of the mobile.

For example, in a typical experiment, the infant was visited at home (see Rovee-Collier & Hayne, 1987, for a review). An attractive mobile was erected on the side of their crib, and a second empty mobile stand was also erected (see Figure 1.2).

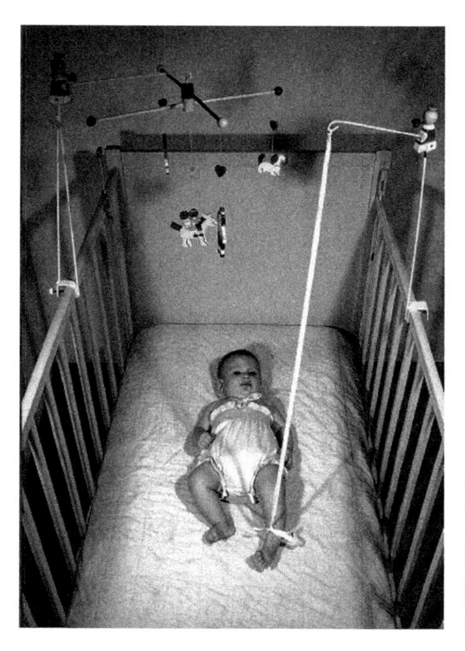
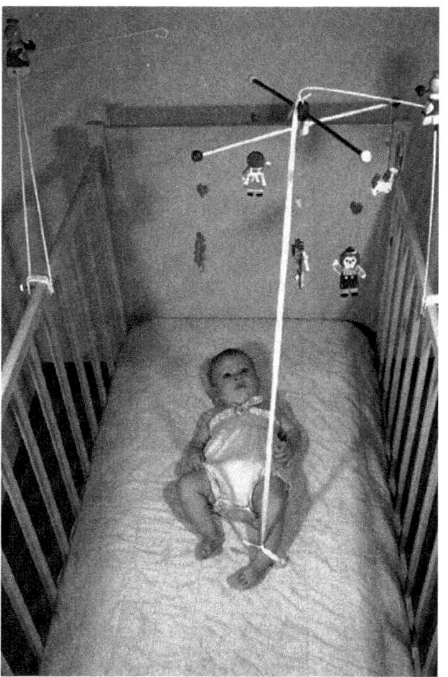

FIGURE 1.2
An infant in Rovee-Collier's causal contingency paradigm (left) during baseline, when kicking cannot activate the mobile and (right) during acquisition, when the ankle ribbon is attached to the mobile. From Rovee-Collier et al. (1980). Copyright © 1980 AAAS. Reprinted with permission.

The ribbon was first tied to this empty stand, to measure the baseline kick rate in the absence of reinforcement with the mobile. After approximately three minutes, the ribbon was attached to the correct mobile stand, and the infant was allowed to kick for about nine minutes for the reward of activating the mobile. The ribbon was then moved back to the empty stand for a final three-minute period. The difference in kick rate between this second three-minute period and the initial baseline period provided a measure of the infant's short-term retention of the contingency. The infant was then visited a second time some days after the original learning phase, and the ribbon was again tied to the empty stand. Long-term retention of the cause–effect relation was measured by comparing kicking in the absence of reinforcement during this second visit with the original baseline kick rate.

Rovee-Collier and her colleagues have found that three-month-old infants show little forgetting of the mobile contingency over periods ranging from two to eight days. By 14 days, however, forgetting of the contingency appears to be complete. Furthermore, as the time between the learning and test periods increases, the infants forget the specific details of the training mobile (its colours and shapes), and respond as strongly to a novel mobile as to the original. Twenty-four hours after learning, the infants remember the objects on the mobile, and will not respond to mobiles containing more than one novel object. By four days, however, they will respond to a novel five-object mobile. This suggests that infants, like older children and adults, gradually forget the physical characteristics or attributes of what they have learned, retaining only the gist or the associations between specific attributes and the context of learning.

Interestingly, at the same time as memory for the mobile itself declines, memory for the surrounding context (e.g., the pattern on the crib bumper) becomes more important in reactivating the infant's memory of the contingency. Infants show perfect retention of the contingency at 24 hours, whatever the pattern on the

KEY TERM

Reactivation paradigm
A procedure in which the participant, usually an infant, is given a reminder of an earlier learned, but apparently forgotten, memory that enables this memory to become accessible again.

crib bumper. By seven days, infants who have been trained with a distinctive crib bumper show apparently complete forgetting if they receive a different crib bumper at test, while infants who receive the distinctive crib bumper at test remember the contingency. The different cues on the crib bumper, such as its colours and the particular shapes in its pattern, appear to be forgotten at different rates (Rovee-Collier, Schechter, Shyi & Shields, 1992). It is difficult to escape the conclusion that details of the learning context, such as details of the pattern on a distinctive crib bumper, are acting to *cue* recall.

If the crib bumper indeed provides an appropriate 'reminder' cue for recall, then we can examine whether 'forgotten' memories become accessible again when appropriate retrieval cues are provided. Rovee-Collier and her colleagues have developed a *reactivation* paradigm to study this question. The retrieval cue that they have studied most intensively is a *reminder* of the mobile contingency, namely showing the infants the moving mobile for three minutes prior to measuring kick rate. During the reminder phase, the mobile is activated by a hidden experimenter pulling on the ribbon, and the infants are prevented from kicking by a special seat that also precludes 'on-the-spot' learning. The infants are then retested in the crib procedure 24 hours after the reminding event. With a reminder, three-month-old infants demonstrate completely intact memories for the mobile contingency 14 and 28 days after the training event. Two-month-old infants show excellent memories after a 14-day delay, but only a third of this age group show intact memory after 28 days. By six months of age, the retention period is at least three weeks (Rovee-Collier, 1993). Thus very young infants can develop long-term memories for causal events, and memory retrieval appears to be governed by the same cues that determine retrieval in adults.

Another way of examining infants' long-term memory for causal events is to use delayed imitation, a technique pioneered by Meltzoff in his studies of learning (see Chapter 2). Mandler and McDonough (1995) used delayed imitation to examine 11-month-old infants' retention of causal events over a three-month period. The events were two-step action sequences, namely 'make a rattle' (by pushing a button into a box with a slot), and 'make a rocking horse' (by attaching a horse with magnetic feet to a magnetised rocker). Imitation of the events was measured on the following day (24-hour retention period), and three months later. On each occasion the infants were simply presented with the materials (the horse, the rocker), and were then observed. To check that the older infants were not simply more likely to discover the sequences without having seen them being modelled, a control group of 14-month-old infants were also given the materials at the three-month follow-up.

Mandler and McDonough found that recall was good at both the 24-hour and the three-month retention intervals, and that there was little forgetting over the three-month period. In contrast, retention of non-causal events (e.g., 'put a hat on the bunny, and feed him a carrot'), was poorer than that of causal events at 24 hours, and non-existent after the three-month interval. Mandler argues that retaining causal relations provides one of the major ways of organising material that is to be remembered in a coherent and meaningful fashion. The importance of causal relations for memory development is covered more fully in Chapter 8.

However, young infants can also show good memory for related events that are not causal. More recently, Rovee-Collier and her colleagues have devised a deferred imitation task using puppets that is a reliable measure of memory in infants as young as six months of age.

The puppet deferred imitation paradigm involves showing infants two puppets, A and B. The puppets are first experienced in a 'sensory pre-conditioning' phase, and then the experimenter demonstrates actions with one puppet, puppet A. For example, Puppet A might be wearing mittens, and so the experimenter demonstrates to the watching infant taking off one of the mittens, waving it around for a bit, and then putting the mitten back onto the puppet's hand. At the delayed imitation test, the six-month-old infants are given the mitten and puppet B. The outcome measure is whether the infant will copy the mitten actions with puppet B. Using this paradigm, Barr, Rovee-Collier and Campanella (2005) demonstrated delayed imitation as long as ten weeks after the initial exposure event for six-month-olds. Meanwhile, Campanella and Rovee-Collier (2005) found that three-month-old infants who experienced the sensory pre-conditioning phase also put the mitten on puppet B when tested at the age of six months, as long as they were given reminder cues (namely seeing puppet A having the mitten put on) in the intervening three months. Clearly, there is learning and recall of ordered events even at three months of age, even when there is no necessary causal relationship to support the memory.

Interestingly, if the infants did not experience puppets A and B simultaneously in the sensory pre-conditioning phase, but experienced them consecutively, they did *not* later put the mitten on puppet B. This appears surprising, but may offer important clues regarding the simple neural learning mechanisms that could underpin these early memories. For example, associative learning may play an important role. Cuevas, Rovee-Collier and Learmonth (2006) experimented with a combination of the kicking/mobile and puppet paradigms, simultaneously exposing six-month-olds to puppets A and B, and then training them to kick to activate a mobile in a setting with a distinctive context (crib buffer) a day later. Subsequently the infants experienced a third phase of the experiment, in which they saw puppet A and the crib buffer. The experimental question was whether this third associative experience would result in an associative connection between puppet B and the mobile, even though neither object had been experienced simultaneously.

This turned out to be the case – the infants demonstrated memory for puppet B and the mobile when the distinctive crib buffer was present during deferred imitation. Note that if the networks of brain cells that encode the visual experience of seeing puppets A and B, the crib buffer and the mobile became interrelated via these experimentally selective pairings, for example by co-activation of adjacent networks of cells, then there would indeed be a 'memory' for puppet B with the crib buffer – via spreading activation. This 'memory' would result from the neural mechanisms used by the brain to encode the original experiences. Such mechanisms would create emergent neural networks that are interlinked, one useful basis for developing a cognitive system such as memory. It is now technically possible to track the development of specific neural networks of this kind using a technique called multi-voxel pattern analysis (MVPA). In MVPA, high-resolution fMRI is able to track the activation of individual voxels (small cubic areas of brain tissue, corresponding to thousands of cells) in a particular network in response to particular objects or to memories of these objects (i.e., when participants are instructed to bring an object to mind). So far, to my knowledge, this difficult technique has not been attempted for infant memory, but it could yield interesting data (see Emberson, Zinszer, Raizada & Aslin, 2017). Note that such a relational cell network would not require language in order to first develop – although it may become far more flexible, connected and enriched as language is acquired.

KEY TERMS

Delayed imitation Imitation of a previously seen behaviour after a delay.

Implicit or procedural memory Memory that is not available to conscious report.

Explicit or declarative memory Memories for earlier experiences that can be readily brought to mind and thought about.

Procedural versus declarative memories?

It is notable that most of the studies discussed above have measured infant event memory in terms of the infants' *behaviour*. Rovee-Collier measured the amount of kicking that was produced to the mobile or whether the infant put the mitten on the puppet, Mandler measured the number of action sequences that were reproduced with the props, and Clifton measured children's reaching behaviour in the auditory localisation paradigm. Paulus et al. measured the mu rhythm, but as a neural correlate of memory rather than as an insight into mechanism. Indeed, for a long time it was believed that infant memories were somehow different *in kind* to the type of memory in which we bring an aspect of the past to conscious awareness (e.g., Mandler, 1990). Most researchers in the field of memory did not regard infant memory as an active remembrance of things past, but conceptualised infant memory more as akin to conditioned responses of the type studied in animals.

This has changed recently (see Mullaley & Maguire, 2014, for a review). It is widely accepted in cognitive psychology that there are *two* types of memory system in humans. One is automatic in operation, and is not accessible to verbal report. This kind of memory is usually called implicit or procedural memory and is measured by behaviour. The second involves bringing the past to mind and thinking about it. This kind of memory is usually called explicit, declarative or episodic memory. One popular definition is that explicit, declarative or episodic memory involves information that has been encoded in such a way as to be accessible to consciousness. The 'what, where and when' of facts and events can be brought deliberately to mind. Infants were generally assumed not to encode explicit or declarative memories until they become verbally competent, a phenomenon that has been called 'infantile amnesia'. This assumption is probably incorrect, and is discussed more fully in Chapter 8.

It is very difficult to measure whether a pre-linguistic infant can bring past events consciously to mind, as we cannot ask them to tell us about their memories. Nevertheless, since the variables that affect declarative memory performance in adults (such as context and length of retention interval) also affect infant memory, available data suggest that these paradigms tap into declarative memories, not implicit memories. Indeed, in their comprehensive review Mullaley and Maguire (2014) argue that declarative memory is online from early infancy, and that it develops rapidly and monotonically. Their argument would support the theoretical view proposed here, that the way in which the brain processes information about events *in itself* forms the basis for the development of cognitive systems, in this case memory. The development of implicit and explicit memories is discussed more fully in Chapter 8.

PERCEPTION AND ATTENTION

Learning and memory in infants and neonates would be impossible if infants lacked adequate perceptual skills and adequate attentional mechanisms. Although there are some important immaturities in the visual system at birth (see Atkinson & Braddick, 1989), recent research has shown that the perceptual abilities of babies are much more sophisticated than was once supposed. We have already seen that visual recognition memory emerges early, as defined by responsiveness to novelty. Attention is clearly a prerequisite if visual recognition memory is to function effectively.

Adequate attentional mechanisms appear to be available shortly after birth. However, it is not clear whether these mechanisms are under the infant's volitional control. It can be very difficult to attract an infant's attention, particularly to a stationary visual stimulus, as many infant experimenters will tell you! At one point it was believed that infants were passive in their selection of visual stimuli. The idea was that attention to certain stimuli was obligatory, and that visual 'capture' by these stimuli controlled infant attention (e.g., Stechler & Latz, 1966).

This view is no longer widely held. The visual world of the baby is an active one, characterised by a dynamic flow of perceptual events over which the babies themselves have no control. In order to deal with this dynamic flow of events, infants need to develop expectations of predictable visual events, around which they can then organise their behaviour (Haith, Hazan & Goodman, 1988). Thus one way to study when attentional mechanisms in infants come under volitional control is to study their *expectations* about visual events. The development of visual expectancies requires the volitional control of visual attention.

Attention in infancy

In order to find out whether babies as young as 3.5 months of age can develop visual expectations, Haith and his colleagues devised a paradigm that involved showing babies a series of stimuli to the left and to the right of their centre of gaze. In Haith et al. (1988), the stimuli used included pictures of checkerboards, bull's-eyes and schematic faces in different colours (the kind of stimuli used by Fantz, 1961, to examine visual perception in babies, see below). Sixty stimuli were used in all. Thirty of these were presented in a left-right alternating sequence, which was thus predictable, and the remaining 30 were presented in a random left-right order. The movements of the babies' eyes were observed during both the predictable and the random presentation sequences. Haith et al. argued that, if the infants could detect the alternation rule governing the appearance of the predictable stimuli, then they should develop expectations of the left-right alternation, and should make anticipatory eye movements to the location of the next slide. Such eye movements should be less common during the random presentation sequences. This was exactly what happened. The infants showed more anticipatory fixations to the predictable (alternating) sequence than to the unpredictable (random) sequence of pictures, and also showed enhanced reaction times, meaning that they were developing expectations for the visual events quite rapidly. This shows that, at least by the age of 3.5 months, babies can control their own perceptual (attentional) activity.

Using a somewhat different task, Gilmore and Johnson (1995) have shown that, by the age of six months, infants can also control their visual attention over delays of at least 3–5 seconds. Gilmore and Johnson's paradigm involved showing the infants an attractive geometric display presented centre-screen, in order to encourage fixation at the centre (see Figure 1.3). Once the infants were reliably looking at the central fixation point, a blue triangle ('cue stimulus') was flashed briefly either to the left or to the right of the centre. The screen then stayed dark for a set time period, until two rotating, multi-coloured cogwheel shapes (which were highly attractive to the infant) appeared, one to the left and one to the right of centre. The experimenters then scored whether the infants showed a preference for looking at the cued location during the delay period, prior to the onset of the cogwheel targets.

Gilmore and Johnson found strong preferences for the cued location at each of the three different time intervals that they studied, which were 0.6 seconds, three seconds and five seconds. They argued that this showed that the infants were maintaining a representation of the spatial location of the cue, and were using it to plan their eye movements several seconds later. In a follow-up study, Gilmore and Johnson cued the eventual left or right location of the target stimulus by presenting

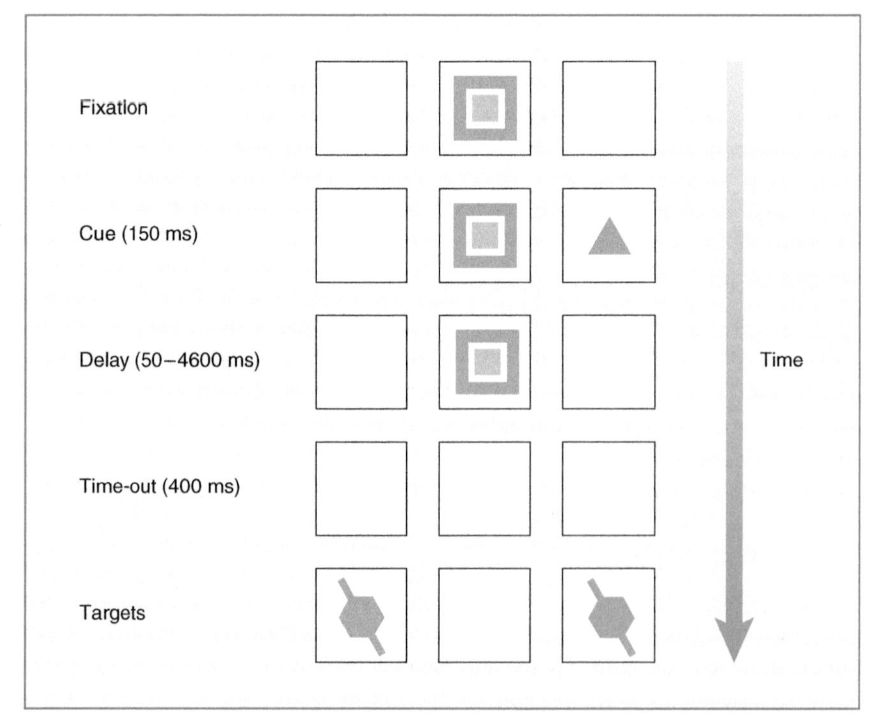

FIGURE 1.3 Example of one of the stimulus presentation sequences used by Gilmore and Johnson to study infant control of visual attention. Each box represents one of the three computer screens. From Gilmore and Johnson (1995). Copyright © 1995 Elsevier. Reprinted with permission.

different geometric displays at the central fixation point, and omitting the blue triangle. For example, if the centre-screen stimulus was a pattern made up of four shifting light- and dark-blue circles, then the target would appear on the right three or five seconds later, whereas if the centre-screen stimulus was a pattern made up of small red and yellow squares that spiralled around each other, then the target would appear on the left three or five seconds later. The infants quickly learned this contingency, and again showed strong preferences to look to the cued location. Gilmore and Johnson argue that their expectation paradigm also shows the early operation of 'working memory' in the infant.

More recent studies of infant working memory and attention, using indices such as heart rate and using EEG, have supported this idea. Heart rate measures have been used to show that during a single look at an object, infants cycle through four phases of attention: stimulus orienting, sustained attention, pre-attention termination, and attention termination (Reynolds & Richards, 2008). During sustained attention, for example, there is a significant and sustained decrease in infant heart rate. Using EEG, two markers of infant visual attention to objects have been found. The first is a negativity of large amplitude (Nc) around 400–800 ms following stimulus onset, typically on central electrodes, and possibly reflecting prefrontal cortex activity (Reynolds & Romano, 2016). The second is a late slow wave (LSW) about 1–2 seconds after stimulus onset. The LSW is thought to index recognition memory, as its amplitude is much greater for novel than for familiar objects. The cortical sources of the LSW are currently unknown. Note that although the Nc and LSW are useful *correlates* of infant memory and attention, by themselves they do not tell us anything about the mechanisms underlying infant memory and attention. This requires the development of further paradigms.

Alternatively, some of the mechanistic measures used in adult cognitive neuroscience could be adapted to study infant attention. In adult studies, brain oscillations (rhythmic fluctuations in electrical activity reflecting the signalling of cell networks) in the alpha band (around 10 Hz) are related to performance in visual attention tasks. For example, when detecting visual targets, adults are unaware of visual stimuli that occur during the trough of an alpha oscillation (the least excitable phase, when fewest cells are firing) in parietal cortex. Adults are most likely to detect visual targets at the oscillatory peak of the ongoing brain rhythm, when the maximum number of cells in the network are discharging action potentials (Mathewson, Gratton, Fabiani, Beck & Ro, 2009). For adults, visual events that arrive 'out of phase' (during the oscillatory trough) do not reach conscious awareness. Furthermore, when adults monitor two visual spatial locations in turn, then a theta-rhythmic process (oscillations around 4–5 Hz) appear to alternately sample each spatial location, with detection benefits correspondingly alternating in a 4 Hz rhythm (Landau, Schreyer, van Pelt & Fries, 2015). These adult studies suggest neural mechanisms that could be studied developmentally in relation to individual differences in children's visuo-spatial attention and their visuo-spatial working memories. Alpha and theta oscillations offer the promise of a deeper understanding of mechanism, and are now beginning to be studied in infants.

Xie, Mallin and Richards (2017) provide an illustration of this approach. Xie et al. studied the relationship between three aspects of infant attention (stimulus orienting, sustained attention and termination of attention) and oscillations in the infant alpha (6–9 Hz) and theta (2–6 Hz) bands. Infants aged six, eight, ten and 12 months watched clips of the TV programme *Sesame Street*, in which a single character such as Elmo appeared singing and dancing in different locations, and occasionally disappeared from view. Alpha power (signal strength) during sustained attention decreased significantly for the older infants (ten- and 12-month-olds), an effect that is also found in adult sustained attention. Theta power during sustained attention increased with age from eight months onwards (theta power in adults usually increases when attention is taxed by working memory demands). Xie et al. concluded that the infant theta rhythm is an earlier index of sustained attention than the infant alpha rhythm. Unfortunately, Xie et al. were not able to conduct the phase analyses used in the adult literature (phase analyses establish whether visual attention is more efficient during the peaks of the alpha oscillation). However, they did find that the infant alpha oscillation was located in the same neural networks utilised by the adult brain (in essence, the 'resting state' network or 'default mode' network). If supported by other studies, this would suggest that the infant brain has essentially the same attention *structures* (localised neural networks) as the adult brain, and that these structures and networks are carrying out essentially the same *functions* via the same *mechanisms*. We will return to this possibility in later chapters.

Visual search

One aspect of visual attention that has been a focus of more recent research with infants is visual search. In the adult attention literature, the ability rapidly and accurately to detect odd-one-out target elements in visual displays (for example, a circle in a field of crosses) has been shown to be an important index of visual attention. One reason that researchers became interested in visual search behaviour in infancy was because it was thought that enhanced visual search could be an early indicator of autism spectrum disorder (ASD). For example, it was suggested that a superior ability to discriminate individual visual features in a display could be a byproduct of an impaired ability to attend to higher-level, 'big picture' information (Happe & Frith, 2006). To investigate attention to individual features in infancy, Cheung and colleagues (2018) studied a sample of 116 infants at family risk for autism (as part of a larger longitudinal study of ASD). The infants were tested in a visual search task when aged nine and 15 months, and their visual search performance was compared to that of no-risk control infants. All infants were shown arrays of eight letters arranged in a circle, with seven identical letters (e.g., X), and one target letter (e.g., O). Examples are shown in Figure 1.4. Infants sat on their mothers' laps during stimulus presentation and their eye movements were recorded with an eye tracker.

At three years of age, the family risk infants were tested for autism spectrum disorder, and 17 were found to meet criteria for ASD. When Cheung et al. retrospectively analysed their visual attention at nine and 15 months, they found that the percentage of 'first looks' that accurately detected the oddball target was significantly greater in the infants later diagnosed with ASD compared to *both* the no-risk

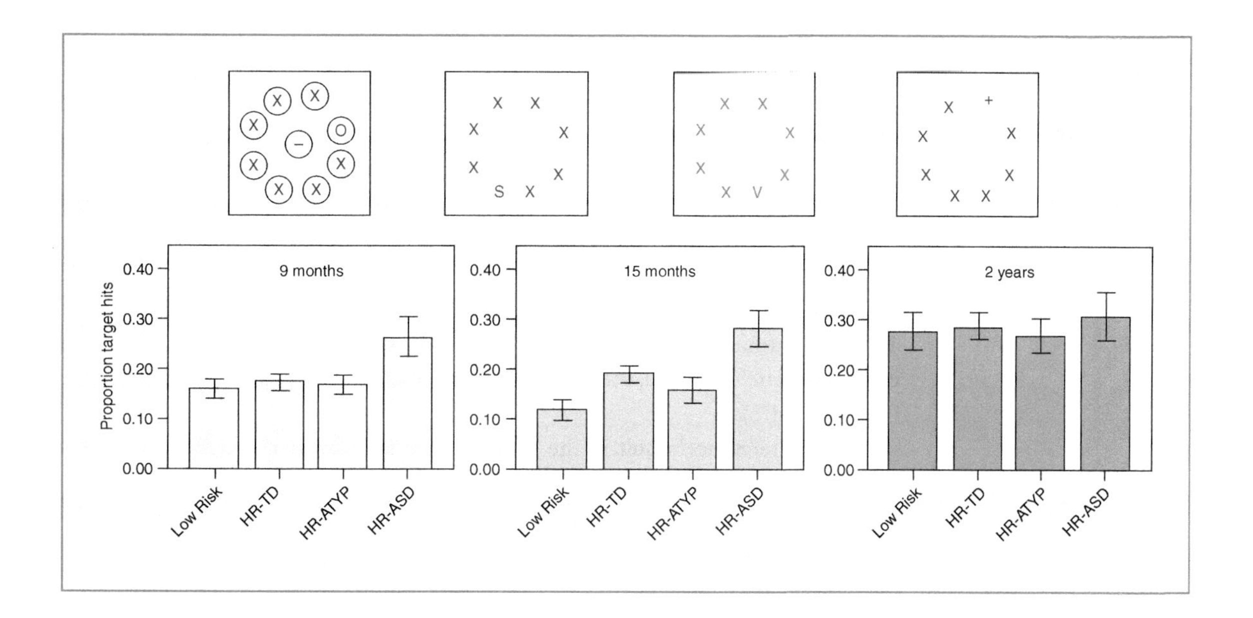

FIGURE 1.4
Examples of the stimuli used by Cheung et al. (2018) to test visual attention in infants and the data patterns at different ages. Groups were either at low family risk of developing ASD, high family risk but turned out to show typical development (HR–TD), high family risk and turned out to show atypical development (HR–ATYP), or high family risk and later qualified for an ASD diagnosis (HR-ASD).

controls and the family risk infants who did not later meet the diagnostic criteria for ASD. This was the case at both the nine-month and 15-month measurement points. The infants later diagnosed with ASD also showed better visual attention during the task than the other infants, looking longer at the stimuli and generating more valid trials (valid trials required a saccade to the centre of the visual display within 100 ms of its appearance).

The developmental mechanisms underlying the association between superior visual search and later ASD are not yet known. Indeed, when tested again with the visual search task at two years of age, the group differences found earlier were no longer present. Interestingly, this was largely because the non-autistic participants had 'caught up' in developmental terms with the ASD infants – they improved their performance in visual search between 15 and 24 months, while the ASD infants did not show further improvement. Cheung et al. (2018) proposed that this could reflect reduced progression of structural connectivity in development in ASD, suggestive of less neural plasticity in this group. While this idea is purely speculative, the data clearly support the importance of documenting early sensory processing when seeking to understand developmental trajectories. In the current example, the operation of a basic sensory mechanism, visual search, is actually superior in early development, yet implicated in the cognitive disorder of ASD. Individuals with ASD have social cognition impairments, for example displaying reduced initiation of social interaction and reduced eye contact with other people. Accordingly, it is possible that the sensory mechanisms required to develop knowledge supporting the emergent domain of naïve psychology are atypical in affected infants, with lifelong consequences. Developmental longitudinal studies, such as that being undertaken by Cheung and her colleagues regarding ASD, are hence critically important not only for documenting *what develops*, but also in helping us to understand why development pursues its observed course.

KEY TERM

Visual preference technique
Infants are shown pairs of stimuli and a preference for looking at one indicates the ability to discriminate between the two.

KEY TERM

Habituation paradigm
Infants are presented a stimulus, usually visual or auditory, until it no longer attracts attention: recovery of attention to a new stimulus (dishabituation) indicates discrimination between familiar and new.

Visual preference and habituation

The existence of visual preferences in infancy provides a useful index of infants' perceptual abilities as well as of their attentional skills. Suppose that we want to discover whether an infant can make a simple visual discrimination between a cross and a circle. One way to find out is to show the infant a picture of a cross and a picture of a circle, and to see which shape the infant prefers to look at. The existence of a preference would imply that the infant can *distinguish* between the different forms. The 'visual preference' technique was first used by Fantz (1961, 1966), who found that seven-month-old infants showed *no* preference between a cross and a circle. Instead, they looked at both shapes for an equal amount of time (see Figure 1.5).

A 'no preference' result in the visual preference paradigm is difficult to interpret. It could mean that the infants were unable to distinguish between the two shapes being tested. Alternatively, it could mean that they found both shapes equally interesting (or equally dull!) to look at. One way to find out whether infants can in fact distinguish two equally preferred visual stimuli is to use the *habituation* paradigm. This has now become one of the most widely used techniques in cognitive research with infants (as noted above, habituation is now conceptualised as an infant form of declarative memory, a cognitive system). Habituation provides a tool for the experimenter to measure infants' conceptual (cognitive) representations.

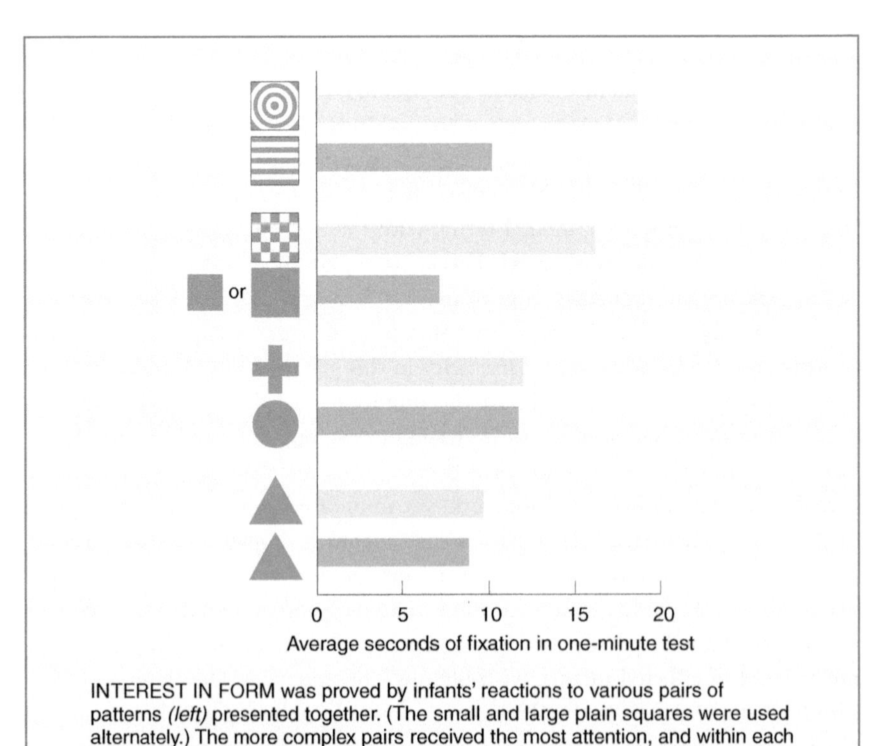

or
0
5
10
15
20
Average seconds of fixation in one-minute test
INTEREST IN FORM was proved by infants' reactions to various pairs of patterns *(left)* presented together. (The small and large plain squares were used alternately.) The more complex pairs received the most attention, and within each of these pairs differential interest was based on pattern differences. These results are for 22 infants in ten weekly tests.

FIGURE 1.5
Examples of the visual preference stimuli adapted from Fantz (1961) to study infant form perception, showing the average looking time for each stimulus.

In simple habituation studies, the infant is shown one stimulus, such as a circle, on repeated occasions. Typically, the infant's interest is at first caught by the novel stimulus, and a lot of time is spent in looking at it. Following repeated exposures of the same stimulus, the infant's looking time decreases. This is quite understandable – seeing the same old circle again and again is not that exciting. Once looking time to the stimulus has fallen to half of the initial level, the old stimulus is removed and a new stimulus is introduced – such as a cross. This is a novel stimulus, so if infants can distinguish between the cross and the circle, renewed looking to the cross should be observed. Renewed looking to a novel stimulus is called 'dishabituation'. When dishabituation occurs, we know that the cross is perceived as a novel stimulus, and this tells us that infants can distinguish between a cross and a circle.

Research with neonates by Slater and his colleagues has shown that infants can indeed discriminate a cross from a circle (Slater, Morison & Rose, 1983). In Slater et al.'s experiment, the cross and the circle were *both* presented during the dishabituation phase, thereby combining the habituation method with the preference technique. Slater et al. showed that when the cross and the circle were presented after habituation to the circle, then the cross was preferred. When the cross and the circle were presented after habituation to the cross, then the circle was preferred. As neonates in a habituation paradigm can distinguish a cross from a circle, we can conclude that the absence of a preference in seven-month-old infants in Fantz's experiments did not arise out of an inability to distinguish between crosses and circles.

Cross-modal perception

The ability to match perceptual information across modalities (cross-modal perception) also appears to be present from early in life. Infants seem to be able to connect visual information with tactile information, and auditory information with visual information, from soon after birth.

Linking vision and touch

One of the most striking demonstrations of infants' ability to make cross-modal connections between vision and touch comes from an experiment by Meltzoff and Borton (1979). They gave one-month-old infants one of two dummies to suck that had different textures. The surface of one of the dummies was smooth, whereas the other had a nubbled surface (see Figure 1.6). The infants were prevented from seeing the dummy when it was placed into their mouths, and so in the first phase of the experiment their experience of the dummy was purely *tactile*. In the second phase of the experiment, the infants were shown enlarged pictures of both dummies, and the experimenters measured which visual stimulus the infants preferred to look at. They found that the majority of the babies preferred to look at the dummy that they had just been sucking. The babies who had sucked on the nubbled dummy looked most at this picture, and the babies who had sucked on the smooth dummy looked most at this picture. This suggests an early understanding of cross-modal equivalence.

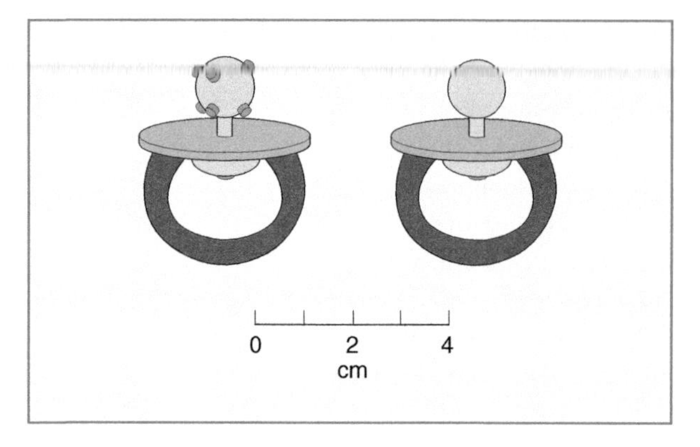

FIGURE 1.6
The two dummies used to study intermodal connections between vision and touch by Meltzoff and Borton (1979). Copyright © 1979 Macmillan Publishers Limited. Reprinted with permission.

Linking vision and audition

Infants also appear to be able to make links between the auditory and visual modalities soon after birth. For example, Spelke (1976) showed four-month-old infants simultaneous films of two rhythmic events, a woman playing 'peek-a-boo', and a baton hitting a wooden block. At the same time, the soundtrack appropriate to one of the events was played from a loudspeaker located between the two screens. Spelke found that the infants preferred to look at the visual event that matched the auditory soundtrack. Again, this preference for *congruence* across modalities suggests an understanding of cross-modal equivalence. Dodd (1979) has found similar results in experiments that required infants to match voices to films of faces reading nursery rhymes. When the soundtrack was played 'out of sync' with the mouth movements of the reader, the infants got fussy. They preferred to look at faces whose mouths were moving in time with the words in the story. Adults also get fussy when they experience this phenomenon – think of being in the cinema when the soundtrack is out of time with the film. Clearly, we have a strong perceptual preference for *congruence* across different perceptual modalities, and this preference is present from early in life. Regarding how neural information processing mechanisms could constrain learning, it is plausible that this preference for congruence reflects the activation of interconnected cell networks. For example, the networks of visual cells that respond to encode the temporal information of the baton hitting the block, and the networks of auditory cells that respond to encode the temporal information of the beat, may be oscillating in synchrony. Accordingly, association networks may be more active when there is temporal congruence in the visual-auditory response patterns. A possible neural method for investigating this possibility comes from a study by Hyde, Porter, Flom and Stone (2013).

Hyde and his colleagues chose to study links between space, time and number, following initial work by Srinavasan and Carey (2010). Neural data from adults suggests that there is a form of analogue coding for these domains in the human brain (see Chapter 10). In a particular area of cortex called the *intraparietal sulcus*, for example, more neurons are electrically active for large numbers than for small numbers. Thus the physical amount represented by the numbers is coded by more neurons firing: this is analogue coding (see also Chapter 10). Hyde et al. wondered whether infants as young as five months of age might demonstrate faster learning of cross-modal magnitude relationships across space (visual information) and time (temporal information) when the magnitude relations were relationally congruent (e.g., 'large' goes with 'long'). Hyde et al. also took care to ensure that the information in the different modalities was *asynchronous*, so that the information received by participating babies was not redundant across modality. Infants were shown cartoon caterpillars that were either short (body made of two sections) or long (body made of ten sections) for 1,000 ms, and simultaneously listened to tones that were

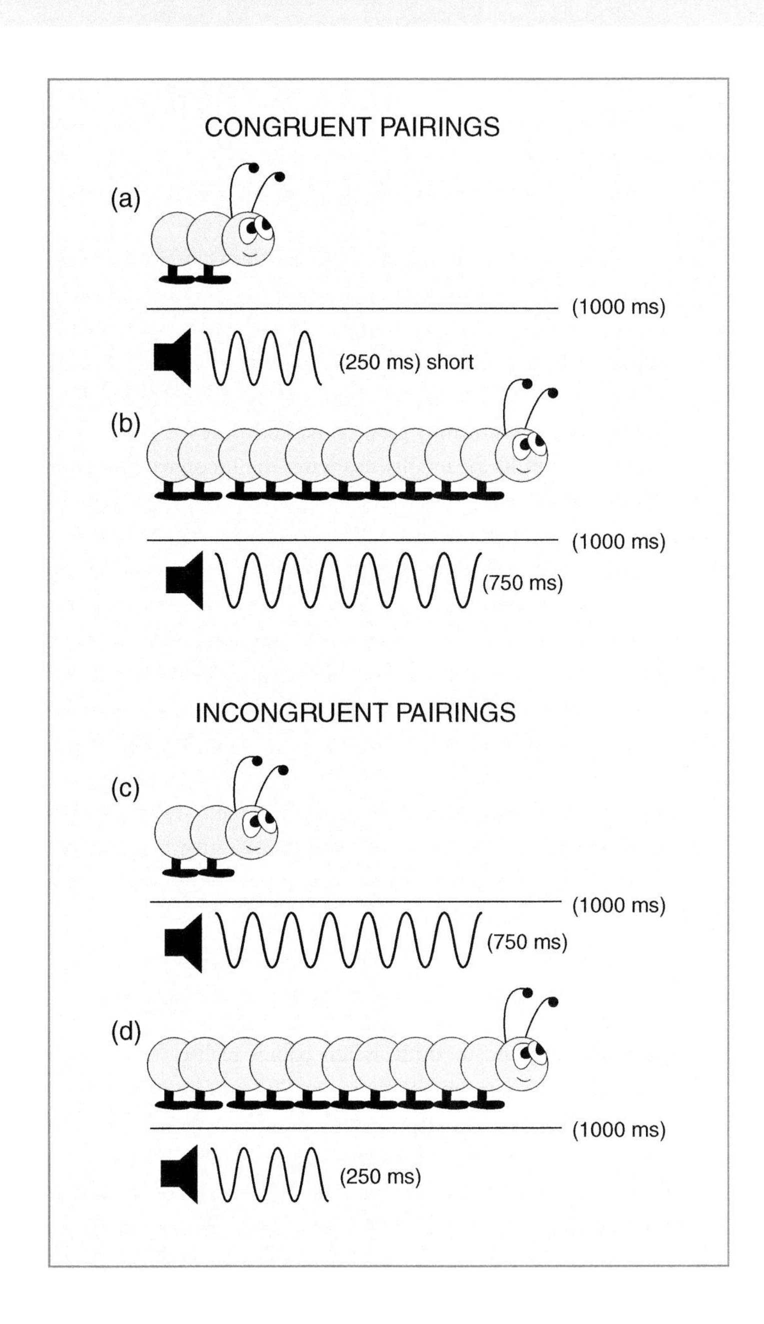

FIGURE 1.7 Examples of the stimuli used by Hyde et al. (2013) to test congruency between long tones and long caterpillars, or short tones and short caterpillars.

either short (250 ms) or long (750 ms) while EEG was recorded. Examples of the caterpillars are shown in Figure 1.7.

Prior to the experimental trials, infants received one of two kinds of familiarisation training, either seeing 20 congruent trials or 20 incongruent trials. At test, all infants received the same pairings of stimuli, both congruent (e.g., short caterpillar and short tone) and incongruent (e.g., long caterpillar and short tone). Hyde et al. (2013) reported a difference in brain response between infants who had received the congruent training and those who had received the incongruent training, which occurred during processing of the test stimuli (recall that the test stimuli were identical for both groups). Sets of electrodes over visual cortex (back of the head, posterior) and frontal cortex (front of the head, anterior) showed different effects by group in three time windows, early (~275 ms following stimulus onset), middle (~600–700 ms following stimulus onset) and late (~1130 ms following stimulus onset).

In the early window, significantly more negative amplitudes were observed for infants who had experienced the congruent familiarisation, while in the middle window, significantly more negative amplitudes to the congruent (familiar) pairings compared to the incongruent (unfamiliar) pairings were observed for this group. For the late window, which measured slow wave activity after the caterpillars were no longer visible, activity was only significantly greater than baseline for the familiar test stimuli (here, incongruent stimuli) for infants receiving the incongruent familiarisation condition. Hyde et al. concluded that the EEG data suggested enhanced processing of and attention to congruent stimuli cross-modally by infants who had been familiarised with congruent caterpillar-tone pairings. While the data strictly show different brain activity by familiarisation condition rather than enhanced processing per se, this is an exciting result. The differences in neural activity between the two groups cannot be explained by the nature of the stimuli themselves, as the stimuli were equivalent for both groups during the test phase. However, this experiment is only a first step. In order to interpret these differences mechanistically, for example in terms of enhanced processing of congruency, we need converging EEG data from other congruency paradigms. Nevertheless, Hyde et al.'s study provides a new methodology for studying the preference for congruence in young infants, and, potentially, for studying why older infants sometimes show the opposite effect, of looking longer at *incongruent* stimuli.

Organising perceptual information into categories

Habituation methods can also be used to study when babies realise that visually distinct objects belong in the same conceptual category. This can be achieved by varying the stimuli that the infant sees during the *habituation* phase of the experiment. The variation of exemplars during habituation requires the infants to *categorise* what they are being shown in some way in order to remember it. At test, we can present the infants with a *new* exemplar of the familiar category that they haven't seen before, as well as a new exemplar from a contrasting category. If the infants have formed a representation of the familiar category, then they should prefer to look at the exemplar from the new category, even though both items presented at test are novel stimuli.

Slater and Morison (1987, described in Slater, 1989) used this categorisation technique with three- and five-month-old babies. During the habituation phase of their study, they showed the babies a variety of types of circle (or of squares, triangles or crosses – see Figure 1.8). At test, they showed the 'circle' babies a new exemplar of a circle, and an exemplar of another shape, such as a cross. The infants preferred to look at the novel shape (the cross). This suggests that the babies had formed a 'prototype' or generalised representation of the familiar shape, to which they appeared to be comparing all subsequently presented stimuli.

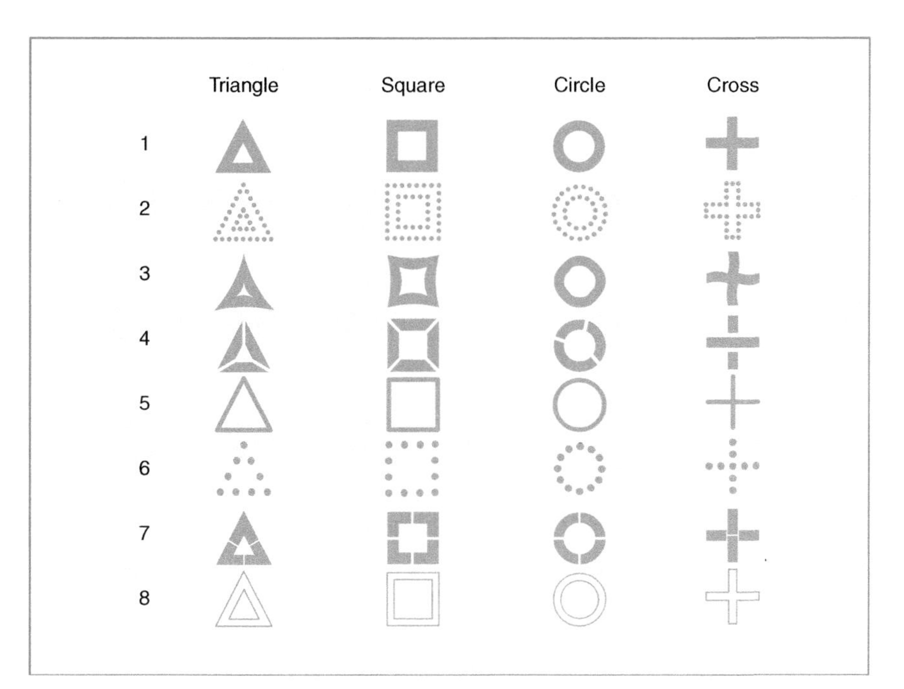

FIGURE 1.8
The different exemplars of triangles, squares, circles, and crosses used during habituation by Slater and Morison. From Slater (1989). Copyright © 1989 Psychology Press.

The ability to categorise exemplars as similar is an important process for cognition. The categorisation of exemplars as similar suggests that a generalised representation or *prototype* has been formed, to which subsequently presented stimuli can be compared. One idea prevalent in adult cognition is that the use of prototypes enables an organism to store maximal information about the world with the minimum cognitive effort (e.g., Rosch, 1978; see Chapter 5). If we were unable to impose categories on the perceptual world, then every percept, object or event that occurred would be processed as if it were unique. This would produce an overwhelming amount of information. The ability to organise incoming information into categories is thus essential for cognitive activity. Habituation studies have used a variety of stimuli to discover whether babies can form prototypes of objects. As noted in the Foreword, prototype formation may be the automatic outcome of neural mechanisms for encoding information. For example, the neural activation patterns corresponding to features that *repeat* across examples, such as the vertex of the cross, would be strengthened via statistical learning.

As another example, suppose that you showed a baby a number of pictures of different stuffed animals. You might show a picture of a stuffed frog, a picture of a stuffed donkey, a picture of a stuffed alligator, a picture of a stuffed bear, and so on. Although these exemplars would differ in numerous features, some features would be common to all of the exemplars. Accordingly, infants may be able to abstract a category like 'stuffed animals' from seeing these different instances, in which case they should eventually habituate to these changing exemplars. By the time they saw their fifteenth stuffed animal, even if it was a novel stuffed octopus, they might find the 'stuffed animals' category rather *too* familiar, and show habituation of looking.

Cohen and Caputo (1978) carried out an habituation experiment that was very similar to the one just described. They used three different groups of babies,

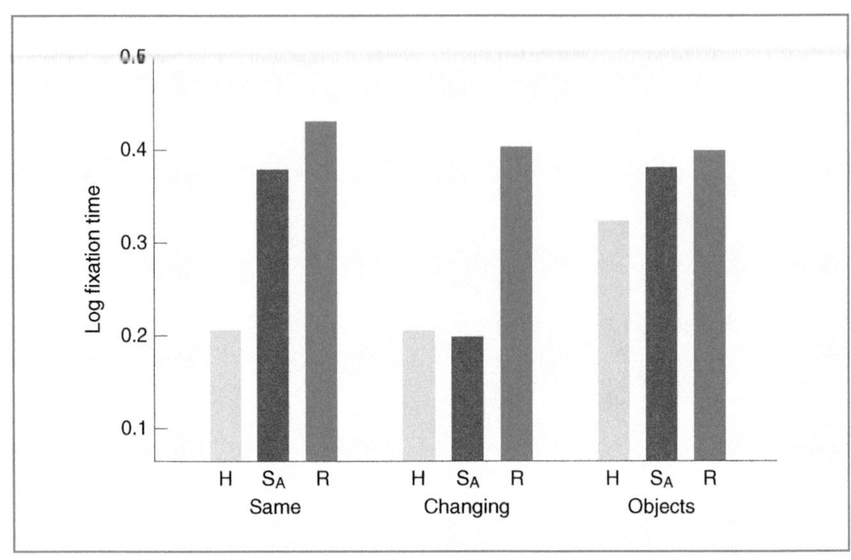

FIGURE 1.9 Looking time on the last habituation trial (H) and the first dishabituation trials with the novel stuffed animal (SA) and the rattle (R) in Cohen and Caputo's (1978) experiment with three groups of babies: those shown the same stuffed animal (Same), those shown different stuffed animals (Changing) and those shown totally unrelated objects (Objects). Figure from Younger and Cohen (1985). Copyright © 1985 Academic Press. Reproduced with permission.

all aged seven months. The first group saw the same stuffed animal on each trial of the habituation phase of the experiment. The second group saw a different stuffed animal on each trial and the third group saw a set of totally unrelated objects (e.g., a toy car, a ball, a stuffed animal, a telephone). At test, the infants were shown a novel stuffed animal and a rattle. The first group showed dishabituation to both the novel stuffed animal and the rattle. The second group showed dishabituation to the rattle only, and the third group (who in any case had shown little habituation) showed no dishabituation. This pattern of results is shown in Figure 1.9. Cohen and Caputo argued that the second group had abstracted a category of 'stuffed animals'.

Processing interrelations between features: The differentiation of prototypes

In order to argue that the infants were abstracting a prototypical 'stuffed animal' from all of these instances, we would need evidence that they were attending to the *interrelations* between the different features of each stuffed animal (the statistical co-occurrence patterns), rather than habituating to a single recurring feature, such as the animals' eyes. If infants can encode the perceptual structure of objects in terms of the correlational structure between different features, then this would be good evidence for representation on the basis of perceptual prototypes. In fact, Rosch (1978) has argued that humans divide the world into objects and categories on just such a correlational basis. Certain features in the world tend to co-occur, and this co-occurrence specifies natural categories such as trees, birds, flowers and dogs (see also Chapter 5). For example, birds are distinguished from dogs partly because feathers and wings occur together, whereas fur and wings do not. According to Rosch, this process of noticing co-occurrences between sets of features results in a generalised representation of a prototypical bird, a prototypical dog and so on, and it has been argued that these perceptual prototypes provide an important basis for *conceptual* representation.

FIGURE 1.10
Examples of the cartoon animals used by Younger and Cohen (1983). Reproduced with permission from Blackwell Publishing.

Younger and Cohen (1983) carried out some of the first experiments on statistical learning, framed as prototype theory. They examined whether infants were able to attend to the interrelations between features as prototype theory required. Younger and Cohen designed an habituation study based on 'cartoon animals' to study this question (see Figure 1.10). The cartoon animals could vary in five attributes: shape of body, shape of tail, shape of feet, shape of ears and shape of legs. There were three different forms of each attribute (e.g., the feet could be webbed feet, paws or hooves). During the habituation phase of the experiment, the babies were shown animals in which three critical features varied. Two of them varied together, and the third did not. For example, long legs might always occur with short necks, but tails could be any shape. Following habituation, the babies were shown three different cartoon animals. One was an animal whose critical features maintained the correlation. The second was an animal whose critical features violated the correlation. The third was an animal with completely different features. Younger and Cohen found that ten-month-old babies showed dishabituation to the second and third animals, but not to the first. This result suggested that the babies were sensitive to the relationship between the different critical features. They had formed a prototype of an animal with a short neck and long legs.

One way to test whether the infants really were encoding the correlational structure between the different features is to show different infants different sets of correlations between features, and then see whether they form different prototypes. Younger (1985) devised an ingenious method to enable such a test. She reasoned that if babies were shown cartoon animals in which all possible lengths of necks and legs could co-occur, then they should form a prototype of the *average* animal. As the different features would be uncorrelated with each other, the infants should abstract a prototypical animal with an average-length neck and average-length legs. However, if they were shown animals in which neck and leg length co-varied in two clusters, for example long legs and short necks and vice versa, then they should form two different prototypes. One would be of animals with long legs and short necks, and one of animals with short legs and long necks.

In order to test her hypothesis, Younger used cartoon animals whose leg and neck lengths could have one of five values (e.g., 1 = short and 5 = long). Infants in a *broad* condition saw animals in which all possible lengths co-occurred except for length 3 (the average value), and infants in a *narrow* condition saw animals in which short legs went with long necks (1, 5) and vice versa. At test, Younger found that the infants in the broad group preferred to look at cartoon animals with either very short legs and very long necks, or very long legs and very short necks (see Figure 1.11). In contrast, the infants in the narrow group preferred to look at cartoon animals whose legs and necks were of average length (3, 3). This suggests that the infants in the broad group found the average familiar, even though they had never seen those particular attributes before. They had abstracted a prototypical animal with an average-length neck and average-length legs. The infants in the narrow group had formed *two* prototypes, and thus found the average animal novel. Younger (1990) went on to demonstrate that babies were also sensitive to correlational structure when stimuli were based on features taken from real animals ('natural kinds').

The use of more natural categories and real features to study prototype formation is important, as the correlational structure of objects in the real world is quite complex. Recently, developmental psychologists have begun to study whether infants can form prototypes of natural kinds, such as cats, horses, zebras and giraffes. This work is relevant to infant understanding of the core domain of biology, and will be considered in Chapter 5.

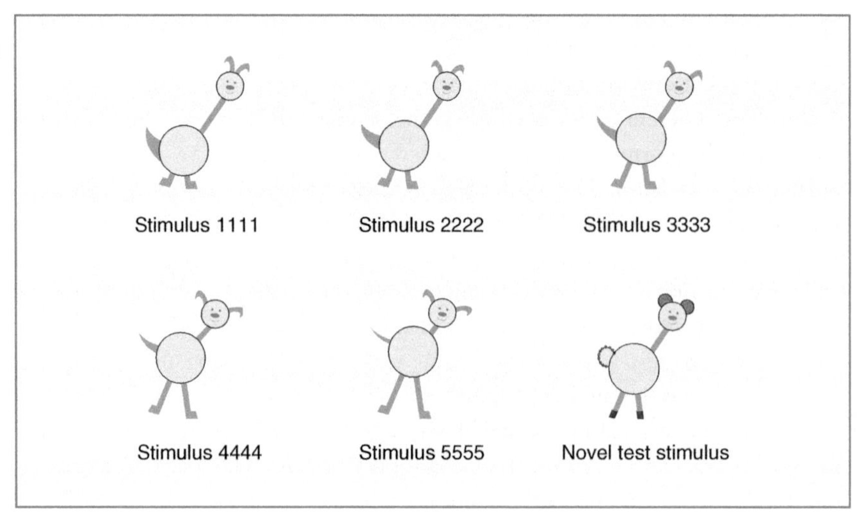

FIGURE 1.11
Some of the cartoon animals used by Younger (1985). Reproduced with permission from Blackwell Publishing.

Prototypes and statistical learning in infancy

The cartoon animal experiments demonstrated that infants could encode the correlational structure between the different visual features being manipulated by the experimenters. This suggests a form of statistical learning. In effect, the infants were learning about statistical patterns. They were learning which features co-occurred together. Recently, there has been an explosion of interest in infants' ability to track statistical patterns, particularly in the auditory domain (the auditory work is discussed in Chapter 6). However, the same questions can be asked in the visual domain. If the ordering of certain objects in the visual world follows a pattern, will infants track this pattern and show dishabituation when it is violated? This question was studied by Scott Johnson and his colleagues, with infants as young as two months of age.

Kirkham, Slemmer and Johnson (2002) created a visual habituation task based on simple coloured geometric shapes. These were presented as a continuous stream by a computer monitor. The participating infants were aged two, five and eight months of age. For example, they might see a blue cross for one second, followed by a yellow circle for one second, followed by a green triangle for one second, and so on. Visual attention to the stream of objects was maintained by having the objects 'loom' at the infant (essentially this means that the objects increased in size from 4 to 24 cm in height during presentation). The order in which the shapes were seen by a particular infant was varied, so that certain pairs of objects always followed each other. For example, a blue cross might always be followed by a yellow circle. Hence the transitional probability that when a blue cross was on screen, the next shape would be a yellow circle was 1.0. Each infants saw a stream of six shapes, with three pairings. This meant that the transitional probability of the next shape after the yellow circle was 0.33. For example, if this particular infant was also seeing the pairs 'green triangle, turquoise square' and 'pink diamond, red octagon', then the likelihood that the yellow circle would be followed by a green triangle was 0.33, the likelihood that it would be followed by a pink diamond was 0.33 and the likelihood that it would be followed by a blue cross was 0.33. The stream of shapes continued for up to 90 seconds per trial for the two-month-olds, and for up to 60 seconds per trial for the older infants.

Following habituation to the stream of shapes, the infants saw six test displays. Half of these comprised the familiar sequence, and half were a novel sequence of new orderings produced randomly by the computer. The only difference between the familiar and novel sequences lay in the transitional probabilities between the shapes. This ensured that any looking time differences at test would depend on the statistical structure governing the sequence. Kirkham et al. (2002) found that all groups looked significantly longer at the novel sequence. The two-month-olds were as good at detecting novelty as the older infants. As there was no *a priori* relationship between the geometric shapes to provide information for co-occurrence, Kirkham et al. argued that they had demonstrated a true sensitivity to transitional probabilities in very young infants.

Again, we see that infants have impressive abilities regarding computing the statistical structure in the input (see also Fiser & Aslin, 2002). The visual input

KEY TERMS

Prototype formation The formation of an internal prototypical or generalised representation of a class of stimuli.

Statistical learning Using the regularities in input to learn which features co-occur together.

structure in this experiment is quite arbitrary. It is not supported by rudimentary conceptual relations such as 'instance of a cartoon animal'. This experiment with geometric shapes suggests that infants have brains that are able to compute environmental structure at a fairly *abstract* level. Indeed, in later work, Kirkham and Johnson have referred to this visual statistical learning as 'learning without the intention to learn', a form of implicit learning that just happens to have a cognitive pay-off (see Jeste et al., 2015). In terms of developmental cognitive neuroscience, all early learning could be conceptualised as *learning without the intention to learn* (although note that infants actively choose which events to pay attention to). The superlative pattern learning mechanisms used by the brain across sensory systems indeed have a cognitive pay-off. Statistical learning mechanisms appear to be fundamental to the development of cognitive systems. The automatic nature of these mechanisms also carries a potential cognitive cost, however. If the sensory input to these neural statistical mechanisms (e.g., the visual features processed by the infant's visual cortex) is degraded or atypical in some way, the resulting cognitive system is also likely to be degraded or atypical in some way. This may affect unexpected areas of cognition. For example, as in the infant ASD paradigm discussed earlier in this chapter, early visual processing may affect social cognitive development (Cheung et al., 2018).

Clearly, if the infant brain automatically encodes and computes statistical patterns in the environment, particularly patterns relating to spatio-temporal structure (i.e., features designating movement and change in the environment), this would be a powerful mechanism for learning about the world. When does this ability come online? The answer is that it is probably online from birth. For example, Bulf, Johnson and Valenza (2011) adapted the geometric shapes paradigm to test newborn infants. Bulf and colleagues created a 'low demand' version of the geometric shapes paradigm, in which newborn infants were shown four geometric shapes in a probabilistic visual sequence. Following habituation to a particular sequence of covariation, infants were shown both a novel instance of this structured visual input along with a randomly ordered sequence of the same shapes. Although both stimulus streams were novel, the infants showed a significant looking preference for the random sequence. Bulf et al. (2011) concluded that visual statistical learning was functional at birth. The infants had no prior knowledge of the geometric shapes nor of the prior probabilities by which they were related (indeed, a computer randomised the particular shape pairings learned by each infant). Nevertheless, the newborn infants' brains automatically learned the statistical structure inherent in the pairings of the shapes. When six shapes were used instead of four shapes, the neonates studied by Bulf et al. (2011) were not able to discriminate the random sequence from the structured sequence of shapes. This suggests that limited cognitive resources (i.e., the ability to discriminate six shapes and store information about the sequential pairings between these six shapes, so perhaps infant working memory limitations) will also play a role in the effectiveness of this innate statistical learning mechanism. This is unsurprising, and does not reduce the importance of this study regarding theories about infants' innate expectations/innate constraints on learning.

One plausible interpretation of Bulf et al.'s data is that the neural mechanisms for visual information processing operate in ways that automatically learn statistical patterns as soon as visual information is available (i.e., outside the womb). For example, groups of cells that respond to each shape in a geometric sequence will be activated each time that shape occurs, and if that shape always occurs with another shape, then this information will also be recorded by either these or by adjacent cell networks (neural association networks). So neural mechanisms for processing the sensory world (here, the visual world) will by their very nature of operation automatically learn probabilities of co-occurrence. Indeed, some recent theories of adult consciousness (based on computational modelling studies) have proposed that the brain automatically encodes *trajectories* with spatio-temporal structure to represent experiences (in our example, the infant brain would be encoding the statistical temporal structure in the input as a neural *pattern of activity*, rather than only encoding each shape piecemeal and separately encoding the probabilities of its occurrence with each of the other shapes). In the adult consciousness literature, it is these 'dynamical activity space' *trajectories* that are the basis for our conscious experience of the world (Fekete, 2010; Fekete & Edelman, 2011). The core idea is that neural representation may rely upon transient activity patterns extended in time that *inherently* carry causal constraints regarding their instantiation (such as the causal constraint that shape A is always followed by shape B). Note that there are neural processes, for example chemical transmitters such as calcium cascades, that can operate at much longer timescales than a neuron which fires an action potential and then recovers. Accordingly, neural encoding mechanisms *in themselves* could impose causal structure on perceptual experiences.

Clearly, this has yet to be shown directly in adult cognitive neuroscience studies, let alone in studies with infants. Nevertheless, the trajectories idea is important. As we will discuss later, the auditory system has similar statistical learning capabilities to the visual system, which are also online from birth, and which by necessity record patterns experienced over time (as auditory statistical information can only be experienced over time). Hence in both the visual and auditory domains (and possibly also the motor domain, as we will see in Chapter 4), the sensory world of the infant is given causal structure simply by the mechanisms of sensory information processing inherent to the human brain. These mechanisms rapidly encode a great deal of structured information, thereby creating 'expectations' or 'constraints' regarding what is learned next. One of the main jobs of the brain is *prediction*, working out what is likely to happen next to prepare us for action. Again, new techniques in cognitive neuroscience could be used developmentally to document such learning with considerable precision (for example, multi-voxel pattern analysis, MVPA). MVPA is however a challenging methodology, and no such studies are yet available. Instead, researchers have recorded EEG during visual statistical learning in the geometrical shapes paradigm (albeit with four-year-olds rather than infants, see Jeste et al., 2015). A reliable neural correlate of learning has been demonstrated, related to the amplitude of the event related potential called the N1. However, the N1 is an averaged measure across the brain, and in itself does not reveal anything specific about underlying neural mechanisms.

Clearly, the brain's ability to track statistical patterns provides a very powerful domain-general learning mechanism for extracting information about structure from the physical world. We can now consider direct evidence regarding infant encoding and representation of *classes of events* in the physical world. Many such events have predictable structure. How much of this predictability is recognised by the infant?

THE PERCEPTUAL STRUCTURE OF THE VISUAL WORLD

Events in the visual world are usually described by *relations* between objects (such as football *collides with* goalpost, child *pushes* truck). The ability to detect and track these structural regularities in the object relations that are inherent in events, which occur over time, would confer great cognitive power. In part, this is because the perceiver of such events is frequently receiving *causal* structural information.

The detection of regularities in causal relations like *collide*, *push* and *supports* between different objects can also be described in terms of *classes* of event, such as 'occlusion', 'containment' and 'support' (see Baillargeon, 2001, 2002). Similarly, other types of relations, such as spatial relations (*above* and *below*) and quantitative relations (*more than* and *less than*) may also be apparent during physical events. One way of measuring infants' ability to process and represent spatial, numerical and causal relations is to introduce *violations* of typical regularities in the relations between objects, which then result in physically 'impossible' events. This is known as the *violation of expectation* paradigm, and has been widely used to study infant cognition (Baillargeon, 2004). For example, an object with no visible means of support can remain stationary in mid-air instead of falling to the ground. The experimental investigation of infants' ability to detect such violations provides an important way of measuring their ability to encode relations between events and to represent the causal structures underlying these relations.

KEY TERM

Violation of expectation paradigm Infants are shown a physical event and then on test trials shown events that are either incompatible (thus, violating expectation) or compatible with the event. Longer looking at the impossible event is taken as evidence that the infants understand the physical principle involved.

Encoding spatial relations

One way to test whether infants are sensitive to spatial relations is to use habituation. For example, if an infant is shown a variety of stimuli which are all exemplars of the same spatial relation, and if the infant shows habituation to these stimuli, then the infant must be sensitive to relational information. If the infant is then shown an example of a *new* spatial relation, dishabituation should occur. This method was used in an experiment by Quinn (1994). He familiarised three-month-old infants to the spatial relations *above* and *below*. This was achieved by showing half of the infants repeated presentations of a black horizontal bar with a dot above it in four different positions, and half of the infants a black horizontal bar with a dot below it in four different positions. These patterns provided exemplars of the spatial relation *above* and the spatial relation *below* respectively. At test, the infants were shown a novel exemplar of the familiar relation (a dot in a new position above or below the bar, depending on the habituation condition), and an exemplar of the unfamiliar relation (a dot on the other side of the bar). Both groups showed a visual preference for the *unfamiliar* relation. This finding suggests that infants can categorise perceptual structure on the basis of spatial relations.

Experiments based on the spatial relations between dots and lines may appear to provide rather impoverished tests of relational encoding and processing. However, there is evidence that infants show the same abilities with far more complex stimuli. For example, Baillargeon and her colleagues investigated whether infants of 5.5 months realised that a tall rabbit should be partially visible when it passed behind a short wall. During the habituation phase of the experiment, the infants saw a display of a tall painted 'wall' (Baillargeon & Graber, 1987). A rabbit appeared at one end of the wall, passed along behind it, and reappeared at the other end. This 'habituating' rabbit could either be tall or short, but as both the tall and the short rabbit were too small to be visible when they were behind the wall, the infants watched the rabbits disappear and reappear as they moved from left to right. At test, the mid-section of the wall was lowered. The wall now had two tall ends and a short middle (see Figure 1.12). The short rabbit could still pass behind the entire length of the wall without being visible, but the tall rabbit could not. The tall rabbit's head would appear as it passed behind the middle section of the wall.

Both groups of infants then again watched the habituating rabbit (tall or short) passing behind the wall. In fact, they saw the *same* event to which they had been habituated. For the 'small rabbit' group, the failure of the rabbit to appear in the mid-section of the apparatus was perfectly acceptable in terms of the spatial relations involved, and accordingly there was no dishabituation. For the 'tall rabbit' group, the test event was not acceptable in terms of the spatial relations involved – in fact, it was physically impossible. The tall rabbit's head should have appeared behind the mid-section of the wall, but it did not – just as in the habituating event. Baillargeon and Graber found that the babies in the 'tall rabbit' group spent much longer staring at the experimental apparatus than the babies in the 'short rabbit' group. The infants' increased looking time at the non-appearance of the tall rabbit suggests that they had encoded the spatial relations between the wall and the rabbit. Later work (Baillargeon & DeVos, 1991) has shown that 3.5-month-old infants

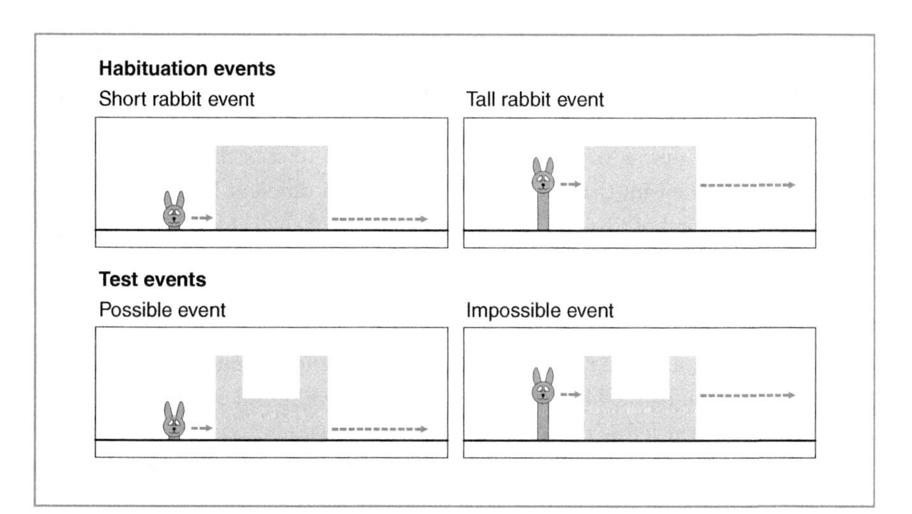

FIGURE 1.12 The habituation and test displays in the tall and short rabbit experiment devised by Baillargeon and Graber (1987). Copyright © 1987 Elsevier. Reprinted with permission.

behave in the same way (this was demonstrated in a modified version of the experiment, which used a tall and a short carrot). Thus very young babies appear to be able to represent spatial relations such as relative height, at least in an occlusion paradigm.

Baillargeon and her colleagues have also used habituation to measure infants' *memory* for spatial locations. This is a strong test of representation, as the infants must retain the spatial relations defining location *over time*. In one experiment, Baillargeon and Graber (1988) showed infants a display that had two possible locations in which a toy could be placed, A and B. The two locations were marked by identical mats. As the infants watched the display, an attractive object was placed at location A (in fact, the object used was a plastic styrofoam cup with matches stuck into its sides, an object that the infants found far more visually interesting than the toys that were used when the experimenters tried to pilot the experiment!). Two screens were then slid in front of the two locations, hiding the mats. As the infants continued to watch the display, a hand wearing a silver glove and a bracelet of bells appeared and the fingers danced around – this was also visually interesting, and was designed to keep the infants attending to the display. The hand then reached behind the screen at location B, and retrieved the styrofoam cup.

Of course, this retrieval was an 'impossible' event. Location B had been visibly empty when the screens slid in front of the mats, and the styrofoam cup should only have been retrievable at location A. Baillargeon and Graber argued that if the babies could remember the location of the object during the delay, then they would be perturbed at this event, and should show increased looking at the display. This was exactly what they found. The babies stared at the impossible retrieval, and looked at the display for a long time. Increased looking time did not occur in a control event, which was a 'possible' event. In this event, the hand retrieved the cup from behind the correct screen, and the infants were not particularly interested. The fact that their attention was caught only when the cup was retrieved from the wrong spatial location suggests that they were able to represent the location of the cup even when it was out of view. Baillargeon, DeVos and Graber (1989) went on to demonstrate that eight-month-old infants could retain these spatial memories for up to 70 seconds. So 'out of sight' is not necessarily 'out of mind' for the infant.

A different test of spatial learning and memory was devised by McKenzie, Day and Ihsen (1984). They seated six- to eight-month-old babies behind a kind of semi-circular 'newsdesk' (see Figure 1.13). The babies sat on their mothers' laps in a central position (like a 'newsreader'), enabling them to scan the entire desk. The shape of the desk meant that there were a number of different locations at which events could occur, both to the left and to the right of the babies. The location at which an event was about to occur was always marked by a white ball. The events were visually exciting to the babies – an adult appeared from behind the desk and began playing 'peek-a-boo'.

KEY TERM

Coding of spatial position
Can occur either in relation to one's own position in space (egocentric) or to external landmarks (allocentric).

McKenzie et al. found that the babies quickly learned to anticipate an event at the spatial location marked by the white ball. As the white ball could either appear to the right or to the left of the midline, the babies could not have learned a specific motor response, such as turning their heads to the right. Instead, they were learning to *predict* the spatial location of the visual events by using the white ball. McKenzie

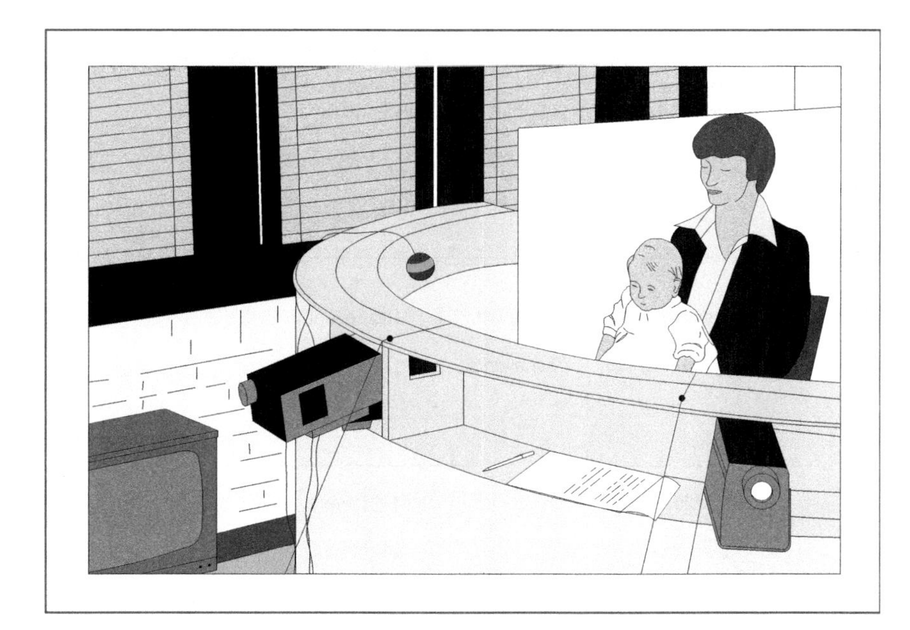

FIGURE 1.13
The experimental set-up used in the 'newsreader' experiment to study spatial learning and memory by McKenzie et al. (1984). Copyright © The British Psychological Society. Reproduced with permission.

et al. argued that this showed that babies did not always code spatial position in memory *egocentrically*, with respect to a motor response based on their own position in space. When given the appropriate opportunity, they could also code spatial location in memory *allocentrically*, with respect to a salient landmark such as the white ball. The representation of spatial relations in eight-month-olds thus involves landmark cues, just as it does in adults.

Encoding occlusion relations

So far, we have considered evidence that babies can use the perceptual structure of events in the visual world as a basis for encoding and representing relational knowledge about space. However, perceptual events can also provide infants with knowledge about the continued existence of objects when they are out of view. When an object is occluded by a second object, we as adults believe that it still exists. Even when one object totally occludes another, we assume that the hidden object continues to exist and to occupy the same location in space behind the occluder.

Babies seem to make similar assumptions about the existence of occluded objects. One of the most ingenious demonstrations of their belief in 'object permanence' comes from an experiment by Baillargeon, Spelke and Wasserman (1985). Baillargeon et al. habituated five-month-old babies to a display in which a screen continually rotated through 180° towards and away from the baby, like a drawbridge (see Figure 1.14). Following habituation, a box was placed in the path of the screen at the far end of the apparatus. As the screen began its 180° rotation, it gradually occluded the box. When it reached 90°, the entire box was hidden from view. For babies who were shown a 'possible event', the screen continued to rotate until it had passed through 120°, at which point it came to rest, apparently having made contact with the box. For babies who were shown an 'impossible

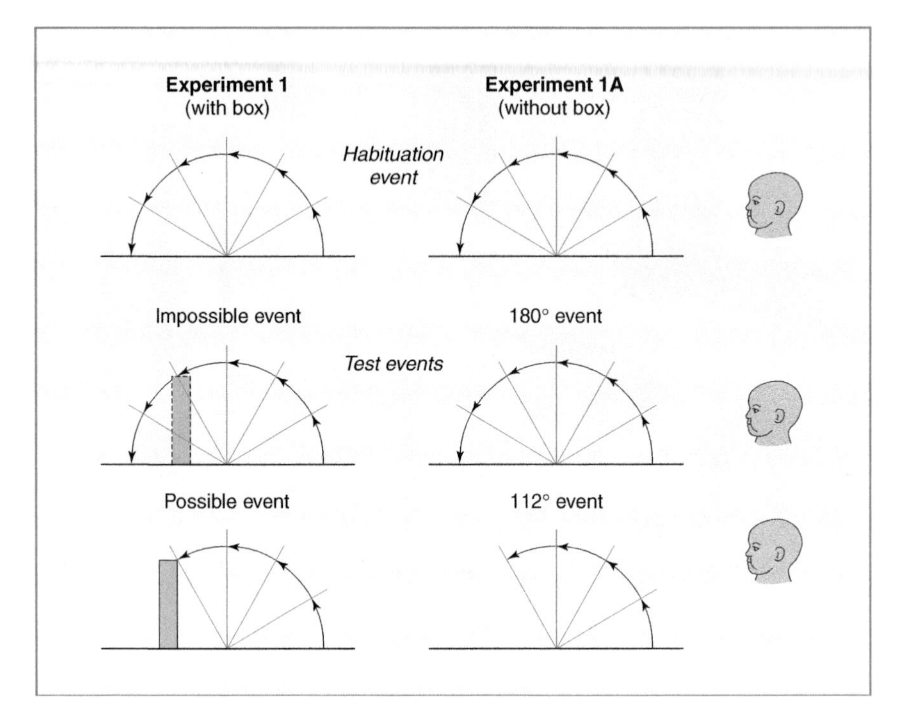

FIGURE 1.14
Diagram of the habituation and test events in the rotating screen paradigm. From Baillargeon et al. (1985). Copyright © 1985 Elsevier. Reprinted with permission.

event', the screen continued to rotate until it had passed through the full 180° rotation. In the physically 'impossible' condition, the box had apparently caused no obstruction to the path of the screen's movement. Although the 180° rotation was the familiar (habituating) event, the babies in the impossible condition spent much longer staring at the experimental display than the babies in the possible condition (who were seeing a novel event). This finding suggests that the babies had represented the box as continuing to exist, even when it was occluded by the screen. They looked longer at the display when the screen passed through an apparently solid object.

In later work, Baillargeon has shown that babies as young as 3.5 months of age look reliably longer when the screen passes through the box, particularly if they are 'fast habituators' (Baillargeon, 1987a). She has also shown that infants can represent some of the physical and spatial properties of the occluded objects, such as whether an object is compressible or not (e.g., a sponge versus a wooden block, see Figure 1.15), and whether it is taller or shorter than the height of the screen (e.g., a wooden box measuring 20 × 15 × 4 cm standing upright versus lying flat; Baillargeon, 1987b). These experiments suggest that not only can young infants represent the existence of hidden objects, they can also represent some of the specific properties of the objects that are hidden. They can then use these physical and spatial characteristics to make predictions about how the drawbridge should behave as it begins to rotate.

Despite the many variations of the 'drawbridge' paradigm that Baillargeon has devised, her use of a rotating screen to demonstrate infants' belief in object permanence has proved to be a controversial one. For example, it has been argued that

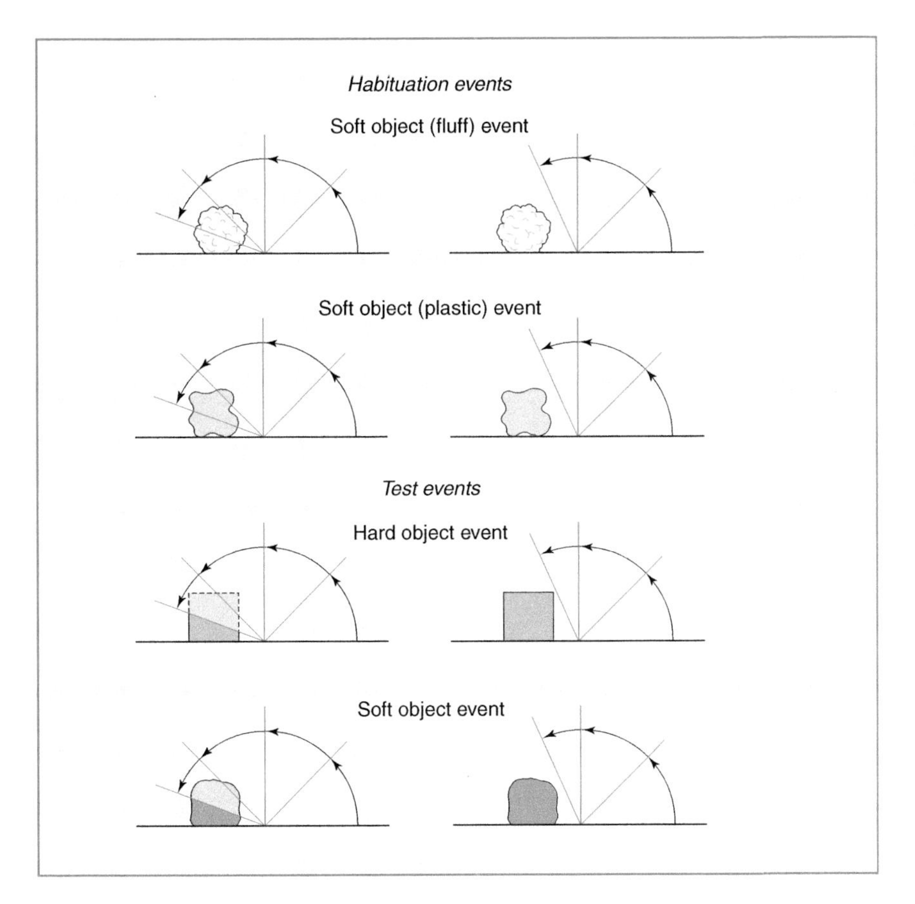

FIGURE 1.15 Diagram of the habituation and test events in the rotating screen paradigm with soft versus hard objects. From Baillargeon (1987b). Copyright © 1987 Elsevier. Reprinted with permission.

the perceptual structure of events in the 'drawbridge' paradigm leads the infants to form a strong expectation that the drawbridge should stop, an expectation that does not necessitate a representation of the occluded object. This criticism is weakened by the demonstration that infants' expectations about the behaviour of the drawbridge differ depending on the nature of the object that is hidden (e.g., Baillargeon, 1987b). Furthermore, the series of drawbridge studies that Baillargeon and her colleagues have conducted are only one piece of evidence that babies represent hidden objects as continuing to exist. A different paradigm, also devised by Baillargeon (1986), tests the same understanding, and does not seem vulnerable to an 'expectation' criticism at all.

This paradigm was based on a toy car and a ramp. During the initial phase of the experiment, 6.5-month-old infants were shown a display in which a toy car was poised at the top of a ramp. A track for the car ran down the ramp and along the base of the apparatus. When the infants were attending to the apparatus, the middle section of the track was hidden by lowering a screen, and the habituation phase of the experiment began. The car ran down the ramp, passed behind the screen, and reappeared at the end of the apparatus. Following habituation to repeated presentations of this event, the screen was raised and a box was placed either *on* the car's track, or behind it. The screen was then lowered again, hiding the box, and the car began its journey. The apparatus used is shown in Figure 1.16.

KEY TERM

Continuity principle
Objects exist continuously in time and space.

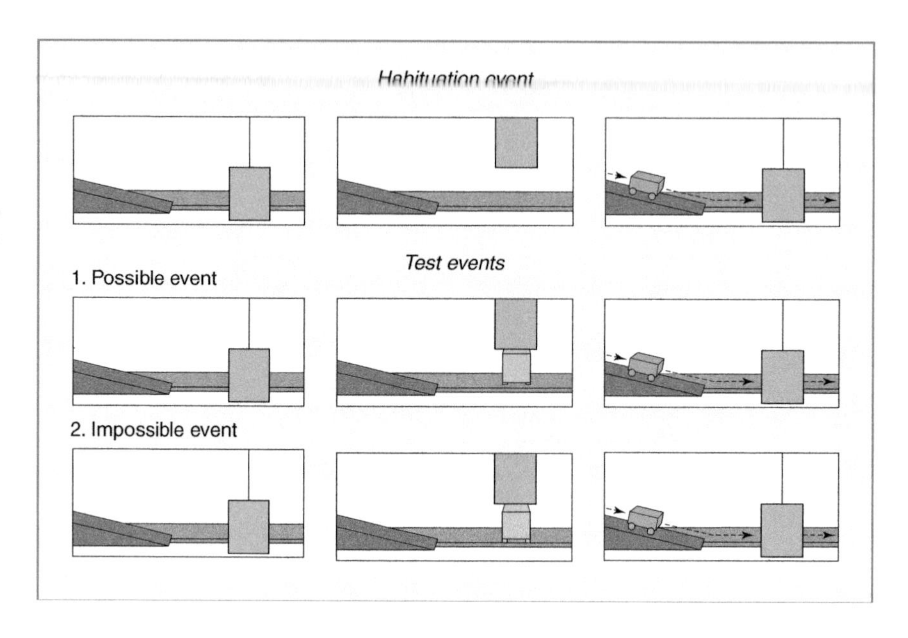

FIGURE 1.16
Depiction of the habituation and test events in the car on the ramp paradigm. From Baillargeon (1986). Copyright © 1986 Elsevier. Reprinted with permission.

All the babies then saw exactly the same set of events as during the habituation phase of the experiment. For babies in the 'possible' condition, the box was behind the track and out of the car's path, and so the reappearance of the car was not surprising. For babies in the 'impossible' condition, however, the box was *on* the track, directly in the path of the car – and yet the car *still* reappeared in the familiar way! The babies in the impossible condition spent much longer staring at the apparatus than the babies in the possible condition. The only explanation was that they had represented the box as continuing to exist and as therefore blocking the car's path, and so they were intrigued by the reappearance of the car. Baillargeon and DeVos (1991) later demonstrated that babies as young as 3.5 months looked for a reliably longer time when the car reappeared despite the fact that a hidden object (a Mickey Mouse doll) was blocking its path.

This paradigm was subsequently used by Kotovsky and Baillargeon (1998) to investigate babies' understanding of collision events more directly. They explored infants' expectations in situations where a stationary object was hit by a moving object. The moving object was a cylinder that varied in size, and the stationary object was a bug on wheels. When the cylinder collided with the bug, it set the bug in motion. As adults, we would expect the distance travelled by the bug following this collision to depend on the size of the cylinder. The impact of a larger cylinder should cause the bug to roll further. Kotovsky and Baillargeon explored whether babies aged from 5.5 months also expected there to be a proportional relation between the size of the cylinder and the distance travelled by the wheeled bug. During habituation trials, the infants were shown the ramp, and a stationary wheeled bug sitting at the bottom of the ramp on the track. No occluders were used. As the infants watched, a medium-sized cylinder ran down the ramp, collided with the bug, and set it in motion. The bug ran halfway along the track and then stopped. In the novel event, either a larger or a smaller cylinder was used. Both cylinders propelled the bug to the *end* of the track. While this was a possible event for the larger cylinder, it should have been impossible for the smaller cylinder. The size of the cylinder should have affected the bug's trajectory. Kotovsky and Baillargeon (1998) found that 6.5-month-old babies and 5.5-month-old female babies looked reliably longer at the small cylinder event than at the large cylinder event. They argued that the babies were engaging in calibration-based reasoning about the size/distance relations in the perceptual display.

Another type of occlusion event has also been the focus of investigations by Baillargeon's group. Hespos and Baillargeon (2001) studied the looking behaviour of babies when potential containers were used as occluders. For example, either a tall or a short container made of PVC piping was used to occlude a brightly coloured cylindrical object with a knob on the top. The tall container completely concealed the object, with just the knob on the top remaining visible. However, the short container was only about half as high as the object. Hence the top half of the object should have remained visible behind this short occluder. In fact, via the surreptitious use of two objects, the visual events seen by the babies were the same. When the object was lowered behind the short occluder, it also became completely hidden, apart from the knob on the top. Of course, this violated the expectation that the object could not become fully hidden by the short occluder. Hespos and Baillargeon (2001) reported that babies as young as 4.5 months looked reliably longer at the impossible event. They apparently realised that the height of an object relative to the height of an occluder will determine whether the object will be fully or only partly hidden behind the occluder.

Encoding support relations

Another set of perceptual relations that are commonly encountered in the physical world are the relations involved in support. Adults are well aware that if they put a mug of tea down on a table and the mug protrudes too far over the edge, then the mug will fall onto the floor. However, if only a small portion of the bottom surface of the mug is protruding over the edge of the table, then the mug will have adequate support and the tea can be drunk at leisure. Baillargeon, Needham and DeVos (1992) investigated similar intuitions about support in young infants. They studied 6.5-month-old infants' expectations about when a box would fall off a platform.

In Baillargeon et al.'s experiment, the infants were shown a box sitting at the left-hand end of a long platform, and then watched as the finger of a gloved hand pushed the box along the platform until part of it was suspended over the right-hand edge (see Figure 1.17). For some infants, the pushing continued until 85% of the bottom surface of the box protruded over the platform, and for others the pushing stopped when 30% of the bottom surface of the box protruded over the platform. In a control condition the same infants watched the box being pushed to the right-hand end of the platform, but the bottom surface of the box remained in full contact with the platform. The infants spent reliably longer looking at the apparatus in the 85% protrusion event than in the full-contact control event. This suggests that they expected the box to fall off the platform (the

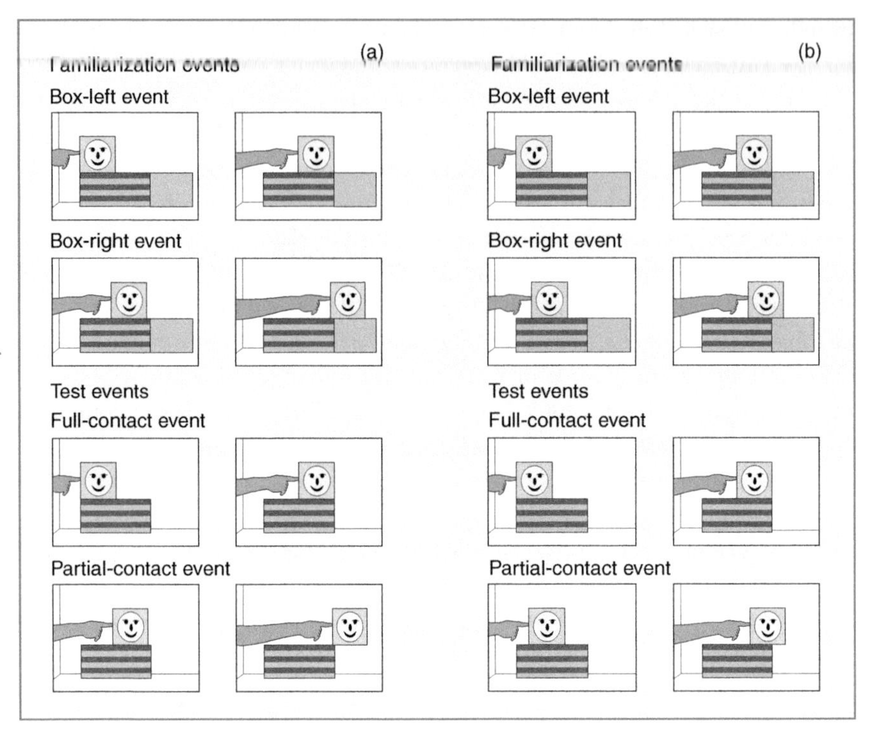

FIGURE 1.17 Depiction of the familiarisation and test events in the 'box on a platform' paradigm devised by Baillargeon et al. (1992). Panel (a) depicts the 85% protrusion event, and panel (b) the 30% protrusion event. From Baillargeon et al. (1992). Copyright © John Wiley & Sons, Ltd. Reproduced with permission.

box was able to remain magically suspended in mid-air via a hidden hand). The infants in the 30% protrusion event looked equally during the protrusion event and the control event. Baillargeon et al. argued that the infants were able to judge how much contact was required between the box and the platform in order for the box to be stable.

Interestingly, younger infants (5.5- to six-month-olds) appeared unable to make such fine judgements about support. They looked equally at the 85% and 30% protrusion events compared to the full-contact control event. Baillargeon et al.'s interpretation of this finding was that younger infants perceive *any* amount of contact between objects to be sufficient to ensure stability. They operate with a simpler causal rule that *no contact = object falls* and *partial contact = object is supported*, even when the partial contact is very partial indeed. In fact, Baillargeon argues that much physical causal reasoning develops according to this all-or-none pattern (see Baillargeon, 2001, 2002). Infants begin with representations that capture the essence of physical events (e.g., contact versus no contact), and then gradually develop more elaborate representations that identify variables that are relevant to the events' outcomes (such as degree of support). Experience of the physical world has an important role to play in this developmental sequence. For example, by around six months of age most babies have become 'self-sitters'. They are able to sit up with support, and for the first time they can be seated in high-chairs etc. in front of tables, and can deposit objects on surfaces and watch them fall off. Baillargeon has suggested that these experiences help infants to refine their understanding of the cause–effect relations underlying support. The idea that specific action experiences of this kind help in the development of specific aspects of physical reasoning will be discussed further in the next chapter.

Encoding containment relations

Another way of investigating what Spelke (1994) has termed the 'continuity principle', that objects exist continuously in time and space, is to explore infants' understanding of containment events. When an object is placed inside a container, it leaves the field of view. However, as adults, we know that the object continues to exist inside the container. Do babies share this understanding? In a series of experiments, Baillargeon, Luo, Wang, Paterson and Hespos studied infants' understanding of containment events (e.g., Hespos & Baillargeon, 2001, 2006; Wang, Baillargeon & Paterson, 2005; Luo & Baillargeon, 2005). For example, Hespos and Baillargeon (2001) designed a containment study that was analogous to the occlusion study with the cylindrical object discussed earlier. The container was a piece of PVC tubing identical in perceptual appearance to the PVC tubing occluder. During the containment events, the infants watched as the cylindrical object was lowered inside the container (rather than behind the occluder). This comparison is shown in Figure 1.18. The tall container was large enough to contain the entire object, leaving just the knob on the top visible to the infants. However, the short container was not. In this condition, the top half of the object should have remained visible when it was lowered into the container. Instead, the object disappeared until only the knob was visible, violating expectations based on object continuity. Both the tall and the short container appeared able to contain the entire cylindrical object.

Intriguingly, Hespos and Baillargeon (2001) reported that infants did not show increased looking times for the short container event until around 7.5 months of age. They did not appear to realise that the height of the container relative to the height of the object determined whether the object would be fully or only partially hidden by the container. This was surprising, as infants aged 4.5 months were able to use the relative heights of the object versus the container as a cue in the highly similar occlusion condition. In the containment condition, infants of 4.5 months, 5.5 months and 6.5 months of age did not appear aware of the importance of the height of the container. They did not look reliably longer at the short container event. Baillargeon and Wang (2002) suggested that infants treated containment events as distinct from occlusion events. They did not generalise their knowledge of a variable like height from one type of event to the other.

In a separate series of studies, Luo and Baillargeon (2005) studied infants' understanding of transparent containers. The infants were shown a plexiglass box, which was then occluded by a screen. An attractive object was then lowered into the box, although its entry into the box occurred out of view of the infant. When the screen was lowered, the infants either saw the object inside the box (possible event) or the empty transparent container (impossible event). Infants did not look for reliably longer at the empty container until ten months of age. However, if the plexiglass was used as an occluder rather than as a container (via a plexiglass screen), then infants looked longer at the physical violation at 7.5 months of age. Again, this

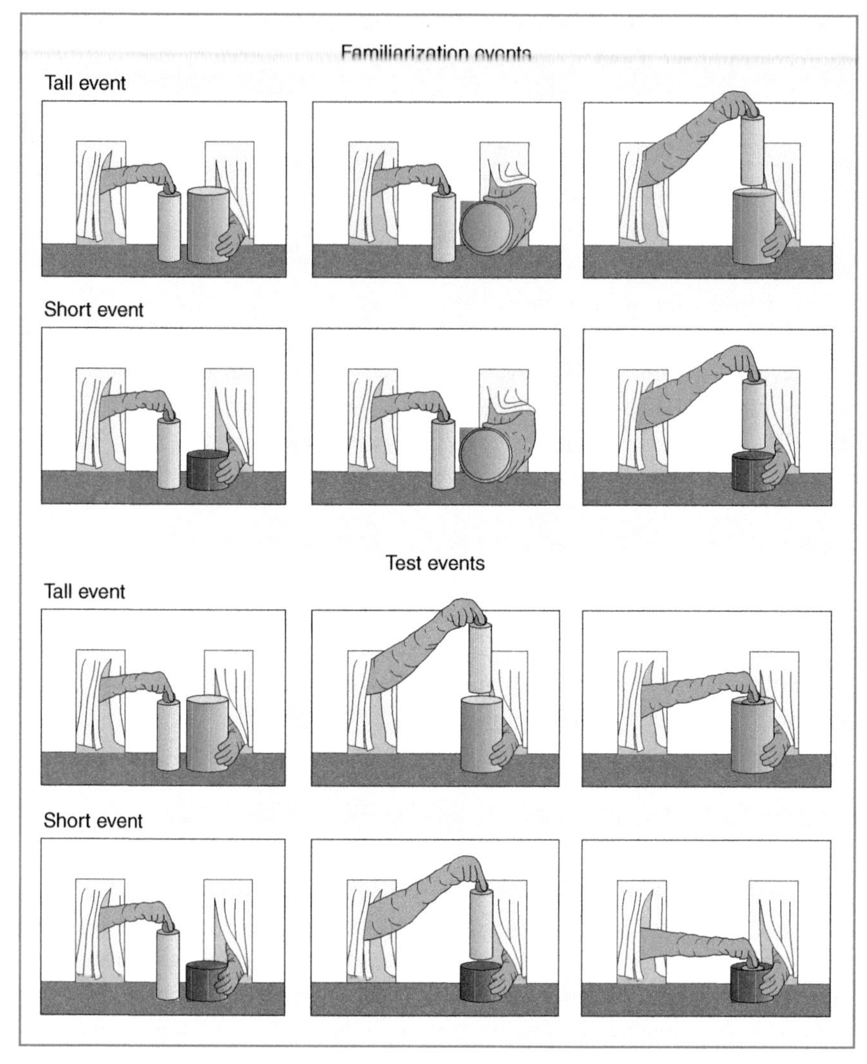

FIGURE 1.18
Familiarisation and test events in the container condition of Hespos and Baillargeon's (2001) study. Reproduced with permission from Blackwell Publishing.

difference in looking behaviour was interpreted in terms of *event categories*. Infants were apparently treating containment events as distinct from occlusion events, and were working out the perceptual variables relevant to a more elaborate representation of occlusion events earlier than they were working out the perceptual variables relevant to a more elaborate representation of containment events – even when these were the same perceptual variables. An alternative possibility is that neural encoding of the visual features of occlusion events (the neural activity patterns extended in time that *in themselves* carry causal constraints regarding their instantiation) is distinct from the neural encoding of the visual features of containment events. Accordingly, coherent causal structure could be given by the nature of the encoding for occlusion events earlier than for containment events. Again, a way forward could be to record fMRI during the two paradigms and then compare neural encoding patterns using multi-voxel pattern analysis. This would enable precise imaging of the voxels in the brain that are active during each kind of event to be compared, allowing any degree of overlap in the representations to be revealed.

Meanwhile, support for Baillargeon's interpretation based on event categories comes from a separate series of studies comparing containment with covering. When a cover is lowered over an object, the same principle of continuity applies as when an object is lowered inside a container. The object continues to exist beneath the cover or beneath the container, and various physical attributes of the object or the cover/container will determine whether any parts of the object remain visible. Wang et al. (2005) used identical tubes as covers or as containers. For example, in one experiment nine-month-old infants watched as a tall tube was used to cover a tall object, or as a tall object was lowered into a tall cylindrical container (see Figure 1.19). These were possible events. In the impossible events, the tall object was covered by a short tube, or was lowered into a short cylindrical container. In all of the events, the object became fully hidden. Wang et al. reported that the infants looked reliably longer at the unexpected event in the containment condition, but not in the covering condition. Only infants aged 12 months detected the violation in the covering condition, 11-month-old infants did not. Again, these discrepancies in behaviour were explained in terms of event categories. Covering seemed to be treated by infants as an event distinct to containment. Therefore, identical variables (such as tube height) were treated as relevant to one event category before they were treated as relevant to another. This supports Baillargeon's idea that infants sort physical events into categories, and then learn separately how each category operates. Again, this idea is amenable to investigation using developmental cognitive neuroscience multi-voxel pattern analysis methods.

Although all the experiments discussed so far have used looking behaviour as the outcome measure in the violation-of-expectation paradigm, more recently Baillargeon and her colleagues have been testing her model of physical reasoning in infancy using action-based tasks. For example, Hespos and Baillargeon (2006) exploited search behaviour to compare infant actions in containment versus

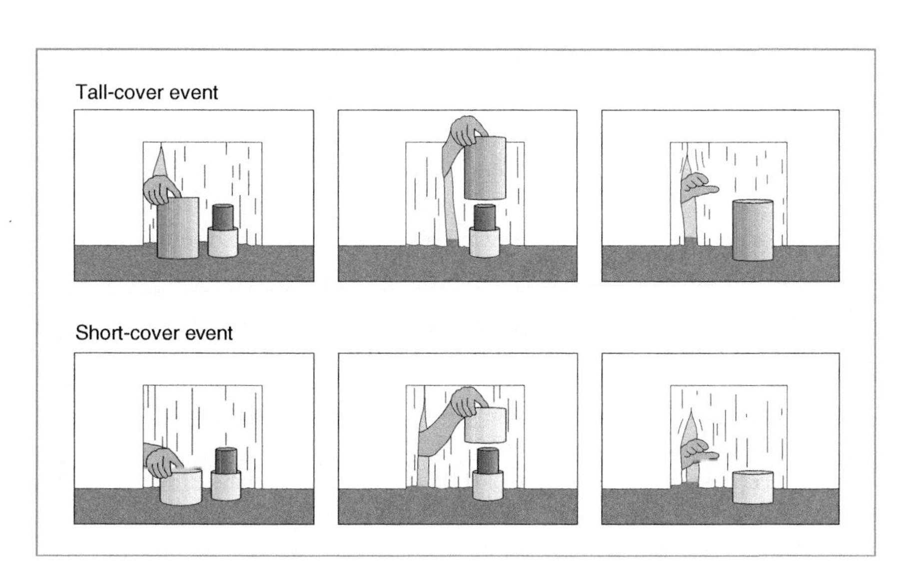

FIGURE 1.19 Possible and impossible events used in Wang, Baillargeon and Paterson's (2005) study. Top: a tall tube was used to cover a tall object (possible); bottom: a tall object was covered by a short tube (impossible). From Wang et al. (2005). Copyright © 2005 Elsevier. Reprinted with permission.

occlusion events. They studied the ages at which infants would search consistently for a toy that was either hidden behind an occluder or that was hidden inside a container. The occluders and containers were the same PVC tubes used in previous experiments. This time, infants were given an engaging tall frog toy to play with. The frog was then removed. The infants were shown a screen, and when the screen was lowered two containers or occluders were revealed, one tall and one short. Frog legs were sticking out of the bottom of each, and the infants were encouraged to find the frog. Six-month-olds and 7.5-month-olds were tested.

Based on their findings in the violation-of-expectation paradigm, Hespos and Baillargeon reasoned that infants identify the variable height as relevant at about 4.5 months in occlusion paradigms and at about 7.5 months in containment paradigms. They therefore argued that the younger infants should search for the frog behind the tall occluder in the occlusion paradigm, but should not preferentially search inside the tall container in the containment paradigm. The older infants were expected to search for the frog successfully in both paradigms. This was exactly what they found. In the occlusion condition, 12 out of 16 7.5-month-olds and 14 out of 18 six-month-olds reached for the tall occluder on three or four of the four search trials. In the containment condition, only the older infants were successful, with 12 out of 16 reliably picking the tall container compared to only two out of 18 of the younger infants. Different control conditions were used to rule out alternative explanations. Hespos and Baillargeon (2006) concluded that evidence from action tasks was consistent with evidence from the violation-of-expectation paradigm concerning physical reasoning by infants. They argued that infants form distinct event categories (occlusion, containment, support), and that they learn about each category separately. Perceptual variables that are identified in one category may not be generalised to another category, even when the variables are equally relevant.

Hespos and Baillargeon suggest that infants' physical reasoning systems are designed to acquire event-specific expectations rather than event-general principles. As noted earlier, a complementary interpretation is offered by computational modelling. Theorists in this field have proposed that the brain automatically encodes 'dynamical activity space' *trajectories* to represent each event that is experienced (Fekete & Edelman, 2011). A series of neural responses over time (a trajectory) will reliably occur when encoding containment events, and a different series of neural responses over time will reliably occur when encoding occlusion events. If these 'dynamical activity space' *trajectories* are the basis for automatic generalisation, then the neural trajectories for containment, occlusion, etc., would be distinct, explaining the lack of generalisation by the infant brain. This explanation is consistent with a basic aspect of brain function revealed by cognitive neuroscience, that of *distributed representation*. The neural networks active during the encoding of an event are typically distributed across the entire brain. The distributed representations brought to mind (re-activated) by new instantiations of events of containment or occlusion will thus always differ to some extent.

Further, Fekete and Edelman (2011) have argued that these neural activity trajectories *in themselves* carry causal constraints regarding their instantiation. Accordingly, neural activity during infants' experiences of different physical events would *in itself* create an emergent physical reasoning system, which would have inherent causal structure based on the particular experiences of the infant. This automatic outcome of neural encoding could then provide the basis for developing *cognitive* knowledge about objects and their relations: a cognitive physical reasoning system. In developing a cognitive reasoning system, motor development and the ability to carry out increasingly sophisticated actions on objects may play critical roles. Becoming able to act oneself upon objects, and being the agent who is causing different relations between them (such as putting one object inside another), may be important for developing a cognitive system from these automatic neural trajectories, and for enriching and enlarging such a system. At the time of writing, this alternative neurally driven model is consistent with the data, but still needs to be tested via experiments in developmental cognitive neuroscience. However, it is relevant to the historical debate that we will consider next, which concerns exactly what is being measured in the violation-of-expectation paradigm.

What is measured in the violation-of-expectation paradigm?

In the past two decades, a number of criticisms have appeared of studies that use habituation, visual preference and violation-of-expectation techniques to study cognitive processes in infancy. Some of these critics have been highly resistant to the notion that young infants engage in physical reasoning (e.g., Haith, 1998). Critics such as Haith point out correctly that looking paradigms were developed in order to study sensory and perceptual questions, not cognitive questions. We saw some examples of these perceptual questions at the beginning of this chapter. An important part of these critiques is the point that it is simply not possible to generate perceptually identical but conceptually distinct stimuli for habituation paradigms (see Sirois & Mareschal, 2002). Critics such as Haith argue therefore that scientists like Baillargeon must be able to discount every possible *perceptual* interpretation of differences in looking time before proposing cognitive interpretations of infant looking behaviour. In a similar vein, Bogartz, Shinskey and Speaker (1997) argued that simple perceptual mechanisms such as novelty, scanning and tracking may explain longer looking times by infants in some perceptual conditions versus others. These points are important: it is not clear that behavioural experiments will ever be able to devise conditions that unambiguously demonstrate *cognitive* representations in infants.

KEY TERM

Cognitive representations
Traditional term for presumed amodal conceptual information held 'in the mind'.

In fact, a theoretical point frequently made by such critics regards what it means to attribute *cognitive* representations to infants at all. For example, Haith (1998) assumed that to count as *cognitive*, the re-computation and transformation of sensory information into an amodal cognitive entity was required. He suggested that infants may have lingering sensory information about objects that have been (for example) occluded, and that it was this lingering sensory information rather than a conceptual representation of the object that yielded the changes in looking behaviour. Perceptual mechanisms such as familiarity, novelty and discrepancy, which operate when objects are visible, may also be operating to create longer looking times to 'impossible' events simply because infants are still operating on actual sensory information, albeit degraded information.

Indeed, related work in adult visual perception has shown that object representations in adults ('object files', or mid-level visual mechanisms for treating part of the visual field as the same object over time) persist over several seconds (e.g., Noles, Scholl & Mitroff, 2005). Object files in adults are also affected by violations of key principles of infant perception, such as cohesion (that an object must always maintain a single bounded contour, see Mitroff, Scholl & Wynn, 2004). A more detailed exploration of how the visual system constructs and maintains object files may throw light on some of the curious discrepancies in infants' use of perceptual variables such as height in the experimental paradigms devised by Baillargeon and her colleagues. In particular, the precise measurement of neural 'trajectories' in terms of which neural networks are involved (e.g., via multi-voxel pattern analysis) could be very informative. Rather than reflecting degraded sensory representations, ongoing network activity may be a necessary component of the statistically learned trajectory.

Furthermore, as noted earlier in this chapter, cognitive neuroscience is rejecting the very notion of 'amodal' cognitive entities as the basis for conceptual representations. Adult cognitive neuroscience studies suggest instead that conceptual representations necessarily depend on the activity of sensory neurons relevant to experiencing that concept (Barsalou, Simmons, Barbey & Wilson, 2003). Reprisal of neuronal activity in sensory systems, rather than redescriptions of these sensory states into an amodal representational form, appear to be how the human brain represents conceptual knowledge (*distributed representations*, see also discussion in Chapter 5). If the brain is representing the concept of a cup, for example, motor neurons activated when the cup is grasped and visual neurons activated when seeing a real cup will become active, along with neurons in various association areas that might link cups to specific contexts like breakfast or specific emotions like the pleasure one gets from drinking coffee. There may be no set of amodal 'cup' neurons that are activated by themselves. Most recently, Barsalou (2017) has discussed conceptual representation in terms of 'situated conceptualisation', exploring how this gradual learning of concepts may occur in different brain regions over time. He has also reviewed adult neuroimaging studies that use MVPA (multi-voxel pattern analysis) to measure exactly which sets of neural areas are active during the processing of any specific concept. In adult studies, it is now possible to know which concept a participant is thinking about simply by measuring which voxels are active in the brain. As noted earlier, this technique (MVPA) could be very powerful for tracking conceptual development in infants and young children. The new experimental techniques offered by cognitive neuroscience also provide us with a way forward in tackling the serious questions about the nature of infant representations raised by scientists like Haith. Meanwhile, interesting work by Stahl and Feigenson (2015) has shown that infants who observe violations of expectations regarding relations such as 'support' are more likely to investigate the objects that have violated their expectations if given the opportunity. In Stahl and Feigenson's study, 11-month-old infants who had seen a toy car suspended in mid-air were more likely to choose to manipulate it rather than a new object when given the opportunity, typically by dropping it to see if it would fall. Similarly, infants who had seen a ball apparently roll through a solid wall were more likely to choose to manipulate the ball rather than a new object when given the opportunity, typically by banging it. Stahl and Feigenson argued that these differential actions were evidence that the infants were trying to work out why the objects had behaved in ways that did not accord with their expectations.

COGNITIVE NEUROSCIENCE AND OBJECT PROCESSING IN INFANCY

A series of EEG experiments with six-month-old infants, reported by Kaufman et al. (2003a), provides a nice example of how difficult questions in cognitive development are amenable to a cognitive neuroscience approach. Kaufman et al. used EEG imaging as a way of examining what infants' 'representations' of occluded objects were actually like. Kaufman et al. recorded the EEG from electrodes placed over the whole scalp when infants were watching different disappearance events, which were either expected or unexpected (the violation of expectation paradigm). The habituation event was a toy train going into a toy tunnel. The train was shown entering the tunnel, and then reversing back out again. Following habituation, the infants watched the train enter the tunnel, and then saw either (a) a hand lifting the tunnel to reveal the train (expected appearance event); (b) a hand lifting the tunnel to reveal no train (unexpected disappearance event); (c) the train leaving the tunnel and the visual field, and a hand subsequently lifting the tunnel to reveal the train (unexpected appearance event); or (d) the train leaving the tunnel and the visual field, and a hand subsequently lifting the tunnel to reveal no train (expected disappearance event). Behaviourally, the infants looked significantly longer at the unexpected disappearance event compared to the expected disappearance event. The two appearance events were not distinguished by looking time.

KEY TERM

EEG (electro-encephalogram) The recording from the scalp of changes in brain electrical activity caused by the oscillating rhythmic electrical activity of neural networks.

Kaufman et al. (2003a) then compared the EEG signal in the gamma band (a relatively fast oscillatory response at around 40 Hz) during the lifting of the tunnel in the expected versus unexpected *disappearance* events. Much higher electrical activity at right temporal sites only was found when the train was occluded. In addition, when the tunnel was lifted to reveal no train (unexpected disappearance), there was sustained right hemisphere gamma band activity that peaked around 500 ms after the lifting of the tunnel. Kaufman et al. argued that this increased activity showed the brain attempting to maintain its representation of the train despite the competing visual evidence that the train was not under the tunnel. This higher activity could also represent the brain's response to an unexpected event. However, Kaufman et al. argued that the fact that higher activity during object occlusion occurred only in right temporal areas ruled out a general explanation of brain activity based on degraded sensory input. It is not clear why degraded sensory representations should be maintained on only one side of the brain and not on the other.

To investigate further whether the sustained EEG activity shown when the tunnel was lifted was related to the representation of non-visible objects, a second experiment was conducted using only *appearance* events. As will be recalled, behaviourally the infants did not distinguish between expected and unexpected appearance in terms of increased looking time. As in the first EEG experiment, it was found that occlusion of the train led to increased EEG activity in the right temporal electrodes during the period that the train should have continued to exist beneath the tunnel. However, there was no change in the EEG when the tunnel was lifted. As the train was always revealed in the appearance conditions, Kaufman et al. argued that there was no need to maintain a representation of the object independently of visual input. However, as the unexpected appearance was unexpected, the higher activity found for unexpected *disappearance* was unlikely to represent the brain's response to an unexpected event. Note that the finding that *increased* EEG activity is associated with the representation of hidden objects again appears to rule out explanations based on degraded sensory inputs (as here, activation should decrease).

Kaufman, Csibra and Johnson (2005) then developed a convergent paradigm to examine the neural activity associated with object occlusion in more detail. This time, they measured gamma-band EEG in six-month-old infants who either watched an object being slowly occluded (e.g., a picture of a toy ball was shown on a screen, and a black screen slowly moved over the ball to conceal it) or an object disintegrating (e.g., a picture of a toy ball gradually dissolved away, leaving a grey screen). Note that while an occlusion event suggests that the occluded object continues to exist even though it is now out of view, a disintegration event suggests that the object no longer exists. As in the train/tunnel study, it was right hemisphere activity in temporal areas of the infant brain that showed significant differences between conditions. Gamma band activity was significantly greater when the object was occluded than when it disintegrated. The effect was strongest 200 ms after the object had disappeared from view, again suggestive of the infant maintaining a representation of an object that has vanished from their view. Kaufman et al. (2005) argued that the sustained gamma band activity was not simply a neural correlate of occlusion, but formed one basis for object representation by the infant.

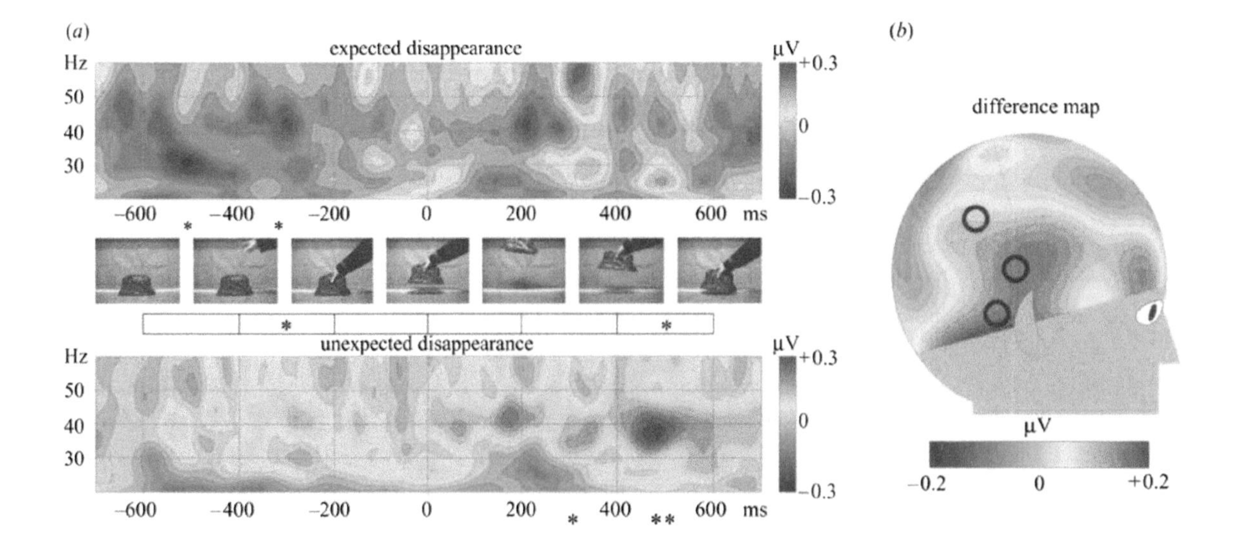

Gamma band activity during expected versus unexpected disappearance events. The time-frequency plots show gamma band (~40 Hz) activity during each event, with significant differences marked by asterisks. Activity differs between conditions both as the tunnel is lifted and also as the infant sees nothing where there should be a train. The electrodes yielding the result were right-lateralised, shown in the head plot.

Future studies, for example using multi-voxel pattern analysis, are required to assess whether this is the case.

Research on how the brain processes objects has also shown that visual objects are processed via two neural pathways (Milner & Goodale, 1995; Ungerleider & Mishkin, 1982). One pathway, the *dorsal* route, is used for spatial and temporal information, and is thought to be important for processing information that may be needed to guide action. This is called the 'where' pathway. The second pathway, the *ventral* route, is used for processing information that is useful for identifying unique objects, such as colour information. This is called the 'what' pathway. The ventral route is also used for processing information about faces. Both pathways appear to process information about the size and shape of objects. Indeed, it has been argued that this partial separation of the neural processing of different features of the *same* objects needs to be taken into account when interpreting experiments with infants (e.g., Mareschal, Plunkett & Harris, 1999; Mareschal & Johnson, 2003). Neuroimaging studies with infants suggest that the object processing pathways are organised in the same way in infants as in adults from the early days of life (Wilcox & Biondi, 2015).

For example, it seems quite plausible that younger infants may be poor at integrating information that is processed separately in the dorsal and ventral pathways. They may not yet have sufficient neural connections. This idea was tested in an ingenious experiment reported by Mareschal and Johnson (2003). They exploited the preference of the ventral route for faces and colour by examining four-month-old infants' memory for surface features of objects versus the spatial location of objects in an occlusion paradigm. The infants were habituated to five repetitions of two objects appearing sequentially from behind two occluders on a computer display. In each trial, the infants watched as one object moved out from behind the first occluder for five seconds, and then moved back behind it, to be followed by the second object moving out from behind the second occluder and returning behind it. In the test trials, the two occluders moved upwards to reveal the objects behind them. In the *feature change* condition, one of the objects (a face or a cartoon asterisk) changed colour or identity. In the *location change* condition, the colour of the cartoon asterisks or the identity of the faces remained constant, but the location of each was switched. In parallel conditions using pictures of familiar graspable toys, either the identity of one of the toys was changed, or the location of the toys was switched. Faces and coloured asterisks were selected because they were expected to be processed by the ventral route. In the parallel conditions, graspable toys were selected because they afforded the infant potential actions and hence should be processed by the dorsal route. In a final baseline condition, the occluders lifted to reveal the expected objects.

Mareschal and Johnson (2003) argued as follows. If four-month-old infants have difficulties in integrating object information processed by the ventral and dorsal pathways, then they should show increased looking time in the feature change condition when the objects were asterisks and faces, but they should show increased looking time in the location change condition when the objects were graspable toys. In each of these cases, one visual route can yield sufficient information for dishabituation. When infants had to process changes in the *conjunction*

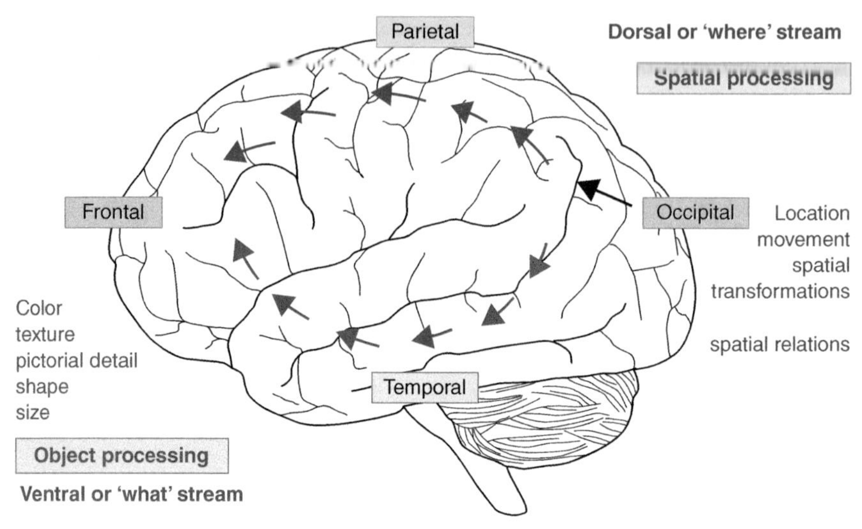

Schematic depiction of the dorsal and ventral visual pathways. The dorsal route is thought to be important for processing information that may be needed to guide action. The ventral route is thought to be important for processing information that is useful for identifying unique objects.

of features and location (e.g., when the object [toy] afforded action but changed in features rather than moving, or when the faces or coloured asterisks preserved their features but changed location), looking time should not differ from baseline. Difficulties in integrating information from the two pathways would prevent the infants from keeping track of conjunctions.

This was exactly the pattern of results found. When the objects in the occlusion paradigm were faces or asterisks, the babies looked significantly longer in the feature change condition only. When the objects were graspable toys, on the other hand, the infants showed increased looking time in the location change condition only. They did not show dishabituation when the faces or coloured asterisks moved location, or when the graspable toys changed in appearance. This is a remarkable result, and illustrates the importance of understanding the neural processing that is underpinning infant looking behaviour. As argued by Mareschal and Johnson (2003), it also has implications for the design of studies probing infants' knowledge about objects. Depending on the kind of object used, infants may process features of that object differently. In particular, for real objects that can be grasped, location information appears to be retained at the expense of surface features like colour. For objects that do not afford action, information about surface features is retained at the expense of information about location. This does not imply that infants cannot process changes in either surface features or location at the same time. Rather, there appears to be a selective loss of information when aspects of the object that are processed by the ventral stream must be integrated with aspects processed by the dorsal stream while an occlusion occurs.

Kaufman et al. (2003b) extended this argument to suggest that the potential *graspability* of the stimuli typically used in infant studies influences how the infant brain will process these stimuli. They propose that the ventral/dorsal route framework could explain apparently conflicting results in the infant literature on object segregation, on the processing of surface features of moving objects, and on object individuation. They proposed that any experiments using small, familiar and

KEY TERMS

Ventral stream
Visual 'what' pathway that processes information about unique objects.

Dorsal stream
Visual 'where' pathway that processes spatial and temporal information.

moving objects as stimuli are likely to activate the dorsal route. Any experiments using larger, stationary objects are likely to be processed by the ventral route. Further evidence relating to these ideas is considered in Chapter 2. For the moment, note that evidence for an *inability* to integrate information from both routes depends on a negative result (no difference in looking time compared to baseline). Negative results can have many causes. A direct measure of neural processing would provide a better test for this plausible hypothesis.

SUMMARY

Clearly, there is considerable continuity between measures of learning, memory, perception and attention in infancy and later in cognitive development. Indeed, some of the neuroimaging studies that we considered suggest that the infant brain has essentially the same *structures* (localised neural networks) as the adult brain, and that these structures are carrying out essentially the same *functions* via the same *mechanisms*. Babies as young as three weeks show learning and memory of simple objects, and three-month-olds can learn causal contingencies and retrieve them 28 days later with appropriate retrieval cues. Six-month-olds can form event memories that can be retrieved two years later in the presence of the right reminders, and six-month-old babies have working memories. Although both learning and memory are demonstrated via infant *behaviour*, more recent analyses suggest that these event memories show declarative memory in action (Mullaley & Maguire, 2014). Infant memories involve bringing some aspect of the past to conscious awareness.

Perceptual and attentional abilities are also impressive early in life. Neonates can discriminate between simple visual forms such as crosses and circles. Work with older infants suggests that the basis of these discriminations is fairly abstract perceptual information, as habituation also occurs to different exemplars of the same shape, novel exemplars being seen as familiar. Habituation at an abstract level suggests that rudimentary categorisation is occurring, with babies forming a generalised conceptual representation or 'prototype' of a particular visual form. Work on prototypes and sequential learning has shown an impressive capacity for tracking conditional probabilities. This provides infants with a very powerful domain-general learning mechanism: statistical learning. Statistical learning about the physical world enables infants to construct a world of predictable objects that behave in predictable ways. Infants appear to organise the physical world in terms of events like occlusion and containment.

We also discussed the possibility that automatic neural encoding mechanisms record spatio-temporal information about events via 'dynamical activity space' *trajectories* (Fekete & Edelman, 2011). Accordingly, the brain encodes event streams in a way that has inherent causal structure. Encoding trajectories of information across time would mean that the brain will rapidly extract meaningful structure, thereby helping infants to predict what happens next and to prepare for future action. Neural encoding mechanisms also adapt automatically to changes in the statistical patterns presented by environments because they learn *conditional* probabilities. The use of conditional probabilities as a basic encoding mechanism means that the brain is also creating *predictive* knowledge structures. The brain automatically computes the probabilities of events based on its prior sensory experiences. This process is described mathematically by Bayesian learning and is discussed further in Chapter 7. Put succinctly, sensory systems are taking their prior knowledge, weighting it in terms of accuracy, and then combining it with new sensory input in a process of continual updating that enhances the perceptual prediction of the most probable outcomes. These automatic perceptual learning processes are rapidly supplemented as the infant becomes able to manipulate objects themselves, to sit up and to act deliberately upon the world. New abilities for action can then drive representational knowledge. The infant as agent can test out different predictions, thereby elaborating early perceptually based representations. The active infant will eventually develop a cognitive system for physical reasoning that encompasses fine-grained knowledge about causes and their effects. This process is discussed further in Chapter 2.

CHAPTER 2

CONTENTS

Infancy: The physical world 2

2

In the last chapter, I argued that much of the early knowledge that we think of as cognitive may develop initially via the automatic operations of our neural and perceptual systems. The automatic neural encoding of perceptual information, including 'trajectories' of information across time, may be the basis of infant 'constraints' on learning. At the same time, the rich repertoire of statistical learning mechanisms used by the brain would very quickly organise initially incomprehensible event streams into meaningful structures with inherent causal information. The infant brain may thus naturally be equipped to learn complex environmental structures in familiar scenarios that will enable it to predict what happens next. Such predictive learning is vital for the infant brain to prepare the appropriate actions. In Chapter 1, I proposed that although these neural encoding mechanisms are automatic, via repeated sensory experiences, agency and action, infants can then develop *cognitive* understanding of physical causal systems.

This neurally driven analysis offers a new way of thinking about the links between perception and cognition. Historically, perception has usually been conceptualised as hard-wired aspects of sensory systems, such as colour and depth perception in the visual system. Cognition has been conceptualised as higher-level judgements or deductions based on that perceptual information, such as making causal inferences about visual events. Yet for both adults and infants, cognitive neuroscience is showing that the line between perception and cognition is actually quite blurred. In particular, the idea of dynamical activity space *trajectories* means that basic causal information, about temporal sequence, for example, may be part of the neural *encoding* of perceptual events. Accordingly, causality would be directly observed, and does not need to be consciously inferred via reasoning. Within this neural explanatory framework, it is interesting to consider perceptual displays that are experienced by adults as causal or animate, even if the perceptual information consists of simple moving geometric shapes (Scholl & Tremoulet, 2000). It seems likely that infants, too, will experience such moving visual displays as causal or animate, suggesting that perception and cognition are intimately linked from the get-go. One very early link between perception and cognition is demonstrated by studies of infant imitation.

PERCEPTUAL STRUCTURE AND CONCEPTUAL INFORMATION

Imitation

Even neonates can imitate the facial and manual gestures of adults. For example, Meltzoff and Moore (1983) showed that babies aged from one hour to three days could imitate gestures like tongue protrusion and mouth opening after watching an adult produce the same gestures (see Figure 2.1). In Meltzoff and Moore's experiment, the babies were supported in a seat in a darkened room. A light then came on for 20 seconds, illuminating an adult's face. The adult demonstrated a gesture such as tongue protrusion for the entire 20-second period, and the light was then extinguished. The babies were then filmed in the dark for the next 20 seconds. Following this imitation period, another gesture was modelled by the adult, and so on. An experimenter who was 'blind' to the experiment later scored the behaviour of the baby in each 20-second segment of video. Significantly more tongue protrusion was scored in periods following modelling of tongue protrusion, and significantly more mouth opening was scored in periods following modelling of mouth opening. The conclusion that the babies were imitating the adults seems inescapable, although it proved controversial both at the time (e.g., Hayes & Watson, 1981; McKenzie & Over, 1983) and up to the present day (e.g., Oostenbroek et al., 2016; Meltzoff, 2017).

KEY TERM

Mirror neurons
Multi-modal neurons in the ventral premotor and parietal cortices that are activated when we carry out an action, see someone else carrying it out, or think about carrying it out.

Despite some failures to replicate imitation in neonates, many experimenters have confirmed Meltzoff and Moore's findings, using a variety of different gestures

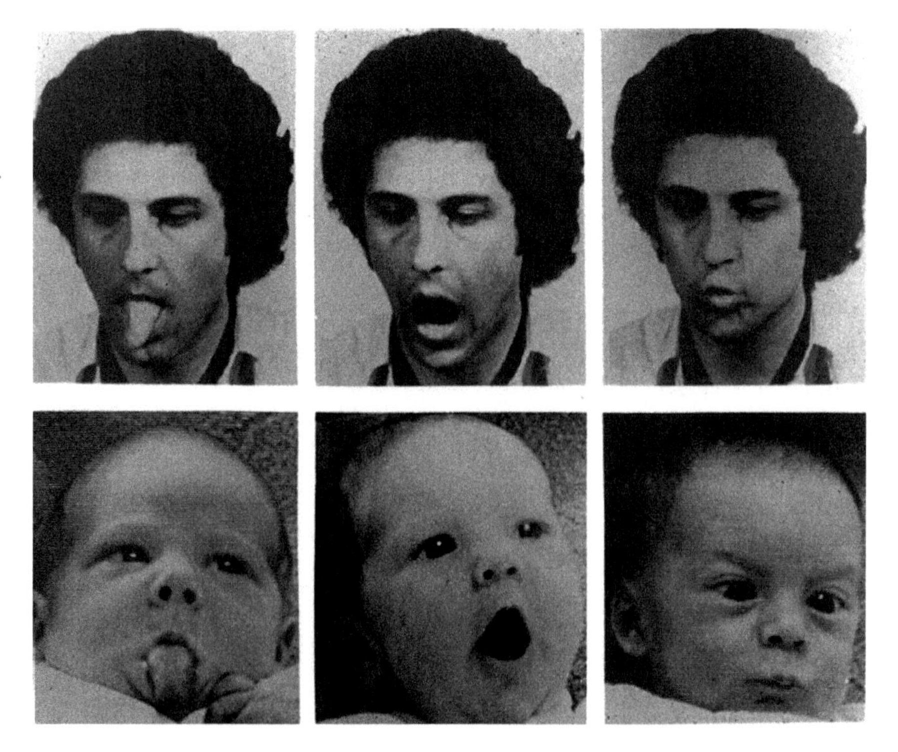

FIGURE 2.1
Babies imitating Meltzoff demonstrating tongue protrusion, mouth opening and lip pursing. From Meltzoff and Moore (1977). Copyright © 1977 AAAS. Reprinted with permission.

and testing babies from a variety of different countries. For example, Vintner (1986) investigated whether newborn Italian babies could imitate tongue protrusion and hand opening-closing. She contrasted imitation in two conditions, an 'active' condition in which the gesture was continuously modelled for a 25-second period, and a 'static' condition in which a protruding tongue or an open hand was maintained for 25 seconds. Vintner found imitation behaviour in the 'active' group only, even though the babies in the 'static' condition spent a lot of time looking at the experimenter. She suggested that movement may be a fundamental property for eliciting imitative responses at birth. Nagy, Pal and Orvos (2014) investigated whether newborn Hungarian babies could selectively imitate the raising of one, two or three fingers, using a sample of over 200 infants. Nagy and colleagues reported differential and accurate imitation of all three modelled gestures. They suggested that the rapid copying shown by the infants indicated that neural representations of certain motor 'schemas' or action patterns could be innate. Recent evidence that newborn monkeys living in research facilities can also imitate the facial gestures of human adult carers makes this suggestion highly plausible (Simpson Miller, Ferrari, Suomi & Paukner, 2016).

Indeed, it was Meltzoff and Moore's (1983) proposal that successful imitation by the neonate necessitated *representational capacity* that generated much of the subsequent controversy. For successful imitation, Meltzoff and Moore argued that infants need to (i) represent the action of the adult, (ii) retain this representation during the period that the adult is invisible in the dark, and (iii) work out how to reproduce the gesture using their own musculature. In 1983, the idea that neonates might have a representational system that allowed them to match their own body transformations to those of others was seen as rather incredible. Thirty-five years later, it seems much more plausible. These early imitative behaviours may be subserved by a neural system designed to differentiate self- versus other-generated movement. Another neural candidate is the so-called 'mirror neuron' system. Meltzoff favours the latter system, and has argued that neural mirroring mechanisms enable a 'like me' analogy, by which infants are able to recognise similarities between themselves and others (Marshall & Meltzoff, 2014). This idea is discussed in detail in Chapter 4. Certainly, neuroimaging studies with adults suggest that 'mirror neurons' (multi-modal neurons) in the ventral premotor and parietal cortex are crucially involved in kinaesthetic-visual matching (Jackson, Meltzoff & Decety, 2006; see also Chapter 4). The role of the infant 'mu' rhythm mentioned in Chapter 1, which is reduced in amplitude when infants observe the motor actions of others, is discussed along with further social imitation studies in Chapters 3 and 4.

Motionese

It is now understood that adults also modify their actions in important ways when they interact with infants. These modifications to action appear to facilitate learning. 'Infant-directed action' is characterised by greater enthusiasm on the part of the adult, closer proximity to the infant, greater repetitiveness and longer gaze to the face when compared to interactions with another adult. Infant-directed action is based on simplifying actions and incorporates more turn-taking. For example,

Brand and Shallcross (2008) allowed infants aged 6–8 months to choose to look at either adult-directed actions or at infant directed actions. They did this by filming a mother demonstrating an object to her baby versus filming her demonstrating the same object to her partner. Other babies were subsequently given a choice of which film to watch. The infants showed a systematic preference for watching the films of infant-directed actions. In a follow-up study, Brand and Shallcross blurred the faces of the adult actors so that there was no emotional or direct gaze information. They found that the infants still preferred to watch the infant-directed actions. It was only when shown still shots of the adult actors in each type of video that the infants demonstrated no preference. These data suggest that it was characteristics of the actions themselves, rather than of the actors, that generated the preferential looking behaviour.

'Motionese' serves to increase the baby's attention to what is taking place during goal-directed scenarios. Motionese seems to help the babies to process the *meaningful structure* underlying the actions – the causal structure. The extra pausing, the exaggeration of the actions and the repetition all help the infant to understand the meaning of the actions and/or the goals of the actor. Motionese signals that 'this behaviour is relevant to you', possibly making it more likely that infants will imitate. The motionese studies also show that infants are not passive learners, simply processing everything in the visual field. Rather, they *select* which actions to watch, as they prefer to watch a particular action in motionese rather than the same action performed for another adult. Typically, the parents are not aware of using infant-directed actions. Rather, motionese appears to be an unconscious way of modifying behaviour when interacting with an infant in a natural setting (just like infant-directed speech – see Chapter 6).

The perception of causality

We noted earlier that some types of perceptual event promote a causal interpretation, in adults as well as in infants. For example, many of the violation-of-expectation experiments discussed in Chapter 1 involved causal relations as well as relations like space, occlusion and containment. In order for most of the physical violations underpinning these experiments to be *unexpected*, a representation of the cause–effect relations underpinning the events was necessary. For example, in the study based on the car running down the ramp (Kotovsky & Baillargeon, 1998), it was the expected *collision* between the car and the box that made the reappearance of the car unexpected. In the study with the rotating screen (Baillargeon et al., 1985), it was the expected *contact* between the screen and the box that made the continued rotation of the screen unexpected. However, recall that all of the causal events in these paradigms occurred *out of view*. This means that we have to infer that, for example, the babies in the car experiment expected the car to collide with the box and thus to stop (they might have looked longer at the display because they expected the car to reappear shunting the box in front of it). In order to assess infants' ability to perceive causal events *directly*, we need to study causal events that occur in full view.

Collision events provide a useful set of events for such experiments. For example, when one billiard ball collides with another, the second ball is launched into motion. This is a pure example of a cause–effect relation, and Michotte (1963) showed that adults always have an impression of causality when they view 'launching' events, even if they are watching patches of light moving on a wall (see Scholl & Tremoulet, 2000, for more recent examples of adult impressions of causality, using very impoverished displays). This impression of causality even in the absence of a mechanical connection was taken by Michotte to show that adults perceive causality *directly*, in the same way as they perceive other perceptual features such as colour or motion. The automatic perception of causality would mean that our perceptual systems *assume* cause–effect relations in the absence of contradictory evidence. Theorising in adult cognitive neuroscience is supportive of this kind of interpretation. As noted in Chapter 1, Fekete and Edelman (2011) argued that the neural mechanisms that encode perceptual experiences may *in themselves* impose causal structure on experiences. Their idea was that phenomenological experiences could be recorded in the brain by *trajectories* of neural activation over time and cortical space. These trajectories would inherently carry constraints regarding the sequence of instantiation. In simple language, the trajectory would determine who fires, where firing occurs, and when it occurs, thereby carrying inherent information about cause and effect.

Young infants certainly appear to possess perceptual mechanisms that assume causality. Further, accumulating data make it more and more plausible that these sensory/neural mechanisms enable the direct perception of causality and lie at the root of the development of causal understanding. An example regarding causal relations comes from experiments devised by Leslie and Keeble (1987) to study six-month-old infants' understanding of *launching* events. In a typical experiment, infants were shown one of two films. In one film, a red block moved towards a green block and then collided with it, directly setting the green block in motion. In the other film, the red block again moved towards a green block and made contact with it, but the green block only began to move after a delay of 0.5 seconds. While the first launching event gave an impression of causality to watching adults, the second did not. Following habituation to one of the films (either direct launching or delayed launching), the infants were then shown the *same* film in reverse. Although the change in the spatio-temporal relations in the films was the same for both groups, the reversal of the 'direct launching' film resulted in a *novel* causal event (green launches red). Leslie and Keeble argued that if the infants in the direct launching condition were perceiving a causal relation (red launches green), then they should show more dishabituation following reversal (green launches red) than the infants in the delayed launching condition.

This was exactly what they found. Leslie (1994) argued that this effect showed that the infants were sensitive to the mechanical causal structure of launching events. In the 'direct launching' film, there was a change in the mechanical roles of the two billiard balls. The 'pusher' was now the 'pushed'. In the 'delayed launching' film the billiard balls did not have roles such as 'pusher', and so the roles could not be reversed. Leslie's description of the launching event in terms of mechanics entails a notion of *agency*, an idea that is discussed further below. Meanwhile, recent

KEY TERMS

Launching event
When one object appears to strike another and seems to cause it to move.

Agency
An understanding of instrumentality in events, whether mechanical or human.

experimental work with adults provides converging data. For example, Moors, Wagemans and de Wit (2017) presented adults with perceptual displays of causal versus non-causal relations such as launching versus non-launching events using a perceptual technique called Continuous Flash Suppression (CFS). The use of CFS meant that the events were initially invisible to the watching adults. The suppressed stimuli were then gradually increased in visual contrast until the participants became able to see them. Moors et al. reported that the causal events used, such as launching, entered conscious awareness *significantly more quickly* than the non-causal events. Accordingly, low-level aspects of the visual system appear to be sensitive to the causal structure of perceptual events and they automatically foreground this perceptual information for the brain. This is a very interesting finding, and it also supports Michotte's theoretical observations.

The perception of animate relations

Our perceptual systems also appear to assume *animacy* in the absence of contradictory evidence. Our perceptual systems thereby accord the status of *agents* to the actors/objects in certain classes of perceptual events. Again, this was first noticed by Michotte (1963), who suggested that simple motion cues may provide the foundation for social cognition (this idea is discussed in more detail in Chapters 3 and 4). Adults who are shown simple displays of moving geometric shapes will describe the displays as animate ("it's trying to get over there", see Heider & Simmel, 1944). In an illustrative experiment, Tremoulet and Feldman (2000) created displays in which a dot moved in one direction for 375 ms, and then changed direction and speed for another 375 ms. In other displays, a rectangle moved in one direction for 375 ms, and then changed orientation when it changed direction and speed (see Figure 2.2). In a control condition, a rectangle did not change orientation when it changed direction and speed. Experiments with adults showed that the first two displays were more likely to be perceived as animate than the control condition. The strongest perceptions of animacy occurred for the rectangle that changed orientation as it changed direction, as in this condition the principal axis of the rectangle was aligned with the direction of motion. Thus very simple motion cues, such as change in speed or change in direction, can give adults an impression of animacy. The rectangle is seen as an animate agent, 'choosing' where to go.

Exactly the same effects are found with infants. In fact, infants as young as three months of age show a preference for 'animate' motion. For example, Rochat and his colleagues gave infants aged 3–6 months two visual displays involving dots (Rochat, Morgan & Carpenter, 1997). In one display, one dot appeared to be chasing another around the screen, giving an impression of a social interaction. This dot systematically approached the second dot at a constant velocity, until the latter accelerated away to a comfortable distance. In the other display, the two dots moved randomly and independently on the screen. Except for the relative spatial-temporal dependence of the dots' movements, all other dynamic aspects of the two displays were controlled and maintained equal. For example, the average relative distance between the two dots was comparable. Both displays were presented concurrently side by side. Rochat et al. found that in general the infants preferred to watch the 'chasing'

KEY TERM

Intentional stance
The attribution of mental causes such as beliefs, desires or goals as the basis for action.

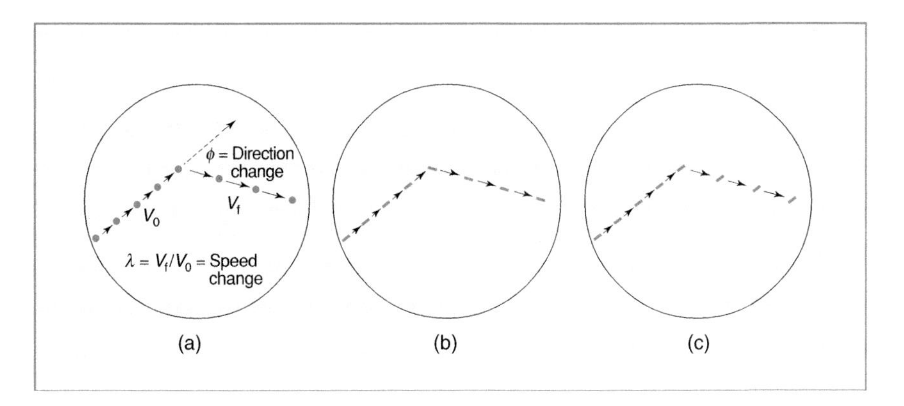

FIGURE 2.2
The shape/alignment conditions used by Tremoulet and Feldman: (a) dot condition (also shown are the motion parameters); (b) aligned condition; (c) misaligned condition. From Tremoulet and Feldman (2000). Copyright © 2000 Pion Ltd, London. Reproduced with permission.

display. Accordingly, they argued that even three-month-old babies were sensitive to movement information that specified social causality for adult observers.

The social causality claim is also given support by a recent brain imaging study using EEG. Galazka, Bakker, Gredebäck and Nyström (2016) designed a 'chasing' paradigm similar to Rochat and colleagues' paradigm for nine-month-old infants. In their experiment, the infant watched as one geometric shape 'chased' another geometric shape during a short video. In the *animate* motion 'chasing' videos, the chaser consistently moved towards the chased, which consistently moved away. In control *inanimate* motion videos, the same shapes moved randomly at similar speeds, but on independent linear trajectories. Galazka et al. (2016) then recorded EEG in a subsequent period when the infants were shown the geometric shapes in a static display. The research question was whether the neural response to the shapes that apparently had agency (i.e., when they were in the roles of 'chaser' and 'chased') would differ from the neural response to the exact same shapes when they had been experienced in an experimental setting that did not suggest agency.

The data showed that two responses called the N290 and P400 (respectively a large negativity in amplitude occurring about 290 ms after stimulus onset, followed by a large positivity in amplitude occurring around 400 ms after stimulus onset, also called the N290/P400 complex) were different depending on whether a shape had been the 'chaser' or had been part of the neutral motion display. The authors noted that the infant N290/P400 complex may be the infant equivalent of the adult N170 response to faces (this is a negative deflection in the averaged event-related potential (ERP) that occurs around 170 ms after stimulus onset, that is reliably associated with face processing). They hence argued that the N290/P400 complex was an indicator of social-cognitive processing. If this social-cognitive claim is supported by independent evidence, the study would provide a nice example of using brain imaging to go beyond brain-behaviour correlations. Here the N290/P400 complex (a neural correlate) is used as an ingenious measure of infants' processing of static geometric shapes whose *previous behaviour* can be potentially designated as social. The identification of a neural marker that is active specifically when previous experience of motion cues has identified a geometric shape as an agent ('chaser') supports the conclusion that simple dynamic motion displays of this nature are processed by infants as 'social' events involving *animate agents*.

Intriguingly, Japanese macaques also appear to interpret dynamic motion displays of this nature as goal directed (Atsumi, Koda & Masataka, 2017). In this study the monkeys were first taught to discriminate between 'chasing' displays and random movement of geometric objects via food rewards. They were then presented with novel chasing or control displays, in which the degree of correlation between the motion trajectories of the two objects, or their proximity, was systematically varied. The monkeys responded to variations in the physical cues in the novel 'chasing' events in the same way as adults who were shown identical chasing versus control displays. Atsumi et al. concluded that the monkeys were using overall motion information rather than specific physical cues to attribute goal-directedness to the movements of the geometric shapes. If supported by further work, this would suggest that the primate brain automatically interprets certain kinds of dynamic motion as specifying animate agents. Like causality, therefore, animacy may be a primary property for the visual system, just like colour or shape.

Meanwhile, Gergely and his colleagues have shown behaviourally that the attribution of agency by infants on the basis of perceptual analyses of geometric shapes also provides evidence for early attribution of *intentionality* to animate agents. In an innovative study, Gergely, Nádasdy, Csibra and Bíró (1995) showed that 12-month-old infants could analyse the spatial behaviour of an agent in terms of its actions towards a goal. They argued that the infants applied an 'intentional stance' to this behaviour when it appeared *rational*. In other words, the infants were attributing a mental cause for the goal-directed behaviour that they observed. Gergely et al. (1995) also argued that when there is no basis for attributing rationality to spatial behaviour that appears goal-directed then an intentional stance is not adopted.

In their experiment, Gergely et al. took as their starting point the fact that the prediction and explanation of the behaviour of agents requires the attribution of intentional states such as beliefs, goals and desires as the mental causes of actions. This is the adoption of an 'intentional stance'. The adoption of an 'intentional stance' towards agents entails an assumption of rationality – that the agent will adopt the most rational action in a particular situation to achieve his or her goal. Gergely et al. were interested in whether 12-month-old infants would generate expectations about the particular actions that an assumed agent was likely to perform in a new situation to achieve a desired goal. They designed a visual habituation study to find out, in which a computer display gave an impression of agency to the behaviour of circles. The infants saw a display in which two circles, a large circle and a small circle, were separated by a tall rectangle (see Figure 2.3). During the habituation event, each circle in turn expanded and then contracted twice. The small circle then began to move towards the large circle. When it reached the rectangular barrier it retreated, only to set out towards the large circle a second time, this time jumping over the rectangle and making contact with the large circle. Both circles then expanded and contracted twice more. Adult observers of this visual event described it as a mother (large circle) calling to her child (small circle) who ran towards her, only to be prevented by the barrier, which she then jumped over. The two then embraced.

Following habituation to this event, the infants saw the same two circles making the same sequence of movements, but this time without a barrier being present. In

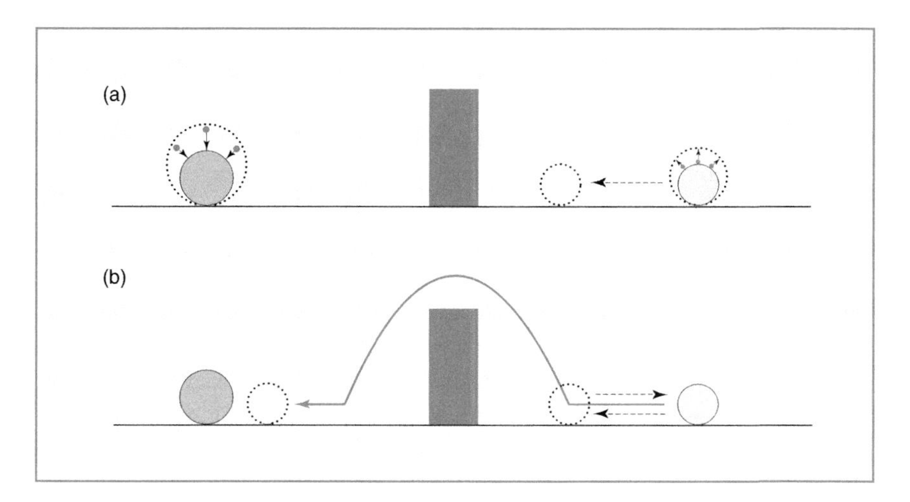

FIGURE 2.3 Schematic depiction of the habituation events shown to the rational approach group by Gergely et al. (1995), with (a) depicting the expansion and contraction events and the first approach, and (b) depicting the retreat, jump and eventual contact. From Gergely et al. (1995). Copyright © 1995 Elsevier. Reprinted with permission.

this 'old action' event, there was no rational explanation of the retreat-and-jump action. In a 'new action' event, the small circle simply took the shortest straight path to reach the large circle. Gergely et al. predicted that, if the infants were making an *intentional* causal analysis of the initial display, then they should spend more time looking at the old action event than the new action event. This was because even though the former event was familiar, it was no longer rational. This was exactly what they found. A control group which saw the same habituating events without the rectangle acting as a barrier (it was positioned to the side of the screen) showed equal dishabituation to the old and new action events. Gergely et al. argued that the control group had not considered the small circle to be a rational agent, as the behaviour of the small circle during the habituation event did not lead the infants to adopt an intentional stance towards it.

Intriguingly, infant chimpanzees appear to make similar causal analyses when shown the same perceptual displays involving geometric shapes (Uller, 2004). Uller used a looking time measure with infant chimps to see whether they also adopted a rational stance when watching the same animated displays of rectangles and circles shown to infants by Gergely et al. (1995). Like the human infants, the infant chimps spent significantly more time looking at the old action event than the new action event once the 'barrier' had been removed. In the control condition, the chimps spent longer looking at the new action event. This result appears to show that the chimps were making an *intentional* causal analysis of the initial geometric display. However, Uller (2004) does not herself make this claim. The data are certainly suggestive concerning the idea that the primate brain automatically interprets certain kinds of dynamic motion as specifying animate agents. They are also suggestive of these neural differences arising at the point of encoding. However, such data cannot prove that the visual system has made an inference that the circle's behaviour is intentional.

More recently, it has also been pointed out that this 'rationality' paradigm (and indeed many other experimental demonstrations of the rationality principle) have depended on scenarios showing infants infrequently experienced or even 'odd' actions, such as the jumping action of the small circle (Scott & Baillargeon, 2013). Accordingly, Scott and Baillargeon proposed that a novel critical test of the *efficiency principle* underlying the intentional stance was required. The efficiency principle is that agents pursue their goals as efficiently as possible. In this critical test, 16-month-olds watched an agent retrieving a desired object (a toy pig) which was either easily accessible (because it had been placed in a transparent container), or not easily accessible (because it had been placed underneath a transparent cover). Less effort was required to reach into the container and take out the pig than to reach to remove the cover, then reach again to get the pig. Scott and Baillargeon (2013) found that infants looked at the scene reliably longer when the agent chose to pursue the less efficient goal. The infants had apparently worked out which was the shorter action sequence required to obtain the desirable pig, and so expected that shorter action sequence to be the rational choice of the agent in that particular scenario.

This intriguing result shows just how powerful the information conveyed by simple dynamic displays may be for the young infant. Infants watch many examples of efficient goal-directed actions in their daily lives, long before they can perform equivalent actions themselves. Observing dynamic interrelations in the everyday world of objects and events may initially be encoded by trajectories of neural activation over time and cortical space, and repeated observations of such events appear to give rise to *cognitive* causal analyses by the infant (e.g., regarding what is rational). Such causal analyses would supply important information about the physical world of inanimate objects and also information about the mental world of animate agents, enabling better future prediction. These repeated perceptual experiences seem to lead to the eventual development of two separable cognitive frameworks which both represent causality, one for explaining the behaviour of objects (physical reasoning) and one for explaining the behaviour of people (psychological reasoning). The foundation of both frameworks appears to be the same: neural sensory processing mechanisms that encode perceptual experiences in structured ways, imposing causal structure or information about agency/animacy on these experiences *at the point of encoding* (via the temporal/spatial sequence of neural responses, Fekete & Edelman, 2011). As noted earlier, neuronal networks in the brain can utilise signalling cascades extended over time via a variety of mechanisms (for example, calcium cascades, or hormonal cascades, see Fekete, 2010). The attribution of psychological causal mechanisms such as belief and desire may thus develop from the same kinds of everyday observations as the attribution of physical causal mechanisms such as collision and support – namely from automatic neural perceptual-structural encoding of the dynamic spatial and temporal behaviour of objects and agents.

Cross-modal cues to causal structure

Of course, in the real world infants are seldom exposed to events in a single modality at a time. Most perceptual experiences are of unified and unitary objects and events. We have already seen that infants experience cross-modal equivalence from birth. However, as pointed out by Scheier, Lewkowicz and Shimojo (2003), modalities like vision and audition can be related in other ways that do not necessarily specify equivalence. For example, hearing something can change the way that we perceive a visual event. Sekuler, Sekuler and Lau (1997) demonstrated that a sound can cause perceptual reorganisation of an ambiguous motion display. Adults who watched a computer display in which two identical disks moved from opposite sides of the screen towards each other and then past one another at a constant speed perceived them as streaming through one another. However, if a sound was presented at the moment when the two discs coincided, many adults perceived them as bouncing off one another. This perception occurred even though there was no actual collision. Scheier et al. (2003) asked whether babies, too, would experience this inter-sensory illusion.

Scheier et al. (2003) tested three groups of infants aged four, six and eight months. The infants watched as two yellow discs appeared at opposite edges of the screen, travelled towards one another, coincided without stopping, and continued on until they reached the other's starting point. At the coincidence point, a beep was heard. The tone was expected to cause a percept of bouncing. This event was repeated until the infants habituated to it. At test, the infants saw either another instance of the tone sounding at the coincidence point (no dishabituation expected), or of the tone sounding either 1.3 seconds before or after the coincidence point. Scheier et al. argued that if the infants were perceiving illusory bouncing, then they should dishabituate in the latter two test trials, but not in the former. This was exactly what they found, but for the six- and eight-month-old infants only, not for the four-month-olds. In a follow-up experiment, the habituation event was visual streaming (the tone did not sound at the coincidence point, but 1.3 seconds before or after it), and the critical test event was visual bouncing (here the tone sounded at the coincidence point). This time, six- and eight-month-olds dishabituated to the bounce event only. Scheier et al. argued that sounds can induce the perceptual reorganisation of ambiguous visual events from six months of age.

The idea of using ambiguous perceptual events with infants is a good one, particularly in a longitudinal design. The bounce/stream illusion is a very intriguing result. It suggests that information from *all relevant modalities* is considered in conjunction by our perceptual systems, at least by six months of age. Such cross-modal information would yield a lot of data about the laws governing physical interactions between objects, which the brain may learn very rapidly indeed (recall Noesselt et al.'s 2007 adult cross-modal study discussed in the Foreword). On such data, sensory systems are taking their prior knowledge, including their prior cross-modal knowledge, and combining it with new sensory input to support the perceptual prediction of the most probable outcomes. Scheier et al. (2003) attribute the apparent failure of the four-month-olds to perceive the bouncing illusion to immaturities in the attentional systems of young infants. They point out that attentional behaviours become more flexible and voluntary at around six months (see Chapter 1). An alternative possibility is that some experience of objects is required in order for infants to associate a complex tone with a collision between two two-dimensional discs on a computer screen. In the real world of objects and events, most collisions will be accompanied by sounds, but the sounds will be meaningful in terms of the objects colliding (e.g., if a child throws a rigid toy at the wall, there will be a predictable thumping sound, not a complex tone). In order for an infant to infer that the complex tone is the sound of two cartoon discs colliding, some experience of such collision events may be required.

Causal frameworks for mechanical agency and human agency

The ability to represent causal relations between objects may appear to be a simple extension of infants' ability to represent relational information in general, but attention to causality is actually a critical cognitive tool. Causal relations are particularly powerful relations for understanding the everyday world of objects and events. Regarding understanding human agency (psychological reasoning), we will consider this in detail in Chapter 3. In other chapters we will see how a sensitivity to causal information underlies conceptual development (Chapter 5), the development of memory (Chapter 8) and logical development in general (Chapters 9 and 11). As just discussed, sensitivity to causal relations also appears to support the development of a notion of *agency*. Agency may be considered in terms of an understanding of mechanical agency or an understanding of human agency. As we will see in Chapter 7, in mechanical scenarios young children appear to give priority to establishing the *agent* of causal events. An early analysis of infant development regarding the understanding of mechanical agency was put forward by Leslie (1994), and this will be considered here. Human agency is considered in Chapters 3 and 4.

In 1994, Leslie argued that causal analyses of motion in infants serve the important purpose of generating mechanical *descriptions* of the events. He argued that such descriptions are more than a perceptual description of what has occurred, and so the perception of cause and effect cannot be said to be purely visual. Instead, this perception is the basis of a *mechanism* for understanding the mechanical properties of agents. Inanimate objects move in the real world as a result of the redistribution of energy, and these mechanical forces differ in important ways from animate sources of motion. Things that move on their own are agents, and things that move because of other things obey certain cause–effect, or mechanical, laws. Leslie proposed that infants' interest in things that move helps them to sort out the source of different cause–effect relations in the physical world.

Evidence that infants make a distinction between mechanical forces and animate sources of motion comes from an experiment reported by Spelke, Phillips and Woodward (1995). They compared seven-month-old infants' use of the principle of contact as a force that causes motion for inanimate objects versus people. The inanimate objects were meaningless patterned shapes five to six feet high, that were moved from behind by hidden people walking at a normal pace. The experiment was based on an occlusion paradigm (see Figure 2.4). During habituation, the infants saw either a person or an inanimate object appear to one side of the display and move behind a central screen. A second object or person then appeared from the other side of the screen and exited the display. These events were then repeated in reverse. The timing of the events was identical for objects and people, and for the objects it was consistent with the first object setting the second object in motion via contact behind the screen.

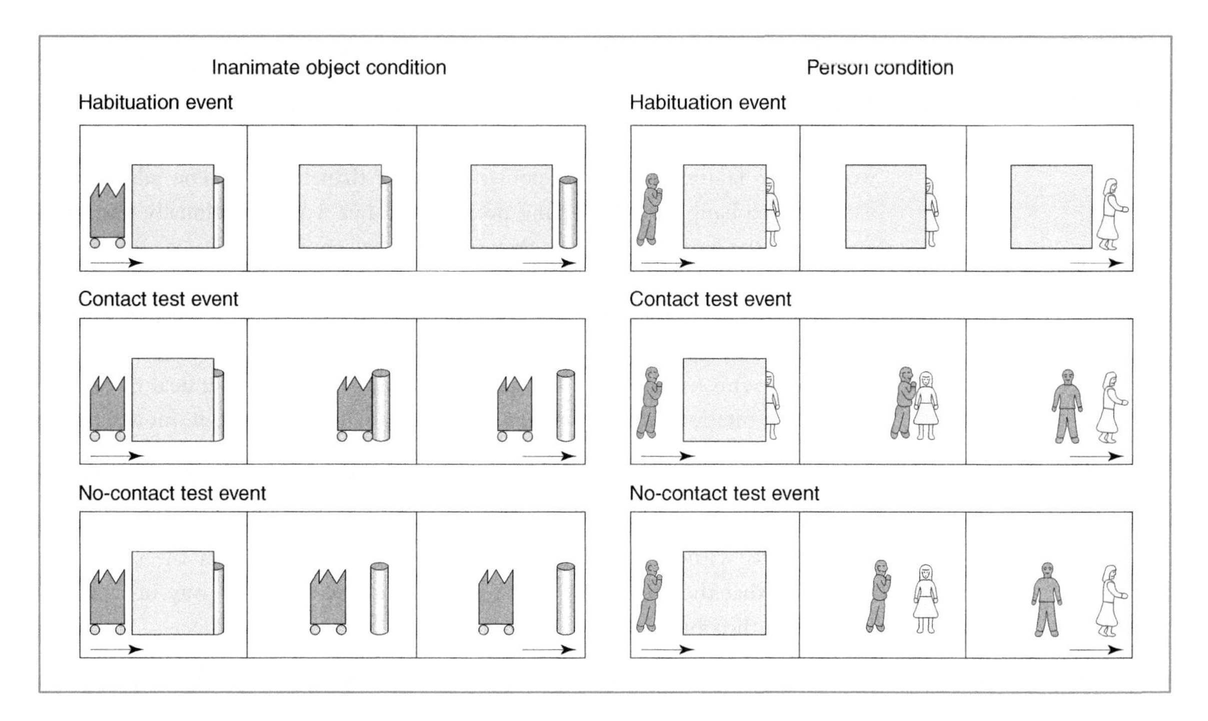

FIGURE 2.4
Schematic depiction of the events used to study infants' understanding of the forceful contact principle. From Spelke, Phillips and Woodward (1995). Reproduced by permission of Oxford University Press.

At test, the screen was removed and the infants saw either a contact event or a no-contact event. In the contact event, the two inanimate objects or people followed the same movement paths as in the habituation event, the second object or person moving once the first had made contact with it. In the no-contact event, the two objects or people never made contact, the first stopping a short distance from the second, which began moving after a suitable pause. Both events were then shown in reverse, and were repeated as long as the infants kept looking at them. If the infants were able to distinguish mechanical forces from animate sources of motion, then they should find the no-contact event more visually interesting for the inanimate objects, where the absence of contact should not cause motion. For the people, there should be no difference in the visual interest of the contact and the no-contact events, as people are agents who are capable of self-propelled motion. This was exactly what Spelke et al. found. A control condition established that there was no difference in the intrinsic attractiveness of the contact and no-contact events. Spelke et al. thus concluded that, at least by seven months of age, infants do not apply the forceful contact principle to people. They reason differently about people and objects, appreciating that people are agents who can move on their own.

Infants' interest in things that move helps them to distinguish between the sources of different cause-effect relations in the physical world.

Converging evidence that infants reason differently about people and objects comes from the deferred imitation paradigm. For example, Meltzoff

(1995a) showed 18-month-old infants either an adult who demonstrated an intention to act in a certain way, or a mechanical pincer device that went through identical motions. In each case, the intention was never fulfilled. For example, the adult/pincer would try to put a string of beads into a cylindrical container, but the beads would keep falling outside the container rather than inside it. The adult/pincer would try to hang a loop of string over a hook, but would continually under- or over-shoot the target. The infants then received the beads, string, etc., to manipulate by themselves.

Meltzoff found that production of the target acts (putting the beads inside the container, hanging the loop of string over the hook, etc.) was very frequent for the infants who had watched the adult agent. Indeed, it was identical to the frequency of imitation in a control group of infants who saw the adult modelling the same actions, but in each case completing the entire action successfully. Given that the first group of infants had never actually seen the target acts being performed, this is a very striking result. A further control group of babies who were given the same objects without any demonstrations produced very few of the target acts, suggesting that the target acts were not in themselves a natural way in which to manipulate the objects.

Most interestingly, however, babies did not behave in the same way when the inanimate pincer device performed the same movements in space as the human hand. Babies who watched the pincer device performing slipping motions matched to those performed by the human hand when failing to complete one of the target acts did not imitate very much. Meltzoff found that, although the babies' attention was riveted by the mechanical device, they were much less likely to imitate the target acts when the inanimate object failed to demonstrate an intended act than when a human agent failed to demonstrate the same act. In fact, babies who watched the human failing were six times more likely to produce the target act than babies who watched the inanimate device failing. Accordingly, the babies were not responding solely to the physics of the situation, they were responding to the human agent. Meltzoff concluded that a psychological understanding of the intention of the human actor led the babies in the experimental group to produce the target acts. They were aware of the acts that the adult had intended to produce, even though the acts themselves had failed.

Again, the evidence suggests that infants represent the behaviour of people in a psychological framework involving goals and purposeful intended acts, and not in terms of purely physical movements and motions (mechanical forces, see also Leslie, 1994). Further investigations of the development of the understanding of psychological causality are discussed in Chapter 3. Note that all the experiments discussed to date for separable frameworks for understanding people versus objects have involved *observable* events. However, understanding causation also involves the understanding of *unobservable* events. For understanding people, these include unobservable mental events such as desires, beliefs and goals, studied in child psychology via the development of a 'theory of mind'. For understanding objects, there may be unobservable causes for changes of state, such as magnetic forces or gravity. In fact, data from children with developmental disorders can be used to support this idea of separable causal frameworks. For example, autistic children appear to understand mechanical cause–effect relations extremely well, yet lack a theory of mind. Developmental disorders such as autism provide evidence consistent with the idea that the development of a 'theory of mind' occurs to some extent independently from the development of an understanding of mechanical agency (see Leslie, 1994, for further discussion).

FROM NEURAL/PERCEPTUAL REPRESENTATIONS OF CAUSALITY TO COGNITIVE REPRESENTATIONS

Traditionally, the ways in which infants might go beyond the perceptual categorisation and representation of events to form concepts and schemas (defined as *meaning*-based knowledge representations) has been hotly debated (e.g., Mandler, 2004). Meaning-based knowledge representations are usually defined in adult cognition as representations that encode what is significant about an event, omitting the unimportant perceptual details (e.g., Anderson, 1990). Concepts and schemas are thought to be sets of ideas held in the mind that explain the world, rather than perceptual copies of how the world is. Cognitive concepts and schemas can be complex units of knowledge that represent what is typically true of a *category*, for example *birds* (has wings, has beak, can fly, etc.) or an *event*, such as *going to the doctor* (report to receptionist, wait a long time, enter surgery, etc.). Schemas for events are also called *scripts* in cognitive psychology.

KEY TERM

Scripts
Schemas for events such as going to the doctor, going to a restaurant.

Previously, it was assumed that infants developed categories that were purely sensory or perceptual in nature rather than being schema-based (e.g., Quinn & Eimas, 1986). Clearly, if this were the case, the problem of conversion to a more cognitive or schematic format that omits perceptual details would be considerable. However, recent cognitive neuroscience studies with adults suggest instead that the activation of a concept always produces neural activity in the sensory modalities associated with those concepts (Barsalou, 2017; see Chapter 5). Hence there may not be a 'conversion problem' to overcome in cognitive development. In terms of brain activation, perceptual details may never be omitted, and so cognitive representations would naturally build upon perceptual representations. This is a core feature of *distributed representations*.

Further, we have already seen that infants store a remarkable amount of information that is not purely sensory in nature. For example, infants store information about launching or chasing events that include encoding of animacy or agency. Older infants store information about the movements of simple shapes that assume trajectories based on rational action (the intentional stance). Traditionally, notions of agency and intentionality have been considered to develop from cognitive analyses of sensory information, but we have been considering the alternative possibility that they may arise directly during perception via the nature of neural encoding mechanisms. We have also seen (in Chapter 1) that infants can retain two-step causal event sequences like 'make a rattle' for long periods of time (three months), and can even retain non-causal or arbitrary event sequences like 'put a mitten on the puppet' over the same period. Research such as this suggests that what is being remembered is all the experienced aspects of an event, which are stored and are reactivated together when reminder cues are given. In Chapter 1, we suggested that 'memory' would thus result from the neural mechanisms used by the brain at the point of encoding the original experiences. On this view, it is unlikely that infants have to wait until they have acquired language to represent knowledge in a meaning-based fashion, as was once thought. In any case, such a proposal begs an important question, namely what kind of cognitive structures underpin language acquisition.

Infant representations in at least some domains thus incorporate more than the sensory information experienced during a particular event. Although early representations of events such as occlusion events or support events may only encode some of the important variables (e.g., presence/absence of support rather than degree of contact with supporting surface), the capacity for independent action leads infants to augment these variables very rapidly into schemas. Meaning-based knowledge representations or schemas are thus *experience-dependent*. With multiple experiences, what is common across many instances is activated in the brain more strongly than what is not. Hence the 'gist' or most significant aspects of a particular concept or schema are eventually encoded more strongly than the variable perceptual details, akin to developing 'prototypical' concepts or schemas.

Specialised modules or distributed representations?

KEY TERMS

Domain-general
An ability or skill that applies across several situations or domains.

Domain-specific
An ability that applies only to one specific area or domain.

An alternative to this domain-general proposal for how perceptual encoding leads to meaning-based knowledge representations has been the modular view. For example, Leslie (1994) argued that there are specialised information processing systems in the brain that provide the basis for cognitive development. According to Leslie's *domain specificity* view, there are mechanisms in the brain that, by virtue of the position that they occupy in the overall organisation, receive inputs from particular classes of objects in the world and end up representing certain kinds of domain-specific information. One example he gave was the mechanism that acquired the syntactic structure of natural language. Other proposed modules included *number* and *music*. In addition, Leslie proposed two core domains that he argued were central to infants' initial capacities for causal conceptual knowledge. These were *object mechanics* and *theory of mind*. In these two core domains, the central organising principle was the notion of cause and effect.

For example, infants' processing of the physical world seems to organise itself fairly rapidly around a core structure representing the arrangement of cohesive, solid, three-dimensional objects which are embedded in a series of mechanical relations such as *pushing*, *blocking* and *support*. Leslie argued that this organised processing was the result of a specialised learning mechanism adapted by evolution to create conceptual knowledge of the physical world. In Leslie's view, the modular organisation of the brain *itself* allowed the infant to acquire rapid and uniform knowledge about object mechanics (and also about psychological causality). One way of describing this position is to say that babies have rudimentary 'theories', such as a 'theory' of mechanics. We have already described objections to this viewpoint using data from cognitive neuroscience. We have also discussed data from cognitive neuroscience suggesting that amodal or modular representations appear unlikely. However, as we will see in later chapters, the position that babies and young children have emergent theories has been used to explain cognitive development in many areas, such as biological knowledge. In my view, the notion of modules has long been in decline (Karmiloff-Smith, 1992). For example, the notion of a specialised 'module' for grammar, the 'language acquisition device', has been challenged on many fronts (see Chapter 6). Cognitive neuroscience has also shown that there is no number 'module', as important numerical information is stored in the language system (see Chapter 10). Nevertheless, Leslie's idea that certain cortical areas, by virtue of the position that they occupy in the overall organisation of the brain, receive inputs about particular classes of objects in the world is likely to be correct. Current research in cognitive neuroscience suggests that, rather than these cortical areas reflecting discrete modules of knowledge, they are part of large and distributed sets of activation patterns across the whole brain that represent knowledge in a distributed fashion. If we re-class a 'module' as a 'distributed representation', that is probably a more faithful description of current insights into the representation of knowledge from cognitive neuroscience.

In this book, therefore, I will follow a non-modular view of cognitive development (including, perhaps controversially, for language development). The assumption will be one of common learning mechanisms, primarily statistical learning (including associative learning), learning by imitation, explanation-based or causal learning and learning by analogy. Using these relatively simple learning mechanisms, some of which appear to be intrinsic mechanisms of neural encoding, the brain appears to build up complex representations about how the world is. Repeated experience of the perceptual/causal structure of objects and events strengthens certain neural pathways and particular trajectories of neural activation, leading to distributed neural activity that captures 'abstract' or 'generalised' world knowledge. As an example, the connections between neurons responding to specific details of a perceptual experience like 'covering' will be strengthened less across multiple instances of observing 'covering' events than the connections between neurons responding to more general aspects of 'covering', such as a cover occluding the covered object, that recur frequently. As a second example, the specific perceptual details of two objects that participate in a launching event may vary, but causal structure (the fact that object A causes object B to move) will not. This viewpoint suggests that meaning-based representations will automatically develop on the basis of the infant's experience, because intrinsic sensory/neural information-processing algorithms that are the same across brains are recording both what is general and what is specific about each experience. These intrinsic methods of sensory/neural processing can be described as simple learning mechanisms (such as statistical learning) that operate in a domain-general fashion.

KEY TERM

Learning by analogy Finding correspondences between two events, situations, or domains of knowledge and transferring knowledge from one to the other.

Evidence for the action of these sensory/neural learning mechanisms with respect to different kinds of representations (such as human versus mechanical forces, or biological versus non-biological kinds) is considered further in later chapters. The way in which an early focus on agency, in particular human agency, contributes to the development of a theory of mind is discussed in the next two chapters, Chapter 3 and Chapter 4. The relationship between spatial/mechanical analyses of the movements of non-human objects and agents and the biological/non-biological distinction that aids conceptual development is discussed further in Chapter 5. Communicative acts and language events are very interesting to babies, and the statistical learning of auditory input is discussed further in the chapter on language, Chapter 6. Causal learning across domains is discussed in Chapter 7. Finally, the evidence that even very young children are developing event schemas is discussed in the chapter on memory development, Chapter 8.

Reasoning and problem-solving about the physical world

Two other hallmarks of cognitive activity in adults are reasoning and problem-solving. Defining reasoning and problem-solving is not straightforward. One popular definition of reasoning is that it denotes those processes in information retrieval that depend on the *structure*, as opposed to the *content*, of organised memory (Rumelhart & Abrahamson, 1973). However, this definition leaves us with the problem of deciding which parts of organised memory are structural and which are content. According to Anderson (1990), reasoning and problem-solving usually involve three ingredients. First, the reasoner wants to reach a desired end state, which usually involves attaining a specific goal. Second, a sequence of mental processes must be involved in reaching this end state. The involvement of a sequence of mental processes is intended to distinguish reasoning from goal-directed behaviour such as opening your mouth when you see a feeding bottle. Third, the mental processes involved should be cognitive rather than automatic. An automatic or routine sequence of behaviour, such as playing a 'peek-a-boo' game, does not qualify as cognitive. Although we will only discuss a few examples of reasoning and problem-solving in infancy here, experiments investigating infants' understanding of the physical world have produced many more. A recent review is provided by Baillargeon, Scott and Bian (2016).

Reasoning about objects and events

One ingenious problem-solving experiment concerning infants' understanding of the physical world asked whether infants understand that, when they have seen a toy bear sitting under an inverted plastic cup, it is impossible to retrieve the *same* bear from a previously empty toy cage. Baillargeon, Graber, DeVos and Black (1990) examined this question by showing infants a display in which a clear plastic cup and a small cage were sitting side by side on a table, with the cage to the right of the inverted cup. A toy bear was visible inside the plastic cup, but the cage was empty (see Figure 2.5). A screen was then raised so that the two containers were hidden from the infant's view.

As the infants watched, a hand appeared to the right of the screen, reached behind it, and reappeared holding the cage. The hand then reached behind the screen for a second time, and reappeared holding the toy bear. This was an impossible event, and the infants looked significantly longer at this event than at the same

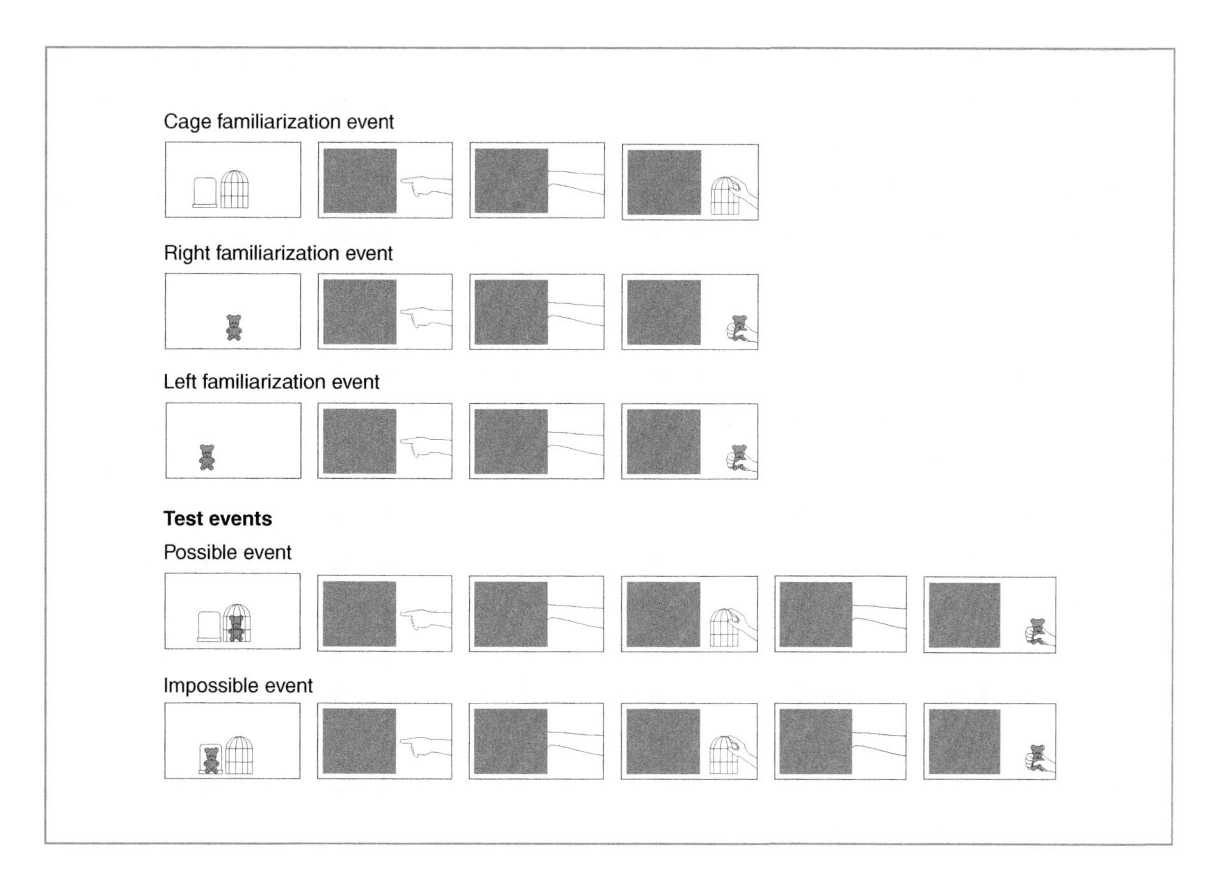

FIGURE 2.5
Depiction of the familiarisation and test events in the 'bear in the cup' paradigm.
From Baillargeon at al. (1990). Copyright © 1990 Elsevier. Reprinted with permission.

event in a control condition in which the bear was first shown to be inside the cage rather than inside the cup. In order to look longer at the impossible event, the infants had to believe that the bear, cup and cage continued to exist behind the screen and also had to retain a representation of their locations once they were hidden by the screen. On the basis of these premises, they then had to infer that it was impossible to retrieve the bear from the empty cage.

Another example of reasoning about physical events in infancy concerns infants' ability to judge the size of a hidden object (Baillargeon & DeVos, 1994). In Baillargeon and DeVos's 'hidden object' studies, the infants' task was to work out the size of an object hidden beneath a cloth. In a typical experiment, the infants were first shown a lump covered by a soft, fluid cloth. This array was then occluded by a screen. As the infants watched the display, a hand reached behind the screen and first reappeared holding the cloth cover, then reached again and subsequently reappeared holding a very large toy dog. This toy dog was so big that, to an adult, it was obvious that it could not have been the object causing the original lump beneath the cloth. However, the infants (12.5-month-olds) did not look longer at the appearance of this impossibly large 'hidden' object.

Subsequent experiments showed that one reason that the infants did not show increased looking was that they found it difficult to retain a memory of the *absolute* size of the lump once it had been occluded by the screen. When Baillargeon and DeVos provided a second, identical protuberance under a cloth as a memory

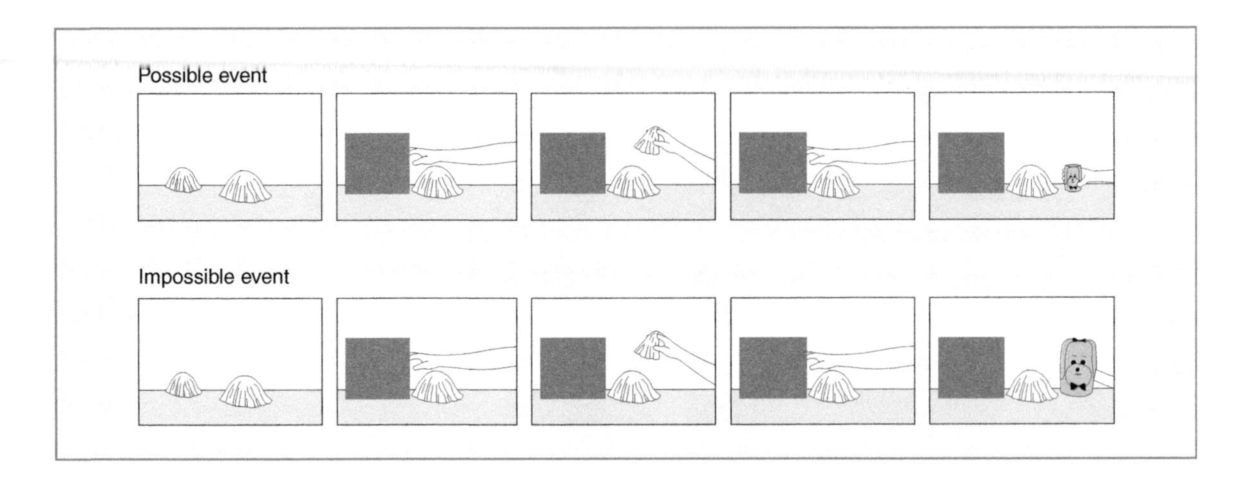

FIGURE 2.6
Depiction of the events used in the 'dog beneath the cloth' experiment when a relational comparison was possible. Redrawn from Baillargeon and DeVos (1994). Reproduced with kind permission from the author.

cue which remained visible when the screen concealed the lump, then 12.5-month-old infants were able to reason about the plausible size of the hidden object (see Figure 2.6). In this study, the infants were able to make a direct comparison between the very large toy dog that emerged from behind the screen and the rather small lump as the hand reached first for the cloth and then for the toy dog. When they could make a relational comparison, the infants looked significantly longer at the emergence of the impossibly large toy dog than at the emergence of a smaller toy dog that was of a suitable size to have caused the lump beneath the cloth. The infants were thus able to use the visible lump to reason about the size of the hidden object.

How can we be certain that infants were indeed *reasoning* about the physical parameters of the situations in the paradigms just discussed? One way is to see whether they would stop being 'surprised' at impossible physical events if the mechanism behind the 'trick' was revealed. This would be good evidence for problem-solving skills in infancy, as it would involve a sequence of mental processes that were non-automatic.

One experiment in which the 'trick' behind an impossible physical event was eventually revealed to the infants was the study by Baillargeon and Graber (1987) concerning spatial relations (discussed in Chapter 1). In this study, infants were habituated to either a tall or a short rabbit passing behind a wall. The mid-section of the wall was then lowered, and the impossible physical event was that the tall rabbit passed behind the short mid-section of the wall without the top half of his body becoming visible. This impossible event generated considerable looking among 5.5-month-old babies. The impossible event was actually produced by using two rabbits. In a follow-up study, Baillargeon and Graber showed a new group of infants how the 'trick' had been produced. These 'informed' infants received two pre-test trials, during which first two tall rabbits and then two short rabbits were shown standing motionless at each side of the windowless screen. The experiment then proceeded as before. This time, neither group showed extended looking when the rabbit failed to appear behind the short section of the wall, whether they had been habituated with the tall rabbit or the short rabbit. The infants had apparently used the information from the pre-test trials to work out that the trick had been produced by using two rabbits.

The disappearance of the dishabituation effect in this experiment is particularly remarkable when it is remembered that the infants did not see the rabbits *in motion* during the pre-test trials. This would appear to rule out a lower-level explanation based on the encoding of dynamic spatial-temporal trajectories. The infants simply saw two static short rabbits, and two static tall rabbits. They then used this information to work out that while one rabbit disappeared behind the fence at the beginning of a habituation trial, a *second* rabbit most probably 'reappeared' at the other end. In this paradigm, the infants were apparently using the information about two rabbits to make sense of a surprising phenomenon. They used the information about the static rabbits to solve the problem of how an impossible event of this nature could have been produced.

Experiments such as these seem to provide convincing evidence of reasoning and problem-solving behaviour by young infants. All of the criteria set by Anderson (1990) for defining reasoning and problem-solving behaviour appear to be fulfilled. The desired end state is to generate an explanation for an impossible event, which is a specifically cognitive goal. This is achieved by a sequence of mental processes, requiring the combination of a number of premises that are represented in memory. These processes are cognitive rather than automatic as the premises involve information that is not directly observable. It is difficult to disagree with Baillargeon's conclusion that the infants in these experiments are engaging in a knowledge-based, conceptual analysis of the physical world (Baillargeon, 2004).

Numerical relations

Another way in which babies might use the perceptual structure of events in the visual world as a basis for conceptual analysis is in the representation of numerosity. The understanding of relations such as 'greater than' and 'less than' are a crucial aspect of the number system. Equally important is the understanding that a quantity remains the same unless something is added to it or taken away from it. Habituation studies have been used to investigate whether babies can represent quantitative relations fairly early in life, and also some numerical relations.

For example, Cooper (1984) devised a habituation paradigm in which infants were habituated to pairs of arrays of coloured squares (see Table 2.1). These arrays either depicted the relation 'greater than' or the relation 'less than'. For the 'greater than' relation the infants might be shown a pair made up of four squares in array 1 versus two squares in array 2, then a pair of four squares in array 1 versus three squares in array 2, and then two versus one square. At test, the infants received either a reversed relation ('less than'; three squares in array 1 versus four squares in array 2), an 'equal' relation (two versus two), or a novel exemplar of the same relation (three versus two). At ten months, the infants dishabituated to the 'equal' relation only, showing that they could differentiate equality from inequality By 14 months, the infants dishabituated to the 'less than' relation as well, showing an appreciation of relational reversal. Similar paradigms have also been used with younger babies (e.g., Starkey & Cooper, 1980).

TABLE 2.1 Examples of numerosity arrays used by Cooper (1984)

Condition	Numerosity of array 1	Numerosity of array 2	Trial type
Less than:			
Habituation	3	4	
	2	4	
	1	2	
Test	3	4	Old
	2	3	New
	4	3	Reversed
	2	2	Equal
Greater than:			
Habituation	4	2	
	4	3	
	2	1	
Test	4	3	Old
	3	2	New
	3	4	Reversed
	2	2	Equal
Equal:			
Habituation	4	4	
	2	2	
	1	1	
Test	4	4	Old
	3	3	New
	2	4	Less than
	4	2	Greater than

There have also been reports of *cross-modal* understanding of number fairly early in life. Starkey, Spelke and Gelman (1983) used a paradigm similar to that used by Spelke (1976, see Chapter 1) to examine whether infants equated three sounds with three objects, and two sounds with two objects. In their experiment, the infants were given a choice of two visual arrays, one of which contained two objects and the other three objects. The objects in the arrays were changed from trial to trial, but one array always contained *two* and the other *three*. The infants also heard a soundtrack from a speaker placed in between the two arrays. The soundtrack was either repeated pairs of drumbeats or repeated triples of drumbeats. Starkey et al. found that the infants preferred to look at arrays of two objects when they were listening to a pair of drumbeats, and at arrays of three objects when they were listening to triples of drumbeats. This preference for cross-modal *congruence* was also found in the cross-modal work discussed in Chapter 1. In order to recognise cross-modal congruence in Starkey et al.'s paradigm, it was argued that the infants must have been representing the *numerosity* of the drumbeats. This striking result has proved difficult to replicate, however (Moore, Benenson, Reznick, Peterson & Kagan, 1987).

One of the most famous demonstrations of an apparent understanding of numerosity in infancy is by Wynn (1992). She studied the ability of five-month-old babies to add and subtract small numbers, using a looking-time procedure. All babies first viewed an empty display area. Once they were attending, a hand appeared in their field of view and placed a Mickey Mouse doll in the display area. Next, a small screen rotated up from the floor of the apparatus, hiding the doll from view. The hand then reappeared and placed a second Mickey Mouse doll behind the screen. When the screen dropped, it either revealed two Mickey Mouse dolls (possible event) or a single Mickey Mouse doll (impossible event). This sequence of events is shown in Figure 2.7. Wynn found that the babies looked significantly longer at the single Mickey Mouse doll, the impossible outcome. A group of babies who initially saw two Mickey Mouse dolls in the display area and then saw one being removed after the screen came up showed the opposite pattern: they looked significantly longer when the screen dropped to reveal two Mickey Mouse dolls (impossible event) than when it dropped to reveal a single Mickey Mouse doll (possible event). Wynn argued that this showed that the infants could compute the numerical results of simple arithmetical operations.

Support for her view comes from a replication of her study by Simon, Hespos and Rochat (1995). Simon et al. pointed out that Wynn's results could be explained on the basis of violations of infants' knowledge about the physical world, rather than on the basis of an innate possession of arithmetical abilities. In the 'impossible addition' condition, objects seen placed behind the screen ceased to exist, and in the 'impossible subtraction' condition, objects that did not previously exist magically appeared. Simon et al. noted that this alone could explain increased looking time. However, if Wynn's results depended on the recognition of physically impossible outcomes regardless of arithmetic, then infants should also show increased looking time in 'possible arithmetic' conditions in which the identity of Mickey Mouse was changed to someone else. Simon et al. therefore included 'impossible identity' and 'impossible identity and arithmetic' conditions in their replication of Wynn's study.

Rather than using Mickey Mouse dolls, however, Simon et al.'s 'impossible identity' conditions involved a switch between the two Sesame Street characters Ernie and Elmo. This is shown in Figure 2.8. For example (impossible identity), if the infants saw a second Elmo doll being added to a first Elmo doll which was concealed by the screen, the screen would drop to reveal two dolls, Elmo and *Ernie*. Similarly, if the infants saw an Elmo doll being removed from behind a screen known to conceal two Elmo dolls, the screen would drop to reveal a single doll, *Ernie*. Both of these outcomes are arithmetically correct but physically impossible. In the arithmetically impossible conditions ($1 + 1 = 1, 2 - 1 = 2$), the identity of the dolls was also switched (Elmo + Elmo = Ernie, 2 Elmos − Elmo = Elmo + Ernie). Thus in this condition the outcome was *both* arithmetically incorrect and physically impossible.

Simon et al. found that five-month-olds in the 'impossible identity' condition behaved just like the infants in the 'possible arithmetic' condition used by Wynn, whereas the infants in the 'impossible identity and arithmetic' condition behaved just like the infants in her 'impossible arithmetic' condition. The infants thus looked longer at arithmetically incorrect outcomes, but not at physically incorrect outcomes (a control condition had established that they *could* distinguish between Elmo and Ernie). Simon et al. argued that the spatio-temporal information in

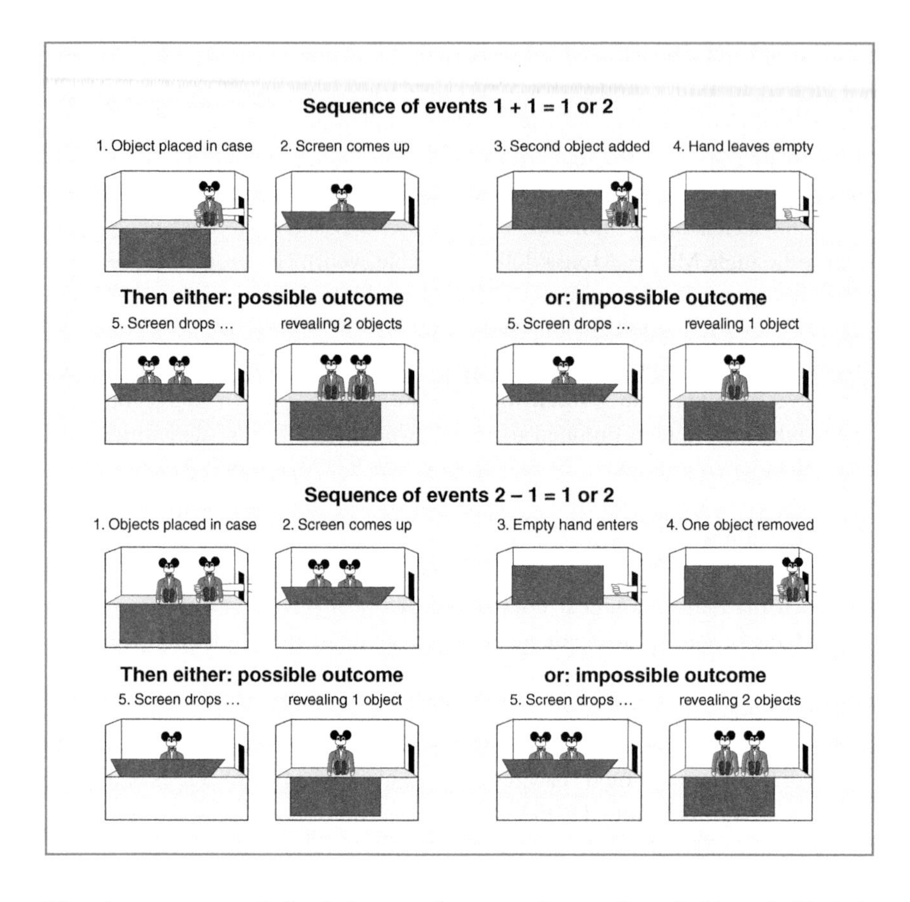

FIGURE 2.7
The addition and subtraction events used by Wynn (1992). Copyright © Macmillan Publishers Limited. Reproduced with permission.

Wynn's set-up caused the infants to focus on the *number* of objects behind the screen rather than on their identity.

Other authors have pointed out that in most of these 'number' paradigms, the change in number also involves changes in basic perceptual variables. For example, two objects have a larger surface area than one object, and infants could be responding to this feature of the displays rather than to numerosity per se. For the cross-modal representation of numbers reported by Starkey et al. (1983), it has been pointed out that the rate of presentation of the acoustic stimuli differed for two beats versus three beats, and so infants could have been responding on the basis of rate rather than of number. When number versus perceptual variables like total surface area and contour density are varied systematically in visual paradigms, then infants do not appear to respond on the basis of number (see Clearfield & Mix, 1999; Feigenson, Carey & Spelke, 2002). More recently, experiments on number have controlled perceptual variables more carefully in the displays, and have reported that infants aged six months can for example discriminate eight from 16, but not eight from 12 (see Xu & Spelke, 2000). Similarly, they can discriminate 16 from 32 when perceptual variables are controlled, but not one from two (Xu, Spelke & Goddard, 2005).

These surprising findings have led to the proposal that infants deal differently with large versus small numbers. For small numbers, namely one, two and three, infants are influenced by perceptual variables in the displays. For larger numbers

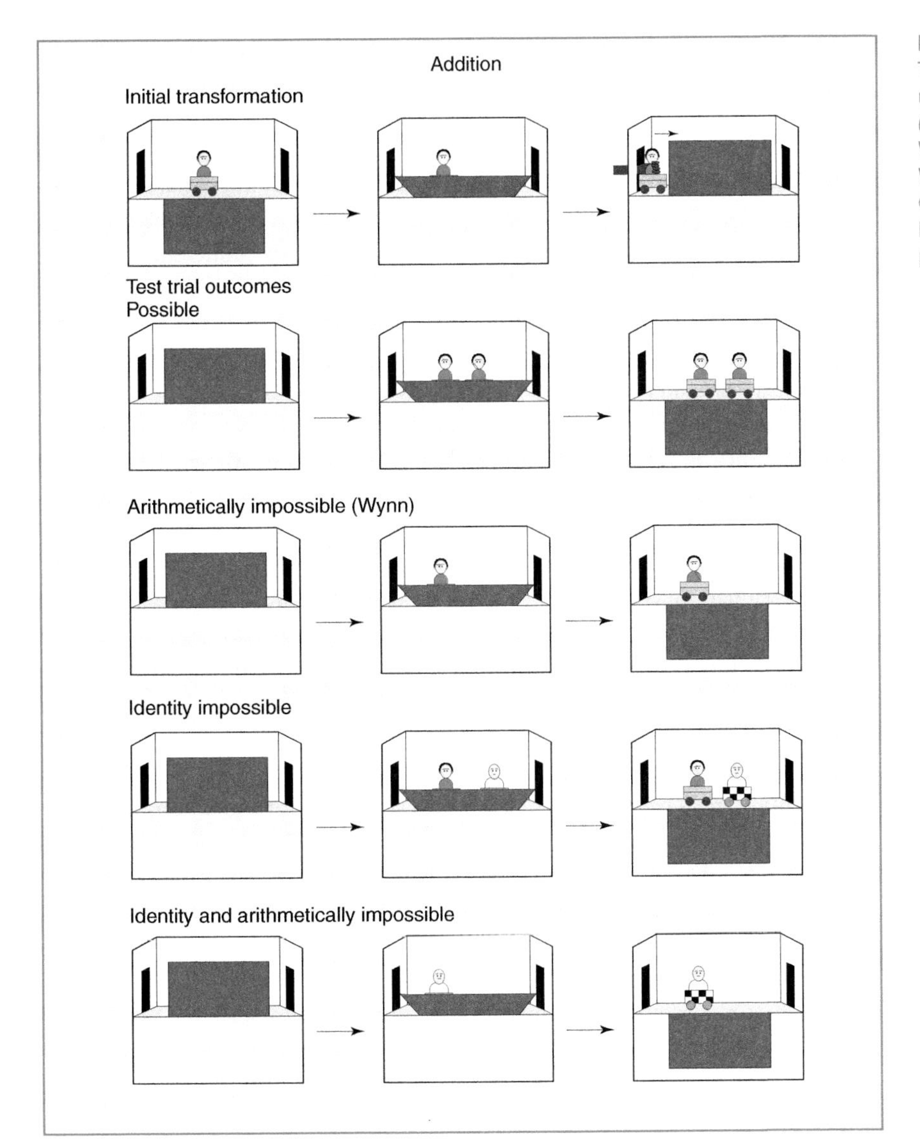

FIGURE 2.8
The addition events used in Simon et al.'s (1995) replication of Wynn (1992). From Wynn (1992). Copyright © Macmillan Publishers Limited. Reproduced with permission.

above three, infants rely on a ratio-sensitive, approximate analogue magnitude representation that is evolutionarily given and that is also found in other species (e.g., Dehaene, Dehaene-Lambertz & Cohen, 1998). This view is discussed further in Chapter 10. For present purposes, it is important to assess these experiments about infant representations of small numbers in the light of the cognitive neuroscience explanation offered by Mareschal and Johnson (2003). This explanation concerned the preference of the dorsal route for information that guides action, and the preference of the ventral route for information that identifies unique objects.

If we apply this knowledge about neural processing to the switch from Elmo to Ernie in the paradigm devised by Simon and his colleagues, we can see that in order for the infant to pick up the switch in identity, ventral processing is required. On the other hand, if the infants were more interested in the number of graspable objects present in the display, then dorsal processing would be dominant. In this situation, the ventral route may not be engaged, and so information about unique identity would be lost. The spatio-temporal information in Wynn's (1992) paradigm also seems likely to selectively engage the dorsal route. Wynn used small attractive dolls that could be manipulated by babies, and the dorsal route is interested in information that will guide action. Hence the apparent discrimination of number in these paradigms could reflect the infant's response to the fact that different actions are required when there is one desirable graspable toy present compared to when there are two. If this interpretation is correct, then experiments that do not use attractive graspable toys but rather large non-manipulable objects to alter the numerosity of visual displays should not yield 'numerical' processing.

Another example of the relevance of this neurocognitive analysis concerns an experiment about infant numerosity that has been frequently cited, by Xu and Carey (1996). They argued that even though babies as young as three months seem to make similar assumptions to adults concerning the continued existence of occluded objects, infants were unable to set up representations of numerically distinct occluded objects until the end of the first year. On the basis of a series of experiments using an occlusion paradigm with two perceptually distinct objects, a toy elephant and a toy truck, Xu and Carey (1996) suggested that infants have a fairly generalised representation of objects until the age of at least ten months. They argued that infants do not represent the identity of *individual* objects until slightly later in development.

Xu and Carey's basic paradigm contrasted two occlusion conditions, a *property-kind* condition and a *spatio-temporal* condition (see Figure 2.9). In the property-kind condition, the infant was shown a single screen. A toy truck was brought out from the right side of the screen and returned behind it. A toy kitten was then brought out from the left side of the screen, and returned behind it. These successive events were repeated three more times. This set of repetitions comprised the habituating event. At test, the screen was removed to reveal either one object or two objects. Xu and Carey reasoned that infants who could set up representations of numerically distinct objects should find the single object outcome surprising, and so look longer at the single object than at two objects. In the spatio-temporal condition, the same sequence of successive emergences of the truck and kitten was preceded by a single trial in which the two toys were brought out *simultaneously* from behind the screen, one to each side. Infants in this condition were again expected to look longer at the single object outcome. Finally, a *baseline* condition assessed infants' intrinsic preference for looking at one object versus two objects in the absence of any occlusion events.

Xu and Carey found that infants looked longer at the two-object outcome in the baseline and property-kind conditions, but looked longer at the single-object outcome in the spatio-temporal condition. This basic finding was then replicated a number of times. Xu and Carey concluded that infants have an intrinsic preference for looking at two objects, and that this preference is overridden in the spatio-temporal condition but not in the property-kind condition. From this, they concluded that ten-month-old infants were unable to use the perceptual differences between the toy kitten and the toy truck to infer that there were two distinct

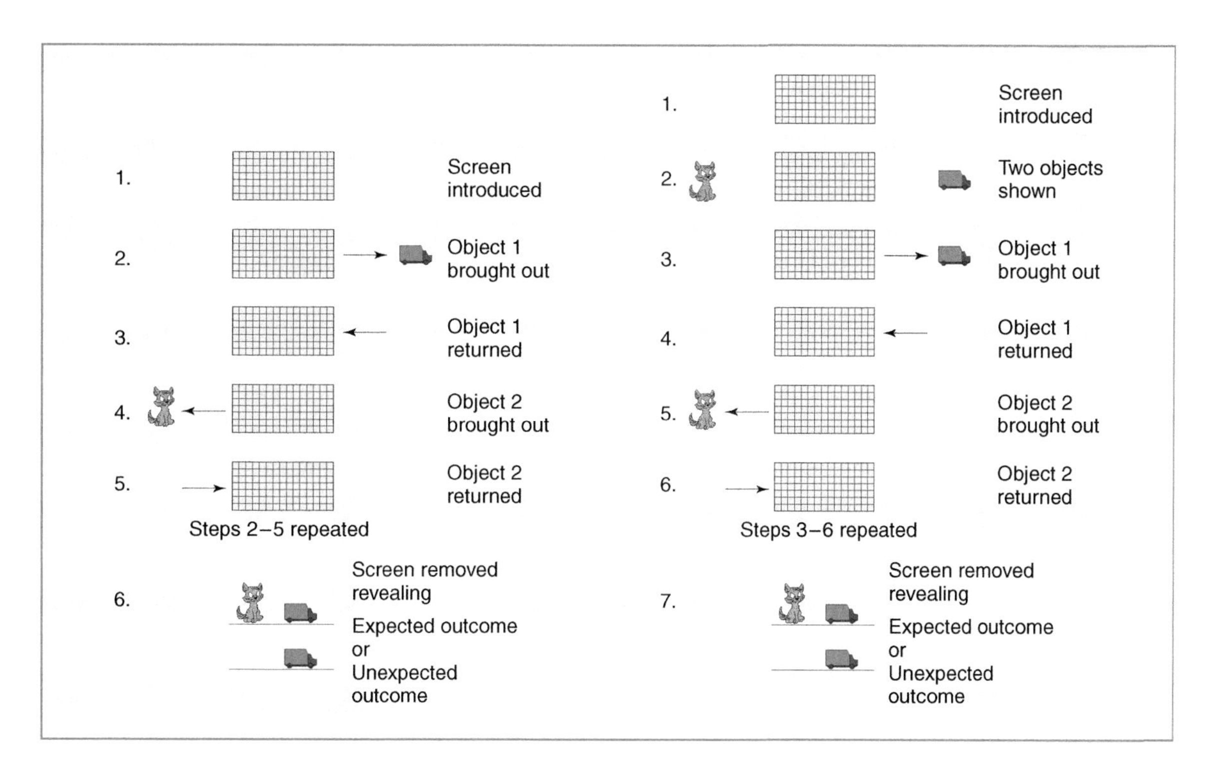

FIGURE 2.9
Schematic representations of the property-kind condition (left) and the spatio-temporal condition (right) used in the experiments by Xu and Carey (1996). Copyright © 1996 Elsevier. Reprinted with permission.

objects behind the screen. Note that this conclusion has been drawn on the basis of a negative result – that there was no difference in looking times. As we have seen, it is risky to conclude that an infant or young child lacks a certain cognitive competence on the basis of a negative experimental result. The explanation may be instead that the chosen paradigm is not sensitive enough.

Indeed, an alternative explanation for Xu and Carey's findings is that both conditions were selectively activating the dorsal visual route, because the stimuli being used were small graspable toys (see Kaufman et al., 2003b). The property-kind condition needed to activate the ventral visual route in order to process surface features, but the nature of the toys being used conspired against this. The infants were more interested in what actions they could perform on the toys, and so dorsal processing was prioritised. Kaufman et al. (2003b) argued that objects that are small, within reach and/or moving are likely to elicit dorsal processing by infants. A consequence of this is that infants will not encode surface features such as colour and texture. Only if objects are large and cannot be grasped will ventral stream processing come into play.

A converging explanation with this action-based view is that younger infants tend to differentiate objects on the basis of their *functions*. The toys used to test object individuation by Xu and Carey (1996) both had similar functions – they were toys. Stavans and Baillargeon (2018) devised a comparable experimental paradigm to Xu and Carey (1996) for four-month-old infants, using two visually distinctive kitchen implements, a potato masher and a pair of tongs. The infants were first shown a demonstration of the function of each object, via a small stage and a display of sponges. As the watching infants sat on their parents' laps, a human hand appeared through a window at the back of the stage, and demonstrated using the masher to squash each sponge in turn, and the tongs to pick up each sponge in turn In the critical test trials, a screen was then lowered. The infants then watched the stage as a hand brought out the tongs from one side of the screen, and the masher from the other side of the screen. The screen was then lowered to reveal either a single object or both objects. Stavans and Baillargeon (2018) found that the infants looked significantly longer when only one object was present. They concluded that object individuation was available to infants as young as four months of age if categorical information concerning *function* was available.

Recently, the nature of the development of object individuation during infancy has been explored using cognitive neuroimaging, and the overall conclusion from such studies has been that younger infants are indeed able to individuate objects by feature (Wilcox & Biondi, 2015). Wilcox and Biondi (2015) reviewed recent infant object processing studies employing fNIRS as a measure of ventral versus dorsal route object processing. The fNIRS studies showed that infants can attend to perceptual features of objects such as colour and use these features as a basis for object individuation by as young as 3.5 months of age. The fNIRS literature has also enabled new insights into the role of dorsal versus ventral streams of information processing during early object processing by infants. Visual acuity is poorer in younger infants, and so they appear to rely more on motion information to identify objects, which is processed by the dorsal pathway. As their visual acuity improves, and possibly as they develop more 'embodied' knowledge about objects (e.g., about the actions that the objects afford), the ventral pathway becomes more dominant. For example, different patterns of neural activation are observed in the anterior temporal cortex (ventral pathway) in response to function versus non-function events with objects, and in response to human versus robotic events with objects, in infants aged 4–9 months. This differential processing is consistent with the cognitive differentiation of objects with different functions shown by Stavans and Baillargeon (2018). Nevertheless, systematic developmental neuroimaging data regarding the ventral object-processing pathway and changes in activation over the first year of life are currently absent from the literature.

LEARNING

As we have already seen, babies learn about the world very quickly indeed: every experience that they have every day constitutes an opportunity for learning. Infant brains automatically encode perceptual experiences in ways that *in themselves* appear to impose causal structure on these experiences, thereby enabling further learning to be more complex. Many researchers have noted that the kinds of information that infants learn appears to be 'constrained', since some types of information are learned more easily than others. This idea is then extended to argue that certain aspects of the external world are *prioritised* for learning. Innate 'constraints on learning' are said to govern these priorities, helping to determine the objects and events that babies give their attention to. However, as discussed in the Foreword to this book, studies in developmental cognitive neuroscience in my view question such descriptions of infant learning. Instead, it can be argued that it is the mechanisms of information-processing that the brain uses to encode sensory information that are constraining what is learned. Accordingly, these intrinsic neural ways of encoding and processing sensory information may in themselves be the building blocks for developing cognitive systems. Rather than nascent cognitive theories driving the infant's learning, the activity of neural sensory systems as these systems acquire and process information, aligned with unconscious behavioural adaptions by adults that facilitate encoding (like *motionese*), guide early learning. As infants become able to act on objects by themselves, this early learning will be enriched and extended into meaning-based knowledge representations – schemas or concepts. It is also important to remember that the infant is an *active* learner. Infants do not simply process everything in the visual field in a passive fashion, infants *select* which actions and events to watch, or which objects to manipulate.

The dorsal/ventral neural architecture of the human visual system provides a good example of this new way of thinking about 'constraints on learning'. If infants prioritise action, and if it takes time developmentally to become efficient at integrating the information yielded by the dorsal and ventral visual routes, then learning that requires attention to the unique identities of objects will be constrained by the nature of the objects in the visual field. A simple way to define learning is to say that it is the modification of behaviour in the light of experience. Even simple organisms such as *aplysia* learn according to this definition. Research with animals has focused on a number of different kinds of learning, including habituation and associative learning. These types of learning are an inherent part of how animal brains learn and, as we have seen, can also be identified in infants. In cognitive psychology, however, learning is usually measured in terms of what has been *remembered* as a result of learning, either via measures of *recognition* or via measures of *recall*. We will examine the apparently more cognitive forms of learning in the rest of this chapter. These are learning by imitation, learning by analogy, and explanation-based learning, which are rarely found in animals. Explanation-based learning is a form of causal learning. Learning by infants also takes off rapidly once they can operate on the world by grasping and examining objects, and then by moving their bodies around their environments, by crawling and then by walking. Once you can walk upright, you can use your hands to carry objects. The impact of self-locomotion on learning is particularly interesting, as we will see.

Even very simple organisms, such as *aplysia*, are able to learn and develop by modifying their behaviour in light of experience.

Learning by imitation

Learning by imitation can be defined as 'B learns from A some part of the form of a behaviour' (Whiten & Ham, 1992). One example is learning the use of a novel tool by imitating the actions of another user with that tool. Most definitions of imitation require that something new is learned, and such learning has proved remarkably difficult to distinguish in animals, even though until quite recently the contrary was believed to be true. In fact, many psychologists now believe that learning by imitation is beyond the cognitive abilities of all animals, even animals like monkeys and chimpanzees, who are arguably the most human-like members of the animal kingdom (e.g., Tomasello, 1990; Tomasello & Call, 2018). Tomasello has argued that humans differ profoundly from apes in their skills of imitation and imitative learning, because the ability to learn novel behaviours via imitation depends on the ability to understand the *intentions* of others.

Despite its cognitive sophistication, learning by imitation is present by the age of at least nine months (Meltzoff, 1988a). As we saw earlier, older infants (18 months) can even imitate an adult who demonstrates an *intention* to act in a certain way, although the action is never completed, suggesting that imitation may indeed involve the ability to understand the intentions of others (Meltzoff, 1995a). In fact, most of our knowledge about imitative learning in infants comes from the pioneering work of Meltzoff, who has expanded his early research on the imitation of adult facial gestures in a number of interesting ways. Many of his more recent experiments depend on the use of *deferred imitation* as a test of learning. Meltzoff's test is to see whether infants can reproduce a novel action that they have observed previously even if they were not allowed access to the critical materials at the time of learning. This is a strong test of imitation, as the ability to duplicate actions that have been absent from the perceptual field for some time makes it more likely that the infant is actively reconstructing what he or she has seen and therefore actively imitating. The involvement of 'recall' rather than 'recognition' memory in deferred imitation also makes it a useful test of memory (see the research by Mandler and her colleagues, discussed in Chapter 1).

In one of his first studies of deferred imitation, Meltzoff (1985) devised a paradigm suitable for 14-month-old infants based on the manipulation of a novel toy. The toy, which was specially constructed for the experiment, was a kind of wooden dumbbell made up of two wooden blocks joined by a short length of rigid plastic tubing. The rigid tubing gave it the appearance of a single object. However, the wooden blocks could in fact be pulled apart, but only if sufficient pressure was applied. In the experiment, the experimenter sat across from the baby at a small table, and produced the toy. In the *imitation* condition, he then pulled it apart three times in succession, using very definite movements. In the control condition, the experimenter moved the toy in a circle three times, pausing between each rotation. Each action (pulling versus circling) lasted for the same amount of time. In the *baseline* condition, the experimenter simply gave the infant the toy to manipulate, without any behavioural modelling. In all three conditions, the infants were then sent home.

Twenty-four hours later the infants returned to the laboratory, sat down at the same table, and were given the toy to manipulate. The critical question was whether the infants in the imitation condition would be more likely to pull the toy apart than the infants in the baseline and the control conditions. This was exactly what happened. Forty-five per cent of the babies in the imitation condition immediately pulled the toy apart themselves, compared to 7.5% of the control and baseline groups, whose data were added together. The babies in the imitation group were also much faster at producing the pulling-apart behaviour than any control babies who managed to do so.

In a later study, Meltzoff (1988b) expanded the novel modelled behaviours to six, and the delay before allowing imitation to one week. In addition to the dumbbell used in the 1985 study (target act = pulling apart), 14-month-old babies were shown a flap with a hinge (target act = hinge folding), a box with a button (button pushing), a plastic egg filled with metal gravel (egg rattling), a bear suspended on a string (bear dancing) and a box with a panel (head touching of the panel, which then lit up). The control group observed six different modelled behaviours with these novel objects, and the baseline group observed no actions. Imitation was again measured in terms of the production of the target acts by each group. Meltzoff found that the infants in the imitation group produced significantly more of the target behaviours than the infants in the control group and the infants in the baseline group, who did not differ from each other. The infants in the imitation group were also significantly faster at producing the target acts. Once again, therefore, Meltzoff found clear evidence for learning by imitation, even after a delay of a week. The fact that six different novel behaviours had to be retained in memory makes this demonstration particularly impressive. In a related paper, Meltzoff (1988a) demonstrated retention of three novel acts by nine-month-olds over a delay of 24 hours, and subsequently he demonstrated deferred imitation in 14-month-olds over delays as long as 2–4 months (Meltzoff, 1995b). Learning by imitation appears to be crucial for infant cognition.

What if infants could learn not only by observing people acting on objects in real life, but by observing people acting on objects in films and videos? Given the ubiquity of the television in the modern home, this would expand infants' potential learning experiences by a huge degree. Meltzoff (1988c) has evidence that infants of 14 months of age can indeed learn novel actions from watching television. In his television study, the target action was again pulling apart (the dumbbell), but this time the infants watched the experimenter model the action on a 22-inch television set and never saw him live. The experimenter was filmed separately for each infant, enabling him to see their reactions on a video monitor so that he could wait until the infant was fixating on the dumbbell before pulling it apart. The experimenter also gained the infant's attention if necessary by calling "Look", or "Can you see me?" to babies who did not immediately begin to watch the television (this was apparently quite rare). Following the demonstration of the pulling-apart action, the infants were sent home. Control and baseline groups were also tested, as in previous studies.

When the infants returned to the laboratory the following day, they were given the toy to manipulate. As in real TV viewing, the infants did not see the experimenter 'live', but were handed the toy to manipulate by their parents, who followed instructions from the televised experimenter. Meltzoff again found that the imitation group were significantly more likely to produce the target behaviour than the control and baseline groups, just as in the 'live' version of his study. Forty per cent of the babies in the imitation group immediately pulled the dumbbell apart, compared to 10% of control and baseline subjects. This demonstration of deferred imitation from the TV is a rather sobering one, as Meltzoff himself points out. If

Meltzoff (1988c) found that young infants can reproduce behaviours they witness on TV and proposed that older children would be similarly susceptible to deferred imitation.

such young infants can reproduce behaviours that they see on TV and incorporate them into their own routines, then it is unlikely that TV viewing leaves older children unaffected. On the other hand, Meltzoff's demonstration is also an exciting one, as it means that TV can be used in a constructive way to enhance learning in target infant groups. Perhaps we should leave the last word to a toddler quoted by Meltzoff, who, on seeing his father pick up a bottle of beer, pointed to the bottle and exclaimed "Diet Pepsi, one less calorie!"

Learning by analogy

Learning by analogy involves finding certain correspondences between two events, situations or domains of knowledge and then transferring knowledge from one to the other (e.g., Keane, 1988). So far, learning by analogy has only been demonstrated in one member of the animal kingdom, the highly unusual ape, Sarah, who had learned a limited language (Gillan, Premack & Woodruff, 1981). As put memorably by Winston (1980), in learning by analogy 'we face a situation, we recall a similar situation, we match them up, we reason, and we learn'. We may decide whether a dog has a heart by thinking about whether people have hearts (see Chapter 5), or we may solve a mathematical problem about the interaction of forces by using an analogy to a tug-of-war (see Chapter 7). Reasoning by analogy has usually been measured in children aged three years or older (see Goswami, 1992, 2001, for reviews). Recent research, however, has shown that learning by analogy is available in infancy.

Chen and his colleagues devised a way of studying learning by analogy in infants as young as ten months of age, following a procedure first developed by Brown (1990) for 1.5- to two-year-olds. Brown's procedure depended on seeing whether toddlers could learn how to acquire attractive toys that were out of reach. Different objects (such as a variety of tools, some more effective than others) were provided as a *means* to a particular *end* (bringing the desired toy within grasping distance). The analogy was that the means-to-an-end solution that worked for getting one toy in fact worked for all of the problems given, even though the problems themselves appeared on the surface to be rather different. Brown and her colleagues used this paradigm to study analogical reasoning in children aged 17–36 months. Chen, Campbell and Polley (1997) were able to extend it to infants.

In Chen et al.'s procedure, the infants came into the laboratory and were presented with an Ernie doll that was out of their reach. The Ernie doll was also behind a barrier (a box), and had a string attached to him that was lying on a cloth (see Figure 2.10). In order to bring the doll within reach, the infants needed to learn to perform a series of actions. They had to remove the barrier, to pull on the cloth so that the string attached to the toy came within their grasp, and then to pull on the string itself so that they could reach Ernie. Following success on the first trial, two different toy problem scenarios were presented, each using identical tools (cloths, boxes and strings). However, each problem appeared to be different to the problems that preceded it, as the cloths, boxes and strings were always dissimilar to

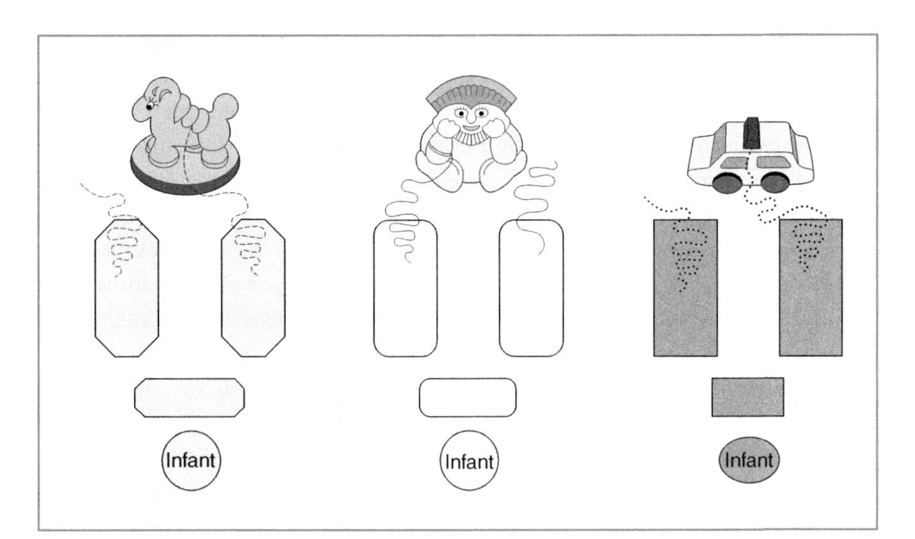

FIGURE 2.10
Depiction of the problem scenarios used to study analogical reasoning in infants by Chen et al. (1997). Copyright © American Psychological Association. Reprinted with permission.

those encountered before. In addition, in each problem *two* strings and *two* cloths were provided, although only one pair could be used to reach the toy.

Chen et al. tested infants aged ten and 13 months in the Ernie paradigm. They found that although some of the older infants worked out the solution for getting Ernie on their own, others needed their parents to model the solution to the first toy acquisition problem for them. Once the solution to the first problem had been modelled, however, the 13-month-olds readily transferred an analogous solution to the second and third problems. The younger infants (ten months) needed more salient perceptual support in order for learning by analogy to occur. They only showed spontaneous evidence of using analogies when the perceptual similarity between the problems was increased (for example, by using the same goal toy, such as the Ernie doll, in all three problems).

Analogical reasoning may be a particularly important form of learning for cognitive development, as it involves reasoning about *relations*. As we will see throughout this book, a focus on relations, particularly *causal* relations, is very important in children's cognition. Analogies, especially analogies involving causal relations, may provide a critical cognitive tool for knowledge acquisition and representation and for conceptual development. More detailed treatment of the view that analogies play a fundamental role in cognitive development can be found in Carey (1985), Goswami (1992) and Halford (1993).

Explanation-based or causal learning

Explanation-based learning in infancy has been explored in detail by Baillargeon and her colleagues, who have proposed that explanation-based learning is the core mechanism used by infants to identify new variables as they build their knowledge of the physical world. As we saw in Chapter 1, Baillargeon (2001, 2002) has proposed a detailed model of the development of physical reasoning in infancy (Baillargeon, Li, Ng & Yuan, 2008, for review). According to this model, infants identify event categories such as containment and support, and these categories

KEY TERM

Core knowledge
Fundamental notions or principles that guide understanding of events, claimed to be innate.

are initially understood in terms of a few fundamental notions deemed to be 'core knowledge'. This core knowledge provides a shallow causal framework for understanding perceptual events, and relies on simple principles such as persistence (that objects exist continuously in time and space). Basic information about physical events such as support is represented, such as *no contact = object falls* and *partial contact = object is supported*. As infants experience more and more events, more elaborate representations are developed in which variables that are relevant to the events' outcomes are identified and represented, such as *degree of contact* for support events. The process whereby infants identify new variables in event categories is thought to be explanation-based learning.

In machine learning, explanation-based learning depends on constructing causal explanations for phenomena on the basis of specific training examples. The aim is to provide machines with ways of applying domain-specific knowledge to formulate valid generalisations from single instances of a phenomenon. The underlying idea is that the ability to generalise from a single example depends on the machine's ability to explain to itself why the training example is an instantiation of a concept that is being learned. However, to develop this explanation, the machine is also given a set of rules and facts that can be used to explain why the target example is an instance of the target concept. Infants are faced with similar problems in learning. They see a variety of instantiations of a particular phenomenon, such as different objects falling from different surfaces. The infants need to extract the generalisation that objects fall when they are inadequately supported. In order to make this generalisation, infants need to work out the relevant rules and facts related to the concept of support, by attending to variables such as the degree of contact between relevant surfaces of the object and the supporting surface, the nature of the object (e.g., is it unevenly weighted?), the nature of the supporting surface (e.g., is it sticky?), and many other variables. Baillargeon has argued that in order to identify these variables, infants engage in three sub-processes that together constitute explanation-based learning. First, infants notice contrastive outcomes (e.g., when an object falls and when it does not). Second, infants search for the conditions that determine the contrastive outcomes (e.g., degree of support). Third, they use core knowledge to supply an explanation for this condition-outcome relationship. Condition-outcome relationships that can be explained via causal explanations are then used to identify new variables relevant to a particular event category.

Baillargeon has argued that the experience-dependent nature of explanation-based learning can explain why infants will learn about certain perceptual variables in some event categories before others. For example, we saw in Chapter 1 that infants identified height as a relevant variable for occlusion events at around four months, as a relevant variable for containment events at around seven months, and as a relevant variable for covering events at around 12 months. One reason for this might be that infants have more experiences of the relative heights of occluders and the objects that they occlude than of containers and the objects that they contain. Caretakers are very likely to lower objects directly into containers, whereas many occlusion events may occur when one object (say a cereal packet) is moved in front of another (say a cereal bowl). In the latter event, the infant has the opportunity to observe the relative heights of the object and the occluder.

If infants are really engaged in explanation-based learning, then manipulating the frequency with which they experience key variables in different events should affect the age at which they identify these variables as relevant to the events. Baillargeon and her colleagues reported a series of 'teaching' experiments designed to test this possibility. In one such series of experiments, Wang and Baillargeon (2006, 2008) explored whether infants aged nine months could be taught to take account of the variable *height* in covering events. As will be recalled, height is not usually taken into account in covering events until the age of around 12 months. Wang and Baillargeon (2008) devised a series of ways of teaching nine-month-old infants to attend to height. For example, in one experiment infants were given three pairs of 'teaching trials' regarding the importance of height for covering events. At the beginning of each trial, a cover (tall or short) stood next to a tall object on the apparatus floor. Whichever cover was not going to be used in that pair of trials stood against the back wall of the apparatus. A hand then rotated the cover through 90° to demonstrate that it was hollow, and returned it to the floor. Next it lifted the cover, lowered it over the object, and released it. In the tall cover trial, the cover completely concealed the object. In the short cover trial, a portion of the object remained visible. After a pause, the hand removed the cover and a new event cycle began.

After the infants had seen three pairs of covering trials, they were shown the test event used in the original covering experiments (Wang & Baillargeon, 2005, see Chapter 1). Both a short cover and a tall cover completely hid the tall object once they were lowered over it. The first was an impossible event. The nine-month-old babies who had received the teaching trials looked reliably longer at this 'short cover' event. They detected the violation three months earlier than babies who had not received any teaching trials. In a second teaching experiment with fewer teaching trials (two pairs of trials only), similar results were achieved. Again, nine-month-old babies looked reliably longer at the 'short cover' event. A third teaching experiment then introduced a delay between the teaching trials and the test trials. The infants received the teaching trials on one day, and came back to the laboratory 24 hours later for the test trials. Again, the infants looked reliably longer at the 'short cover' event. Wang and Baillargeon (2008) argued that exposure to the contrastive outcomes with the tall and short covers during the teaching trials was triggering explanation-based learning. Consequently, infants were identifying relevant variables for covering events on the basis of specific experiences. Further evidence for this account was gained by preventing the infants from experiencing contrastive outcomes. For example, small objects were covered in the training trials, so that the entire object was concealed by *both* tall and short covers. In these conditions, the infants did not learn about the importance of height in covering events.

To check that the infants were engaging in causal learning, Wang and Baillargeon (2008) conducted a final experiment. In this experiment, they created a situation that was perceptually very similar to the teaching events, but for which no causal explanation could be provided. This was done by putting false tops into the tall and short covers, so that both were only 2.5 cm deep. When the covers were rotated at the beginning of the teaching trials, the infants could hence see that they were extremely shallow. It thus made no sense in causal terms when the short cover only concealed a portion of the object, while the tall cover concealed the entire object. There was no plausible causal explanation for the contrastive outcomes. In this final experiment, the nine-month-old infants did not appear to learn about the height variable, as they did not look reliably longer at the 'short cover' event. Wang and Baillargeon concluded that explanation-based learning is distinct from associative learning in infants. Even nine-month-old infants were making causal interpretations of condition-outcome regularities.

These demonstrations of explanation-based learning in infants can also be explained neurally in terms of dynamic encoding via *trajectories* of neural activation over time and cortical space (Fekete & Edelman, 2011), which would make explanation-based learning an automatic outcome of neural mechanisms for encoding and processing information. These extended neural trajectories would by their very nature result in causal learning, as the temporal sequence of neural activation in itself would specify the causal relations. These neural trajectories would also by their very nature be contrastive – as they would differ at the point of encoding. Accordingly, the particular training examples experienced would enable the automatic detection of 'violations'. Therefore, explanation-based learning could result from automatic processes (as, of course, is the case for machine learning). Once the infants became able to manipulate the tall and short covers for themselves, however, then learning may become explanation-based in the cognitive sense originally intended by Baillargeon. Infants would themselves determine the actions on objects that were required to learn the relevant rules and facts related to a concept like 'cover'. Active infants would understand *why* (for example) short covers cannot conceal tall objects. Subsequently, they should be able to reason about height information in *any* novel covering event, even if this event was very different in terms of perceptual features from the learning events. The infants, like the machines, would be able to formulate valid generalisations from single instances, because via a mixture of observational and causal learning, they had worked out the relevant rules and facts related to that domain for themselves.

Grasping, crawling and walking facilitate quantum leaps in causal learning

This viewpoint regarding causal learning would mean that motor development plays a critical role in cognitive development. Once babies are able to grasp efficiently, important developments in learning about objects become possible. Once babies can manipulate objects by themselves, then causal learning can really take off. This was demonstrated in a clever set of experiments using 'sticky mittens' with very young babies (three-month-olds, see Needham, Barrett & Peterman, 2002; Sommerville, Woodward & Needham, 2005). The sticky mittens manipulation revealed that being able to handle objects significantly accelerated learning. Usually, the ability to grasp objects unaided begins at around four months. In these experiments, however, the 'sticky mittens' had Velcro on them, and so soft toys stuck to the three-month-old babies' hands. This enabled these younger babies to experience making a range of different actions on the toys. The 'sticky mitten' babies subsequently showed earlier understanding of the actions of an adult who was reaching for objects. This was demonstrated in comparison to other babies of three months, who simply watched the same toys being manipulated by someone else. So carrying out actions *oneself* is important for effective learning.

Another important development in motor terms is being able to sit up unsupported. This usually occurs between four and six months of age. Being able to sit upright enables babies to expand their range of actions on the world. For example, babies can now turn objects around, and can turn objects upside down. They can feel the textures of objects, see the objects from different angles, and pass objects from hand to hand (note that 'baby gym' toys offer prone babies similar learning opportunities). As might be expected, becoming a 'self-sitter' has been shown to be related to babies' understanding of objects as 3D (Soska, Adolph & Johnson, 2010).

An even more important developmental motor milestone is becoming able to move around by yourself. Crawling and then walking enables the baby to get to the places where he or she wants to go. Crawling makes it difficult to carry objects with you on your travels, but learning to walk frees up the hands, enabling infants to carry things. Indeed, walking babies spend a lot of their time selecting objects and taking them to show their carer (Karasik, Tamis-LeMonda & Adolph, 2011). Adolph's studies show that infants spend on average 30–40 minutes in each waking hour interacting with objects. Consider that even though an expert crawler can move more quickly and more efficiently than a novice walker, babies persevere in learning to walk. Walking usually develops between 11 and 12 months in Western cultures, and babies practise and practise. Research by Adolph and her colleagues showed that newly walking infants take over 2,000 steps per hour, which means that they cover the length of approximately seven football pitches every hour (Adolph et al., 2012). The average distance travelled of 700 metres each hour means that during an average waking day, and allowing for meals, bath-time etc., most infants are travelling over five kilometres!

Self-generated movement is clearly critical for cognitive development. Being able to crawl and then to walk enables babies to go to the people and the places that *they* choose, and to initiate object-focused social interactions when they arrive. As we all know, a number of places that babies choose include places that adults do not want them to visit, like staircases, fireplaces and plug sockets. Also, initial understanding of what is physically possible for the infant to accomplish appears quite all-or-none. For example, research shows that novice walkers can have very bad judgement even with respect to choosing which path to follow. Work by Adolph and her colleagues has shown that babies will hesitate at the top of a steep slope for ages, but then plunge down headfirst nonetheless (Adolph & Robinson, 2013). Similarly, babies will dangle their foot into a gap that they cannot possibly cross, then try anyway, and fall. These all-or-none efforts are reminiscent of Baillargeon's idea concerning the simple all-or-none rules that infants first appear to employ during physical reasoning about events such as support, when, for example, they appear to assume that any degree of surface contact will allow one object to be supported upon another. Nevertheless, Adolph's research shows that most falling is *adaptive*, as it helps infants to gain expertise. This enables them to elaborate their all-or-none understanding of physical possibilities with more nuanced understanding of which variables are relevant. Indeed, Adolph's studies show that newly walking infants fall on average 17 times per hour. But they always get up again and carry on.

From the perspective of cognitive development, one important aspect of 'motor milestones' like crawling and walking is that they enable greater *agency* on the part of the infant (self-initiated and self-chosen behaviour). Like the scientist, the baby can now intervene in events, and see what happens next. Examples of the 'baby as scientist' that impinge on other family members include unplugging the Hoover during hoovering, and manipulating the buttons on the TV or DVD unobserved, thereby changing the settings. Nevertheless, the sheer enormity of experiences accumulated by infants who spend their waking hours exploring and investigating their environments is quite staggering.

Repetitive behavioural routines and early motor immaturities

It should be noted, however, that some surprising gaps have been documented in infants' cognitive abilities, gaps that have frequently been explained in terms of cognitive immaturities. Research in cognitive neuroscience has suggested that these gaps may arise for non-cognitive reasons. Rather than reflecting cognitive confusions, they appear to reflect constraints in the flexibility of neural processing, often related to the immaturity of the frontal cortex and/or its relation to action. The frontal cortex is important for the planning and monitoring of action.

One definition of cognitive activity was considered earlier in this chapter, taken from Anderson's (1990) discussion of reasoning and problem-solving. Anderson noted that the characteristics of reasoning and problem-solving situations are that the reasoner wants to reach a desired end state, that a sequence of mental processes are involved in reaching this end state, and that the mental processes are cognitive rather than automatic. He pointed out that routine sequences of behaviour did *not* qualify as cognitive. When we examine the documented gaps in infants' cognitive abilities in some detail, most of them turn out to involve repetitive, or perseverative, motor routines.

KEY TERMS

Frontal cortex
Part of the brain typically activated by higher thought processes such as the planning and monitoring of cognitive activity and the inhibition of action.

A-not-B error
When the infant attempts to retrieve an object from a previously used hiding location (A) despite having seen the object newly hidden at location B.

Search errors in reaching

The best known of these apparent gaps in cognitive ability is a search error that emerges at around nine months of age, which was first documented by Piaget (1954). This search error occurs in simple hiding-and-finding tasks that involve more than one location. Imagine that an object is hidden at one location, location A, for a number of trials. The infants retrieve the object without difficulty. The hiding location is then moved to another location, location B. Although this switch in hiding location occurs in full view of the infants, the infants persist in searching at location A. This is the 'A-not-B' error, and is shown in Figure 2.11. The A-not-B error was originally thought to stem from a cognitive confusion. Piaget (1954) argued that infants might initially believe that the location of objects was dependent upon their own actions. He proposed that infants might initially rely on egocentric spatial codes, and might therefore link the location of objects to

KEY TERM

Perseverative behaviour
The repetition of behaviour that is no longer appropriate.

places associated with previously successful actions. In order to find objects, therefore, they would search perseveratively at locations that had provided them with previous successes. A growing number of studies suggest that the A-not-B error may have a more prosaic explanation. The A-not-B error, and other surprising gaps in infant performance that depend on perseverative behaviours, may be connected with immaturities in the prefrontal cortex.

The frontal cortex of the brain is active during a number of cognitive functions. It is thought to play a central role in the planning and monitoring of action and cognitive activity. The frontal lobe is said to be the site of higher thought processes, abstract reasoning and motor processing, and also contains the primary motor cortex, which is thought to be important for the planning and control of movements. Adult patients with lesions to the frontal cortex exhibit a tendency to persist in certain motor actions. For example, if a frontal patient is asked to discover the rule (colour, shape or number) for sorting a pack of cards (with feedback), the patient will do so very successfully. If the rule is then switched, however, for example from colour to shape, the patient is unable to sort the cards according to the new rule. Instead, the patient continues sorting the cards according to the rule that was previously correct (e.g., Milner, 1963). The patient is aware that the rule has changed, but is unable to use this knowledge to guide action.

Monkeys with frontal cortex lesions demonstrate similar *perseverative* behaviours. The classic test for prefrontal cortex function in non-human primates is 'delayed reaching'. In delayed reaching tasks, monkeys retrieve a desired object from one of two identical hiding wells after a short delay. The hiding location is varied randomly over trials. Monkeys with lesions to the frontal cortex fail the delayed reaching task following delays as brief as 1–2 seconds, although they succeed if there is no delay (Diamond, 1988). Similarly, nine-month-old infants show perseverative searching in the A-not-B task following a delay of 1–2 seconds, but succeed if there is no delay. Older infants *also* show A-not-B errors if they are subjected to *longer* delays. In fact, the delay needed to produce the A-not-B error increases at an average rate of two seconds per month (Diamond, 1985). Eight- to 12-month-old infants also fail the delayed reaching task following short delays (Diamond, 1990).

FIGURE 2.11 The sequence of events in the A-not-B error. The experimenter hides an object at location A (1), and the infant reaches successfully (2). The object is then hidden at location B (3), whereupon the infant again searches at A (4). From Bremner (1988). Reproduced with permission from Blackwell Publishing.

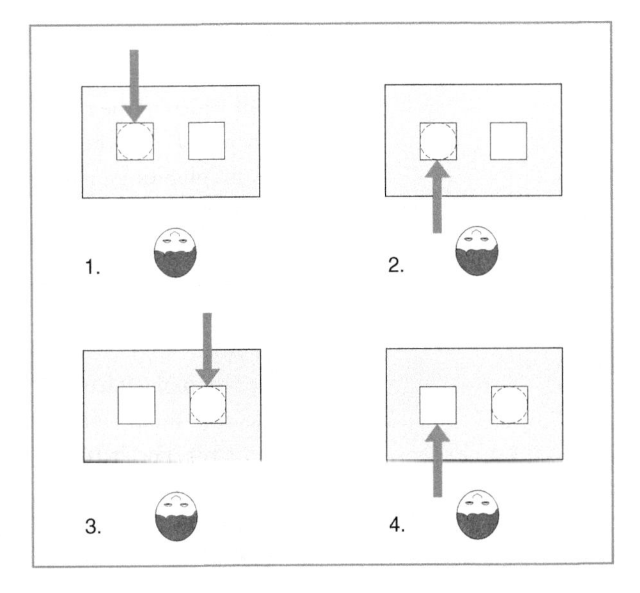

Diamond (and others) have suggested that babies who make 'perseverative' errors such as continuing to search in location A when the object has been hidden in location B are doing so because of immaturities in the frontal cortex. Her argument was that perseveration is a symptom and not an explanation of the problem faced by these infants when they are searching for desired objects. Their underlying problem is an inability to *inhibit* a predominant action tendency. The predominant action tendency in the A-not-B task is to search at A. When the hiding location is moved to B, infants find it difficult to inhibit their tendency to search at location A, and thus show perseverative errors. Cognitively, they know that the location of the object has changed, but they are unable to use this cognitive knowledge to guide action. The fact that they show these errors even when the object is *in full view* at B supports this explanation (Butterworth, 1977). In the same way, frontal patients who are unable to sort a pack of cards according to a new rule are finding it difficult to inhibit their prepotent tendency to sort by the old rule. They may actually tell you this themselves. Some patients say, as they are sorting the cards by the old rule, 'This is wrong, and this is wrong...' (see Diamond, 1988).

On the other hand, a study exploring the role of inadvertent social cueing in the classic A-not-B search paradigm showed that many ten-month-old infants can successfully find the object at location B if the change in hiding location does not depend on a visible human agent (Topal Gergely, Miklósi, Erdohegyi & Csibra, 2008). Topal and colleagues argued that the face-to-face interactive nature of the hiding trials at location A, during which the experimenter maintains eye contact with the infant and talks to the infant to maintain attention, could lead infants to mistakenly infer that they are being taught that 'this kind of toy is to be found in container A'. In a control condition, Topal et al. thus presented the same number of hiding events at location A as in the standard face-to-face interaction condition, but with the experimenter and her hands fully concealed by a sheet. This change in social context led to a dramatic improvement in successful search when the hiding event was shifted to location B. While 86% of infants in the standard face-to-face condition showed perseverative searching at location A during B-location trials, only 36% of infants in the non-social condition showed perseverative searching at location A during B-location trials. Topal et al. concluded that there was still a role for inhibition in explaining the A-not-B error, as the perseverative error was reduced rather than eliminated. However, they noted that as infants are highly social creatures, their predisposition to learn from the intentional actions of adults can be misleading in certain experimental situations.

Search errors in crawling

We will return to the role of inhibition in cognitive development later in this book (in Chapter 9). For the time being, it is interesting to note that infants show difficulties in inhibiting prepotent action tendencies in crawling as well as in reaching. Perseverative crawling has been observed both in studies using random switching of the location-to-be-crawled-to (analogous to the random switches in delayed reaching) and in studies requiring infants to crawl to a consistent location which is then switched following a series of successful trials (analogous to the A-not-B paradigm).

For example, Rieser, Doxey, McCarrell and Brooks (1982) examined whether mobile nine-month-olds could crawl around a barrier to reach their mothers (see Figure 2.12). The barrier went across the centre of the experimental room, and was too high for the infants to see over it. However, they could hear their mothers calling to them from the far side. Prior to being allowed to crawl to their mothers, the infants were carried around the apparatus, and were shown that the barrier was open at one end. On the first trial, 85% of the infants crawled successfully to their mothers. However, on subsequent trials 75% of the infants crawled to the *same* side as before, even though the 'open' side of the barrier was then varied randomly. In fact, these infants crawled perseveratively to the same side on *every* trial, whether this led to success or failure, despite being shown the open end of the barrier on *each* trial. The infants seemed incapable of inhibiting the previously executed motor pattern. In a second study of infants aged from nine to 25 months, Rieser et al. showed that the perseverative crawling response dropped out slowly with increasing age. Perseverative crawling was shown by 80% of the nine- and 13-month-olds, 44% of the 17- and 21-month-olds, but only 6% of the 25-month-olds.

McKenzie and Bigelow (1986) carried out a similar 'detour crawling' task with ten-, 12- and 14-month-olds. However, instead of varying the open end of the barrier at random, they kept the open end of the barrier consistent across a series of trials and then changed it. The open path lay either to the right or to the left side of the barrier for four trials at a time. The crucial measure was the direction of crawl on the fifth trial (which is similar to the first B trial in an A-not-B search task). McKenzie and Bigelow found that 75–80% of the younger babies showed perseveration of the now inappropriate motor response on the fifth trial, crawling to the same side as before. In contrast, only 25% of the 14-month-olds made a perseverative response. As the second block of four trials progressed, however, the younger babies, too, were able to learn to crawl to the correct side of the barrier. This suggests that the babies could learn to correct their perseverative errors. It would be interesting to measure the effect of imposing increased delays on infants of different ages in detour crawling tasks. With longer delays, perseverative crawling may be observed in later trials as well.

The observation that infants have difficulties in inhibiting prepotent motor tendencies in crawling as well as in reaching supports the view that the A-not-B search error does not arise from shortcomings in basic cognitive abilities. Instead of reflecting a difficulty in cognition, it seems to reflect a difficulty in gaining control of one's behaviour so that one's behaviour can reflect what one is thinking (Diamond, 1988). This may or may not reflect immaturities in the frontal cortex.

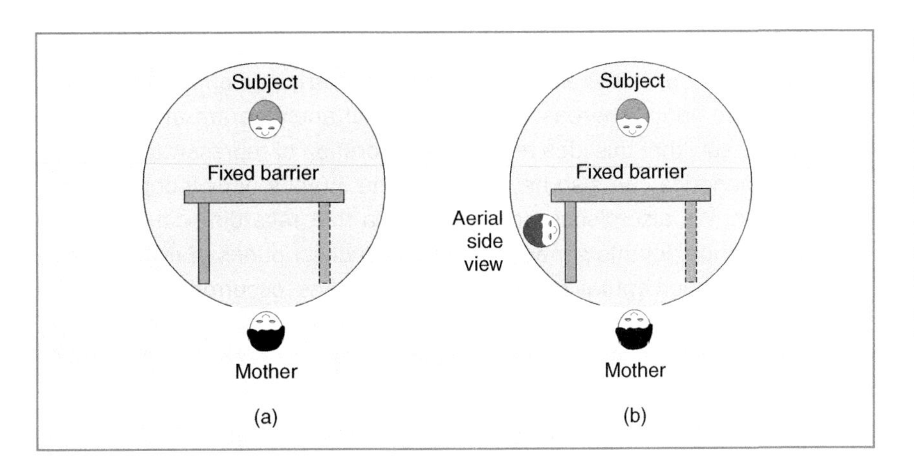

FIGURE 2.12 The experimental set-up used by Rieser et al. to study perseverative crawling, seen (a) from the viewpoint of the infant, and (b) depicting the aerial view shown to the infant. From Bremner (1988). Reproduced with permission from Blackwell Publishing.

The conclusion that knowledge representation, reasoning, problem-solving and causal learning are progressing well in infancy is therefore not undermined by errors such as the A-not-B error. Instead, it seems that we are justified in concluding that basic cognitive abilities are well-established by the end of the first year of life.

SUMMARY

This chapter set out to examine the development of cognitive knowledge about the physical world in infancy. Knowledge representation, reasoning and problem-solving about the physical world proved easy to demonstrate. We argued that knowledge representation is rooted in the automatic processes of neural encoding of the causal structure of perceptual events, encoding processes which are extended over neural time and cortical space. Accordingly, aspects of this knowledge rapidly become *conceptual*, because similarities in the perceptual/causal structure of experienced events will cause stronger encoding of core physical features than of variable perceptual details, resulting in schema-based representations. Schemas encode what is significant about the causal structure of events as well as encoding related perceptual and sensory details. On the basis of repeated experience, what is common across many instances of events is activated more strongly than what is not. This simple fact of neural learning means that the 'gist' or even 'essence' of a particular concept or schema is eventually encoded more strongly than the (variable) perceptual details.

Further, the processes by which perceptual experience yields conceptual representations are becoming increasingly amenable to investigations utilising cognitive neuroscience methods (Fekete, 2010). Historically, we saw that Baillargeon suggested that initial perceptually based descriptions of concepts such as support, and core descriptions like 'has contact', are gradually refined in the light of active experience of the physical world into more elaborate representations that identify variables that are highly relevant to outcome, such as 'degree of contact with supporting surface'. This description still captures the process of development, but it was suggested here that early perceptual analyses may be automatic, whereas more conceptual analyses are afforded by action. We saw that this idea regarding elaboration of representations via active experience can also be applied to the infant's early independent locomotion. We also discussed Leslie's idea that regarding causal analyses of motion, infants generate mechanical descriptions of events that go beyond a perceptual description of what has occurred, eventually enabling them to understand the mechanical properties of agents. Using more recent experiments involving simple dynamic perceptual displays, it was argued that these perceptual causal analyses are probably automatic. Perceptual encoding of dynamic spatio-temporal events might yield meaning-based representations simply because the ways in which the brain processes different types of spatio-temporal information automatically encode causal or agentive relations. Certain types of motion, for example, specify agency, and this is considered further in Chapter 3. Finally, it is important to recognise that perceptual information may never be discarded in favour of an amodal conceptual schema. Instead, as the actively learning infant gains deeper knowledge about causal structure, the distributed representations of initially perceptual information may be increasingly elaborated to include conceptual knowledge.

An analogy used by Fekete (2010) is useful here. Fekete noted that someone who enjoys drinking wine may decide to take a course in wine tasting in order to become a connoisseur. During the tasting course, this person would repeatedly sample different wines, and gradually educate his or her palate. However, the intricacies of flavour noted by the connoisseur (but not the novice) were always present in the wine, even when the drinker was a novice. What has changed is the *concept* of the wine, which has been increasingly enhanced and elaborated. Via repeated learning experiences, the representation of the environment (here, the taste of the wine) has been irrevocably transformed. Neurally, this translates to the multi-dimensional distribution of wine-related activity in the visual, tactile and olfactory cortices being differently organised. As noted earlier, cognitive neuroscience studies with adults suggest that the activation of a concept always produces neural activity in the sensory modalities associated with those concepts (Barsalou et al., 2003; Barsalou, 2017). What changes with expertise is the structure of the cortical activity space.

Reasoning and problem-solving have traditionally been seen as dependent on conceptual representations. We saw that reasoning and problem-solving can be defined as requiring at least three ingredients. The first is a desired end state or specific goal, the second is that a sequence of mental processes are involved in reaching the goal, and the third is that these mental processes are cognitive and not automatic. A series of experiments by Baillargeon and others based on impossible physical events, in which the 'tricks' behind impossible physical events were subsequently revealed, supports the view that young infants are indeed capable of reasoning and problem-solving behaviour defined in this way. Infants' capacities for learning by imitation and learning by analogy, cognitive skills almost universally absent in other species, also support the view that infant cognition is highly sophisticated. The ability to imitate an action seen only on a video is a particularly remarkable demonstration of this sophistication.

Finally, we considered what infants can't do. The conclusion reached was that tasks that suggest immature cognition in infancy typically involve motor perseveration. Frontal cortex continues to develop throughout childhood and adolescence, and patients with frontal cortex damage have difficulties in the planning and execution of action. Like frontal patients, infants find it difficult to inhibit predominant action tendencies in certain contexts. Infants are poor at monitoring their activities, especially when these activities involve perseverative behavioural routines. They find it difficult to stop themselves searching at a previously successful location or crawling along a previously travelled route. These 'cognitive' failures appear to provide further evidence for the intimate connection between cognitive activity and the underlying neural substrates. Although our understanding of this intimate connection is still very preliminary, developmental cognitive neuroscience could transform our understanding of infant cognition in the coming decades. So far, however, developmental cognitive neuroscience techniques have been applied to understanding early social-cognitive development rather than early physical reasoning. We consider some relevant studies in the next chapter.

CHAPTER 3

CONTENTS

Infancy: The psychological world 1

3

The means by which we come to know others as persons like ourselves has long fascinated philosophers. To explain how we become able to understand that others have unobservable things called minds is quite difficult. Somehow, we come to understand that the actions of other agents will depend on the knowledge and beliefs held in their minds, and that this knowledge and these beliefs may differ from the knowledge and beliefs that we hold in our minds. This is a sophisticated cognitive feat. We have already seen that one possible mechanism supporting naïve psychology is the infant's capacity for imitation. Infants appear to be born with a propensity to attend to and interact with other people, finding people intrinsically more interesting than other kinds of entities. Meltzoff has argued that the first common code between self and other is the ability to map the actions of other people onto the actions of our own bodies. As we saw in Chapters 1 and 2, he suggests that this imitative ability is innate, and he argues that basic aspects of the representation of action must therefore also be innate. We have considered evidence that imitation is not simply an immediate mimicry of another's actions, without any representation of those actions. We know that it is not due to arousal, and that it cannot be dismissed as a biologically based response that is released in the presence of certain 'releasing' stimuli. Rather, infant imitation appears to be creative and open to modulation, and can occur hours or even days after the modelled event.

We also saw that infants prefer to imitate people, not machines. In the last chapter, we saw that Meltzoff argued that this might be evidence for two separable causal frameworks by the age of 18 months. Infants appear to use physical causality for explaining the behaviour of things, and may be able to use psychological causality for explaining the behaviour of people. Over 20 years ago, it was proposed that a 'revolution' in social understanding in infancy occurred at around nine months of age (Tomasello, 1995). At nine months, babies begin to engage in measurable joint attention activities. They begin to make communicative gestures, and they show social referencing, using the behaviour of others as a guide to how to respond to novel objects or events in the world. There is also evidence for a growing understanding of intentionality, another mechanism related to developing a 'theory of mind'. Reading the intentions of others is critical for predicting their unobservable mental states. Further, infants appear to begin developing an understanding of the self as a causal agent. The traditional notion (stemming from Freud) that young infants are initially unable to distinguish themselves from their environments appears to be quite wrong. As we will see, very simple cognitive skills such as the ability to detect causal contingencies between events may underlie the development of an understanding of self and agency in infancy. As new methodologies transform our abilities to carry out cognitive experiments with infants, new data

Infants appear to be born with a natural propensity for finding interaction with other people more interesting than with other entities.

make it unlikely that there is a sudden revolution in development at nine months of age. Infants appear to be able to predict the intended goals of actors as early as three months of age, if they have relevant motor experience (e.g., the 'sticky mittens' studies, see Chapter 2). In fact, joint action on objects with caretakers plays a key role in the development of joint attention and social referencing (Rodriguez, 2007). Accordingly, the 'triad' of *infant, caretaker and object* is central in very early social development, with shared actions on objects acting as a crucial cultural support for psychological understanding.

THE PROPENSITY FOR SOCIAL INTERACTION

Many theorists have commented that infants enter the world predisposed for social interaction. For example, Bowlby (1971) pointed out that newborn babies seem to come equipped with behavioural mechanisms for ensuring proximity to the mother or primary caretaker. Innate actions such as rooting, grasping, crying and smiling effectively play a role in social interaction. It is very difficult to ignore the cries of a baby! Acoustic evidence suggests that this is no evolutionary accident: these cries are at a particular pitch and amplitude that is highly stressful for the caretaker, prompting immediate attention (e.g., Zeskind, Sale, Maio, Huntington & Weiseman, 1985). Infants also show strong preferences for the caretaker, preferring the mother's smell and the mother's voice (e.g., Cernock & Porter, 1985). Some of these preferences may be established while in the womb. This was demonstrated in an ingenious study by DeCasper and Fifer (1980), who tested babies' memory for their mother's voice approximately 12 hours after they had been born.

DeCasper and Fifer's (1980) experiment was based on the finding that infants can hear noises while they are inside the womb from at least the third trimester onwards (6–9 months). One sound that they hear a lot is their mother's voice. She may be talking to other people during her daily routines, talking on the telephone or even talking to the infant in her womb. If infants can recognise and remember the sound of their mother's voice, then they should be able to distinguish her voice from the voice of a female stranger. In order to see whether infants were able to do this, DeCasper and Fifer first measured how strongly the infants sucked on a dummy in the absence of any auditory stimulus. They then introduced two tape recordings, one of the infant's mother reading a story, and one of a strange woman reading the same story. For some infants, every time their suck rate increased compared to baseline, they were rewarded with the tape of their mother's voice. Every time their suck rate fell below the baseline measure, they heard the tape of the voice of the

KEY TERMS

Reversal learning
The ability to reverse a learned contingency, an ability that is present from birth.

Contingency learning
The ability to detect and learn conditional relationships between events and/or actions, such as between one's own actions and events in the environment.

stranger. For other infants, the contingencies were reversed. A low suck rate relative to baseline was rewarded with their mother's voice, and a high suck rate relative to baseline was rewarded with the voice of the stranger.

Both groups of infants rapidly learned to suck at the appropriate rate to hear their mother's voice. This shows that they remembered the sound of their own mother's voice, and that it was a familiar and comforting stimulus. Even more impressive, they could remember the contingency in a second test session given on the following day. Babies who had learned to suck strongly to hear their mother began by sucking strongly on the dummy, and those who had learned to suck slowly began by sucking slowly. The experimenters, however, had reversed the contingencies. Babies who had learned to suck strongly for their mother's voice were now meant to suck slowly, and babies who had learned to suck slowly were now meant to suck strongly. Around 80% of the babies learned to reverse their suck rate. As well as demonstrating the strength of infants' preference for their mother's voice, this is good evidence for learning and memory in these extremely young babies. In fact, the ability to reverse a learned contingency is considered to be a strong test of cognition in animals. The rapid reversal learning demonstrated in these day-old babies shows that they are already cognitively sophisticated as compared with other species.

In order to provide a strong test of the idea that memory for the mother's voice does indeed occur via learning in utero, rather than from very rapid learning during the first few hours after birth, DeCasper and Spence (1986) conducted a further study in which participating mothers read three stories onto a tape. The mothers then selected one of the three stories and read it every day during the last six weeks of their pregnancies. Following birth, the infants' baseline suck rates were established, and the infants were then rewarded for sucking either above or below baseline by their mother's voice reading the familiar story. If sucking fell to baseline, the infants heard their mother's voice reading an unfamiliar story. DeCasper and Spence found that the infants consistently sucked at the rate that was appropriate to produce the familiar story. Interestingly, a second group of infants showed the same pattern of preferences when tested with another mother's voice reading the stories. DeCasper and Spence argued that the target stories were preferred because the infants had heard them before birth. The babies had apparently learned something about the acoustic cues specifying a particular target passage as foetuses, and could recognise these cues even when a strange female voice was reading the story. As we will see in Chapter 6, auditory learning mechanisms in babies are extremely powerful, and provide them with a lot of information about environmental structure.

Infant preferences for the mother's face, voice and smell are part of the constellation of attachment behaviours that serve to foster and strengthen the relationship with the primary caregiver. This relationship is very important for the healthy psychological development of the infant. However, a discussion of the attachment literature is beyond the scope of this book (see Fearon & Roisman, 2017, for a recent overview). We will focus instead on the more cognitive aspects of understanding the self and the minds of others. We begin by considering the ability to learn contingencies, as demonstrated by the sucking behaviour of neonates. Contingency learning appears to be a crucial aspect of the development of social cognition.

The role of detecting contingency in awareness of self and other

The early ability to detect and learn contingencies between one's own actions (e.g., sucking) and the auditory environment can also be found in the visual environment. As we saw in Chapter 1, three-month-old babies can learn a causal contingency involving a relationship between their own actions and visually rewarding objects such as mobiles. Rovee-Collier et al. (1980) showed that babies will learn to kick to activate an attractive mobile at three months of age. In these experiments, the mobile's movement was dependent upon a ribbon tied to the baby's ankle. The babies learned the contingency between their action and the rewarding event. Following suggestions by Watson (1994), Gergely (2001) has argued that this ability to detect the contingencies of their own motor actions is the mechanism whereby babies first develop a primary representation of the bodily self. He argues that babies can identify the consequences of the motor actions of their own bodies by 2–3 months of age, and that they do this most easily when there is a perfect contingency between their actions and certain consequences.

An interesting study of young babies' ability to notice contingencies based on their own bodily actions was carried out by Bahrick and Watson (1985). They videoed babies aged three and five months of age as they sat in a seat kicking their legs. The infants could see two video monitors as they kicked. In one monitor, they saw their legs moving in real time. The display showed the current live image, and so there was a perfect contingency between the legs on the screen and the babies' own movements (as they were watching their own legs). The second monitor showed earlier footage of the babies' legs. They were still seeing their own legs, but now there was no contingency between the movements they were in the act of performing and the movements of the legs on the screen. The older babies looked much longer at the non-contingent image, as did half of the younger babies. The other half of the three-month-olds reliably preferred to watch the contingent image. All infants had a reliable preference, indicating awareness of the contingencies.

Mirror self-recognition

If babies can recognise the contingency between their own actions and the actions of a live video-recording of themselves by three months, it seems likely that they would also be able to recognise themselves in a mirror. However, the literature on mirror self-recognition ascribes this ability to rather older babies, aged on average two years. The critical test of mirror self-recognition is the 'mark test'. A noticeable mark is made on the baby in a location that cannot be viewed naturally, such as the face. For example, a spot of rouge is placed on the child's nose. The test of mirror self-recognition is whether the child, while looking in the mirror, notices and touches the spot on their nose. The 'mark test' is intended to provide an objective test of mirror self-recognition.

KEY TERM

Mark test
A mark is made (usually in red) on the infant's nose and if the child touches the spot when looking in a mirror this is a sign of mirror self-recognition. Sometimes called the 'rouge on the nose' test.

Observing an infant testing out different actions in the mirror is one way of measuring mirror self-recognition.

Amsterdam (1972) used the mark test with babies aged from three to 24 months. The babies were given a mirror in their playpens, large enough to view their entire bodies. Their mothers drew their attention to the mirror, pointing at the babies' faces and saying 'See! See! See! Who's that?' The babies were then left to explore the mirror. Amsterdam reported that over half of the babies tested out the mirror, even in the youngest age group (3–5 months), for example by observing their image as they moved different body parts. Some also looked behind the mirror or tried to reach into it. However, only the older babies showed mirror self-recognition in terms of touching the dot of rouge, or saying their own name in response to the image (42% of 18–20 month olds and 63% of 21–24 month olds). Amsterdam concluded that objective evidence for mirror self-recognition was only present from 18 months, although she noted that there was data indicative of emergent awareness (e.g., testing out the mirror) much earlier than this. More recently, longitudinal work has established that girls tend to pass the mirror self-recognition test earlier than boys, suggesting a link with general pro-social development (Kristen-Antonow, Sodian, Perst & Licata, 2015).

Interestingly, it has been shown that animals like elephants, apes and dolphins also show mirror self-recognition. For example, Plotnik, de Waal and Reiss (2006) demonstrated mirror self-recognition in an Asian elephant in New York Zoo. The researchers mounted an enormous mirror in the enclosure occupied by three elephants, Patty, Happy and Maxine. They then examined whether the four stages of mirror self-recognition found to be typical in other species would be displayed. These are (1) social responses to the mirror; (2) physical mirror inspection, such as looking behind the mirror; (3) repetitive mirror-testing behaviour; and (4) self-directed mirror behaviour (passing the 'mark test'). All of the elephants displayed stage 2 and 3 behaviours during the first four days of having the mirror. They touched and sniffed the mirror, checked behind it with their trunks, and tried to climb over the wall on which the mirror was mounted. They also tested the mirror repetitively by making non-species-typical trunk and body movements. However, when a large white mark was placed on the face on day five, only Happy passed the mark test. She checked the mark 47 times. Although Patty and Maxine appeared to ignore their marks, they continued to show self-directed behaviour at the mirror. Plotnik et al. argued that awareness of the self–other distinction is present in a number of species who show complex sociality and co-operation.

Adult contingent behaviours

Another reason that infants may develop psychological understanding relatively early in life is that their caretakers treat them as social partners. When caring for infants, adults usually make their behaviour contingent upon, rather than ignoring of, infant attempts to communicate. In fact, caretakers may treat their

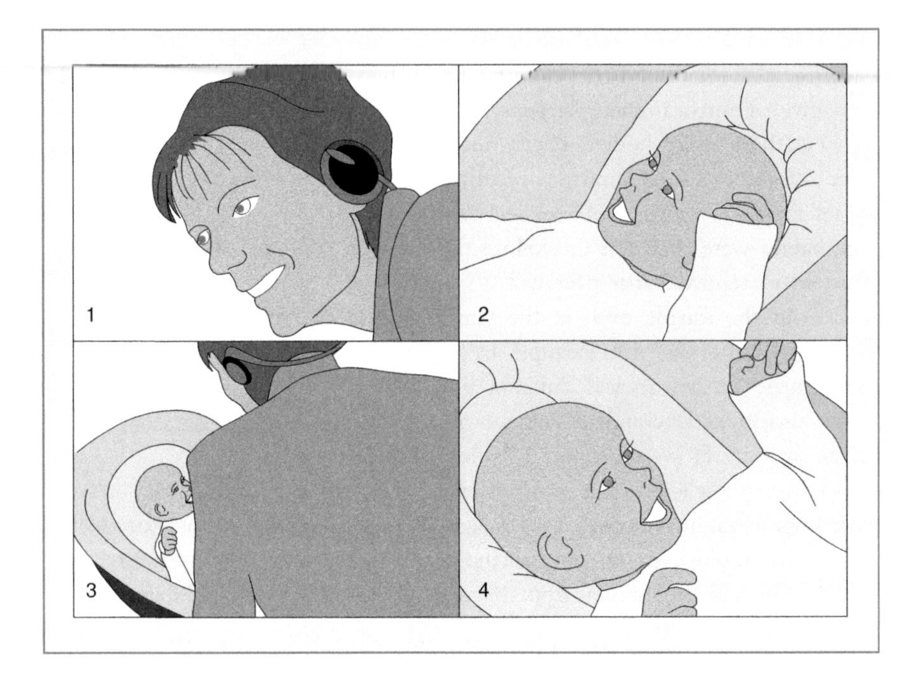

FIGURE 3.1
Set up for studies 1 and 2

infants as acting communicatively even before infants are intentionally acting in this way (Meins et al., 2002; Meins, 2013). Striano, Henning and Stahl (2005) explored infants' sensitivity to social contingencies, by studying babies aged one and three months during face-to-face interactions with their mothers (see Figure 3.1).

In Striano et al.'s study, maternal contingencies were manipulated by the experimenters. All mothers were first instructed to interact with their infants 'as they normally did'. The infants and mothers faced each other with eyes 50 cm apart, and three minutes of normal interaction were recorded. Mothers and babies then went home. A week later, they came back to the laboratory and three different contingencies were experienced. The one-month-olds experienced these different contingencies for one minute each, and the three-month-olds for three minutes each. *Normal* interaction was as on the previous visit. In a second condition, *Non-contingent* interaction, the mothers wore headsets that played them one minute of their interaction from the prior week. They were required to reproduce this interaction as they heard it. The third *Imitation* interaction required the mother to mirror their infants' facial expressions, arm/hand gestures and vocalisations. The experimenters recorded the gazing and smiling shown by the infant during each of the contingencies. Striano and her colleagues found that the one-month-olds did not distinguish between the different contingencies in terms of their gazing and smiling behaviour. By three months of age, however, the infants were behaving differently in response to the different contingencies. They showed more gazing in the Imitation condition, and more smiling in the Normal interaction condition. Striano et al. argued that infants were sensitive to social contingencies by three months of age.

This sensitivity to contingency and awareness of the bodily self may play an important role in the development of social cognition. Reactive social partners tend to provide high but not perfect contingencies (for example, parents usually 'talk back' to a cooing baby, but not always). Gergely (2001, see also Gergely, 2010) argues that by three months of age babies have adapted to this aspect of the social environment, and now prefer imperfect contingencies. They gradually realise via contingency detection that their own states and behaviours exert control over the caretaker's behaviours, and thus pay more attention to these states and behaviours, thereby beginning to develop emotional self-awareness and control. According to Gergely, the very young infant engages in social interactions because these actions have important evolutionary advantages, such as maintaining proximity with the caregiver. Gradually, via mechanisms such as contingency detection, the infant becomes aware of the self as a separate (bodily) self, with its own intentional and affective states. This helps to lay the foundations for an understanding of others as having their own (subjective) mental states and intentions.

A slightly different position to Gergely's is taken by Meltzoff (2002). He argues that the crucial thing about early interactions with other agents such as caregivers is that the infant comes to recognise the other person as 'just like me'. The important mechanism for this is imitation, which as we have seen is operational from birth (just like contingency detection). Meltzoff points out that imitation shows that, at some primitive level, infants are mapping the actions of other people onto the actions of their own bodies. They are connecting the visible bodily actions of others with their own internal states. This cross-modal knowledge of what it feels like to do the act that was seen then provides a privileged access to people as special kinds of entities. Meltzoff suggests that the infant experiences his or her own internal desires (e.g., wanting their bottle) and experiences the actions (concomitant bodily movements) required to fulfil these desires or goals (reaching for the bottle). This helps the infant to make sense of the object-directed movements of others. When another person is seen reaching for an object, the action can be imbued with goal-directedness, because of the infant's own experience with similar acts.

THE CENTRAL ROLE OF THE ACTIONS OF OTHER AGENTS

Viewing the actions of others as goal-directed

When we as adults observe the actions of others, we assume that they have a purpose. We assume that the actor has certain goals, and that he or she is acting to achieve these goals. Meltzoff's suggestion is that infants' ability to map from self to other (as documented by imitation studies) serves as a catalyst for understanding their own behaviour and the behaviour of others in terms of goals. For example, infants imitate successful yet arbitrary acts performed by the experimenter, such as turning on a light panel by bending forward and pressing it with the forehead (see Chapter 2). They also imitate unsuccessful acts that they infer that the experimenter had intended to perform successfully, such as putting a string of beads into

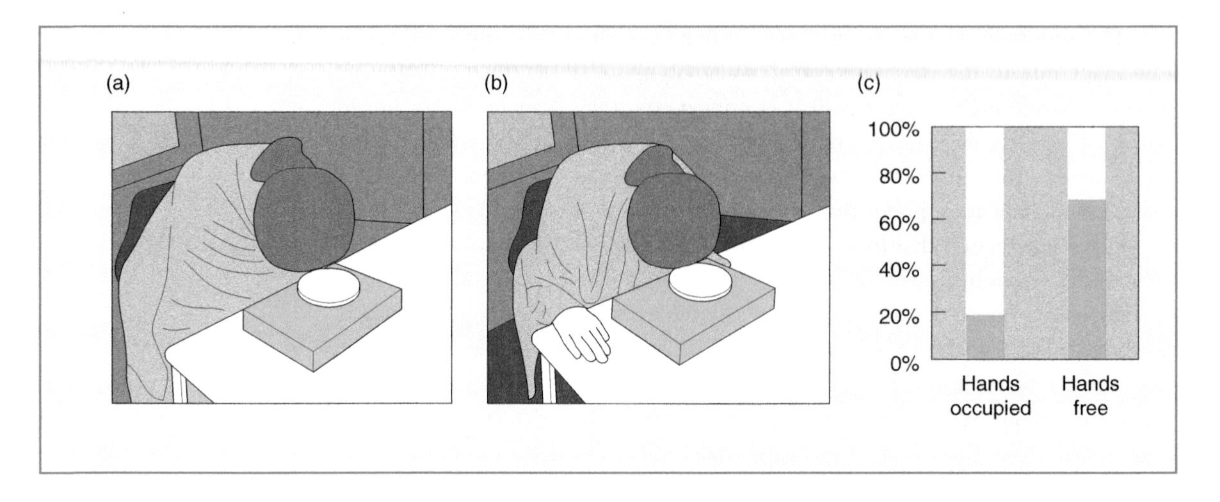

FIGURE 3.2
The light panel procedure showing (a) the Hands Occupied condition, and (b) the Hands Free condition. (c) shows the methods used by infants to switch on the light box after watching the demonstrator under these two conditions. (Blue: head was used, yellow: hands were used.) From Gergely et al. (2002).

a cylindrical container (see Meltzoff, 1995a, 1988b). These imitations suggest that certain goals are imputed to the actor. In the first example, infants appear to believe that there must be a reason for the actor using their head rather than their hand to activate the light, and accordingly they follow suit. In the second example, the infants infer the intended goal of the actor and execute this goal, rather than imitating the unsuccessful act. However, neither experiment provides *direct* evidence for these inferences. A study by Gergely, Bekkering and Király (2002) has filled this gap, by demonstrating that infants who imitate are indeed imputing goals to the actors that they are copying.

For their demonstration, Gergely et al. (2002) extended the light panel procedure invented by Meltzoff (1988b, see Figure 3.2). In this procedure, the infant watches as an experimenter switches on a light panel with her head. The panel is set into the top of a box, which sits on a table in front of the experimenter. The experimenter illuminates the light by bending forwards and touching the panel with her forehead. This is a rather odd act, but the infants in Meltzoff (1988b) still imitated it. In Gergely et al.'s (2002) experiment, a rationale was provided for this odd behaviour. The experimenter complained of feeling cold, and wrapped herself in a blanket. Her hands were thus occupied with the blanket, and were not free to press the light panel on the box. Consequently, she used her head to press the panel. The 14-month-old infants who observed this did not use their own heads to illuminate the light panel. When given the box to operate, 79% of them chose to use their hands to illuminate the light, and only 21% used their foreheads. However, in a second condition, other 14-month-old infants saw the same event when the blanket was just draped around the shoulders of the experimenter, whose hands were thus free. In this 'hands free' condition, 69% of the infants pressed the light panel with their foreheads, thereby replicating Meltzoff's original findings (Meltzoff, 1988a).

This experiment shows neatly that understanding the goals of another transforms the bodily motions of another into purposive behaviour. Although the action in Gergely et al.'s (2002) experiment was identical in each scenario, the goals of the agent were assumed to differ. When the experimenter was not clasping the blanket and had her hands free, her use of her forehead was assumed to be intentional and therefore important *vis-à-vis* the activation of the light panel. This is a strong experimental method, because differential intentions are suggested by simple changes in context. Differential imitation of the same action because of changes in context is therefore strong evidence for attributing an understanding of goal-directed action to infants.

Carpenter, Call and Tomasello (2005) devised a different paradigm based on simple contextual changes in order to explore how early young infants use context to determine goals. They demonstrated hopping and sliding actions made by an actor with a toy mouse to infants aged 12 and 18 months in two different contexts. In each context, the infants watched an experimenter make a toy mouse take a distinctive hopping or sliding journey across a mat. In the hopping journey, the experimenter made the mouse cross the mat in a series of eight jumping actions, accompanied by suitable hop noises ('bee', 'bee', 'bee'...). In the sliding journey, the mouse was made to cross the mat in one long slide ('beeeeeeeee'). In one condition, the mouse ended its journey by being put into a little house on the other side of the mat (see Figure 3.3). The journey followed a straight line into one of two possible goal houses. In the second condition, no houses were present. In this condition, the experimenter appeared to be making the mouse hop or jump just for the fun of the action. The question was what the infants would imitate in each condition.

Carpenter et al. predicted that the infants would put the mouse straight into the goal house in the first condition, when the little houses were present. In this condition, the apparent goal of the adult was to get the mouse into its house. In the second condition, however, the infants were expected to jump or slide the mouse across the mat. Without any goal houses, the differential actions were presumably being made for their own sake. In the test trials, the infants were handed the mouse and told 'Now you'. Carpenter et al. (2005) found that the infants in the House condition were significantly less likely to imitate the action style of the adult, at both ages, in comparison to the infants in the No House condition. In the House condition, most of the infants simply put the mouse into the correct house. In the No House condition, in contrast, the majority of the older infants made the mouse hop or slide across the mat. Carpenter et al. argued that the infants were analysing the ends and means of the adult's actions, and were thus choosing to imitate either 'putting the mouse in the house' or 'making the mouse hop up and down'. The infants were interpreting the actions of the adult in terms of her assumed goals. They were also making causal analyses of her actions based on context.

Another way of examining whether infants understand that the actions of other agents are goal-directed is to show them actions that fail to achieve their intended outcome. If infants can distinguish successful from unsuccessful actions, this must involve an understanding of the mental goals of the actors. A study of successful versus unsuccessful actions was reported by Brandone, Horwitz, Aslin and Wellman (2014), with infants aged eight and ten months. Eye tracking was used to measure infant looking. During the study, infants watched a video event in which an actor either did or did not achieve his intended goal of picking up a ball. The actor was sitting at a table which had a barrier on the top, and was gazing

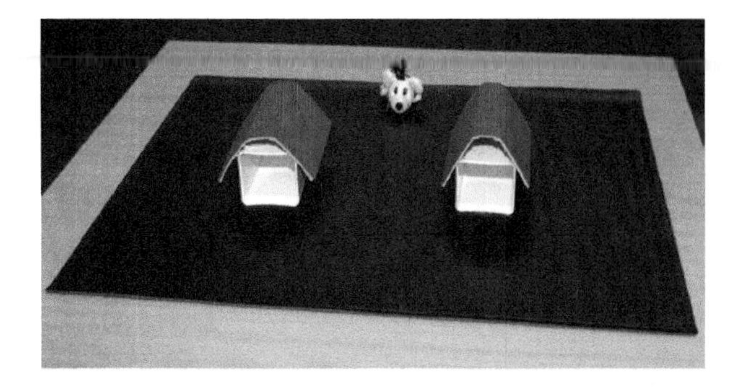

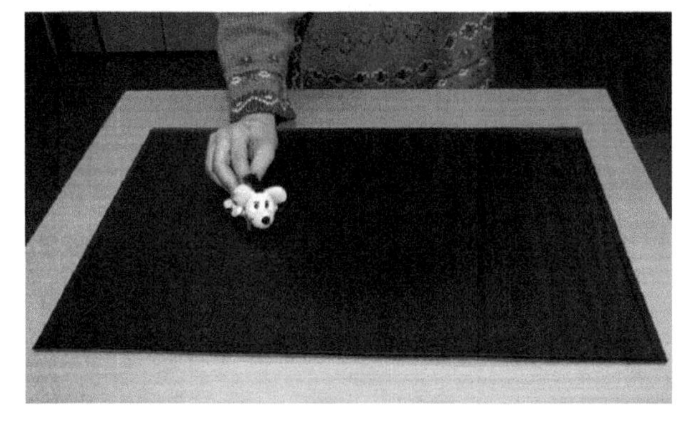

FIGURE 3.3
Top: the mouse at the start location in the House condition of Carpenter et al.'s (2005) study; bottom: the mouse at the end location in the No House condition. From Carpenter et al. (2005). Reproduced with permission from Blackwell Publishing.

at a ball on the other side of the barrier. In the *successful* reaching event, the actor reached over the barrier with an arcing motion, grasped the ball and brought it back towards him. He looked down at the ball in his hand and the video then froze. In the *unsuccessful* reaching event, the actor reached over the barrier with an arcing motion, hovered his hand above the ball and brought his hand back again empty. He looked down at his empty hand with a disappointed expression, and the video froze. Infants were assigned to either the successful or unsuccessful event conditions, and watched the video ten times. The key measure of interest was anticipatory looking to the ball by the infant, which would indicate an understanding of the actor's goal (obtaining the ball). Also of interest was whether infants would cease looking to the ball in the unsuccessful action condition, having watched repeated failures.

Brandone et al. (2014) reported that no infants showed anticipatory looking on the first trial, for either condition. For the infants viewing the *successful* reaching condition, however, both eight- and ten-month-olds increasingly produced anticipatory looking over trials 2–10. For the infants viewing the *unsuccessful* reaching condition, the ten-month-olds produced anticipatory looking until around trial 8, when this behaviour stopped. The eight-month-olds viewing the unsuccessful reaching condition did not show evidence for anticipatory looking. Brandone et al. (2014) concluded that by ten months of age, infants predict the intended outcome of an action even when the action is repeatedly unsuccessful. Note that *predictive* looking, which occurs before the action event takes place, is good evidence that infants (here, as young as eight months of age) understand that specific goals usually underlie human actions.

Meanwhile, Carpenter, Akhtar and Tomasello (1998a) have shown that infants aged 14 to 18 months will *imitate* the acts of others differentially depending on whether they perceive these acts to be intentional or accidental. Studying infants' responses to accidental acts is another way of investigating their understanding of the goals of the actor. When something is done accidentally, it was not the goal of the actor to carry out this act. The actor's goal was to carry out a different act. This provides an interesting contrast to Meltzoff's (1995a) paradigm involving unsuccessful acts that *were* intended (such as trying but failing to place a string of beads into a container). In Meltzoff's work, the infants imitated the acts that the actor had intended, but had not actually modelled. In Carpenter et al.'s study, the question was whether they would imitate modelled acts that the actor had *not* intended.

Carpenter et al. (1998a) designed eight different objects that afforded two distinct actions by an actor (see Figure 3.4). Each object had two parts that could be moved, and an 'end result'. For example, one object was a bird feeder. The

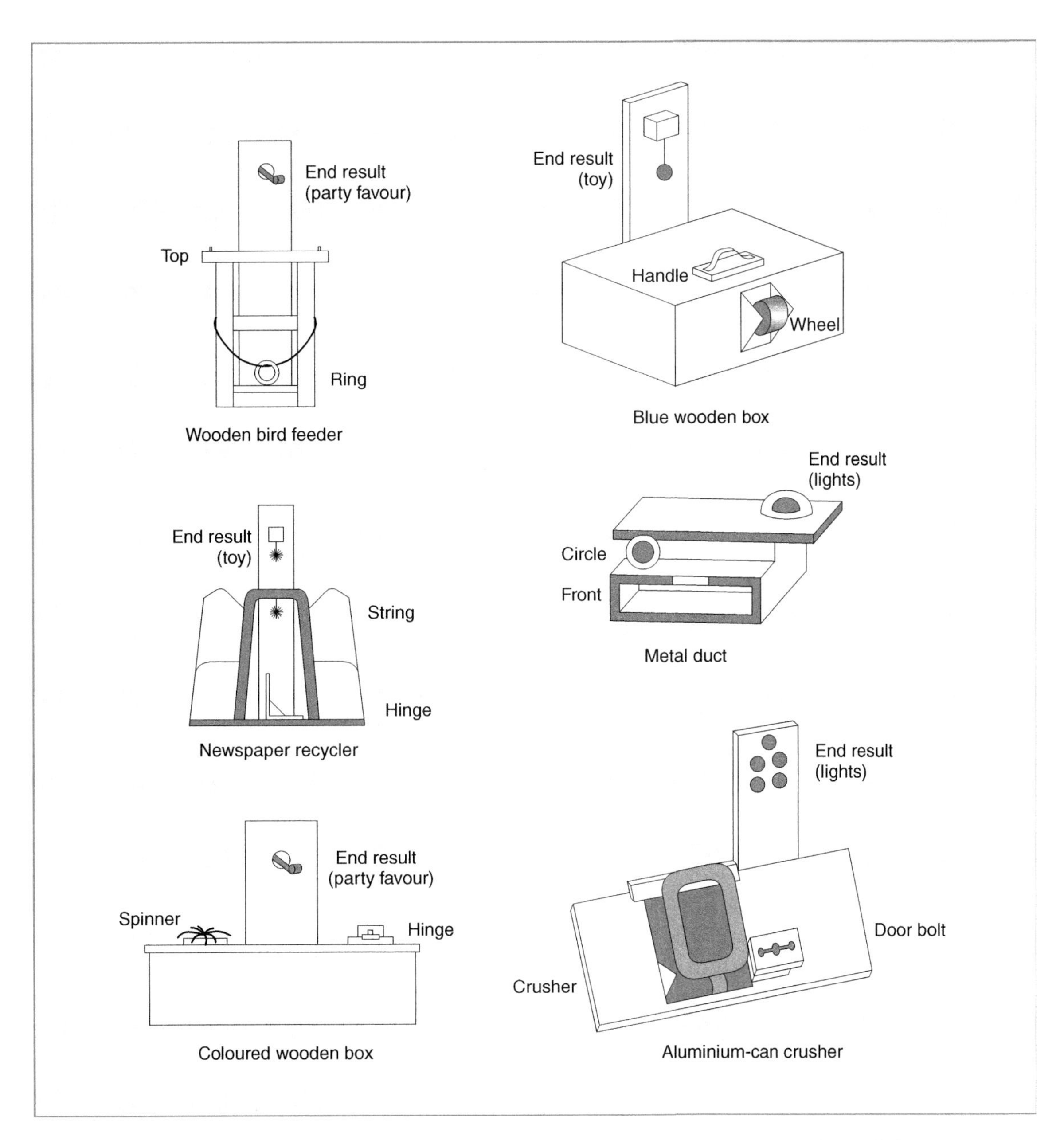

FIGURE 3.4
The test objects used by Carpenter et al. (1998a). From Carpenter et al. (1998a). Copyright © 1998 Elsevier. Reprinted with permission.

bird feeder had a top that could be moved up and down and a string attached to its middle with a ring on it. Either pulling the ring or moving the top resulted in a 'party favour' being activated – the type of favour that you blow into to make a long paper tongue shoot forwards. The party favour was not activated by the moving parts on the bird feeder, but by a hidden experimenter with a small pump. This enabled the experimenters to correlate the end result with certain actions on the object and not with other actions. The experimenter serving as the actor always carried out *both* possible actions on the object. The experimenter said 'Watch! I'm going to show you how this works!' She then modelled two actions.

For example, for the bird feeder she always both pulled the ring and moved the top. However, one action was done intentionally, and the other accidentally. The intentional action was accompanied by the experimenter saying "There" in a satisfied voice, and the accidental action by the experimenter saying "Whoops!" The order of these actions was counterbalanced across infants. Infants were then allowed to interact with the objects themselves, the experimenter asking "Can you make it work?"

Carpenter et al. (1998a) found that the infants were much more likely to imitate the intentional acts. Intentional acts (as marked by "There") were imitated on 78% of occasions, and accidental acts (as marked by "Whoops") on 43% of occasions. The acts themselves, of course, were identical whether they were intentional for some infants or accidental for others. Carpenter et al. argued that the infants were interpreting the adult's overall behaviour in terms of the goal of producing the 'end state'. They were interpreting the adult's behaviour as intentional, screening out the accidental and unintended actions. As Carpenter et al. point out, the ability to recognise intentional actions provides a powerful boost to the infant's capacity for imitative learning. An infant who selectively imitates only the intentional acts of others will thereby acquire many significant cultural skills.

More recently, this broad cultural explanation has been refined into a theory of 'natural pedagogy' (Csibra & Gergely, 2009). Csibra and Gergely argued that the tendency to imitate the novel actions of others reflects infants' innate assumption that the intention of the 'teacher' is to impart some new and relevant knowledge. Typically novel goal-directed actions are accompanied by 'ostensive' (manifestly demonstrative) communicative signals on the part of the social partner. These include the use of Parentese and making direct eye contact. On this *natural pedagogy* view, actions such as using one's forehead to touch the light panel or making the mouse hop are manifestations of a communicative intention. The ostensive signals accompanying the demonstration of these novel goal-directed acts tell the infant that the action is important even though it may appear to be inefficient.

To test the natural pedagogy view, Király, Csibra and Gergely (2013) repeated the light panel procedure with the demonstrator having a blanket (cf. Gergely et al., 2002), but added an *incidental observation* condition. In this condition, the demonstrator did not interact directly with the infant, nor complain of being cold. The incidental observation manipulation meant that infant imitation could be compared when ostensive cues were present versus absent, and when the demonstrator's hands were either not free (as she was holding the blanket) or were free. Király et al. (2013) reported that 65% of infants imitated the head movement when the demonstrator's hands were free in the ostensive communication condition, compared to only 29% in the incidental observation condition. This suggests that the communicative signals were indeed important regarding infant intention-reading behaviour Although the natural pedagogy view requires converging evidence from other paradigms, selective imitation of pedagogical demonstrations by infants and young children would be a fast and efficient way of acquiring culturally relevant knowledge.

Distinguishing different types of intentional action

Of course, not all goal-directed actions towards infants are underpinned by communicative intentions. The actions of others can also stem from very different intentions. The same gesture (e.g., not giving an infant a desirable toy) could occur because the adult does not wish the infant to have the toy, or because the adult is unable at the time to secure the toy. Carpenter and her colleagues have investigated whether infants can distinguish between different *kinds* of intentional acts. They have focused on the distinction between an actor who is *unwilling* to hand the infant a desirable toy, and an actor who is *unable* to hand the infant the same toy. In both cases, the infant does not receive the desired toy. However, in the first case this was the intention of the actor, and in the second case it was not. In a third condition, the infant does not receive the toy because the experimenter was distracted from the game. The research question is whether the frustrated infants will behave differently in these three conditions.

In order to create a context for these different intentional acts, the study was based on a game of handing each other toys (Behne, Carpenter, Call & Tomasello, 2005a). Each time the experimenter was about to pass a new toy to the infant, she said "Oh, look!" Then she handed over the toy. In some trials, however, the infant did not receive the toy. In the Unwilling condition, the experimenter held out the toy and then withheld it in a teasing fashion. In other Unwilling trials, she refused to hand it over, leaving the toy sitting in front of her in full view of the infant, or continuing to handle the toy herself. In the Unable condition, the experimenter either offered the toy but then clumsily dropped it, or appeared unable to extract the toy from its transparent container. In some Unable trials, she seemed unable to remove the lid from a different transparent container holding the toy. In the third Distracted condition, the experimenter was either about to hand over the toy when the telephone rang, or she got distracted by talking to her assistant, or she appeared to forget to hand the toy over while she searched in a bucket for the next toy. The question in each case was how the infant would react to the withholding of the toy. Groups of both 12- and 18-month-old infants were tested.

The results showed that the infants behaved very differently in the Unwilling, Unable and Distracted conditions. They reached more for the toy in the Unwilling condition, and tended also to look away from the experimenter for longer in this condition. When the experimenter was trying to give them the toy but was unable to, they reached less for the toy and looked away from the experimenter less as well. They reacted in a similar way in the Distracted condition, reaching less for the toy than when the experimenter was unwilling to hand it over. They appeared to be adapting their behaviour appropriately given the social situation. Behne et al. argued that the infants were basically impatient when the adult refused to give them the toy, but patient when the adult was unable to give them the toy. They knew that the goals of the actor in each case were different, and responded accordingly.

Younger infants aged nine months showed a similar pattern of performance. They reached more for the toy and banged the table in frustration more in the Unwilling condition than in the Unable condition. Six-month-old infants, however, showed no differential behaviours. Behne et al. thus argued that their results supported the idea that a 'revolution' in infant social cognition begins at around nine months of age. At this age, the first clear evidence for the understanding of the psychological states of others is emerging. However, the negative result in this experiment for the six-month-olds cannot be taken as evidence that younger babies lack any understanding of differential intentions. As usual in psychology, a negative result simply means that we cannot draw any strong conclusions.

Younger infants also perceive human actions as goal-directed

Looking time data in contexts that do not vary intentions suggest that younger infants are also able to perceive the actions of others in terms of goals. An important paradigm was developed by Woodward (1998), who compared the looking times of babies aged nine months in a paradigm contrasting a human agent with a non-human agent. The infants watched as either a human actor grasped a toy, or a poster tube topped with a sponge made contact with the toy by following the same reach trajectory as the hand (see Figure 3.5). The babies watched these events on a small stage. The stage contained two toys, a teddy and a ball. For the human actor condition, the habituation trials comprised an arm appearing at the side of the stage, approaching one of the toys, and grasping it. The actor then remained still. For the mechanical condition, the poster tube (which was similar in size to the actor's arm, and was coloured the same as the actor's t-shirt) approached the toy, taking the same path as the actor's arm. When the sponge made contact with the toy, it rested there, just as the hand rested in the grasp condition. Babies were either habituated to the mechanical event or to the human actor event. At test, the position of the toys were reversed, and two events were shown. In the 'new path' event, either the actor or the poster tube approached the same toy as before, thereby following a new trajectory. In the 'new goal' event, the actor or the poster tube made the same action as previously, which led to the novel toy being 'grasped'.

Woodward (1998) found that the babies were most interested when the human actor had a new goal. The babies who were habituated to the human actor looked significantly longer at the 'new goal' (same path) event than at the 'new path' (same goal) event. The babies who were habituated to the poster tube showed no preference. However, the poster tube babies did show dishabituation to both events, indicating that they were aware that they were seeing something novel. In a follow-up experiment with six-month-old babies, Woodward showed that even younger infants were also interested in the goals of human actors. The younger babies were shown the hand/grasp event on the toys, contrasted with a mechanical pincer performing the same actions. The babies who watched the human hand grasp one of the toys during habituation were only interested in the 'new goal' event at test (same path, new goal). The infants habituated to the mechanical pincer were not, showing reliable dishabituation only when the pincer followed a new path but grasped the same toy. Woodward concluded that her six- and nine-month-old participants were selectively encoding the aspects of actions that were relevant to the goals of the human agent, thereby construing human actions as goal-directed.

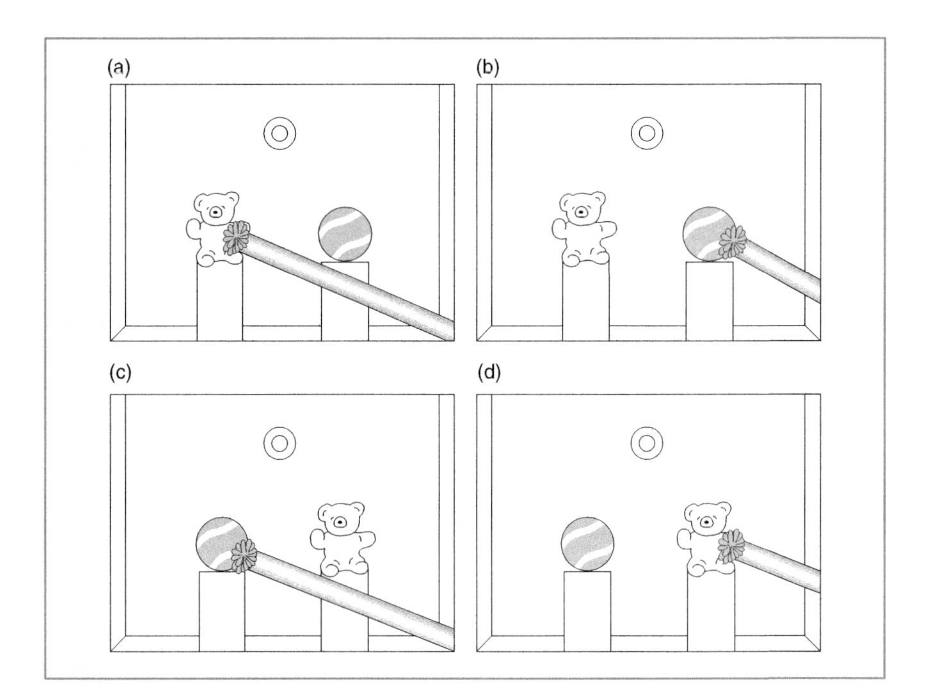

FIGURE 3.5 The events in the mechanical condition of Woodward's (1998) study. From Woodward (1998). Copyright © 1998 Elsevier. Reprinted with permission.

Southgate and Begus (2013) also utilised a mechanical pincer, in a study of *motor prediction* by nine-month-old infants. Southgate and Begus recorded EEG (the mu rhythm, the putative electrophysiological signature of action-effect learning) as their measure of the infants' prediction of action. Infants watched videos involving goal-directed actions on objects, in which either the pincer or a human hand moved towards one of two objects visible at the end of a table. The hand or pincer grasped one of the objects, the *target* object, and moved it to the centre of the table. This occurred four times. In a control *self-propelled* condition, the target object moved by itself to the centre of the table four times. The other *distractor* object never moved during these demonstrations. At test, the infants were either shown the target object or the distractor object, with either the hand or the pincer visible at the bottom of the video screen. After a static period, an action occurred. In the control condition, either the target or the distractor object was visible on screen, and nothing else. Southgate and Begus reasoned that on the basis of prior experience, the infants would expect a goal-directed action to be coming next for the target object, but not for the distractor object. Thus they predicted selective desynchronisation of the mu rhythm in the human hand condition only.

Consistent with their expectations, Southgate and Begus found significant suppression of the mu rhythm in the watching infants for the target object trials, but not for the distractor object trials, consistent with action prediction. Intriguingly, however, this neural evidence for action prediction was found for all three conditions, human hand, pincer, and self-propulsion. Southgate and Begus (2013) argued that this was evidence that suppression of the mu rhythm is a neural signature of action prediction that is not dependent on whether the infant can themselves produce the expected action. Hence terming the mu rhythm *motor resonance* (as found in some of the literature) is incorrect. The suppression of the mu rhythm seems to relate to an expectation of action by the motor system, irrespective of whether the observing infant can themselves perform this action. As Southgate and Begus point out, an action-anticipation mechanism that is independent of motor skills would enable infants to learn rich causal information simply by watching the actions of agents and of self-propelling mechanical objects.

Kim and Song (2015) developed a similar paradigm for younger six-month-old infants, and succeeded in obtaining *predictive* evidence that infants can infer object preferences in a human agent. In an adaptation of the reaching paradigm, the infants watched on video as an actor repeatedly reached to one of two objects placed in front of them, and grasped it. One object was on the left of the actor and one object was on the right, and the selected object was the goal object. The actor was wearing a visor (like a baseball cap) so that the infants could not use her gaze as a complementary cue. After six reaches to the goal object, the actor disappeared and the location of the two objects was exchanged. The actor then reappeared and sat quietly. Kim and Song (2015) measured where the infants now looked. As in the study by Brandone et al. (2014) discussed earlier, *predictive* looking behaviour suggests a *mentalistic* interpretation of the agent's reaching behaviour. This was exactly what Kim and Song (2015) found. The infants looked longer at the new location containing the goal object. Kim and Song argued that their data provided evidence for a mentalistic interpretation of the viewed reaching event. The infants had inferred that the actor *preferred* the goal object, and thus they expected her to reach for it again, even when its location had moved.

Sommerville et al. (2005) have extended the reaching paradigm to three-month-olds, using habituation of looking. In their study, infants sat in front of a stage containing two toys, a ball and a teddy bear, which were shown side by side. As the infants watched, an actor reached for and grasped one of the toys, but did not move it. This event was repeated until the infants habituated to it. At test, the infants either saw a 'new goal' event, in which the actor reached for the other toy, or a 'new path' event, in which the location of the toys was swapped and the actor again reached for the familiar toy. The infants dishabituated to both novel events, but looked equally in the 'new goal' and 'new path' conditions. This behaviour was in marked contrast to a second group of three-month-olds, who were allowed to manipulate the toys before the habituation and test events. This group of babies, the 'action' group, were allowed to play with the toys for three minutes, and were then given a further three minutes wearing Velcro mittens that caused the toys to adhere to their hands. This gave them experience of different actions on the same toys (in fact, there was significantly more handling of the toys when wearing the Velcro mittens). When this 'action' group watched the two test events, they showed significantly longer looking for the 'new goal' event compared to the 'new path' event. They also showed significantly greater looking on the first habituation trial than the group who had experienced passive watching first (this latter group were also allowed to manipulate the toys prior to the test trials, but only after watching the agent grasping the toys in the habituation trials). Sommerville et al. (2005) argued that receiving the action experiences *first* caused the infants to be more attentive to reaching events that they saw performed by another agent. Their ability to detect the goal structure of actions was improved by their initial experience of object-directed behaviour, which led them to interpret the habituation trials in terms of the agent's goals. Even three-month-old babies thus appear to be aware of goal-directed actions, at least when they have had the opportunity to perform these actions themselves.

GOAL-DIRECTED ACTION AND THE ATTRIBUTION OF MENTAL STATES

Teleological versus psychological analyses of actions

How sure can we be that more recent studies, such as that by Kim and Song (2015), involve mentalistic reasoning (naïve psychology) on the part of the infant? The assumption that the ability to *detect* the goal structure of the actions of other agents or to *predict* their actions involves an ability to *understand* their goals has been controversial. For example, regarding violation-of-expectation paradigms with infants, researchers such as Heyes have objected to a 'rich' interpretation of infant competence, arguing that attention to 'low-level novelty cues' may underpin apparent false belief responding (Heyes, 2014). Should we credit a baby who can detect the goal structures of the actions of others with an emergent understanding of psychological causality? According to a rich interpretation, the infants are able to understand the goals of other agents because they understand that others have mental states that cause them to act to achieve certain goals. This assumption that very young infants have an emergent understanding of mental states has only recently represented the consensus view (Baillargeon, Scott & Bian, 2016). For example, a strong alternative position was previously proposed by Gergely and his colleagues. Gergely and Csibra (2003) proposed that prior to achieving a mentalistic understanding of others' actions on objects, infants adopted a 'teleological stance' to the representation of action. In a teleological stance, goal-directed actions are represented in terms of (a) the goal state, (b) the action as a means to the goal state, and (c) the relevant aspects of reality as constraints on possible actions. The mental states of the agent are not considered. On a teleological view, the ability to draw inferences on the basis of goal-directed actions is *not* the same thing as attributing mental states to others.

KEY TERM

Teleological stance
The view that it is possible to understand goal-directed actions without the need to attribute mental states to agents.

One line of evidence supporting the idea that infants did not evoke mental states in order to understand goal-directed actions came from demonstrations that infants also attribute goals to *non-human* agents. Relevant experiments have been carried out by Johnson and her colleagues (see Figure 3.6). For example, Shimizu and Johnson (2004) adapted Woodward's (1998) looking time paradigm to incorporate non-human objects that behaved as though they were agents. In the critical experimental condition, a novel faceless green oval object behaved in self-propelled ways that appeared intentional. For example, it acted as though it could interact with its social environment, beeping contingently in response to small talk from the experimenter. It also acted as though it could choose between the relative merits

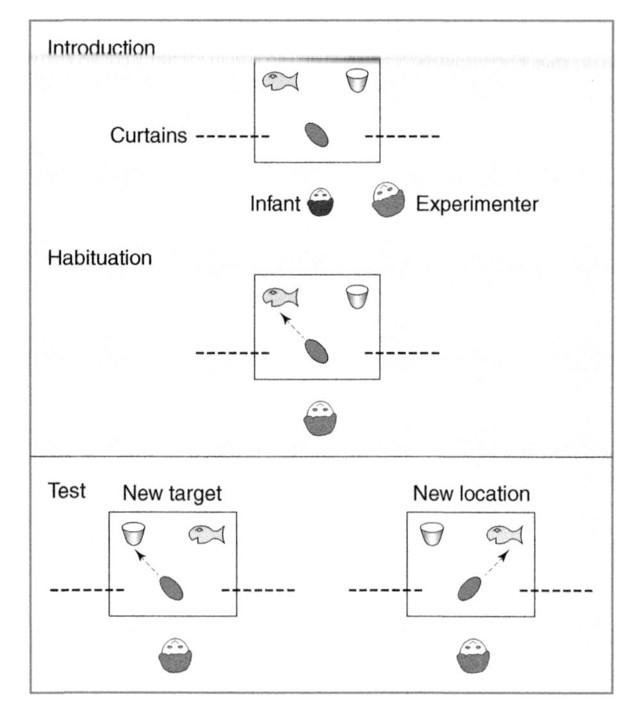

FIGURE 3.6
Top view of the stage for the Agent and Non-Agent conditions, showing the introduction, habituation and test phases of Shimizu and Johnson's (2004) experiment. From Shimizu and Johnson (2004). Reproduced with permission from Blackwell Publishing.

of two toys, deliberately turning away from one and approaching the other. Infants aged 12 months participated in one of three conditions. In one condition, a human hand performed grasping actions on one of two toys. In the second, the Agent-Object beeped to the infant, and then repeatedly approached one of the toys. In the third condition, the same green object was first shown behaving in a random and non-intentional way, and then repeatedly approached one of the toys exactly as in the Agent-Object condition.

Following habituation, the human hand or the green object either performed the same action as before, but to a new toy (as the location of the two toys had been switched), or performed a novel action by reaching to the old toy in the new location. Woodward (1998) had reported increased looking by infants when the human hand performed a familiar action with a novel goal, but not when a cardboard tube did the same thing. Shimizu and Johnson replicated this effect for the human hand, but found the *same* effect for the Agent-Object. When the green oval object had been introduced as though it was capable of intentional behaviour, infants watched longer when it performed a familiar action with a novel goal than when it performed a novel action with a familiar goal. Infants who saw the same green oval object in the Non-Agent condition did not differentiate between the two events. Shimizu and Johnson argued that infants do attribute goals to non-human agents, thereby supporting a lean rather than a rich interpretation of infant sensitivity to goal-directed action.

The theoretical position put forward by Gergely and Csibra (2003) also follows directly from the computer animation experiments with the rectangle and the large and small circles by Gergely et al. (1995) discussed in Chapter 2. As shown in Chapter 2, the results of these experiments were interpreted in terms of an 'intentional stance'. Adoption of an 'intentional stance' towards agents entails an assumption of rationality – that the agent will adopt the most rational action in a particular situation to achieve his or her goal. This rationality assumption was also demonstrated in the current chapter, by the imitation experiment concerning the activation of a light panel with the forehead (Gergely et al., 2002).

Explanations that are teleological depend on the relevant aspects of the actual situation rather than on causal explanatory analyses of the mental states of agents. For example, in the perceptual display of Gergely et al. (1995) discussed in Chapter 2, it was mentioned that adult observers of the movements of the shapes described the event as a mother (large circle) calling to her child (small circle) who ran towards her, only to be prevented by the barrier (rectangle), which she then jumped over. This is a psychological causal explanatory analysis of the perceptual display, which is mentalistic in nature. In contrast, according to a teleological explanation of the same event, the interpreter would be 'mind blind'. He or she would still attribute

KEY TERMS

Mind blindness
An apparent inability to attribute mental states to others.

False belief
An understanding that others may have beliefs that do not reflect current reality.

rationality to the actors, and therefore would expect the novel 'straight line' pathway shown in the dishabituating event, whereby the large ball took the most direct path to the small ball. This would be expected because it was the rational way to reach the small ball. However, when the large ball took the familiar 'jumping' path, which now appeared to be an inefficient means to the goal of reaching the small ball, the teleological interpreter should be surprised. Nevertheless, this surprise would be based on the perception of the situational constraints concerning direct and indirect paths, rather than on a mentalistic interpretation concerning what the large ball 'wanted'. As will be recalled from Chapter 2, young monkeys showed similar behaviour when viewing this paradigm to human infants.

To distinguish between these 'rich' and 'lean' interpretations of infant behaviour, we need empirical tests. The critical empirical test requires a situation in which the infant must attribute rational actions to the agent on the basis of *false beliefs* rather than true beliefs. If the infant is 'mind blind', then the assumption of rational action must always be based on visible context, not on hidden context (such as a belief). As long as goal-directed actions are occurring in contexts where all the variables are visible or known (e.g., visible barriers or hidden barriers whose existence is known to the infant, see Csibra, Bíró, Koós & Gergely, 2003), then according to the teleological view, infants should be able to interpret actions as a means to goals. In such circumstances they should be able to evaluate the relative efficiency of different available means (via the principle of rational action) and generate the relevant inferences (e.g., that it is no longer rational to take a 'jumping' pathway). However, if the actor's action is driven by a *false* belief, infants should not be able to generate the relevant inference. This should be impossible on the basis of teleological reasoning. If infants turn out to be able to evaluate actions based on false beliefs, then the mental states of the actor would have to be represented in order to interpret the actions as rational. We consider some infant false belief studies later in this chapter.

Meanwhile, an intriguing series of studies reported by Kuhlmeier, Wynn and Bloom (2003) provide some basis for preferring a mentalistic interpretation of infants' abilities concerning goal-directed action. Rather than using a violation-of-expectation paradigm with infants, Kuhlmeier et al. (2003) created a situation in which one perceptual animation was preferable to another given a specific prior context. The displays were based on squares, triangles and circles animated on a computer screen that depicted a pair of 'hills' to be climbed (see Figure 3.7). The infants (12-month-olds) watched as a small red ball set out to 'climb' the first, small hill. Once at the top, it expanded and contracted, and then began to climb the second, larger hill. When it reached the halfway point, it slid back to the base of this second hill. The ball then began a second climbing attempt. In one movie, a green triangle moved down behind the ball, and pushed it to the top of the hill. The ball then expanded and contracted once more. This first movie was the 'Help' movie, in which the triangle 'helped' the ball. In a second movie, a yellow square moved down in front of the ball, and pushed it downwards, so that it 'fell' all the way back to its original starting position. This second movie was the 'Hinder' movie, as the square 'hindered' the ball. Infants saw both movies until they had habituated to them (for half of the infants, it was the triangle who 'hindered' the ball and the square who 'helped' the ball).

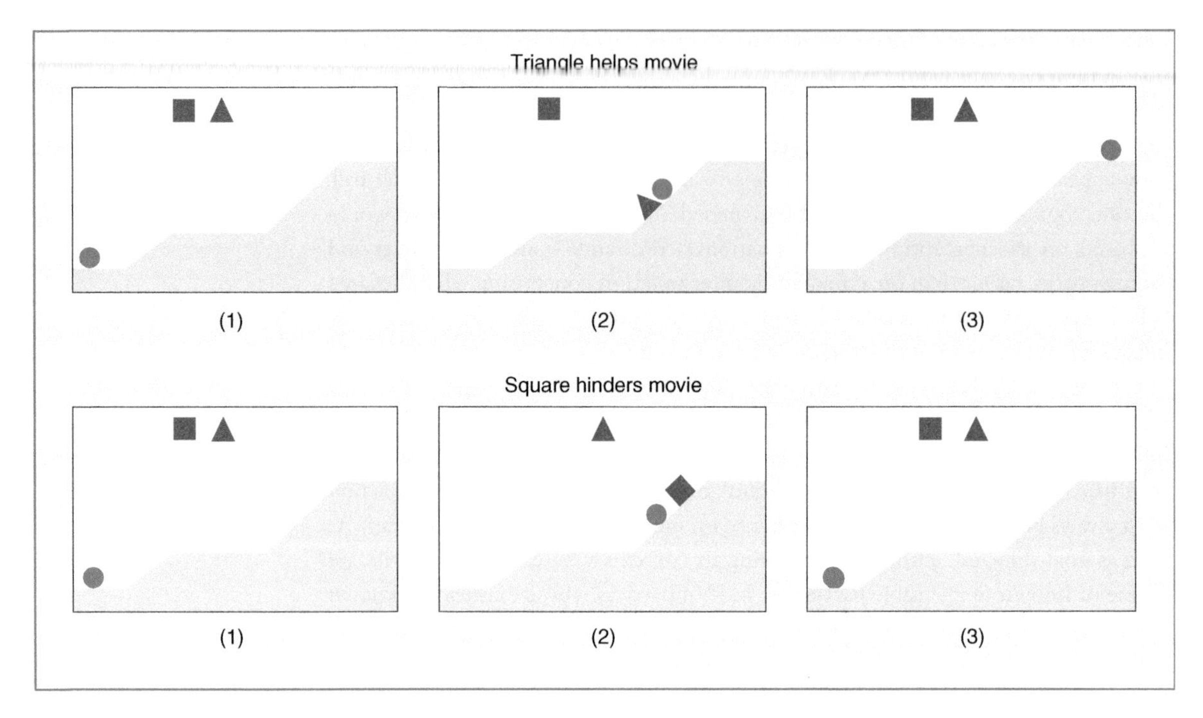

FIGURE 3.7
The habituation movies used by Kuhlmeier et al. (2003). From Kuhlmeier et al. (2003). Reproduced with permission from Blackwell Publishing.

The infants were then shown a second animation. The same three objects were present, with the triangle and square at opposite sides at the top of the display, and the ball in the centre at the bottom. There was no 'hill'. The ball rose to the middle of the screen, paused, wiggled from side to side (as though deciding which direction to move in), and then either moved to rest next to the triangle (its helpmate) or next to the square (its hindrance). Prior piloting with adults had shown that the animation in which the ball 'chose' its helpmate was seen as a more coherent continuation of the previous movies. Kuhlmeier et al. (2003) found that the infants shared this view. They looked significantly longer at the movie in which the red ball approached its helper. This result held whether the helper had been the triangle or the square during the habituation trials. This was seen as being consistent with an interpretation of the animations based on mental states. The red ball was seen as 'preferring' or 'liking' its helper more than its hinderer, and so the movie in which the ball 'wanted to be with' its helpmate attracted the infants' attention more. Differential attention was not shown when the square and triangle approached the red ball rather than the other way around. By 12 months of age, therefore, the spatio-temporal characteristics of dynamic motion displays such as this appear to result in mentalistic reasoning. More recently, Hamlin, Wynn and Bloom (2010) have reported that infants as young as three months of age show distaste for the 'hinderer' when shown these same helping/hindering movies. So even very young babies can evaluate the 'social' behaviour of the geometric shapes that move in certain ways in these visual scenarios.

Intriguingly, bottlenose dolphins shown similar displays of moving geometric shapes appear to make similar mentalistic interpretations. Johnson et al. (2018)

showed dolphins 'helping' versus 'hindering' scenarios comparable to those used by Kuhlmeier and her colleagues in which one (target) oval would be following an upward path, and then falter and wobble downwards. At this point a second oval would appear and either 'help' the first oval on its way by moving beneath it and lifting it upwards again, or 'hinder' it by moving into its path and pushing it downwards towards the floor. In additional novel 'hit' and 'caress' sequences, the motion of the second oval either showed gentle and repeated contacts, suggestive of stroking, or the second oval made abrupt and strong contacts with the target oval, that briefly deformed its shape. The 'helping' and 'caressing' ovals were visually distinctive, as were the 'hindering' and 'hitting' ovals. At test, the dolphins were shown all three ovals, the target oval, the helper and the hinderer (or the caresser and the hitter), along with an occluder. The target oval was floating above the occluder, with the 'friendly' and 'unfriendly' ovals to each side of the occluder. The unfriendly and friendly ovals then moved off screen at their respective sides, and the target oval moved down behind the occluder as though to follow them. The measure of interest was whether the dolphins would show anticipatory looks to the side exited by the friendly oval, or to the side exited by the unfriendly oval.

Eighty per cent of the dolphins first looked to the side exited by the friendly geometric shape. Some of the dolphins also spontaneously displayed high-arousal social behaviours like blowing bubbles and making tail slaps. This remarkable study suggests that simple motion displays involving geometric shapes specify pro-social versus anti-social behaviour to the brains of other mammals as well as human infants. The dolphins' *predictive* looking to the display in the absence of any reinforcement (e.g., food rewards) is strong evidence for the automatic encoding of goal-directed causal information by the brain, in ways that enable a mentalistic (psychological reasoning) interpretation of observed events.

Using brain imaging (EEG), Southgate and Vernetti (2014) were recently able to show that even young infants (six-month-olds) can make mentalistic *predictions* about the actions of an agent based on that agent's belief. In Southgate and Vernetti's paradigm, infants watched an experimenter sitting opposite to them (behind a small stage) as she observed one of two events involving a ball and a box. In event 1, a ball jumped out of the box and rolled off-stage, while in event 2, a ball rolled onto the stage and jumped into the box. The box then closed, and a curtain came down, obscuring the view of the experimenter. The infant then watched further as either event 1 continued via the ball reappearing and jumping into the box, which had reopened, or as event 2 continued via the box opening again, the ball jumping out, and disappearing off-stage. The box lid closed again, and the curtain lifted to reveal the experimenter gazing at the closed box. Note that in event 1, the experimenter believed that the box was empty, whereas the infant knew that it contained the ball. In event 2, by contrast, the experimenter believed that the ball was in the box, whereas the infant knew that the box was empty. Southgate and Vernetti (2014) were interested in whether changes in the infant mu rhythm (the assumed electrophysiological signature of action effect learning, an alpha band response) would indicate an expectation of action in event 2, but not in event 1.

Southgate and Vernetti (2014) indeed found differential neural activity as indexed by the infant alpha rhythm for the two events (recall that when the EEG was measured, the infants were watching the same scene in both conditions, namely the experimenter gazing at the closed box). Suppression of alpha activity was only found for event 2, in which the infants were anticipating an action (based on a false belief). Hence the differential brain responses depended on what the *agent*, rather than the *infant*, believed to be the actual physical state of affairs. The six-month-old infants observing event 2 knew that the box was empty, but they also knew that the agent thought that the ball was in the box. An automatic index of this knowledge, a particular neural activation pattern associated with action-effect learning, suggested that the infants expected the agent to open the box to retrieve the ball. This paradigm is a nice example of using brain imaging to investigate cognition in a way that goes beyond the possibilities offered by behavioural studies. Further, the use of a *predictive* measure provides converging evidence for a mentalistic interpretation of goal-directed action on the part of the participating infants.

Gaze following and gaze monitoring

KEY TERMS

Gaze monitoring Following another's gaze to work out what is capturing that person's attention.

Still face paradigm Where the mother (or interacting other) stops interacting with her child and adopts a fixed 'still face' during a face-to-face interaction.

Social referencing Using feedback from others in order to appraise a current situation and determine how one should respond.

Another foundational mechanism for understanding psychological causation that is based in action may be infant gaze monitoring behaviour. The information from another person's eyes is very important for social cognition. For most of us, it is second nature to monitor another person's gaze. We follow gaze in order to work out what is capturing the attention of another agent, and we look into their eyes to try and infer their emotions, their intentions and their likely future actions. If someone deliberately avoids or prevents eye contact, it makes us feel uncomfortable. Babies, too, feel uncomfortable when another person stolidly refuses to meet their gaze. This is known as the 'still face' paradigm, in which mothers adopt a 'still face' during a playful face-to-face interaction, suspending interpersonal contingency (see for example Toda & Fogel, 1993). Babies react to a still face by avoiding gaze and getting upset. During the 'still face' periods, they tend to show a decline in visual attention to the mother, and show negative affect. Some infants attempt to restart engagement by vocalising, gazing at the mother while making motor movements (e.g., 'pick me up' movements), smiling and so on. Infant responsiveness in a 'still face' task has been shown to predict mirror self-recognition at 24 months, with infants who engage in more re-engagement behaviour (specifically by gazing to the still face) at nine months later showing better self-recognition in the mirror (Kristen-Antonow et al., 2015). In an important book on the development of the mind, Rochat (2009) suggested that infants' upset reaction to a 'still face' is early evidence for the importance of social interaction in forming a concept of the self. Put simply, he argued that if others are positive to us and interact warmly with us, we feel good about ourselves. If others are hostile to us or ignoring, we feel bad about ourselves. Accordingly, Rochat (2009) suggested that the need to *avoid social exclusion* or rejection determines much of our pro-social behaviour. Gaze following and gaze monitoring behaviour has been studied in order to find out how early in development babies begin to use the gaze of another as a clue to internal mental states.

One of the first experimental investigations of infants' ability to follow changes in adult gaze direction was carried out by Scaife and Bruner (1975). They were interested in the capacity for joint visual attention in the infant. When the visual attention of the mother-infant pair is directed jointly to objects and events, this facilitates learning. In fact, joint attention episodes have been described as 'hot spots' for learning (Tomasello, 1988). For infants to engage in joint attention behaviours, they must be able to follow their mother's gaze. Scaife and Bruner investigated gaze following by infants aged from two to 14 months. The infants were tested in a plain room, and were supported in an infant seat. After the mother had settled them in, an unfamiliar adult experimenter played with them for a while, and if they appeared content, the mother left the room. The experimenter then established eye contact with the infant, and then looked away, silently turning her head through 90° and gazing at a point on the wall for seven seconds. The experimenter then turned back to interact with the infant, before repeating the head turn, but this time looking to the other direction.

Gaze following was attributed to the infant if they (a) looked in the same direction as the experimenter, (b) did not look elsewhere first, (c) looked within seven seconds, and (d) appeared to be looking for or at something. Using these criteria, the percentage of infants showing gaze following was found to increase steadily with age. Around 30% of the youngest infants followed the experimenter's gaze, compared to around 65% of those aged 8–10 months and 100% of those aged above 11 months. Scaife and Bruner noted that the older infants also showed social referencing, looking back to the experimenter and then again at the wall, as though looking for something to look at. In other non-experimental trials using the mother, this referencing behaviour was even more marked, and more infants showed gaze following when alerted by language ("Oh look!").

Gaze following appears to become robust between six and 18 months of age. However, not everyone agrees that robust gaze following is evidence that infants are representing the other person as volitionally choosing to 'look at' an object in the environment. It has been argued that acts like gaze following could be the result of conditioned learning. For example, infants could learn that when an adult turns their head, something interesting is likely to be occurring at the location that they are now gazing at (see Moore & Corkum, 1994, for this type of analysis). However, if gaze following is a kind of conditioned learning based on adult head turning, then infants should still follow the gaze of an adult who turns their head when their eyes are closed. Brooks and Meltzoff (2002) investigated whether infants aged 12, 14 and 18 months would follow gaze in these circumstances.

Brooks and Meltzoff (2005) suggested that infants who were advanced in recognising the connection between looker and object would in turn benefit from facilitated language acquisition.

In their experiment, Brooks and Meltzoff (2002) contrasted infant looking in two conditions, *open-eyes* and *closed-eyes*. In each condition, an experimenter first made eye contact with the infant, who was sitting opposite on their mother's lap. She then silently turned her head towards a target, and gazed at it for 6.5 seconds. After this time, she returned her head to midline, made eye contact with the infant and resumed play. In the closed-eyes condition, she first closed her eyes before performing this movement. Four trials were given with two-minute intervals. Looking by the infant was credited if the infant's head and eyes were aligned with the target for at least 0.33 seconds. Brooks and Meltzoff found that the looking scores were significantly higher in the open-eyes condition than in the closed-eyes condition. Infants also looked significantly longer at the target in this condition (for almost two seconds), pointed at it more frequently and vocalised more. Exactly the same results were found for the older infants when the adult either wore a headband as a blindfold (hence could not see the target) versus had the headband around her head (hence could see the target). Brooks and Meltzoff pointed out that not only did these results rule out a conditioned learning explanation of gaze following, infant pointing and vocalising in addition to head turning was suggestive of the importance of gaze following for joint attention. Infants were interpreting the adult's look as an object-directed act.

Subsequently, Brooks and Meltzoff (2005) replicated the open-eyes/closed-eyes conditions with younger babies aged nine, ten and 11 months. They found that the ten- and 11-month-old infants almost only looked at the target when the experimenter's eyes were open. They were also more 'talkative' in this condition, producing many spontaneous vocalisations. The nine-month-olds did not distinguish between the two conditions. Furthermore, correct interpretation of gaze + vocalisation at 10–11 months predicted language comprehension at 14 and 18 months. Brooks and Meltzoff suggested that infants who were advanced in recognising the connection between looker and object might have an advantage in using gaze to disambiguate the referent of linguistic utterances. Understanding the line of regard of another person should facilitate language acquisition, as verbal labels usually refer to objects that are being looked at. The act of looking thus has referential meaning for both language and behaviour (following gaze can disambiguate both another person's emotional behaviour and their linguistic behaviour).

In neuroimaging studies of adults, it has been found using fMRI that a specialised area of the fusiform gyrus (the 'fusiform face area' or FFA) responds selectively to faces. Meanwhile, EEG studies with adults show that faces elicit a distinctive brain electrical potential, the N170 (a negative potential occurring most prominently over occipital-to-temporal scalp regions). These adult studies provide two neural markers that can be applied to studying face processing by infants. Tzourio-Mazoyer et al. (2002) studied whether two-month-old infants would show activation in the fusiform face area identified by adult studies when viewing unfamiliar female faces. The infants were shown coloured slides of women's faces, with a headscarf covering the hair so that only facial information was visible. The women were instructed to show gentle neutral expressions. The faces were interspersed with geometric patterns of circles matched with the faces for luminance. Tzourio-Mazoyer et al. reported that when viewing the female faces, the infants activated a network of brain areas belonging to the core system for the fusiform face area in adults. Even though this neural region is still immature in two-month-old infants, it appears to be selectively active during a relatively specialised activity, namely human face processing.

Converging evidence that face processing is specialised from early in development comes from an EEG study by Farroni, Csibra, Simion and Johnson (2002).

They studied whether four-month-old infants would display an N170 when processing human faces, and compared faces with direct eye gaze to faces where eye gaze was averted. Clearly, for psychological understanding, eye contact (mutual gaze) is essential for establishing a communicative context. Farroni et al. (2002) predicted that an early preference for eye contact would facilitate the neural processing of faces with direct gaze. Again, female faces without a hair contour and wearing a neutral expression served as stimuli. The females were either directing their gaze straight-on to the viewers, or were looking to one side. Faces with direct or averted gaze were presented in random order and remained on display for as long as the babies were willing to look at them. Analysis of the EEG recordings showed that the 'infant N170' (which actually occurred later than in adults, at around 240 ms) showed a significantly greater amplitude (i.e., more negativity) to the faces with direct gaze compared to the faces with averted gaze. Farroni et al. concluded that the presence of direct gaze *facilitated* the neural processes associated with face encoding. They noted that this early facilitation could be important for the interpretation of eye gaze signals as referential communicative acts. The assumption that greater neural activity reflects facilitation may not be correct, as expertise is often accompanied by reduced neural activation. Nevertheless, it is clear that the infant brain is responding differently to direct gaze. Similar specialised neural processes in infants and adults thus appear to be activated when faces are being viewed.

Brain imaging using fNIRS with five-month-old babies has revealed that infants are also sensitive to when an adult follows their gaze during the first year of life. Grossmann, Lloyd-Fox and Johnson (2013) contrasted the infant hemodynamic (blood oxygenation) response for two outcomes, both contingent upon the infant's attention being drawn to a novel object. During the experiment, the infant watched a person's face in the middle of a computer screen. There were also two cars visible on the screen, one at each side. After the infant was gazing at the person's face, one car moved briefly, attracting the infant's gaze. In the *congruent* condition, the social partner raised her eyebrows, smiled, and then followed the infant's gaze to the object. In the *incongruent* condition, the social partner again raised her eyebrows and smiled, but instead gazed at the other object. Grossmann et al. reported that the neural response was significantly different in the two conditions. The neural response was also left-lateralised in both conditions (compared to baseline). These neural data suggest that the infants were interpreting the same behaviour (the adult gazing at a car) differently in the two different contexts. The hemodynamic data reveal that infants can keep track of a social partner's eye gaze at younger ages than previously realised. Accordingly, infant social cognition can benefit from triadic social interactions (infant, partner, object) during the first six months of life.

Seeing as a 'mental act'

If infants follow the gaze of another because they want to see what she is seeing, then infants must interpret adult gaze as volitional and intentional. They must understand the looking behaviour of adults as a mental act of seeing some *particular* thing. In order to see whether infants aged from seven to 12 months would interpret adult gaze as intentional, signifying a relationship between actor and object,

Woodward (2003) devised an experiment using the visual habituation paradigm. Her method involved an extension of the 'teddy and ball' paradigm that she has also used to investigate infants' understanding of goal-directed action (Woodward, 1998).

In Woodward's gaze experiments, the infants were shown a stage containing two objects, a teddy bear and a ball. The two objects were sitting on pedestals a short distance apart. An actor was sitting at the back of the stage, with her upper body and head visible. At the beginning of an habituation trial, she made eye contact with the baby, saying "Hi". When she had the baby's attention, she then said "Look", and turned her head to gaze at one of the objects. She then kept looking at this toy for the duration of the trial. Following habituation, the position of the toys was reversed, and two types of test trial were given. In the 'novel trajectory' test, the actor turned her head to gaze in the opposite direction, thereby looking again at the same toy (as the toy locations had been reversed). In the 'novel object' test, the actor gazed in the same direction as before, thereby looking at the other toy. The results showed that both the seven-month-old and the nine-month-old infants tested followed the actor's gaze. Thirteen out of 16 seven-month-olds and 18 out of 19 nine-month-olds looked at the same toy as the actor, irrespective of whether this toy was novel or not. Infants aged 12 months tested in the same paradigm also showed overwhelming gaze following behaviour, but additionally registered the distinction between the old toy and the new toy. They spent longer overall looking at the object holding the actor's gaze when the test trials involved a new toy. Woodward argued that gaze was a powerful attentional signal for young infants. Note that ostensive cueing is also a feature of this paradigm.

If infants follow an adult's gaze because they want to see what she is seeing, then they should also move their location if this is necessary to obtain a good viewing angle. Moll and Tomasello (2004) investigated whether 12- and 18-month-old infants would move their viewing position if an adult was gazing at something behind a barrier. The barrier blocked the child's own line of sight. In their experiment, four different situations were studied. In one, the barrier was a dividing wall, made of wood and cardboard. In the second, the barrier was a cardboard box lying on its side. The third barrier comprised a movable wooden panel. The fourth barrier was the bottom drawer of a filing cabinet which was open, thereby blocking the child's line of sight. All of the barriers had fixed positions in the testing room.

In the experimental trials, an attractive toy was hidden behind one of the barriers. The experimenter looked behind the barrier, and said "Oh!" in an excited fashion. She gazed behind the barrier for about three seconds. She then looked back at the child. In the control trials, the experimenter gazed at a toy that was in full view of the child on a wall, in the opposite direction to the barrier. She again said "Oh" in an excited fashion. The dependent measure was whether the child moved to look behind the barrier.

Moll and Tomasello found that all the babies were significantly more likely to crawl around to look behind the barrier in the experimental trials than in the control trials. The 18-month-olds did this in more trials overall than the 12-month-olds. Both age groups also followed the experimenter's gaze in the control trials, looking at the toy on the wall. For the control condition, in which the object of the experimenter's gaze was clearly visible, the 12-month-olds followed the experimenter's gaze almost as frequently as the 18-month-olds. Gaze following to a visible target appeared to be developmentally easier for infants. In a second study, Moll and Tomasello checked that infants in the experimental condition were not simply having their attention drawn to the barrier by the adult's gaze, and then deciding to crawl around it by themselves. This time, the adult gazed at the barrier in both the experimental and control conditions. In the control condition, the experimenter gazed at a sticker on the side of the barrier. The infant could see this sticker without moving their position. Again, babies were found to be more likely to crawl around the barrier in the experimental condition, at both ages. Although the objects that the adult was gazing at were initially out of the infants' view, even the 12-month-olds wanted to see what the adults were seeing. Moll and Tomasello concluded that 12-month-olds understood that others were intentional agents, just like themselves.

Infant understanding of the *communicative intent* of adult gestures was addressed further in a study by Behne, Carpenter and Tomasello (2005b). They investigated whether infants could use communicative gestural cues, namely gazing and pointing, as clues to the location of hidden objects. The infants were aged 14, 18 and 24 months, and the experiment was introduced as a hiding game. In the hiding game, an attractive toy was hidden in one of two identical opaque boxes placed on a table in-between the experimenter and the infant. The experimenter then indicated the baited box either by gazing or by pointing. In the gazing condition, she repeatedly alternated her gaze from the baited box to the child and back again. She also raised her eyebrows and used other facial gestures to express communicative intent. In the pointing condition, she did these same things while also pointing at the baited box. Behne et al. (2005b) found that the infants could find the toys by using the gestural cues in both conditions at all ages. There were no differences in finding for the 14- and 24-month-olds whether the hiding place was indicated by gazing or by pointing. The 18-month-olds found significantly more toys from the pointing gesture. Behne et al. concluded that even infants of 14 months of age understand gazing as an expression of communicative intent in certain contexts.

Social referencing

Another way of using gaze to understand the internal mental states of others occurs during social referencing. Social referencing means appraising a current situation on the basis of the emotional expressions and behaviours of others and then regulating your own behaviour to it accordingly. The infant modulates his or her reaction to an object or event *by reference to* information gained from the actions of another. An interesting question is why infants regulate their own behaviours in social referencing situations. One possibility is that they modulate their behaviour because of a mentalistic interpretation of the reaction of another. For example, they may have an interpretation like 'she is reacting like that because she is scared, this is a potentially dangerous toy'. A second possibility is that they modulate their behaviour simply because the emotional display acts as a signal, telling them what to do

KEY TERM

Visual cliff
Apparatus that has a solid glass-topped surface but which appears to have a drop (the 'cliff' side) on one half. Originally designed to study depth perception in infants.

(for example, 'that expression means that I should stop'). Clearly, only the former possibility implies understanding of the internal mental states of another agent.

The classic study on social referencing used an apparatus called the visual cliff. A 'visual cliff' is created by building an apparatus that has a solid transparent surface that allows infants to crawl freely. Beneath the transparent surface, different cues to depth can be manipulated, so that one area of the apparatus appears to fall away like a cliff. This is usually done with geometric shapes, such as black and white squares, which generate powerful depth cues. The visual cliff was originally used to study depth perception in infants. It was found that babies of crawling age (around nine months) would not crawl over the edge of the 'cliff', even though the solid transparent surface enabled them to do so (Gibson & Walk, 1960). This was interpreted as showing that infant depth perception was well-developed, enabling them to avoid crawling over dangerous drops. Sorce, Emde, Campos and Klinnert (1985) demonstrated that crawling behaviour on the visual cliff could be modulated by the emotional expression of the mother. They put one-year-old infants on a visual cliff apparatus with an apparent drop of 12 inches. The infants typically crawled to the edge of the 'cliff' and then looked at their mothers. When the mother made a fearful face, 17 out of 17 infants did not cross the cliff. When the mother made a happy face, 14 out of 19 infants crossed the drop.

More recently, social referencing experiments have used procedures based on novel toys. Typically, the toy is faintly alarming, being mechanical and noisy and possibly able to move. The question is whether the infant will approach and play with the novel toy. In studies in which the social referencing provided by the mother or by another female adult is successful in modulating infant behaviour, both facial and vocal signals tend to be given. For example, Hornik, Risenhoover and Gunnar (1987) showed infants aged 12 months three novel toys, one intended to be pleasant (a musical Ferris wheel), one aversive (a mechanical monkey that clapped cymbals together incessantly) and one ambiguous (a stationary toy robot that recited facts about outer space in a mechanical voice). The infants' mothers were trained to show either positive, negative or neutral affect when the toys were presented. As the mothers found difficulty in expressing fear in a convincing manner to the toys, disgust was used as the negative emotion. Both facial expression of the emotions and vocalisations were used (e.g., the toys were described as either yucky or fun). Hornik et al. reported that the mothers vocalised almost all the time in the positive and negative affect conditions. The results showed an effect of negative affect only. Infants whose mothers expressed disgust played less with the target toy and also stayed further away from it compared to the neutral condition, irrespective of which toy it was. In contrast, positive affect did not seem to influence the infants' behaviour with the toys in any systematic way.

Mumme, Fernald and Herrera (1996) set out to investigate whether both facial and vocal signals were required for social referencing to be successful, or whether information from the face alone could be sufficient. If infants can read the emotional expressions of another person in the absence of any vocal cues, then this would suggest that they are interpreting the emotional expressions as reflecting internal mental states with predictable antecedents and consequences. This would be consistent with the idea that social referencing is another building block in acquiring a theory of mind. Vocal emotional signals can also reflect internal states, but with vocal signals intrinsic acoustic properties of the signals may induce emotions directly. For example, properties like loudness and pitch may induce fear (see Fernald, 1993).

Ninety 12-month-old infants were tested in Mumme et al.'s (1996) experiment, using novel mechanical toys that made noises and moved (like Magic Mike, the golden mechanical robot). The infant, mother and toy were placed so that they formed a triangle. For some infants (Face Only condition), the mother's face was fully visible, whereas for others (Voice Only condition) she was facing away with her back to the infant. In the key experimental trial, the novel toy appeared and the mother either showed fear, happiness or modelled a neutral reaction. For the Face Only condition, mothers were trained to show the target emotion using facial cues only, producing a 'fear' face or a 'happy' face. For the Voice Only condition, mothers were trained to show the target emotion using only their voice, saying either "Oh, how frightful!" or "Oh, how delightful", using the appropriate intonation. The phrase "Oh, how insightful" was spoken in a monotone for the neutral condition. The phrases were deliberately chosen to be meaningless to the infants, so that intonation would be the main acoustic cue. The mothers' efforts were later rated, and only successful emotional productions (face or voice) were used to calculate any experimental effects.

Infants were scored for how often they looked to their mother, how close they got to the toy or to their mother, and for their own affect (positive or negative). The results showed that when social referencing depended on facial information alone, there were no systematic effects. The only effect found was for girl infants, who looked at their mothers more when she was making a fearful face compared to a neutral face, and actually also approached the toy more. For the vocal cues, however, infants of both sexes looked longer at their mothers when she was vocalising fearfully, and also approached the novel toy less compared to the neutral condition. They also expressed more negative affect themselves. The happy vocalisations had no systematic effects on infant behaviour. The data suggest that 12-month-old infants may not have a fully referential understanding of facial emotional expression, even though in the visual cliff apparatus infants *were* able to modulate their behaviour by reference to the mother's face. Mumme et al. (1996) suggest that facial expression is a more powerful cue in the visual cliff apparatus, where clear danger is apparent. With the novel toy setting used in their own experiment, there was no particular danger, and this could explain the difference in results.

A visual cliff experiment by Vaish and Striano (2004) suggested that Mumme et al. (1996) were correct in emphasising the potential importance of the context in which social referencing takes place. They compared the contribution of facial versus vocal cues in social referencing on the visual cliff. In their study, 12-month-old infants were placed on a visual cliff with a 28 cm 'drop' (see Figure 3.8). All infants tested looked up to their mother after looking down at the drop, and the mother was then instructed to act according to one of three possible conditions. In the Face Plus Voice condition, the mother faced the infant across the cliff, smiled, and vocalised to the infant to encourage them to cross the drop. In the Face Only condition, the mother faced the cliff and smiled and nodded to her infant. In

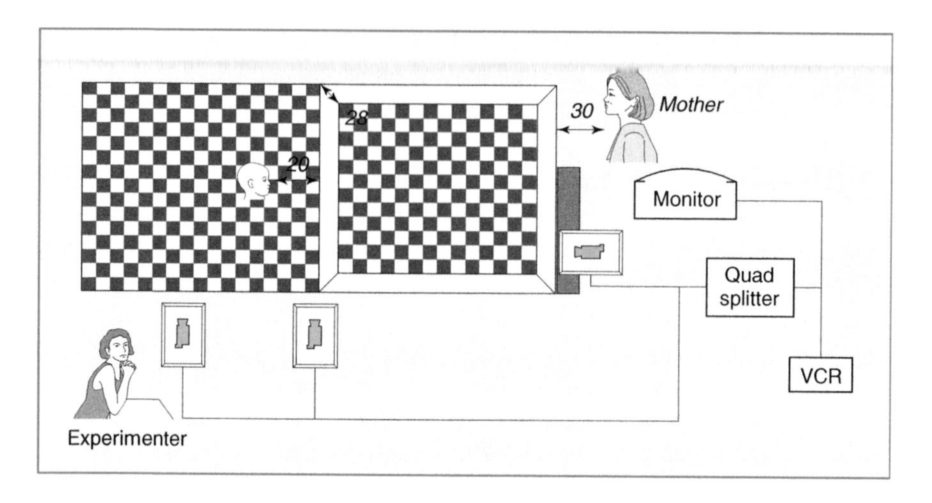

FIGURE 3.8 The visual cliff experimental set-up used by Vaish and Striano (2004). From Vaish and Striano (2004). Reproduced with permission from Blackwell Publishing.

the Voice Only condition, the mother had her back to the cliff, but vocalised to encourage the infant to cross the drop. Vaish and Striano measured the time it took the infant to cross the visual cliff. They found that infants in the Face Plus Voice condition crossed fastest, taking around a minute. Infants in the Voice Only condition crossed next fastest, taking just under two minutes. Crossing time in these two conditions did not differ significantly. Infants in the Face Only condition were slowest to cross, taking almost four minutes. This was significantly slower than the other two conditions. However, all infants did cross the cliff in the Face Only condition, suggesting that the face is an important social referencing cue in a potentially dangerous situation.

Note that in terms of social referencing, vocal cues can be experienced even when the mother's face cannot be seen (e.g., if she is carrying you on her back). Vaish and Striano pointed out that infant sensitivity to vocal cues deserved greater attention in studies of the ontogeny of human social cognition. Obviously, from the point of view of the mother, vocal interaction is often the most direct way of guiding the responses of her infant. Adopting the framework of natural pedagogy, vocal interaction is also a very effective ostensive cue.

It seems that further experiments using more ecologically valid paradigms are needed, experiments that systematically vary the presence or absence of ostensive cueing. This would enable the isolation of the effects of facial cues per se, and would enable emotional reactions to be natural rather than feigned, creating genuine ambiguity about whether the infant should approach or avoid. Such experiments would give us better insights into whether infants can use the emotional facial expressions of others as a basis for developing an understanding of psychological causation.

KEY TERMS

Protodeclarative pointing
When the infant's goal is to solicit joint attention by making another person attend to or recognise something that is of interest to the infant.

Protoimperative pointing
When the point is used to obtain an object (e.g., 'I want that').

ACTIONS BY INFANTS

Protodeclarative versus protoimperative pointing

As well as considering infant reactions to the actions of other agents as a source of information about psychological development, it is important to consider what we can learn from the infant's own actions. Pointing is an infant behaviour that has interested researchers for decades. A distinction has been made between two kinds of pointing: 'protodeclarative' pointing and 'protoimperative' pointing. When a point has a *protoimperative* function, it is used to obtain an object. The infant points in order to communicate 'I want that', or 'Get me that' (see Baron-Cohen, 1989a). This type of pointing has traditionally been conceived of as an instrumental behaviour. Protoimperative pointing does not necessitate an understanding of the mental states of others. The infant could simply be pointing because the usual outcome is getting the desired object (this would be stimulus-response or instrumental learning). When a point has a protodeclarative function, the point is used to 'remark on' the world to another person. This is thought to involve a higher, more mentalistic level of communication: the infant appears to want to influence the mental state of another person. The infant is communicating something along the lines of 'Look over there!' In protodeclarative pointing, the infant's goal is to make the other person attend to or recognise something of interest to the infant. The goal is joint attention.

Protodeclarative pointing is thought to be a precursor for, or early indication of, a 'theory of mind' (Baron-Cohen, 1989a). The use of protodeclarative pointing suggests that the infant knows about 'thinking', and is aware that another person may be thinking about something else. Pointing is used to initiate joint attention, which is necessary for the turn-taking behaviours that are the foundation of social interaction. The average age at which protodeclarative pointing emerges is around 12 months.

This was shown in a longitudinal study of 24 mother-infant dyads by Carpenter, Nagell and Tomasello (1998b). They found that for some babies, protodeclarative pointing emerged as early as ten months. Butterworth (1998) cites data for protodeclarative pointing in 33% of eight-month-old babies, with an advantage for girls.

In order to elicit proto-declarative pointing in the laboratory, Carpenter et al. devised a situation in which it was thought likely that the infants would want to engage the attention of an adult. For example, the infant was given a relatively uninteresting toy to play with, and then a much more interesting toy was produced by an experimenter. This was done behind the back of a second experimenter. Carpenter et al. measured when the infant gestured to the second experimenter about the more interesting toy. These 'distal' protodeclarative gestures had a mean age of emergence of 12.6 months. The mean age of protoimperative pointing was slightly later, at 14 months. There appeared to be a significant developmental change in the likelihood of producing a protodeclarative point between 11 and 12 months, and in producing a protoimperative point between 12 and 13 months. The ability to follow the point gesture of someone else emerged slightly earlier. Carpenter et al. (1998b) reported that the average age of the emergence of point-following was 11.7 months.

Protodeclarative pointing is used to initiate joint attention, and suggests that the infant is aware that the other person could be thinking about something else.

An EEG study of protodeclarative pointing was carried out with 14-month-old infants by Henderson, Yoder, Yale and McDuffie (2002). They measured EEG power in the 4–6 Hz and 6–9 Hz ranges at 40 electrode sites when the infants were 14 months of age, and correlated these neural 'resting state' measures with protodeclarative versus protoimperative pointing at 18 months. Henderson et al. argued that contrasting the neural activity related to two acts of pointing enabled them to control for the influence of active motor movements on the brain, so that the neural activity associated with the pragmatic function of the points could be isolated. Once motor responding was equated, they found that EEG power in frontal regions at 14 months was correlated with protodeclarative pointing at 18 months, but not with protoimperative pointing. Hence brain resting state activity in frontal regions at 14 months was correlated to pointing four months later only when pointing was being used to initiate shared attention and not when pointing was being used to regulate behaviour.

Indeed, brain activity is also different when participants are *observing* pointing actions versus instrumental actions. As noted earlier, protoimperative pointing is an instrumental behaviour, it has the instrumental goal of obtaining the item being pointed at. Pomiechowska and Csibra (2017) argued that grasping is also an instrumental behaviour, as grasping an item usually serves the instrumental goal of obtaining that item. Protodeclarative pointing, on the other hand, is a communicative or social behaviour, with the goal of sharing attention and mental states. Pomiechowska and Csibra thus recorded EEG (the mu rhythm, the putative electrophysiological signature of action-effect learning) while adult observers watched two types of actions, pointing and grasping. They reasoned that as grasping was an instrumental action, viewing grasping should cause suppression of the mu rhythm. By contrast, as pointing was a communicative action, viewing pointing should not result in mu suppression.

The EEG data showed that mu suppression did not occur when the adult observers were watching pointing acts, but that mu suppression did occur when the observers were watching instrumental acts. Pomiechowska and Csibra (2017) argued that the data suggested that while grasping was an instrumental action, pointing was a communicative action. Interestingly, if the grasping action was preceded by ostensive cueing (saying "Look!" to the observers), then mu suppression was not observed. The ostensive cueing changed the *context* for the action into a communicative one. This experiment suggests that suppression of the mu rhythm is specific to an interpretation of observed actions as *instrumental*, a point that we will return to in Chapter 4. Most recently, Pomiechowska and Csibra (2018) have provided eye-tracking data supporting the same distinction for infants aged 12 months. The infants' eye movements discriminated between observing pointing at an object (a communicative, referential act) and observing grasping an object (an instrumental, object-directed act). It would be interesting to repeat this study while recording the mu rhythm in these young infants.

Joint attention

As noted, both protodeclarative and protoimperative pointing are object-directed, in that the objects being pointed at are available in the physical and social environment rather than being unobservable entities in someone's head. Similarly, in order to follow someone's point, the infant must understand that pointing is intended to highlight or signal the presence of an object or event worthy of remark. They must understand that the actor is attending to the object that they are pointing to, and that the actor's intention is to bring another person (the infant) into shared attention on the object. Woodward and Guajardo (2002) set out to investigate infants' understanding of the relation between the object being pointed to and the person who points, using a habituation procedure. They argued that understanding the 'object-directedness' of pointing would be important for a general understanding of pointing as an intentional act.

Woodward and Guajardo (2002) studied babies aged nine and 12 months of age, in order to capture the period during which it was generally believed that a change in infants' understanding of pointing occurs. They used a variant of the stage apparatus used by Woodward (1998), in which a teddy bear and a ball were displayed each sitting on a pedestal 14 cm apart. During habituation, the infants watched as an actor pointed at one of the two toys, touching it with her finger (see Figure 3.9). Only the actor's hand was visible, and the hand either pointed to the object nearest to the right side of the stage, or passed in front of the first pedestal to point to the further object. For some infants, the actor's upper body was visible during habituation as well. For these infants, the actor pointed from above, and said "Look" at the same time as she pointed. After the infants had habituated to the actor pointing at the chosen toy, the position of the toys was switched and two types of test trial were given. In the 'new object' trials, the actor pointed to the same side as before, which meant that she pointed to a novel object. In the 'new side' trials, the actor pointed to the opposite side to before, which meant that she pointed to the familiar object. The 12-month-olds looked significantly longer at the 'new object' event. The nine-month-olds did not.

Woodward and Guajardo argued that the 12-month-olds were responding selectively to the change in relation between actor and object. The actor was now pointing at a novel object. This selective response was found whether the infants saw only a hand pointing during habituation, or saw both the hand and the upper body of the actor. During the habituation trials, the nine-month-olds habituated to the pointing events in the same manner as the 12-month-olds. Woodward and Guajardo hence argued that the younger babies were able to attend to the events of interest. In fact, further analyses showed that the point gesture acted as an effective 'spotlight' for attention for the younger babies as well as for the older babies. Thirty-seven of the 39 younger babies tested looked longer at the toy being pointed at during the test trials. However, these younger babies did not react selectively according to whether this toy was a novel object of attention for the actor. In a second study, Woodward and Guajardo (2002) found some evidence that babies who themselves pointed at objects were more likely to encode the habituation events as object-directed. This appears to indicate that the production and comprehension of object-directed points are developmentally linked.

By at least 12 months of age, therefore, infants appear to understand pointing as an intentional act. These data support the assumption that protodeclarative pointing is an early indication of a nascent 'theory of mind', indicating an effort by the infant

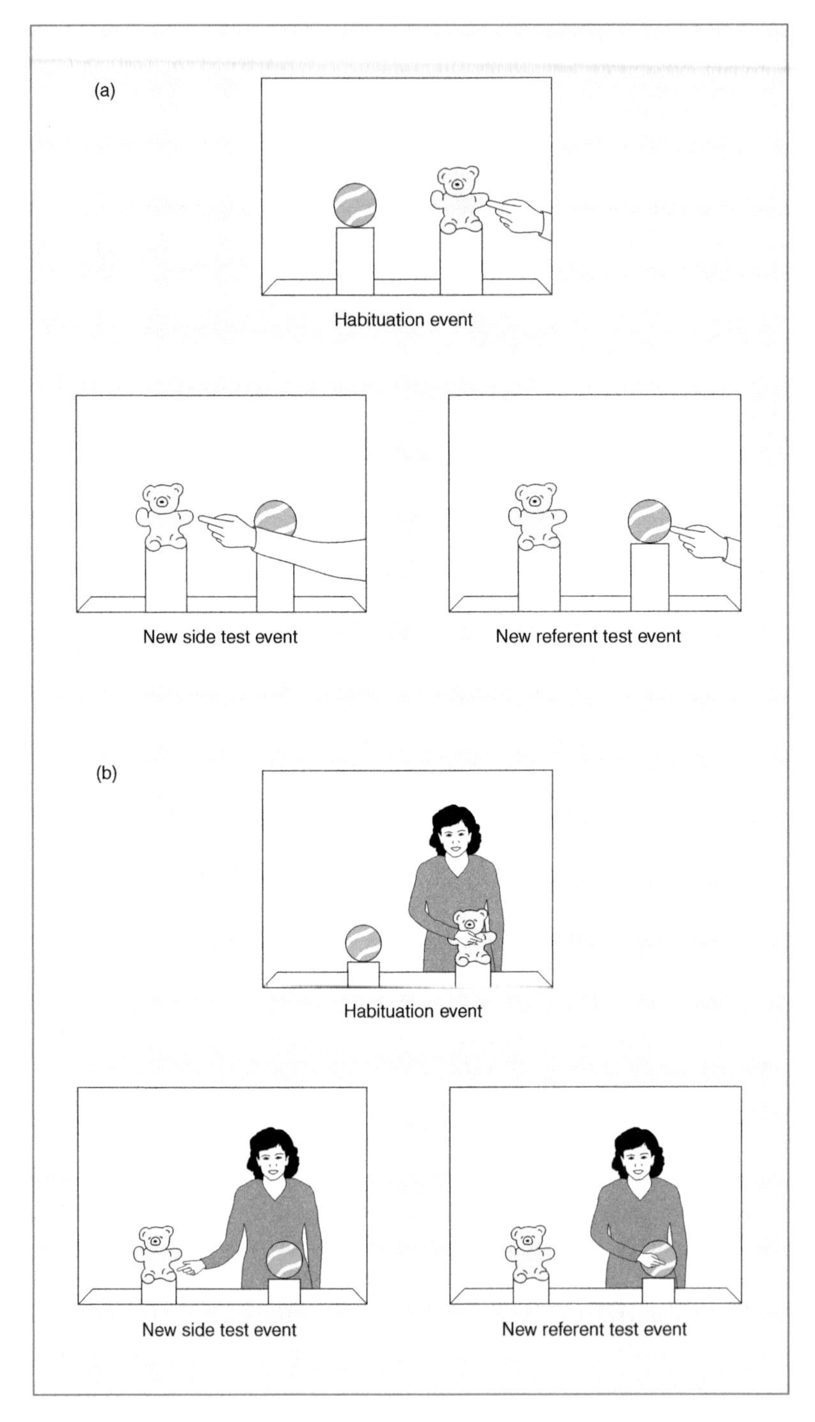

FIGURE 3.9
Woodward and Guajardo's (2002) habituation and test events for (a) the hand-only condition, and (b) the face-and-hand condition. From Woodward and Guajardo (2002). Copyright © 2002 Elsevier. Reprinted with permission.

to establish joint attention (Baron-Cohen, 1995). Bates, Camaioni and Volterra (1975) have also suggested that protodeclarative points are intentionally communicative acts, noting that the pointing infant usually also alternates his or her gaze between the adult's face and the object being pointed at. Bates et al. argued that this gaze alternation indicates a desire to affect the adult's behaviour. In the studies discussed above by Brooks and Meltzoff (2002, 2005), the authors reported that their infants pointed at the target object much more frequently when the adult had his or her eyes open. This also suggests an intention to communicate with the adult by pointing – pointing is only useful if the other person is able to move their gaze to the referent.

FIGURE 3.10
Staged photograph of the testing situation in Liszkowski et al.'s (2004) study showing the puppet appearing from behind the curtain. From Liszkowski et al. (2004). Reproduced with permission from Blackwell Publishing.

The possibility that infants who display protodeclarative pointing *intend* to establish joint attention was examined empirically by Liszkowski, Carpenter, Henning, Striano and Tomasello (2004). They compared what happened when adults either rewarded infants' protodeclarative pointing with shared attention and interest, or did not. In their study, 75 12-month-old babies interacted with an adult experimenter in one of four conditions. In all the conditions, the experimenter first established herself as a social partner for the infant, who was seated next to her. The infant was facing some curtains, and the experimenter was at 90° to the curtains but could easily see them by turning her head. The experimenter chatted to the infant, made eye contact and engaged in joint play with a bead toy. Once the infant was relaxed, the experimenter gradually withdrew from the interaction. She then signalled to a second experimenter to produce an interesting event from behind the curtains. For example, a hand puppet might appear from the curtains and dance about (see Figure 3.10).

Typically, the infant pointed to this exciting event, and the experimenter then responded in one of four ways. In the Joint Attention condition, she showed shared attention and interest, looking backwards and forwards from the event to the infant's face, talking excitedly about the puppet, and commenting that they were seeing it together. In the Face condition, she continually gazed at the infant's face, talking excitedly but not about the puppet ("Oh! I see you are in a good mood! Did you sleep well?"). In the Event condition, she looked at the puppet, but did not look at the infant and did not vocalise. In the Ignore condition, she took no notice of the infant's protodeclarative pointing, simply gazing at her hands and picking at her nails.

Liszkowski et al. measured how often and how long the infants pointed during a given trial, and how many times they looked at the first experimenter. They expected more pointing and looking when the experimenter was refusing to share her attention with the infant, and this was essentially what they found. Infants were significantly less likely to point frequently in the Joint Attention condition, pointing more within a given trial when the experimenter was not engaging in joint attention behaviours. Their enjoyment of sharing the novel event with a responsive adult in the Joint Attention condition was shown by the fact that their points were of significantly longer duration when shared attention was being engaged. The infants also looked at the experimenter significantly more frequently in the Event condition, when the adult appeared interested in the event but was not sharing her interest with the infant. This condition produced the most gaze alternation between the infants, the adult and the puppet. Liszkowski et al. argued that protodeclarative pointing indeed had a social motive. The social context within which the pointing took place, in terms of the partner's reaction, had a significant effect on the infants' behaviour. Infants point protodeclaratively to share attention and interest with other people. Their pointing behaviour has communicative intent.

Kovács and her colleagues have made the complementary point that protodeclarative pointing by infants could also serve as an 'epistemic request' for new information. A 12-month-old infant who points to an object may wish to learn information about the object, as well as to share a mental state with another person. Kovács, Tauzin, Téglás, Gergely and Csibra (2014b) devised a variation of Liszkowski et al.'s (2004) procedure in which adults systematically responded to pointing by infants either by sharing attention, or by providing new information. In the Shared Attention condition, when the puppet appeared from behind the curtain and the infant pointed, the adult would smile and nod and share her attention. In the New Information condition, when the puppet appeared from behind the curtain and the infant pointed, the adult would provide the name of the puppet and some emotional information about it (e.g., saying "Juj" in a fearful voice, or "Aah" in a delighted voice). Eight trials with novel puppets were delivered overall. Hence while joint attention was established in both conditions, the infants also received a learning opportunity in the New Information condition. Kovács et al. (2014b) found that the infants pointed significantly more frequently in the New Information condition. This is an interesting result, as it shows that pointing can be used deliberately by infants to learn new information. The infants in this case are active learners, using pointing to choose *themselves* what to learn about next.

Indeed, Csibra has made the important claim that infants interpret objects as *symbols* of object kinds, rather than as one piecemeal object after another. Csibra and Shamsudheen (2015) argued that when someone demonstrates something to an infant using an object as a pedagogical tool, then infants make the assumption that the object stands for or exemplifies the *whole class* of objects of the same kind. So an individual object used in a joint attention situation of natural pedagogy will be used by the infant to learn general knowledge about the whole class of objects that this single object represents. If correct, this rudimentary symbolic capacity of individual objects (that is, symbolic when in the presence of ostensive cues in a natural pedagogic context) would enable rich learning about the environment. Put simply, Csibra's idea is that for infants, caretaker demonstrations with objects in contexts denoting natural pedagogy create a learning situation comparable to that of adults of watching an advertisement on TV. When someone on TV shows a box of pills and says "This is what I take when I have a headache", we understand that the person is demonstrating not that particular box of pills, but that brand of medicine. Similarly, if infants watch a demonstration with a novel object, they may view this as a learning opportunity relevant to that *category* of objects. If infants indeed routinely learn generic knowledge in this way (that is, learn about a whole class of objects rather than about the single object being demonstrated), then powerful culturally relevant symbolic learning would result.

THE UNDERSTANDING OF FALSE BELIEF

The evidence discussed so far shows that many of the mechanisms underpinning social cognition are developing rapidly during the first year of life. As new technologies such as EEG and eye tracking are used more widely with infants, it seems likely that the ages at which infants first show reliable evidence of mentalistic understanding of adult actions like gazing and facial expression will be pushed even earlier. The (largely) behavioural data discussed here have shown that infants detect contingencies between their own behaviour and events in the world from birth. They rapidly come to view the actions of others as goal-directed, they follow others' gaze, they engage in joint attention and they ascribe actors with communicative intent. However, in the studies discussed so far, infant understanding of the mental states of others has usually been correlated with reality. The actual state of affairs in the real world has typically been represented by both the other agent and by the infant (the exception was the EEG study by Southgate & Vernetti, 2014). Even in the experiments where the adult used gestures to communicate the location of a hidden toy to an infant who did not know where it was hidden, both agents shared a belief about the location of the toy. The adult had a true belief that she knew where the toy was, and the infant had a true belief that the adult knew where the toy was. There was no contradiction in mental states from the perspective of veridical reality in the world.

One aspect of mental state understanding that must be purely representational, however, is understanding *false belief*. In fact, the philosopher Dennett has argued that successful reasoning about false beliefs is the *only* convincing evidence for the attribution of mental states to others (Dennett, 1978, see also Chapter 4). This is because a person who acts on the basis of a false belief acts in a way that would not be predicted by the real situation in the world. Until a few years ago, it was thought that a fundamental change in children's ability to understand the minds of others occurred at around four years of age. This was thought to be the median age at which children became able to respond successfully in false belief tasks (see Chapter 4). The ability to pass false belief tasks was taken as evidence for the acquisition of a *representational* 'theory of mind'. Passing the false belief task was thought to be proof that the child had come to understand that the minds of others held beliefs that were not necessarily direct reflections of reality. Another person might believe something to be true that you yourself knew to be false.

KEY TERM

Attribution of mental states
Explaining the behaviour of others by detecting and understanding the intentions underlying their goal-directed behaviour.

The idea of a developmental 'watershed' in understanding at around four years of age means that infants were not expected to 'pass' false belief tasks. Yet many studies are now appearing that provide tests of false belief understanding in infants (see Scott & Baillargeon, 2017, for a recent survey). We will consider a selection of relevant studies here. One popular paradigm has been an adaptation of the *false location* task used with young children (see Chapter 4). The infant watches an actor hide an object at a particular location. Then, while the protagonist is absent, the hiding place is changed. Consequently, when the protagonist returns and seeks the object, the rational act is to look in the location at which the protagonist *believes*

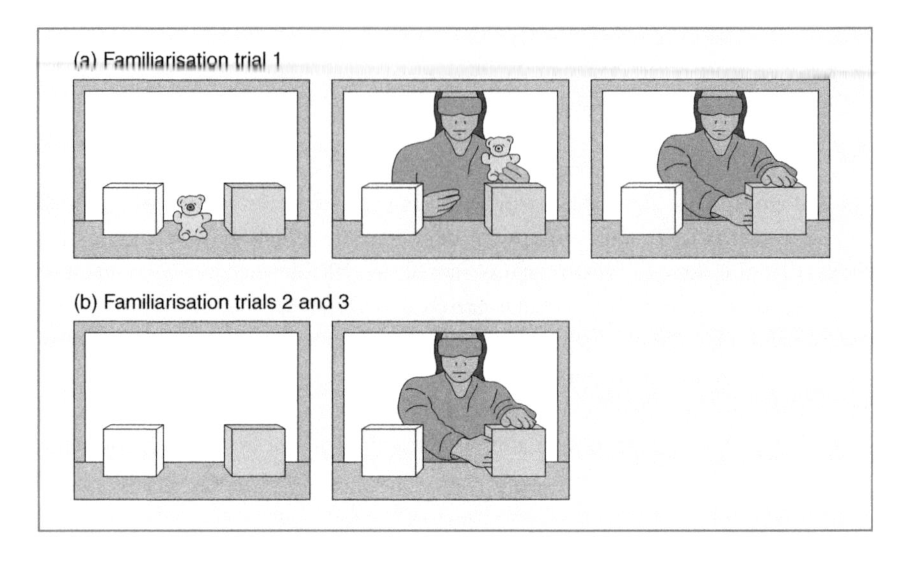

FIGURE 3.11
The events shown during (a) the first familiarisation and (b) the second and third familiarisation trials of Onishi and Baillargeon's (2005) study. From Onishi and Baillargeon (2005). Copyright © 2005 AAAS. Reproduced with permission.

the object to be hidden. But this belief of the protagonist is now false – the object is in a new location. Hence in order to find the object, the protagonist must discover this new location. If infants/children show that they understand that the protagonist will *first* seek the object at the wrong location, this is an indication that they have an understanding of false beliefs. As will be recalled, Southgate and Vernetti's (2014) EEG experiment discussed earlier used a variant of the false location paradigm.

We already know that infants can keep track of an object's location across changes in spatial position, even when the object is hidden (e.g., Baillargeon & Graber, 1988). We also know that they view the actions of others as goal-directed, and attribute intentionality to the behaviours of others. Onishi and Baillargeon (2005) devised one of the first false location tasks for infants, using a version of the hidden object paradigm developed by Baillargeon and Graber (1988). The basis of their paradigm was a search task. During familiarisation trials, 15-month-old infants watched as a protagonist hid a toy in one of two possible locations (a yellow box or a green box, see Figure 3.11). The openings of the boxes were at 90° to the infants, so that the infants could not see into the boxes, and were concealed by fringing. In the first familiarisation trial, the actor put the toy into the green box. In the next two familiarisation trials, she put her hand into the green box as though to grasp the toy, and rested her hand there. The infants then saw a belief induction trial. In the *false belief* condition, they watched as the toy moved location into the yellow box. The actor did not see this event, and the infants observed that she could not see it. In the *true belief* condition, the actor and infant both watched as the toy moved location into the yellow box. A test trial was then given, during which the actor simply placed her hand into one of the boxes. In the false belief condition, infants were expected to look longer when this was the yellow box, as the actor believed that the toy was in the green box. In the true belief condition, infants were expected to look longer when this was the green box, as the actor believed that the toy was in the yellow box. Further conditions checked the analogous predictions when the first hiding place was the yellow box.

Onishi and Baillargeon (2005) reported that the infants looked significantly longer when the box that the actor chose to search was *inconsistent* with her belief about the toy's location (violation of expectation). This was the case whether the actor's belief was false or true. The infants appeared to expect the actor to reach to where she *believed* the toy to be. Onishi and Baillargeon argued that this was evidence for a rudimentary representational theory of mind in 15-month-old infants. Infants not only attribute mental states such as goals and intentions to other agents, they also attribute beliefs. These beliefs may be true or false, and are used to make sense of others' actions.

Using a different violation of expectation paradigm with 17-month-old infants, Scott, Richman and Baillargeon (2015) have also produced evidence that infants understand *strategic deception*, namely that the goal of some actions may be to implant a false belief in another person. In the strategic deception paradigm, the infants watched events on a small stage, during which two experimenters interacted with rattling toys. The toys belonged to one experimenter ('owner'), who during six familiarisation trials tested out six toys while being watched by the second experimenter ('thief'). Three of the toys tested by the owner rattled, and three were silent. Each time the owner left the stage to fetch a new toy, the thief covetously rattled the rattling toys, but did nothing for the silent toys. Each time the owner returned with a new toy, she placed the previous toy in her box if it had rattled, and discarded it into a trash can if it had been silent. The key manipulation was a test trial during which a seventh toy, identical in appearance to the last silent toy, was joyfully demonstrated as a rattling toy by the owner. She then left the scene as usual. At this point the other experimenter 'stole' the new rattling toy, by putting it in her pocket. She then substituted the silent toy of identical appearance, which she retrieved from the trash can.

The important measure was how long infants looked at this scenario compared to a control scenario, in which the thief stole the new rattling toy in the test trial, but replaced it with a silent toy from the trash can that looked quite different to the toy that rattled. Scott et al. argued that if infants have an emergent understanding of false beliefs, then they should look longer at the scenario where the substitute toy does not match the stolen toy in appearance. In this situation, the theft would become obvious as soon as the owner returned. This was exactly what they found. The infants watching the deception scenario with a non-matching toy looked for significantly longer than the infants watching the deception scenario with a matching toy. A series of further control studies ruled out lower-level visual feature-based explanations. Scott et al. (2015) argued that they had shown a mentalistic interpretation by infants of the goal-directed actions viewed during the strategic deception scenario. The ingenious use of deception in this scenario appears to provide strong evidence for 'mindreading' infants, and an early (rich) understanding of psychological causality.

However, as noted in Chapter 2, violation of expectation paradigms have difficulty in matching all of the spatio-temporal information in the displays being watched by the infant. Further, as also discussed in Chapter 2, spatio-temporal information, particularly involving motion, can specify agency and trigger causal explanations of physical events (such as 'chaser' and 'chased'). The spatio-temporal motion of geometric shapes can thus also be used to test false belief understanding in infants. Surian and Geraci (2012) devised a false belief paradigm to explore whether infants can generate explanations based on hidden psychological causes. In Surian and Geraci's task, infants watched a triangle 'searching' for a circle. Infants aged 17 months viewed a video screen on which a red triangle followed a blue circle around the screen in several directions. Eventually, the circle entered a Y-shaped tunnel leading to two boxes, one at each exit of the Y. The circle could be seen going through the tunnel and up one of the arms of the Y, exiting the arm, and settling into the relevant box ('hiding' – once it entered the box, the circle disappeared from view). The waiting triangle then entered the tunnel, followed the same route, and entered the appropriate box.

Following two of these events (familiarisation trials), a test trial occurred. This time, after hiding in one box, the circle changed location and moved to the other box. For some infants, this change of location occurred in the presence of the waiting triangle, while for other infants, the triangle went off screen and so was absent when the hiding place changed. In the latter scenario, the triangle thus held the 'false belief' that the circle was still in the first box. The key measure was which box the infant made *anticipatory* looks to as the triangle entered the tunnel in the test trial. Surian and Geraci found that 17 out of 23 infants in the *false belief* condition showed anticipatory looking to the box that the circle had entered when the triangle was still present (the 'empty' box). Meanwhile, 18 out of 24 infants showed anticipatory looking to the new hiding location in the true belief condition, the box that the triangle 'knew' that the circle had moved into.

This is a very interesting result. As with the studies by Kim and Song (2015), Brandone et al. (2014) and Southgate and Vernetti (2014) discussed earlier, anticipatory or predictive behaviour is a good index of a mentalistic interpretation of physical events. What is remarkable here is that Surian and Geraci have evidence for a mentalistic interpretation of a spatio-temporal scenario involving geometric shapes, not human agents. This is suggestive evidence that the automatic processes operating during the neural encoding of events *in themselves* give rise to causal analyses, in this case, an analysis of *psychological* causality. The use of a 'false belief' scenario makes this particular demonstration even more noteworthy. According to traditional philosophical analyses, babies' behaviour in these paradigms is evidence for a representational theory of mind.

On the other hand, infants aged 11 months tested in the same scenario showed ambiguous anticipatory looking behaviour, with no clear evidence for false belief understanding. This null result is difficult to interpret, but may suggest that a certain amount of *experience* is required for infants to ascribe psychological causes in spatio-temporal scenarios where hidden beliefs must be inferred in order to comprehend goal-directed actions. In such cases, it is not simply the spatio-temporal event sequences that matter, but prior experience of similar spatio-temporal event sequences which *have* had psychological causes. The idea that experience enriches the interpretation of events was also discussed in earlier chapters, and must necessarily be true. The most parsimonious interpretation of these data regarding psychological reasoning is that while the spatio-temporal dynamics of viewed events might impose a certain level of causal structure on experience at the point of encoding, real-world experience will gradually enable enriched understanding of such viewed events, thereby supporting the development of meaning-based knowledge representations (schemas). This may happen fairly rapidly. As in the metaphor of the connoisseur of wine (see Chapter 2), the representation of features in the environment thus becomes transformed irrevocably by experience. For understanding psychological causation, ostensive cueing and language are also likely to play key developmental roles.

Converging evidence for false belief understanding in infancy has used an adaptation of the 'false contents' task that is also used to measure false belief in young children. In the false contents task, children are shown a distinctive tube of Smarties (candies) and then when the tub is opened they find out that the tube actually contains a pencil (Hogrefe, Wimmer & Perner, 1986). Buttelmann, Over, Carpenter and Tomasello (2014) devised a false contents task for 18-month-olds in which infant understanding was measured by an active response. The infants first met two experimenters who wondered what to play, and suggested playing with 'block boxes' that contained building blocks. One experimenter then fetched a block box from a shelf (the box had a picture of building blocks on it), opened it with the infant, and took out the block. This was repeated two more times, with two more block boxes being fetched from the shelf, opened, and found to contain a building block (the blocks were also used to build a tower). On the fourth trial, the first experimenter briefly left the room and the second experimenter fetched the fourth block box. This was opened and found to contain a spoon. This block box was then returned to the shelf. The first experimenter then reappeared, carrying a bowl. She put the bowl by the tower of blocks and said "Now I'm looking for a...". The second experimenter appeared occupied, so the infant was prompted to help. The second experimenter put down a block and a spoon, and said "Go and get it for [first experimenter's name]". The question was whether the infants would fetch the block or the spoon.

Buttelmann et al. (2014) found that 67% of the infants gave the first experimenter the block and not the spoon. By contrast, in a true belief control condition in which the first experimenter was present when the spoon was revealed inside the fourth 'block box', 67% of infants gave the first experimenter the spoon. This 'unexpected contents' paradigm provides a nice test of infants' ability to keep track of the false beliefs (and true beliefs) of others. As Buttelmann et al. point out, their task is also more complex than the false location task, as the infants had to infer that the first experimenter was expecting a block to be in the fourth 'block box' on the basis of what block boxes typically contain, rather than on direct experience of the fourth 'block box' containing a block. This task is similar to the 'Smarties' false belief task used with older children (a false contents task), and provides strong converging evidence for an understanding of false beliefs in older infants.

Finally, returning to our metaphor of the wine connoisseur, we noted in Chapter 2 that machine learning analyses that have modelled neural trajectories over multiple learning experiences have proposed that what changes with expertise is the structure of the neural activity space. This makes it interesting to consider whether the neural activity space engaged during infant false belief tasks shows any similarity to the neural structures that are engaged when adults reason about false beliefs. In order to make such comparisons, we need studies in which infants and adults are reasoning about the exact same scenario. This is important, as the brain will automatically respond in specific ways to specific visual features of a scenario, which could also lead to group differences in neural imaging indices.

Kovács and her colleagues devised a false belief paradigm that could be used with both seven-month-old infants and with young adults (Kovács, Téglás & Endress, 2010). They then used the same false belief paradigm in an fMRI study with adults only (Kovács, Kuehn, Gergely, Csibra & Brass, 2014). In their false belief paradigm, observers (infants or adults) watched a cartoon video involving a table, a ball and an occluder. At the beginning of the video, an agent placed the ball on the table, and the ball rolled behind the occluder. The ball then rolled out again and back again, either in the presence of the agent, or after the agent had left the scene. Alternatively, the ball rolled out from the occluder and left the scene completely, either in the presence of the agent or after the agent had previously left the scene. This enabled the comparison of four belief states plausibly held by the agent: a *true belief* that the ball was either behind or not behind the occluder, and a *false belief* that the ball was either behind or not behind the occluder, However, all of these belief states were irrelevant to the task required of the adult observers, which was to press a button when the occluder was lowered if the ball was revealed to be present. Intriguingly, adult reaction times were faster both when they themselves believed the ball to be behind the occluder and when the agent falsely believed the ball to be behind the occluder. Kovács et al. (2010) argued that humans *automatically* compute the beliefs of other agents when they observe visual events. A parallel set of studies with seven-month-old infants showed that the infants looked significantly longer when only the agent falsely believed the ball to be behind the occluder.

Kovács et al. (2010) concluded that the mere presence of social agents automatically triggered online belief computations in young infants. Further, the infants could hold these computations in mind along with alternative representations, such as their own veridical knowledge of where the ball was actually located. Kovács and her colleagues (Kovács et al., 2014) then investigated the neural structures that were activated during this implicit false belief task. Of interest was whether the same neural structures would be activated as during explicit false belief reasoning. A specific set of neural structures focused around the right temporoparietal junction is frequently activated during explicit 'theory of mind' tasks such as the false belief task (Saxe, 2006). Kovács et al. (2014) were interested in whether their implicit false belief task would activate the same neural structures. They reported that their automatic belief tracking task did indeed activate the right temporoparietal junction, just like explicit false belief tasks. As seven-month-old infants showed (via looking time) the same pattern of performance as adults showed (via reaction time) in the automatic belief tracking task, it would clearly be very interesting to run the parallel neuroimaging study with infants.

SUMMARY

The development of psychological understanding in the infant, including knowledge of self and agency and the development of social cognition, seems to begin with simple mechanisms for ensuring proximity to the caretaker and facilitating attention to human faces, voices and odours. Human infants are born with a propensity for social interaction, and they notice social contingencies as well as contingencies between their own actions and the environment from the first weeks of life. Contingency detection plays a role in developing a primary representation of the bodily self, and also in fostering social interactions. Infants reward their caregivers with the most smiles when the behaviour of their caregivers is contingent on their own responses. These early interactions also enable infants to recognise that other agents are 'just like me'. In particular, via imitation, infants come to understand that the visible bodily actions of others can provide clues to their internal mental states. This is hypothesised to be worked out by analogy to the self. The infants' own acts are related to their internal desires, and by analogy the object-directed movements of others are related to their internal desires. The ability to interpret actions as goal-directed is present by at least three months of age.

Goal-directed action has a special role to play in the development of psychological understanding. The interpretation of the goals of other agents is thought to require some insight into intentionality. Although it is possible to interpret individual experiments regarding goal-directed behaviour teleologically, by reference to the relevant aspects of reality as constraints on possible actions, infants appear to go beyond teleological analyses. Infants are more likely to imitate the intentional acts of other agents than their accidental acts, for example, even if the acts are identical. They can also distinguish between actors who are unwilling versus unable to act in certain ways. Depending on the context in which modelled acts are performed, infants will infer the intended goal of the actor. They will then use the most efficient actions themselves to achieve that goal, rather than simply imitating the action sequences modelled for them. By the age of six months, they will expect an actor to reach for a desired object when that actor *believes* that the object is in a box, even though the infants themselves know the box to be empty (Southgate & Vernetti, 2014).

Gaze following and gaze monitoring are also relatively simple mechanisms that can confer psychological understanding of others. The information from the eyes is very important for social cognition, and we look into people's eyes to try to infer their intentions and emotions. Typically-developing infants are attracted to direct gaze from early in life, and follow changes in the direction of gaze because they interpret seeing as a mental act. They do not follow changes in gaze if another agent is blindfolded or has their eyes shut. They will also use the facial expressions of others as clues to how to behave themselves, for example inhibiting crawling in potentially dangerous situations if the mother's emotional expressions suggest fear. In social referencing, information from the voice is as important if not more important as information from the face. Indeed, recent work concerning 'natural pedagogy' shows the importance of ostensive cueing, notably via the voice, in helping infants to identify intentionality when viewing goal-directed actions.

Joint attention and pointing also play a special role in the development of social cognition. When an infant points to 'remark on the world' to another person, pointing has communicative intent. Such protodeclarative pointing emerges as early as eight months in some infants, with an advantage for girls. Pointing in order to establish joint attention has been seen as an important indicator of the development of the understanding that others have unobservable mental states. Protodeclarative pointing is usually accompanied by gaze alternation between the face of the social partner and the object being pointed at, and also by increased vocalisation. Research also shows that babies point as an 'epistemic request' for new information. This constellation of behaviours suggests that babies point protodeclaratively with the intention of affecting the mental states of their social partners, actively choosing what to learn about. Indeed, babies who are skilled at recognising the connection between looker and object at ten months have better language development at 14 and 18 months. As discussed, the act of looking has referential meaning for both language and mental states.

Finally, the possibility that babies are able to represent false beliefs was discussed. Successful reasoning about false beliefs has been argued to be the only type of evidence that shows convincingly that humans attribute mental states to others. This is because someone who holds a false belief will act in ways that would not be predicted by the actual state of affairs in the world. The ability to predict these actions demonstrates that the infant or child understands that the contents of the mind of another are not necessarily a direct reflection of reality. A series of remarkable infant studies that have adapted the 'false location' and 'false contents' versions of false belief tasks suggest that infants are able to attribute false beliefs to a human agent. Indeed, infants during the second year of life also appear able to understand strategic deception, when a false belief is deliberately used by a thief to confer an advantage upon themselves. Such experiments provide intriguing evidence that the development of psychological understanding is proceeding well in the first years of life, before the acquisition of language gives it a major boost. The intimate relationship between language development and the development of social cognition will be discussed further in later chapters.

CHAPTER 4

CONTENTS

Social cognition, mental representation and theory of mind: The psychological world 2

4

Theory of mind was a term first used by Premack and Woodruff (1978) to refer to the ability of a person to impute mental states to the self and to others. Clearly, having a theory of mind is very important for social cognition. Understanding the mental states of others enables you to predict their behaviour on the basis of their beliefs and desires. Essentially, having a theory of mind enables an analysis of psychological causation. However, analyses of psychological causation will only work if we generally attribute to others beliefs that are true and actions that are rational. In this chapter, I will summarise experimental work on the development of the ability to impute mental states, both to the self and to other agents, including a brief survey of the relevant mirror neuron literature.

As we saw in Chapter 3, research on the psychological world of infants suggests that many mechanisms that support the development of social cognitive understanding are at work early in life. These mechanisms include gaze following, joint attention, the monitoring of goal-directed actions and the monitoring of intentions. Further, infants can make predictive mentalistic inferences from social-cognitive information, suggestive of an emergent understanding of the mental states of others. For example, predictive looking studies show mentalistic (e.g., desire-based) interpretations of the goal-directed actions of other people for infants as young as six months of age (Kim & Song, 2015). Converging evidence is provided by EEG studies in which the infant is viewing the same outcome, but in one case was expecting a particular action on the basis of the actor's prior *beliefs* (Southgate & Vernetti, 2014). Infants expect other people to act on the basis of their beliefs, and this expectation supports their development of a theory of mind. Accordingly, even during the first year of life, some form of metarepresentational ability appears to be present. Even very young infants appear to have some ability to represent the knowledge states (mental representations) of others, and to make predictions about their behaviours based on causal analyses of their likely goals and desires – analyses based on *psychological causation*.

KEY TERMS

Theory of mind
The ability to impute mental states to the self and to others.

Metarepresentational ability
The ability to take a representation itself as an object of cognition.

Pretend play
To pretend that one object or event is another, often called symbolic play.

This capacity for 'metarepresentation' is at the heart of our ability to understand the mental states of others and to use knowledge about their mental states to make predictions about their behaviour. However, young children can sometimes fail false belief tasks that appear to be similar in their underlying structure to tasks that are passed by infants. This is because metarepresentational abilities seem to develop from at least three sources in addition to the early mechanisms discussed in Chapter 3. These sources are the capacity for imitation, the development of pretend play and language. The importance of metarepresentational abilities for developing a 'theory of mind' was first discussed in detail by Leslie (1987).

PRETEND PLAY, IMITATION AND METAREPRESENTATION

In a seminal paper, Leslie (1987) suggested an important link between the development of a theory of mind and pretend play. His idea was that pretend play was one of the earliest manifestations of a child's ability to characterise and manipulate their own (and others') cognitive relations to information. In order to pretend that (for example) a banana is a telephone, the child must decouple the *primary* representation of the banana given by the sensory systems (yellow, edible object) from the pretend representation (telephone receiver) that is necessary for playing at answering the telephone. The primary representation is a direct representation of the object – the child is seeing and touching a banana. The pretend representation is part of the game. It is crucial for cognition that our primary representations are veridical. In order to pretend that the banana is a telephone, therefore, this primary representation must be marked off or 'quarantined' from the pretend representation of a telephone receiver. This quarantining suspends the true semantics of being a banana. The pretend representation is not a representation of the objective world, rather it is a representation of a representation from that world. It is a *meta*representation. Leslie argued that the fact that early pretence is usually shared with others (e.g., a pretend tea party) shows that young children also have some understanding of pretending by others. Hence the act of pretending suggests some understanding that others also have metarepresentations.

Leslie (1987) suggested that pretend play was one of the earliest manifestations of a child's ability to characterise and manipulate both their own and others' cognitive relations to information.

According to this analysis, the emergence of pretence is another marker of the beginning of a capacity to understand cognition itself – to understand thoughts as entities. Social partners frequently use language to help young children to understand pretend representations. For example, they may say "Do you want a turn with my telephone?" As we will see later, language plays other roles as well in helping children to understand mental states. For example, a theory of mind emerges earlier in children who live in families who talk explicitly about their emotions and feelings (Dunn, Brown & Beardsall, 1991a).

An important aspect of Leslie's (1987) analysis of metarepresentation was the explicit connection that he drew with pretend play. Prior to this analysis, pretend play had been analysed largely in terms of the development of a symbolic capacity. Therefore, pretend play had been considered largely in relation to language development. Both language and pretending were thought to require a general capacity for creating 'symbols in thought'. In pretend play, objects are used for symbolic purposes irrespective of their appearance. For example, a baby bottle may be used as a comb (Ungerer, Zelazo, Kearsley & O'Leary, 1981). Similarly, in language, words are used for symbolic purposes. The sound patterns that constitute words have no direct link to the objects or to the events that they refer to. Language and pretend play were hence thought to depend on the same cognitive capacities. It was thought, for example, that the ability to sequence pretend acts (e.g., cook a pretend meal, and then feed it to a doll) and the ability to sequence words into sentences depended on the same symbolic mechanism. Certainly, a large number of studies produced evidence consistent with a joint timetable for the development of language and of pretend play.

One example comes from a detailed analysis of the sequence of pretend play behaviours that develop between eight and 30 months, carried out by McCune-Nicolich (1981). McCune-Nicolich was able to document a number of parallel developments in pretend play and in language. Young children's pretending begins by being closely tied to the veridical actions that people make on objects. For example, below 12 months, pretending might constitute 'drinking' from an empty cup while making playful slurping noises. During the second year, pretending gradually becomes more abstract, and divorced from the immediate context. For example, a doll might be made to drink from the empty cup, or the child might play at 'cleaning' something that isn't dirty. Initially, pretending seems to depend on the child's knowledge of the structures and functions of real objects. Gradually, action schemes are combined, so that, for example, a doll might be given a drink, do some cleaning and then be put to bed. Late in the second year, the child becomes able to plan games mentally before embarking on them. At this stage, the child will search for the particular objects that are required for the planned pretend game, or will substitute different objects (e.g., a stick becomes a horse). Planned pretend is clear when the child announces the planned pretend act, or searches deliberately for a required prop. While two years is the median age for planned pretend, longitudinal studies reveal its emergence as early as 18 months, or as late as 26 months (Nicolich, 1977).

The shared requirement of pretend play and language for 'symbols in thought' was assumed on two grounds. First, both activities involve the communicative function of sharing objects with others. Second, children use both pretend play and language for 'trying out' various representational equivalences. Accordingly, McCune-Nicolich (1981) went on to document some structural developments in language which appeared to parallel different levels of pretend play. For example, first words tend to co-occur with first pretend behaviours. McCune-Nicolich suggested that this signalled a separation between the means of signifying a meaning and the meaning itself. A second parallel was that the combination of words tends to co-occur with the sequencing of pretend behaviours. This coupling was suggested to arise from a fundamental cognitive development that allowed the child to construct *relations* between symbols. As a third example, the production of rule-based (syntactically structured) utterances appeared to emerge at the same time as planned pretending. The essential common element for this coupling was thought to be an internal cognitive structure that allowed symbolic elements to be related to one another directly.

The apparent correspondences between pretend play and early language were also appealing because of the view that a general capacity for symbolic representation set human cognition apart from cognition in other animals. The symbolic capacity was thought to account for representation across many domains, for example play, language, drawing and mental imagery. Leslie's (1987) paper pointed out that the development of 'symbols in thought' was better understood as a 'metarepresentational' capacity. He argued that this 'metarepresentational' capacity was also at the core of the developing understanding of the *internal states of others*. One reason that it might be useful to separate pretend play from language was demonstrated in a study by Fenson and Ramsey (1981). They showed that the developmental sequence of pretending could be influenced when adults modelled pretend behaviours for the child to imitate. Fenson and Ramsey demonstrated that pretend play following modelling was more advanced than pretend play in the absence of modelling. Fenton and Ramsey focused on combinatorial pretend acts, which are typically performed spontaneously at around 24 months. An example is pouring imaginary liquid from a teapot into a cup, and then giving a doll a drink from the cup.

Fenson and Ramsey studied infants aged 12, 15 and 19 months. Examples of the modelled combinatorial behaviours included stirring a spoon in an empty pan and then feeding a doll pretend food, and placing a doll into a bed and tucking her in with a blanket. A substantial number of both the 15-month-olds and the 19-month-olds in the study were found to imitate the combinatorial pretend acts, and a couple of the 12-month-olds did so as well. Two subsequent studies confirmed this developmental pattern. Fenson and Ramsey concluded that the general ability to imitate sequences of pretend acts was present by 19 months of age rather than by 24 months of age. In language development, single words are usually present by about the same age (see Chapter 6). Fenson and Ramsey speculated that in both domains (pretend play and language) the appropriate cognitive components were in place before combinations began to occur. An alternative way of thinking about these results is that imitation is a powerful tool for social cognition. Children will imitate pretend acts modelled by others as a way of sharing communication, and they will imitate pieces of language for the same purpose (see Chapter 6). In neither case is it necessary to propose a particular cognitive development enabling such behaviours. Learning by imitation has been present from birth. The child may be utilising imitation in the realms of both language and action as a means of gaining a deeper understanding of representational activities. Initially, the imitation may occur for purely social purposes. Eventually, it may give the child fundamental insights into human social cognition.

Converging evidence that imitation enables a deeper understanding of representational activities comes from a study reported by Bigelow, MacLean and

Proctor (2004). They were interested in pretend play during episodes of sustained joint attention, where an adult and an infant share attention to an object. As we saw earlier, joint attention can be considered a 'hot spot' for learning. Bigelow and her colleagues expected infants to show increases in more advanced play with objects during episodes of joint attention. The infants studied were aged 12 months, the youngest age group studied by Fenson and Ramsey (1981). Twelve months is the age at which infants begin to be intentional in their play with objects: rather than simply sucking or banging objects, infants now begin to attend to the specific functional uses of their toys, and play gradually becomes symbolic. Bigelow et al. provided some enticing objects for the infants to play with, including a tea set, a doll with her bed, and a toy telephone.

Following a period of free play by the infants alone, mothers and infants were encouraged to play as they normally would. Infants' activities with the toys were then scored in both conditions. Bigelow et al. found that the infants had significantly more advanced play within joint attention episodes, and significantly more stereotyped play outside joint attention episodes. Bigelow et al. argued that mature play hence depended on more than simply the presence of the mother. Adult scaffolding of play within joint attention episodes had a particular effect on symbolic development. This supports social learning theories about the potential importance of 'collective intentionality' in pretend play and psychological development. The view that psychological development is dependent on cultural imitative learning and 'collective intentionality' has been proposed by Rakoczy, Tomasello and Striano (2005). Rather than conceiving the individual mind as the main driver in the child's attainment of an understanding of mental states, Rakoczy et al. argue that the developmental process is a more social one. Carpendale and Lewis (2004) and Rochat (2009) also offer social interaction views of the child's developing understanding of mind.

Imitation, metarepresentation, mirror neurons and the 'like me' analogy

The argument that imitation is crucial for gaining a deeper understanding of mental states and mental representations has been made in a stronger form by Meltzoff and Decety (2003). They pointed out that infants not only imitate the actions of others, but are also aware when others are imitating the actions of the infant. On this basis, Meltzoff and Decety argued that when infants see others acting 'like me', they project that others have the same mental experience that is mapped to those behavioural states in the self. The evidence that infants know when they are being copied comes from a series of experiments reported by Meltzoff (1995b). In these studies, Meltzoff asked adults to become 'social mirrors' for babies aged 14 months. The adults simply had to copy every action that the infant made, behaving like a kind of 'shadow'. In fact, during the experiments, two adults were sitting opposite to the infants. One shadowed the infant's behaviour with a novel toy. If the infant banged her toy three times, the shadow banged her toy three times. If the infant mouthed the toy, the shadow mouthed her toy. The other adult served as an active control, and imitated the infant's past behaviour with the toy, which was shown on a video screen out of the infant's line of view. Meltzoff found that the infants preferred to look at the adult who was imitating them. They smiled more at this adult, and also tested out her reactions, for example suddenly stopping an action to see whether the shadow would follow suit. Meltzoff argued that caretakers who act as social mirrors take infants beyond the initial starting point of behaviour. Caretaker reflections of infant behaviour capture not only the motor aspects of this behaviour, but add in intentions and goals. For example, if an infant waves a toy, the caretaker might interpret this as waving in order to shake and make a sound. Hence intentionality and desire become part of the imitative episode, reflecting psychological attributions back to the infant. In this way, imitation serves to aid the development of both a notion of the self and a notion of other minds.

On this view, action and imitation are critical to developing a theory of mind. Meltzoff and Decety (2003) argued that human acts are especially relevant to infants, because other humans look like the infant feels himself or herself to be. The infant recognises 'that seen event is like this felt event'. Human acts are also events that the infant can intend. This enables infants to make inferences about the visible behaviour of others, and the underlying mental states of others, such as what they intend. Infants thus infer the internal states of others through an analogy to the self, the 'like me' analogy discussed in Chapter 2. Recently, Meltzoff's perspective on imitation and metarepresentation has been supported by neuroimaging studies based on the mu rhythm, the putative electrophysiological signature of action-effect learning, and also by studies based on mirror neurons, neurons that are activated not only when a person is performing a certain action, but also when the person watches *someone else* perform the same action. Indeed, early mirror neuron studies were interpreted as evidence that the mirror neuron system was a specialised neural substrate for reasoning about beliefs, perhaps a 'theory of mind module'.

The ideas about modular cognition discussed in Chapter 2, that there are specialised information processing systems in the brain that provide the basis for cognitive development, included the proposal of a 'theory of mind module' in the human brain (Leslie, 1994). The idea that the mirror neuron system could be this neural module in fact became popular for a time (see Frith & Frith, 2003; Gallagher & Frith; 2003, Rizzolatti & Sinigaglia, 2010, for reviews). Neuroimaging studies of adults indeed identified an extensive network of neural regions that were typically active when a person was engaged in false belief reasoning, involving the temporal poles, posterior superior temporal sulcus and medial prefrontal cortex. However, such neuroimaging data simply demonstrate structure-function *correlations*, leaving the underlying mechanisms for psychological reasoning unclear. For example, part of the neural activity in these extensive regions may simply reflect the complex demands made by false belief tasks on general abilities like memory and language.

For more specific information regarding underlying mechanisms, we need neural studies of how actions are represented in the brain, and how action, imitation and understanding intentionality may be linked. As we saw in Chapter 3, both EEG measures and predictive looking measures suggest that the understanding of goal-directed action can already be mentalistic during the first year of life. A decade earlier, groundbreaking research related to the neural representation of goal-directed action and its potential role in metarepresentation was conducted by

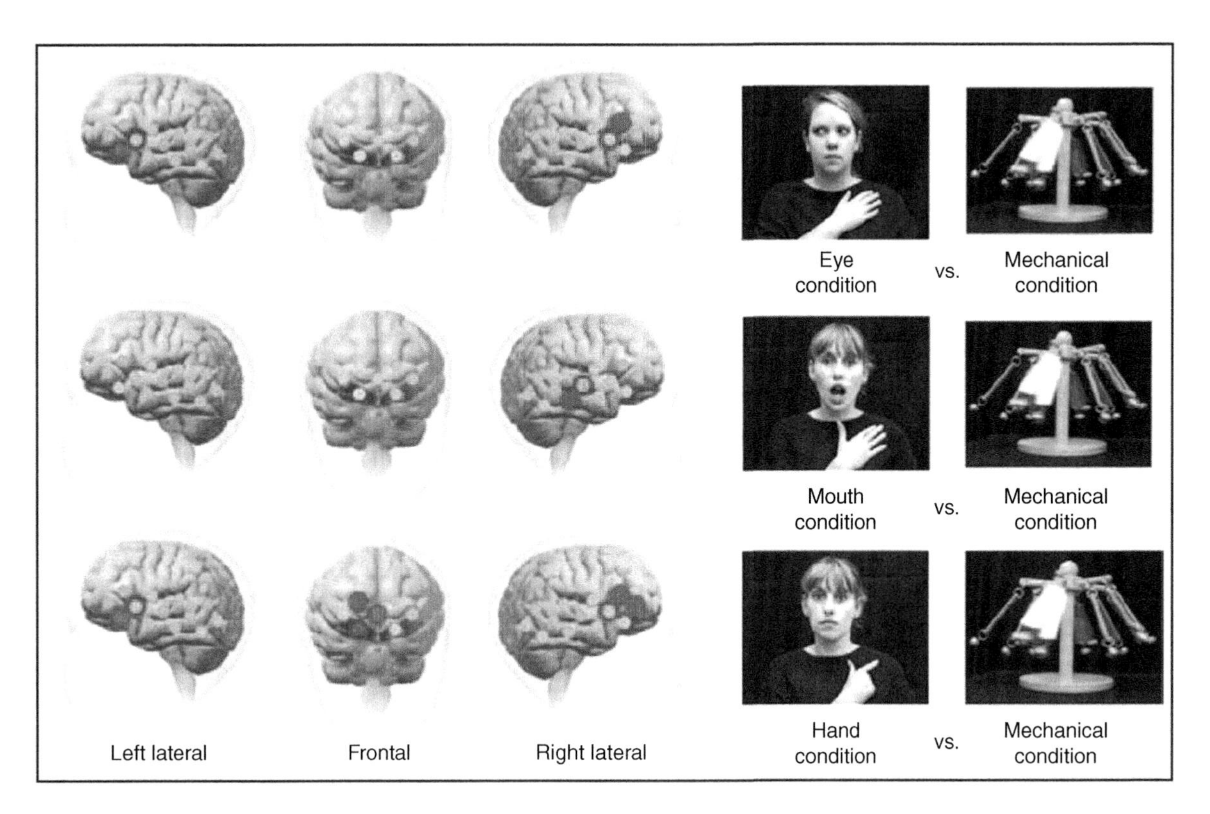

Infant data consistent with the view that the mirror neuron system is biologically tuned. The three biological conditions were each contrasted with mechanical motion, and the corresponding head plots show sites of differential activation. As can be seen, the fNIRS channels showing significant effects were different for the three kinds of biological motion.

Rizzolatti and his colleagues. They were the first researchers to identify the 'mirror neuron' system in monkey and man, a system for representing action (Rizzolatti & Craighero, 2004). Rizzolatti and his colleagues produced evidence showing that the mirror neuron system was active when the self made a certain action, and also when someone else made the same action. As the actions of both self and other usually have *intentions*, they suggested that the brain will thereby also represent the likely intention of another person who is performing a certain action, using the *same neural system*. It was thus argued that the mirror neuron system may be the neural substrate for *imitating* the actions of others, and possibly also for understanding their *intentions*.

Mirror neurons were first discovered in primate premotor cortex. It was found that mirror neurons were activated when a monkey performed object-directed actions such as tearing, grasping, holding and manipulating. Furthermore, the *same* neurons also became active when the animal observed someone else performing the same class of actions, and were even activated by the *sound* of an action, such as paper ripping or a stick being dropped (Rizzolatti & Craighero, 2004). The researchers pointed out that action recognition has a special status, as action implies a goal and an agent. Hence action recognition may involve an understanding of the agent's *intentions*. To find out whether the mirror neuron system had a role in coding intentions, Iacoboni et al. (2005) devised a paradigm in which the context for performing an identical action varied. Adults were shown grasping a mug in two different contexts, when partaking in a tea party, or when clearing up after a tea party. The visual display of teapot, jam, biscuits, plates, etc., was identical in both scenarios, as was the mug and the grasping action. The contexts were differentiated by the presence of crumbs and partially consumed cakes. The brain activity in the mirror neuron system was different for the two contexts, despite the grasping action being identical. Iacoboni et al. argued that the mirror neuron system does not simply provide a substrate for action recognition, but also for intention coding.

Iacoboni et al.'s ideas about action recognition and intention coding require that the mirror neuron system is activated only by biological actions. For example, we saw in Chapter 2 that studies of the development of intention-reading in infants have shown that infants will imitate a human hand trying (although failing) to pull apart a dumbbell apparatus, but will not read the 'intentions' of a mechanical hand that fails to perform the same actions on the dumbbell (Meltzoff, 1985). Tai, Scherfler, Brooks, Sawamoto and Castiello (2004) have demonstrated that for adults, the mirror neuron system is indeed activated only by biological actions. They measured mirror neuron activity in adults who observed either a human hand performing manual grasping actions or a robot arm performing analogous actions. Mirror neuron activity was observed for the manual grasping actions performed by the human hand, but not for the actions performed by the grasping robot hand. Tai et al. (2004) argued that the mirror neuron system is biologically tuned. Nevertheless, it is possible that this 'biological tuning' is dependent on experience. As will be recalled, in Chapter 3 we saw that infants aged nine months show neural suppression of the mu rhythm both when they are expecting a human hand to make a goal-directed action, and when they are expecting a mechanical pincer to make a goal-directed action (Southgate & Begus, 2013). Nevertheless, an fNIRS study with infants aged five months was able to show selective activation to biological motion when it was contrasted with motion from mechanical toys (Lloyd-Fox, Blasi, Everdell, Elwell & Johnson, 2011). The infants watched biological motions such as a hand opening and closing, eyes opening and closing, or a mouth opening and closing, and also watched mechanical motions such as a merry-go-round toy. Lloyd-Fox et al. found differential neural activation to the three types of biological motion in widespread frontal and temporal regions, a much larger area than the mirror neuron area per se.

The data are contradictory, and the location and existence of a mirror neuron system in human infants have indeed been widely debated. Typically developmental studies have used EEG and have measured the mu rhythm, the desynchronisation of alpha band activity that is thought to be a neural signature of action-effect learning. Marshall and Meltzoff (2011) pointed out that early studies of the mu rhythm called it a 'motor resonance' response, which can be recast as 'neural mirroring' (although we saw in the previous chapter that Southgate and Begus disputed this in their 2013 study). Nevertheless, the presence of the mu rhythm and its desynchronisation in certain experimental conditions has been used as a way of measuring the neural processes underpinning action perception, action production and imitation in pre-verbal infants. Some researchers argue that mu rhythm suppression may provide an index of the development of metarepresentations as well.

For example, Meltzoff has produced infant neural data that support his claims about imitation, neural mirroring mechanisms and the 'like me' analogy, using EEG and the mu rhythm. Testing older infants (14-month-olds), Meltzoff and his colleagues contrasted the neural activity elicited by watching an agent press a button using either her hand or her foot. They predicted that desynchronisation of the mu rhythm would be greater over electrodes overlying 'hand' areas of sensorimotor cortex when the infant was watching the hand press the button, while desynchronisation of the mu rhythm would be greater over electrodes overlying 'foot' areas of sensorimotor cortex when the infant was watching the foot press the button. This was exactly what they found (Saby, Meltzoff & Marshall, 2013). In subsequent studies, they established that the mu rhythm was selectively desynchronised over the same electrode sites when the infant herself was pressing the button using either her hand or her foot (Marshall, Saby & Meltzoff, 2013). Marshall and Meltzoff (2014) argued that their data supported a shared neural mirroring system underpinning the 'like me' analogy. Further, they made the important point that while this neural substrate could provide a functional basis for the development of intersubjectivity (understanding what it is for others to be like us and vice versa), social learning during development (for example, via pretend play and language) was likely to transform the computations carried out in this neural activity space.

A complementary approach to measuring the mu rhythm is provided by studies by Turati and her colleagues (Turati et al., 2013; Natale et al., 2014), who have studied infants aged three, six and nine months of age. They have used electromyography (EMG) to measure infant motor activation and thereby document mirror-like neural motor mechanisms. EMG measures the electrical potentials generated by muscle cells. By using EMG to record from the mouth-opening muscles of young infants, Turati and colleagues were able to compare motor responses as infants watched a video of an actress reaching for an object (a dummy or a piece of Lego), and then bringing it to either her mouth or to her head respectively. Turati et al. (2013) showed that the neural motor responses of six-month-olds were reliably different in the two conditions. The EMG recorded greater muscular activity when the infant was viewing the mouth action compared to the head action, suggesting recruitment of the motor system while viewing the object-to-mouth actions. Three-month-olds did not show a differential EMG response to the two conditions, while nine-month-olds tested subsequently (see Natale et al., 2014) showed similar differential effects to six-month-olds.

As EMG rather than EEG was used to measure electrical responses directly from the relevant muscles, this differential response at six and nine months of age provides clear evidence for motor resonance effects during the first year of life. The automatic simulation of observed actions in muscle cell activity also showed a developmental difference between six and nine months of age. For the younger infants (six-month-olds), the EMG response became different between the two conditions only as the reached-for object was brought to the mouth. For the older infants (nine month olds), neuromotor activity was already different as the female grasped the dummy, suggestive of activity related to her *intention* to take the dummy to her mouth (activity predictive of the expected act). Natale et al. (2014) concluded that they had demonstrated a functional role for the infant motor system in perceiving the intentions of other agents. While the EMG data actually provide evidence for action prediction rather than understanding of intentions per se, they are consistent with Meltzoff's claims about the functional role of the 'Like me' analogy during imitation.

In a similar vein, although the 'like me' analogy provides a potential mechanism for the development of metarepresentations, the infant neural data reviewed thus far do not actually document that metarepresentations (representations of what the actor intends or believes) are being generated in the different experimental paradigms. Even the activation of mouth muscle electrical potentials by nine-month-olds when watching the agent grasp the dummy could be argued to reflect the application of learned associations rather than an understanding of intention. To tackle this issue, Kampis, Parise, Cisbra and Kovács (2015) devised a novel paradigm for eight-month-old infants, based on their prior findings of enhanced gamma band activation when infants were maintaining an object representation of an occluded object (see Chapter 2). The metarepresentation that they studied using this gamma band signature was false belief. Infants viewed a video showing a rotating box containing an object. The box only had two sides, so that depending on its orientation during rotation, either both the infant and a watching adult could see the object, or only the adult could see it, or only the infant could see it. The adult had previously been observed to be interested in the object in the box. In a control condition, the same rotations were presented to the infant, but first the object disintegrated in full view of both the infant and the actor. In this case, the rotation of the box did not present different viewpoints of the object, as the box was now empty.

In the first part of the experiment, the infant gamma response was compared for when only the infant could view the object and when only the adult could view the object. Kampis et al. established that the infants showed an increased gamma band response both when only they could see the object, and when only the adult could see the object. There was no change in gamma band response to the same box rotations in the control condition, where no object was present. This pattern of neural activity suggested that the infants were successfully computing the visual perspective of the adult. In a second part of the experiment, once the object was occluded from the actor, it also disintegrated. In this scenario, the watching adult had the false belief that the object was still present behind the occluder, whereas the watching infant knew that it had disappeared. Kampis et al. reported that the infant gamma band response in this *false belief* condition was significantly higher than in the matching true belief condition from the first part of the experiment (when both the infant and the adult knew that the object was no longer present). Kampis et al. argued that their EEG data provided evidence for infant metarepresentation. The infants' neural response suggested that they were encoding that the adult was still representing the object, even though the true state of affairs in the world was

Bigelow et al. (2004) found that the infants had significantly more advanced play within joint attention episodes, and that adult scaffolding of play (for example by the mother) had a particular effect on symbolic development.

that the object no longer existed. The neural data are consistent with the view that the infants were ascribing a *representation* of the (absent) object to the adult. In other words, the infants were forming a *metarepresentation* of the belief content of the adult's brain, by using their own neural object representation system. Although the *motor* neural mirroring system is not being activated in this paradigm, the infants are ascribing to others the representations that they themselves can form. This is functionally the same mechanism as that used by the 'like me' analogy framework to explain why the capacity for imitation is linked to the development of metarepresentational abilities. As noted by Kampis et al. (2015), such a metarepresentational capacity is not dependent on language. However, learning language is likely to enrich and deepen its function considerably. Further studies are required to support this 'rich' interpretation of infant gamma band activity.

Early desire-based psychology?

Inferring the internal states of others through an analogy to the self might be easier for some internal states than for others, however. For example, it has been proposed that the ability to understand the internal state of desire might develop earlier than the ability to understand the internal state of belief. It may be cognitively easier to infer the desires of others from their actions than to infer the beliefs of others from their actions. This view, that early psychological understanding is based on desires, and that later psychological understanding is based on the interaction between beliefs and desires, was proposed for example by Wellman and his colleagues (Wellman, 2002; Wellman & Woolley, 1990).

KEY TERMS

Simple desire psychology
Construing of human actions in terms of desires or internal dispositions.

Belief-desire psychology
Construing of human action by predicting the actions and reactions of others on the basis of their beliefs.

For example, Wellman and Woolley (1990) argued that young children (two-year-olds) first construe human action in terms of desires. This 'simple desire psychology' was thought to precede 'belief-desire' psychology because desires motivate behaviours, and desire-behaviour action sequences are relatively easy to map. Beliefs frame behaviours, and are less easy to map. It is easier to predict the actions and reactions of another person that are related to desires than to predict the actions and reactions of another person that are related to beliefs. Wellman and Woolley hence argued that younger children will be simple 'desire psychologists' and older children (from three years on) will be 'belief-desire psychologists'. For example, if you are hungry, that is a physiological state. Hunger is a drive that can be overcome by eating anything that is edible. However, if you want an apple, that is a psychological state. The desire for an apple cannot be overcome by eating a banana. Via examples such as these, Wellman and Woolley showed that desires are intentional constructs, whereas basic physiological drives (hunger, fear, pain) are not. They then argued that whereas beliefs always entail metarepresentations, desires may not. To understand the beliefs of another person, it is necessary to understand thoughts as entities – to understand representations of representations. To understand the desires of another person, it is only necessary to understand objects and events – primary representations.

Wellman and Woolley characterise simple desire psychology as attributing to other agents internal dispositions towards or against certain primary representations (i.e., actions or objects). An actor might not want a drink of milk, or might want a certain toy. These desires then cause the actor to do certain things. Usually, they engage in goal-directed actions, e.g., they refuse the milk, or they go and fetch the toy. These goal-directed actions can be predicted by understanding desires. Simple desire reasoning can proceed without a conception of belief, and hence without a metarepresentation. The child can predict that the other person will go and get the desired toy from wherever the child knows the toy to be. The child can make this prediction without attributing knowledge of where the toy is to the actor (a belief-representational state). Simple desire reasoning does not allow a child to explain why an actor may engage in an action that is contrary to prediction (for example, going to fetch the toy from a wrong location – this might occur because the actor *believes* that the toy is in that location). However, although simple desire reasoning does not require a conception of belief, it does require a conception of intention or internal disposition.

In order to test their claim that two-year-olds engage only in desire-based reasoning about the actions of others, whereas three-year-olds engage in belief-desire reasoning about the actions of others, Wellman and Woolley devised experiments contrasting desire-reasoning with belief-desire reasoning. For example, children were asked to predict the actions of story characters in different situations. The children might be told "Here's Johnny. He wants to find his dog to take it to the park, because that's what he really wants to do. His dog might be in the house, or it might be in the garage. So, he's looking for his dog to take it to the park. Watch, he's looking for his dog in the garage. Look. He doesn't find his dog. What will Johnny do next?" This scenario can be solved by simple desire reasoning. If Johnny doesn't find his dog in the garage, the child can predict that he will next look for his dog in the house. The two-year-olds were extremely good at making these desire-based predictions.

In order to contrast predictions based on desire psychology with predictions based on belief-desire psychology, Wellman and Woolley changed their scenarios slightly. In their second experiment, the correct prediction in each case required the child to go against their own desires or beliefs. This was done so that the child could not give the correct prediction on the basis of happening to have that desire or belief herself. For example, the child might be told "At Betsy's school, they can play with puzzles in the classroom or they can play with sand on the playground." The child was then asked which option they would choose. If the child said the sand, the experimenter would say "Betsy wants to play with puzzles today, she doesn't want to play with sand." The child was then asked to predict where Betsy would go. This experimental design enabled a parallel scenario to be constructed requiring belief-desire reasoning. Here the child might be told "Sam wants to find his puppy. His puppy might be in the garage or under the porch. Where do you think Sam's puppy is?" If the child said under the porch, the experimenter would say "Sam thinks his puppy is in the garage, he doesn't think it is under the porch." Again, the child was asked to predict where Sam would go.

Wellman and Woolley predicted that the two-year-olds in their study would be good at desire reasoning and poor at belief-desire reasoning. In fact, the children taking part in the study were quite good at both. As a group, the 20 children tested were 73% correct on the belief stories, and 93% correct on the desire stories. Both performance levels were significantly above chance. However, when a stricter criterion was adopted requiring correct responding in all three story scenarios based on either belief or desire, then 85% of the two-year-olds passed criterion for the desire stories and only 45% passed criterion for the belief stories. Based on this and further experiments, Wellman and Woolley (1990) concluded that two-year-olds understand others' mental states in intentional terms, but not in representational terms. Two-year-olds understand that others can want certain objects (have desires), but not that others can believe certain states of affairs (have beliefs). Simple desire psychology grants young children a beginning awareness of how people's internal intentional states can guide their behaviour. However, it does not grant young children awareness of other people's cognitive relations to information. Hence by this account, children's first understanding of the mind is non-representational. Very young children's analyses of psychological causation do not depend on an understanding of belief.

It is interesting to reconsider the false belief experiments with infants discussed in Chapter 3 from a simple 'desire psychology' perspective. Both the 'block box' paradigm devised by Buttelmann et al. (2014) and the strategic deception paradigm with rattling toys devised by Scott et al. (2015) involved the infant being aware of the *desires* of the agent with the false belief. On the other hand, in order to compute that the agent had a *false belief* about the blocks or the rattling toys, the infants required an awareness of that agent's cognitive relations to information. This was also demonstrated nicely via the object occlusion paradigm used by Kampis et al. (2015) discussed earlier, in their false belief paradigm using EEG with eight-month-old infants. Therefore, more recent data contradict the idea that younger children have a non-representational understanding of the mind. Nevertheless, many studies since Wellman and Woolley (1990) have confirmed an early-developing desire psychology in very young children. Demonstrations of an early-developing desire psychology do not rule out the co-existence of an early-developing belief-desire psychology.

For example, Repacholi and Gopnik (1997) devised a paradigm for exploring reasoning about desires in children aged 14 and 18 months. This paradigm entailed a food-request procedure, in which the toddlers were asked to select one of two foods to give to an experimenter. One of the foods had previously been preferred by the same experimenter. For example, in a contrast between crackers and broccoli, the experimenter might have tasted the crackers and pulled a face of disgust saying "Eew! Crackers! I tasted the crackers! Eew." The same experimenter might have tasted the broccoli and shown happiness, saying "Mmm! Broccoli! I tasted the broccoli! Mmm." The experimenter then placed her open hand mid-way between the two foods and said "Can you give me some?" The experimental measure was whether the toddler gave the experimenter the food that was the target of her positive affect (for some toddlers, this was the crackers and not the broccoli). Repacholi and Gopnik found that of the 14-month-olds, only 54% of children offered the preferred food. For the 18-month-olds, 73% offered the preferred food. As almost all the children themselves preferred the crackers to the broccoli, adjustments were necessary depending on whether the child was in the matched condition (both child and experimenter prefer crackers) or in the mismatched condition (child prefers crackers, experimenter prefers broccoli). These analyses indicated that while the 18-month-olds reliably offered the food preferred by the experimenter, the 14-month-olds reliably offered the food preferred by themselves. Repacholi and Gopnik concluded that, by 18 months of age, children have a psychological understanding of desire. They understand that desire is an internal psychological state, and that two people can have different internal dispositions towards the same entity (i.e., the broccoli).

More recently, Repacholi, Meltzoff, Rowe and Toub (2014) adapted this paradigm to study imitation by young children (15-month-old infants) when faced with a different internal psychological state, anger. In their paradigm, the infants were 'bystanders' to the interactions between two adults. The infants watched while one adult performed actions on novel toys, and the other adult responded by expressing anger ("That's aggravating! That's so annoying!"), as though the actions were forbidden and were social transgressions. The infants were then given access to the toys. For some infants, the angry actor remained present, either reading a magazine (distracted) or facing the infant with a neutral expression. For others, the angry actor left the room or turned her back on the infant. Repacholi et al. found that the infants were significantly less likely to touch the toys or to imitate the 'forbidden' acts while the actor was looking at them or was able to look at them but was distracted by her reading. The infants were using their *prior observation* of the reactions of the angry adult to regulate their own behaviour with the toys. As the adult was not currently showing anger, the infants were *simultaneously* maintaining a representation of the adult's likely future mental state if they touched the toy, and a representation of her current mental state. Repacholi and her colleagues argued that these young infants were using their representation of her prior state to modulate their behaviour accordingly. In terms of the development of the understanding of psychological causation, of course, belief and desire can be equally important in determining human action (Astington, 2001). The most likely explanation is that the *same* developmental mechanisms – for example, intention reading, imitation and interpreting goal-directed actions – underpin the development of the psychological understanding of *both* desires and beliefs.

MENTAL REPRESENTATIONS, BELIEF AND FALSE BELIEF

So far, we have seen that pretend play is an early manifestation of the infant's growing representational ability, and that studies of neural mirroring suggest that infants activate the same neural substrates when they or another human agent perform a certain action (such as bringing a dummy to the mouth, or activating a button with the foot), or when they or another human agent are representing the continued existence of an occluded object. The substitution of objects in pretend games (as in using a banana as a telephone) or the maintenance of a representation about both the veridical state of affairs and another's false belief about the state of affairs (as in the object occlusion paradigm used by Kampis et al., 2015) require the ability to conceive *simultaneously* of two contradictory models of reality, thereby requiring metarepresentational ability. Yet traditionally, the metarepresentational understanding required for false belief reasoning was thought to be more complex than that required for pretending. For example, the philosopher Dennett argued that successful reasoning about false beliefs was the *only* convincing evidence for the attribution of mental states to others (Dennett, 1978).

In Chapter 3, we noted that for a long time it was believed that children did not have a representational understanding of false belief until the age of 3–4 years (Wimmer & Perner, 1983). This was because passing the false belief task was equated with having a 'theory of mind'. As the classic false belief tasks involving change of location or change of contents were failed by younger children and passed by older children, it was argued that important changes in social cognitive development occurred between the ages of three and four years. During this period, children were thought to acquire a previously absent representational concept of belief, and they were thought to become able to represent the contents of other people's beliefs. We will now consider briefly the evidence base for this 'developmental watershed' assumption. Note that even though the 'watershed' idea turned out to be incorrect, it is likely that mechanisms such as pretend play, imitation and language considerably enrich children's understanding of beliefs during this key period of early childhood.

The copious developmental literature using false belief tasks began with a classic paper by Wimmer and Perner (1983). They devised a simple and ingenious paradigm for measuring children's understanding of false belief based on a change of location. A story character Maxi puts chocolate into cupboard X. He then goes out to play. While he is absent, his mother moves the chocolate into cupboard Y. The children are then asked where Maxi will look for his chocolate when he returns. In order to answer correctly, children need to distinguish their own true belief concerning the location of the chocolate (cupboard Y) from Maxi's false belief (cupboard X). Wimmer and Perner argued that children need to have an explicit and definite representation of Maxi's wrong belief in order correctly to select cupboard X. When children aged 4–6 years were tested with toy scenarios, 50% of those aged 4–5 years wrongly chose cupboard Y, whereas 92% of those aged 5–6 years correctly chose cupboard X. This dramatic difference occurred despite evidence that the younger children remembered where Maxi had left his chocolate. This was shown by the inclusion of a memory control question in the experimental paradigm ("Do you remember where Maxi put the chocolate in the beginning?"). This memory question was usually answered correctly. Wimmer and Perner concluded that a novel cognitive skill emerged between the ages of four and six years. This was the ability to represent wrong beliefs. More generally, Wimmer and Perner suggested that it was only between the ages of four and six years that children became able to represent the relationship between two people's epistemic states.

In fact, when this form of the 'false location' task is used, the ability to represent false beliefs involving location indeed appears to emerge between the ages of four and six years in children in all cultures so far studied. For example, Callaghan and her colleagues (2005) tested false belief understanding in 267 children aged between 30 and 72 months in five different cultures. The participating countries were: Peru, where children from a rural Andean town were tested; Samoa, where children from traditional Polynesian agrarian villages were tested; Canada, where children from a rural town were tested; India, where children from a large city were tested, and Thailand, where disadvantaged children attending a Buddhist temple school in a large city were tested. All children were tested by local female experimenters using a change of location false belief task. Callaghan et al. (2005) found that in all settings, a majority of three-year-olds failed the false belief task, while a majority of five-year-olds passed it. The four-year-olds were usually evenly split – around half demonstrated understanding of false belief by passing the task, and around half did not. Callaghan et al. concluded that a shift in understanding false belief was a universal milestone of development, and that it occurred between three and five years of age.

Despite many replications finding similar results, Wimmer and Perner's original claim that their change of location task measured children's understanding that beliefs can be false did not go unchallenged. For example, it was argued that the context within which the children were questioned in the false belief task could mislead children into thinking that they were being asked where Maxi would *need* to look to find his chocolate. Siegal and Beattie (1991) suggested that a more child-friendly way of asking children about Maxi's false belief would be to change the question that they were asked to "Where will Maxi look *first* for his chocolate?" Siegal and Beattie tested 40 three-year-olds and 40 four-year-olds in a false belief paradigm involving lost puppies and kittens. For example, the children were told "Sam wants to find his puppy. Sam's puppy is really in the kitchen. Sam thinks his puppy is in the bathroom. Where will Sam look *first* for his puppy?" Children tested with the 'look first' question tended to give the correct answer. For the three-year-olds, 85% were correct in answering on the basis of a false belief in at least one of the story scenarios used, compared to 40% in the standard paradigm. For the four-year-olds, the figures were 75% and 50%. Siegal and Beattie argued that it was unlikely that conceptual limitations prior to the age of four explained children's performance in Wimmer and Perner's false belief task. Rather, the successful performance shown by three-year-olds in the 'look first' paradigm indicated that the nature of the conversational environment, namely its social context, was critical to explaining children's response patterns.

A different suggestion was that developmental 'deficits' in executive function lay at the root of the wrong answers given by younger children in the false belief task. It was argued that younger children might fail the false belief question because they were worse than older children at inhibiting themselves from giving an answer based on their very salient current mental representation of reality. Russell and his colleagues developed the 'windows task', a measure of young children's capacity for strategic deception, to explore this alternative explanation (Russell, Mauthner, Sharpe & Tidswell, 1991). In the windows task, children learned to play a competitive game with the experimenter in order to win chocolate for themselves. The game was based on two little boxes, one of which always contained chocolate. The child's job was to point to one of the two boxes on each trial, to tell the experimenter 'where to look' for the chocolate. If the competitor opened the empty box, the child got the chocolate. If the competitor chose the box containing the chocolate, the competitor kept the chocolate. Thus the winning strategy was to point to the empty box.

When the child was learning the game, the boxes were opaque. Once the child had learned how to play (after 15 trials), however, the opaque boxes were replaced with boxes that had windows facing the child (and facing away from the competitor). The child could now see via the windows which of the two boxes on a given trial contained the chocolate. It was explicitly pointed out to the child that it was now easier to make the competitor go to the wrong box. Twenty trials with the windows then commenced, and three- and four-year-olds were tested. The results showed that, on the first windows trial, most of the three-year-olds pointed to the box containing the chocolate, while most of the four-year-olds pointed to the empty box. The four-year-olds hence won the chocolate for themselves, while the three-year-olds lost the chocolate. Remarkably, 65% of the three-year-olds continued to point to the baited box for all 20 trials, thereby never winning the chocolate. No four-year-olds showed this pattern. Russell et al. argued that the younger children's difficulty lay in inhibiting pointing to a salient object (i.e., the chocolate). Their physical knowledge concerning the location of the chocolate was so salient that it controlled their behaviour.

Russell et al. went on to argue that younger children faced similar problems in the false belief task. Here, the younger children's physical knowledge of the new location of Maxi's chocolate controlled their verbal responding. Indeed, Russell et al. (1991) demonstrated that the same children who failed in their windows task were likely to fail a traditional false belief task as well. As we will see in Chapter 9, developmental deficits in executive function can account for many instances of younger children failing to inhibit salient information when initiating action. The large variety of situations in which this does and does not occur suggests that executive deficits do not provide a parsimonious explanation of younger children's performance in false belief tasks.

What about Wimmer and Perner's original interpretation, that younger children are poorer at co-ordinating different *mental* representations? This idea received apparently strong support from a related series of experiments on children's understanding of appearance versus reality, conducted by Flavell and his colleagues. Flavell, Flavell and Green (1983) were interested in situations in which appearance and reality differed; for example, in children's understanding of a sponge that looked like a lump of rock. They thought that young children might not be able to keep the difference between appearance and reality clear in their minds when appearance and reality differed. Younger children might not yet be able simultaneously to hold two representations of the same object in mind, the same claim made by Wimmer and Perner (1983). To investigate this question, Flavell et al. gave children aged 3–5 years a series of tasks involving contrasts between appearance and reality. The objects used included the rock/sponge, an imitation pencil made out of rubber, a hand puppet covered with a white handkerchief so that it looked like a ghost, and a piece of white card placed beneath a pink transparency, so that it appeared pink. The children were then asked a series of questions, including "What is this really really? Is it really really (a rock) or really really (a piece of sponge)?", and "When you look at this with your eyes right now, does it look like (a rock) or does it look like (a piece of sponge)?" Flavell et al. reported that the four- and five-year-olds rarely confused appearance and reality, whereas about half of the three-year-olds

KEY TERM

Appearance–reality distinction
Awareness that appearance and reality may differ, both in terms of physical objects (e.g., a sponge that looks like a rock) and psychological states (e.g., appears interested but is actually bored).

did so. They concluded that the younger children had a general metacognitive difficulty in dealing with mental representations. Although young children have and use mental representations, Flavell et al. suggested that they are less able to 'stand back' from their representations and reflect on their veridicality. What develops is thus the ability to reflect on and index one's own representations, tagging their internal source so that both appearance and reality can be kept in mind together. This conclusion implies a role for metacognition, as source monitoring (from where did I gain this knowledge?) is an important aspect of metacognitive behaviour. This is considered further in Chapter 9.

Meanwhile, some of the infant studies reviewed in Chapter 3 suggest that infants are able to hold two representations simultaneously in mind in certain circumstances, casting doubt on this aspect of Flavell's explanation for young children's confusions regarding appearance and reality. A novel paradigm for investigating younger children's (18-month-olds) ability to hold both appearance and reality in mind at the same time has been developed by Buttelmann, Suhrke and Buttelmann (2015). Their experiment used a rock-sponge, a book-box (a box that looked like a book), a branch-pencil (a pencil that looked like a tree branch) and a duck-brush (a brush that looked like a duck). There were also eight additional objects, respectively a real rock, book, branch and duck, and a real sponge, box, pencil and brush. During the experiment, the infants were first primed to 'help' experimenter A by giving her objects. During experimental trials, the infant and experimenter A were shown one of the objects (for example, the rock-sponge) by a second experimenter, experimenter B. Experimenter A showed her interest in the rock-sponge, and then left the room. Experimenter B then showed the infant the real function of the rock-sponge – that it was a sponge, not a rock – and placed it on a shelf. Experimenter A then came back into the room, and tried to reach the object on the shelf. She asked experimenter B to help her, but experimenter B turned away, inadvertently moving a piece of cardboard on the floor that had been hiding two more objects (in the current case, a real rock and a real sponge). The infant then was asked to help get experimenter A what she wanted. In this scenario, the infant should get her the real rock rather than the real sponge. By choosing the rock, the infant was showing that she could hold two representations in mind at the same time, her own representation that the rock-sponge was a sponge, and experimenter A's representation that the rock-sponge was a rock.

In a control condition, both the infant and experimenter A watched a demonstration of the object before experimenter A left the room. In this scenario, experimenter A had a true belief that the rock-sponge was a sponge, rather than a false belief that it was a rock. Hence on re-entry, when experimenter B refused to help her and accidentally knocked the cardboard, the infant should hand experimenter A the real sponge. Buttelmann et al. (2015) found exactly this cross-over effect, which suggests that 18-month-old infants can hold two representations in mind at the same time. The infants were more likely to hand experimenter A the rock in the false belief condition, and the sponge in the true belief condition. They behaved similarly with the other appearance-reality objects, and they did this from the very first trial onwards. This clever paradigm provides evidence that even 18-month-olds are able to 'stand back' from their representations and reflect on the veridicality of those representations, contrary to the conclusions drawn by Flavell and his colleagues. In non-verbal tasks with a low cognitive load (i.e., in which the two possible response options are both physically present), therefore, infants are able to hold two representations in mind simultaneously. This is supportive of metarepresentational capacity in 18-month-olds, and also provides converging evidence for an early understanding of false belief, this time using an 'unexpected identity' paradigm.

Pictorial versus mental representations

A different method for exploring children's ability to hold two competing representations in mind at once was developed by Zaitchik (1990). She decided to develop a task that did not depend on mental representations, but on non-mental representations. Her solution was a 'false photograph' task. In this task, a photograph was taken of an object A at one location, location X. The object was then moved to location Y. The child was then asked "In the picture, where is object A?" The correct answer was of course location X. Three-, four- and five-year-olds were tested, and all received prior experience with a Polaroid camera. This type of camera takes real pictures and develops them instantaneously. The child took pictures of familiar soft toys (e.g., Ernie from Sesame Street) and watched them develop. During the experimental test, the child watched a little show in which Bert and Ernie were taking a picture of Rubber Duckie with a Polaroid camera. Bert took a photo of Rubber Duckie on the bed, and developed the picture. Ernie then moved Rubber Duckie to the bathtub, so that he could go to bed. The child was asked "In the picture, where is Rubber Duckie?"

Zaitchik found that both the three-year-olds and the four-year-olds were at chance in working out where Rubber Duckie would be in Bert's picture. In fact, they were worse in the false photograph task than in a standard false belief task given at the same time. Zaitchik (1990) suggested that mental representations are difficult not because they are mental, but because they are representations. When children must reason about representations that are supposed to describe the real state of affairs in the world, be these photos or beliefs, they run into difficulty when the representations in fact do *not* describe the veridical state of affairs in the world. Zaitchik suggested that it is misrepresentations rather than mental representations that cause younger children difficulties.

Zaitchik's (1990) 'false photograph' task explored children's ability to hold two competing representations in mind at once, using the familiar figures of Bert and Ernie from *Sesame Street*.

Zaitchik's conclusions have been called into question by more recent studies. For example, Slaughter (1998) gave three-year-olds five different tasks involving representations, including a false belief task, a false photograph task and a false drawing task. The false photograph task involved a Polaroid camera, which was used to take a photo of a toy frog on a chair. The frog was then removed, and a teddy bear was put onto the chair instead. The child was asked "What is on the chair in this picture?" The false drawing task was similar. A doll was placed on the chair, and the experimenter drew a picture of the doll on the chair. The child checked the picture, which was then turned face down, and a car was put on the chair in place of the doll. The child was asked "What is on the chair in this picture?" Slaughter reported that the false photograph and false drawing tasks were significantly easier than the false belief task for her three-year-olds. For example, 76% of her participants passed the false photograph task, compared to 32% passing the false belief task. Slaughter suggested that children's understanding of pictorial versus mental representations was not developmentally related.

Mental representation in the deaf

This view of the developmental independence of mental versus pictorial representations has been supported by a quite different line of investigation. This innovative research examines deaf children's understanding of mental states. Siegal (e.g., Peterson & Siegal, 1998) pointed out that conversations about mental states were likely to be important in the normative development of an understanding of mind, and that deaf children were unlikely to have such conversations. The reason for this is that most deaf children are born to hearing parents, and so miss out on many rich early communicative experiences. Typically, deafness is not diagnosed early, and once it is diagnosed both the hearing parents and the deaf child need to learn sign language before adequate communication can take place between them. Surprisingly, most deaf children do not learn sign language prior to schooling, and many hearing parents with deaf children lack fluency in signing. Accordingly, most deaf children experience a relatively isolated early social environment. This is bound to have effects on the development of social cognition and the understanding of mental representation. As we saw when considering pretend play, social partners are very important for the early development of an awareness that others have their own minds and representations. For example, regarding representational understanding, caretakers frequently use language to help young children to understand thoughts as entities.

Peterson and Siegal (1998) studied signing deaf children's performance in two representational tasks, the false photograph task and the standard false belief task involving the change of location of a desired object. The deaf children ranged from five to 11 years in age, and were compared with typically-developing three- and four-year-olds. Peterson and Siegal found that the deaf children had no difficulty with the false photograph task, but were at chance in the false belief task. The deaf children were significantly poorer in the false belief task than four-year-old children, being comparable to typically-developing three-year-olds (who also failed the false belief task while passing the false photograph task). Peterson and Siegal concluded that deaf children have a special difficulty with the concept of false mental representations. They do not have a difficulty in conceiving of pictorial representations, because the latter are not facilitated by conversations. Peterson and Siegal suggested that the absence of pervasive family talk about abstract mental states lay at the root of deaf children's poor performance in theory of mind tasks.

Clearly, family talk about abstract mental states should not be absent in deaf families where the parents are fluent signers. Indeed, it has been shown that native deaf signers (deaf children who learn to sign from birth) acquire a theory of mind at the same developmental rate as other children. Woolfe, Want and Siegal (2002) recruited two groups of deaf children aged four to eight years to their study, a group of native-signing deaf children being raised by deaf signing parents, and a group of late-signing deaf children being raised by hearing parents. The children were tested by a native deaf signer who tested them in sign language. The tasks used to test the children's understanding of mental representations were false and true belief tasks, shown pictorially. For example, in the false belief task a boy was shown fishing, and then shown looking happy and excited when he felt a weight on his rod. However, the child was shown that he had actually caught a boot. The child was then required to select an item to go into a new picture of the boy showing a 'think bubble' above his head (i.e., the child had to choose a picture to represent the boy's belief). The child also had to select the picture of the item that was really at the end of the rod. In order for the child to be credited with having a theory of mind, they were required to pass both the belief and the reality measures successfully.

Woolfe et al. (2002) found that the native deaf signers performed at a similar level to hearing four-year-olds in the belief tasks, showing an understanding of false belief. The late deaf signers did not. Both native and late deaf signers succeeded on the false photograph test, which was also administered. This success demonstrated that both groups understood pictorial representations. Woolfe et al. argued that the late deaf signers were likely to be cut off from communication about mental states with their parents and siblings by their lack of a shared language. The native deaf signers were not. The importance of conversations about mental states for a normative development of an understanding of mind was emphasised.

More recently, as discussed in Chapter 3, paradigms have been designed to test false belief understanding in pre-verbal infants, and this is also the case for deaf infants. In order to assess the importance of family conversations about mental states in the earliest years, Meristo et al. (2012) compared a group of deaf infants aged on average 23 months with a matched group of hearing infants on the false belief paradigm designed by Surian and Geraci (2012). As will be recalled, in this paradigm infants watch as an entity being chased (for Meristo et al.'s study, this was the mouse Jerry from the 'Tom and Jerry' cartoon about a cat and a mouse) enters a Y-shaped tunnel which leads to two boxes, one at each exit arm of the Y. As infants watched, Jerry was seen going through the tunnel and up one of the arms of the Y, exiting the arm, and settling into the relevant box ('hiding'). Tom the cat then entered the tunnel ('chasing' Jerry), followed the same route, and found Jerry in the appropriate box. In the 'true belief' test version of the scenario, Jerry swapped hiding boxes in full view of the infant and of Tom. In the 'false belief' test version of the scenario, only the infant saw Jerry swap his hiding box. The key experimental measure was which box the infant would make *anticipatory* looks to as Tom entered the tunnel. Meristo et al. (2012) found that both the deaf and the hearing infants were significantly more likely to make anticipatory looks to the correct box on the true belief trials. However, in the false belief trials, *all* the deaf infants made anticipatory looks to the wrong box – the box in which Jerry was now hiding. They did not look at the empty box in which Tom *thought* that Jerry was hiding. This striking group difference could not be explained by other factors in the experiment, such as lack of visual attention to the scenarios.

This is an extremely interesting result. The fact that all the deaf infants made predictive looks to the 'wrong' box on the false belief trials, while all the hearing infants made predictive looks to the correct (false belief) box, indeed suggests that communicative experiences play an important role in the early understanding of other minds. As Meristo et al. (2012) note, further research is required to document which exact communicative experiences are the key developmental drivers. Likely factors include general family interactions around emotions and other mental states, and overhearing conversations containing mental state language. The negative result reported in this study (the failure of *all* the deaf infants to make predictive looks consistent with Tom's false belief) is certainly consistent with an important role for conversations about mental states to the development of children's mentalistic reasoning. The negative result is also interesting with respect to our discussion in Chapter 3, about the extent to which spatio-temporal information involving motion can specify agency and trigger causal explanations of physical events based on hidden mental processes. Clearly, for the viewing deaf infants in the Tom and Jerry scenario, the spatio-temporal characteristics of the events were insufficient by themselves to generate mentalistic explanations. From this perspective, it would be interesting to also give deaf infants a version of the 'helping' and 'hindering' scenarios used by Kuhlmeier et al. (2003), scenarios that appear to cause even bottlenose dolphins to make mentalistic interpretations (Johnson et al., 2018, see Chapter 3). The key difference in the spatio-temporal events signifying helping and hindering is, of course, that a false belief is not involved. Therefore, deaf infants should do better, as the availability of family conversations about mental states should be less important for psychological understanding. Accordingly, deaf infants should behave like hearing infants in the 'helping' and 'hindering' scenarios.

THE ROLE OF LANGUAGE AND DISCOURSE IN METAREPRESENTATIONAL DEVELOPMENT

The idea that conversations and communication about the mind, even communications that the infant eavesdrops on rather than being directly involved in, might play a pivotal role in acquiring an understanding of other minds is an idea that has been around for a long time. Bretherton and Beeghly (1982) suggested that the exchange of information via language played a central role in the development of social cognition, and theorists such as Piaget (1962) suggested that pretend play and language development were linked, as both reflected an emerging ability to manipulate symbols. In fact, the kind of discourse found during pretend play might have a special role in explaining why conversations help to develop young children's mindreading abilities. There are a number of converging sources of evidence for these links between language as a symbol system, language as a communication system, pretend play and the development of metarepresentational understanding.

Here we consider some of this evidence, focusing on the development of children's spontaneous use of mental state language, the kinds of communicative experiences that seem to matter within families, and the role of siblings and peers in developing social cognition, particularly as focused around pretend play.

Piaget (1962) suggested that pretend play and language development were linked, as both reflected an emerging ability to manipulate symbols.

The use of internal state terms

Children begin using mental state terms in their everyday conversations some time during the second year. Internal state language – that is, the use of labels for one's own states such as fatigue, disgust, pain, distress and affection – is found in children aged as young as 20 months. Although some of these internal states are emotions, other internal states such as fatigue are not emotional states. Between 20 and 28 months, there is a large increase in the use of mental state terms in children's discourse. This was demonstrated in a study by Bretherton and Beeghly (1982). They asked mothers of children aged 20 and 28 months to keep a record of their child's internal state language in six different categories. The first category was perception (sight, hearing, touch), for example "Hear wind Daddy? Blowing, blowing" and "I'm going to be a cloud in the sky so that you can't see me". This category also incorporated skin senses (touch, pain, temperature), for example "I'm too hot, I'm sweating!" The second category was physiology (hunger, thirst, states of consciousness), as in "Are you awake?" and "I not hungry now". The third category was positive and negative affect (joy, surprise, love, anger, fear, distress, kindness, disgust), for example "You boo-hoo all better?", "Don't be mad, Mummy!" and "Daddy surprised me!" The fourth category was volition and ability (desire, need, ability to do something difficult), as in "My baby needs me" and "Do you think I can do this?" The fifth was cognition (knowledge, memory, uncertainty, dreaming, reality versus pretending), for example "Jim knows where it is", "Those monsters are just pretend, right?" and "I had a dream about a dog". The final category was moral judgement and obligation (moral transgression, permission and obligation), as in "Matthew won't let me play!", "Was he naughty?" and "If I'm good, Santa will bring toys".

Bretherton and Beeghly found large individual differences in the children's utterances. Nevertheless, 90% of the children at 28 months produced labels for pain, fatigue, disgust, love and moral conformity. Distress was the most commonly labelled emotion, and knowing was the most commonly labelled cognition. The scores for cognition labels (knowing, remembering, thinking, pretending, dreaming) lagged significantly behind the scores for affect and morality labels. As Bretherton and Beeghly pointed out, it is much more difficult to infer processes like thinking than to infer emotions and intentions, as the latter have more explicit behavioural correlates. Unasked, the mothers also collected examples of their children's causal utterances involving mental states. These included utterances like "I scared of the shark. Close my eyes", "I give a hug. Baby be happy", "I'm hurting your feelings, 'cause I was mean to you" and "You sad Mommy. What Daddy do?" Most causal utterances were emotion-related, indicating some grasp of the causes and consequences of mental states even at 28 months. Bretherton and Beeghly concluded that linguistic evidence suggested that the ability to analyse the goals and motives of others was already well-developed in the third year. They argued that the exchange of information via language played a central role in the development of social cognition. Theoretically, they predicted that the acquisition of psychological knowledge about the self and about others should be greatly facilitated by intentional communication (see also Chapter 3).

The role of communicative experiences

Subsequent work by Dunn and her colleagues suggested that this idea was correct. For example, Dunn et al. (1991a) investigated the links between family dialogue about feelings and children's later understanding of the emotions of others. The children were observed at home when aged 36 months in conversation about feeling states with their mothers and older siblings. They were then followed up at six years of age using an affective-perspective-taking task. The naturalistic observations revealed that a number of different emotional themes were discussed in family talk, with positive themes (pleasure, affection, sympathy) and negative themes (fear, anger, distress) discussed about equally. No gender differences were found in the frequency of children's references to feeling states, and mothers did not discuss feelings more frequently with girls than with boys. However, discussions about feeling states were most common when the child was engaged in a dispute, either with their mother or with a sibling.

At six years, the children's ability to identify the emotions of others was measured using video vignettes of acted emotion scenarios between adults (happiness, anger, anxiety, sadness). In these vignettes, the emotions portrayed changed from the beginning of the scenario to the end. The children were asked to identify how the protagonist was feeling at the beginning of the interactions, and at the end. Dunn et al. found that there were highly significant associations between early differences in family talk about feelings (measured in terms of frequency of discussions, causal feeling-state discussions, diversity of themes and disputes) and the children's ability to identify emotions in others at age six. These associations were independent of the verbal ability of the children and the total amount of mother–child talk in different families. Dunn et al. concluded that there was a continuity between early family discourse about feelings and children's understanding of the emotions of others. It seems likely that this continuity reflects the linguistic exchange of information about emotions and their causes, particularly in situations that may be highly emotionally charged for the child, such as a dispute with a sibling.

What about possible links between family dialogue and children's understanding of the *beliefs* of others? Family discourse about the causes of behaviour and events might well be linked to individual differences in the later ability to understand the connections between another's beliefs and another's behaviour. Dunn, Brown, Slomkowski, Tesla and Youngblade (1991b) suggested that family conversations about causality offered young children opportunities to enquire, argue and reflect about why people behave in the ways that they do. Dunn et al. (1991b) tested their hypothesis in a study using methods analogous to those used by Dunn et al. (1991a). Participating children (this time aged 33 months) were observed at home in conversation with their mothers and older siblings during spontaneous interactions. At a later time point (at age 40 months), the children were given both false belief and affective-perspective-taking tasks. The latter involved emotion-inducing situations enacted by puppets (e.g., having a frightening dream). The false belief tasks involved puppets believing that a Band-Aid box would contain sticking plasters, and finding out that it did not. Instead, the sticking plasters were in a plain unmarked box. The child was then asked to predict where a new puppet (who had a cut) was likely to look for sticking plasters. Dunn et al. found that their different measures of feeling state talk in family conversations when the children were aged 33 months (total amount of talk, mother-to-child talk, causal talk) were related to both the children's emotional understanding and to their understanding of false belief when aged 40 months.

The existence of significant correlations over time between explicit family conversation about mental and feeling states at time 1, and later emotional and false belief understanding at time 2, is suggestive of a causal link. Indeed, rich documentation of such links is available in Hughes (2011), who has followed a cohort of toddlers and their families in an extended longitudinal design. However, even longitudinal correlations are still correlations. A *training* study is required to demonstrate whether a *specific* causal relationship in fact exists. Lohmann and Tomasello (2003) provided an example of such a study. Lohmann and Tomasello highlighted the potential role of requests for clarification and discourse concerning misunderstandings as being particularly important for developing an understanding of family members' different perspectives about situations. Two training conditions were then devised. These conditions enabled the comparison of the effects of perspective-taking discourse that also involved the frequent use of mental state words, with the effect of perspective-taking discourse alone. The basic training procedure was based on deceptive objects. For example, an object might appear to be a plastic flower, but turn out to be a pen. In the mental state condition, the deceptive nature of the objects was highlighted and discussed using mental state verbs like 'think' and 'know' ("What do you think this is? ... You thought it was a flower"). In the contrasting condition, the experimenter only used phrases like "What is this?" and "A flower". The effects of training were assessed using (1) further deceptive objects, (2) the traditional Wimmer and Perner false belief task involving change of location, and (3) Flavell's appearance-reality tasks.

Lohmann and Tomasello (2003) reported that the three-year-olds tested showed significantly more improvement in the metarepresentational post-tests if they experienced training with perspective-taking discourse involving mental state terms. Lohmann and Tomasello concluded that mental state language was necessary for young children to make progress in false belief understanding, with this language used by other persons to structure the children's perspective-taking experiences. Lohmann and Tomasello linked their findings to those of Siegal and his colleagues with deaf children, discussed earlier. They argued that it seems to be difficult for children to construct an understanding of the representational nature of mental states from visual scenes alone. Rather, rich linguistic communicative experiences are required in order for children to develop adequate social understanding. This is consistent with our earlier conclusion for infants, that communicative *experience* is important, even during the pre-verbal stage. Rich linguistic communicative experiences help infants to ascribe psychological causes in situations where hidden beliefs (particularly false beliefs) must be inferred in order to comprehend visual scenarios depicting goal-directed actions (as in the Tom and Jerry scenario devised using the Surian and Geraci (2012) experimental paradigm with the Y-shaped maze).

KEY TERM

Mind-mindedness
An interactional stance in which parents and caregivers treat young children as individuals with minds.

'Mind-mindedness' talk by mothers, security of attachment and metarepresentational understanding

Meins (1999, 2013) has developed an interesting idea concerning how children internalise representations of the self and others via family discourse about mental states. She has suggested that a crucial factor in explaining individual differences is whether parents and other caregivers treat young children as individuals with minds. She terms this parental stance 'mind-mindedness', and links mind-mindedness to security of attachment. Although an analysis of the attachment literature is beyond the scope of this book, theories linking aspects of cognitive development with aspects of attachment are very welcome. Originating with Bowlby's (1969, 1973) analysis of security of attachment as a key variable in explaining social and emotional development, the attachment literature offers an alternative way of analysing how children internalise the notion of self and come to characterise others as distinct mental agents expected to act in certain ways. A core notion in attachment theory is the 'internal working model' of the self, which is thought to be developed from parenting experiences. Via experiencing either consistent or inconsistent responding from caretakers, the infant develops a notion of the self as an entity that is more or less deserving of loving attachment from others. The internal working model therefore involves understanding other people and their psychological characteristics, particularly in terms of their likely behaviour towards the self.

Meins made the important point that individual differences in parenting experiences in infancy might have effects on later metarepresentational understanding. In particular, ways in which caregivers initiate, maintain and control interactions with their infants might be important, in particular how they choose to interpret their infants' acts. The proclivity to treat one's child as an individual with a mind from an early age might well have a positive effect on the child's developing understanding about other minds. Terming this proclivity 'mind-mindedness', Meins and Fernyhough (1999) set out to explore whether maternal mind-mindedness assessed when infants were aged 20 months would be predictive of maternal mind-mindedness at age three, and also predictive of the children's performance in theory of mind tasks at age five. Maternal mind-mindedness at 20 months was assessed using two measures. The first was whether the mother reported that their infants used non-standard vocalisations to reliably intend certain meanings, systematically replacing a given English word with such a vocalisation. Mothers who interpreted infant vocal behaviour as *intending meanings* were deemed more mind-minded. The second was whether the mother reported that their infants made many vocalisations that they could not understand (such vocalisations were frequently referred to by the mothers as "double Dutch" or "gobbledegook"). Mothers who did not report that their infants used meaningless speech were deemed to be more mind-minded.

Continuity in mind-mindedness when the children were three years was assessed by asking the mothers to describe their children, and then scoring the use of mental descriptors as a proportion of the total descriptors produced. For example, a mother who described her child in terms of height, weight, hobbies, position in the family, etc., would score low on mind-mindedness ("He's a typical lad"). A mother who described her child in terms of the child's emotions, desires, mental life and imagination would score high on mind-mindedness ("She knows what she wants").

Hughes et al. (1998) found that the hard-to-manage preschoolers were more likely than controls to snatch toys and to engage in rule-breaking behaviour. Moreover, these 4-year-olds went on to display deficient moral understanding as 6-year-olds.

Meins and Fernyhough (1999) found that mothers of securely attached infants were more likely to be mind-minded. Maternal attribution of meaning to their infant's early vocalisations was significantly related to the likelihood that the mothers would describe their children in terms of mental characteristics 16 months later. All three maternal mind-mindedness measures were significantly related to the children's performance in the false belief tasks at age five. Meins and Fernyhough suggested that the proclivity to treat one's child as an individual with a mind was important for the child's own developing understanding of other minds. Relating this to the family discourse work of Dunn and her colleagues, they pointed out that mothers in those studies who employed mentalistic strategies to diffuse disputes, explaining why a sibling had behaved in a certain way using mental state terms, were similarly treating their children as mental agents. Meins and her colleagues have also published longitudinal evidence showing that mothers' mental state talk to their *six-month-old* infants predicts future theory of mind performance on verbal tasks (Meins et al., 2002). Adopting a 'mind-mindedness' stance to your pre-verbal infant may be more common in Western cultures, however. This idea is suggested by a cross-cultural study of parental mind-mindedness and preschoolers' theory of mind that compared the United Kingdom and Hong Kong (Hughes, Devine & Wang, 2017). Hughes and her colleagues found that the UK children showed superior theory of mind performance to the children from Hong Kong, and that the UK parents showed greater levels of mind-mindedness. In both cultural settings, however, parental mind-mindedness was related to the development of children's theory of mind.

Pretend play/mental state discourse between siblings and peers and metarepresentational understanding

Of course, it is not only parents who treat infants as mental agents. Older siblings are also likely to treat younger children as mental agents, and a child with an older sibling is likely to be involved in many episodes of pretend play. Indeed, there is some evidence that children with siblings acquire a theory of mind somewhat earlier than children without siblings, and that older siblings may cause larger developmental effects (e.g., Perner, Ruffman & Leekam, 1994). In Perner et al.'s study of

76 children aged between three and four years, those with two siblings were found to be almost twice as likely to pass false belief tasks as those who were only children. There was also a trend for older siblings to have a greater effect on false belief development than younger siblings, although this trend was non-significant. However, Perner et al. did not collect information about the children's verbal and cognitive abilities, and so the association between family size and false belief understanding could be mediated by children from larger families having higher cognitive or verbal abilities. Jenkins and Astington (1996) therefore investigated the potential association between false belief understanding and family size by systematically measuring these variables as well.

Jenkins and Astington studied a cohort of 68 children, 22 of whom were only children. The others had either one sibling (32 children), two siblings (13 children), or three siblings (one child). All of the children were given four different false belief tasks, including versions of the standard Wimmer and Perner (1983) task, as well as comprehensive language and memory assessments. Jenkins and Astington reported that general language ability and verbal memory were significantly associated with false belief understanding after controlling for age. General language ability was also related to family size, with children from larger families generally having more advanced language skills. The central question was whether family size would be related to false belief understanding once individual differences in language ability were controlled. Jenkins and Astington found that even with age, language ability and birth order controlled, family size contributed a significant amount of unique variance to individual differences in false belief understanding. They also found that the effects of having siblings were actually stronger for children with lower language abilities, and that whether the sibling was older or younger than you did not matter. Hence it seems to be having a sibling per se that is important. A sibling provides a ready playmate as well as a competitor for parental attention, and the intensity of children's interactions with their siblings seems likely to provide a large number of opportunities for reflecting upon both one's own and another's desires, beliefs and emotions.

However, the presence of siblings does more than provide rich linguistic communicative experiences that may contribute to the development of understanding other minds. Siblings also offer different kinds of play experiences compared to mothers. Regarding pretend play, Youngblade and Dunn (1995) pointed out that pretend play with siblings differs from pretend play with the mother, as siblings are more likely to be actors in the drama themselves. Pretend play with siblings is thus more likely to be social pretence, and may be highly charged emotionally. It may thus have particular effects on the development of metarepresentational understanding. To investigate this hypothesis, Youngblade and Dunn observed naturally occurring pretend play episodes between children aged 33 months and their older siblings. The pretend play episodes were rated for factors such as diversity of themes, role enactment and role play. The experimenters then measured the same children's performance when aged 40 months on the Band-Aid false belief and puppet affective-perspective-taking tasks used by Dunn et al. (1991b). The median age gap between the target children studied and their siblings was three years.

Youngblade and Dunn (1995) found that pretend play with sibling differs from pretend play with the mother, as siblings are more likely to get involved in the drama themselves.

Youngblade and Dunn (1995) found that the target children spent much more time in pretend play with their siblings than with their mothers. Further, conversations about feelings were much more frequent during pretend play with the sibling than with the mother. In order to examine predictive relations with understanding of false belief and emotional states, longitudinal correlations were computed for four measures of pretending, namely total amount of pretend play, diversity of themes, role enactment during pretend and role play during pretend. Of these four measures, only the role enactment measure was found to be significantly related to later performance in the false belief task. For the affective understanding task, only total amount of pretending showed a significant longitudinal correlation. Youngblade and Dunn concluded that certain aspects of children's interactions with their siblings were particularly closely linked to developments in understanding other minds.

Interactions with siblings are, of course, only possible if you have siblings, and interactions with friends may also be important in developing an understanding of other minds. There may also be important developmental changes in when and why children talk about mental states. Brown, Donelan-McCall and Dunn (1996) reported that at 33 months of age children's talk about mental states was predominantly within their families, whereas by 47 months mental state talk was more frequent with siblings and friends. Many children begin nursery in their fourth year of life, which gives them access to multiple friendships. The role of play with their friends in children's growing understanding of the beliefs, desires and intentions of others is becoming an increasing focus of developmental studies. Pretend play between friends makes high demands for imaginary and co-operative interaction.

For example, Hughes and Dunn (1998) recruited 25 pairs of friends (mean age three years 11 months) who met daily at their nurseries, and studied whether dyadic play was characterised by mental state talk. They video-recorded each dyad during 20 minutes of pretend play on three occasions spread over the course of a year. The 25 pairs comprised ten boy-boy dyads, ten girl-girl dyads and five boy-girl dyads. Hughes and Dunn also studied the children's understanding of mental states and emotions at the same three time points. To ensure that rich interactions would occur during the sessions being recorded for the study, Hughes and Dunn (1998) supplied novel and exciting dressing-up materials and role-play toys at each recording session. These props included walkie-talkies, police clothes and handcuffs, and an extensive set of toy cooking equipment including a blender and a cooker. Children's mental state talk during the dyadic play sessions was measured, and related to the different outcome measures of theory of mind and emotion understanding development. These measures included variants of the false belief and puppet affective-perspective-taking tasks used in prior studies, along with additional measures of false belief including deception.

Hughes and Dunn found that the rate of mental state talk between the dyads and performance on the false belief and emotion understanding tasks was highly correlated at all three measurement points. Within individual dyads, individual differences in the frequency of mental state talk remained reliable across the year of the study. These individual differences between dyads were significantly associated with theory of mind performance a year later: higher rates of talk at session 1 led to better theory of mind performance when measured at session 3. Interestingly, mental state talk was both more advanced and more frequent in pairs of girls than in pairs of boys. Hughes and Dunn concluded that children's friendships provided a rich social context for learning about others' minds.

Individual differences in friendships and in topics selected for pretending

Of course, children's friendships can vary markedly in quality. Dunn and Cutting (1999) explored whether the nature of interactions with friends might have differential effects on the development of a theory of mind. For example, while some friendships are characterised by amity, shared amusement and positive affect, others are characterised by frequent dramatic quarrels, conflicts and then making up. Cutting and Dunn studied the friendships of 128 four-year-olds from both middle-class and deprived backgrounds. The children's theory of mind development was measured in terms of false belief and deception understanding. Their understanding of emotions was measured using puppet vignettes and other measures, and assessments were also made of their language development, family background and temperament. Each friendship pair was videoed for two separate 20-minute periods of playing by themselves, in a separate room supplied with a set of props similar to those used by Hughes and Dunn (1998). Dyadic play was then rated for the frequency of co-operative pretend play, co-ordinated play, conflict, communication and amity.

Dunn and Cutting reported that the nature of the friendships between children differed dramatically between the dyads. While some children shared an imaginary world together with great skill and enjoyment, others created shared pretence rarely, preferring to engage in boisterous games or even 'shared deviance' (e.g., killing flies together). Children with more shared pretend play were those who scored more highly on the theory of mind measures. These were also the children who talked more to their friends, showed fewer 'failed bids' at communication and showed less dyadic conflict. When both friends had higher mindreading scores, there was more pretending within the dyad. Maternal education and family background also made important contributions to better performance on the theory of mind measures. The amount of connected conversation between children did not in itself contribute unique variance to the understanding of other minds. In fact, this kind of conversation often focused on competitive discussions (e.g., who had the better toys at home) rather than on sharing thoughts and feelings. Clearly, it is mental state discourse rather than discourse per se that is important for the development of mindreading skills.

There is also evidence that skilled mindreading does not always go hand in hand with amicable play characterised by low conflict. Studies of older children have shown that those who bully others may show advanced performance on theory of mind tasks. For example, Sutton, Smith and Swettenham (1999) studied possible links between social cognition and bullying in a cohort of 193 children aged between seven and ten years. They pointed out that whereas the popular stereotype of a bully combines physical power with intellectual backwardness (the 'oaf'), bullies may actually be manipulative experts who can organise gangs of other children and use subtle and indirect methods for control. The latter kind of bully would require superior mindreading skills. Sutton et al. characterised the children in their sample into six groups on the basis of peer and self-nomination. These were Bully (13%), Victim (18%), Assistant (helps bully, 6%), Reinforcer (encourages the bully indirectly via watching and laughing, 8%), Defender (sticks up for the victim, 44%) and Outsider (resolutely refuses to become involved, 11%). Theory of mind development was measured using a comprehensive assessment of 11 types of social cognition (e.g., understanding deception, 'double bluff' and emotion).

Sutton et al. reported that the Bullies scored higher in terms of total social cognition scores than any other group. They scored significantly more highly than Victims, Assistants, Reinforcers and Defenders, but not significantly more highly than Outsiders. Initiative-taking, ringleader behaviour by bullies was the type of bullying most highly correlated with total social cognition score. Sutton et al. argued that possessing theory of mind skills that were superior to those of their followers and victims put bullies at an advantage. Bullies appear to perceive and interpret social cues very accurately. However, Sutton et al. also noted that their study was a correlational one. Hence they could not distinguish between the possibility that having advanced theory of mind skills enables a child to become a bully and the alternative explanation that the experience of bullying itself aids children's social cognitive development.

For younger children, it appears that pretend play with peers has effects on the development of a theory of mind because it entails frequent discussion of mental states. For older children, other developmental factors may also come into play, such as a propensity to manipulate others. Younger children who engage in anti-social play are not necessarily good mindreaders. This was shown in a study of 'hard-to-manage' preschoolers reported by Hughes, Dunn and White (1998). They studied 40 children nominated by their parents as being difficult to manage, and compared their performance on theory of mind and emotion understanding tasks with 40 age- and gender-matched four-year-olds from the same urban area. Family background, language development and non-verbal ability were also measured.

Hughes et al. found that the hard-to-manage preschoolers showed delayed understanding of emotion, showing poorer affective perspective-taking skills even when family background, language and cognitive ability were controlled. These children were basically poorer at understanding how others might feel in particular situations. Once language abilities were controlled, the hard-to-manage preschoolers were equivalent to their controls on most of the theory-of-mind tasks. An exception was the emotion false belief task, which involved a protagonist experiencing either a nice surprise or a nasty surprise. The hard-to-manage preschoolers found it easier to pass the false belief task involving a nasty surprise. As Hughes et al. (1998) pointed out, rather than showing a bias in understanding nasty intentions before nice intentions, this difference could simply reflect the home lives of the hard-to-manage children, which may have included more hostile interactions. Hughes et al. found that the hard-to-manage preschoolers were more likely than controls to snatch toys, to call their friends names and to engage in rule-breaking behaviour. Hughes et al. speculated that these permissive attitudes towards social transgressions could reflect norms within the families (see also Hughes, 2011). Children who are developing in family contexts that do not facilitate an understanding of social norms are likely to show impairments in pro-social behaviour. Everyone feels anger at an act that harms the self, and it is normal to feel aggression towards the perpetrator. However, many apparently provocative acts are *neutral* in intent. Families who discuss the emotions and intentions around such acts foster the development of an understanding that many such acts were not intended to provoke and harm – facilitating the development of a *benign attribution bias* in the child. However, if provoking experiences occur in families that do not talk about the intentions and emotions of others and that do not explicitly discuss social norms, then children are likely to exhibit *reduced* social understanding and may develop a *hostile attribution bias*.

In fact, the hard-to-manage preschoolers studied by Hughes and her colleagues were also significantly more likely to engage in violent pretend play at the age of four years. Violence that involved killing or inflicting pain on another was particularly frequent, and in fact the friends of the hard-to-manage preschoolers often refused to continue with these games (child brandishing sword "Kill! Kill! Kill me!"; friend drops his sword "No"). When the hard-to-manage preschoolers were followed up as six-year-olds, Hughes and Dunn (2000) found that they showed deficits in moral awareness and in social understanding. Language abilities, theory of mind performance and moral sensitivity were all related in this sample, suggesting developmental continuity in mentalising and language abilities. However, there was also an independent relationship between the violent pretend play measure and individual differences in later moral understanding (see also Dunn & Hughes, 2001). The children showing more violent pretend play as four-year-olds had deficient moral understanding as six-year-olds.

As noted, these differences regarding violent play could again reflect the home lives of the hard-to-manage children, which may have included more violent interactions. It is an unfortunate fact that in some families, parent–child relationships are quite hostile, and parental control strategies are inconsistent and ineffectual, or rely on harsh discipline. In such families, children are more likely to engage in bullying behaviour or physical aggression against other children once they reach school. For example, in one study children from families in which parents reported using control strategies like hitting, grabbing and shoving the child were more likely to exhibit behaviours like starting fights, disrupting classroom discipline and being defiant to the teacher (Dodge, 2006). As family interactions are central to the development of pro-social understanding, families that are characterised by sustained (not occasional) violence and aggression to children, a hostile attribution bias on the part of parents and punitive child-control methods, tend to produce children with a hostile attribution bias, in whom the development of social understanding is impaired. A fuller account of the developmental mechanisms involved in aggression and impaired social-cognitive development is available from Dodge (2006).

Pretend play, mental state discourse and metarepresentational understanding: What are the causal connections?

Regarding normative development, the consistent finding that *language* development plays an important role in the development of metarepresentational understanding deserves deeper consideration. From the evidence discussed so far, it seems that language development facilitates metarepresentational understanding because play (particularly pretend play and role play) provides a context in which discussions about feelings, thoughts and desires take place. This discourse about mental states enhances understanding, leading to further development of theory of mind skills. In fact, as noted by Lillard (2002), as children get older less time is spent in actual play, whereas more and more time is spent in negotiating the plot in planned pretend play and in negotiating what each other's roles will be. This negotiation requires sophisticated language abilities and seems likely to engender further metarepresentational insights.

It is also worth pointing out that language and pretend play can facilitate the development of mental state understanding even when the child is playing with an *imaginary* friend. Studies suggest that between 20% and 50% of preschool children have an imaginary friend (Trionfi & Reese, 2009). First-born children are those most likely to invent an imaginary friend, with more girls inventing imaginary friends than boys. Most children invent a friend who is the same gender as they are, and some children have more than one imaginary friend. Research shows that children who have imaginary friends are no more shy or anxious than other children, and have as many live friends as children who do not invent imaginary companions (Trionfi & Reese, 2009). However, those children who do have imaginary friends tend to have richer language skills and tend to be better at constructing narratives than other children. Indeed, creating an imaginary companion requires the child to create a detailed story about the imaginary friend's name and appearance, likes and dislikes, and actions and intentions. Sharing imaginary worlds, reading the intentions of your friends and discussing co-operatively how to fit everyone's actions and mental states to the game at hand, appears to have beneficial effects on the development of a 'theory of mind', even when the friends are imaginary. Taylor, Mottweiler, Aguiar, Naylor and Levernier (2018) reported that older children who create 'paracosms', imaginary worlds which are often highly elaborated (e.g., with their own languages and geographies), had usually had imaginary friends when they were younger (83%). The children with paracosms did not have better language skills than children who did not create such imaginary worlds, but they did show better performance on creativity measures that had social content and involved narrative skills, such as providing an ending to a story begun by the experimenter. For adults, the extension of experience through narrative (e.g., reading a novel, watching a play) is thought to promote insight into psychological causation. Taylor et al. (2018) proposed that the creation of paracosms may provide similar insights for children.

Russell (2005) has suggested a complementary reason to narrative skills for thinking that language should play a central role in the development of representational understanding. He points out that even the most basic aspects of language acquisition entail a conception of beliefs. This is because language acquisition involves becoming a labeller of objects. First, a true labeller of objects seeks the *correct* word for the object – there is a requirement for veridicality. Second, the true labeller realises that the source of this veridicality is label use by other language users. It is only correct to label a cat as 'cat' if there is a social consensus that 'cat' is the correct label for small furry animals. Russell argues that this insight entails a notion of intersubjective truth – and thereby also of intersubjective error. If two people are fixating the same object, but using different labels, one is mislabelling the object.

Koenig and Echols (2003) demonstrated that infants as young as 16 months will correct a human mislabeller, and will spend more time looking at a mislabeller than at a human who labels objects correctly. Koenig and Echols pointed out that words carry information both about the world and about the people who use the words. The choice of label made by human speakers who use labels is a reflection of their internal mental life. If someone says assertively "That's a cat", then it is likely that his or her utterance reflects a belief that the object being viewed is indeed a cat, and that he or she is informing an audience of that fact about the real world. Koenig and Echols termed this type of assertion a 'belief-report', and pointed out that even a simple statement such as "That's a cat" is a belief-report, as it reflects the speaker's beliefs about the world (as well as the state of affairs in the world itself, i.e., that there is indeed a cat present).

To explore this experimentally, Koenig and Echols showed 16-month-old infants colour slides of five different familiar objects (shoe, ball, duck, cat, chair) while the infants were sitting on their parent's lap. The parents wore eye shields so that they could not influence their infant's looking behaviour. As each slide appeared, an adult experimenter who was sitting looking at the slides with the infant provided a label, for example "That's a cat". On half of the trials, the label provided was false. Koenig and Echols reported that the infants looked significantly longer at the pictures when the label was true, but looked significantly longer at the experimenter and at their parent when the label was false. The infants also corrected the false labels, produced other vocalisations or shook their heads to the false labels, and attempted to get support from their parents by trying to remove their eye shields. In fact, 15 of the 16 infants in the study showed corrective labelling behaviour, and some infants began crying at the false labelling. Koenig and Echols argued that infants expect truthful labelling from human labellers who have visual access to the objects being named. Infants take labels to reflect the *intentional states* of human speakers, and expect veridicality in labelling. In an attempt to ensure veridical labelling, some infants became very active, for example pointing at their own shoes when the experimenter was mislabelling a picture of a cat as a "shoe".

Just as language may facilitate metarepresentation in ways additional to the need to use language to discuss the feelings, thoughts and desires of others, for example of characters in pretend play, so pretend play is also likely to facilitate metarepresentation in ways additional to this requirement for mental state discourse. Two other ways in which pretending might facilitate metarepresentational development are via the need to interpret the intention of the playmate in pretending, and the need for active quarantining of what is pretend and what is real. Researchers have explored both of these demands of pretence in empirical studies. Tomasello, Striano and Rochat (1999) investigated whether children aged 18, 26 and 35 months could interpret adult pretend gestures in the context of a game about giving the adult desired objects. For example, the adult would gesture wanting a hammer by making a hammering motion with their fist, or gesture wanting a book by opening their hands in a cupped form like a book. Tomasello et al. reported that children in all three age groups were able to hand the adult the desired object in this game. The pretence was then made more abstract, by involving substituted objects. For example, a cup was substituted for a hat, with the adult demonstrating putting the cup on her head. A ball was substituted for an apple, with the adult demonstrating biting motions on the ball and pretending to chew. The child was then again asked to play the object-giving game. This time, the adult gestured putting a hat on her head when she wanted the cup, and gestured biting into an apple when she wanted the ball. Now only the 26-month-olds and the 35-month-olds were able to interpret the intention of the gestures at above-chance levels. These findings are consistent with the earlier work of McCune-Nicolich (1981). She showed that on average the ability to substitute one object for another during pretend play emerges at around two years of age.

Children's capacity for active quarantining during pretend play has been explored by investigating the occasions on which young children do confuse the pretend and the real. An example is scary pretence. Games involving scary monsters can be actively upsetting for children. Harris, Brown, Marriott, Whittall and Harmer (1991) asked children aged four and six years to play a pretend game in which they had to imagine a scary creature such as a monster, and then to pretend that it was inside a large black box that was actually present in the room. In fact, two black boxes were present in the room, and the children were asked to pretend that a little puppy was inside the other black box. Each box was about one metre square, and had a small hole in the front. The children were asked to imagine that if they put their finger in the hole, the puppy would lick it, whereas the monster would bite it off. The children were then asked to choose which box they would put their finger into, and whether they would prefer to poke a stick through the hole rather than their finger. Harris and his colleagues found that the majority of the children chose to put their finger into the box with the imaginary puppy. Situations involving scary pretence hence seem to present children with some difficulties in quarantining the pretend from the real. However, even adults show pretend-real boundary problems for certain pretend entities such as witches and the Devil. Lillard (2002) has suggested that in cases where emotions colour real-world behaviour, there will be failures in quarantining. However, she argues that these failures bear little relation to children's everyday pretending.

SUMMARY

Social cognition, language and metarepresentation are linked in numerous intimate ways. Whereas cognitive development in the foundational domains of naïve physics and naïve biology depends on veridical primary representations (e.g., of objects), cognitive development in the foundational domain of naïve psychology depends on representations of representations, or *metarepresentations*. In order to develop metarepresentational understanding, the child must 'quarantine' the primary representation and take the representation itself as an object of cognition. As we have seen in this chapter, metarepresentational development depends in important ways on shared social activities such as imitation, pretending and language, activities that may be less available to children who are born deaf. Pretend play and family discussion of feelings and emotions help the young child to take desires and beliefs as entities that can be reasoned about and used to predict behaviour. However, despite the critical role for communicative activities involving sharing objects with others ('shared intentionality') in social cognitive development, metarepresentational development depends also on mechanisms within individual minds. The ability to imitate is one example of a social mechanism which can also be individual. Intention-reading and the analysis of goal-directed actions are others.

Metarepresentations are not observable, and the development of an understanding of the mental states of others undergoes extensive development. One idea has been that early insights about mental states are based on an understanding of desire, which is then supplanted by belief-desire psychology. Desires are more closely tied to observable motivational states and to observable emotional expressions, whereas beliefs are not. However, it was argued here that belief-based analyses may also be available early in development. It may simply be more difficult to observe successful understanding by children of beliefs compared to desires, in which case the *same* developmental mechanisms would underpin the development of the psychological understanding of *both* desires and beliefs. Plausible candidate mechanisms include intention reading, imitation, language development and family dialogue, and experience-based interpretation of the goal-directed actions of others. The representation of action in the brain is linked to imitation and possibly also to intention-reading, for example via the neural mirroring and object manipulation systems. Action recognition may involve an understanding of the actor's *intentions* from early in development, and the representation of action may also be linked to *metarepresentation*. Some EEG studies, such as Kampis et al.'s (2015) rotating box study, are consistent with the view that infants compute both the visual perspective of another agent with respect to an object, and also that agent's *beliefs* about the object's existence, and hold both in mind simultaneously. Kampis et al.

argued that infants were forming a *metarepresentation* of the belief content of the adult's brain, by using their own neural object representation system. Accordingly, mental state understanding requires more than neural mirroring systems.

A classic task in terms of measuring children's understanding of beliefs has been the false belief task. This is because making a correct prediction about the behaviour of another on the basis of their *false* belief requires the child to make a prediction about behaviour that goes against current reality. The development of infants' and children's understanding of false belief has been investigated very intensively, and is still the subject of controversy. For example, infants appear competent in scenarios in which young children fail to show false belief understanding. However, it has also been demonstrated that children have difficulties with *true beliefs* when these no longer reflect current reality. One key developmental mechanism may thus concern keeping track of whether mental representations are veridical or non-veridical. What develops may be the ability to reflect on and index one's own representations, tagging their internal source so that both current reality and past reality can be kept in mind together. This would imply an important developmental role for metacognition, and the development of source monitoring is considered in Chapter 9.

Language, communication and discourse also play important developmental roles in children's growing understanding of metarepresentation. Language enables the exchange of information about beliefs and desires, and thereby contributes to social understanding, as well as giving children experience of mental state terms. Deaf children who lack early communicative experiences show delays in mental state understanding, and deaf toddlers behave differently from hearing toddlers in some paradigms thought to show automatic social-cognitive processing (e.g., Tom and Jerry chasing each other in the Y-shaped maze, Meristo et al., 2012). Pretend play with siblings or peers is an especially rich source of discussion about mental states, and children from larger families show earlier development of a theory of mind. This suggests an additional role for pretend play in the development of understanding mental states. Language acquisition also depends on veridicality (the label should be the *correct* word), and the use of language labels by other users is the source of this veridicality (correct labels depend on social consensus). Theoretical analyses such as these suggest that language may play multiple roles in the development of metarepresentational insights. We consider language development in more detail in Chapter 6.

CHAPTER 5

CONTENTS

Conceptual development and the biological world

5

Knowledge about the kinds of things in the world, or conceptual knowledge, was the third of the 'foundational' domains in cognitive development proposed by Wellman and Gelman (1998). Although originally conceptualised as 'naïve biology', I will expand the term here to cover the development of conceptual knowledge about the world in general. Conceptual knowledge covers not simply the objects in the environment (such as the distinction between animate and inanimate entities, the development of knowledge about natural kinds such as animals versus plants, and the understanding of the properties of the different classes of inanimate objects, such as vehicles versus furniture), but also knowledge about other aspects of experience such as actions, events and mental states. From very early on, infants are using their abilities to detect associations between features, to extract statistical regularities in the environment, to learn conditional probabilities, to connect causes and effects and to perceive relational similarities and make analogies as a basis for developing their understanding of the conceptual world around them.

Conceptual development has traditionally been treated in terms of inductive learning and categorisation. Generalising on the basis of a known example is one of the most common forms of inductive reasoning, and is the basis of categorisation. Because generalising on the basis of an *object* is easiest to study empirically, the majority of studies of conceptual development concern children's knowledge about objects in the world. For example, a child may never have seen a robin before, but may have seen many sparrows. The likely behaviour and properties of robins can be inferred from the usual behaviour and properties of sparrows. In their everyday worlds, children and adults are frequently required to 'go beyond the information given' in this way, and to make inductive inferences. When there are gaps in our knowledge, we have to reason by induction. The ability to reason by induction is widely accepted to be present very early in development. Because inductive reasoning is assumed, the focus of research has been on the extent and organisation of the knowledge that determines the ability to categorise entities as instances of the same concept. As noted in the Foreword, these core principles that determine induction have been referred to as innate 'constraints on learning' (Gelman, 1990). I have been arguing instead that developments in cognitive neuroscience suggest that these constraints are better understood as arising from inbuilt mechanisms of neural information coding and processing, and inbuilt mechanisms for the neural transmission of information. An example of a core principle that may arise initially from intrinsic neural mechanisms for encoding information about the motion of objects is the animate/inanimate distinction. It seems plausible that cognitive systems such as biological knowledge emerge because our neural sensory systems learn information about the world in particular ways that entail assumptions about animacy and agency. Via mechanisms such as active experience, language and communication, these initial assumptions are developed into *explanatory frameworks* or naïve theories. These explanatory frameworks involve *beliefs* about the world.

KEY TERM

Categorisation
Categories have different levels of inclusiveness; e.g., animal (superordinate), cat (basic level), Siamese (subordinate).

SUPERORDINATE, SUBORDINATE AND 'BASIC LEVEL' CATEGORIES

Categorisation is thought of as a cognitive activity because inductive inferences depend on more than purely perceptual similarity. Historically, concepts have been thought to be internalised amodal representations that are relatively abstract. For example, concepts have been described as mental structures that are more than the sensory perceptual representation (Haith, 1998; Quinn, 2002). Robins will be categorised as similar to sparrows because they share many perceptual features such as beaks and wings. However, if the child has seen only muddy sparrows, the feature 'being muddy' will not be automatically projected to robins. The child knows that 'being muddy' is not an enduring property of a sparrow in the way that 'having wings' is, and irrelevant aspects of the sensory perceptual representation are *not* generalised (Gelman, 1988). Neisser (1987) defines categorisation as the ability "to treat a set of things as somehow equivalent, to put them in the same pile, or call them by the same name, or respond to them in the same way" (p. 1). For example, a person may indicate the category 'bird' by naming a robin, an eagle, a chicken and an ostrich 'bird', or may indicate the category 'things that can be sat on' by sitting on a chair, a couch, a stool and a tree stump (Mervis & Pani, 1980). Categorisation is seen as an essential cognitive activity because the world consists of an infinite number of discriminably different stimuli, and each object or event cannot be treated as unique. Recognising novel objects or events as familiar because they belong to a known category enables us to know more about those objects or events than is possible just from looking.

Basic-level category exemplars contain the greatest number of features in common with one another, and the fewest with members of contrasting classes. The robin and the sparrow have more features in common with each other (as birds) than, for example, with dogs.

Categorising has thus been treated as more than another form of perceiving, even though perceptual and conceptual processes are intertwined. They are intertwined both because of the intimate connections between perceptual information and cognition (for example, between modes of movement and inferences about animacy), and because categories involve *beliefs* about the world. These beliefs can be thought of as cognitive constraints, and are discussed later in this chapter. Nevertheless, initially the perceptual structure of the world must be an important source of information for the development of these beliefs. Perception of the attribute structure of the world provides a reasonable basis for the assignment of objects to particular categories. This point was first made by Rosch (1978). She argued that the world comes naturally bundled into sets of attributes, and that this attribute structure is most accessible at the so-called 'basic level'. The 'basic level' involves both seeing directly what things are, and having a theory that tells us how they should be classified. This theory is prototype theory.

Here I will discuss some of the key elements of prototype theory as related to cognitive development. However, it should be borne in mind that evidence from visual neuroscience suggests that perceptual prototypes are an automatic product of the neural processing of visual information. Via experiments into visual search and 'change blindness' (when a change is introduced in a visual scene and observers do not notice it), neuroscience researchers are showing that perceptual encoding mechanisms adapt automatically to changes in the statistical patterns presented by environments, even though the conscious observer is not aware of these changes nor of these adaptations (Hochstein, Pavlovskaya, Bonneh & Soroker, 2015). In cognitive neuroscience studies of perceptual learning, average feature values are automatically computed by the brain whenever a set of novel exemplars is experienced, for example a set of colours or a set of facial expressions. When shown brief presentations of, say, facial expressions, either simultaneously or in succession, the brain automatically *computes summary statistics* for a variety of relevant features. Hochstein and his colleagues argue that conscious perception *begins* with these summary statistics. These summary or higher-order statistics are automatically encoded by populations of neurons ('population coding'), and category prototypes may correspond to these summaries. Again, new insights from cognitive neuroscience suggest that neural mechanisms of information coding may *in themselves* give rise to representations that are treated by cognitive psychology as requiring development or learning. In the case of perceptual category formation, such computations would automatically yield category encoding via *prototypes*, even for totally novel sets of exemplars. In other words, the perceiver's brain is always extracting summary statistics of visual scenes, and these summary statistics are the basis for conscious perception. Such data suggest that higher-order or global information is perceived first. If this view turns out to be correct, then it would throw new light upon some of the debates concerning prototype formation by infants and young children that we will consider here, such as whether global categories precede basic level categories or vice versa.

'Basic level' categories and prototypes

As we saw in Chapter 1 when considering perceptual development, the categorisation of exemplars as similar implies that a generalised representation or *prototype* of a category has been formed. Subsequently presented stimuli are then compared to this prototype. Experiments showing that infants can code the correlational structure between different features of cartoon animals showed the key role of statistical learning in prototype formation (e.g., Younger, 1985). The infants in these experiments were learning which features co-occurred together: their brains were extracting conditional probabilities and constructing prototypes. Rosch pointed out that many naturally occurring categories in the world can be distinguished by perceptual similarity. Rosch argued that at the so-called 'basic level' of category abstraction, concepts such as 'cat', 'bird', 'cow', 'tree' and 'car' were perceptually 'given' by covariations in the constituent features of category members (Rosch & Mervis, 1975; Rosch, Mervis, Gray, Johnson & Boyes-Braem, 1976).

According to this account, the world contains intrinsically separable things because features co-occur in regular ways. For example, feathers reliably co-occur with wings, with flight, and with light body weight, and this co-occurrence helps the child to distinguish the category 'bird'. Basic level category exemplars are thought to have the greatest number of features in common with one another, and the fewest number of features in common with members of contrasting classes. For example, birds share more features in common with each other than they share with dogs. Rosch further proposed that the basic level was the level of categorisation that offered the greatest *psychological* utility. She argued that at this level, an organism could obtain the most information about a category with the least cognitive effort. Her views have also been widely interpreted to imply that the most efficient means of storing conceptual information will be in terms of *prototypes*, or highly typical basic level objects (in 1978 Rosch herself argued that prototypes do not constitute a theory of representation for categories). In terms of conceptual *development*, it has been assumed that if children indeed categorise the world around them at the basic level, distinguishing between objects such as cats, birds, cows, trees and cars, then this will result in the development of a conceptual system that codes categories by prototypes.

As we have seen, the perceptual abilities of infants are certainly sophisticated enough to respond to the featural co-occurrences that distinguish a prototypical category member (e.g., Younger & Cohen, 1983). Although the experiments discussed in Chapter 1 involved man-made categories such as stuffed animals and cartoon animals, similar experiments have been done with pictures of real animals. For example, Eimas and Quinn (1994) studied whether infants can form prototypes of natural kinds such as cats, horses, zebras and giraffes. They habituated three- and four-month-old infants to coloured photographs of horses, using 12 different photographs. Following habituation, the infants were given three different kinds of test trial in a looking preference paradigm: a novel horse paired with a cat, a novel horse paired with a zebra and a novel horse paired with a giraffe. The infants reliably preferred to look at the photograph of the new animal on each type of test trial, showing that they could distinguish between horses, zebras, giraffes and cats. Research such as this shows that the information-processing systems of young infants can cope with the natural featural variation found among exemplars of real species, and can form categorical representations of these species which are perceptually based. In the world of real objects as opposed to photographs, other perceptual features such as relative size, characteristic sounds and movement patterns would also help to differentiate between these natural kinds.

In fact, babies aged three months can categorise objects just as effectively on the basis solely of motion cues as on the basis of rich visual perceptual features. Arterberry and Bornstein (2001) showed infants photographs of real vehicles and real animals on natural backgrounds, or showed them point light displays of vehicles and animals in motion (see Figure 5.1). Point light displays remove all attributes except motion. They are usually created by putting points of light on key joints (animals) or intersections (vehicles) of objects and filming them as they move. When the points of light are then viewed against a plain background, a strong impression of an animal moving or a vehicle moving is experienced by the perceiver. The stimuli created by Arterberry and Bornstein (2001) were computer-generated, but were judged by adults with 100% accuracy as either animals or vehicles.

The infants then viewed either the static pictures of animals and vehicles or the point light displays. Half of the infants in each condition were habituated to animals, and the other half to vehicles. At test, they were shown either new novel exemplars from the familiar category, or an exemplar from the contrasting category. For example, infants habituated to point light displays of different animals either saw a point light display of a new animal, or of a vehicle. Arterberry and Bornstein reported that all infants distinguished the novel category from the familiar category,

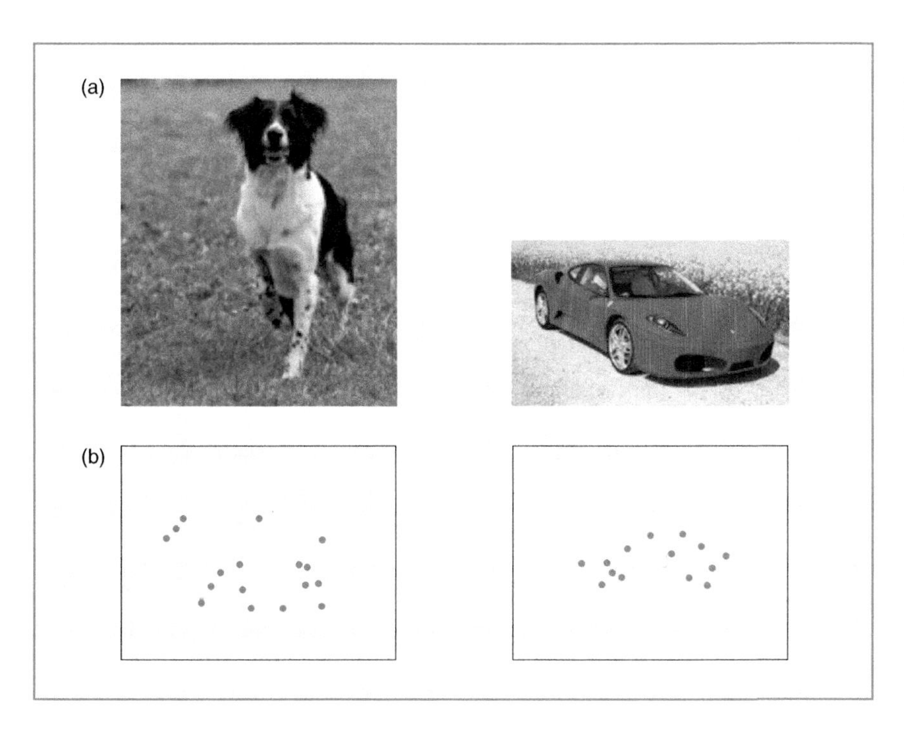

FIGURE 5.1
(a) Examples of the type of photos used as colour static images of exemplars in the animal and vehicle categories used by Arterberry and Bornstein (2001). (b) Single frames from the dynamic point light displays of the dog and the sports car. Adapted from Arterberry and Bornstein (2001).

irrespective of whether they had seen static pictures or point light displays. All infants also failed to dishabituate to the novel exemplar from the familiar category. Categorisation of animals versus vehicles was thus equivalent whether rich perceptual feature cues were provided, or sparse information about motion. These data may appear surprising, but they fit well with the data from visual neuroscience discussed earlier, showing that the brain first encodes the higher-order properties or 'gist' in a visual event (Hochstein et al., 2015; 'high level first' theory). This means that prototype extraction is automatic for any given visual experience via the neural computation of summary statistics. As *type* of motion very clearly distinguishes cars from animals, the development of prototypes or internal models for each category is as accurate when based on motion cues as when based on rich perceptual features.

Historically, the prototype effects demonstrated by experiments such as these were argued to provide evidence for *perceptual* rather than conceptual categories (Mandler, 2004). In order to provide evidence for basic level *conceptual* categories, it was argued that developmental psychologists needed to demonstrate that these perceptual differences have *conceptual* significance for children. One way of investigating this has been to study sorting behaviour in pre-verbal children. If children reliably group objects by their basic level category, for example by putting toy dogs with other toy dogs and toy cars with other toy cars, then this may imply conceptual significance.

KEY TERM

Sequential touching
Children from around 18 months of age tend to touch objects from the same category in sequence more than one would expect by chance, and this provides a measure of categorisation.

Sequential touching as a measure of basic level categorisation

One way to examine sorting behaviour is to measure touching. Infants and children like to handle objects that interest them, but spontaneous grouping of objects is rarely seen prior to 18 months of age. However, sequential touching is often observed: children tend to touch objects from the same category in sequence more often than would be expected by chance, and this systematic touching behaviour gradually develops into systematic sorting. Children's tactile behaviour was used to measure emergent categorisation in a set of studies designed by Mandler and Bauer (1988). They used sequential touching to see whether there was any evidence that children would sort toys into basic level categories.

The toys used in Mandler and Bauer's studies were (a) dogs versus cars (a basic level contrast), and (b) vehicles versus animals (a superordinate level contrast). In the basic level task, the children were given a toy poodle, collie, bloodhound and bulldog, and a toy sports car, sedan, station wagon and Volkswagen Beetle. Sequential touching of the cars or the dogs was taken as evidence for differentiation of these basic level categories. In the superordinate level task, the children were given a toy horse, spider, chicken and fish, and a toy airplane, motorcycle, truck and train engine. A set of kitchen things versus bathroom things was also used (mug, spoon, plate, pan versus soap, toothbrush, toothpaste, comb) to explore so-called 'contextual categories'. Contextual categories were defined as categories of objects that are associated because they are found in the same place or are used in the same activities. The sequential touching behaviour of infants aged 12, 15 and 20 months who were faced with these pairs of categories was then measured, and any spontaneous grouping of the objects was also noted.

Spontaneous grouping of objects is indicative of conceptual organisation, but is rarely witnessed in children prior to 18 months of age.

Mandler and Bauer found that the touching behaviour of the 12- and 15-month-old infants as a group indicated categorisation at the basic level only, as their touching was non-randomly sequential only for the dogs versus cars. Sequential touching within the contextual and superordinate categories was only reliably non-random at 20 months. However, there were large individual differences within this overall pattern. For example, 25% of the 12-month-olds were responsive to the superordinate categories. In a second experiment, Mandler and Bauer compared sorting behaviour with basic level categories when the objects either came from the *same* superordinate class or *different* superordinate classes. Objects from the same superordinate classes (animals) shared a high degree of within-category similarity while *minimising* between-category differences (e.g., dogs versus horses), while objects from different superordinate classes (animals versus vehicles) shared a high degree of within-category similarity while *maximising* between-category differences (e.g., dogs versus cars). Only older children were tested (16- and 20-month-olds). The results showed that basic level categories were clearly distinguished when animals were compared with vehicles, but not when animals were compared with animals. Mandler and Bauer argued that the role of superordinate classes in the development of categorisation may be greater than was previously supposed. In their studies, so-called 'basic level' sorting apparently occurred only when the basic level coincided with *different* superordinate categories.

Another way of describing these results, however, is to argue that the children's sorting behaviour reflected the differing amounts of *perceptual* similarity evident in the stimuli in a particular experiment. The contrast between basic level objects from the same or different superordinate categories does not hold perceptual similarity constant. For example, the distinction between dogs and cars is an easy one to make in terms of perceptual similarity, because dogs are more similar to each other than they are to cars (all dogs have heads, tails, four legs, etc., whereas all cars have wheels, no legs, seats, and so on). The distinction between dogs and horses is less easy to make in terms of perceptual similarity, as both dogs and horses have heads, tails, four legs, and so on, and so it is more difficult to differentiate between them (at least, when one is looking at *toy* dogs and horses, which are similar in size!). Mandler and Bauer's negative result does not necessarily imply an *inability* to differentiate at the basic level within the same superordinate category.

Furthermore, features that suggest membership at the basic level also suggest membership at the superordinate level. For example, feathers suggest 'animal' as well as 'bird', 'fur' suggests 'animal' as well as 'cat' (see Murphy, 1982). This is recognised in Rosch's theory, as she argues that perceptual similarity *correlates with* structural similarity. The perceptual similarity between dogs and horses reflects a deeper underlying structural similarity, namely that both dogs and horses are natural kinds. Accordingly, automatic extraction of perceptual similarity will provide the learner with information about this underlying structural similarity 'for free'. Similarly, cars have wheels and no legs because they are artefacts, and are thus more perceptually similar to aeroplanes and trains, with which they share a deeper underlying structural similarity. Of course, in the real world dogs and horses are easily distinguished at the basic level by a striking difference in size, texture of coat, barking versus neighing, smell, and so on, distinctions that are not a feature of plastic toy animals. In this sense, both basic level and superordinate concepts originate from perceptual knowledge. The differing degrees of perceptual similarity between horses, dogs and cars are an intrinsic part of Rosch's hypothesis.

Pauen (2002) has also produced touching-based evidence that infant categorisation involves conceptual insights. She used an object examination paradigm to show that infant generalisations were not based simply on the variability in the perceptual characteristics of the particular set of objects being used in a particular experiment, but could draw on pre-existing conceptual insights. In Pauen's object examination task, infants were given a series of objects from one category to examine manually, one at a time, until they were familiar. For example, they were given a series of toy animals. Following familiarisation, they were then given a novel exemplar from the familiar category (a new toy animal) and a novel exemplar from a new category (a toy chest of drawers) to manipulate in turn. If the infants spent longer manipulating the chest of drawers, Pauen took this as evidence for categorical discrimination. To explore the question of whether infants form on-line perceptually based representations in categorisation tasks, Pauen systematically varied the degree of perceptual similarity between the objects from different superordinate categories in her experiment. Her categories were animals and furniture.

Pauen's idea was to create toy furniture that was matched with toy animals for overall perceptual similarity. The animals were made of wood but did not appear naturalistic. The furniture was also made of wood, and each piece had black-and-white dots that could be interpreted as eyes, had the same global shape as one of the animals, had the same colours as another animal, and had the same surface pattern as a third animal. For these artificial stimuli, adult raters judged that there was more perceptual similarity *between* the animals and furniture than within each category. Hence perceptually, the individual animals were more different from each other than they were from the individual pieces of furniture. Different, more naturalistic toy animals and toy furniture were also used. These toys differed markedly in perceptual similarity. Examples of Pauen's stimuli are shown in Figure 5.2.

Pauen was interested in whether categorisation would be better with the realistic toy replicas of animals and furniture than with the artificial replicas. If infant categorisation in the object manipulation task depended on the similarity of the perceptual attributes of particular objects, then categorisation should be very difficult with the artificial replicas. This was because the artificial toy animals and furniture were more similar to each other than the animals were to other animals and the furniture was to other pieces of furniture. However, if infants use pre-existing knowledge to group objects in this task, then there should be no difference in responding across the artificial and naturalistic conditions. Pauen found the latter result. Infants spent much more time examining an object when it was from a new category than when it was from a familiar category, irrespective of perceptual similarity. Pauen therefore argued that infant behaviour in the object manipulation task was *knowledge-based*.

FIGURE 5.2
The stimuli used by Pauen (2002). Top: natural-looking models of toy animals and furniture with low between-category similarity. Bottom: artificial-looking toy models of the same animals and furniture items with increased between-category similarity. Adapted from Pauen (2002) with permission from Blackwell Publishing.

The matching-to-sample task

Another way to look at category knowledge is to use a matching-to-sample task. In matching-to-sample tasks, children are given a sample or target object, and are asked to select the correct match for the target from a pair of alternatives (this task is often used with monkeys!). Bauer and Mandler (1989) used a matching-to-sample task to contrast superordinate and basic level matches in children aged 19, 25 and 31 months. The distinction between superordinate matches and basic level matches was created by varying the objects in the choice pair. For example, triads like *bird*, bird, nest or *toothbrush*, toothbrush, toothpaste, enabled basic level matches. Triads like *chair*, table, person, or *monkey*, bear, banana enabled superordinate matches (see Table 5.1). The children were told that they were going to play a 'finding' game. The experimenter would indicate the target object (e.g., the toothbrush or the monkey), and then say "See this one? Can you find me another one just like this one? Can you show me the other one like this?"

Bauer and Mandler found that, although basic level sorts were slightly easier overall, the children were highly successful at categorical sorting in both tasks at all ages. For the basic level sorts, correct responses by age were 85% for the 19-month-olds, 94% for the 25-month-olds and 97% for the 31-month-olds. For the superordinate level sorts, the figures were 91% correct for the 19-month-olds, 81%

KEY TERM

Matching-to-sample task
In which children are given a sample or target object and asked to select the correct match for the target from a pair of alternatives.

TABLE 6.1 Examples of triads of stimuli from Mandler and Bauer (1988)

Column 1	Column 2	Column 3
Bird	Nest	Bird
Toothbrush	Toothpaste	Toothbrush
Mug	Juice can	Glass
Brush	Mirror	Comb
Pear	Knife	Apple
Chair	Person	Table
Baby	Bottle	Adult
Flower	Vase	Plant
Bed	Pillow	Crib
Spoon	Plate	Measuring scoop
Coat	Umbrellla	Sweatshirt
Pot	Spatula	Skillet
Hammer	Nail	Pliers
Wagon	Child	Trike
Monkey	Banana	Bear
Pail	Shovel	Flowerpot
Lion	Cage	Elephant
Shirt	Hanger	Pants
Sink	Soap	Bathtub

Basic level triads comprised two non-identical objects from Column 1 and one object from Column 2. Superordinate level triads comprised one object each from Columns 1, 2 and 3. The first three rows of stimuli formed the example triads.

for the 25-month-olds and 93% for the 31-month-olds. We can conclude that a sensitivity to *both* basic level and superordinate level categories exists by at least 19 months.

The core developmental role of the superordinate level?

Rosch's assumption that the basic level of categorisation would have developmental primacy has been challenged by some researchers. For example, Mandler and her colleagues have proposed that sensitivity to superordinate level categories may *precede* sensitivity to basic level categories, thus reversing the sequence of development proposed by Rosch (Mandler, 2004). In a similar vein, Quinn and his colleagues have proposed that the development of categorisation proceeds from more general to more specific representations (Quinn, 2002). Both proposals assume that more global category representations are differentiated with development into narrower ones (such as basic level ones). However, whereas Mandler made a distinction between perceptual (infant basic level) categorisation and conceptual (older infant object examination) categorisation, Quinn assumed that perceptual learning processes were sufficient to account for the 'global to basic' sequence. As noted earlier, more recent experiments in visual neuroscience suggest that Quinn was correct. The brain automatically encodes the summary statistics of the visual field or its 'gist', with specific information about certain visual features never reaching conscious awareness.

To test her idea, Mandler carried out a series of studies contrasting basic level objects within the *same* superordinate categories, such as horses and dogs (both animals, e.g., Mandler, Bauer & McDonough, 1991; Mandler & McDonough, 1993). For example, Mandler et al. (1991) used the sequential touching technique to investigate basic level distinctions between toy dogs and toy horses (which have a low degree of perceptual contrast) toy dogs and toy rabbits (which have a medium degree of perceptual contrast), and toy dogs and toy fish (which have a high degree of perceptual contrast) in 19-, 24- and 31-month-olds. They found that only the 31-month-olds could differentiate the dogs from the horses with any degree of reliability ($p < .08$). For the medium degree of perceptual contrast (dogs versus rabbits), both the 24- and 31-month-olds could reliably differentiate the animals, and for the high degree of perceptual contrast (dogs versus fish), all groups could differentiate the animals reliably. All groups could also differentiate animals from vehicles (a superordinate level contrast with high perceptual dissimilarity). From these data, Mandler et al. argued that categorisation proceeds from the differentiation of a global domain (such as animals) through successively finer distinctions until the basic level of abstraction is approximated. On the other hand, the failure of infants to distinguish toy dogs from toy horses in this task need not imply the *absence* of the conceptual distinctions. This is because a negative result cannot be used to argue for the *lack* of knowledge, as it may reflect other aspects of the experimental task.

Further evidence thought to go against Rosch's view that categorisation begins at the basic level and then differentiates upwards to the superordinate level and downwards to the subordinate level was reported by Mandler and McDonough (1993), working with seven-, nine- and 11-month-old infants. They again compared the basic level contrasts of toy dogs versus toy fish and toy dogs versus toy rabbits, but this time they used an object examination task rather than the object manipulation task used in their studies with older children. As will be recalled, an object examination task was also used by Pauen to argue for knowledge-based representations. To use the object examination task to explore basic level distinctions, infants were first presented with a series of exemplars at the basic level. For example, they may be given a series of toy fish. Following familiarisation, they were given a novel exemplar from the familiar category (a new toy fish) and a novel exemplar from a new category within the same superordinate (a toy dog) to manipulate in turn. If the infants show increased examination time for the object from the new basic level category (dog) compared to the novel object from the familiar basic level category (fish), then they were assumed to differentiate the two categories. Using this technique, Mandler and McDonough found no evidence that dogs were discriminated from fish, nor that dogs were discriminated from rabbits. However, animals were discriminated from vehicles (a superordinate level contrast). On the basis of these data, Mandler and McDonough argued that infants have a fairly undifferentiated concept of animals that does not include basic level information. Again, however, a clear weakness is that the study depended on *toy* animals, not on real animals. Toys only preserve some of the features of the objects that they represent (such as overall appearance, number of legs and number of eyes), and omit many others (such as relative size, smell, sound and texture of skin). These other features may make a key contribution to basic level distinctions between, for example, dogs, horses, fish and rabbits (real fish and real dogs look, feel and smell very different!). Again, the failure to distinguish toy dogs from toy fish in this task need not imply the *absence* of the relevant conceptual distinctions

Quinn's view of a perceptually based global to basic sequence of category development was derived from connectionist modelling. Quinn and Johnson (1997, 2000, see Quinn, 2002, for a summary) essentially gave a computer model information about different environmental features of mammals and furniture. For example, the model was given information about the different leg lengths associated with tables, chairs, dogs and cats. This information was given to input nodes in the model, and had to be mapped to output nodes via a layer of hidden units. The hidden units had the job of forming representations (i.e., connection strengths) from the featural information given to the input nodes that would yield the relevant concepts at the output nodes. Such computational models learn via progressive extraction of statistical regularities in the input, just like babies. A variety of different simulations converged on the same result: global categories emerged prior to basic level categories.

Quinn argued that connectionist modelling had the advantage of showing what kinds of representations are possible *in principle*, given perceptual features as input. He reported that the connectionist networks initially tended to devote large numbers of hidden units to coding the global level. Large differences in a small number of attributes were focused on (e.g., faces and tails being present in mammals and absent in furniture). Subsequently, more and more hidden nodes were allocated to coding more subtle distinctions in attributes that were relevant to the basic level (e.g., leg length of different types of furniture). As learning progressed, the models overall were dedicating more nodes to basic level distinctions, and fewer to global level distinctions. Hence as more and more exemplars were encountered, the 'encoding' developed by the model favoured the basic level. This fits the conclusions in visual neuroscience discussed earlier, which were drawn on the basis of adult and patient data ('high level first' theory, see Hochstein et al., 2015). Perceptual distinctions that would be termed global-level distinctions in the categorisation literature could equally result from neural visual encoding mechanisms, which automatically compute and represent higher-level statistics in visual scenes and present those summary statistics to conscious awareness.

KEY TERMS

Global to basic sequence
The view that the development of categorisation during childhood proceeds from the initial formation of global categories to the later appearance of basic level categories.

Connectionist modelling
The computational modelling of learning via 'neural networks'. Each unit in the network has an output that is a simple numerical function of its inputs. Cognitive entities such as concepts or aspects of language are represented by patterns of activation across many units. Connectionist models have been applied to many areas of child development, e.g., perception, attention, learning, memory, language, problem-solving and reasoning.

Cognitive neuroscience studies of infant categorisation

More recently, infant researchers have turned to EEG to measure early categorisation behaviour. In this literature, infant ERPs are used to provide an index of discrimination between stimuli, offering an alternative to object examination procedures, and enabling younger participants to be tested. ERP studies have produced evidence consistent with both the basic level being discriminated first, and with the global level being discriminated first.

For example, in the first-ever study of neural markers of infant categorisation, Quinn, Westerlund and Nelson (2006) contrasted pictures of cats with pictures of dogs, a basic level distinction. EEG was recorded as six-month-old infants watched a series of 18 different pictures of cats, and then were shown either more cat pictures, or some dog pictures. Quinn and his colleagues reported a novel ERP, the Nc (a negative central component), related to viewing the novel category (dogs). For the basic level of categorisation, therefore, a distinct neural marker appears to be associated with categorical discrimination.

Subsequent studies have confirmed the Nc component as relevant to infant categorisation at the subordinate level as well (e.g., Saint Bernard dogs versus beagles; Quinn, Doran, Reiss & Hoffman, 2010). Meanwhile, Pauen and her colleagues have used the Nc to investigate infant discrimination of superordinate categories, using the animal versus furniture paradigm also used in their object exploration work (Elsner, Jeschonek & Pauen, 2013). Elsner et al. showed seven-month-old infants repeated pictures of a standard global level item, such as a rabbit, with occasional pictures of 'oddball' items, such as a giraffe (same category oddball) versus a chest of drawers (different category oddball). The standard picture was seen on 60% of trials, and each oddball was seen on 20% of trials. In order to ensure that the infants were responding on the basis of the global categories (animals versus furniture), the category mismatch (the different oddball, for example the chest of drawers when the rabbit was the standard item) was designed to have more perceptual features in common with the standard rabbit picture than the category match picture (in this case, the giraffe). Elsner et al. argued that if the infants focused mainly on perceptual differences, they should show a later Nc amplitude peak for the giraffe oddball, whereas if they were making a category discrimination, they should show a later Nc amplitude peak for the chest of drawers oddball.

The data showed a later Nc peak for the chest of drawers oddball. Given the perceptual similarity control, this study provides nice evidence for global level categorisation. Although this early ERP evidence for *both* basic and global level categorisation may appear confusing, as discussed previously, neural studies of adult behaviour in perceptual categorisation studies have found that average feature values are automatically computed by the brain for *any set of images* that is viewed (Hochstein et al., 2015). Accordingly, the ERPs recorded by researchers in infant categorisation paradigms are likely to depend heavily on the choice of images for a particular experiment and on the nature of the experimental controls for perceptual similarity. Indeed, it is possible that the Nc amplitude peak that is observed reliably in these infant studies may reflect the output of this automatic system for computing average feature values, rather than constitute a distinct neural marker for categorisation per se. Future EEG studies with infants could be devised to test these possibilities.

Evidence from sorting and match-to-sample paradigms with preschoolers

Meanwhile, Poulin-Dubois and her colleagues have produced evidence consistent with Rosch's original developmental proposals, using both a sorting task and a match-to-sample paradigm. Wright, Poulin-Dubois and Kelley (2015) argued that while a range of *implicit* paradigms such as those discussed above have been used with infants and toddlers, there is little *explicit* evidence showing that children are forming categories that involve naïve theories. To remedy this, Wright et al. (2015) asked children aged four and five years to complete both sorting tasks and match-to-sample tasks requiring different types of categorisation, and also to explain the basis for their behaviour. The children were given replica (toy) objects at both the basic level and the superordinate level, with the replicas also representing either animate or inanimate entities. Wright et al. argued that the animate/inanimate distinction may develop later than basic level and superordinate level categorisation.

The toys given for the sorting task included figurines of people, animal replicas (e.g., dog, elephant, bird, fish), furniture replicas (e.g., bed, desk, table, chair) and vehicle replicas (e.g., car, motorcycle, bus, helicopter). The sorting task was introduced by giving the children two bowls, each containing a different replica object (e.g., ice cream and grapes). The child was then given another replica object (e.g., another ice cream), and told "If I gave you this one, you would put it here, because it is the same kind of thing". The ice cream was then placed in the 'ice cream' bowl. For the match-to-sample task, the children were shown detailed colour drawings of entities on a computer with a touchscreen. The children were introduced to one centrally located image with two comparison pictures below it, and were asked to "touch the same kind of thing". Practice trials were given prior to the experimental trials for each task.

Children's sorting of the replica objects and their matching-to-sample behaviour was explored for either basic level objects (e.g., dogs versus birds), superordinate level objects (e.g., animals versus furniture) or with objects representing the inanimate-animate distinction (e.g., people versus cars). The results showed that both the four-year-olds and the five-year-olds were very good at sorting or at matching at both the basic and the superordinate levels, scoring 80–90% correct in each. The children had more difficulty with the animate versus inanimate level, although the five-year-olds were above chance at this level also, performing at around 60% correct. As might be expected, the children were less good at explaining verbally why the replica objects went together. They were best at the basic level, with taxonomic explanations given 63% of the time, and next best at the superordinate level, with taxonomic explanations given 46% of the time. Wright et al. (2015) concluded that children's spontaneous categorical decisions followed Rosch's core proposal. Categorical knowledge was acquired at the basic level first, then at the superordinate level, and finally at the animate/inanimate level.

Typicality effects in sequential touching tasks

Highly typical basic level objects are called *prototypes*, yet the role of typicality in conceptual development has been relatively under-researched. A prototype is an exemplar of a category that is considered highly representative of that category. Prototypes have been conceived of either as the category member with *average* values on the features or attributes associated with the category (Rosch & Mervis,

1975), or as an individual exemplar that is judged as highly typical in terms of the number of other exemplars of the category that it resembles (Medin & Schaffer, 1978). In practice, the two measures usually generate the same prototype: the category member with average values is often the exemplar that is most typical.

Bauer and her colleagues have tackled the question of whether typicality facilitates category formation by using the sequential touching measure. Bauer, Dow and Hertsgaard (1995) tested 13-, 16- and 20-month-old infants' categorisation of sets of objects that either consisted entirely of prototypical exemplars, or of non-prototypical exemplars. The categories that they used were *animals* and *vehicles*, at either the superordinate or the basic level. For example, a set of prototypical animals at the superordinate level might include a cow, a dog, a pig and a cat. A set of prototypical vehicles at the superordinate level might include a bus, a motorcycle, a truck and a car. A set of non-prototypical animals at the superordinate level might include a snail, a rhinoceros, an alligator and an ostrich, and a set of non-prototypical vehicles at the superordinate level might include a canoe, a tank, a space shuttle and a battleship. The basic level sets contrasted prototypical and non-prototypical fish, dogs, cars and aeroplanes (e.g., trout, salmon, bass, pike; or sunfish, eel, fancy goldfish, nurse shark; see Table 5.2). If typicality is important in category formation, then prototypicality should affect sequential touching at *both* the superordinate and basic levels. For example, children should sequentially touch the cow, pig, dog and cat as frequently as they sequentially touch the trout, salmon, bass and pike.

TABLE 5.2 Examples of the stimuli used by Bauer et al. (1995)

Stimulus type/ Category	**Level of contrast**		
	Global level	**Basic level**	
	Animal vs. vehicle	**Dogs vs. fish**	**Cars vs. aeroplanes**
Prototypical			
Category 1	pig	German shepherd	Mercedes-Benz
	house cat	collie	Mustang convertible
	dog	Labrador-retriever	Renault sedan
	cow	brown mongrel	Thunderbird
Category 2	school bus	bass	KLM airliner
	motorcycle	trout	Pan Am airliner
	pick-up truck	walleyed pike	Comanche prop
	four-door sedan	salmon	airforce jet
Non-prototypical			
Category 1	alligator	bulldog	Indy racer
	snail	Chihuahua	drag racer
	rhinoceros	spitz	Lotus
	ostrich	terrier	3-wheeled roadster
Category 2	wooden canoe	sunfish	glider
	armoured tank	eel	stealth bomber
	battleship	fancy goldfish	WWI bomber
	space shuttle	nurse shark	X-wing fighter

Bauer et al. found that the categorisation of the prototypical object sets was indeed superior to that of the non-prototypical object sets. However, for the 13 month-olds, categorisation of the prototypical object sets occurred at the basic level only, whereas for the 16- and 20-month-olds, categorisation of the prototypical objects sets occurred at *both* the basic and the superordinate levels. Categorisation of the non-prototypical object sets was more variable. For the 16-month-olds, categorisation of non-prototypical objects (e.g., sunfish, eel, fancy goldfish, nurse shark) was found only at the basic level. Sequential touching of all the other non-prototypical object sets did not differ from chance. Again, it is difficult to draw strong conclusions from this negative result. For the 24-month-olds, categorisation of the non-prototypical object sets was only found at the superordinate level, while for the 28-month-olds categorisation of the non-prototypical object sets was found at both levels (basic and superordinate). Finally, prototypicality accounted for more variance than either age or categorical level (basic versus superordinate). This finding demonstrates that typicality is likely to play a developmental role in categorisation. As young children gain more knowledge about categories and kinds, stronger typicality effects would be expected. More studies exploring the links between typicality and different levels of categorisation are obviously important, particularly as in adults inductive inferences can be governed by typicality.

Cognitive neuroscience studies and child-basic categories

As cognitive developmental neuroscience methods are utilised more widely by cognitive developmental researchers, a deeper understanding of how concepts are represented by infants and young children should emerge. As noted in earlier chapters, techniques like multi-voxel pattern analysis (MVPA) have yet to be used to study infant conceptual development, but could provide interesting data. In such studies, the voxels (small cubic regions of brain, typically containing thousands of neurons) that are reliably activated by certain concepts are studied. The spatial activation patterns across voxels differ for different concepts, and so brain activity can be used to classify cognitive activity – activation of one group of voxels may reliably signify thinking about a cat, while activation of another group of voxels may reliably signify thinking about a dog Some of the behavioural data discussed so far has been quite contradictory, for example regarding the primacy of basic level versus global level concepts, and infant ERP data have not resolved this contradiction.

Of course, one reason for contradictory findings may be that the infant brain is attending to aspects of the stimuli that have not been considered by the experimenters (and, accordingly, have not been controlled for in a stimulus set and have not been built into a connectionist simulation). The features of a stimulus which a child or an infant considers to be important and therefore assigns the greatest weight in computing similarity and making inductive inferences may differ from those features used by an adult. For example, in Chapter 1 we saw that Mareschal and Johnson (2003) were able to demonstrate that four-month-old infants treated 'being graspable' as an important feature of toys. 'Being graspable' was an important aspect of the stimuli to be categorised for the infants, because it meant that these toys afforded the infant the *same potential actions*. Using data from cognitive neuroscience, Mareschal and Johnson argued that any studies using small, familiar and moving objects as stimuli were likely to activate the *dorsal* route of visual processing, while any studies using larger, stationary objects were likely to be processed by the *ventral* route. In Chapter 2, we saw the likely importance of *function* for early infant categorisation. Infants can individuate objects which are observed to have different functions from four months of age (Stavans and Baillargeon, 2018). In Chapter 2, we also discussed infant neuroimaging data reviewed by Wilcox and Biondi (2015), which suggested the importance of *motion* and of *embodiment* for infant categories. Embodiment is the idea that characteristics of one's physical body play a role in cognitive processing, including an active conceptual role. Clearly, given their different motoric abilities, these embodied characteristics would differ for infants and adults. Features that afford actions by the infant seem likely to be particularly salient for early categorisation.

Indeed, in adult cognitive neuroscience embodiment is playing an increasingly important role in explaining categorisation and concepts. We have already discussed some adult studies on the mirror neuron system (see Chapter 4), showing that goal-directed actions on objects are coded by a special neural system. Mirror neurons are activated both when a person makes an action themselves, such as picking up a cup, and also when a person is sitting passively watching someone else pick up a cup (Rizzolatti, Fogassi & Gallese, 2001). Mirror neurons are found in frontal, parietal and premotor cortex, and are also active when a person *imagines* making a movement. Remarkably, when reading *words* for action concepts such as 'kick', 'pick' and 'lick', the motor areas in the adult brain that are used when moving the feet, fingers and tongue become activated respectively (Hauk, Johnsrude & Pulvermuller, 2004). Similar findings have been reported for preschool children listening to words like 'jump' and 'clap' ('hand' verbs and 'leg' verbs, see James & Maouene, 2009). The children (four- to six-year-olds) showed activation of motor cortex when listening to hand/leg verbs but not when listening to adjectives like 'furry' or 'small', thereby showing embodiment effects. When imagining the colours of objects, the visual cortex region related to colour perception is activated (Barsalou, Simmons, Barbey & Wilson, 2003). Cognitive neuroscience studies with adults and children thus essentially show that the activation of particular concepts reliably produces neural activation in the sensory and motor brain regions associated with those concepts.

Barsalou (2017) provides a recent review of what is now known about adult conceptual behaviour from cognitive neuroimaging studies. He argues that while a lot has been learned regarding the neural areas involved in conceptual processing, including voxel-level precision regarding which neural networks are active during the processing of specific concepts, neuroscience studies have so far thrown little light on the *mechanisms* that implement conceptual processing. For example, he points out that while adult neuroimaging studies employing MVPA are now sophisticated enough to be able to *predict* which concept has just been activated in the mind on the basis of which voxels in the brain are activated, mapping such correlations does not in itself illuminate the mechanisms of conceptual knowledge. As argued in this book, more studies of the developing brain, including multi-modal MVPA studies, along with studies of how language acquisition may change and enrich conceptual knowledge, may be more useful than adult studies with respect to understanding underlying mechanisms. The adult is already a cognitive expert, and accordingly studying the growth of conceptual expertise in infants and young children may throw greater light on underlying mechanisms.

Barsalou's (2017) survey also reached conclusions consistent with the view proposed here concerning the distributed nature of conceptual knowledge. The brain codes objects, actions, events and other concepts in terms of the sensory modalities that are active when the concepts are being directly experienced, and automatically keeps track of which areas are typically activated together. Object categories have very distributed representations, and any category is represented by activation in multiple modalities at the same time. When a child activates the concept 'robin', knowledge about how robins look, sound, move and what they feel like to touch, as well as the emotions associated with seeing robins, are simultaneously active and together constitute the child's conceptual knowledge. Neurally, there is no evidence for an amodal or abstract 'concept' for 'robin' that is held in a separate conceptual system (such as semantic memory, see Chapter 8). Barsalou (2017) chooses the example of the concept of 'bicycle' (p. 19). Across interactions with bicycles, the adult brain will have established aggregate information in the relevant sensory systems as a person perceives, evaluates and interacts with bicycles. On encountering future bicycles, this vast distributed network of 'bicycle knowledge' will become activated and produce multimodal inferences which will support effective goal-directed action. The concept 'bicycle' is stored as a *distributed representation*.

Accordingly, the brain is conceived of as activating conceptual knowledge in order to predict the likely next state of affairs in the real world, in order to act in the most effective manner. In Chapter 2, we discussed this idea about prediction in terms of the rich repertoire of statistical learning mechanisms used by the infant brain to organise experienced event streams into meaningful structures. The infant brain uses statistical learning to predict what will happen next in order to prepare the appropriate actions. We also discussed Fekete and Edelman's (2011) theoretical ideas about 'dynamical activity space' trajectories, whereby basic causal information is an intrinsic part of the neural encoding of experienced events. Accordingly, the process of neural encoding in itself would record *core conceptual information* such as object function and object embodiment (like 'being graspable'). We also noted Fekete's (2010) metaphor of the wine connoisseur, for whom via repeated learning experiences (wine tastings), the representation of the environment (here, the taste of the wine) has been irrevocably transformed. In my view, this metaphor nicely captures the likely processes underlying conceptual development by infants and children. The challenge for developmental cognitive neuroscience is to go beyond metaphor and to specify mechanism. To understand mechanisms, methods such as MVPA are likely to be most helpful when used in conjunction with methods such as EEG, which enable the precise temporal determination of dynamic network interactions.

As also noted in Chapter 2, a neural conceptual system based on aggregating information would lend itself naturally to development. As infants and children gained more conceptual knowledge about the world, both via direct experience and via language, more and more information would be aggregated to a specific concept, which would become more 'adult-like' over accumulated experience. In fact, children's early concepts could naturally differ from those of adults, depending on the sensory, motor, causal and emotional information being attended to. Similarly, behaviour in experiments measuring knowledge of subordinate, basic level and superordinate categories could depend on the overlap between the particular patterns of neural activation resulting from the particular exemplars of those categories chosen for a given experiment. Because two superordinate categories like vehicles and animals, or two basic level concepts like car and dog, are likely to activate different aggregated neural networks, these distributed representations also lend themselves naturally to the simultaneous development of categorical knowledge at more than one level. We already know from MEG studies that if adults are asked to make simple decisions (man-made or natural?) about different superordinate categories such as animals, flowers, clothes and furniture, spatially distinct neural networks are activated by different basic level exemplars of these categories. For example, seeing pictures of a bear, wolf, deer and fox all activate a particular neural network distributed widely across the scalp, and this network of neurons is geometrically partially distinct from the network activated by seeing pictures of a rose, sunflower, orchid and tulip (Löw et al., 2003). Depending on the particular exemplars that have been experienced, these networks may differ at different ages. Clearly, the field of conceptual development is ripe for exploration using cognitive neuroscience methods.

The idea that what is core conceptual knowledge for a child may differ from what is core conceptual knowledge for an adult has of course been proposed before. Regarding Rosch's original theorising, this idea was proposed in terms of 'child-basic' categories, which were thought to be distinguished from 'adult-basic' categories, for example by Mervis and her colleagues (e.g., Mervis, 1987). Mervis pointed out that children may notice or emphasise different attributes of the same object than adults do, for example because of their different experiences and their different knowledge of the culturally appropriate functions of objects. They may thus form somewhat different basic level categories compared to adults. Children's categories may thus be broader than, narrower than, or overlap with the corresponding adult categories.

KEY TERM

Child-basic categories Categories that can differ from 'adult-basic' categories because children might notice or prioritise different attributes of objects than adults.

To illustrate her theory about 'child-basic' categories, Mervis kept a detailed record of the development of her son Ari's first category, which was *duck*. The objects that he was first prepared to countenance as members of this category are shown in Figure 5.3, which also shows the objects that he excluded. Ari's category boundaries were tested by giving him sets of four objects, and asking "Can Ari get the duckie?" Over time, Ari's duck category evolved to include first pictures of ducks, then the plastic duck rattle shown in Figure 5.3, and finally the plush duck head rattle and the Donald Duck head. At this point, he began to spot instances of ducks that his mother failed to notice, such as a picture of a duck inner tube in a magazine that she was reading and a swan soap dish in a shop. The fact that Ari initially considered some toys to be ducks and not others is obviously important for the design of future studies of basic level distinctions that rely on toys.

FIGURE 5.3 Examples of the objects included in and excluded from Ari's initial duck category. Top: objects included immediately (plush mallard, carved grebe, porcelain snow goose, wind-up chicken). Second row: objects that were included as soon as they were available for testing (plush Canadian goose, swan, great blue heron, ostrich). Bottom: objects initially excluded (plastic duck rattle, Donald Duck head, porcelain song bird, plush owl). From Mervis (1987). Copyright © Cambridge University Press. Reproduced with permission.

Perceptual similarity as a guide to structural similarity in categorisation

In the studies of categorisation at the basic, superordinate and subordinate levels discussed so far, a recurring factor used to explain children's categorisation behaviour has been the degree of perceptual similarity between different category exemplars. As noted above, one reason for this is that perceptual similarity is correlated with structural similarity. Perceptual similarity can act as a guide to structural similarity, providing an indication that objects share deeper characteristics (such as the non-observable features shared by biological kinds: has a heart, has blood inside, and so on). However, we can also study how children behave when perceptual similarity and conceptual similarity are *not* correlated. When perceptual similarity is pitted against category membership, children prove surprisingly adept at categorising on the basis of deeper structural characteristics. However, younger children perform better when language helps their intuitions about category membership (two-year-olds), whereas slightly older children do not (three-year-olds). This has been demonstrated in a series of experiments by S. A. Gelman and her colleagues. They devised a picture-based technique that allowed category membership and category appearance to be independently manipulated (e.g., Gelman & Coley, 1990; Gelman & Markman, 1986, 1987).

For example, Gelman and Coley (1990) asked two-year-old children questions about the properties of typical and atypical members of familiar

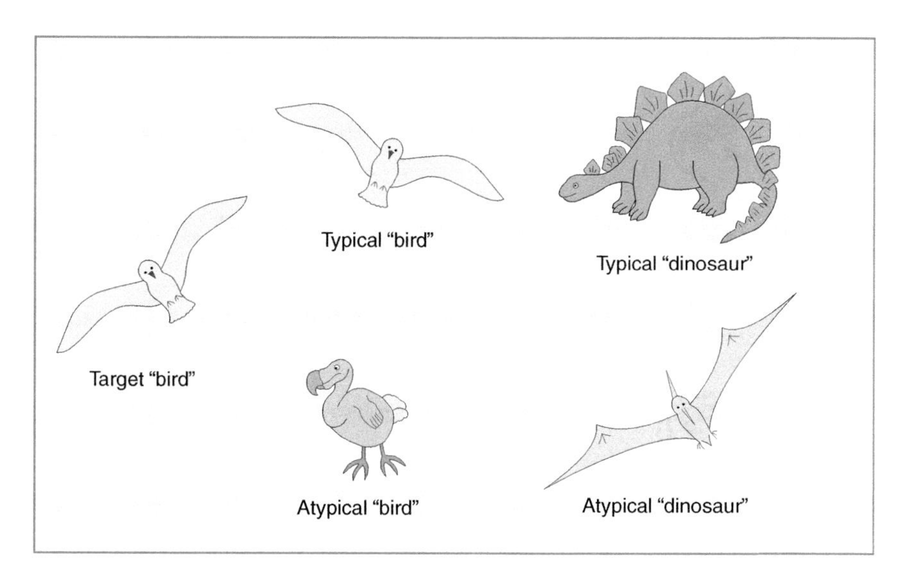

FIGURE 5.4
Examples of the stimuli used to test the 'bird' category by Gelman and Coley (1990). Copyright © American Psychological Association. Reprinted with permission.

categories like birds and dinosaurs. The children were shown pictures of birds and dinosaurs, and were asked questions such as "This is a bird. Does it live in a nest?", and "This is a dinosaur. Does it have big teeth?" Each category comparison was introduced by showing the children a target picture of a typical category member, such as a typical bird. The children were then asked one question about the target picture ("Does it live in a nest?"), and this picture remained in view during the presentation of the test pictures. Pictures of other category members and of the members of a contrasting category were then shown to the child one by one. The same question ("Does it live in a nest?") was asked for each picture in turn.

The key manipulation was that one of the members of the contrasting category looked highly similar to the target picture, and one of the members of the same category looked highly dissimilar to it (see Figure 5.4). For example, for the bird category just mentioned, the highly dissimilar test picture was a *dodo* (atypical category member) and the highly similar test picture was a *bluebird* (typical category member). In the contrasting category of dinosaurs, the highly dissimilar test picture was a *stegosaurus* (typical category member) and the highly similar test picture was a *pterodactyl* (atypical category member). If the children answered the questions on the basis of overall appearance, then they should have judged that the bluebird and the pterodactyl both live in a nest, while the stegosaurus and the dodo do not. However, if they were sensitive to the deeper structural properties that specify category membership, then they should have judged that the bluebird and the dodo both live in a nest, while the stegosaurus and the pterodactyl do not.

Gelman and Coley found that the two-year-olds correctly ascribed the different properties to the atypical category members only 42% of the time (dodo), compared to 76% for the typical category members (bluebird). The 42% level of responding was significantly *below* chance, suggesting that appearance does seem to control judgements about category membership in the absence of linguistic

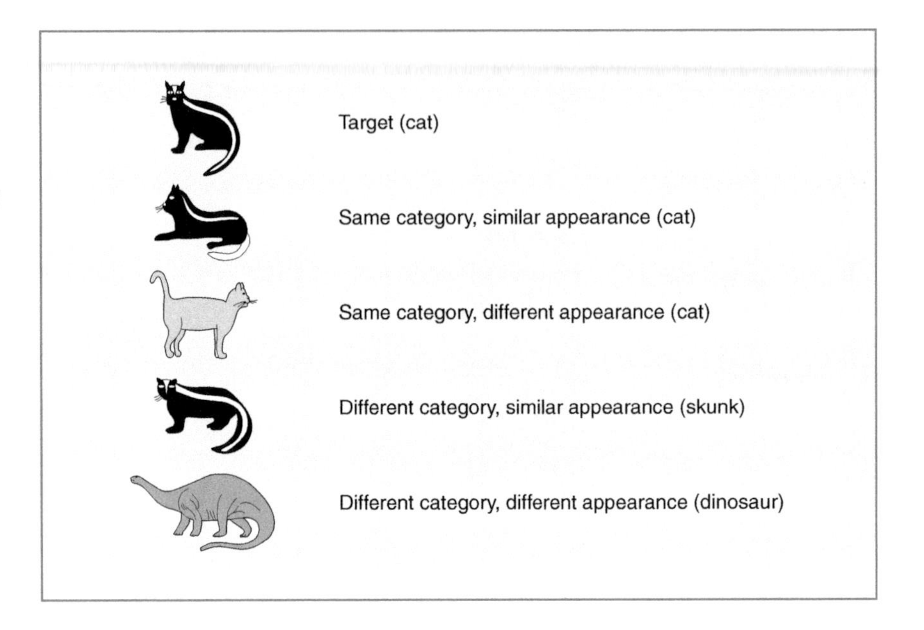

FIGURE 5.5
Examples of the stimuli used to test the 'cat' category. From Gelman and Markman (1987). Reproduced with permission from Blackwell Publishing.

support in this age group. However, if category membership labels were provided during questioning, for example by saying "This is a bird/dinosaur. Does this bird/dinosaur say 'tweet tweet'?", then the two-year-olds correctly ascribed the different properties to the atypical category members 69% of the time, and to the typical category members 74% of the time. Adults given the same task succeeded even without labels. Of course, this is unsurprising as all the atypical members were unfamiliar to the children, but not to the adults. Nevertheless, Gelman and Coley's experiment demonstrates clearly that, when children are provided with category labels, category membership rather than perceptual appearance guides their inferences about the extension of category properties.

In related work, Gelman and Markman (1986, 1987) have demonstrated that three- and four-year-olds succeed in a similar categorisation task, even without labels. For example, in their work with three-year-olds the children were shown a picture of a target object, such as a *cat*, and were told a new fact about it, such as that it "can see in the dark". They were then shown pictures of four more animals (see Figure 5.5), one that looked like the target picture and was of the same category (another cat), one that looked like the target picture but was of a different category (a skunk), one that did not look like the target picture but was of the same category (a cat with different colouring), and one that did not look like the target picture and was of a different category (a dinosaur). The children were asked in each case whether the animal shared the property ascribed to the target picture ("can see in the dark"). Gelman and Markman found that the children consistently assigned properties on the basis of category membership rather than perceptual appearance. Therefore, three- to four-year-old children can use category information *alone* as a basis for drawing inductive inferences about biological kinds.

THE ROLE OF LANGUAGE IN CONCEPTUAL DEVELOPMENT

The potentially critical role of language in conceptual development is clearly illustrated by the finding of Gelman and her colleagues that category membership labels promote accurate conceptual distinctions in two-year-olds. As noted, while active exploration of the environment by the child is one developmental mechanism for enriching perceptually based conceptual knowledge, being able to ask questions and to receive knowledge verbally about categories and kinds via natural pedagogy is another. In fact, there is a lot of experimental work on the relation between language and conceptual development (a detailed discussion of the relation between language and thought is beyond the scope of this book, although some examples are discussed in Chapter 6). Here I give a brief flavour of the kind of research that has been carried out into the role of language in conceptual development. A key finding in such work is that children seem to have *linguistic biases* that guide their conceptual organisation at the different hierarchical levels identified by Rosch.

Learning new words teaches children about conceptual relations between objects and *classes* of objects. In Chapter 3, we saw that Csibra has suggested that when someone demonstrates an object to an infant in a situation of *natural pedagogy*, using ostensive cues like eye contact and labelling the object, then the infant learns general knowledge about the *whole class* of objects that this single object represents (Csibra & Shamsudheen, 2015). By this view, naming informs infants about *categories* of objects in situations of natural pedagogy. Names in natural language can designate relations between basic level objects and superordinate and subordinate relations. The provision of a common label like 'animal' for multiple referents like dogs, horses and fish acts *in itself* to classify these referents as members of the same superordinate class. Indeed, it can be shown using connectionist modelling that naming choices (e.g., using only global level names) utilised with the *same set* of perceptual inputs will change the nature of the category prototypes generated by the connectionist model in its hidden layers (Westerman & Mareschal, 2014). In particular, work in language acquisition has shown that children interpret the introduction of novel *nouns* as highlighting superordinate categories, but the introduction of novel *adjectives* as highlighting subordinate categories.

One researcher whose studies have led to this conclusion is Waxman. She devised an experimental technique that involved teaching children *novel* labels (Japanese words) for objects from familiar categories which had been sampled at different hierarchical levels. For example, Waxman (1990) taught three-year-olds novel labels for superordinate categories like *animal*. Here the basic level objects were photographs of dogs, cats and horses, and the subordinate level objects were photographs of collie dogs, Irish setters and terriers. The children were shown some Japanese dolls, who were introduced as being unable to speak English and also as being "very picky" (choosy). Each doll only liked a certain kind of thing, and three examples of the thing that the doll liked were given by the experimenter. If the doll liked *animals*, the experimenter placed photographs of a dog, bird and fish by the doll (superordinate level), and if the doll liked *dogs* then the experimenter placed photographs of a setter, a bulldog and a poodle by the doll. The children were then given a variety of other photographs to assign to the dolls (e.g., *superordinate sort*: horse, elephant, duck, pig, mixed in with photographs of clothing and food; *basic level sort*: photographs of four other varieties of dog mixed in with photographs of varieties of cats and horses).

Some of the children were given the sorting task in the context of novel nouns, and others were given the sorting task in the context of novel adjectives. For example, in the novel noun condition the experimenter would say "This doll likes only *suikahs*, and these are the *suikahs*". In the novel adjective condition, the experimenter would say "This doll only wants *sukish* ones, and these are the ones that are *sukish*". Waxman found a striking cross-over effect in sorting behaviour, with the three-year-olds in the novel noun condition classifying more pictures correctly at the superordinate than at the subordinate level (e.g., doing better with animals than with dogs), and the three-year-olds in the novel adjective condition classifying more pictures correctly at the subordinate than at the superordinate level (e.g., doing better with dogs than with animals). As the pictures being sorted were the same in each case, this indicates that nouns are interpreted as indicating superordinate categories, and adjectives are interpreted as indicating subordinate categories. The perceptual cues were equivalent in both conditions, as the children were sorting the same sets of photographs.

Furthermore, classification at the basic level was close to ceiling in both conditions, with linguistic cues neither facilitating nor inhibiting performance. Waxman argued that, at non-basic levels, children used syntactic cues to aid the establishment of taxonomic classes. Recent experimental studies show that children as young as 18 months can use syntactic structure to induce whether novel labels are nouns or verbs (Carvalho, He, Lidz & Christophe, 2019). In doing so, they are helped by prosodic cues, as we will see in the next chapter. Waxman argued that children have a linguistic bias to behave like this, because they are sensitive to the powerful links between conceptual hierarchies and the language that we use to describe them. Novel labels can thus promote classification at precisely those levels that are most subject to cultural influence and variation – the non-basic levels.

THE BIOLOGICAL/NON-BIOLOGICAL DISTINCTION

As noted earlier, however, categories also involve *beliefs* about the world. At the same time as they are developing conceptual hierarchies *within* categories like animal and vehicle, children are also developing knowledge about some fundamental conceptual distinctions. One of the first and most important of these conceptual distinctions is that between biological and non-biological entities. Biological entities engage in certain distinctive processes. They can move on their own, they can grow taller, fatter or (in some cases) change their colour or form, and they can inherit the characteristics of their forebears. They also share certain core properties, such as blood, bones or cellulose. Non-biological entities do not engage in self-generated movement, and do not exhibit growth, metamorphosis or inheritance, although they can also share certain core properties (e.g., they may be made of plastic). Infants and young children are aware of some of these differences between animates and inanimates at a surprisingly early age. This basic understanding of the biological/non-biological distinction is then enriched by the child's growing experience of the world.

Evidence from studies of biological movement

One way to examine whether infants and young children are sensitive to the animate/inanimate distinction is to see when they distinguish biological from non-biological movement. An ingenious study by Bertenthal, Proffitt, Spetner and Thomas (1985) suggests that this distinction is already present by 36 weeks of age. In their study, Bertenthal et al. showed infants displays of 'point-light walkers'. Point-light walker displays were first created by Johansson (1973), who placed small lights on the major joints and head of a person, dressed them in black and then filmed them walking in the dark. Johansson found that adults easily recognised these 10–12 points of moving light as a person walking. Adults could also recognise people doing push-ups, people dancing and people riding a bicycle. Later work showed that the gender of the person could also be determined just from seeing the moving points of light (Cutting, Proffitt & Kozlowski, 1978).

KEY TERM

Point-light walker displays
Displays resulting from the movements of small lights that are placed on the major joints and head of a person in dark-coloured clothes who is filmed walking in the dark.

One of the key cues in recognising the points of light as a human form turns out to be the patterns of occlusion created by the act of walking. Imagine a person walking past you. Each time the lights on the limbs on the far side of his or her body (e.g., wrist, knee, elbow, ankle) pass behind the near-side limbs or torso, they will be briefly occluded. Bertenthal et al. (1985) used this occlusion cue as a test of infants' recognition of the point-light displays as human walkers. They created computer displays of points of light, enabling occlusion to be manipulated experimentally, and then showed the babies point-light displays with and without occlusion. Babies were either habituated to an occluded display and then shown the non-occluded version, or vice versa. Additional control groups of babies saw scrambled point-light displays with and without occlusion, which tested their detection of occlusions that did not specify biological motion.

Bertenthal et al. found that the babies dishabituated to the point-light displays that specified biological motion ('canonical' displays), but not to the scrambled displays. This suggested that they were indeed preferentially sensitive to the occlusion information characteristic of biological motion. In a later experiment, Bertenthal et al. demonstrated that the babies' sensitivity was due to their implicit detection of the body of the point-light walker. The babies discriminated the canonical walker from a random occlusion display, but did not show the same discrimination when the walker was presented upside-down. Bertenthal and his colleagues argued that this implied that the detection of biological motion is due to rapid learning on the part of the infants, who detect biological motion only in the familiar upright position (as do adults). Consistent with this idea, younger babies of 20 and 30 weeks did not show dishabituation to the canonical point-light walker displays.

However, as we have seen before, it is risky to use a negative result in developmental psychology to assume a lack of cognitive competence. In fact, more recent studies suggest that even two-day-old infants can differentiate point-light displays showing biological motion from inverted control displays, or from random point-light motion (Simion, Regolin & Bulf, 2008). Simion et al. (2008) used a point-light display created from a walking hen, which they had used previously to test sensitivity to biological motion in newborn chicks. Two-day-old infants tested could discriminate the walking hen display from the inverted control display and from random point-light motion. In a later study, Bardi, Regolin and Simion (2011) compared the point-light display of the walking hen to a point-light display generated by a rotating rigid object. Two-day-old infants tested again preferred to watch the walking hen display, despite the fact that this time both point-light displays contained intrinsic structure. Such data suggest that the detection of biological motion is an inherent capacity of the visual system.

Indeed, a recent fMRI study employing point-light displays of people engaged in acts like walking found that children aged seven to 13 years showed no age-related changes in the neural activity generated by the biological motion displays (Kirby, Moraczewski, Warnell, Velnosky & Redkay, 2018). Individual differences in sensitivity to the biological motion displays (which were compared to random point-light motion) were, however, related to differential activation of core 'social brain' areas, such as the amygdala and superior temporal sulcus. Meanwhile, eye tracking data show that older infants can use biological motion to detect human *interaction*. Galazka, Roche, Nystrom and Falck-Ytter (2014) used eye tracking to show that infants aged 14 months preferred to watch point-light displays created by two people interacting by pushing each other, or by one falling into the arms of the other, compared to inverted versions of the same point-light displays. The preference in infants for watching motion that specifies humans interacting, along with the demonstration that watching biological motion activates 'social brain' networks, suggests that the automatic processing of biological motion information may be an important basis for social-cognitive development (see also Chapters 3 and 4).

Other basic visual cues also seem to be used to distinguish biological from non-biological motion. Lamsfuss (1995) suggested that the predictability and regularity of motion was another useful cue to the biological/non-biological distinction, as while we can usually make fairly accurate predictions about the movement patterns of non-biological kinds such as cars and other machines, we cannot make accurate predictions about the movements of biological kinds such as house flies. In order to test her idea, Lamsfuss showed four- and five-year-old children different pairs of 'tracks' that had purportedly been left by either an animal or a machine. The tracks were simple dot patterns (see Figure 5.6), one of which was always more regular than the other. Dot patterns were used so that no additional perceptual information about the object that produced the motion would be provided by the displays. Lamsfuss found that, when asked which of the two tracks looked more like it could have been left by a machine, the children chose the regular track significantly more often than when asked the same question about animals. She argued that this indicated that children expect animals to move in unpredictable ways, regular as well as irregular, whereas they expect machines to produce highly

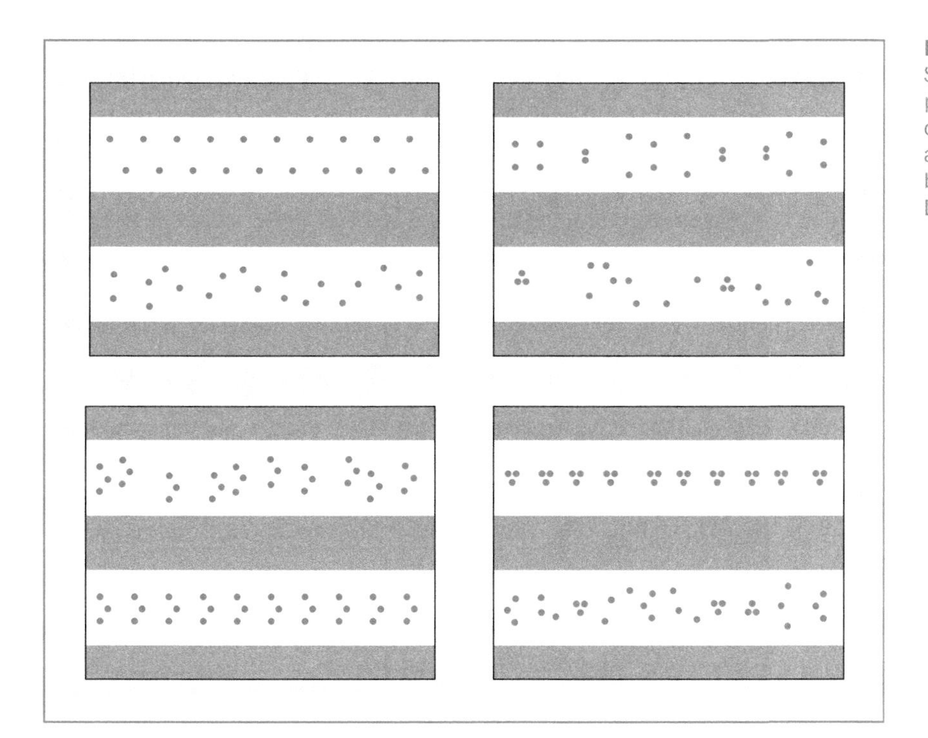

FIGURE 5.6
Some of the simple dot patterns used to test children's intuitions about biological and non-biological movement by Lamsfuss (1995).

predictable, regular movements. Adults and biology experts who took part in the study showed the same response pattern as the young children. Thus movement cues seem to be an important source of information about the biological/non-biological distinction.

Evidence from studies of self-initiated movement

Another way to assess young children's ability to make a conceptual distinction between biological and non-biological entities is to see whether young children recognise that biological entities can move on their own. Massey and R. Gelman (1988) investigated three- and four-year-olds' understanding of self-initiated movements by showing them photographs of unfamiliar objects and asking them whether the objects could move up and down a hill on their own. The photographs depicted unfamiliar exemplars of two animate categories, mammals (e.g., tarsier, marmoset) and non-mammals (e.g., tarantula, lizard), and exemplars of three inanimate categories, statues of animals, wheeled objects (e.g., golf caddy, bicycle), and complex rigid objects (e.g., camera, exercise machine). Notice that the golf caddy and the bicycle can move *down* a hill on their own. This choice of categories was intended to pre-empt responding on the basis of shared perceptual features, such as linear versus non-linear edges. Only the animate objects could move *uphill* on their own.

Massey and Gelman introduced the children to the task by showing them a picture of a hill, and asking them whether photographs of some practice items (a man, a little girl, a fork and a chair) could go up and down the hill "all by itself".

The target pictures were then shown in a randomised order, and the child was asked to decide whether these objects, too, could go up and down the hill by themselves. Overall, the three-year-olds made correct decisions about 78% of the photos, and the four-year-olds made correct decisions about 90% of the photos. Even though the animals were not depicted as moving and feet were seldom evident, the children's comments often focused on the feet and legs. For example, (echidna) "It can move very slowly … it has these little legs. Where's the legs? Underneath?" and "It has feet and no shoes". [*Experimenter* – can you point to the feet?] "I can't see them". In contrast, where feet were depicted for inanimate objects (e.g., the statues), the children denied that they were feet, because they were not movement-enabling. Most of the errors made by the three-year-olds concerned the unfamiliar animate non-mammals, such as the tarantulas and lizards. The younger children who made errors with these pictures nonetheless appeared to be basing their decisions on the animate/inanimate distinction. For example, they would say that 'bugs' (the spider) could not go up the hill by themselves because they were too little to go up such a big hill.

Similar results concerning young children's assumptions about self-generated movement have been reported by S. A. Gelman and Gottfried (1993, discussed in Gelman, Coley & Gottfried, 1994). They showed four-year-old children videotapes of animals, wind-up toys and household objects moving across a surface. The animals and toys were deliberately chosen from unfamiliar categories, for example a chinchilla and a wind-up toy sushi. The household objects (e.g., a pepper mill) were all transparent, in order to see whether this would make it more difficult for the children to assume an internal cause for movement. In a control condition, all of the animals, toys and objects were moved manually, and the hand doing the moving was clearly visible on the video.

The children were asked three critical questions about each animal and object. These were: Did a person make this move? Did something inside this make it move? Did this move by itself? Gelman and Gottfried found a clear distinction between the animals and the toys and objects in the children's responses to these questions. In the manual carrying condition, the children said that the animals moved on their own, whereas the toys and the objects were moved by a person. In the self-generated movement condition, the children attributed movement to internal mechanisms, even for the transparent objects. However, they were unable to explain how this movement occurred. Many children suggested the involvement of a supernatural agent, hidden persons or invisible natural causes such as electricity. So even children as young as four were quite clear that animals are different. Animals can move on their own, whereas toys and objects can only move with the help of an external agent.

Indeed, even newborn infants seem sensitive to the cues of self-propelled motion and change of trajectory as specifying biological entities. Di Giorgio, Lunghi, Simion and Vallortigara (2017) showed newborn infants videos which either specified the self-propulsion of an entity or the self-determined change of trajectory of an entity. They reported newborn sensitivity to self-propulsion (the control condition was non-self-propelled motion), but not to self-determined change of trajectory (the control condition was a change of trajectory determined by contact with an occluder). Again, the negative result (infants' lack of a visual *preference* for watching the motion specifying a self-determined change of trajectory) does not mean that they are *unable* to discriminate this cue at birth. Indeed, a follow-up study by the same authors showed that the infants *could* discriminate between self-determined and not-self-determined changes of trajectory, but that they did not prefer to view one over the other.

Accordingly, some of the visual motion cues specifying 'animal' are available even to newborn infants. This is consistent with some of the studies reviewed in Chapters 2 and 3 concerning *goal-directed actions*. Goal-directed movement is particularly informative about animacy, as we saw in experiments like the 'chasing' studies. Young children, too, treat goal-directed movement as a core property of living things. In an experiment by Opfer (see Gelman & Opfer, 2002), children aged four, five, seven and ten years observed blob-like shapes in motion. In one condition, the blobs were apparently pursuing a goal, while in another condition the same blobs made the same motions without a goal being present. From age five onwards, the children decided that the blobs in the goal-directed movement condition were living things. They attributed life, biological properties and psychological properties to these blobs, describing them as jellyfish or bugs. In the control condition, where no goal was present, the blobs were not attributed any biological or psychological properties, and were described as clouds or meteors.

Evidence from the assumption of shared core properties

A third way of probing children's understanding of the biological/non-biological distinction is to ask children to make judgements about similarities and differences in the 'insides' and 'outsides' of objects. 'Insides' are more important than 'outsides' for understanding the true nature of an object. For example, biological kinds share key internal properties (dogs and birds both have blood and bones), and these differ from the key internal properties of non-biological kinds (chairs and doors may be wood or metal). In order to examine young children's understanding of the 'inside-outside' distinction, S. A. Gelman and Wellman (1991) asked three- and four-year-old children to make a series of judgements about which pictured objects shared insides or outsides.

In their study, the children were asked "Which has the *same kinds of insides* as x?" (insides), or "Which *looks most like* x?" (outsides). The pictures were always presented in threes, two of which were similar in their insides, and two in their outsides. For example, in the triad *orange, lemon, orange balloon*, the orange and the lemon had the same insides, while the orange and the balloon had the same outsides (appearance). In the triad *pig, piggy bank, cow*, the pig and the cow had the same insides, and the pig and the piggy bank had the same outsides. Gelman and Wellman found that both the three-year-olds and the four-year-olds performed at levels above chance, although the four-year-olds were correct on more trials than the three-year-olds (73% correct and 58% correct, respectively). Additional analyses showed that errors were not due to a reliance on similarity of appearance (a perceptual error, such as saying that the pig and the piggy bank had the same insides).

Although such errors occurred, an equal number of errors were made on the basis of using insides to assess appearances (saying cows and pigs looked similar). Gelman and Wellman suggested that what developed with age was the ability to deal with conflicts between insides and outer appearances (as in the piggy bank), rather than the ability to distinguish insides from outsides.

This conclusion is supported by the results of a study by R. Gelman and Meck (cited in Gelman, 1990). Gelman and Meck asked three-, four- and five-year-old children about the insides and outsides of animate and inanimate objects. The animates were person, elephant, cat, bird and mouse, and the inanimates were rock, ball, doll and puppet. The children's answers showed a clear distinction between the animates and the inanimates. The animates were said to have blood, bones and hearts inside, while the inanimates had hard stuff (rock, ball) or material and cotton inside (dolls and puppets). All the inanimates were said by some children to have 'nothing' inside, while none of the animates was ever thought to contain 'nothing'. The children also tended to say that the animates would have different insides from outsides (the outsides included skin, hair and eyes), whereas the inanimates were judged to have the same outsides as insides ('hard stuff', 'material'). Accordingly, Gelman (1990) proposed an 'innards' principle that she suggested acted as an innate 'constraint on learning'. The 'innards' principle is that self-propelled agents have insides that make possible their behaviour. More recently, Setoh, Wu, Baillargeon and Gelman (2013) have used the violation-of-expectation paradigm to show that eight-month-old infants indeed expect novel entities that they classify as animate to have insides.

Setoh et al. (2013) introduced the infants to two novel objects, one of which was capable of self-propulsion and agency and one of which was not. The novel objects were a visually distinctive can and a visually distinctive box. For example, if the can was the 'animate', it would move in a bouncing manner to and fro as if self-propelled, and then initiate a 'conversation' with an experimenter, by making quacking sounds, to which the experimenter would talk back. If the box was the 'animate', it would move in a zig-zag manner to and fro as if self-propelled, and then initiate a 'conversation' with the experimenter by making beeping sounds, to which the experimenter would talk back. When the objects were 'inanimate', they did not move, and the experimenter did not talk back to quack and beep noises. Following these displays, the can and box were both turned upside down, so that the infant could see inside them. Depending on the condition, the can/box was either revealed to be hollow and empty, or to be closed. The results showed that the infants looked significantly longer when the can/box had appeared to be animate, yet was revealed as hollow and empty. Setoh et al. (2013) argued that when the novel object had identified itself as animate (via autonomous motion and holding a conversation), the infants expected it to have 'insides'.

Concrete or abstract knowledge?

Although these responses appear to indicate that young children have fairly concrete ideas about the insides and outsides of animates and inanimates, Simons and Keil (1995) have argued that the basis of young children's judgements about insides and outsides are *abstract expectations* about the sorts of things that should differentiate the two. Rather than having concrete knowledge about which insides are appropriate for biological versus non-biological kinds, Simons and Keil argued that children have an abstract framework of *causal* expectations about natural kinds and artefacts, and that this abstract framework then guides their search for concrete differences. To test their idea that younger children lack concrete knowledge about these differences, Simons and Keil conducted a series of experiments designed to examine the kinds of things that three-, four- and five-year-old children expected to be inside biological and non-biological kinds. They argued that their hypothesis was consistent with the findings discussed above, as the tasks used by Wellman and S. Gelman and by R. Gelman and Meck actually required abstract knowledge about category membership.

KEY TERM

Shared core properties Properties that are important for categorisation but may be unobservable, such as the insides of biological objects that differ from the insides of non-biological kinds.

In their experiments, Simons and Keil introduced children to a toy alligator, Freddy, who had the ability to "see right through the outsides of things into the inside". The children were told that Freddy had never been to Earth before, so he sometimes got confused about what was inside different sorts of things. The children were asked to help Freddy to decide which of a pair of things had the real insides. For example, in one study the children were shown two pictures of either a natural kind (such as a sheep, a frog or an elephant) or an artefact (such as a clock, a telephone or a bus). Each picture had a computer-generated inside depicted in its middle (see Figure 5.7). One inside was always animal-like, and the other was machine-like. For example, one of the two pictures of a sheep had cogs and gears inside it, and the other had some internal organs. The children were asked to show Freddy which picture showed 'a sheep with real sheep insides'. In another study, the children were just shown a single picture of a sheep without any depicted insides, and had to choose the appropriate insides from a set of three glass jars. One jar contained gears, dials and wire (machine insides), a second contained the preserved abdominal organs of two cats (biological insides) and the third contained some small white rocks suspended in gelatine (representing a mixture of biological and non-biological insides, or 'aggregate insides'). The children were asked to point to the sort of insides that Freddy would see if he looked right through the outside of each animal or machine.

The findings across these different studies were quite consistent. The younger children had more difficulty in selecting the correct insides than the older children, but they did not err randomly. Instead, they showed a clear distinction between the

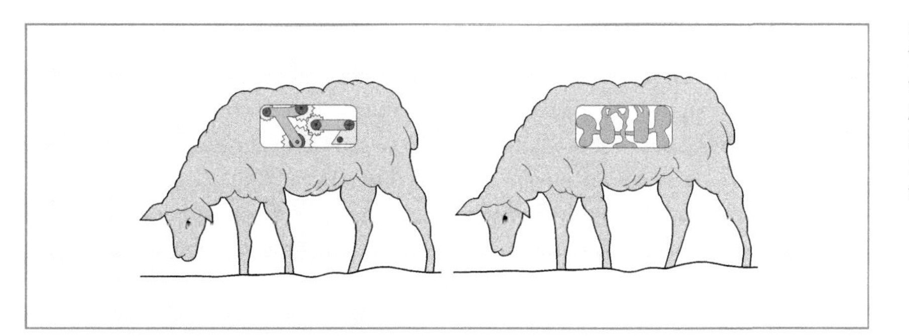

FIGURE 5.7 The sheep with animal versus machine insides. Adapted from Simons and Keil (1995). Copyright © 1995 Elsevier. Reprinted with permission.

natural kinds and the artefacts. Even the youngest children were highly accurate at selecting the correct insides for the machines, but for the natural kinds they tended to choose the aggregate insides as frequently as the biological insides. Simons and Keil argued that the younger children did not know what insides are like, but that they did know that some things were more likely to be inside animals than inside machines, and vice versa. These general ideas about what insides should look like were taken to indicate that even the youngest children had *abstract* expectations about the sorts of things that can be inside animals and machines. However, they lacked experience with *concrete* examples of insides.

Another way of examining children's knowledge about shared core properties is to ask them to make *verbal* judgements about the internal properties of biological and non-biological kinds. Without the aid of pictures, children are forced to reason about categories as abstract wholes. Gelman and O'Reilly (1988) asked five- and eight-year-old children whether different biological and non-biological kinds, such as dogs, horses, snakes and tractors, had "the same kinds of stuff inside". These comparisons were made sequentially. The children were asked, in a random order, (1) whether all dogs had the same kinds of stuff inside, (2) whether dogs and horses had the same kinds of stuff inside, (3) whether dogs and snakes had the same kinds of stuff inside, and (4) whether dogs and tractors had the same kinds of stuff inside. Gelman and O'Reilly found that the children knew that animals had the same kinds of internal parts, and differentiated the animals from the artefacts such as the tractor. For example, they told the experimenters "Every dog has the same stuff unless they're missing a tail or something", and "All chairs aren't the same. Some of 'em have metal, some of 'em have wood. Some of 'em have iron."

Structure versus function in categorising natural kinds and artefacts

The research on children's intuitions about 'insides' and 'outsides' rests on the assumption that the shared core properties that are important for categorisation are similar for biological and non-biological kinds. However, Keil (1994) has pointed out that, while children may judge shared *structure* (insides and outsides) as important for categorising living kinds, they may judge shared *function* as more important for categorising artefacts. While shared function in artefacts does not necessitate similarity of appearance, shared structural similarity in animals frequently does. For example, the handles of bags can look quite different. They may be rigid or flexible, thick or thin, and long or short. Nevertheless, we do not categorise bags according to the appearance of their handles as long as these differences have no *functional* implications. Different varieties of rodent, however, may have tails that vary in appearance as much as the handles of bags (thick or thin, bushy or hairless, long or short), and yet these variations in appearance may be very important for classification purposes. Such differences do not necessarily affect the function of tails, but they may indicate important differences between species (e.g., squirrels versus rats). In the case of animals, differences in the appearance of parts thus often imply other underlying differences, such as differences in specific genetic structure.

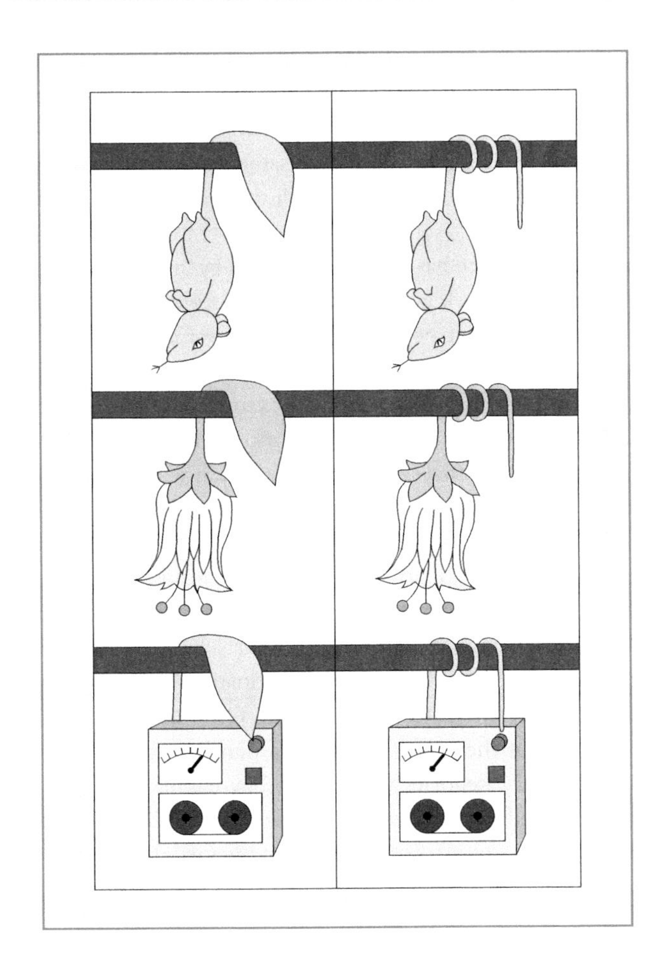

FIGURE 5.8
The same perceptual feature functioning as a mouse's tail, a plant stem or a tape recorder's handle. From Pauen (1996a).

In order to test the idea that children judge shared *function* as important for categorising artefacts and shared *structure* (insides and outsides) as important for categorising living kinds, Pauen (1996a) created pictures of pairs of artefacts and pairs of biological kinds which shared a key part. Her idea was that the perceptual similarity of this key part could be manipulated across the pairs. The function of the key part was the same within each biological or non-biological pair, but its appearance differed. For example, a pair of mice had either a wide tail or a narrow tail, and a pair of tape recorders had the same perceptual feature forming either a wide handle or a narrow handle (the two mice or tape recorders were otherwise identical in appearance, see Figure 5.8).

Four- to five-year-old children were shown the different matched quadruples of pictures, and were told a cover story about needing to tidy a room. Tidying required putting the pictures together "that were the same kind of thing", and the experimenter began this by separating the artefacts from the living kinds. An 'expert' then appeared, and told the children that this was not the proper way to organise things, as some of the pairs of pictures that had been put together were "not really the same kind of thing". It was necessary to separate one of the two pairs (either the artefacts or the living kinds). The children were then asked which of the experimenter's pairs could be separated. Pauen found that the majority of the children said that the biological kinds could be separated rather than the artefacts. This supports the idea that perceptual dissimilarities are taken to specify different subcategories within biological kinds. The same does not appear to apply to artefacts, at least as long as the function of the dissimilar feature remains the same.

More recently, perceptual appearance has been manipulated by 'morphing' one image gradually into another. Diesendruck and Peretz (2013) morphed the images of either artefacts or living kinds, so that perceptual similarity varied statistically over different exemplars. They then asked children aged three and five years to make forced-choice categorisation decisions about novel exemplars of each kind, and also supplied them with some extra conceptual information. The extra conceptual information contradicted the perceptual similarity information with respect to the required categorisation decision. The extra information could be about internal properties ("has green blood", "has a small motor") or about the intention of a boy to create a new category member ("I want to create a Muso just like these ones"). Diesendruck and Peretz reported that information about internal properties affected categorisation decisions for animals more than artefacts, but only for the five-year-olds, not for the three-year-olds. In contrast, information about a boy's intentions affected categorisation decisions for artefacts only, for both age groups. Diesendruck and Peretz argued that their data supported a domain-specific view of categorisation behaviour. Animal categories were defined by intrinsic 'essences' and artefact categories were defined by varying features such as function and the creator's intent. When children were not given conflicting conceptual information, perceptual similarity governed category assignment for both the living kinds and the artefacts.

Evidence from studies of growth

A fourth way of examining children's understanding of the difference between biological and non-biological kinds is to study children's understanding of growth. As time goes by, biological kinds change in their appearance. They may grow bigger (a tree), they may change colour (a tomato) and they may even change their appearance (a caterpillar changing into a butterfly). Artefacts do not alter as time goes by. They may become scuffed or worn, but they cannot grow, change their shape or change their colour.

In a series of studies examining young children's understanding of growth, Rosengren and his colleagues have shown that children as young as three are aware of these distinctions. For example, Rosengren, Gelman, Kalish and McCormick (1991) showed three- and five-year-old children pictures of baby animals and brand-new artefacts, and then asked them to choose which of two other pictures showed the animal or the artefact after it had been around for a very long time (see Figure 5.9). In some example pairs the children had to make a choice between a picture showing the target the same size and a picture showing the target as larger (*same size-bigger* condition), and in others they were given a choice between

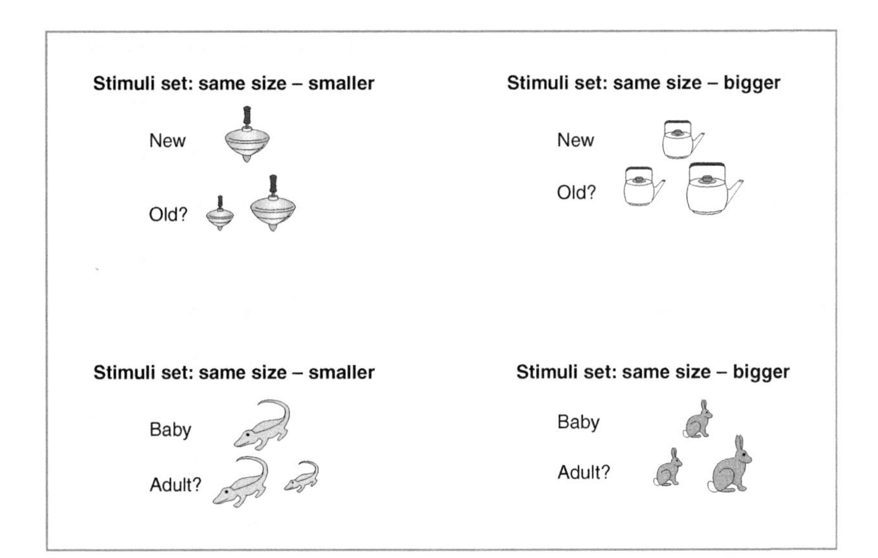

FIGURE 5.9
Examples of the artefact (top row) and natural kind (bottom row) stimuli used to study children's understanding of growth. From Rosengren et al. (1991). Reproduced with permission from Blackwell Publishing.

a picture showing the target the same size and a picture showing the target as smaller (*same size-smaller* condition). In the case of the artefacts, the same-size pictures were drawn to show the passing of time, with cracks and scuff marks. The animals depicted included alligators, bears and squirrels, and the artefacts included mugs, lightbulbs and televisions. If children understand that animals grow but that artefacts do not, then they should consistently select the picture of the artefact that is the same size in each type of pair, and they should never select the picture of the animal that is smaller.

Rosengren et al. found that the five-year-olds' performance was at ceiling level for the animals, being 100% correct for the same size-smaller comparison, and 97% correct for the same size-bigger comparison. The three-year-olds also performed at high levels in the animal task, at 78% correct for the same size-smaller comparison, and 89% correct for the same size-bigger comparison. Performance with the artefacts was also at ceiling for the older children. In contrast, the three-year-olds performed at 78% correct for the artefacts in the same size-smaller comparison, and were at chance in the same size-bigger comparison. Both age groups thus expected animals to change in size over time, and knew that they got larger and not smaller. However, the three-year-olds seemed uncertain as to whether artefacts grew over time, occasionally selecting the larger artefact rather than the aged and scuffed one in the same size-bigger condition.

KEY TERM

Principle of growth
Understanding that animals undergo changes over time that do not affect their identity and that they only grow bigger and not smaller over time.

Rosengren's work suggests that the principle of growth is understood first in the biological domain. Even young children expect animals to undergo changes over time that do not affect their identity, understanding that biological kinds only grow bigger and not smaller over time. Indeed, younger children themselves are getting bigger as time passes. Artefacts are less well understood. Although an emerging understanding of the fact that artefacts do not grow with the passing of time was clear in these studies, the younger children did not seem to have fully grasped the kind of changes that artefacts actually undergo. Nevertheless, by the age of five the children were drawing a principled distinction between animate and inanimate patterns of transformations.

Analogy as a mechanism for understanding biological principles

Convergent evidence for the idea that five- to six-year-old children have grasped that the principle of growth applies only to biological kinds comes from a study by Inagaki and Hatano (1987). They were interested in how often children base their predictions about biological phenomena on analogies to people. As human beings are the biological kinds best known to young children, and as we already know that analogical mappings can be made quite early in development (see Chapter 2), it seems plausible that children may use their biological knowledge about people to understand biological phenomena in other natural kinds. This has been termed the 'personification analogy'. In order to study personification analogies, Inagaki and Hatano asked five- and six-year-olds to make biological predictions about a person, a rabbit, a tulip and a stone. The growth question was "Suppose someone is given a baby X and wants to keep it forever the same size because it's so small and cute. Can he or she do that?" Inagaki and Hatano found that 89% of the children said that s/he couldn't do that for the person, 90% said that s/he couldn't do that for the rabbit and 81% said that s/he couldn't do that for the tulip. Eighty per cent of the children also said that he or she could keep the stone the same size. The understanding that growth is inevitable for biological kinds thus appears to be present in this age group. Inagaki and Hatano also found that the children had some idea about the biological mechanism underlying inevitable growth. They tended to make statements like "No, we cannot keep the baby the same size forever, because he takes food. If he eats, he will become bigger and bigger and be an adult."

In fact, analogies to people appear to provide an important source of preschoolers' understanding of a variety of biological phenomena. Inagaki and Sugiyama (1988) asked four-, five-, eight- and ten-year-olds a range of questions about various properties of eight target objects, including "Does X breathe?", "Does X have a heart?", "Does X feel pain if we prick it with a needle?" and "Can X think?" The target objects were people, rabbits, pigeons, fish, grasshoppers, trees, tulips and stones. Prior similarity judgements had established that the target objects differed in their similarity to people in this order, with rabbits being rated as most similar and stones being rated as least similar. The children all showed a decreasing tendency to attribute the physiological properties ("Does X breathe?") to the target objects as the perceived similarity to a person decreased. Apart from the four-year-olds, very few children attributed physiological attributes to stones, tulips and trees, and even four-year-olds only attributed physiological properties to stones 15% of the time. A similar pattern was found for the mental properties ("Can X think?"). This study supports the idea that preschoolers' understanding of biological phenomena arises from analogies based on their understanding of people. This pattern of responding can also be designated *anthropocentrism*, that children's understanding of biology emerges out of their understanding of people. Anthropocentrism is discussed further later in this chapter.

Evidence from studies of inheritance

Another biological principle is that living things transmit some of their properties to their offspring. Baby kangaroos have the properties of adult kangaroos, and baby goats have the properties of adult goats. Artefacts are different. They do not reproduce, and so they cannot transmit their properties. A coffee pot cannot transmit its shape or colour to a smaller coffee pot, as coffee pots are created by man. Young children appear to know certain facts about biological inheritance from quite an early age.

One important fact about inheritance is that 'genes will out'. If you are a baby kangaroo, you will grow up to be an adult kangaroo, even if you live with goats. S. A. Gelman and Wellman (1991) investigated young children's understanding of this essential fact by telling four-year-olds about baby animals that were raised among members of a different species. For example, the children were shown a picture of a baby kangaroo, which looked like a shapeless blob, and were told that it was taken to a goat farm as a baby and raised with goats. A picture of the goat farm was then shown to the children, and they were asked how the baby kangaroo behaved when she grew up. For example, was she good at hopping or good at climbing? Did she have a pouch? The children were almost all sure that the grown-up kangaroo was good at hopping, and had a pouch.

Another important fact about inheritance is that identity is maintained over transformations in appearance. For example, if a doctor bleaches the hair of a tiger and sews a mane onto its neck, it looks like a lion, but it is still really a tiger. If the same doctor paints the skin of a zebra to conceal its stripes, it is still a zebra. However, if a doctor saws off the handle and the spout of a coffe epot, seals the top and then attaches a bird's perch, a little window at the side and some birdseed, it not only looks like a bird feeder, it can function as a bird feeder (see Figure 5.10). Keil (1989) has shown that younger children behave as though such transformations in appearance change identity for both the natural kinds (e.g., tiger-lion) and the artefacts (coffee pot-bird feeder), whereas older children (seven- and nine-year-olds) only accept identity changes for the artefacts. Keil's explanation is that the older children are operating on the basis of a biological *theory*, in which natural kinds are identified by underlying essences and deep causal relations. In contrast, artefacts are identified by virtue of the functions that they serve, by children of all ages.

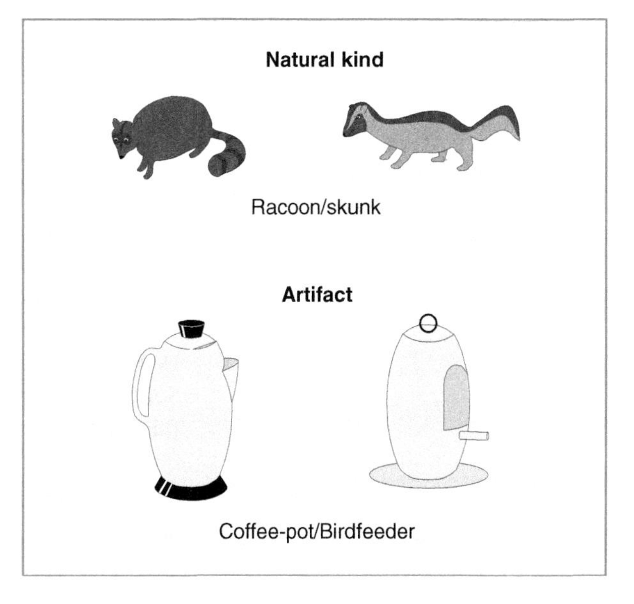

FIGURE 5.10
A racoon transformed to resemble a skunk, and a coffee pot transformed into the bird feeder. Two of the examples used by Keil (1989). © Massachusetts Institute of Technology, by permission of the MIT Press.

For example, when the younger children were asked "After the operation, was the animal a tiger or a lion?" for the bleached tiger with a mane, they would say things like:

Child: "I think he changed it into a real lion."
Exp.: "OK. Even though it started out as a tiger, you think now it's a lion?"
Child: "Um hmm."
Exp.: "Why do you say that?"
Child: "Because a tiger doesn't have long hair on his neck."

In a similar paradigm, older children would say things like:

Child: "It looks like a lion, but it's a tiger."
Exp.: "Why do you think it's a tiger and not a lion?"
Child: "Because it was made out of a tiger."

In contrast, a typical response to the coffee pot-bird feeder example at all ages was:

Child: "I think they made it into a bird feeder because it doesn't have a spout, and coffee pots need spouts, and it doesn't have a handle… and how are you supposed to hold onto it if it doesn't have a handle?"
Exp.: "Can it be a bird feeder even though it came from a coffee pot?"
Child: "Yes."

However, even the youngest children seemed to realise that transformations that appeared to change an object from a natural kind into an artefact or vice versa were impossible (e.g., porcupine to cactus). This fits with Keil's notion that biological knowledge is theory-driven, with natural kinds being identified on the basis of deeper structural characteristics like being alive and having offspring. For example, the same young child who had argued that a tiger could be changed into a lion denied that a porcupine could be changed into a cactus:

Child: "I think he's still really a porcupine."
Exp.: "And why do you say that?"
Child: "Because he started out like a porcupine."
Exp.: "Oh, OK. And even though he looks like a cactus plant, you think he's really a porcupine?"
Child: "Um hmm."
Exp.: "OK. Can you think of any other reasons why he's still a porcupine? Something you know about him?"
Child: [Shakes head.]

It is also possible however that the questioning technique used by Keil and his colleagues underestimated the knowledge of the younger children. Meyer, Gelman, Roberts and Leslie (2016) asked children aged four to seven years about heart transplants. The children were told stories in which a heart was transplanted from a pig or a monkey into a child, and were asked to make inferences about the transfer of pig and monkey attributes, such as rolling in the mud and eating bananas. They were also asked whether the child receiving the heart transplant would *become* a pig or a monkey. Meyer et al. reported that the children were willing to believe that attributes like rolling in the mud were transferred by the heart transplants, but not that category membership changed. The children did not agree that the child receiving a pig's heart became a pig. Thus in certain paradigms younger children, too, can operate on the basis of a biological theory. According to this theory, natural kinds are identified by underlying essences and deep causal relations.

Finally, whereas some bodily characteristics such as eye colour and gender are inherited and so cannot be changed or modified, others such as running speed and body weight can be modified by training or diet. Inagaki and Hatano (1993) investigated whether Japanese children aged four and five years were aware of such distinctions. For example, the children were told "A boy, Taro, has black eyes. He wants to make his eyes blue like a foreigner's [Caucasian]. Can he do that?", and "Taro is a slow runner. He wants to be a fast runner. Can he do that?"

Inagaki and Hatano found that the children were very good at distinguishing between whether Taro could change his eye colour if he wanted to (no), and whether Taro could become a faster runner if he wanted to (yes). For the modifiable characteristics, they gave explanations like "He can run fast if he practices more." Inagaki and Hatano concluded that even rather young children understand biological phenomena like inheritance. Perhaps the younger children studied by Keil demonstrated a poorer understanding of inheritance because of particular aspects of his paradigms. Alternatively, the young American children studied by Keil and his colleagues may have shown poorer understanding because they were imbued in a particular cultural depiction of inheritance that is common in storybooks for young children in the USA. Geerdts, Van De Walle and LoBue (2016) analysed American storybooks for children aged three to six years that were about animals and inheritance. The books were all on sale in the USA in 2012. Geerdts et al. found that *none* of the storybooks used the correct biological mechanism of inheritance to explain the transmission of physical properties between parents and their offspring. The animals were always presented anthropomorphically, that is, as surrogate people. Moreover, the storylines were typically about social influences and psychological properties. Anthropomorphism is discussed further at the end of this chapter.

Evidence from studies of natural cause

Finally, we can examine children's understanding of natural cause, which is not quite the same as inheritance, but taps the same underlying conception that features can be inborn. For example, a rabbit may not hop at birth, but the ability to hop is inborn. In contrast, the behaviour of artefacts is not a result of natural cause. A ball can bounce, but this is because someone made it that way – the cause is man-made.

S.A. Gelman and Kremer (1991) asked four- and seven-year old children about the behaviours of a variety of natural kinds and artefacts (e.g., rabbits hopping, birds flying, leaves changing colour, salt melting in water, balloons going up into the sky, cars going up hills, telephones ringing, guitars playing music, crayons drawing).

For example, the children might be shown a picture of a rabbit, and told "See this? It's a rabbit. It hops." They were then asked "Why does it hop?" This open-ended question was followed by two direct questions, such as "Did a person make it hop?" and "Is there anything inside it that made it hop?" For the balloon, the equivalent questions would be "Did a person make it go up into the sky?" and "Is there anything inside it that made it go up into the sky?"

Gelman and Kremer found that the children tended to overgeneralise the involvement of man-made causes to the less familiar natural kinds. For example, human influence was attributed to the dissolution of salt in water (42%), but not to the colour change of leaves (0%). However, the children were extremely accurate at identifying man-made causes in the case of the artefacts, knowing for example that guitars couldn't play music on their own. Although natural causes were ascribed to artefacts in some cases, the ascription of internal cause depended on the artefact. Internal causes were largely attributed to *self-sustained* properties of artefacts, such as a telephone ringing, a balloon going up into the sky or a car going up a hill. Internal causes were seldom attributed to properties that were not self-sustained, such as guitars playing music or crayons drawing.

Gelman and Kremer concluded that children as young as four realised that natural causes existed independently of human influences. The children applied *different causal mechanisms* to natural kinds and to artefacts, and realised the importance of internal causes, which were applied to all of the natural kinds: "The leaf just makes itself change colours", "Rabbits are made to hop", "Flowers open up theirselves", "It grew that way". Gelman and Kremer argued that children can develop a core understanding of natural cause for objects, their properties and their behaviours before knowing the precise origins of such natural causes. Also, children appreciate that causal mechanisms can be inferred rather than directly observed. As we will see, the idea that young children can go beyond information that can be directly observed and can grasp the significance of non-obvious properties for surface appearances is becoming increasingly important in explaining conceptual development. This idea is called *psychological essentialism*, the belief that a causal internal essence gives rise to shared attributes or behaviours of category members. Conceptual development seems governed by children's beliefs about the world, beliefs which (as we have seen) are based on their direct experience of things moving, growing, dying and having shared core properties.

THE REPRESENTATION OF CATEGORICAL KNOWLEDGE: A HISTORICAL PERSPECTIVE

Before we discuss psychological essentialism, however, we need to address the question of how conceptual knowledge is represented in memory. Despite the recent work in cognitive neuroscience discussed earlier, suggesting that conceptual knowledge is stored in a distributed form across a range of neural systems, the dominant view is still that adults organise their semantic memories on the basis of *categorical knowledge*. At one time, it was thought that young children did not share this categorical organisation of memory. Instead, it was thought that young children organised their conceptual knowledge in terms of *thematic* relationships.

The role of thematic relations in organising conceptual knowledge

The belief that young children organise semantic memory according to thematic associations arose from some experiments that suggested that younger children were more inclined to learn about thematic relationships than about categorical relationships. A thematic relationship is an associative one: dogs go with bones and bees go with honey. As the young child tends to experience instances of different categories along with associated instances of other categories, the notion that categories are first represented in terms of thematic relations seemed quite plausible.

KEY TERM

Thematic relations Associative relationships such as 'dogs go with bones' and 'bees go with honey'.

For example, a picture-sorting study carried out by Smiley and Brown (1979) found a preference for thematic over categorical relations in four- and six-year-old children. Only ten-year-olds appeared to prefer categorical relations. In Smiley and Brown's study, children were given a matching-to-sample task using triads of pictures. The triads used included *bee*, honey, butterfly; *dog*, bone, cat; and *bird*, nest, robin (see Table 5.3). The children were asked "Which one goes best with the [bee], the [honey] or the [butterfly]?" Children who chose *honey* were scored as preferring a thematic match, and children who chose *butterfly* were scored as preferring a category match. Smiley and Brown found that the conceptual preferences of the younger children were consistently for the thematic match.

However, the instructions used in this task were very open-ended. More recent work has shown that even one-year-olds are able to sort objects by category relations rather than by thematic relations when they are given more direct sorting instructions. For example, Bauer and Mandler (1989) used the matching-to-sample task with 16- and 20-month-olds in a paradigm similar to that used by Smiley and Brown, except that triples of real objects were presented rather than triples of pictures. For example, the toddlers were shown a toothbrush, and were asked to select the correct match from another toothbrush (category relation) and some toothpaste (thematic relation). Alternatively, they might be shown a hammer, and asked to make a choice from some pliers (category relation) and a nail (thematic relation). Bauer and Mandler changed the verbal instructions given to the children to "find the other one just like this one". They also checked that the thematic relations were familiar to their young subjects.

TABLE 5.3 Examples of the stimulus sets used by Smiley and Brown (1979)

Standard	Thematic	Taxonomic
Bee	Honey	Butterfly
Cow	Milk	Pig
Crown	King	Hat
Spider	Web	Grasshopper
Dog	Bone	Cat

Under these circumstances, a preference for thematic selections was shown on only 26% of trials by the 16-month-olds, and 15% of trials by the 20-month-olds. Although it could be argued that the children were matching on the basis of object identity, a follow-up study using slightly older children found similar results, even though this time the triads were at the superordinate level (*monkey*, banana, bear; *hammer*, nail, pliers; *bed*, pillow, cot; this study was discussed earlier in this chapter). Bauer and Mandler's work suggests that children organise semantic knowledge in the same way as adults do, in other words, on the basis of categorical relations. They only show a preference for thematic relations under the influence of certain task instructions.

However, the demonstration that very young children can organise conceptual knowledge according to *either* thematic or categorical relations is again consistent with distributed representations. If conceptual knowledge is represented in both the sensory systems first used to experience the concept and associated neural areas that represent the conjunction of this particular set of sensory information with other sets of sensory information, then knowledge about both categorical and thematic relations between concepts would be developing at the same time. Accordingly, children, like adults, can organise their conceptual knowledge according to categorical knowledge when required. The next question is how this categorical knowledge is stored in semantic memory. Classical work on adult concepts suggested that there were at least two sets of features that could be important. Concepts could be stored on the basis of their *defining* features, or on the basis of their *characteristic* features.

Representing categories in terms of characteristic versus defining features

KEY TERM

Characteristic-to-defining shift
The view that the basis of categorisation changes during development from being based primarily on well-known characteristic features to being based primarily on more sophisticated defining features.

A characteristic feature is a feature that is typically associated with a concept. For example, a characteristic feature of grandmothers is that they are old. A defining feature is a feature that applies to 100% of all the instances of a concept. A defining feature of a grandmother is that she is the mother of your parent. One possibility that has interested developmental psychologists is that children initially represent concepts in terms of *characteristic* features, which tend to be perceptually salient. As they learn more about the world, children pass through a period of conceptual reorganisation, developing conceptual representations that take account of *defining* features. According to this hypothesis, the basis of categorisation changes developmentally from being based on well-known characteristic features to being based on more sophisticated defining ones. This hypothesised reorganisation was called the 'characteristic-to-defining shift' (Keil, 1991). Earlier investigators have talked in similar general terms of a 'concrete to abstract' shift in conceptual development, a 'perceptual to conceptual' shift, and a 'holistic to analytic' shift (see Keil, 1987).

One way to examine the possibility that children's conceptual representations pass through a 'characteristic-to-defining' shift is to pit characteristic features against defining ones, and then to examine whether younger children prefer characteristic features and older children prefer defining ones. Keil and Batterman (1984) used this technique with five-, seven- and nine-year-old children. They told the children pairs of stories about familiar concepts like *uncle*, *robber* and *island*. The first of the stories in each pair had no information about the characteristic features of being an uncle, a robber or an island, but did include a defining feature. The second story in each pair had no information about the defining features of being an uncle, a robber or an island, but included a number of characteristic features. The children were then asked "Could X be an uncle/robber/island?"

Examples of the 'defining feature' stories include:

> *"Suppose your mommy has all sorts of brothers, some very old and some very, very young. One of your mommy's brothers is so young that he's only two years old. Could that be an uncle?"*
>
> *"This very friendly and cheerful woman came up to you and gave you a hug, but then she disconnected your toilet bowl and took it away without permission and never returned it. Could she be a robber?"*

Examples of the 'characteristic feature' stories include:

> *"This man your daddy's age loves you and your parents and loves to visit and bring presents, but he's not related to your parents at all. He's not your mommy or daddy's brother or sister or anything like that. Could that be an uncle?"*
>
> *"This smelly, mean old man with a gun in his pocket came to your house one day and took your coloured television set because your parents didn't want it anymore and told him he could have it. Could that be a robber?"*

Keil and Batterman reported that the five-year-olds relied on characteristic features in making their judgements, whereas the nine-year-olds relied on defining features. Although the children did not shift at the same time for all concepts, the younger children usually said that the 'characteristic feature' stories were instances of the concept, while the older children chose the 'defining feature' stories. Keil and Batterman concluded that the children seemed to represent the concepts in different ways at different ages. Of course, this could reflect increasing knowledge. As children learn more about defining features, these could replace characteristic features as the basis for representation.

On the other hand, cognitive neuroscience studies enable a different interpretation. If concepts and categories have distributed representations, including activation of the sensory modalities that were active when the concepts were being directly experienced, then any concept will be represented by activation in multiple modalities at the same time. These distributed representations would mean that even apparently trivial aspects of experimental situations would be crucial for guiding which particular conjunctions were reactivated in experimental scenarios and therefore deemed relevant. For example, as real-world experience increases, children's 'robber' concept might still incorporate aspects like 'looking mean' and 'having a gun', but the fact that a parent had told this person that they could take the television would activate conjunctions associated with charity and giving away old things. The evidence for multi-modal representational systems is very strong, and is a developmentally appealing one, as discussed earlier. Lakoff (1986) argued a long time ago that the properties that are relevant for the characterisation of human

KEY TERMS

Distributed representations Concepts and categories are represented in the brain in terms of activity in multiple modalities at once and therefore are highly distributed across networks of neurons.

Essentialism Children's tendency to search for hidden, non-obvious features that make category members similar.

categories do not exist objectively in any case. Instead, what we *understand* as properties depends on our interactive functioning with our environment (Barsalou, 2017, made a similar point with his 'bicycle' example). Thus our theories about the world are important for our decisions about what is categorically similar. This notion has been incorporated into developmental psychology by reference to the importance of 'essences' for children's conceptual understanding.

CATEGORIES AND BELIEFS ABOUT THE WORLD: 'ESSENCES' AND NAÏVE THEORIES

Some researchers in adult cognitive psychology have argued that category membership is defined not only in terms of characteristic and defining features, similarity to prototypes, etc., but in terms of 'essences'. One of the major proponents of this view is Medin (1989). His view can be summarised by the following quote: "People act as if things (e.g., objects) have essences or underlying natures that make them the thing that they are" (p. 1476). In other words, people have implicit assumptions about the structure of the world, and about the underlying nature of categories, and these beliefs are represented in the categories that they develop. This view is sometimes called 'psychological essentialism'. According to this view, categories are not discovered via the passive observation of correlations between features. Rather, they are created by 'carving nature up at its joints'.

We can illustrate how category membership can go beyond clusters of characteristic features by returning to our example of birds. It is true that feathers reliably co-occur with wings, with flight and with light bodyweight, and that these co-occurrences help to distinguish the category 'bird'. But adults also have a 'theory' about why these features go together. This theory involves the causal relations necessary to enable flight. Adults believe that low bodyweight, feathers and wings facilitate flight, thereby imposing a degree of *causal necessity* on the covariation of these features in birds. This tendency to create causal explanatory constructs may not be limited to adults. Children, too, may create intuitive theories to understand conceptual structure. These 'theories' would correspond to core sets of interconnected beliefs about category membership.

The essentialist bias

Such sets of causal beliefs about the co-occurrence of core properties apply to a great many concepts (although not all). They apply particularly to categories of natural kinds, such as animals, birds and plants. A number of developmental psychologists have suggested that children's growing understanding of the category of natural kinds is partly governed by their implicit appreciation of the causal/explanatory relations that explain featural clusterings within this category (e.g., Carey & Spelke, 1994; Gelman et al., 1994; Keil, 1994). For example, S. A. Gelman et al. suggested that young children have an essentialist bias, and that this bias constrains the ways in which they reason about natural kinds. Young children's early understanding of living things is theory-like, leading them to search for invisible causal mechanisms to explain object actions (Gelman, 2004). Indeed, Gelman and her colleagues have described psychological essentialism as a 'folk theory'.

Gelman (2004) argued that psychological essentialism may be an early cognitive bias. She suggested that young children have an early tendency to search for hidden, non-obvious features that make category members similar. Some of the evidence reviewed in this chapter is consistent with Gelman's idea that even young children go beyond observable features when developing biological concepts. For example, young children appear to assume that living things maintain their identity over superficial transformations and transmit some of their properties to their offspring. Evidence discussed in previous chapters also supports her idea, as we have seen that children have an inherent tendency to search for causal explanations of phenomena in their everyday worlds, and that they show an early ability to go beyond surface features to focus on structural characteristics (e.g., when reasoning by analogy, see Chapter 2). Children also prioritise causality in deciding which properties members of a category should share. One appealing aspect of essentialism is that it does not propose dichotomous development, for example from perceptual to conceptual categories, or from concrete to abstract categories. Rather it assumes that categories have two distinct, interrelated levels, the level of observable reality and the level of explanation and cause. The level of explanation and cause are essentially placeholders for as-yet unknown properties of concepts.

However, a strong version of the 'psychological essentialism' theory argues that the theories and core principles that guide essentialism are *innate*. This innate knowledge is thought to guide cognitive development by setting important limits (or 'constraints') on the information that can and cannot be learned (e.g., Carey & Gelman, 1991). The origins of this innate knowledge are not well-specified. However, we have also considered the view that what appear to be 'innate constraints' may be a byproduct of automatic neural mechanisms for encoding and processing information. New studies in robotics are showing that machine learning systems can extract hidden or latent variables automatically during instance-based learning and can make inductive inferences about category membership successfully on the basis of these 'essences' (e.g., Oved, Cheung & Barner, 2014). Another intriguing theoretical proposal was that in addition to conceiving of causality as either mechanistic or intentional, young children conceive of it as 'vitalistic' (Inagaki & Hatano, 2004). Children assume a 'vitalistic causality', namely a vital life force taken from food and water which makes humans and other animals active. This vital power explains target biological phenomena such as growth and health. As argued previously, as we understand more about the ways in which perceptual structure in itself gives rise to conceptual assumptions, it may turn out that the neural processing algorithms by which our perceptual systems encode information in themselves provide the source of these core principles and causal explanations. At the beginning of this chapter, we discussed Hochstein et al.'s (2015) claim from visual neuroscience that because of population coding by the brain, conscious perception *begins with* summary statistics. Coupled with children's inherent tendency towards explanation-based learning, discussed further in Chapter 7, the perceptual structure of the world thus may *in itself* be enough to support the patterns of conceptual development that have been documented in this chapter. Mechanisms of neural encoding and processing may also support the patterns of inductive inference currently described by 'cognitive biases' such as essentialism, vitalism and constraints on learning. The role of explanation in concepts and intuitive theories has also been discussed at length by Keil (2006), who foregrounded the importance of causal explanations in cognitive development (see Chapter 7).

CONCEPTUAL CHANGE IN CHILDHOOD

Although we have reviewed a large amount of evidence indicating that young children have rich conceptual structures that they have abstracted from their everyday experience of the world, this does not mean that they never experience conceptual change. The level of knowledge that can be abstracted from perceptual causal information about different entities has in many cases been transcended by modern physics and biology. By and large, such information must be taught. It seems likely that conceptual change in such instances (for example, from medieval theories of motion to Newtonian theories, see Kaiser, Proffitt & McCloskey, 1985) depends on direct and focused tuition. However, it has been argued that children do experience spontaneous conceptual change, without direct tuition, regarding the biological world. When such conceptual change occurs, then new principles are said to emerge that 'carve the world at different joints' (see Carey & Spelke, 1994, for a fuller discussion). For example, at some point children may need to distinguish plants as biological entities that are essentially similar to animals, even though plants differ markedly from animals in terms of their capacity for self-generated movement. It seems likely that sufficient real-world experience is enough to enable children to draw this new distinction.

In important work, Carey argued that conceptual change in childhood (and in science) depends on children and scientists making *mappings* between different domains. Such mappings entail relating objects in one system (e.g., people) to objects in another (e.g., plants). If such a mapping is created, then the principles that govern children's understanding of people can be applied to their understanding of plants. We have already seen that children use analogical mappings from people to decide whether animals and plants can be kept small and cute forever or whether they would feel pain if pricked by a needle (see the work of Inagaki and her colleagues discussed earlier in this chapter). Furthermore, there is a growing body of work which demonstrates that analogical mappings are used by children as young as three in other areas of cognition such as causal reasoning (Goswami & Brown, 1989), physical reasoning (Pauen & Wilkening, 1997) and reasoning about natural kinds and artefacts (Goswami & Pauen, 2005). The availability of the mapping mechanism proposed by Carey is thus well-documented.

Carey (1985) herself made a strong case for her view about the importance of conceptual change in the domain of biology (although see Kuhn, 1989). Carey argued that preschool children's understanding of biological phenomena differed radically from that of older children. In her book (Carey, 1985), she documented numerous studies of inductive inference which showed that younger children based their understanding of animals on their understanding of people. The children appeared to project behavioural and psychological properties onto other animals according to how similar these animals were to human beings. For example, the attribution of the property 'breathes' was made to humans by 100% of the four-year-olds studied by Carey, to aardvarks by 78% of the children, to dodos by 67% and to stinkbugs by 33%. The property 'breathes' was never attributed to plants. Only older children showed a coalescence of the concepts *animal* and *plant* into the new concept, *living thing*. Carey thus argued that children's understanding of biology emerged *out of* their understanding of people. This tendency to attribute physiological and mental properties to other objects on the basis of their similarity to people was consistent with the 'personification analogy' perspective (Inagaki & Hatano, 1987; Inagaki & Sugiyama, 1988), and has also been termed *anthropocentrism*.

More recently, however, it has been argued that anthropocentrism, the notion that children's understanding of biology emerges out of their understanding of people, is actually a product of cultural learning. Waxman and Medin (2007) reported that while young urban-living children accorded people a special status when making inductive inferences about biological properties, replicating Carey (1985), young children growing up in rural environments did not show anthropocentrism. Younger rural children (four- to five-year-olds), who had greater direct experience of animals and nature, were equally likely to use dogs and bees as a basis for biological inductive inferences as they were people. Hence rural children did not rely on personification analogies for biological reasoning. Regarding the urban children, Herrmann, Waxman and Medin (2010) pointed out that much media for young children (such as storybooks, cartoons and TV) utilise anthropocentrism. Many books represent non-human animals as surrogate humans, for example wearing clothes, speaking the child's language, having emotions and engaging in human activities like birthday parties. Herrmann et al. (2010) suggested that, particularly for young urban children with limited exposure to the natural world, such media may unintentionally support the development of a personification analogy for reasoning about the biological world.

To test this idea, Herrmann et al. (2010) devised a version of Carey's inductive inference task suitable for three-year-olds who were growing up in a large American city (Chicago). In this task, the children were introduced to two puppets, one of whom was 'silly' and one of whom was not. While the silly puppet would say things like "No! That is not a chair" when shown a picture of a chair, the other puppet would say "Yes! That is a chair." Once the children were comfortable with the task format, they were introduced to a novel biological property, 'andro', which was inside both dogs and humans. For example, they were told "Dogs (or humans) have andro inside them. Andro is roundish, greenish and it goes inside." The children would then colour in pictures of dogs or humans, making their insides green. In the test phase of the study, the children were shown new pictures, and were asked whether the depicted entities had 'andro' inside. For example they were shown pictures of a robin, a bee, a fish, a tree and a wristwatch. In each case, one puppet would say that the depicted entity had 'andro' inside, and the other puppet would say that it did not. The children had to decide which puppet was correct.

Herrmann et al. (2010) found an interesting divergence in the response patterns of the three-year-olds that they tested and a comparison group of five-year olds. While the older children showed anthropocentric reasoning patterns, replicating Carey's (1985) findings, the three-year-olds did not. The five-year-olds were significantly more likely to project 'andro' to other biological kinds like birds and bees if the human had 'andro' inside than if the dog had 'andro' inside. The three-year-olds made as many inductions on the basis of dogs as on the basis of humans. Both age groups were less likely to project having 'andro' inside to the tree and to the wristwatch. Herrmann et al. argued that anthropocentrism is not the young child's initial entry point for reasoning about the biological world. Rather, it is learned from cultural practices.

Indeed, converging evidence for the view that anthropocentrism is culturally learned has come from Waxman and her colleagues, who have shown that the inductions made by five-year-olds concerning who has 'andro' inside can be rapidly changed by reading fact-based picture books. Waxman, Herrmann, Woodring and Medin (2014) used the silly puppet task with five-year-olds who had either been read an animal encyclopaedia book about bears or a popular cartoon-based book in which the bears were depicted as surrogate humans. Waxman et al. (2014) found that while those children reading the cartoon book adopted an anthropocentric stance towards induction, those reading the animal encyclopaedia adopted a more biological perspective. For example, they were more likely to project 'having andro' from a dog to a novel bear (94% projections, compared to 48% projections for children who read the cartoon book). Waxman and her colleagues argued that anthropocentric reasoning in urban children may emerge in direct response to the images portrayed in children's books and other child-focused media. Accordingly, conceptual change does not arise, as proposed by Carey (1985), because preschool children's reliance on anthropocentrism is replaced by older children's use of biology. Rather, as proposed by S. A. Gelman, younger children appear to be essentialists, and anthropocentrism appears to develop in direct response to experience with cultural artefacts like books, videos and other child media.

Accordingly, the most representative view of the mechanisms of conceptual change comes from the notion of foundational domains, summarised by Wellman and Gelman (1998), and adopted in this book. Wellman and Gelman pointed out that young children are probably developing several alternative conceptual frameworks *at the same time*. Rather than developing a monolithic understanding of the world, infants and young children are probably developing distinct yet interlinked conceptual frameworks to describe the 'foundational domains' of biology, psychology and physics. Many concepts will of course be represented in *more than one* of these foundational frameworks. For example, persons are psychological entities, biological entities *and* physical entities. These foundational domains will then engender, shape and constrain other conceptual understandings. At the same time, children will have access to at least two levels of description within any framework, one that captures surface phenomena (mappings based on attributes) and another that penetrates to deeper levels (mappings based on relations like 'essences'). The need to compare, share, merge and create new conceptions is likely to be encouraged by the assumptions of surrounding children and adults, by the technology of the culture and by systematic teaching received in school. These mechanisms of sharing and merging conceptual understandings across foundational domains do not accord with the kind of conceptual change envisaged by Carey and Spelke (1994). Although new conceptions and understandings will be created, these may not 'carve the world at different joints'. The issue of whether children's cognitive development is *ever* characterised by conceptual change is discussed further in Chapters 9 and 10.

SUMMARY

Methods from cognitive neuroscience seem particularly likely to impact the field of conceptual development, a domain in which it is currently barely used beyond ERP studies of the Nc. Many of the historically important debates about conceptual development, such as the potential developmental primacy of the basic level of categorisation, whether there is a meaningful distinction between perceptual versus conceptual categories, and the role of 'constraints' on learning (the core principles that govern induction), may turn out to be natural products of the way that our neural architecture encodes, stores and reactivates knowledge about the world of natural kinds and artefacts. As noted in Chapter 1, the everyday world of the infant may be given causal structure simply by the mechanisms of sensory information processing inherent to the human brain. Previous chapters have documented evidence that these neural mechanisms rapidly encode a great deal of structured information, structure that creates 'expectations' or 'constraints' regarding what is learned next. Here we discussed data collected by visual neuroscience researchers which shows that visual perceptual encoding mechanisms adapt automatically to changes in the statistical patterns presented by environments over very short timescales, even though the conscious observer is not aware of this. It is these summary statistics (population coding) that appear to be the basis for conscious perception. New techniques in cognitive neuroscience thus offer the potential to document conceptual learning in much greater detail. For example, techniques such as multi-voxel pattern analysis (MVPA, widely used for mapping semantic networks in the adult brain) offer considerable precision regarding which brain voxels are active when processing particular concepts, and could be used longitudinally to track changes in neural activation patterns with accrued conceptual learning. MVPA could even be used to index possible conceptual change, thereby testing Carey and Spelke's idea about carving nature at different joints. Although connectionist modelling has suggested similar conclusions to those promised by MVPA, for example that concepts have distributed representations (e.g., Quinn, 2002), connectionist models have only worked on the features that modellers *assume* are used by the infant brain when attending to aspects of the stimulus. Recent developments in machine learning may be able to circumvent this constraint, however, and relevant research is discussed in Chapter 11. A deeper understanding of how the brain processes perceptual information may show that primary features additional to or completely different from those built into existing connectionist simulations are important for encoding, and thereby also important for early conceptual representations. Advances in understanding embodiment appear particularly promising from a developmental perspective.

Nevertheless, the sheer variety of studies discussed here on different aspects of conceptual development demonstrate the complexity of the knowledge that is built up rapidly by the infant and by the young child. Even infants make some assumptions about shared core properties such as 'insides' and 'outsides', as well as about other features such as the gait typical of natural kinds. Even relatively simple perceptual mechanisms, such as the computation of whether a movement originates from a biological or a non-biological entity, yield rich information relevant to distinguishing biological kinds and the latent variables or essences that underlie their behaviour, for example their likely animacy and agency (see also Chapters 3 and 4). The study discussed here by Arterberry and Bornstein (2001) showed that babies of three months can make distinctions between the motion of vehicles versus the motion of animals on the basis of very impoverished cues. As we saw in Chapter 4, cognitive neuroscience studies with adults show that mirror neurons are tuned to respond only to biological motion (Tai et al., 2004), and infants are already showing differential neural activation to biological motions such as a hand opening and closing, eyes opening and closing, or a mouth opening and closing at five months of age (Lloyd-Fox et al., 2011). As more discoveries of this nature are reported, the many important observations about conceptual development yielded by behavioural work may fit into a clearer 'bigger picture' concerning how young children develop conceptual knowledge about different aspects of their experience (natural kinds, artefacts, actions, events), and why effects such as characteristic versus defining features emerge as part of this knowledge.

CHAPTER 6

CONTENTS

Language acquisition

6

So far in this book, we have considered cognitive development largely independently of language. This is not accidental. Language acquisition has traditionally been studied separately from cognition, and as we have seen in previous chapters, there is strong commitment to the idea that basic concepts are pre-verbal. Another reason for the traditional distinction between language and thought was that language acquisition seemed such a remarkable feat for the infant brain that it was assumed that special capacities must be at work. These capacities were thought to be distinct from the capacities underpinning broader cognitive development. For example, it was postulated that infants were born equipped with a 'language acquisition device' or LAD, which had the special job of acquiring the spoken language of whichever culture the infant entered. Chomsky, the original proponent of the LAD, argued that infants are born with innate knowledge of the general rules that all languages obey, along with innate knowledge of permitted variations (e.g., Chomsky, 1957). Hence an infant can as readily acquire a language that makes heavy use of the passive tense ("The boy was bitten by the dog"), like Sesotho, as a language that does not, like English ("The dog bit the boy"; Bates, Devescovi & Wulfeck, 2001).

More recently, it has become clear that language acquisition depends on the same kinds of learning mechanisms that underpin broader cognition. We have already seen in Chapters 1 to 5 that infants acquire a remarkable amount of information simply by looking at and listening to events that occur in their worlds. In the case of the physical, biological and psychological worlds, the primary sense is probably vision. In the case of language, the primary sense is audition. Infants acquire a remarkable amount of information simply by listening to what the people around them say. I will argue that infants use the same abilities to acquire language that they use to acquire knowledge about the physical, biological and psychological worlds, for example statistical and associative learning. The infant brain is automatically tracking statistical dependencies and conditional probabilities in language. Auditory perceptual information is replete with statistical patterns. There are acoustic cues to the phonotactic patterns of the language (the sounds that make up the language, and the orders in which they can be combined), to word boundaries and phrasing (largely carried by speech rhythm and duration cues), and to the emotional content of speech (largely carried by prosodic stress patterns and loudness cues). As in the visual world of objects and events, information that is initially gained passively via automatic neural encoding processes is rapidly supplemented by information gained through direct action by the infant, for example by imitation. In the case of language, infants start babbling and trying out sound combinations for themselves from very early on, and they also initiate verbal interactions by gooing and making comfort or distress sounds – in other words, they attempt to *communicate*.

KEY TERMS

Language acquisition device (LAD)
An innate neural mechanism, first suggested by Chomsky, with the special job of acquiring the language the infant encounters.

Phonotactic patterns of language
Speech sounds that make up language, and the order in which these speech sounds can be combined to make lawful words.

"Motherese" or infant-directed speech (IDS) is an exaggerated prosodic register that emphasises word and phrase boundaries, often with target words receiving primary stress at the end of the utterance: "what a pretty DAISY!"

Social interaction is fundamental to natural language learning, and interpersonal synchrony – the temporal co-ordination between the communicative acts of the infant and their interlocutors – probably plays as important a role in learning language as learning the acoustic statistical patterns described above. Direct gaze may be particularly important. For example, mothers will unconsciously stop speaking in infant-directed speech if their infant is looking at them but can no longer hear them, even though the mother believes that their infant is hearing them clearly (Lam & Kitamura, 2010). The importance of making shared meaning probably explains why no one has yet managed to build a computational system that can learn language and successfully use it interactively (Kuhl, 2004). Auditory statistical information is augmented by other kinds of perceptual information that convey meaning and reference. Parents usually talk to babies about things that are happening right now, in the direct visual field, and the situational context often makes the meaning of their vocalisations highly predictable. The facial expressions of interlocutors, their gestures and touch, the things that they are attending to as they speak as well as the current situation that everyone is in (e.g., a mealtime) all convey cues concerning *reference*, or what is being spoken about. The learning task for the baby is further facilitated by two factors. One is the use by caretakers of a special register to speak to babies, called 'Motherese', 'Parentese' or infant-directed speech (IDS). IDS is an exaggerated prosodic register that emphasises word and phrase boundaries, and appears to make the segmentation of the speech stream easier for the infant. The second is the apparently inborn propensity for attachment, social interaction and communication described in Chapter 3. Infants learn language because of social interactions with partners, and not simply because of exposure to sequences of sounds. The same factors that underpin the emergence of psychological understanding, for example the capacity for joint visual attention and benefitting from natural pedagogy, the recognition of communicative intent and the desire to imitate, thus also underpin the acquisition of human language.

KEY TERMS

Motherese/ Parentese/infant-directed speech (IDS) Special register used by caretakers when speaking to infants; it has an exaggerated prosodic profile that emphasises words and phrase boundaries.

Phonemes A short-hand term for the individual sound elements that make up words in languages. Phonemes are an abstraction from the physical signal.

PHONOLOGICAL DEVELOPMENT

Although you may be a fluent speaker of English, and may be able to get by in a couple of other languages, you are unlikely to speak all of the 6,000-plus languages of the world with fluency. In fact, if someone speaks to you in Chinese, the context might give you a general idea of what they are talking about, but you would probably have no way of understanding the sounds that they are making as individual words. For example, if a Chinese person holding a map, carrying a camera and close to the Tower of London stops you with a questioning air, they are probably asking you directions. You can infer this from the situational context and perhaps from global cues such as whether their utterances end by rising in a questioning fashion. However, you probably have no idea of the individual words in the utterance, because you have no idea of the patterns of sound combinations that constitute words in Chinese. This type of learning is phonological learning. It has two aspects. One is to learn the sounds and combinations of sounds that are permissible in a particular language, so that the brain can develop 'phonological representations' of the sound structure of individual words. The second is to learn to produce these words yourself. Both types of learning undergo protracted development. Although children are usually fluent comprehenders and producers of spoken language by the age of five years, new challenges (such as learning to read) require further development of phonological representations (see Chapter 10).

Categorical perception

Early experimental research on phonological development focused on when infants learn the *phonemes* that make up their particular language. *Phonemes* is the short-hand term used for the individual elements that make up words in languages. For example, words like *bat* and *bit* differ by one phoneme, the middle phoneme. *Bat* and *pat* differ by one phoneme, the initial phoneme, and *bat* and *back* differ by one phoneme, the final phoneme. The notion of a phoneme is an abstraction from the physical stimulus (indeed, it can be seen as a 'cerebral construction', see Mersad & Dehaene-Lambertz, 2016), which is why it is a short-hand term (for example, the 'a' phoneme in *bat* and *back* is not exactly the same sound). Nevertheless, 'phoneme' is a useful term for explanatory purposes. Languages are based on two types of phonemes, consonant phonemes and vowel phonemes. These are selected from a repertoire of around 600 consonants and 200 vowels that are distinctive to the human brain. In practice, most languages use a very small set of all the possible phonemes, for example English uses about 40. An important job for the infant learner, therefore, is to learn the phonemes or speech sounds of their native language. This has been a fertile area of research. It is now apparent that infants come to learn the phonemes of their native language very quickly indeed, within the first year of development. At the same time, they lose their ability to distinguish the phonemes of other languages.

Phoneme perception is *categorical* in adults. In terms of the actual physical sound, there are many similar but non-identical sounds that we would recognise as the phoneme /b/, and many other similar but non-identical sounds that we would recognise as the phoneme /p/. However, there is a measurable point at which sounds that are highly similar physically stop being perceived as /b/, and begin being perceived as /p/. This is called categorical perception. The brain is exposed to a physical continuum of sound, the vocal cords are vibrating to produce the sound, and the airflow of the sound is obstructed at the lips to produce a 'plosive' (for plosives like /b/ and /p/, this obstruction is complete). However, the brain imposes a category of /b/ sounds, and a category of /p/ sounds onto this continuum. At some degree of voicing (i.e., degree of vibration of the vocal cords, which will vary in physical terms with the age, gender and communicational intent of the speaker), the adult brain decides that it is no longer hearing 'bat', but is hearing 'pat'. How quickly does the infant brain reach a similar conclusion?

In a classic study, Eimas and his colleagues investigated the categorical perception of phonemes in infants aged one and four months (Eimas, Siqueland, Jusczyk, & Vigorito, 1971). The infants began the experiment by sucking a dummy to a background sound, for example the syllable /ba/ being repeated over and over again. The rate of sucking gradually declined. The experimental question was whether the rate of sucking would increase to a new syllable /pa/. Six stimuli were used over the experiment, each varying in voice onset time by 20 ms. The voice onset time values were −20, 0, 20, 40, 60 and 80 ms. For an adult, the +20 ms stimulus was heard as /ba/, and the +40 ms stimulus as /pa/. Hence for an adult, the change in voice onset time (VOT) of 20 ms from +20 ms to +40 ms signalled a novel perceptual event: the category boundary between /b/ and /p/ had been crossed. Eimas et al. (1971) reported that both groups of infants showed significant dishabituation in suck rate to the change from the +20 ms stimulus to the +40 ms stimulus. In contrast, changes of the same absolute magnitude (−20 ms to 0 ms VOT, and 60 ms to 80 ms VOT) did not lead to dishabituation in suck rate for either age group. Clearly, the month-old infants had already developed categorical perception for these sounds.

In another classic study, Werker and Tees (1984) showed that young infants had categorical perception for phonemes in other languages as well. Two contrasts were compared, the English contrast /ba/ and /da/, and the American Indian contrast /ki/ and /qi/. Infants were rewarded for turning their heads when the sound changed from /ba/ to /da/, or from /ki/ to /qi/. Adult English listeners had to press a button when they heard the sound change. Werker and Tees found that English babies aged 6–8 months and American Indian adults could perceive both contrasts. However, most English adults tested could not perceive the distinction between /ki/ and /qi/. Older English babies were then tested, aged 8–10 months and 10–12 months. At 8–10 months, 57% of the infants could discriminate the non-native /ki/-/qi/ contrast (eight infants out of 14). By 10–12 months, only one baby out of ten showed categorical perception of this contrast. Comparable patterns were found for Hindi contrasts (see Figure 6.1). Werker and Tees (1984) were also able to test a couple of American Indian infants aged 10–12 months, who had no difficulty in distinguishing the /ki/-/qi/ contrast. A longitudinal study following the same babies from six until 12 months of age confirmed these cross-sectional patterns. Werker and Tees concluded that the ability to discriminate non-native phonetic contrasts declines during the first year of life.

Further work has established that the physical changes where languages place phonetic boundaries are not random (Kuhl, 1986). General auditory perceptual abilities seem to influence where these 'basic cuts' are made, and in fact other mammals such as chinchillas seem to partition sounds in the same ways. This probably explains why infants are sensitive to the acoustic boundaries that separate phonetic categories in all human languages from birth. The choice of sounds that comprise the phonetic repertoire of the languages of the world capitalise on natural auditory discontinuities (Kuhl, 2004). The important point is that these basic cuts are rather rough, so further learning is required. During development, infants

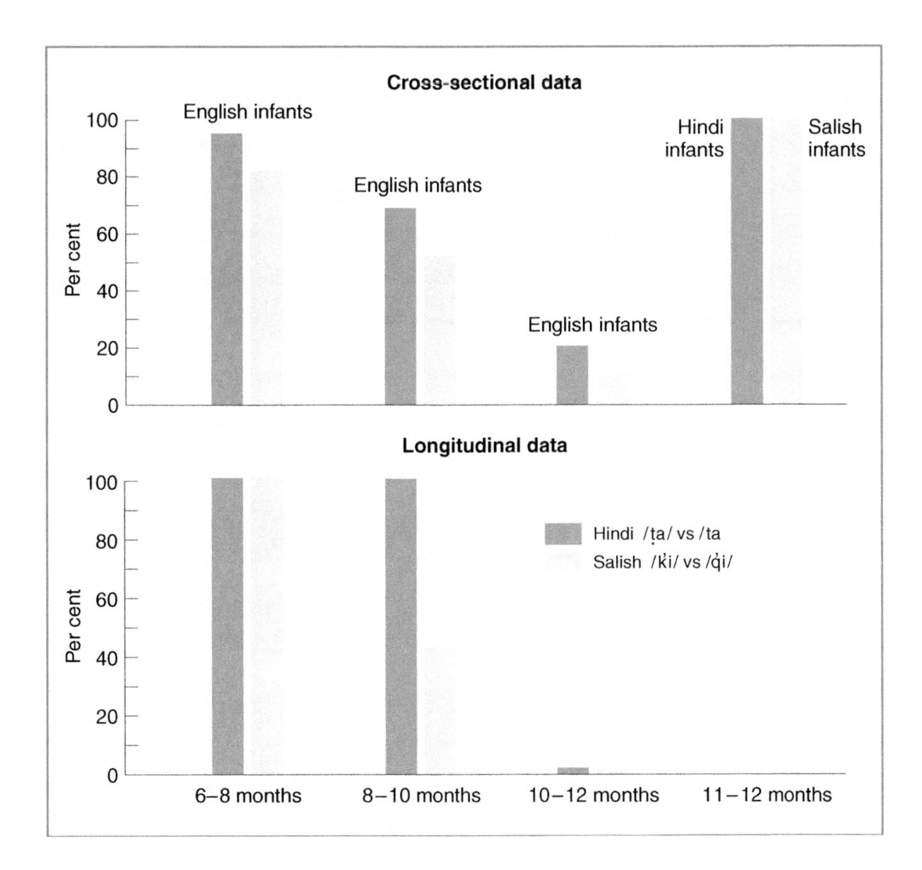

FIGURE 6.1 Proportion of infant subjects from three ages and various backgrounds reaching criterion on Hindi and Salish contrasts. From Werker and Tees (1984). Copyright © 1984 Elsevier. Reprinted with permission.

need to learn the locations of the phonetic boundaries that are important for *their* language. As they continually hear the sounds of, say, English rather than Hindi, or English rather than Chinese, their brains specialise in the sounds of English. Language-specific patterns of listening develop, so that infants become highly adept at discriminating phonemes in their native language and lose the ability to discriminate phonemes in other languages. They develop 'prototypes' of the phoneme categories special to their language (Kuhl, 1991). This specialisation is well under way by one year of age, leading to the idea that there is a sensitive period for language acquisition.

KEY TERM

Perceptual magnet effect Prototypical sounds in a language that come to act as a 'magnet' for perceptually similar sounds, so that similar sounds are perceived as belonging in the same category.

Cognitive neuroimaging studies of categorical perception

Our understanding of speech processing by adults has been transformed by neuroimaging, and our understanding of linguistic development looks set to follow. Studies using fMRI with adults have shown that, as in the visual system, there are two pathways for processing incoming signals in the auditory system. An anterior pathway, analogous to the 'what' pathway for vision, is interested in information about vocalisations and acoustic-phonetic cues ('what are they?'). A posterior pathway, analogous to the 'where' pathway for vision, is more interested in information about sound localisation and how sounds are made, hence articulatory

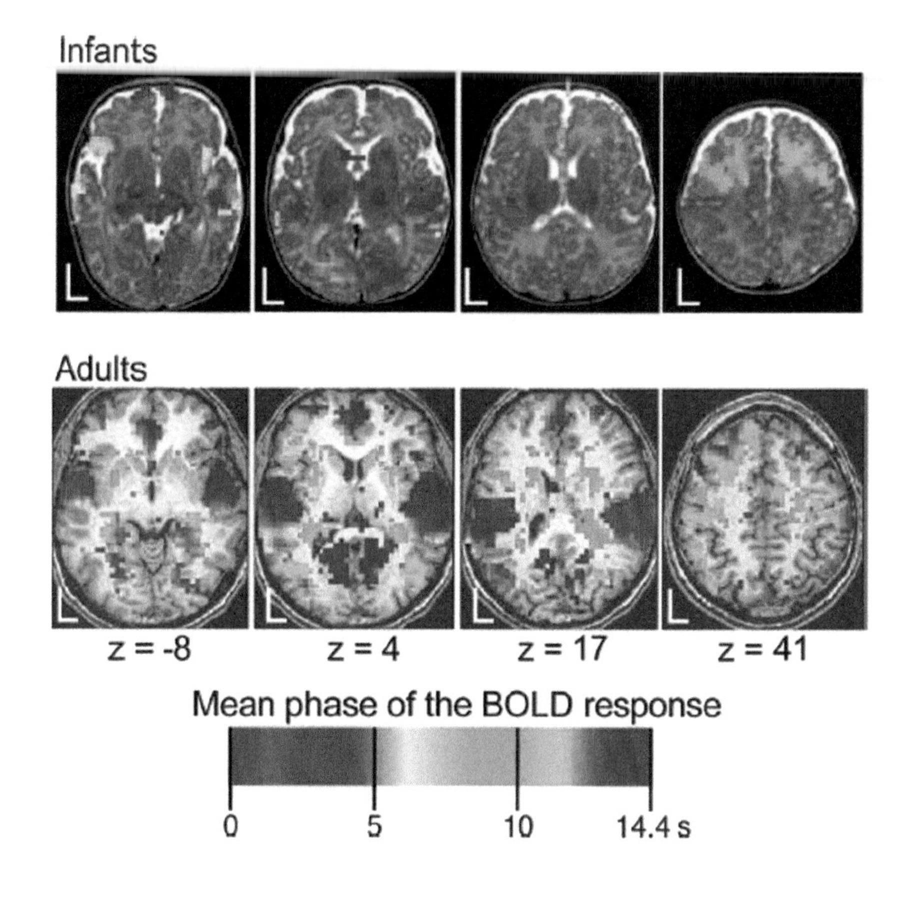

fMRI activation in response to hearing a sentence in infants and adults. Comparison of the cerebral responses shows a remarkable degree of similarity in location and temporal gradient (phase of the BOLD response) in the two groups.

information (Scott & Johnsrude, 2003). Speech perception has been shown to depend on *multiple representations* of the input, for example both acoustic-phonetic representations and motor/articulatory representations. Speech is processed both as a sound and as an action. These multiple and complementary representations appear to help to explain why speech perception is robust across different speakers, noise contexts and variability in accent. Speech processing is also one of the few areas in which cognitive neuroimaging has shown the *mechanisms* used by the brain to encode the signal. An important mechanism is neuronal oscillatory 'entrainment' to the amplitude modulation patterns in speech, discussed further later in this chapter.

Studies using fMRI have shown that speech processing by infants appears to utilise the same areas of the brain that are used during speech processing by adults. In seminal studies, Dehaene-Lambertz and her colleagues used fMRI with very young infants to study the functional organisation of speech processing in the infant brain. For example, Dehaene-Lambertz, Dehaene and Hertz-Pannier (2004) played awake and sleeping infants aged three months a recording of a female voice reading a story in a vivid theatrical manner. The infants heard 20-second excerpts of the story, played either forwards or backwards, and images of the brain were acquired. Dehaene-Lambertz and her colleagues found that both forward- and backward-speech led to specific activation in the left temporal lobe, particularly in Heschl's gyrus, a region of primary auditory cortex which is also typically active during speech processing by adults. There was also significantly greater activation to the forward-speech in some left temporal sites, in particular the angular gyrus and the precuneus. Dehaene-Lambertz and her colleagues argued that the infant cortex was already structured into several regions of functional importance for speech processing by the age of three months. In a subsequent study, Dehaene-Lambertz et al. (2006) played three-month-old infants sentences from these stories which were repeated, and studied the time course of the hemodynamic response. Fast responses to the sentences were observed in Heschl's gyrus, with slower responses in left anterior temporal and inferior frontal regions. In adults, activity in left inferior frontal gyrus (Broca's area) is related to the overt production of speech, silent rehearsal and short-term memory. In infants, the left inferior frontal gyrus was the only brain area sensitive to sentence repetition, producing a stronger response when a sentence was heard for a second time. Dehaene-Lambertz and her colleagues argued that a memory system for speech (and possibly other auditory) stimuli was already active in three-month-old infants, and that it could be localised to Broca's area.

The EEG methodology, which measures the time course of neural processing to millisecond accuracy, is ideal for studying phonemic processing by infants. The best-established response in this EEG literature is the mismatch negativity response, or MMN. When a 'standard' stimulus such as /p/ is repeated many times, there is decreased activity in the network of neurons that respond to auditory /p/ signals (i.e., there is neural habituation). However, if the stimulus is changed to the 'oddball' phoneme /b/, there is renewed neural activity. The brain was expecting to hear another /p/, and so the mismatch /b/ triggers a change in neural response. For adults, oddball detection is typically marked by an increased negative potential at around 270 ms following stimulus onset, the MMN (Naatanen & Picton, 1987; note that sometimes with infants the mismatch response (MMR) has increased *positivity*). Given that sleeping babies show robust MMRs (since the auditory system does not 'switch off' during sleep), the MMR is well-suited to phonological studies.

Dehaene-Lambertz and Gliga (2004) summarised relevant MMR studies carried out with infants, and concluded that EEG responses in phonetic perception tasks with infants are remarkably similar to the EEG responses found with adults. When syllables varying along place of articulation are used as stimuli (e.g., /ba/ to /da/), infants show a large MMR for a change that crosses the phonemic boundary. A change of similar amplitude *within* the phonemic category yields a significantly smaller mismatch response. For example, Cheour et al. (1998) used MMRs to measure the neural response to native versus non-native phonemes in infants aged six months and one year. They recruited Finnish and Estonian infants to their study, and used (a) vowel phonemes present in both languages, and (b) a vowel phoneme that was present in Estonian but not in Finnish. Finnish infants aged six months showed very similar MMNs to this vowel phoneme and to the Finnish vowel phonemes, suggesting that all three vowels were distinguished, probably on the basis of acoustic cues. By the time the Finnish infants were aged 12 months, differential responding was found. The MMN shown by the Finnish infants was significantly smaller to the absent vowel phoneme compared to the vowel phonemes that were part of the Finnish language. The MMN shown by the Estonian infants

KEY TERMS

Broca's area
The left inferior frontal gyrus; typically activated during the overt production of speech, silent rehearsal and short-term memory.

Mismatch negativity (MMN)
A change in averaged neural electrical activity when a repeated sound changes to another.

was similar for all vowel phonemes, as all were phonemes in the Estonian language. Hence neural responding mirrors the behavioural effects documented earlier by researchers such as Werker and Tees (1984). Nevertheless, the Finnish infants still generated a (smaller) MMR for the absent vowel phoneme, suggesting that it was still discriminated at the sensory level. Rivera-Gaxiola, Silva-Pereyra and Kuhl (2005) replicated this finding in a study that followed the same English-exposed infants over time. Rivera-Gaxiola et al. tested the English-exposed infants at both seven and 11 months, for discrimination of both native and non-native phonemes. They also found that the older infants retained neural responsiveness to the non-native contrast. Nevertheless, neural responsiveness to the native phonemes got significantly stronger between seven and 11 months, replicating the Finnish pattern of MMR results.

Indeed, more recent research has shown that the neural response to human speech is unique from birth. May, Gervain, Carreiras and Werker (2017) used fNIRS to track the neural response to normal speech (English versus Spanish) versus backwards speech and non-speech (a 'click' language, a surrogate language in which whistling is used to communicate information over long distances) in newborn English-learning infants. Backwards speech is frequently used as an experimental control in adult studies of speech processing, as it matches normal speech for acoustic complexity while not conveying meaning. May et al. (2017) reported bilateral activation over the anterior temporal lobe that differentiated normal from backwards speech in the language heard in the womb (English), but not for the language not experienced in the womb (Spanish). Moreover, they reported that the newborn infants showed a differential neural response to the unfamiliar real language (Spanish) but not to the whistled surrogate language, with neural differentiation in largely temporal regions. These findings are consistent with those reported by Dehaene-Lambertz and Gliga (2004), who argued that infant studies suggested a common neural basis with adults for phonetic perception in the (posterior) temporal lobe.

More recently, Dehaene-Lambertz has produced neural evidence for early *phonetic normalisation* across co-articulated phonemes in three-month-old infants (Mersad & Dehaene-Lambertz, 2016). As well as being capable of categorical perception, speakers need to be able to recognise a given phoneme independently of its surrounding phonemes, this is termed phonetic normalisation. Yet there is significant acoustic variation in consonant phonemes depending on what sound comes next. For example, the /d/ sound in the syllables /di/ and /du/ is very different acoustically (you can check this by saying the two syllables yourself and feeling the difference). Mersad and Dehaene-Lambertz (2016) used the mismatch response (MMR) to measure whether infants could recognise (for example) that a sound like /b/ was consistent across different consonant–vowel (CV) combinations like 'ba' and 'bee'. Infants would hear a stream of three CVs beginning with /b/, and then another CV with a novel vowel beginning with /b/ versus a CV beginning with /g/. An MMR to the oddball CV with a /g/ onset was taken as evidence for phonetic normalisation. Mersad and Dehaene-Lambertz reported that infants indeed showed a significant MMR to the change of initial consonant. Infants also showed a significant MMR to initial consonant change when they only heard the stimuli rather than both seeing and hearing the syllables. As the infants were too young to produce /b/ and /g/ phonemes, Mersad and Dehaene-Lambertz argued that they could recover consonant identity by perceiving statistical patterns in the acoustic signal itself. They noted that quails also succeed in phonetic normalisation tasks (Kluender, Diehl & Killeen, 1987), further evidence that the physical changes where human languages place phonetic boundaries are not random. Note that such data also go against the classic 'poverty of the stimulus' view of language learning (Chomsky, 1957).

Finally, Kuhl and her colleagues have used neuroimaging (MEG) to show that the ability to discriminate native versus non-native phonetic contrasts that emerges during the first year of life has both acoustic and motor components. Kuhl, Ramírez, Bosseler, Lin and Imada (2014) tested English learning infants aged seven and 11–12 months, contrasting native English phonemes with native Spanish phonemes (via different voicing for /d/ or /t/ sounds). As expected, the MMR showed that the younger infants could discriminate both native and non-native sounds, and the MEG recording showed brain activation in both auditory and motor regions. The MMR also showed, as expected, that the older infants had lost the ability to discriminate the non-native speech sounds. However, the MEG recording revealed an interesting double dissociation between acoustic and motor neural regions for the older infants. For auditory brain areas, there was greater neural activation for the native speech sounds than for the non-native speech sounds, while for motor brain areas, there was greater neural activation for the non-native speech sounds than for the native speech sounds. Adults who were tested in the same paradigm showed the same double dissociation. Kuhl et al. (2014) were also able to replicate these neural patterns in Finnish-learning infants who were tested with Finnish versus Mandarin Chinese sounds. These are interesting data. The neural specialisation for the sounds of the native language documented in the early behavioural studies by Kuhl, Werker and their colleagues appears to be accompanied by differential specialisation of auditory versus motor brain responses as a function of learning. Kuhl et al. (2014) suggested that infants' increasing abilities to form these speech sounds themselves (canonical babbling) could underpin this developmental dissociation.

The onset of canonical babbling could also be important for language learning for a number of other reasons, discussed further below. However, it is interesting to note here that there is a progressive shift of infant attention over the first months of life regarding the eyes and mouth in a talking face, a shift that is coincident with the onset of babbling. Monolingual infants show an early attentional focus on the *eyes* when looking at a talking head, perhaps to optimise intention reading (Lewkowicz & Hansen-Tift, 2012). Later, infants shift to focus attention on the *mouth*. This shift typically occurs at around 8–10 months of age, thus coinciding with the onset of canonical babbling. The onset of babbling may create a greater need for motor information about speech. Although a discussion of bilingualism is beyond the scope of this book, it should be noted that bilingual babies follow a similar pattern to monolingual infants. They specialise their phonemic discrimination skills over the first year of life, in their case specialising for *two* sets of phonemic categories (Werker & Hensch, 2015). Bilingual infants are learning two sets of sound systems, two sets of rules for word order, and two lexicons. Bilingual infants also show more attention to the mouth of a talking face than monolingual infants do, even at four months (Mercure et al., 2018). The increased attention to the mouth shown by bilingual infants may signal an increased need to use visual cues to articulation in developing two different lexicons.

Statistical learning of native speech sounds

We noted above that infants are developing 'prototypes' of the phoneme categories that are specific to their language (Kuhl, 1991). Just as in other areas of cognitive development, the formation of prototypes depends on statistical learning. We saw in the last chapter that prototypes for representing concepts are automatically computed by the brain, as the brain generates summary statistics for variable features present in visual input and these summary statistics are thought to form the basis for conscious perception (Hochstein et al., 2015). Accordingly, summary or higher-order statistics should also be computed automatically for the features that specify different phonemes in speech input. For language learning, we can think of a prototypical /p/, or a prototypical /b/ phoneme. Other sounds which we would categorise as /p/ or /b/, even if they are relatively distant from the prototype in terms of constituent features, are called *allophones* in linguistics. In the visual world, infants track conditional probabilities to create prototypes. They learn which visual features in a given stimulus typically occur together, and perceptual information about correlational structure is used to abstract the prototype. Exactly the same mechanisms appear to govern the acoustic world and language learning. Infants track the distributional properties of the sounds in the language that they hear, and register the acoustic features that regularly co-occur. These relative distributional frequencies then yield phonetic categories.

Once a prototype is formed, we know that non-prototypical members of a category are perceived as more similar to the category prototype than they are to each other. This occurs even though the actual physical distance between the stimuli may be equal. This effect is called the 'magnet' effect. If magnet effects reflect experience with specific languages, then babies should only show the magnet effect for prototypes from their native language. This possibility was tested by Kuhl and her colleagues (Kuhl, Williams, Lacerda, Stevens & Lindblom, 1992). They played six-month-old English-learning versus Swedish-learning babies vowel sounds from both languages. The babies heard prototype vowels from either English (/i/) or Swedish (/y/) as a background stimulus. They were then trained to turn their heads when the prototype changed into a variant; that is, when the prototype changed into an acoustically similar but not identical sound that would still be classified as /i/ or /y/ by an adult. Kuhl and her colleagues found that the English and Swedish babies behaved differently. The English babies perceived the English vowel variants as different from the prototype on 33% of trials, whereas the Swedish babies perceived the Swedish vowel variants as different from the prototype on 34% of trials. For the non-native prototype, performance was 49% and 44% respectively. Hence the English babies grouped together the English vowel variants, and the Swedish babies grouped together the Swedish vowel variants. Vowel variants from the non-native language were not grouped perceptually in the same way.

The prototypical sounds in each language were functioning as a kind of 'magnet' for perceptually similar sounds. However, what was experienced as perceptually similar *depended on the input*. The distributional properties of the sounds comprising English versus Swedish respectively determined which sounds were perceptually assimilated to the prototype. Accordingly, Kuhl et al. (1992) demonstrated that linguistic experience alters phonetic perception by the age of six months. Kuhl, Coffey-Corina, Padden and Rivera-Gaxiola (2006) subsequently produced evidence that infants' performance with native prototype phonemes improved during the period of six months–one year. Kuhl described this as evidence for 'neural commitment' to the speech sounds of the native language. Kuhl has also shown that the increased experience of making speech sounds during the first year enhances the native language magnet effect (Kuhl et al., 2014).

Finally, social interaction plays a critical role in perceptual learning. As noted earlier, infants learn language because they are motivated to interact with partners and to communicate. They do not learn language simply because they are passively exposed to sequences of sounds. The core role of social interaction in language learning was demonstrated in an ingenious study by Kuhl and her colleagues (Kuhl, Tsao & Liu, 2003). This study exploited the decline in sensitivity to the phonetic units of non-native languages that occurs between six and 12 months of age. Kuhl et al. (2003) exposed nine-month-old American infants to native Mandarin Chinese speakers, and then tested their speech perception for Mandarin contrasts. The infants experienced around five hours of spoken Mandarin during play sessions with an adult Mandarin speaker, who entertained them with toys and books. A control group of nine-month-olds experienced identical play sessions, but the language used was English. The adult entertainers spoke in Motherese, made frequent eye contact with the infants, and used their names during the sessions, which on average exposed the infants to 33,000 Mandarin syllables. At test, the infants were given a Mandarin Chinese phonetic contrast that does not occur in English. The group exposed to Mandarin Chinese were significantly more sensitive to the critical contrast, performing as well as infants tested in Taiwan who had heard Mandarin for their entire lives.

Kuhl et al. (2003) then replicated the study using video tapes of the Mandarin Chinese adults. A new group of nine-month-olds got the same amount and quality of Mandarin Chinese input from a video. The toys and books were visible on screen, and the adults appeared to be looking directly at the infants, as the films were taken from the infants' perspective. Analysis of the language used showed that on average 50,000 Mandarin Chinese syllables were heard over the course of the videos. The infants were very interested in the videos, touching the screen and attending avidly to the entertaining adults. However, when their sensitivity to the critical Mandarin Chinese contrast was tested at the end of the experiment, they were indistinguishable from the control infants tested in the first experiment (see Figure 6.2). Watching and listening to films of foreign-language material did not result in any measurable perceptual learning. Kuhl and her colleagues argued that the live speakers provided many subtle social cues that facilitated language acquisition. For example, the speaker's gaze tended to focus on the toys or pictures that they were talking about, and the infants' gaze followed. This conveyed referential

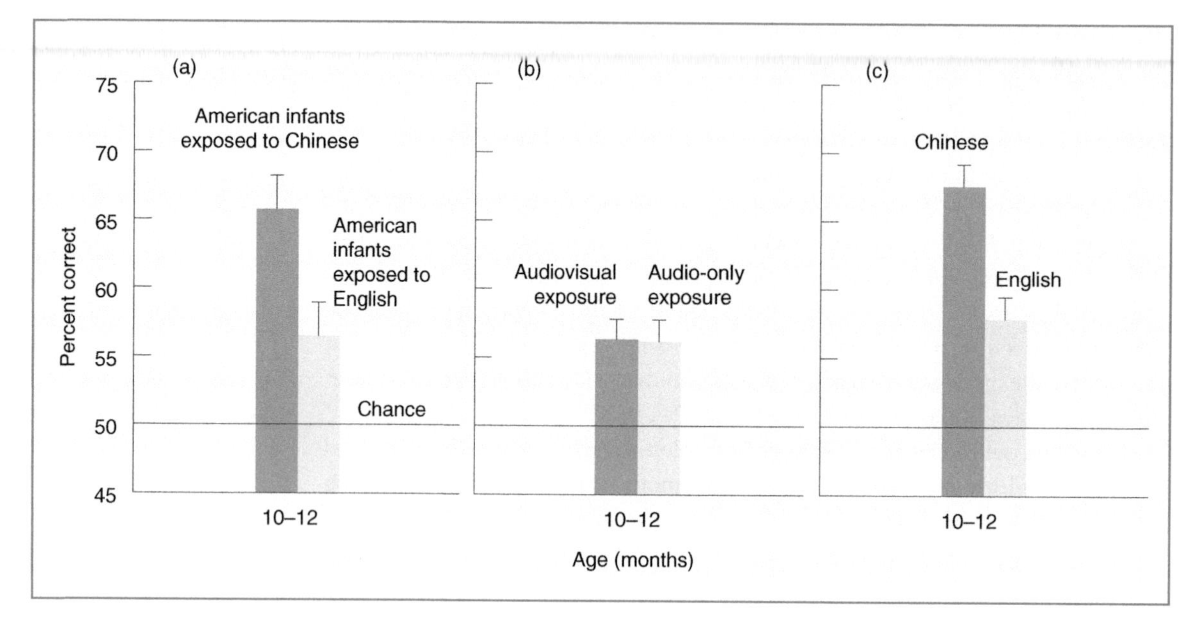

FIGURE 6.2
The results of Kuhl et al.'s (2003) study. (a) Effects of live foreign-language intervention in infancy. (b) Mandarin Chinese foreign-language exposure in the absence of a live person (audio-visual or audio only) shows no learning. (c) The same Mandarin speech discrimination tests on monolingual Mandarin-learning and English-learning infants. From Kuhl et al. (2003). Copyright © 2003 by the National Academy of Sciences. Reproduced with permission.

information, while the live presence of the adult also provided interpersonal cues that attracted attention and possibly motivated learning. The infants were learning language for a purpose, namely communication. Simply being exposed to the raw auditory sensory information is apparently insufficient to trigger perceptual learning.

Statistical learning of language phonotactics

So far we have seen that infants rapidly learn to discriminate between individual phonemes; however, words are composed of sequences of phonemes. The infant needs to group together the phonemes that comprise individual words, and learn which phonemes belong to one word and which phonemes belong to the next word. Some sequences of phonemes are more frequent than others, and some sequences cannot occur at all. The rules that govern the sequences of phonemes used to make words in a particular language are called phonotactics. Information about phonotactic probabilities helps us to determine where one word ends and another word begins. In English, for example, syllables can end in 'ant', but cannot end in 'atn'. If a sequence like 'atn' is heard, it probably crosses a word boundary (as in 'at night').

Infants become able to extract words from continuous speech by about seven months of age (Jusczyk & Aslin, 1995). They seem to do so on the basis of various cues, one of which is phonotactic probability. Phonotactic probability is a good cue, because the sequencing of sounds within particular words is usually heavily constrained. Usually, the transitional probability between two sounds is greatest when the two sounds are within the same word. The probability of sounds following each other tends to dip at word boundaries (as in 'atn'). Hence the ability to track these statistical probabilities is a powerful cue both to the sequences of sounds that make words, and to word boundaries.

The classic demonstration of the fact that infants use transitional probabilities to segment speech comes from research by Saffran and her colleagues (Saffran, Aslin & Newport, 1996). They gave eight-month-old infants novel 'words' to learn on the basis of transitional probabilities. The 'words' were novel three-syllable units like "bidaku" and "padoti". These items were repeated in random order in a monotonous stream (no stress cues or pauses) by a computer speech synthesiser that sounded like a female voice, for two minutes ("bidakupadotigolabubidaku..."). The transitional probabilities were 1.0 for syllable pairs within words (like bi-da), and 0.33 for syllable pairs that crossed word boundaries (like ku-pa; as will be recalled, the same transitional probabilities were used in the visual sequencing experiment by Kirkham et al., 2002, discussed in Chapter 2). These transitional probabilities told infants where the word boundaries were (i.e., bidaku padoti golabu bidaku...). After this two-minute learning trial, a series of 12 test trials were given, half comprising items familiar from the learning phase (like "bidaku") and half comprising new sequences (like "dapiku"). Although the elements of the new 'words' had been experienced during learning, their transitional probabilities had been 0 (i.e., the probability of "da" being followed by "pi" had not been experienced, as "da" was always followed by "ku" during learning). Saffran et al. (1996) reported that the eight-month-olds dishabituated to the novel words like "dapiku", showing that they were computing the necessary sequential statistics.

In a second experiment, Saffran et al. (1996) studied whether eight-month-old babies could also keep track of sequential statistics when the transitional probabilities were relative rather than absolute. For example, learning from the string "bidakupadotigolabubidaku..." was tested with novel items like "kupado". For novel items like these, the serial order of the syllables (ku pa do) has been experienced previously. However, the transitional probability between the first two syllables (ku pa) was 0.33 (as this sequence spanned the boundary between "bidaku" and "padoti"), and the transitional probability between the second two syllables (pa do) was 1.0 (as this sequence was in "padoti"). These 'part words' provided a strong test of statistical learning, as in order to distinguish the 'words' the infants had to pay attention to statistical patterns over the whole learning corpus. Again, the eight-month-old babies showed dishabituation when the novel 'part words' were presented, showing longer listening time to these test stimuli. They were picking out the three-syllable strings with the highest transitional probabilities. In other words, they were learning the structural properties of the input.

While the monotonous diction used in these experiments is not representative of real speech to infants, it has the advantage of isolating infants' ability to learn transitional probabilities between syllables. More recently, brain imaging (EEG) has been used to index infants' statistical learning, in order to study their acquisition of 'long-distance dependencies'. Long-distance dependencies are conditional statistics that relate *non-adjacent syllables* in an utterance. Non-adjacent dependencies are frequent in language, for example in syntax. To test infants' sensitivity to long-distance dependencies, Kabdebon, Pena, Buiatti and Dehaene-Lambertz (2015) created three-syllable non-word stimuli similar to those used by Saffran and her colleagues, but for which the first and third syllables were systematically related. The 'rule' was that for a triple ABC, there was an A_C dependency. Hence for a stream containing items like "kulebi", "kufibi" and "kugobi", the novel item "kunabi" would follow the rule, but the novel item "fibina" would not follow the rule, even though parts of "fibina" were familiar. Eight-month-old infants were tested, as by this age the ability to extract adjacent transitional probabilities is well-established by behavioural measures. Following exposure to the learning stream, the infants were tested with both 'rule words' and 'part words'. Kabdebon et al. (2015) found significantly larger ERP amplitudes to the 'rule words' than to the 'part words', with a left hemisphere bias. They also produced evidence for phase locking effects to both the words (i.e., the trisyllables) and to the individual syllables. Phase locking is an important neural mechanism for successful speech comprehension and is discussed in the next section.

Infant-directed speech, rhythm and prosody

In the real world, the transitional probabilities explored in experiments like those by Saffran and her colleagues are supported by strong prosodic and speech rhythm cues. We speak to babies in a special way, originally called 'Motherese'. When making eye contact with a baby, it is very difficult to stop oneself from speaking in a sing-song rhythmic intonation that is higher in pitch than usual, and that exaggerates certain words by using increased duration and stress. The universal tendency of adults (and children) to talk to babies using infant-directed speech (IDS) suggests that this special prosodic patterning has a developmental purpose. The characteristically higher pitch and exaggerated intonation of IDS may help the infant to pick words out of the speech stream. Although we do not yet know exactly how IDS helps infants to do this, we have extensive evidence that babies prefer to listen to IDS, and that cues such as duration and stress help them to identify words.

KEY TERM

Prosody
The melody, rhythm and stress patterns of language.

Evidence that IDS plays a role in word identification has been gathered by Fernald and her colleagues. Fernald and Mazzie (1991) argued that if the function of IDS was simply to gain the infant's attention and encourage social interaction, then there should be no relationship between prosodic structure and linguistic structure. Similarly, if prosody is simply used to mark new information in all speech, then mothers should use pitch to mark particular words just as frequently when speaking to adults as when speaking to infants. In order to control experimentally which new information was to be marked in natural speech, Fernald and Mazzie designed a picture book called *Kelly's New Clothes*. The book depicted a child getting dressed, with a new item of clothing introduced on each page. Each item was shown as brightly coloured when first introduced, and then appeared in grey on the following page. Mothers of 14-month-old babies were then asked to 'tell the story' to either their infant or to an adult listener on separate occasions. Acoustic measures of prosodic emphasis were made for the target words (the new items of clothing, for example shoes, hat, shirt, etc.).

Fernald and Mazzie found that the target words were those receiving primary stress on 76% of occasions when the mothers told the stories to their infants, compared to 42% of occasions when the mothers told the stories to another adult.

When the new information was mentioned a second time, it was again highly stressed on 70% of occasions to the infants, compared to just over 20% of occasions to the adults. Furthermore, the mothers tended to place the target words at the end of their sentences, increasing their salience further (e.g., "Then he put on his yellow SOCKS"). For the infants, 75% of new words were in utterance-final position, compared to 53% for speech directed to another adult. Hence the mothers were using distinctive prosodic patterns to speak to their infants about new information. Target words were most likely to occur on an exaggerated pitch peak at the end of the utterance. Accordingly, prosodic patterning is used to highlight new words in speech to infants. Fernald and Mazzie suggested that mothers across cultures may converge on this strategy because of its perceptual effectiveness.

Indeed, cross-linguistic research shows that infants are sensitive to prosodic and rhythmic patterning in language from very early indeed. Mehler and his colleagues were the first to show that infants as young as four days use information about rhythm and stress (accentuation) to distinguish their native language from other languages (Mehler, Lambertz, Jusczyk & Amiel-Tyson, 1986). Mehler et al. tested four-day-old babies who had been born in France, and who had thus been exposed to the rhythms and intonations of French while in the womb. For the experiment, recordings were made of a bilingual speaker of French and Russian telling the same story in French and in Russian. Fifteen-second segments of these stories were then played to the babies. The babies showed a clear preference for listening to the native language, in this case, French. The researchers then played the tapes backwards. This meant that whereas the absolute parameters of the signal such as voice pitch were preserved, the relative cues such as intonation and melody were modified. With the reversed speech, the babies could no longer tell the difference between French and Russian. Mehler and his colleagues argued that the infants were relying on rhythmic and prosodic cues to distinguish the two languages. In a control experiment in which the speech was filtered so that only the rhythmic cues were preserved (filtered speech sounds a bit like someone speaking under water), the four-day old infants could again distinguish between Russian and French. Related work by Mehler and his colleagues suggests that the same rhythm-based cues could be important for learning grammar (Bonatti, Peña, Nespor & Mehler, 2005). Vowels in syllables are the main carriers of prosody, and in many languages vowels provide cues to syntax, for example via prominence and lengthening, or via vowel harmony. A relatively high percentage of vowels compared to consonants is found in languages using agglutinative morphology. This means that the duration, intensity and pitch of vowels, which comprise the prosodic information, also carries information about the syntactic structure of a language.

Prosodic cues also carry important information about how sounds are ordered into words when the words are *multi-syllabic*. For example, 90% of English bisyllabic content words follow a strong-weak syllable pattern, with the stress on the first syllable (e.g., *monkey, bottle, doctor, sister*). Jusczyk and his colleagues explored whether infants could use prosodic strategies to segment words from continuous speech. For example, if infants can learn that word onsets are aligned with strong (stressed) syllables, then this would be a useful strategy for picking out words in speech (Cutler & Norris, 1988). Jusczyk, Houston and Newsome (1999) carried out a

KEY TERMS

Head turn preference procedure
In which infants are rewarded, for example with a flashing light, if they turn their head when a sound changes, the reward is located to left or right of midline.

Syntax
Knowledge of how words can be combined into phrases and sentences.

Morphology
The set of rules governing the internal structure of words.

series of experiments to see whether infants were successful in segmenting words following the typical strong-weak pattern from fluent speech, and whether they tended to mis-segment words (such as *guitar* and *surprise*) that followed an atypical weak-strong pattern. In general, strong syllables are louder and longer than weak syllables, and have a higher pitch (frequency). Using a habituation paradigm, Jusczyk et al. found that infants aged 7.5 months could segment words with strong-weak patterns from fluent speech, and treated the words as bisyllables (i.e., they were responding to the total word *doctor* and not just to the strong syllable *doc*). At this age, infants appeared to mis-segment words following a weak-strong pattern, like *guitar*. For example, if they heard a sentence like "her guitar is too fancy", they segmented *taris* as a plausible word (treating *taris* rather than *guitar* as familiar during the dishabituation test). By 10.5 months of age, infants did not make these mistakes with words comprised of weak-strong syllables. Sensitivity to the predominant stress patterns of English words is clearly important for segmentation.

There is also some evidence that when word forms differ by one phoneme but have identical stress, infants do not distinguish between them. Swingley (2005) explored whether the phonological representations of 11-month-old infants showed high specificity for familiar words. For example, would the infants differentiate between *dog* and *bog*, or between *dirty* and *nirty*? Swingley used an infant-controlled head turn preference procedure, in which infants are played speech stimuli for as long as they continue to fixate a flashing light. In these paradigms, infants usually spend longer looking at the light when stimuli are familiar. Swingley compared looking time for familiar words like *dog* and non-words that either varied in the onset consonant, like *bog*, or in the final consonant, like *daub*. His data showed that infants were able to spot the mispronunciations based on onset consonants (as in *dog/bog*), but not in offset consonants (as in *dog/daub*), for stressed syllables. Swingley therefore argued that the 11-month-olds were already building a lexical phonological system that enabled the distinction of minimal pairs (real words differing by a single phoneme). Using a similar paradigm, Vihman, Nakai, DePaolis and Halle (2004) showed that infants' recognition of mispronounced words depended on the stress system of a language. As we have seen, in English stress is usually placed on the first syllable of a bisyllabic word. Vihman and her colleagues showed that English-learning infants could recognise *nirty* as a mispronunciation of *dirty*, but did not recognise *dirny* as a mispronunciation of *dirty*.

Indeed, Curtin, Mintz and Christiansen (2005) showed via computational modelling that stress combines with transitional probabilities in speech in an additive way. They analysed a corpus of phonologically transcribed speech directed to British infants aged between six and 16 weeks, to see whether a connectionist model would be able to learn word representations better when stress provided an additional cue. The addition of stress to the syllable representations led to better segmentation performance by the model. This result suggests that lexical stress makes it easier to distinguish transitional probabilities in the speech stream. Hence it is possible that infants code lexical stress as part of their initial phonological representations. This possibility has been tested experimentally. Using an adaptation of the paradigm designed by Saffran (2001), Curtin et al. (2005) familiarised seven-month-old infants with novel words presented in real English sentences. The novel words either had the lexical stress typical of English (DObita), or atypical stress (doBIta). The question was whether the two types of word, which contained the same phonemes and transitional probabilities, would be represented as distinct by the infants on the basis of whether they contained initial or medial stress. The results showed that the infants preferred the sentences with the words with initial stress. Curtin et al. concluded that lexical stress is retained in the proto-lexical representation. Early phonological representations of potential word forms encode lexical stress as well as segmental information. Therefore, word forms with identical phonemes but differential stress patterns are treated as different words (and in natural language, may indeed have different meanings, as in CONtent and conTENT).

Regarding brain imaging, the MMN has been used to study infant processing of the stress patterns of natural language. The MMN studies suggest that different stress patterns are distinguished early in life, and that the stress patterns characteristic of the ambient language rapidly attain a special status in long-term memory. For example, Weber, Hahne, Friedrich and Friederici (2004) studied German infants' sensitivity to stress patterns using the MMN. Ninety per cent of bisyllabic German nouns are stressed on the first syllable (the trochaic pattern) rather than on the second syllable (the iambic pattern). Weber et al. investigated infant sensitivity to changes in trochaic and iambic patterns. They reported that infants aged four months did not distinguish between two-syllable items stressed on the first versus second syllable. By five months of age, however, infants clearly separated the iambic and trochaic patterns, showing a strong MMN for the trochaic items. These neural findings complement the demonstration by Curtin et al. (2005) that infants are coding lexical stress as part of their initial phonological representations. The pattern of EEG findings reported by Weber and her colleagues suggests that language experience helps the infants to focus on the stress patterns characteristic of their spoken language, a template for which is then stored in the phonological lexicon. This phonological representation of the typical rhythm and stress patterns supports learning and enables the identification of atypical patterns.

Rhythm, prosody, infant-directed speech and neural entrainment

EEG and MEG studies also offer the potential for understanding individual differences in *neural entrainment* to the speech stream. Put simply, neural entrainment refers to the temporal alignment of brain rhythms with speech rhythms. When accurate, this temporal alignment supports the listener in extracting phonological units of different sizes from the speech stream (Goswami, 2018, for review). Neural entrainment, also called *phase locking*, was first identified in adult studies as a fundamental mechanism for speech processing by the brain (Giraud & Poeppel, 2012; Poeppel, 2014). Instead of conceptualising of the speech stream as a linear sequence of successive phonemes, this neural literature focuses on the *amplitude envelope* of speech, which can be thought of as the (relatively) slowly varying changes in signal intensity as someone produces syllables. For amplitude-based analyses, the syllable rather than the phoneme is the core psycholinguistic unit. Modulations or changes in amplitude (intensity) yield speech rhythm, and so, broadly speaking, the amplitude envelope is the major acoustic cue to speech rhythm patterns. Within the overall 'envelope' of sound reaching the ear, however, there are local variations in signal intensity that are related to the production of *particular* syllables and their constituent phonemes. The brain is interested in these local variations in amplitude patterns as well as in overall speech rhythm. In fact, there is a relatively *predictable* variation in signal amplitude as the speaker produces each syllable, and an even more salient and predictable variation as the speaker produces each stressed syllable. 'Rise times', increases in amplitude in the speech envelope as syllables are produced (the amplitude of the speech signal peaks each time that a vowel is produced), are critical perceptual cues to these local amplitude variations. Rise times for stressed syllables are particularly large and salient, and are important for the learning brain. In fact, neuroimaging studies with adults show that the brain uses amplitude rise times as a temporal clue to the different rates of amplitude modulation in the speech envelope, thereby enabling accurate alignment between ongoing neural oscillatory rhythms and speech rhythms. For example, if the rise times associated with syllable-level modulations are removed from speech, adult listeners can no longer comprehend what is being said (Doelling, Arnal, Ghitza & Poeppel, 2014). If simple clicks are then inserted into the signal in place of the rise times, the speech becomes intelligible again. These and other neural data suggest that amplitude rise times are important mechanistically for the automatic parsing of syllables and multi-syllabic words from continuous speech.

Across languages, speakers produce approximately two stressed syllables per second, and five syllables per second (see Goswami, 2018). A good model for understanding how the amplitude envelope supports speech processing by children is the *amplitude modulation phase hierarchy* model based on statistical modelling of the English nursery rhyme (Leong & Goswami, 2015). Rhythmic linguistic routines for the nursery are found in many languages. For English, a nursery rhyme phrase like "Pussycat, pussycat, where have you been?" has regular repetition of the stressed syllable 'pu' in "pussy", followed by two unstressed syllables ('sy-cat'), again repeated. The metrical structure of the nursery rhyme makes the speaker *time the production* of the two unstressed syllables to fit one 'beat' of the underlying rhythmic template, with the stressed syllable getting a full beat of its own. This rhythmic patterning produces a predictable metrical template of speech energy changes, to which the developing brain can entrain (align its endogenous oscillatory rhythms). Indeed, it is likely that nursery rhymes and other cross-cultural forms of rhythmic nursery speech (such as knee-bouncing songs and lullabies) provide input that is optimally structured for accurate neural entrainment (see Goswami, 2018). The changes in amplitude that contribute to the perception of speech rhythm are encoded by the brain via oscillating networks of neurons, primarily in auditory and motor cortex. For example, a cortical network oscillating at ~2 Hz (in the oscillatory delta band) will align itself (entrain or phase lock) to the modulation pattern of the stressed syllables in the input, while another network oscillating at ~5 Hz (in the oscillatory theta band) will align itself (phase lock) to the modulation pattern of the syllables. The neural networks automatically *phase-reset* their activity using amplitude rise times. As discussed previously, ongoing neuronal oscillations are found throughout the brain. Networks of brain cells oscillate or rhythmically alternate between excitatory and inhibitory states, cycling between activation and inhibition at different temporal rates like delta, theta, alpha and gamma. Oscillations occur even if there is no input. For the auditory cortex, when speech is heard, local rise times in the amplitude envelope of speech reset this oscillatory activity so that peaks in neural firing align with peaks in amplitude modulation. For amplitude modulations in speech, when the peaks at delta and theta rates are aligned, the listener perceives a stressed syllable. When a delta trough in amplitude modulation is aligned with a theta peak, the listener hears an unstressed syllable (Leong, Stone, Turner & Goswami, 2014). In fact, there are phase-dependent relations between three temporal rate bands of amplitude modulation in English nursery rhymes: delta, theta and beta/low gamma. These amplitude modulation bands are hierarchically nested in the speech signal, and governed by the slowest (delta) band, so that the phase of slower oscillations in amplitude governs faster oscillations. Hence there are statistical patterns in the amplitude modulation content of speech that help to specify its phonological structure.

Recently, Leong, Kalashnikova, Burnham and Goswami (2017) modelled the amplitude envelope of IDS being spoken to English-learning infants at different ages. They found that the temporal modulation structure of the IDS had highly predictable amplitude modulation patterning, patterning that was similar in its hierarchical statistical structure to that found in children's nursery rhymes. This amplitude modulation patterning is shown schematically in Figure 6.3 for the children's nursery rhyme 'Jack and Jill went up the hill' (middle panel). Leong et al. found that for IDS as spoken to infants aged seven, nine and 11 months, the delta-band speech energy (a band of amplitude modulations centred on ~2 Hz) was significantly *greater* than in adult-directed speech (ADS). This means that IDS foregrounds low-frequency speech energy, shown in the red and yellow colours in Figure 6.3. Furthermore, phase alignment between amplitude modulation bands centred on ~2 Hz and ~5 Hz (delta-theta phase synchronisation) was also significantly greater in IDS than in ADS (Leong et al., 2017). This tighter phase synchronisation between bands would foreground the alignment between peaks and troughs in each band of amplitude modulation. Acoustically, the modelling shows that IDS provides an enhanced and predictable rhythmic template to support the infant brain in successful neural entrainment. In effect, key acoustic landmarks are exaggerated and the nested hierarchical structure has more reliable temporal alignment. For example, more delta-band modulation energy would help infants to identify word and phrase boundaries, which are typically marked by stress and intonation. These exaggerated acoustic statistical patterns were still present in IDS when corrections were made for the different speaker rates that characterise IDS and ADS (IDS is typically spoken more slowly than ADS).

Accordingly, Leong et al. (2017) suggested that IDS was optimally structured to facilitate neural entrainment to prosodic information in speech by the infant brain. The significantly stronger phase alignment between delta and theta bands of amplitude modulation would facilitate the extraction of rhythm patterns (e.g., trochaic versus iambic rhythm), statistical patterns known to be important for language-learning. Indeed, a related EEG study comparing infant neural entrainment at age seven months to IDS versus ADS showed enhanced neural entrainment to the

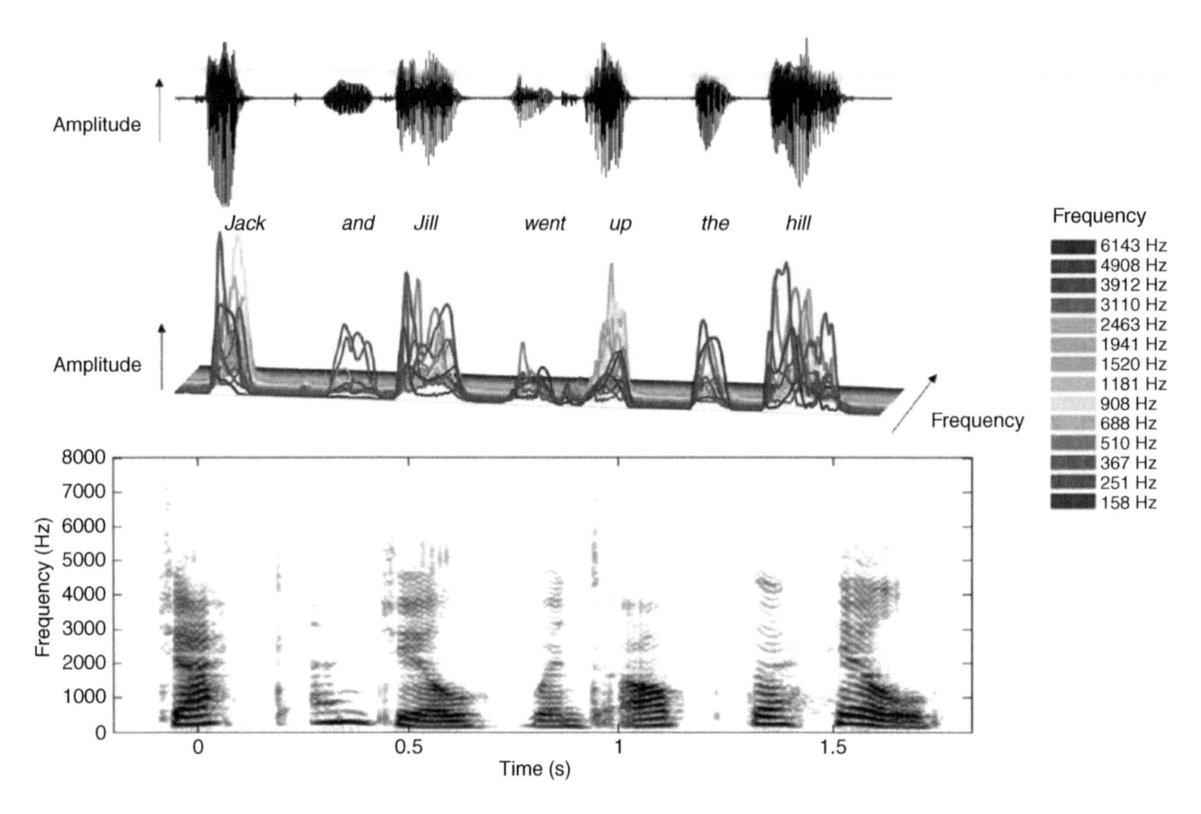

FIGURE 6.3
A schematic depiction of the raw speech signal of the nursery rhyme phrase 'Jack and Jill went up the hill', top panel, with the middle panel showing its amplitude modulation structure using the S-AMPH model and the bottom panel showing the same speech modelled as the speech spectrogram. Reproduced from Goswami (2018).

IDS in a broad low-frequency band encompassing both delta and theta amplitude modulation information (1–8 Hz; Kalashnikova, Peter, Di Liberto, Lalor & Burnham, 2018b). Meanwhile, a study of newborn German-learning infants using a complex amplitude-modulated noise stimulus (which has inherent rhythmic patterning and is matched acoustically to the complexity of the speech signal) has been able to demonstrate successful delta-band neural entrainment at birth (1–3 Hz; Telkemeyer et al., 2011). Given that adult studies have identified a core neural mechanism for understanding speech, more focused studies of neural entrainment by infants are likely to become available in the coming years, hopefully across languages. Indeed, the acoustic statistics in the amplitude envelope have been shown to be relevant to learning syntax as well as phonology (Flanagan & Goswami, 2018). Patterns of acoustic prominence (amplitude modulation patterns) provide cues to syntactic and prosodic phrasing, hence the acoustic statistics nested in the amplitude envelope support grammatical as well as phonological learning (Cumming, Wilson & Goswami, 2015). Accordingly, infant entrainment studies have the potential to identify early neural 'markers' of possible language impairment, long before the production of spoken language has commenced.

Meanwhile, fNIRS has been used to study whether newborn infants can use prosody to learn about word order. Benavides-Varela and Gervain (2017) devised simple four-word utterances by combining function (F) and content (C) words, each word having natural intonation. Infants were habituated to utterances of a certain prosodic format (for example, FCFC), and then heard a block of deviant utterances with a new word order (e.g., FFCC). The newborn infants were able to detect the changes in word order, showing significant changes in the hemodynamic response to the deviant word orders. Benavides-Varela and Gervain found this 'mismatch' response in blood flow in both hemispheres, although it occurred for more channels in the left hemisphere. However, when they created four-word sentences with the same words, but varying the intonational phrasing across the whole sentence rather than word by word, the infants no longer showed a mismatch response. Instead, they showed repetition suppression over right hemisphere channels in the standard trials. This suggests that the infants were sensitive to the repetition of the same prosodic contours during the experiment. In adult neural studies of speech processing, prosodic processing typically shows right-lateralised effects. Benavides-Varela and Gervain (2017) concluded that newborn infants can use prosodic cues to track word order, but that initially it was easier for them to individuate prosodically shorter units.

Early phonological production

Infants begin vocalising from very early on. Infants first signal to their caretakers by crying, and they also make other early non-vocal sounds such as grunting, sneezing and coughing. However, some vocalisations are speech-like, and it is thought that speech-like vocal development goes through a number of stages. During the first stage, from 0–2 months, infants produce 'comfort sounds' that have normal speech-like phonation. These comfort sounds most typically sound like vowel phonemes. By 2–3 months of age, infants enter the so-called 'gooing' stage (Oller, 1980), producing phonetic sequences that are precursors to consonant phonemes. This is followed by an expansion stage, typically from 4–6 months, during which infants produce a variety of new sound types, including trills, squeals, growls, whispers and proto-syllables which are termed 'marginal babbling' (Oller & Eilers, 1988). Mature syllables appear in the canonical stage, which is when canonical babbling begins, at around 7–10 months. Now infants produce the reduplicated sequences of consonant–vowel (CV) syllables that we typically consider babbling, saying /mamamamama/ or /dadadadada/. The onset of the canonical stage is considered critical, as it represents the point in development at which infants produce mature syllables that can function as the building blocks of words.

KEY TERM

Canonical babbling
Babbling that begins around 7–10 months in which infants produce reduplicated sequences of consonant–vowel (CV) syllables that can function as the building blocks of words.

However, new methods for investigating infant competence are beginning to challenge this sequence. For example, Chen, Striano and Rakoczy (2004) showed that neonates can produce the mouth movements appropriate to forming consonant sounds in response to modelling by an adult, even though they cannot produce the sounds themselves. Chen and her colleagues presented newborn infants with an adult making the sounds /a/ and /m/ repeatedly in a face-to-face interaction with the infant, and videoed the infants' faces. They found that infants were significantly more likely to make 'mouth clutching' responses to seeing /m/, and significantly more likely to make 'mouth opening' responses to seeing /a/. Newborn infants can thus imitate the mouth movements needed to produce consonant and vowel phonemes. In fact, even infants who had their eyes closed could produce the two kinds of motor response, just from hearing the speech sounds. This is suggestive evidence that speech is processed as both a sound and as an action from birth.

Nevertheless, the onset of canonical babbling does seem to play a special role in linguistic development. Longitudinal studies have shown that the onset of babbling is a strong predictor of the production of first words. For example, McGillion and her colleagues observed a group of 46 infants monthly between the ages of nine and 18 months, recording both their babbling and their pointing behaviour. McGillion et al. (2017) reported that the onset of canonical babbling preceded the onset of protodeclarative pointing in their sample, and that only babbling was a significant predictor of the onset of the production of first words. The onset of babbling was also a significant predictor of both expressive and receptive vocabulary at 18 months of age; the onset of pointing was not. McGillion et al. argued that their data showed the importance of early phonological production in the creation of the lexicon.

Cross-language work has shown that infants across cultures typically babble the same kinds of sounds in the same order. For example, stops like /b/ and /p/ and nasals like /m/ are easier to produce than fricatives like /f/ and liquids like /l/. The most frequent sounds to be babbled early are /d/, /b/, /m/, /n/, /g/ and /t/. Hence babbling /dadada/ and /mamama/ emerges early in all cultures. However, the relative frequency of easily produced sounds in children's babbling is very similar to the frequency of those sounds in the ambient language. For example, French uses the sounds /p/, /b/, /m/, /f/ and /v/ more frequently than English. Accordingly, French babies babble those sounds more frequently than English babies. Babbling also reflects the rhythmic properties of the adult language. De Boysson-Bardies, Sagart and Durand (1984) took samples of babbling from six-month-old babies who were learning either French, Cantonese or Arabic. They then played them to French-speaking adults. The adults were asked to pick out the babbling of the French babies. They managed to do this very accurately, apparently by relying on the intonational prosodic patterning of the babble.

It is often reported anecdotally that deaf babies produce the same kinds of babbling sounds as hearing infants. This belief has been very important theoretically. If deaf babies babble in the same way and at the same age as hearing infants, it would suggest that babbling is biologically predetermined, and can unfold without auditory experience. The first systematic study of babbling by deaf infants was carried out by Oller and Eilers (1988). They were able to collect vocalisations from nine deaf infants aged from one month. This was a significant achievement, as hearing impairment is not often diagnosed early, and is often accompanied by other learning difficulties. These nine babies were compared to 21 hearing infants, and all infants were followed longitudinally. Oller and Eilers found that all the hearing infants started canonical babbling between six and ten months of age. For the deaf infants, the earliest age for canonical babbling was 11 months, and the latest age was 25 months. Further, the nature of their babbling was quite different. They produced far fewer vocalisations, and the quality of what they produced was restricted. Oller and Eilers concluded that the idea that audition played no role in normal babbling was a myth.

One interesting question concerns whether there is any link between rhythmic babbling and other forms of rhythmic patterning. For example, babies shake rattles and make rhythmic hand movements. Babbling, however, follows the rhythmic, timing and stress patterns of natural language prosody, suggesting that it is not simply another motoric rhythmic behaviour. Rather, it is specifically linguistic. The linguistic hypothesis of babbling assumes that the production of the structured rhythmic and temporal patterns of the ambient language is a crucial part of language acquisition. An alternative, motor hypothesis of babbling attributes the rhythms of babble to the physiological properties of the jaw. Petitto, Holowka, Sergio, Levy and Ostry (2004) contrasted the motoric and linguistic accounts of infant babble by studying the rhythmic hand movements of two groups of babies. One was a group of hearing babies born to hearing parents, and the second was a group of hearing babies born to deaf parents. The latter were learning sign language rather than spoken language, even though they could hear.

If infants are exposed to spoken language, they will babble sounds. If they are exposed to sign language, they will babble signs.

Pettito et al. reasoned as follows. Babies exposed to sign rather than to spoken language might go through a developmental stage of 'babbling' on their hands. In an earlier study of deaf babies, Pettito and Marentette (1991) had indeed found a unique set of hand movements in deaf babies that appeared to contain a reduced subset of the sign units of natural sign language. This 'hand babble' was produced repetitively in accord with the general prosodic contours of natural sign languages, duplicating the rhythmic timing and stress of hand shapes in natural signs. However, if hearing babies who are not exposed to sign language also produce manual babbling, then this would not be specifically linguistic in the sense of indexing production of a reduced form of the ambient language. Rather, it would be suggestive of a universal rhythmic motor activity, but produced with the hands rather than with the mouth. The hearing babies studied by Pettito et al. all produced vocalisations, but the key question was whether they would also produce the same kind of manual hand movements as each other.

KEY TERM

Manual babbling Rhythmic hand movements of babies exposed to sign language that approximate the rhythmic auditory babbling of hearing infants.

In order to get very detailed information about hand movements, Pettito et al. attached infrared emitting diodes onto the babies' hands. Sensitive sensors then tracked the trajectory and location of the babies' hands while the babies played with a parent. During the videotaped play period, the parent offered toys like rattles and soft toys, played peek-a-boo, or talked and signed to the baby. The trajectory and hand locations emitted by the diodes were then calculated independently of the video and blind of whether the babies were being exposed to speech or sign. The resulting movement segments were then matched to the videos. Pettito et al. found that only the hearing babies who were being exposed to sign language produced a low-frequency hand activity that was rhythmic at around 1 Hz (i.e., one hand movement cycle per second). Both groups of hearing babies produced high frequency hand activity at around 2.5–3 Hz (three cycles per second), which was designated as non-linguistic excitatory activity. The low frequency activity was designated as manual babbling. Video matching showed that around 80% of this manual babbling was produced within 'signing space' – the restricted spatial area around the face and chin used for linguistic signs. In contrast, only around 20% of high-frequency hand movements were restricted to the signing space. These data suggest that babbling is not simply another motoric rhythmic behaviour, but is rather specifically linguistic. If infants are exposed to spoken language, they will babble sounds. If they are exposed to sign language, they will babble signs. In each case, the babies are discovering and producing the most rudimentary structures of the natural language to which they are exposed. These rudimentary structures are rhythmic ones. Further, given that oscillatory neural entrainment to spoken language relies in part on rhythmic patterning (discussed above), it would be extremely interesting to explore the loci of neural entrainment to the rhythm patterns in signed languages in deaf infants and children. Given that the rhythmic structure in a signed language is linguistic, it is plausible that oscillatory networks in both auditory and motor cortex may be important in processing signed languages.

Finally, early infant babbling is also related to how much IDS they hear. Ramirez-Esparza, Garcia-Sierra and Kuhl (2014) clipped small recording devices to the clothing of infants aged 11 and 14 months, aiming to record all the language input to each infant over a four-day period, as well as record the infant's own vocalisations. This clever technique enabled the analysis of completely naturalistic language data. Ramirez-Esparza et al. reported that on average the infants heard 30,000 words over the four-day measurement period. However, individual differences in the number of words heard did not relate to infant vocalisations. Instead, the key factor was whether the words were heard in IDS. The percentage of time that an infant spent in hearing IDS in a one-on-one setting (i.e., infant–mother or infant–father dyad) was significantly related to the quantity of infant vocalisations. A follow-up study when the infants were two years of age showed that the quality (hearing IDS) but not quantity (number of words heard) of language input at 11 and 14 months was significantly related to later vocabulary development, even after socio-economic status (SES) was controlled. Ramirez-Esparza et al. (2014) concluded that the *quality* of speech experienced by infants was important for their linguistic development. As discussed earlier, in IDS phonetic sounds are clearer, longer and more distinct. Further, IDS has exaggerated prosody and presents an optimal stimulus for neural entrainment.

LEXICAL DEVELOPMENT

KEY TERM

Lexical development
Building a vocabulary.

Although phonological development is central to language acquisition, the primary function of language is communication. Infants need to learn what words *mean*. In order to learn word meanings, infants must map acoustic patterns to specific concepts. Lexical development (building a vocabulary) is hence intimately tied to conceptual development. Words often name things that are in the world. As we saw in earlier chapters, infants understand a lot about the kinds of things that are in the world. They also understand the kinds of events and actions named by language long before they are fluent speakers of language themselves. Nevertheless, words are part of meaning-making experiences from very early in development. Carers will typically name the objects that are the focus of the child's interest long before the child is talking, and will comment on joint activities or on the child's behaviour or apparent feelings. The situational context in which words are used is an important cue to what they might mean. Indeed, as also noted earlier, the classic situation of *natural pedagogy* is an adult–infant interaction involving an object (Csibra & Gergely, 2009). Adults will typically name and point to the object, using Parentese, and making eye contact. Their use of such *ostensive cueing* makes the infant assume that the adult intends to impart some new and relevant knowledge. How novel objects should be named is one example of such new information.

Carers will typically name the objects that are the focus of the child's interest long before the child is talking, employing highly frequent constructions.

Astonishingly, a study of mothers' utterances (rather than individual words) to their toddlers showed that children heard an estimated 5,000–7,000 utterances a day (Cameron-Faulkner, Lieven & Tomasello, 2003). Around a third of these utterances were questions. In Cameron-Faulkner et al.'s study, more than half of maternal utterances began with one of 52 highly frequent constructions, such as "Look at…", "Are you…", "Let's…" and "Here's…". The utterances that children heard were thus quite repetitive and frequent, with some constructions experienced literally hundreds of times per day. Similar estimates have been made by other researchers. Hart and Risley (1995) estimated that children from high SES families heard around 487 utterances per hour. Children from families on welfare heard around 178 utterances per hour. By the time they were aged four years, it was estimated that high SES children had been exposed to around 44 million utterances, compared to 12 million utterances for the lower SES children. These differences are remarkable. The brain has received far more language input as a basis for learning for the higher SES children. Hart and Risley also found that vocabulary grew faster in those children exposed to a greater quantity of language, irrespective of their SES. As their study tested children rather than infants, Hart and Risley's findings are complementary to the data regarding the *quality* of language input gathered by Ramirez-Esparza et al. (2014) with infants, which found that hearing more IDS led to better vocabulary growth. Indeed, frequency effects in early word learning are found across languages. Young children's phonology is more accurate whey they have heard a larger number of different words in their language containing target phonemes or phonotactic patterns (Edwards, Beckman & Munson, 2015).

Word learning usually takes off some time after the end of the first year. The earliest age for producing your first word is around nine months. By 16 months of age, median spoken vocabulary size is 55 words (Fenson et al., 1994). By 23 months, it is 225 words. By 30 months, median vocabulary size is 573 words, reflecting a tenfold increase in 14 months. By age six, the average child has a spoken vocabulary of around 6,000 words and a comprehension vocabulary of around 14,000 words (Dollaghan, 1994). Clearly, some powerful learning mechanisms are at work.

What are children's first words like? Most children produce words that are highly relevant to their daily lives, such as words for salient individual objects (*Mummy*, *teddy*), words for salient categories of objects (*doggy*, *cup*), words for actions (*up, gone*), words for recurrence (*more, again*) and words for social routines (*bye-bye, night-night*). Word learning is particularly interesting because it is symbolic. Words are linguistic symbols, they *refer* to an object or to an event, but they are not the object or the event itself. This use of linguistic symbols in systematic patterns that in themselves convey meaning (i.e., grammatical patterns, see below) is uniquely human (Tomasello, 1998). Linguistic constructions have meaning independent of the meaning of the individual words making up the utterances. Following the one-word stage at about 16 months, and the two-word stage at about 20–24 months, children from around 30 months onwards rapidly acquire and produce many different kinds of linguistic constructions.

Early in language acquisition, the comprehension of words outstrips the production of words. Benedict (1979) studied eight infants from when they were nine months of age until they were one year and eight months old. She visited the infants at home twice weekly to observe and record their vocabulary, and in between her visits the mothers kept a diary of their children's new words. Benedict reported that the rate of word acquisition for comprehension was twice that of production. On average, she found that the children understood 50 words before they were able to produce ten words. When ten words could be produced, the number of words comprehended ranged from 30 to 182. There was a gap of approximately five months between attaining the 50-word level in comprehension and attaining the 50-word level in production. Thus comprehension developed earlier than production, and it developed much more rapidly. For the first ten words comprehended, over 50% were action words (*give, kiss*). Most of the early words produced were words that referred to things.

KEY TERM

Comprehension precedes production The finding that children can understand more words than they can produce.

Much of our knowledge concerning early lexical development has relied on parental report. For example, Fenson, Bates and colleagues developed the Child Language Checklist to give to parents. The checklist comprised the first few hundred words and phrases typically acquired by American English children. Fenson et al. (1994) asked parents to mark how many of the words they thought their children knew at different ages. This checklist has now been translated into at least 12 languages, enabling cross-language comparisons of lexical development. Three key findings are (a) that word comprehension appears to onset between 8–10 months across the world's languages, (b) that word production onsets between 11–13 months across languages, and (c) that there is huge variation in the lexical growth shown by individual children. For example, at two years, the range in word production is from zero words to more than 500 words. In terms of what the words are, cross-language checklists suggest that all children produce similar first words, words that are salient to the concerns of their everyday lives. However, there are some interesting culturally related differences. For example, the child's word for 'grandmother' is on average the fifth social word produced by an Italian toddler, but the 30th social word produced by an American toddler. This difference probably reflects the fact that Italian children tend to live in closer proximity to the extended family (Bates et al., 2001).

The checklist studies also revealed that a sudden acceleration or 'burst' in vocabulary acquisition did not describe the language development of most children. Rather, vocabulary acquisition was incremental, and was best fit by a smoothly accelerating exponential function (Fenson et al., 1994). This was surprising, as the 'vocabulary spurt' observed in longitudinal studies of individual children was widely held to be a central aspect of early language acquisition. These individual studies had documented a sudden surge in the number of new words acquired just prior to the achievement of a 50-word lexicon, at around 17–19 months of age (Bloom, 1973; Benedict, 1979). Theoretically, this sudden spurt had been interpreted to show that children had achieved a *new level* of referential understanding – they had achieved the *insight* that words can name. It was thought that the vocabulary burst marked an important change in the symbolic status of words. However, other authors had pointed out that this 'naming explosion' may only characterise young children who were focused on learning names for things. In a longitudinal study reported by Goldfield and Reznick (1990), 30% of the children studied never showed a dramatic acceleration in word learning. These children maintained a steady balance of nouns and other kinds of words throughout early lexical development. More recent experimental work (discussed below) shows that the 'insight' that words can name probably develops much earlier than 17–19 months of age, sometime during the first six months of life.

Another aspect of development revealed by the checklist studies was the mediating role of gesture. The first communicative skill to emerge was word comprehension. The second was communication gestures and routines. These gestures were usually elicited in response to verbal input, for example waving 'bye-bye'. Gestures recognising the function of common objects appeared shortly afterwards, such as holding a telephone receiver to the ear. Verbal naming then began. Gesture is a form of production rather than comprehension, and it has communicative intent – children are gesturing in order to convey information. Young children are using action to express or lexicalise meanings that could have been put into words (Volterra & Erting, 1990). Gesture thus provides a kind of cognitive bridge between comprehension and production.

KEY TERM

Vocabulary spurt
An apparent sudden surge around 17–19 months of age in the number of new words acquired, originally thought to reflect achieving the insight that words can name.

The use of gestures to communicate between the ages of ten and 21 months was studied by Zinober and Martlew (1985). They studied four types of gestures: instrumental gestures (e.g., gestures intended to control the caretaker's behaviour, such as pointing to a desired toy); expressive gestures (e.g., gestures conveying emotions); enactive gestures (e.g., imitating the turning of a door handle to convey opening the door); and deictic gestures (for focusing joint attention). These intentional signals by the infant were co-ordinated with emergent vocalisations including babbling, proto-words and single words, in two contexts, free play with the mother and a shared picture-book session. The use of gesture was found to increase between ten and 18 months, and then to decline. Vocalisation showed a steady increase, and by 21 months was the dominant form of communication. Usually, even at ten months, gestures and proto-words were used together to convey the same message. For example, one child always accompanied his gesture for opening and shutting with the proto-word "shuh". In general at this period, the gestures were easier to understand than the vocalisations. Later in development, from around 18 months, vocalisations and gestures were used together to convey more complex meanings than either could convey alone. For example, William conveyed the message "You choose the book, Mummy" by responding to his mother's invitation "Billy choose one" with the word "No" (shaking his head), and then "Book", at the same time placing his mother's hand firmly on the pile of books. Gesture thus plays a key role in enhancing the effectiveness of communication in early language development.

Word learning by infants

What learning mechanisms do children use to acquire all these words? Clearly, the perceptual mechanisms reviewed above play an important role in stabilising the acoustic patterns that constitute words, but these patterns must be linked to concepts in order for a lexicon to develop. In preceding chapters, we have seen that infants are developing a number of physical concepts (such as containment) during the first year of life, and are making distinctions between animates and inanimates as well. These pre-existing conceptual categories can then be linked to words. However, language is primarily part of the social world. Infants first make sounds in order to communicate. Furthermore, the social world provides some of the most salient and important early concepts for the infant – the infant's mother, and the infant's self. In fact, word learning seems to begin with learning one's own name.

Infants as young as 4.5 months of age can reliably recognise their own names, and appear to have a relatively detailed phonological representation of this salient word. This was demonstrated in an habituation study by Mandel, Jusczyk and Pisoni (1995). Each infant listened to repetitions of four different names, their own name and three foils. One foil matched the stress pattern of the infant's name. For example, an infant called Joshua might be tested with the foil Agatha, whereas an infant called Becca might be tested with the foil Aaron. Two foils had a different stress pattern. Here Joshua might be tested with Maria and Eliza, while Becca might be tested with Rumiz and Michelle. A different set of foils was used for each infant, in case some names were inherently more appealing than others. The names were recorded by a female speaker using a lively intonation pattern, as though she were calling to the infant.

During the experiment, the infant sat on their caretaker's lap, and watched a centrally placed flashing green light. Once the infant was fixating at centre, a red light on one side (either left or right) came on, and when the infant turned their head towards this red light, a name was played. The name continued until the infant stopped looking at the red light. Mandel et al. reported that the infants looked significantly longer when rewarded with their own name. They could distinguish their name both from the foil with the same stress pattern, and from the other foils. There was no difference in looking time to the different kinds of foils. Clearly, one's own name is one sound pattern that has been learned by four months of age.

The salient cue of your own name or of your word for 'Mummy' can help in learning adjacent words as well. Bortfeld, Morgan, Golinkoff and Rathbun (2005) played six-month-old infants passages containing novel words, in which the novel word either followed their own names, or an unfamiliar name. For example, an infant named Maggie might hear "The girl rode Maggie's bike" if participating in the 'familiar name' condition, and "The boy played with Hannah's bike" if participating in the 'unfamiliar name' condition. There were six such sentences in each passage. Learning of the novel words (bike, cup, feet, dog) was then tested via a head-turn procedure. When the infant turned his or her head to look at a blinking light, the same (female) voice that had read the passages would begin speaking a particular word. At test, the infant either heard words that had been paired with their own name spoken repeatedly (words like "bike" in the example above), or heard words that had been paired with unfamiliar names (e.g., "cup"). The infants chose to listen significantly longer to the words that had been paired with their own (familiar) name.

In a second experiment, the same effect was found using the name most often used for the infant's mother (e.g., Mama, Mommy). Infants listened significantly longer to words like "dog" after hearing sentences like "Mommy's dog barked only at squirrels" than after hearing sentences like "Lola's dog barked only at squirrels". Bortfeld et al. (2005) argued that the infants were using their own name or their word for mummy as an 'anchor point' in the speech stream, enabling top-down processing that also identified novel words. Such top-down processing would be a potent language-learning device. In such cases infants, like adults, would be using stored lexical knowledge to segment other likely words from speech. The infants must have been matching a stored representation of a phonological form (like 'Mommy') against the input, and hence segmenting out adjacent words as well. So far in this chapter, we have largely considered the 'bottom-up' perceptual and statistical mechanisms available to help infants to learn words. These learning mechanisms are based solely on the perceptual characteristics of the input. By the age of six months, babies are also using *top-down* learning mechanisms as a basis for word acquisition. This top-down processing appears to begin with their own names.

Bottom-up learning processes for word learning are clearly operating at six months, as shown by Shukla, White and Aslin (2011). Shukla and her colleagues devised a looking time experiment to investigate whether six-month-old infants were able simultaneously to segment a novel word from prosodically organised continuous speech and associate it with a visual referent. Infants watched short videos showing simple objects, and listened to the repetition of a short utterance containing a novel name for one (target) object, which moved along a 'table' during the video. The syllables in the utterance were manipulated so that two syllables always had the highest transitional probabilities, but this bisyllabic name was either also cued by a prosodic boundary or crossed a prosodic boundary. During a subsequent test, infants listened to the two syllables of the novel name produced in isolation while their looking was measured via an eye tracker. The data showed that the infants fixated the target object only when the novel word had been aligned with a prosodic phrase boundary. This finding again highlights the key role of speech rhythm and prosody in early language learning by infants.

Complementary data comes from Bergelson and Swingley (2012), who investigated whether six-month-olds could distinguish single lexical referents like 'banana' or 'mouth' when spoken in isolation. They also used a visual task. Infants sat on the parental lap and looked at a screen, which either showed two images side by side (e.g., a picture of a banana and a picture of hair), or a single domestic scene in which various food items could be seen, here including bananas. Their parent was then instructed over headphones to produce a particular word, for example "banana". Overall infants heard eight food or food-related words (*banana*, *apple*, *cookie, juice, bottle, milk, yoghurt, spoon*) and 12 body part words (such as *hair, leg, eyes*). Using eye tracking, Bergelson and Swingley found that the majority of infants were significantly more likely to look at the target item than the distractors, both for the paired-pictures condition and the domestic scenarios. On average the infants knew 75% of the words in the experiment. Interestingly, none of the parents thought that their infants knew any of these words! Clearly, word learning is more precocious than previously thought.

Cognitive expectations about language and labels

Waxman (e.g., Waxman & Gelman, 2009) has proposed that infants are equipped with a broad initial expectation that words are linked to concepts, and that this initial expectation becomes increasingly fine-tuned on the basis of their experiences with the objects and with events. Waxman has argued that in order to acquire *specific* meanings, which she calls 'word-to-world' links, infants must expect that words highlight commonalities among named entities. The early lexicon has more nouns (words for objects) than verbs (words for actions). Waxman argues that the consistent pairings of certain nouns with certain objects will lead the infant to understand a particular noun as the label for that category of objects. As early words also tend to refer to basic level categories (*dog, cup*), these objects will be salient in the environment, facilitating learning of this referential relationship. Once infants understand referential function with respect to nouns and objects, they will evolve more specific expectations about referential function, linking particular types of words (nouns, verbs, adjectives like *pretty*) to particular types of relations among objects (object categories, actions, object properties). Interestingly, content words like nouns, verbs and adjectives also tend to be perceptually salient (they tend to receive greater stress and have more interesting melodic contours). Hence perceptual saliency supports infants in this important task of mapping words to objects and events in the world.

In a classic study, Baldwin and Markman (1989) explored the first steps in acquiring word–object relations. Baldwin and Markman argued that children were unlikely to pay much attention to correlations between labels and objects unless they had a general expectation that the sound patterns uttered by adults were connected to things in the external environment. They proposed that a way to test for the presence of this expectation was to investigate whether infants would pay greater attention to objects that were labelled. They showed infants aged 10–14 months completely novel objects, like snorkels, padlocks and flippers. Some of the novel objects were labelled by adults ("See the snorkel? That's a snorkel"), and others were presented in silence. The question was whether the infants would show more attention to the novel objects that were labelled. During the experiment, the toys were placed within reach of the infants for 60-second periods, and the infants were allowed to play with them. During the labelling condition, the toy was repeatedly labelled during the 60 seconds (resulting in approximately ten repetitions of the label). In the control condition, the toy was presented in silence. Looking time at the toy was measured in each case. The results showed that, even by the age of ten months, the infants showed more looking to the objects that were named than to the objects that were not named. In a follow-up study, Baldwin and Markman (1989) compared the effects of pointing to the novel toy to increase attention to it with both labelling the toy and pointing to it at the same time. This time, two toys were presented in one trial, and the experimenter picked out one of them by either pointing to it repeatedly, or pointing to it while naming it. The time that the infant spent looking at each toy during a subsequent play period was then measured. The infants spent reliably longer looking at the toys that were both labelled and pointed to. Baldwin and Markman argued that labelling sustained infants' attention to objects, facilitating the establishment of word–object relations.

Baldwin and Markman (1989) argued that labelling novel objects helps to sustain the infant's attention, thereby facilitating the establishment of word–object relations.

In fact, infant attention to objects is intimately related to word learning. Yu and Smith (2012) put head cameras on infants aged 18 months and gave them novel toys to play with. The infants' parents were also present, and they were encouraged to name the novel toys in a natural way as part of interactive play. Frame-by-frame analysis of the head camera images revealed that the ways in which the infants grasped the novel toys often meant that they only had one single object in view when the parent was providing the novel label. In such instances of selective attention, the infants successfully learned the novel label. If the parent happened to name a novel toy during an interaction that did not afford the infant such selective attention, the infant did not learn the novel label. This study suggests that learning the intended referent of a novel label may not be as difficult for infants as is sometimes assumed. Infants' own active exploration of objects may naturally lead them to isolate new objects of interest, thereby reducing referential ambiguity. As will be recalled, Csibra has made the important claim that infants interpret objects as *symbols* of object kinds, rather than as one piecemeal object after another (Csibra & Shamsudheen, 2015). In situations of natural pedagogy, infant actions may serve to visually isolate a new object, while interacting adults provide the name for the object and simultaneously use ostensive cues, such as making eye contact, speaking in IDS, and saying the infant's name. Accordingly, object labelling seems likely to offer infants powerful culturally relevant symbolic learning experiences that will affect both their conceptual and linguistic development.

More recently, extensive database studies have enriched these conclusions. For example, Swingley and Humphrey (2018) analysed 130,000 utterances by mothers to their young children taken from the CHILDES online corpus (MacWhinney, 1995), utterances which contained around 455,000 words. Swingley and Humphrey were interested in which features of the words used by mothers could predict children's language comprehension at 12 and 15 months and language production at 15 months. Their analyses showed that the infants were more likely to know and say words that were more frequent, and also that were more frequently uttered *in isolation* (presumably as this removed the need to segment the words from ongoing speech). The infants were also more likely to know and say words that were *concrete* (i.e., that referred to specific identifiable objects or events), for both nouns and verbs. Hence verbs like 'kiss', 'eat' and 'hug' were more likely to be known than verbs like 'hurry' or 'make'. Interestingly, each mother's data predicted learning best in their own children, supporting conclusions from observational studies that the individual learning environments offered by families are related to children's language development (Hart & Risley, 1995). Nevertheless, there was consistency across mothers regarding the properties of the words that they produced, properties that made the words easier for one-year-olds to learn. Word frequency, the isolated production of individual words and the concreteness of those words were all important properties aiding learning, as well as the length of the words and the overall utterance length.

Waxman and Markow (1995) explored how infants make links between a particular kind of word–object relation, that between nouns and categories. Waxman and Markow argued that infants approach the task of lexical acquisition with a bias to interpret words applied to objects as referring both to that object and to other members of its kind. This hypothesis was tested by giving infants aged 12–13 months of age different sets of objects to manipulate, and then measuring how long they attended to each subsequent object. In the Noun condition, the objects were named for the infant. For example, four toy cars might be presented one by one for 30 seconds each. In the Noun condition, the experimenter said "Look, a car" (basic level) for each one, and added "Do you like the car?" as she removed each toy. In a control condition (No Word condition), the experimenter simply said "Look at this" for each toy. Waxman and Markow argued that, if infants interpret nouns applied to objects as referring both to that object and to other members of its kind, then attention to each subsequent object should decline more rapidly in the Noun condition compared to the No Word condition. Basic level nouns were also compared to nouns at the superordinate level. Here, the infants might be handed four different toy vehicles, with the experimenter saying "Look, a vehicle" for each toy.

In a subsequent test phase, infants were then given both another novel object from the familiar category (e.g., another car), and an object from a different category, such as a toy aeroplane (or a toy tool for the superordinate condition). The question was whether they would prefer to manipulate the object from the novel category (here, either the aeroplane [basic level] or tool [superordinate level]). Waxman and Markow (1995) found different results depending on whether the objects were labelled at the basic level or at the superordinate level. At the superordinate level, both predictions were confirmed. The infants showed significantly less attention to the labelled objects ("Look! A vehicle") compared to the objects in the No Word condition, and they showed significantly more attention to the novel object at test (the aeroplane). At the basic level, performance in the two conditions was comparable. Waxman and Markow argued that nouns at the superordinate level do indeed highlight categories of objects for infants.

Experimental work has also shown that labelling distinct objects with distinct names appears to highlight their individuality. Recall the experiment by Xu and Carey (1996), considered in Chapter 2. In that study, infants aged ten months appeared to have difficulty distinguishing perceptually distinctive toy objects. However, work by Baillargeon showed instead that if the objects had distinctive functions, then even four-month-olds produced individuation responses (Stavans & Baillargeon, 2018). Xu (2002) revisited her original paradigm, to see whether providing labels would support object individuation. Xu (2002) contrasted a toy duck with a toy ball, and tested nine-month-old infants. The objects were again shown emerging from behind the screen one at a time, but this time each appearance was labelled with a distinctive noun ("Look, a duck", or "Look, a ball"). In a control condition, each appearance was labelled with the phrase "Look, a toy". At test, the infants were shown either one or two objects. The expectation was that the infants would look longer at the single object. In the experimental condition, ten of the 12 participating infants looked significantly longer at the unexpected, one-object outcome. This did not happen in the control condition, where ten out of 12 infants spent more time looking at the expected two objects. Accordingly, labelling plays an important role in establishing objects as distinct.

Perhaps unsurprisingly, *consistency* of labelling also appears to be important for learning. Waxman and Braun (2005) used the same experimental procedure as Waxman and Markow (1995) to investigate learning new words when a consistent novel noun was introduced to babies compared to a variable novel noun. Small toys were used to make up sets of objects. Infants were then shown a set in one of three different conditions. In the two labelling conditions, the experimenter named each toy with either a consistent novel label ("Look, it's a keeto!... Yes, it's a keeto" for each toy), or with a variety of labels ("Look, it's a –. Yes, it's a –" *keeto, bookoo, dimbee* respectively). In a control condition, no label was provided. The experimenter simply said "Look! Look here!" for each toy. At test, the infants were shown two objects, a new object from the familiar category, and a new object from a new category (e.g., tools). Exploratory behaviour (looking, manipulating) to each object was then recorded. Waxman and Braun found that infants in the consistent noun condition showed a significant preference for the toy from the labelled category. Infants in the variable noun condition did not. Infants in the variable noun condition were in fact indistinguishable from infants in the control condition, suggesting that simply hearing novel labels in the presence of a new set of objects does not lead to the formation of inclusive categories. Applying the *same* name to a set of distinctive objects has the conceptual effect of supporting categorisation. A natural consequence of Waxman's argument about mapping commonalities is clearly that early word learning has a reciprocal effect on conceptual development.

Linguistic influences on early concepts

The influence of language on conceptual understanding is not limited to labels for objects, however. As Clark has pointed out (Clark, 2004), while children's earliest conceptual representations of objects, relations and events underpin early word learning, languages differ subtly in how they encode experience. As children begin to learn particular languages, therefore, it might be expected that linguistic differences will affect which aspects of conceptual categories become most salient. Words will draw attention to some aspects of a category and not to others. A good example is the domain of spatial relations. As we saw in Chapters 1 and 2, infants are aware of spatial relations such as support and containment in the first year of life. Clark notes that different languages partition these spatial concepts differently (see Figure 6.4). In English, we use 'in' to match containment, and 'on' to match support. However, we also use 'on' to match attachment to a surface. We say "the cup is on the table" and "the fridge magnet is on the fridge", even though in the first case the cup is on top of the table (support in the horizontal plane), and in the second the magnet is attached to the door of the fridge (support in the vertical plane). In contrast, Dutch has three different words for these three types of spatial relation. In Dutch, being on the table is denoted by 'op', and being on the door is denoted by 'aan'. In Spanish, the same word 'en' is used for all three types of spatial relation. The word 'en' is used for 'in', 'on' and for attachment in the vertical plane. Clark's argument is that these linguistic differences will lead to differences in the conceptual encoding of space. This interesting idea remains to be tested empirically.

KEY TERM

Over-extension
When a child uses a single label inappropriately to refer to multiple objects.

A different way in which language might shape concepts is suggested by the phenomenon of over-extension in children's language production. Over-extension is when a child uses a single label to refer to multiple objects. For example, the child might use the label 'dog' to name not only dogs, but also lions, cats and horses (Fremgen & Fay, 1980). Over-extension is generally found before around 2.5 years of age. Other examples of over-extensions include using the label 'ball' for apples, grapes, bell-clappers and other round objects, or the label 'tee' for sticks, a cane, an umbrella, a wooden board and other stick-like objects (Clark, 2003). One hypothesis is that over-extensions reflect the fact that children have less well-differentiated conceptual categories than adults do. For example, objects of similar

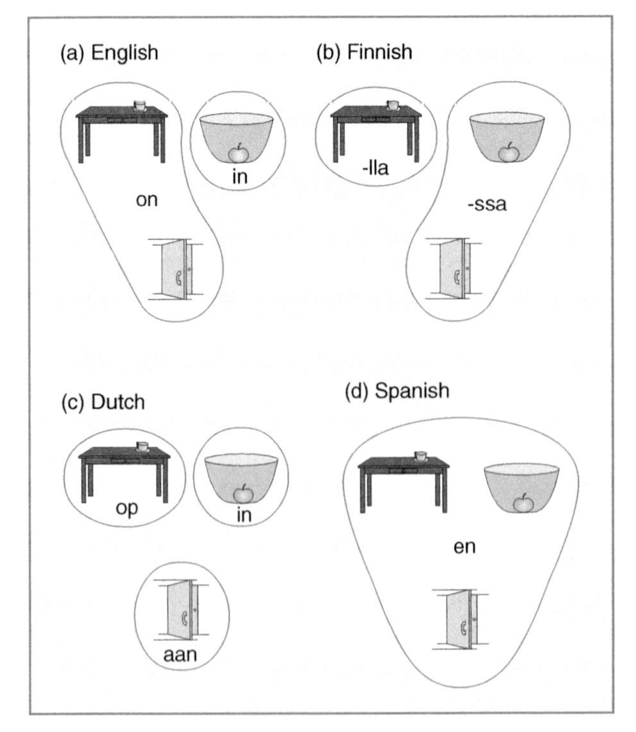

FIGURE 6.4
Linguistic terms for three static spatial relations compared for English, Finnish, Dutch and Spanish, for talking about the locations of the cup, apple and handle in the settings picture. From Clark (2004). Copyright © 2004 Elsevier. Reproduced with permission.

shape may be grouped into the same category of 'long, thin, inanimate things' (see Clark, 1973). According to this hypothesis, as more words are acquired they will cause the child to differentiate objects within these rather general conceptual groupings. The acquisition of more verbal labels will result in meanings no longer being under-specified. Hence language will change conceptual groupings. If over-extension really does reflect the child ascribing a more general meaning to a word than adults do, then over-extensions should be as frequent in comprehension as in production. This does not seem to be the case.

Fremgen and Fay (1980) provided an empirical test of the frequency of over-extensions in comprehension versus production. They studied children aged from one year two months to two years two months, visiting them at home. Following a play period with the experimenter, the child was shown a series of black-and-white line drawings of objects or animals for which the child's mother had previously reported over-extension. The child was asked to name each picture. For example, if the child used the label 'dog' to refer to other animals, the child might be shown pictures of a cow, a cat, a horse and finally a dog. This formed the production test. In a subsequent comprehension test, the child was shown sets of four pictures. Two of these pictures in a given trial were irrelevant to the label being tested, and one of them had been over-extended in production. For example, if the child was being asked to show the experimenter the 'dog', the pictures might be of a dog, a cat, a vase and a car. The results of the experiment were extremely clear. The children in the study never over-extended the meanings of words during the comprehension test, even though the test was based on over-extensions produced by the same child during the production phase of the experiment. Fremgen and Fay (1980) argued that children's knowledge of the meaning of words was much more precise than was suggested by their utterances. They suggested that over-extensions occur because young children are stretching their vocabularies to the limit in their efforts to communicate. When the correct vocabulary item is lacking, the child simply substitutes an item that is present that is similar enough in meaning to what she wants to express. Again, communication is at the heart of linguistic behaviour.

Fast mapping

By the age of around two years, children are acquiring approximately ten new words a day. This high rate of learning novel 'word-to-world' links suggests that a powerful form of exclusion learning must be at work. Children must be rather good at rapidly narrowing down the potential meanings of a new word. This ability to form quick and rough hypotheses about the meaning of a new word has been termed 'fast mapping' (Carey, 1978; Heibeck & Markman, 1987). Using the context in which new words are encountered (non-linguistic information), and their position in a sentence (linguistic information), fast mapping enables the learner to eliminate potential candidates for the meaning of a new word rapidly and efficiently. Even two year olds can quickly infer the meaning of new words for colour, shape and texture on this basis (Heibeck & Markman, 1987).

To explore fast mapping experimentally, Heibeck and Markman taught children aged two, three and four years novel words such as 'turquoise', 'rectangle'

KEY TERM

Fast mapping
Rapidly narrowing down the potential meanings of a new word by excluding other possibilities.

and 'fibrous' using a 'helping' task. The children were asked to help the experimenter by fetching her certain items from a chair in the corner of the room. For example, in the colour condition, the child might be told "Oh, there's something that you could do to help me. Do you see those two books on the chair in the corner? Could you bring me the turquoise one, not the red one, the turquoise one?" The novel colour words were presented via books, and were contrasted with familiar colour words like 'red' and 'blue'. The novel shape words were presented via different trays, and the novel texture words via different little boxes covered with material. Familiar shape and texture words like 'round', 'square', 'fuzzy' and 'smooth' were used to contrast with the unfamiliar shape and texture words in each case.

Heibeck and Markman (1987) reported that the children comprehended the novel words at levels well above chance at all ages. Strongest performance was found for the shape words, followed by the colour words and finally by the texture words. This reflected existing vocabulary entries: in a vocabulary assessment, the children knew 80% of the colour and shape terms tested, but only 36% of the texture terms. Girls comprehended the novel words better than boys. As a further test of word learning, the children were asked to provide a proper contrast for the new word (e.g., to contrast a new texture word with a known texture word). The experimenter would select a novel item, and say "See this box? It's not fibrous, because it's…?" If the child said a texture word like "soft" or "fuzzy", the child was credited with domain comprehension. If the child said a familiar word from a different domain (e.g., "It's not fibrous because it's blue"), domain comprehension was not credited. Heibeck and Markman (1987) found that older children were better than younger children at this task. Sixty-three per cent of two-year-olds answered these questions correctly, compared to 90% of three-year-olds and 96% of four-year-olds. Shape words were again easier than colour words, which were easier than texture words. Finally, the children were asked to produce the novel words that they had learned. Novel word production was much poorer than comprehension, with only 43% of children being able to produce the novel shape words, 4% the novel colour words and 8% the novel texture words. There were no age or gender differences. Clearly, fast mapping is a very efficient comprehension strategy, but more protracted learning may be required for accurate production.

Although first conceptualised as a dedicated language mechanism, fast mapping can be found for other types of learning as well. Markson and Bloom (1997) taught three- and four-year-old children novel names for objects, and novel facts as well. Retention of the novel names and facts was tested either immediately, following a one-week delay, or following a delay of one month. The novel words were taught via a measuring game. Children were given novel objects to measure with the experimenter, such as a plastic tube and a rubber disc. Some familiar objects were also present, such as a pencil and a ruler. During the game, the child was introduced to a novel label such as 'koba'. The child was told "Let's measure the koba. We can count these to see how long the koba is", or "Let's use the kobas to measure which is longer. Line up the kobas so we can count them." The novel facts were also introduced via the game. For example, in the latter scenario the experimenter said "We can use the things my uncle gave me to measure which is longer. My uncle gave these to me." To test retention, all ten objects used in the measuring game were presented together. The children were asked to show the experimenter which one was the koba, and which object had been given by the uncle. Children remembered the correct word–object mapping at levels significantly above chance at all retention intervals, and also remembered which novel object had been given to the experimenter by her uncle. Markson and Bloom argued that because novel facts were retained as well as novel labels, fast mapping was not special to word learning. However, a control condition requiring learning of a novel fact presented visually suggested that there did seem to be a connection between fast mapping and information conveyed through language. In this control condition, the children had to remember where to place a sticker on a novel object ("Watch where this goes. This goes here. That's where this goes"). When asked to put the sticker where it should go, they showed significant forgetting, and by one month performance was at chance levels. Hence fast mapping does not apply to any arbitrary retention task.

Further evidence that fast mapping might be a general cognitive ability of exclusion learning comes from evidence that dogs, too, can learn novel words. As dogs do not develop language, this implicates general auditory learning and memory mechanisms rather than a dedicated language-learning mechanism. The dog 'Rico' was a pet border collie, whose owners claimed that he knew the words for over 200 items (mainly for children's toys and balls). Rico had been learning label-object pairings since the age of ten months, and was rewarded with food or play for fetching named items from around the owners' flat. His abilities to fast map were tested experimentally by Kaminski, Call and Fischer (2004). First, Rico's knowledge of familiar labels was tested by asking him to fetch things ("Fetch the sock"). Rico correctly retrieved 37 out of 40 items. A novel object was then placed in another room along with seven familiar objects. Rico was first requested to bring one of the familiar objects, and was then asked for the novel object. There were ten trials with novel objects overall, and Rico was correct in seven out of ten trials. Retention was tested for these items one month later. Each novel object was placed in a room with four other novel objects and four familiar objects. Rico was correct on 50% of trials. Although Rico may appear to be an exceptional dog, another dog Chaser has learned over 1,000 unique words (Pilley & Reid, 2011). Clearly, therefore, dogs can also acquire the expectation that sound patterns uttered by adults are connected to things in the external environment. Furthermore, dogs can learn specific word–object pairings by fast mapping, and store the novel pairings in memory. Perhaps surprisingly, the perceptual and cognitive mechanisms required to compute word-to-world links are not special to humans. The fact that dogs are also quite good at intention-reading (e.g., Kaminski, Schulz & Tomasello, 2012) may explain why they are able to acquire so many novel lexical items.

Cognitive neuroimaging of lexical development

It is technically challenging to use neuroimaging to explore how infants build a lexicon, as a measure is needed that tracks the integration of the acoustic patterns representing specific concepts into the meaningful contexts provided by other acoustic patterns representing other concepts. That is, words have to be interpreted within sentence or conversational contexts. This means that a brain response like the MMN, which is very sensitive to acoustic differences (for example between phonemes, stress patterns or words), is not in itself a semantic measure. Further, measures of semantic integration will also inevitably measure conceptual development and the organisation of semantic memory (see Chapters 5 and 9), hence such measures will not be language-specific.

In adult studies, the most widely used index of semantic integration in the mental lexicon is an ERP called the N400, measured by recording EEG. The amplitude of the N400 is thought to be associated with the integration of a potentially meaningful stimulus into the current semantic context. For example, adults show an MMN when integrating a word into a sentence, or a picture into a picture story (see Friedrich & Friederici, 2006). The classic technique for eliciting an N400 is to give adults sentences to read. The electrophysiological response to a specific word that is either semantically congruent or semantically incongruent with the rest of the sentence is then measured. For example, the second word in the sentence "the *shirt* has been ironed" makes sense semantically. The second word in the sentence "the *storm* has been ironed" does not. The electrophysiological response to the second word in each sentence can then be compared. Recognition of the semantic incongruency of the second sentence is shown by an increased negativity at around 400 ms when reading this sentence. This increased negativity is the N400.

There is some debate about how to characterise an N400 in the immature brain. Mills and her colleagues measured a significant negativity occurring 200–400 ms after stimulus onset that they called a 'word recognition ERP', but their task was essentially phonological (Mills et al., 2004). Mills et al. (2004) studied 14-month-olds and 20-month-olds. Three categories of words were compared. The first category comprised words already known to the infants, such as 'bear', 'cup' and 'milk'. The second category comprised non-words generated by changing the initial phoneme, as in 'gare', 'tup' and 'nilk'. These non-words shared high phonemic similarity with the familiar words. A third category of non-words were selected to be phonetically dissimilar to the familiar words, for example 'kobe', 'mon'and 'keed'. Each item was presented six times randomly interspersed with other items from the same category, and ERPs were recorded to the different categories. Mills et al. (2004) found age differences in the brain responses shown by the 14-month-olds versus the 20-month-olds. The younger infants showed similar brain responses to the familiar words ('bear') and the phonetically similar non-words ('gare'). These items were distinguished from the phonetically dissimilar non-words ('kobe'), but were not distinguished from each other. The 20-month-olds showed differential brain responses to all three categories. The amplitude of the negative brain responses around 200–400 ms after onset (N200–N400) was significantly larger to the familiar words, while the two types of non-word ('gare' and 'kobe') did not differ from each other. Mills et al. argued that, at 14 months, the neural responses indicated mistaken recognition of items like 'gare' as the real word targets. By 20 months, the children were showing neural responses only to known words that were phonologically correct in all details.

In order to measure *semantic integration* in early language comprehension, Friedrich and Friederici (2004) developed a paradigm linking words to pictures. All words selected for the study were at the basic level (see Chapter 5). During the experiment, the infants, who were aged 19 months, watched a screen which displayed pictures of familiar objects. After each picture appeared, it was named using either the correct or an incorrect label. The same pool of 44 early-acquired words were used as labels and each word appeared twice, so that semantic congruency depended on the picture context. An N400-like semantic congruency effect was then sought in a broad temporal range (200–1500 ms). The response to incongruous words was found to be significantly more negative in these 19-month-olds from around 700 ms onwards. Friedrich and Friederici argued that this broadly distributed negativity to the incongruent words was an infant N400. An earlier negativity between 150–400 ms (similar to that measured by Mills and her colleagues) was also observed, but this negativity was greater to *congruous* words. Friedrich and Friederici argued that this early response reflected a priming effect of the picture (an early 'context' effect). The presence of a picture made the infants expect to hear a real word containing certain phonemes.

Friedrich and Friederici (2005) then extended their paradigm to 12-month-olds. For these younger infants, no effect of semantic incongruency was observed in the picture–word paradigm. The 12-month-olds did however show the early negativity to congruous words shown by the 19-month-olds, between around 100–500 ms. Friedrich and Friederici argued that this early negativity demonstrated that the younger infants did have lexical-semantic knowledge about the words used in the study. Extra experimental conditions including pseudo-words and non-words showed that the 12-month-old infants also had well-developed knowledge of the phonotactics of the language, as they distinguished real words from pseudo-words and non-words. Friedrich and Friederici (2005) argued that mechanisms of semantic integration were not yet mature in 12-month-old infants.

As in behavioural work, of course, the absence of an effect (a negative result) cannot be used to argue that a certain capacity is lacking. Nevertheless, longitudinal investigation (e.g., showing the emergence of an N400 at different ages in different children) could be informative. Friedrich and Friederici (2006) were able to introduce a longitudinal component for the 19-month-olds measured in their 2004 study. When the children in Friedrich and Friederici (2004) reached 30 months of age, they were given a standardised language development test. They were then grouped on the basis of their expressive language abilities. Children with age-adequate skills at 30 months were found to have displayed an N400 at the age of 19 months. For children who had deficits in expressive language skills at 30 months, however, the N400 at 19 months of age was found to be absent. This is an interesting finding. However, the absence of a neural response does not by itself show that semantic integration is not occurring, and hence lack of a neural response cannot provide an unambiguous marker of risk for later language impairments.

More recently, Friedrich and Friederici (2017) have extended their para digm to three-month-olds. In a variant of their picture–word task, these younger infants were taught novel words for pictures of novel objects using an observational learning task. The infants consistently observed pairings of the novel objects and words. The objects were shown first, and then the word was spoken, enabling the assessment of ERPs time-locked to the objects versus the words. Eight specific object–word pairs were presented eight times while EEG was recorded. A memory test was then presented, in which either the correct label was paired with an object, or an incorrect label was given. Friedrich and Friederici (2017) were able to show via the neural measures that the infants learned both the objects and the words. Regarding learning object–word associations, a late negative component over the left hemisphere discriminated between correct and incorrect pairings. However, this component did not show the spatio-temporal characteristics of the 'infant N400', Accordingly, Friedrich and Friederici argued that the three-month-olds were storing 'proto-words' rather than lexical items with referential meaning. Although the infants were learning the statistics of the object–word pairings, the lack of an infant N400 was said to show that these rapidly formed associations did not have lexical (referential) meaning. Further support for the 'proto-word' interpretation of came from the fact that the associations had been forgotten by the following day, when no significant ERPs were found for the three-month-olds. However, six-month-olds tested in the same paradigm did show an infant centro-parietal N400 (Friedrich & Friederici, 2011). This was interpreted to show that the older infants *were* storing lexical information. However, these older infants also failed to show retention on the following day, as the infant N400 was no longer significant (Friedrich & Friederici, 2011). Nevertheless, the older infants did show some ERP evidence for retaining the word–object associations. Again, it is difficult to draw strong developmental conclusions from the absence of various ERPs. Nevertheless, Friedrich and Friederici have been able to show brain activation related to word learning in infants as young as three months of age.

GRAMMATICAL DEVELOPMENT

Grammatical development encompasses syntactic development (the set of grammatical rules that determine how words can be combined into sentences and phrases) and morphological development (the set of rules governing the internal structure of words – we can say 'burglary' but not 'stealery'). The basic unit of morphological analysis is the morpheme. A word like 'dog' is a single morpheme, but the word 'dogs' comprises two morphemes, since the additional phoneme /s/ conveys plurality and hence is also a morpheme. Sometimes morphology is inflectional, as when we add verb endings like *–ing* and *–ed*. At other times morphology is derivational, as when we create new words by adding affixes (*un* + *happy* = *unhappy*) or suffixes (*make* + *er* = *maker*). All of these grammatical rules appear to be acquired by young children on an implicit basis, by listening to the adults around them.

When young children first produce combinations of words, at around 20–24 months, their usual aim is to convey additional meaning. The two-word stage marks a turning point in grammatical development, with many core constructions rapidly acquired between the ages of two and three years. A child who says "No bath" is conveying a different meaning from a child who says "Bath". A child who says "Wet doggie" is conveying a different meaning from a child who says "doggie".

TABLE 6.1 Semantic relations underlying first word combinations in English and Italian (adapted from Braine, 1976)

Semantic functions	English examples	Italian examples
Attention to X	See doggie!	*Gadda bau*
Property of X	Big doggie	*Gande bau*
Possession	My truck	*Mia brum-brum*
Plurality or iteration	Two shoe	*Due pappe*
Recurrence	Other cookie	*Atto bototto*
Disappearance	Daddy bye bye	*Papà via*
Negation or refusal	No bath	*Bagno no*
Actor-action	Mommy do it	*Fa mamma*
Location	Baby car	*Bimbo casa*
Request	Have dat	*Dà chetto*

These first word combinations have various semantic functions (in the examples here, these functions are negation and 'property of X', see Bates et al., 2001). Cross-language data suggest that young children everywhere are trying to get across the same basic stock of meanings. These are possession, location, volition, disappearance/reappearance, and aspects of transitivity (in, on, etc.). Examples of the basic stock of meanings are shown in Table 6.1.

Originally, children's facility in acquiring grammatical rules was interpreted to mean that grammar was a language universal. All children were thought to acquire language on the same schedule and in the same way. It was thought that telegraphic combinations of uninflected words in ordered strings appeared first ("more biscuit", "Mary come"), followed by inflections like *–ing* and *–ed* ("me ironing that"), and then by function words ("Where's the spoon?"). However, cross-language research quickly showed these universals to be non-existent. For example, English has relatively little inflectional morphology, but word order is very important. Hence an English child is just beginning to produce inflected forms like "foots" and "breaked" by around 2.5 years of age. In contrast, each inflection in Turkish is a stressed syllable, and there are almost no irregular forms (like 'break/broke'). This exceptional regularity and phonological salience means that children growing up in Turkey have pretty much mastered inflectional morphology for nouns and verbs before the age of two.

Spoken language errors like "foots" and "breaked" seem to be quite frequent between the ages of two and five. Originally, this was thought to provide good evidence that young children were figuring out the syntactic rules underlying linguistic utterances. Over-regularisation of the past tense construction *–ed* in English was particularly intensively studied. Once children have acquired the past tense morpheme *–ed*, it was thought that they over-applied it to all verbs, thereby displaying so-called 'U-shaped' developmental growth. Development was 'U-shaped' because this over-regularisation followed a period of correct usage. Prior to the age of around 2.5 years, children use the past tense forms that they hear from adults. They say "broke", "came", "went", and so on. However, this correct usage is then followed by a period of over-regularisation. The same children start saying "breaked", "comed" and "goed". Finally, children appear to learn that there

KEY TERM

Over-regularisation
A tendency to over-regularise irregular forms of words such as past tense and plurals (e.g., *goed* and *mans*).

are exceptions to the rule, and begin to display appropriate use of irregular and regular past tense forms. This U-shaped pattern of over-regularisation errors was for a long time taken to support the idea that grammatical development depends on the acquisition of rules.

However, when more systematic studies of over-regularisation were conducted, it was found that over-regularisations were actually quite rare. In a study by Marcus et al. (1992), 11,521 irregular past tense utterances from the spontaneous speech of 83 children were comprehensively analysed. It was found that only 2.5% of irregular verbs were over-regularised, suggesting that over-regularisation errors are quite rare. Once frequency was taken into account, it became clear that irregular forms that were heard frequently by children were rarely over-regularised (e.g., 'goed' was rare). In fact, there was a direct connection between how often a parent used an irregular form and how often a child over-regularised it. The more often a parent used a particular irregular form, the less likely the child was to over-regularise it. Over-regularisations were more likely for verbs that were not frequent in spontaneous conversation (e.g., 'build'), and many individual verbs were used in both the correct and over-regularised forms for a period of some months or even years. However, all children showed a period of correct usage before the first over-regularisations appeared. Marcus et al. suggested that over-regularisations were a consequence of children beginning to mark tense intentionally. When they first did this, they relied on memory of the relevant irregular forms. When memory failed, they treated irregular forms as though they were regulars, and consequently they made errors.

Children's ability to use rules can also be studied directly. In a seminal study, Berko (1958) tested children's implicit knowledge of grammatical rules by using an analogy task based on nonsense words. The children, who were aged 4–7 years, were told that they were going to look at some pictures. The pictures were of objects, cartoon-like animals and of men performing various actions. The experimenter would point to a picture and read some accompanying text (see Figure 6.5). The child was required to fill in the missing word. In the example concerning plural endings shown in Figure 6.5, the children were meant to say "Now there are two wugs." Berko also explored the use of appropriate rules for creating the past tense ("This is a man who knows how to spow. He is spowing. He did the same thing yesterday. What did he do yesterday? Yesterday he? [spowed]"), the rules for generating possessives ("The niz's hat"), and knowledge about the rules for adjectives ("This dog has quirks on him. This dog has more quirks on him. And this dog has even more quirks on him. This dog is quirky. This dog is? [quirkier]. And this dog is the? [quirkiest]").

This is a wug.

Now there is another one.
There are two of them.
There are two ________.

FIGURE 6.5 An example of the text used by Berko (1958) to test children's implicit knowledge of grammatical rules.

Berko reported that the children were in general able to handle plural endings, although performance varied with different nonsense words (e.g., 91% of children tested could complete the analogy "wug-wugs", whereas only 36% succeeded with "gutch-gutches"). Similarly, children could handle the past tense, although again there was variation depending on the phonological form of the nonsense word (e.g., 78% could complete "hing-hinged", whereas 33% completed "mot-motted"). For the possessive, 84% were correct for "wug" ("the wug's hat"), compared to 49% for "niz" ("the niz's hat"). The adjectival inflections were too difficult for most children, with only one child able to generate "quirkier" and "quirkiest". Berko concluded that young children had a good grasp of English morphological rules. The large variation in performance with the phonological patterning required was explored in terms of the child's possible rules for forming plurals, etc., rather than in terms of the number of real-world analogies that the child may have learned (e.g., bug-bugs, jug-jugs, mug-mugs, plug-plugs, rug-rugs; but only hutch-hutches, clutch-clutches).

The invention of new words by children also reveals their knowledge of word-formation paradigms in the language. Becker (1994) documented new words invented by one boy between the ages of two and five years. The boy produced many delightful inventions over this period. He produced different types of novel nouns, as in "He's a cock-a-doodle-doo" (a rooster), "You have good earsight, Daddy" (meaning sensitive hearing), "He has real sneak shoes" (to refer to the quiet shoes Father Christmas uses to tiptoe into houses), and "That's a nose beard" (meaning a moustache). He also produced novel verbs, as in "I'm gonna horn you – do you want to be horned?" (have a horn blown in your face), and "I wanna tennis it" (hit it with a tennis racket). Verb innovations were rarer than noun innovations.

Becker found that most innovations reflected common word-formation devices, such as compounding or combining single words (e.g., 'earsight'). Bowerman (1982) pointed out that children's gradual learning of implicit constraints on word formation can also be revealed by the decline of certain kinds of innovations. For example, the prefix *un–* can only be used with verbs of a specific type. These verbs denote enclosure or surface attachment. Between the ages of four and seven, Bowerman's daughter Christy produced various innovations involving *un–*, but these innovations only reflected inappropriate affixing when Christy was four. At this age, she produced phrases like "I hate you! And I'll never unhate you or nothing!" By age seven, she was producing innovations like "I'm gonna unhang it", innovations which do follow the hidden rule.

How do adults respond to children's innovations or errors? The role of feedback from caretakers in shaping grammatical development has long been debated. The first naturalistic studies suggested that overt correction of grammatical errors was rare. Caretakers did not often use direct correction, and when they did, it was seldom successful, as illustrated in this example:

Ch. "Want other one spoon daddy."
D. "You mean, you want the other spoon."
Ch. "Yes, I want other one spoon please daddy."
D. "Can you say – the other spoon?"
Ch. "Other one spoon."
D. "Say 'other'."
Ch. "Other."
D. "Spoon."
Ch. "Spoon."

D. "Other spoon."
Ch. "Other – spoon. Now give me other one spoon?"

(from Braine, 1971)

More recently, it has been demonstrated that adults in fact provide extensive feedback when their children make errors in language production, but that they do this by reformulating the child's utterance rather than by overtly correcting it. This was nicely demonstrated in a study by Chouinard and Clark (2003). Chouinard and Clark selected all the spontaneous utterances recorded from five children in the child language database CHILDES between the ages of two and four years, and the subsequent utterances made by the parent. They then took random samples of 200 utterances from six-month slices of the data. They reported that almost 70% of children's erroneous productions at age two years were reformulated by their caretakers. All types of erroneous production were likely to be reformulated, whether the errors were grammatical, morphological, phonological or semantic (errors of word choice). There was no prioritising of grammatical errors. Examples of reformulations include:

KEY TERM

CHILDES
Child Language Data Exchange System, an online database of utterances made by children in the age range 2–4 years and their caregivers.

Child: I want butter mine.
Father: Okay, give it here and I'll put butter *on it*.
Child: I need butter *on it*.

Similar patterns were found irrespective of language:

Child: Une petit de lait (= a little of milk).
Mother: Une petite boite de lait (= a little carton of milk).
Child: Petite boite de lait.

Reformulations also occur via adult use of expansion. Adults often repeat an ungrammatical utterance made by a child by expanding it, at the same time providing the correct grammatical form. Again, this is conversationally more natural than simply correcting the child's production. For example, if the child says "Muffy step on that", the mother might say "Who stepped on that?". The child might reply "Muffy", and the mother then says "Muffy stepped on it". An experimental study by Cazden (1970, cited in Peccei, 2005) compared the effects of expansion on the utterances of 2–3-year-old children at a daycare centre. Three types of feedback were given to the children for 40 minutes a day. One group was given extensive and deliberate expansions of their utterances. A second group spent 40 minutes in which the researcher provided models of adult forms by continuing the conversation (e.g., Child "I got apples". Experimenter "Do you like them?"), and a third group received no special treatment. When language progress was assessed at the end of the study, the children in the expansion group had made no more progress than the children in the control group. It was children in the natural conversation group who had made the most progress.

KEY TERM

Mean length of utterance (MLU)
A measure of language development in terms of the number of morphemes in utterances; useful measure of increases in language development in early childhood.

Syntactic development is also measured in terms of a concept called the 'mean length of utterance' or MLU. This term was coined by Brown and Hanlon (1970), who followed the language development of three children, Adam, Eve and Sarah, in great detail from the age of around 18 months. Mean length of utterance is measured in morphemes, or units of meaning. 'Dog' has one morpheme, 'dogs' has two morphemes, 'big dogs' has three morphemes and 'big dogs ran' has five morphemes (because the use of the past tense of 'run' adds two morphemes). These utterances would yield an MLU of 1, 2, 3 and 5 respectively. Eve, Adam and Sarah showed strikingly different rates of grammatical development using the MLU measure. Whereas Sarah was already producing utterances with an MLU of 4 by age two years, two months, Eve and Adam did not reach an MLU of 4 until they were three years and five months old.

Bates made the important point that the MLU measure was dominated by the child's ability to use two aspects of grammar, inflectional morphology (*–s*, *–ed*) and grammatical function words such as *the* and *he*. In the checklist study (Fenson et al., 1994), children first began to use morphemes like *–s* and *–ed* at about 16 months. At this age, rather few children used grammatical suffixes. However, by 22 months, the majority of children were using all four of the morphemes studied (*–s*, *–ed*, *–ing*, and possessive 's', as in "Maggie's dolly"). Fenson et al. (1994) argued that sentence complexity might be a better estimate of grammatical development than MLU, particularly after the age of 30 months. They collected data on sentence complexity by giving parents a forced choice task. The parents were asked which kinds of sentences sounded most like "the way your child talks right now". Sentences were paired to reflect increases in complexity, for example "doggie table" versus "doggie on table", or "I want that" versus "I want that one you got". A steady expansion in the complexity of sentences used was documented, with large individual variation. Furthermore, the best predictor of grammatical complexity was vocabulary size.

It thus seems quite possible that children's early utterances do not in fact reflect the range of grammatical constructions imputed to them by adults. Rather, children are acquiring pieces of language. Consequently, grammatical development is intimately related to vocabulary development. This viewpoint has been argued quite compellingly by Tomasello, who has developed a usage-based theory of grammatical development (Tomasello, 2000, 2008). Tomasello has shown that children's early linguistic constructions are most likely to reflect the types of utterance that they hear around them. Consequently, grammatical development depends on the piecemeal acquisition of particular constructions that are good grammatical forms. Children then build upon these early linguistic constructions by using the same pattern-finding mechanisms that underpin learning in other areas of development (statistical learning, categorisation, induction, analogy). In this way, children gradually create the more abstract dimensions that we recognise as grammar. Tomasello's argument is that early linguistic constructions are learned as concrete 'pieces' of language. Children are learning 'communicatively effective speech act forms', and these may well correspond to whole adult utterances (e.g., "I wanna do it", "Lemme see"). For the child, these utterances represent a single relatively coherent *communicative intention*.

Tomasello (2008) marshalled a variety of experimental evidence in support of this viewpoint. Tomasello, Akhtar, Dodson and Rekau (1997) showed that one word or phrase in a multi-word utterance often acts as a 'pivot' in determining the effective 'speech act' (see also Braine, 1963). This pivot utterance can then be paired with a wide variety of words to create novel utterances (e.g., "More juice", "More milk", "More grapes", "Want cup", "Want cookie", "Want dolly"). Tomasello et al. (1997) showed that 22-month-old children could combine totally novel nouns with their pivot words. When taught the label 'wug' ("Look! A wug!") the children could produce phrases like "More wug!" and "Wug gone". In the case of questions, Rowland and Pine (2000) showed that 2–4-year-old children either produced certain linguistic constructions 100% correctly ("How did…?", "How do…?", "What do…?") or 100% incorrectly ("Why I can…?", "What she will…?", "What you can…?"). Again, the children appeared to be acquiring single 'pieces' of language that had a certain communicative function, and sticking with them. Only later did children appear to extract a more abstract knowledge of language. Tomasello proposed that the main learning mechanism underpinning this change was reasoning by analogy.

By Tomasello's account, the acquisition of grammar depends on learning, and development is consequently slow but steady throughout early childhood. Tomasello argues that more studies using nonsense constructions are required in order to get an accurate picture of development, and that these studies must be done across languages. In his own studies with nonsense words, Tomasello consistently finds evidence for slow and steady progress. For example, he finds that children have difficulties in interpreting totally novel verbs. Akhtar and Tomasello (1997) used novel verbs like 'dack' with children aged two years nine months and three years eight months. The children were shown a variety of toy scenarios to depict 'dacking', and were then told to "Make Cookie Monster dack Big Bird". Only 30% of younger children succeeded at this task, although they could all succeed with familiar verbs. By three years eight months, all children could 'dack' without difficulty. Experiments such as these suggest that children learn from the input, using the same general learning mechanisms that they deploy in other domains. Children learn language by recognising patterns, and using these patterns to determine the set of rules governing the internal structure of words and the ways in which words can be combined into sentences and phrases.

Even quite young children can use syntactic structure alone as a guide to what a novel word might mean. Yuan and Fisher (2009) showed that two-year-olds were aware of the different implications of transitive ("Jane blicked the baby!") versus intransitive ("Jane blicked!") constructions. A transitive construction implies two core participants (here, the 'blicker' and the 'blickee'). An intransitive construction implies one core participant (here, the 'blicker'). In their experiment, two-year-olds watched videos of two people talking, using either a transitive or an intransitive construction:

P1: "Guess what! Jane blicked the baby!"
P2: "Hmm, she blicked the baby?"
P1: "And Bill was blicking the duck."
P2: "Yeah, he was blicking the duck."

P1: "Guess what! Jane blicked!"
P2: "Hmm, she blicked?"
P1: "And Bill was blicking."
P2: "Yeah, he was blicking."

The children were then shown two scenarios, an event with two participants (e.g., one person swinging a baby's leg) and an event with one participant (e.g., a person making circles with their arm), and told "Find blicking!" Yuan and Fisher (2009) reported that the toddlers looked longer at the two-person event after hearing transitive constructions, and longer at the one-person event after hearing intransitive constructions. Accordingly, by at least two years of age, sentence structures *in themselves* are statistical patterns that carry meaning.

In fact, using EEG, Christophe and her colleagues have been able to show that two-year-olds are building expectations *online* about the syntactic category of upcoming words in a sentence. Brusini, Dehaene-Lambertz, Dutat, Goffinet and Christophe (2016) taught toddlers new words during an interactive play session with toy props. For example, for the novel word 'poune', the children played various games with the experimenter during which 'poune' was shown to mean putting a saddle on a horse. The key constructions to be used during EEG recording were never used. A week later the toddlers came back to the lab and were shown videos of the same experimenter playing with the same toys (e.g., "Martin chooses a saddle to poune his horse. Meanwhile, the horse is playing with the cat...") while EEG was recorded. During the short videos the novel word was used both correctly and incorrectly. For a correct example, the experimenter might say "Martin pounes it clumsily". For an incorrect example, the experimenter might say "Now the poune is calm". The toddlers showed the same ERP effects as adults shown the same scenarios. When a novel item was used incorrectly, they showed a late positive component at around 700–900 ms, comparable to the P600 which in adults is associated with detecting a grammatical error. The toddlers also showed a left-lateralised early negativity, which occurred before the novel word had ended. This early effect suggested that the computation of syntactic category expectations was occurring rapidly and online as the toddlers were listening to each successive sentence. Brusini et al. (2016) concluded that early syntactic processing is more robust than had been appreciated from behavioural studies. Toddlers are sensitive to the grammatical usage of newly learned nouns and verbs, a skill that would in turn help with lexical acquisition.

PRAGMATIC DEVELOPMENT

Language is about communication, and the 'pragmatics' of language development are about learning how to communicate competently. For example, in order to have a meaningful conversation with another person, you need to take turns in speaking. You also need to judge your contributions to the conversation, making sure that the other person can understand the information that you are conveying. Anyone who has had a telephone conversation with a young child is aware that it takes time for children to learn these aspects of communication. Young children frequently talk about things that have happened to them, assuming that a listener who was not present at the event will understand them. They seem to assume that adults always know what they are talking about. Young children fail to adopt the perspective of their conversational partner. Young children may also switch conversational

KEY TERM

Pragmatics
Understanding how to use language to communicate effectively and socially appropriately.

topic without warning, breaking the implicit 'rules' of discourse. Finally, pragmatics involves being able to use language both socially and appropriately. Young children may be oblivious of the social aspects of dialogue, for example they may be unaware of what is 'rude' or 'polite' in a given context. They may also be unaware of differences in the social status and familiarity of conversational partners that would lead an adult to modify the language that they use.

One of the first attempts to measure early pragmatic development came from Dale (1980). He argued that early pragmatics had been studied in terms of 'speech acts' or 'pragmatic functions', which measured children's ability to produce declarative or imperative functions (e.g., "That's mine", "Gimme that"). Further dimensions requiring inclusion were affirmation versus negation, requests for objects versus requests for information, and reference to immediately perceivable objects and events versus reference to absent objects and events. Dale (1980) decided to try and devise a measurement tool for pragmatic functions during the second year of life. This tool was based on two sets of structured tasks. The first set was designed to elicit declaratives from the child, words or phrases commenting on a state of affairs in the world. The idea was to set up a repetitive task, and then introduce a new element, to see whether the child communicated about this new element. The declarative tasks included dropping blocks into a bucket and then offering the child a doll, rolling balls to the child and then rolling a baby bottle, and tapping a xylophone with a stick and then offering the child a hammer. The imperative tasks were designed so that the child would have to request assistance from the experimenter. For example, a dancing tiger toy was wound up by the experimenter and then allowed to run down, or some attractive toys were presented packaged in a clear plastic container that required opening. Children aged one year, one year three months, one year six months, one year nine months and two years were tested. The children were credited for utterances that responded to the newness of the objects introduced in the declarative tasks, and for the way in which they communicated (e.g., pointing versus verbalising). They were also credited for any imperatives produced, again whether verbal or non-verbal (e.g., removing the experimenters' fingers from a desired toy).

Dale (1980) found that the number of pragmatic language categories produced was highly correlated with age. Naming emerged first, followed by greetings and ritualised forms, and then comments about objects and their attributes. Requests concerning the here and now emerged next, followed by affirmation and then denial. Reference to past and future and requests for absent objects emerged last. There was no relationship between pragmatic development and mean length of utterance (MLU). Dale concluded that the range of pragmatic functions developing in the second year of life was measurable, and was tapping information about language development that was separate from syntactic development. However, the imperative tasks were more successful at eliciting communicative behaviours than the declarative tasks. This made it difficult to generate an overall measure of communicative performance for an individual child.

How do children acquire the social roles for language use? It is difficult to see how children could acquire social speech and politeness routines without direct input from adults. As pointed out by Gleason (1980), learning the social roles for language use is part of the development of social cognition. The language used in many social routines has no intrinsic meaning, rather the child is required to recognise a certain kind of social situation and apply the appropriate formula. For example, the child may have to say "thank you" for a present that she does not actually like, or say "God bless you" when someone sneezes. Adults are not concerned with the truth value of the routine involved, but with the performance of the child.

As discussed in Chapters 3 and 4, the notion of *communicative intent* is core to the development of both language and social cognition. Successful acquisition of the pragmatics of communication clearly requires insight into the mind of another person. In order for a conversation to be meaningful, an assessment must be made of how much your conversational partner understands. These aspects of pragmatics are as important as obeying conventions like turn taking, and are likely to be the focus of future studies of pragmatic development.

SUMMARY

Language acquisition is a remarkable feat, yet most children are competent speakers of their language within three years of being born. They seem to achieve this via powerful general learning mechanisms coupled with a desire to communicate and to be part of the conversation. Children do not learn language from the television or the radio. They learn language when conversational partners are interacting with them, and they learn the language structures used by those partners. The *communicative intentions* of their conversational partners play a key role in which words are learned, and so does the context of learning. The context within which new words are heard usually provides rich clues to the meanings of those new words, and mechanisms such as joint attention facilitate the use of the relevant context by infants. By the age of two years, children are also using the linguistic structure of utterances to gather clues about the meanings of new words, for example whether they are verbs or nouns.

For phonological learning, the input is crucial. Infants rely on auditory statistical learning, attending to the distributional properties of the sounds in the language to extract the crucial phonetic contrasts. Although the brain is particularly sensitive to the prototypical phonetic contrasts that are used by a particular language early in life, continued exposure to more than one language preserves the ability to make phonetic distinctions that are not used by the dominant language. Nevertheless, within the first year of life, monolingual babies focus on the phonetic contrasts that are important for their native language, showing reduced sensitivity to the phonetic contrasts that are important for other languages. General auditory perceptual abilities seem to influence where the physical boundaries for phonetic contrasts are placed by human languages, and general physiological properties of the mouth and jaw seem to influence which sounds are easiest to make and hence which sounds are frequent across languages. Rhythmic and prosodic patterning are critically important. Infants use stress patterns as a way of segmenting words from the speech stream, prosodic cues determine how individual sounds are ordered into words, and patterns of prosodic prominence offer important acoustic cues for accurate neural entrainment to the speech signal as well as statistical clues to syntactic structure. Producing sounds yourself (babbling) also appears to be important for perceptual learning, and babbling typically reflects the frequent sounds and the intonation patterns of the ambient language.

Measurable word learning takes off around the end of the first year, even though infant studies show secure knowledge of certain words within the first six months. For word learning, the 'input' is more complex. Children need to be aware of the context in which new words are being used by their caretakers, and of the communicative intent of the speaker. Adopting a basic premise that the sound patterns uttered by adults must be connected to things in the external environment, infants rapidly make these 'word-to-world' links. Most children acquire the meanings of hundreds of words during their second year. In fact, it is estimated that two-year-olds acquire around ten new words per day. Children's early words tend to be highly relevant to their daily lives, with a special role for their own name and for extremely salient labels like 'Mummy'. Production tends to lag behind comprehension, and so gesture is used to increase communication. By combining gestures with words, two-year-olds can get quite complex messages across. Language also has a reciprocal effect on conceptual organisation, for example of spatial relations.

At the same time as busily acquiring words, children are learning how to put words together. They are learning the set of grammatical rules that determine how words can be combined into sentences and phrases, and the set of rules governing the internal structure of words. Again, the learning involved appears to depend on general learning mechanisms. Children initially learn 'pieces of language' which may correspond to whole adult utterances. Children want to communicate, and so they use statistical learning, categorisation, induction and analogy to reuse structures that enable them to fulfil certain communicative intentions. In fact, the best predictor of grammatical complexity at any age is vocabulary size. Children's brains do not have a special learning device for learning grammar. Rather, children learn grammar from the phrases used by those around them, termed 'usage-based theories'. When young children produce ungrammatical utterances, their caretakers reformulate these utterances as a way of extending the conversation. Children notice these reformulations, once again learning from the input in a *communicative context*. Children do not seem to learn much from direct correction. At the same time, children must learn the pragmatics of effective communication. The 'pragmatics' of language development are about learning how to communicate competently, and this depends particularly on social cognitive development. As we saw in Chapters 3 and 4, social development and language development are intimately linked.

CHAPTER 7

CONTENTS

Causal reasoning and the human brain

7

In the previous edition of this book, I suggested that causal learning was a domain-general skill that played a core role in cognitive development. I also argued that infants have an apparently innate bias to learn about causal relations and to acquire causal explanations. As we saw in Chapter 2, new experimental evidence, including neuroimaging data, suggests that some forms of causality are perceived *directly*, which would provide an explanation for this innate 'bias'. In the same way as other perceptual features such as colour or motion are perceived directly by the visual system, the brain perceives causality directly, and indeed prioritises learning about causal information (Moors et al., 2017). Consistent with this perspective, research conducted over the past decade has also revealed that perceptual information is replete with causal structure for the brain to learn. We now understand that the brain *automatically* encodes this causal structure, via its intrinsic mechanisms for encoding sensory information. Neural encoding utilises activity patterns that are extended in time (via mechanisms such as calcium cascades) and that thereby *in themselves* impose temporal causal structure on memory and on perceptual representations (see Chapters 1 and 2). In my view, the new challenge for developmental psychology is to discover the basic statistical structures in the sensory information surrounding the infant, the statistical structures that will underpin causal learning in different developmental domains. This perceptual statistical information may differ markedly from what we have previously assumed to be core knowledge for different domains.

For example, for acoustic information relevant to learning language, we saw in Chapter 6 that recent research in auditory cognitive neuroscience has led to the discovery of a novel set of acoustic statistics based on amplitude modulation patterns (local variations in signal intensity) and that these statistics play a crucial role in linguistic learning. These acoustic statistics are complementary to more traditional perceptual analyses of the speech stream, which were based on sequential and rapid changes in frequency (pitch). Yet as we saw, Parentese turns out to selectively enhance those amplitude modulation patterns that are foundational for language learning (Leong et al., 2017), as well as to enhance certain pitch-based characteristics such as fundamental frequency. Thus, human cultural practices facilitate the extraction of the statistical structures that govern early learning.

On the other hand, the sensory information relevant to initial cognitive development in physical, psychological and biological reasoning appears to have been reasonably well-described by previous research. For example, perceptual information about motion yields multiple cues relevant to agency and possibly also intentionality, as well as cues about animates versus inanimates (see Chapters 3, 4 and 5). Again, human

KEY TERM

Causal mechanisms
Neurally hard-wired mechanisms that give rise to automatic causal inferences, enabling us to predict the effects that one object may have upon another.

cultural practices facilitate the extraction of this spatio-temporal statistical information. For example, adults unconsciously employ 'motionese' to enhance infant's perception of the causal structure of actions. If given the choice, infants prefer to watch action scenarios in motionese. The extra pausing, the exaggeration of the actions and the repetition that characterise motionese help the infant to understand the *meaning* of the actions and/or the goals of the actor. This perceptual information provides emergent knowledge structures that provide a basis for deeper understanding, for example via family discourse around the goals and intentions of others. Accordingly, for a domain like agency, encoding the statistical structure of events involving motion, particularly goal-directed actions, provides a rich source of causal information. Neural encoding mechanisms automatically encode the causal *structure* of physical events that involve motion, although representing their meaning may require additional mechanisms (as shown by the 'chasing motion' study with deaf toddlers carried out by Meristo et al., 2012, discussed in Chapter 4).

Accordingly, observing the world yields important clues to its *causal structure*. This causal structure can be inherent in the perceptual structure of the events that are observed and experienced by the infant and young child. The evidence considered so far suggests that the causal structure of events plays an important role in interpreting, representing and remembering from very early in development. The evidence that the adult brain prioritises learning about casual events (like launching events) suggests that this important role is to some extent an automatic result of neural learning mechanisms. Nevertheless, the infant and the young child need to go beyond observing and registering events to making *predictions* about likely future events. As noted, the brain can be described as a prediction machine. Research in cognitive neuroscience, particularly visual neuroscience, shows that inherent neural encoding mechanisms are *in themselves* one source of accurate prediction. Via conditional learning, these encoding mechanisms automatically adapt to changes in the statistical patterns presented by environments, extracting summary statistics (Hochstein et al., 2015). Indeed, some cognitive neuroscientists believe that conscious perception *begins* with these summary statistics. However, it is also important for children to be able to notice associations explicitly and then to generate predictions that they can decide consciously to act on. Explicit notising and predicting provide evidence for cognitive causal reasoning. One way of fostering such explicit noticing is asking children to explain why causal events happen as they do, which necessitates reflection on knowledge that may initially be implicit. Reflection is cognitive rather than automatic, enabling explicit explanation-based learning. Accordingly, in this revised book, I argue that the origins of causal cognition lie in neurally hard-wired mechanisms that give rise to automatic causal inferences. Casuality does not require a conscious inference, as argued by philosophers like Hume (1748), but can be directly observed, as argued by Michotte (1963). Children rapidly enhance their automatic inferences via an early and explicit focus on acquiring causal information. By continually asking adults why things are as they are, children extend and enrich these automatic causal inferences into cognitive systems for causal reasoning and scientific reasoning.

Causal understanding is thus fostered by asking trusted and more experienced informers why things happen. It is also generated by children carrying out *interventions*; that is, acting on their environments themselves in order to test out cause–effect relations. Such additional learning mechanisms enable young children quickly to develop causal explanatory frameworks for making sense of events in the world around them. These explanatory frameworks depend on the infant or young child actively manipulating different causes and observing the effects, having access to trusted sources of information, and also actively *reflecting* on events using language. Once a child has worked out how an event was caused, the child can predict or control that event, causing it to happen by intervening in the world in a principled way. An intervention involves imposing a change on a variable in a causal system from outside the system. In this chapter, we will consider different types of research documenting how children become effective at choosing interventions, at making causal predictions and at engaging in explanation-based learning. The importance of trusted sources of information for early development has recently been reviewed by Harris and his colleagues (Harris, Koenig, Corriveau & Jaswal, 2018).

How do children learn about causal structure? Some causal explanatory frameworks can clearly emerge from consideration of the relevant perceptual variables. For example, when examining the development of naïve physics, we saw that perceptual information about object trajectories and relative size and weight leads infants to infer notions like 'transfer of force' to explain whether the collision of one object with another can be expected to set the other object in motion (e.g., Kotovsky & Baillargeon, 1998; Chapter 1). In other foundational domains, for example naïve psychology, we discussed how key variables for the construction of an explanatory framework may be unobservable, requiring infants to *infer* their existence (e.g., mental states, see Chapter 3, or biological essences, see Chapter 5). In the cases of physical and biological causality, when key variables are unobservable, infants and children have to *infer* causal mechanisms to predict the effects that one object may have upon another. The kind of causal mechanisms that are inferred will also depend on prior experience. In most causal learning situations, we have some *prior knowledge* about basic causal features, and this prior knowledge guides our causal inferences. For infants and young children, this prior knowledge is often gained via intentional action, and may not at first be explicitly available for reasoning. Nevertheless, it may still guide their choice of interventions, as knowledge about action may 'bootstrap' learning about causation, and vice versa (Buchsbaum, Griffiths, Plunkett, Gopnik & Baldwin, 2015).

I will thus adopt a semi-historical perspective in this chapter, first reviewing early studies concerning whether children have the ability to make causal inferences. Foundational work in this domain by Tom Shultz (Shultz & Kestenbaum, 1985) took the form of studying the principles of causal inference defined by the philosopher Hume (1748). Shultz proposed that children would only be able to understand causal mechanisms if the underlying processes of causal reasoning were functioning efficiently. Hume had suggested in 1748 that people are exposed to patterns of data, and infer causality from these patterns. Hume argued that we infer causality (that event X causes event Y) when X precedes Y in time, when X is contiguous with Y in time and space, and when the contiguity is regular (see Koslowski & Masnick, 2010). Accordingly, Shultz set out to measure these underlying processes or 'Humean indices' in young children. He devised experimental methods for studying children's understanding of the principles of causal priority, temporal contiguity and covariation.

More recent research in causal reasoning has focused on causal Bayes nets as a means of representing the statistical relations present in a body of evidence. As described at the end of Chapter 1, patterns of conditional probability between experienced potential causes and their effects can be used by the brain to infer *structured causal relationships* via statistical learning. Indeed, causal Bayes nets have been very effective in research on machine learning and in adult cognitive neuroscience (Griffiths & Tenenbaum, 2009). Bayes nets are attractive because they can represent complex conditional relations, such as relations of conditional independence. Even though Bayes nets are very sophisticated, statistical data alone is usually insufficient for inferring a unique causal model, and so structural constraints are also important. Principles such as temporal order, intervention in situations, and real-world knowledge about likely causes and effects (e.g., that a switch is probably a cause of something) are all important for inferring causal structure. Such principles are taken into account in Bayes' theory, because Bayesian models assume some prior knowledge on the part of the learner. For our purposes, this prior knowledge has likely been extracted automatically by the brain of the infant or child via the statistical learning mechanisms documented in earlier chapters. For example, in Chapter 6, we considered an explicit and mechanistic model of how the brain might extract hierarchically nested structure from the speech signal (hierarchically nested amplitude modulations at different temporal rates) and encode this structure in a way that enabled the segmentation of phonological units. For causal reasoning, the basic computational problem is similar. A temporally continuous stream of events (in this case, actions rather than spoken language) needs to be divided into discrete and hierarchically organised units. Some of the basic units can be segmented by using distributional or statistical clues in the input, such as action goals (see Buchsbaum et al., 2015, for detail). Recent research suggests that even two-year-olds show patterns of causal reasoning consistent with Bayesian analyses.

Finally, causal reasoning is also central to the development of scientific reasoning. Understanding the importance of interventions also involves the ability to recognise the relationship between hypotheses and evidence. A major goal of science is to identify and understand the causes of different phenomena, so scientific reasoning can be considered the consummate form of causal reasoning. However, successful scientific reasoning typically requires the consideration of multiple causal factors at once. Even adults can be poor at reasoning simultaneously about multiple causal factors, as we will see.

REASONING ABOUT CAUSES AND EFFECTS

By the age of around three years, children have experienced many different physical causes and their effects. For example, they have many experiences of *cutting* (apples, paper, hair), of *melting* (chocolate, snow, butter), of *breaking* (a toy, a cup, a chair), of *wetting* (washing clothes, rain, having a bath), and so on. One way to investigate the development of causal reasoning is thus to ask whether children know that causal agents can produce transformations in objects, changing them from one state to another. For example, if children are shown a picture of a cup and a picture of a shattered cup, do they know that a hammer is a more likely agent of the causal transformation than a knife or scissors?

Reasoning about the causal transformations of familiar objects

R. Gelman, Bullock and Meck (1980) used three-picture causal sequences based on such transformations to investigate cause–effect reasoning in three- and four-year-old children. They first trained children to read the picture sequences from left to right. Following this training, they showed the children picture sequences that either depicted an object being transformed from its *canonical* form (an intact cup) to its *non-canonical* form (a broken cup), or from its non-canonical form (a broken cup) to its canonical form (an intact cup). The middle picture in the sequence always depicted the causal agent (the correct agents in this example were a hammer or some glue). As well as familiar causal transformations like cups breaking, the children were shown unfamiliar causal transformations like a cut banana being restored to its canonical form using a needle and thread.

In the experiment, the children were shown the picture sequences in an incomplete form, with one of the three pictures missing. Their job was to select the correct picture to fill the empty slot from three alternatives. For example, if the picture of the agent was missing, then pictures of three possible agents were provided as alternatives. If the first or the final picture in the sequence was missing, then the children were shown pictures of the correct object with either a correct or an incorrect causal transformation, and a picture of an incorrect object with the correct causal transformation. Examples of each type of trial are depicted in Figure 7.1.

Gelman et al. found that 92% of the three-year-olds and 100% of the four-year-olds could select the correct causal agent when the middle picture was missing from the canonical stories (e.g., cup to broken cup), and that 75% of the three-year-olds and 100% of the four-year-olds could select the correct causal agent when the middle picture was missing from the non-canonical stories (e.g., broken cup to cup). Performance fell slightly for the younger children when the missing picture was the final item (83% and 100% respectively for canonical sequences, 58% and 100% respectively for non-canonical sequences), and fell more markedly when the missing picture was the first item (66% and 92% respectively for canonical sequences, 58% and 100% respectively for non-canonical sequences). Nevertheless, performance was significantly above chance in all cases. Gelman et al. concluded that preschool children could predict or infer the states of objects changed by a causal transformation, and could also infer the kind of transformation that related two object states.

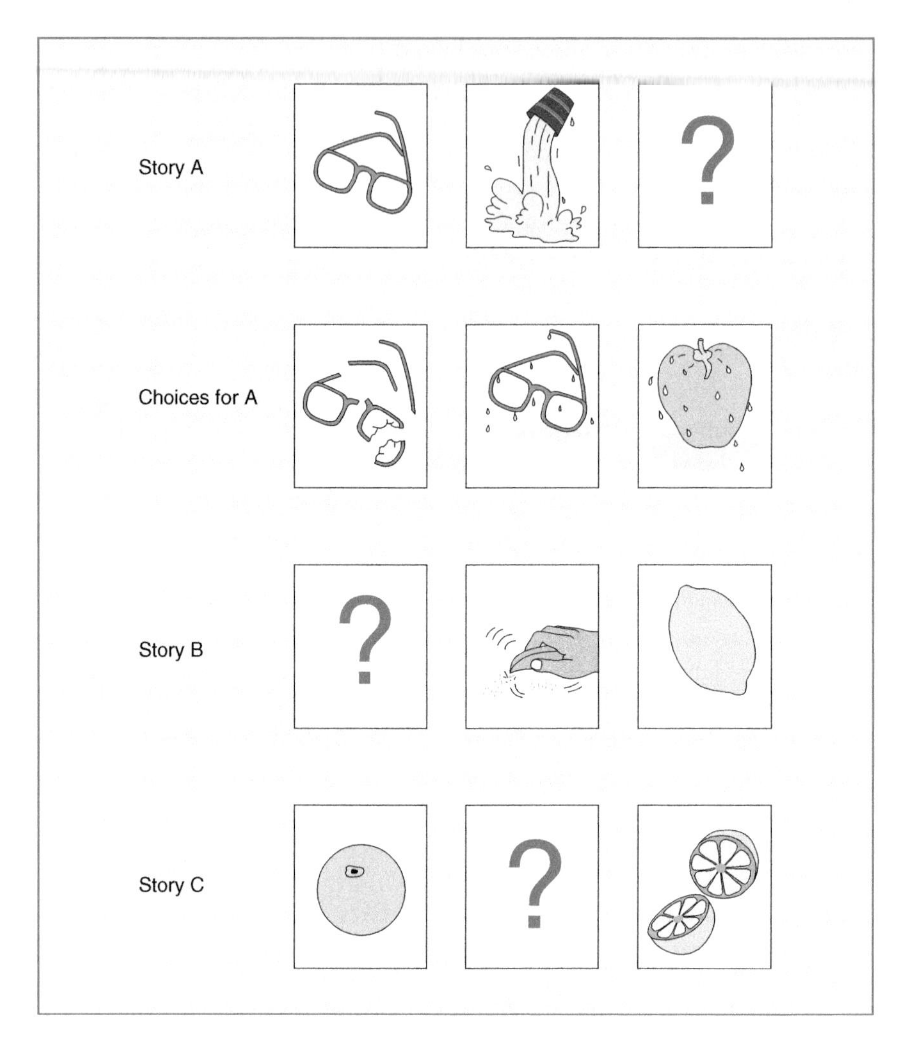

FIGURE 7.1 Examples of the three-picture causal sequences used by Gelman et al. (1980). Story A has the final picture missing, story B the initial picture and story C the picture of the agent. From Gelman et al. (1980). Reproduced with permission from Blackwell Publishing.

KEY TERM

Reversibility of causal reasoning
The ability to think about the same objects in different (reversed) causal relationships

Reversible reasoning about causal transformations of familiar objects

In a second study, the *reversibility* of the same children's causal thinking was investigated. In this study, the missing picture was always the middle picture, and the children were first asked to select a picture of an agent for a *left-to-right* reading of the causal sequence, and then a picture of an agent for a *right-to-left* reading of the causal sequence. This request required the children to think about the same object pairs (e.g., cup and broken cup) in two different (reversed) ways. The three-year-olds found this task quite difficult, and were only correct on 49% of trials. The four-year-olds succeeded on 75% of trials. Gelman et al. argued that the three-year-olds scored poorly because they tended to impose their *own* causal ordering onto the task. They preferred canonical to non-canonical readings over non-canonical to canonical readings, and so were scored as being wrong on half of the trials. Overall, however, Gelman et al. claimed that the three-year-olds' representations of the causal transformations were abstract enough to permit reversibility.

Although Gelman et al.'s work suggests an early understanding of many cause–effect relations and of their reversibility, Das Gupta and Bryant (1989) criticised their studies on methodological grounds. Das Gupta and Bryant argued that it was possible to 'solve' the reversible causal sequences by associative rather than by causal reasoning. This was an important criticism, as associative reasoning is assumed to be a less sophisticated type of reasoning than causal reasoning. Das Gupta and Bryant argued that, rather than considering *both* the initial and the final states of the object when choosing the causal instrument, the children could simply have focused on the *more salient* non-canonical state of the object (e.g., the broken cup). They could then have selected the instrument associated with that non-canonical state. Thus the children could have solved the problems without taking the object's initial state into consideration at all.

Das Gupta and Bryant argued that a *genuine* causal inference depended on the children being able to work out the *difference* between the initial and final states in a causal chain. To examine whether three- and four-year-olds could do this, Das Gupta and Bryant showed them three-picture causal sequences in which the objects *began* as non-canonical in one way (e.g., a broken cup), and ended up as non-canonical in *two* ways (e.g., a wet, broken cup). Trials were paired, so that later in the experiment the children also saw a picture sequence that began with a wet cup and ended with a wet, broken cup. As in Gelman et al.'s procedure, the children's job was to select the missing middle term in the three-picture sequences, which was the causal agent. However, in Das Gupta and Bryant's task, the selection of the agent *most highly associated* with the non-canonical form of the object (a hammer) would be an *incorrect* response in the picture sequence broken cup to wet, broken cup, but the *correct* response in the picture sequence wet cup to wet, broken cup (see Table 7.1). Thus if the same causal agent was chosen on a given pair of trials (e.g., the hammer for broken cup to wet, broken cup *and* for wet cup to wet, broken cup) then a genuine causal inference was unlikely to be being made.

Following Gelman et al., the children were again required to choose from three possible causal agents, in this case *hammer*, *water* and *feather* (an irrelevant agent). Das Gupta and Bryant found that the three-year-olds chose the same causal agent (e.g., the hammer) in 49% of trials for the pairs of sequences, whereas the four-year-olds only chose the same causal agent on 21% of trials. The younger children got the correct answer to both sequences in a pair on 39% of occasions, and the four-year-olds on 78% of occasions. Das Gupta and Bryant concluded that three-year-olds were often distracted by the salience of particular causal effects (such as breaking),

TABLE 7.1 One of the pairs of causal sequences used by Das Gupta and Bryant (1989)

1. Wet cup	[Blank]	Wet broken cup
2. Broken cup	[Blank]	Broken wet cup

which led them to disregard the relation between initial and final states. This made the causal status of their inferences questionable.

The salience of non-canonical states in early causal reasoning

In a second experiment, Das Gupta and Bryant went on to argue that the best test of children's ability to make genuine causal inferences was to use non-canonical (broken cup) to canonical (whole cup) sequences. Such sequences necessitated a causal inference based on the *difference* between the object's initial and final state. In contrast, canonical to non-canonical sequences (cup to broken cup) could be solved on the basis of the departure from canonicality. The salience of the broken cup could lead children to 'correctly' select a hammer as an agent in a canonical to non-canonical sequence, but would probably also lead them to choose a hammer in the non-canonical to canonical sequence. In support of their claim, Das Gupta and Bryant showed that three-year-olds were significantly poorer at reasoning from a non-canonical initial state to a canonical state (broken cup to cup) than from a canonical initial state to a non-canonical one (cup to broken cup), success rates being 47% correct and 88% correct, respectively. Their conclusion was that the ability to make genuine causal inferences developed between three and four years, rather than being already present at age three. However, the use of picture sequences may not be the optimal method for probing the development of causal understanding. Real-world reasoning tasks suggest that children behave as though they are making genuine causal inferences as young as two years of age. This evidence is discussed below, when we consider causal Bayes nets.

KEY TERMS

Causal principles
The principles underlying causal contingencies.

Priority principle
Understanding that an event can be caused by another event occurring prior to it and not by a subsequent event.

Covariation principle
Understanding that the true cause of an effect is likely to be the event that regularly and predictably co-varies with the effect.

Temporal contiguity principle
Causes and effects must be contiguous in time and place.

REASONING ON THE BASIS OF CAUSAL PRINCIPLES

In this section, we ask whether children's causal reasoning follows recognised causal principles. Take the simplest kind of causal contingency, when one event A causes another event B. In order for A to cause B, a number of causal principles must apply. One is that A must either precede B or occur at the same time as B. It cannot occur after B. This asymmetry of causal relations is called the *priority* principle (that causes precede or co-occur with their effects). Other important principles in assigning causality are the *covariation* principle (that causes and their effects must systematically co-vary); the *temporal contiguity* principle (that causes and effects must be contiguous in place and time), and the *similarity* principle (that causes and their effects should have some similarity to each other, for example, that a mechanical effect should have a mechanical cause).

The priority principle

The age at which children develop the understanding that causes precede their effects has largely been measured with action-based tasks. For example, imagine that you see a puppet dropping a marble into an apparatus, and that soon afterwards a jack-in-the-box pops out of the middle of the apparatus. You are likely to attribute the appearance of the jack to the action of the marble. It is unlikely that you will attribute causation to a second marble that is dropped into the apparatus by a second puppet *after* the jack has appeared.

Bullock and R. Gelman (1979) used this 'jack-in-the-box' apparatus to investigate whether three- to five-year-old children understand that the jack-in-the-box can only be activated by an event *preceding* the jack's appearance, and not by an event that occurs after his jump. In their task, the children were shown a long black box. This box was divided internally into three sections (see Figure 7.2; these divisions were not visible to the children). The two outer thirds of the box each had a tunnel for marbles to roll down, with each tunnel running on a sloping path towards the centre of the box. The tunnels were visible through plexiglass windows, so that only their ends (in the middle third of the box) were hidden from view. The central section of the box was opaque, and concealed the ends of the tunnels and the jack-in-the-box. The experimenter could make the jack jump by dropping a marble down either tunnel, although the jack was in fact controlled by a hidden pedal and not by the marbles themselves.

In order to test the children's understanding of the unidirectional order of causes and effects, the experimenter (via two puppets) dropped a marble down one tunnel before the jack jumped, and another marble down the second tunnel after he had jumped. The children's task was to infer which puppet's marble had made the jack jump. Bullock and Gelman found that the majority of children at all ages tested could work out that the first marble had made the jack jump. This was the case for 75% of the three-year-olds, 88% of the four-year-olds, and 100% of the five-year-olds. Bullock and Gelman then made the task more difficult by physically separating the causally appropriate runway from the other two-thirds of the box. Following the separation, one tunnel was in apparent contact with the jack, and the other tunnel was

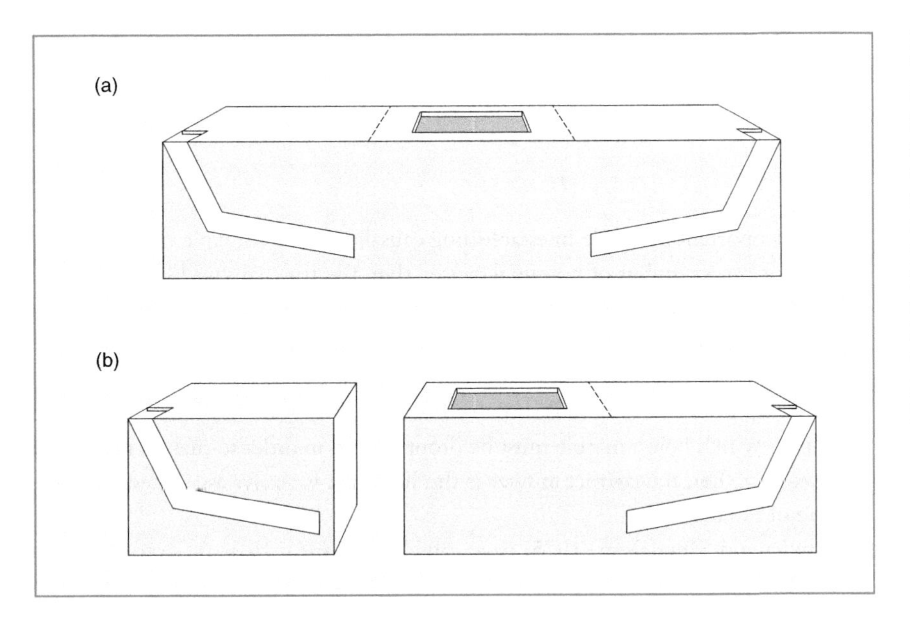

FIGURE 7.2
The apparatus used in Bullock and Gelman's (1979) 'jack-in-the-box' study, showing (a) the complete apparatus, and (b) its appearance when the causally appropriate runway was separated from the remaining two-thirds of the box. From Bullock and Gelman (1979). Reproduced with permission from Blackwell Publishing

not. The experimenter dropped a marble down the detached tunnel before the jack jumped, and a marble down the attached tunnel after he had jumped. Even under these more stringent conditions, causality was attributed to the first marble by 75% of the three-year-olds, 94% of four-year-olds and 100% of five-year-olds. The children in Bullock and Gelman's study appeared to assume that temporal ordering cues were more critical than spatial proximity in determining causality in the jack-in-the-box apparatus. Nevertheless, they were surprised that action-at-a-distance was possible in the detached tunnel condition, and assumed that it needed some explanation ("It's a trick, right?", "Its magic", "When I wasn't looking, the ball slided over").

However, it is not necessary to assume from this data that the children assigned *more* causal importance to temporal priority than to spatial proximity. The simplest conclusion from Bullock and Gelman's data is that the children were basing their judgements on what was *causally relevant* to the experimental set-up. Consistent with this conclusion, Shultz (1982) has shown that children will favour spatial factors over temporal ones when spatial factors are more causally relevant to a particular outcome than temporal ordering factors. Shultz's demonstration was based on an apparatus in which two electric air blowers were directed at a lit candle. The candle flame could be protected from the jets of air by a three-sided plexiglass shield, which could be rotated to field the air emitted by one of the blowers at a time. Shultz showed five-year-old children this apparatus, and then switched on one of the blowers when the shield was in a position to protect the candle. Five seconds later the second blower was switched on, and at the same time the shield turned to protect the candle from this second blower. The flame immediately went out. The children correctly attributed the flame's extinction to the action of the first blower, even though the onset of the second blower was the event that was temporally contiguous with the candle going out. The particular *mechanisms* of causal transmission, rather than spatial or temporal parameters per se, thus appear to determine children's causal attributions. This illustrates the importance of context and background knowledge in children's causal reasoning. The Humean indices cannot be treated independently of the objects and relations involved in a particular causal event. As Shultz showed, children are aware of the importance of the causal *agent* in a given set-up.

The covariation principle

Another important principle in establishing causality is the principle of covariation. If an effect has a number of potential causes, then the true cause is likely to be the one that regularly and predictably co-varies with the effect. For example, if a child is shown a box with two levers, and has to work out which lever causes a light on the lid of the box to come on, the correct answer is the lever that is always activated when the light is on. If a child is shown a box with two holes in the top, and has to determine which hole a marble must be dropped into in order to make a bell inside the box ring, then the correct answer is the hole that is always associated with the ringing of the bell.

Shultz and Mendelson (1975) gave causal problems such as these to three- to four-year-old and older children to determine their ability to use the covariation principle. Covariation information was varied by manipulating the number of times that the cause and the effect were associated with each other. For example, if lever 1 caused the light on the lid of the box to come on, then the children might receive the following pairings of the light and the two levers: lever 1, light; lever 2, no light; lever 1, light; levers 1 and 2, light; lever 2, no light; levers 1 and 2, light. Shultz and Mendelson found that even the three- to four-year-olds could use this kind of covariation information to determine causality, with the majority of children choosing the correct cause across all the different kinds of apparatus used. Shultz and Mendelson thus concluded that the ability to make causal attributions on the basis of covariation information for simple physical phenomena was present by at least three years of age (see also Siegler and Liebert, 1974, for work on covariation with older children).

The temporal contiguity principle

The principle of temporal contiguity, which states that causes and effects must be contiguous in time and place, is intimately related to the covariation principle. In many causal situations, the same cause is implicated by both temporal contiguity and covariation (an example is Bullock and Gelman's jack-in-the-box study, discussed above). Temporal contiguity is also intimately related to the priority principle, which states that causes must temporally precede or co-occur with their effects. However, the temporal contiguity principle refers to the fact that, in addition to co-varying systematically, causes and effects must be linked to each other by an intervening chain of contiguous events (Sedlak & Kurtz, 1981). If there is a physical rationale for a temporal delay between cause and effect, then the principle of temporal contiguity may still hold.

For example, imagine that you are shown an apparatus consisting of a box painted half green and half orange which sits on top of a wooden stand. The box is linked by a piece of rubber tubing 34 inches long to another box, which has a bell inside it. The green and orange box has two holes in it, one on the green side and one on the orange side. If a marble is dropped into the hole on the green side, a five-second delay ensues, and then the bell in the second box rings. If a marble is dropped into the hole on the orange side, then the bell in the second box does not ring. If a marble is dropped into the hole on the green side, and then five seconds later another marble is dropped into the hole in the orange side, the bell rings immediately. Which side of the box is responsible for making the bell ring, the orange side or the green side?

On the basis of covariation information, it seems as though the green side is responsible for making the bell ring. However, on the basis of temporal contiguity, the orange side is a more plausible candidate – except that the marble must pass through the rubber tubing before it can reach the second box which contains the bell. Mendelson and Shultz (1975) showed children this apparatus in two conditions. In one condition the rubber tubing was present, and in the second condition the first box sat directly on top of the second. They found that when the tubing was present, most of the children (who were aged 4–7 years) attributed the ringing of the bell to the green side of the box. However, when the tubing was absent, most of the children attributed the ringing of the bell to the orange side of the box, even though they knew that in some cases dropping a marble into the hole in the orange side failed to make the bell ring. Mendelson and Shultz concluded that, in the absence of a physical rationale for a temporal delay, children assigned more causal importance to information about temporal contiguity than to information about covariation. However, when they could see a reason for the temporal delay, then they attributed causality to the consistent covariate (the green side), despite the lack of temporal contiguity. Again, it seems that prior knowledge of potential reasons for delays between causes and effects, coupled with a focus on the identification of potential causal agents, has an influence on assumed causal mechanisms.

The principle of the similarity of causes and effects

So far, we have seen that young children's causal reasoning follows the principles of priority, covariation and temporal contiguity. When attempting to reason about causality in the absence of any information about temporal contiguity or covariation, however, then the *similarity* of potential causes and effects can be useful. For example, imagine that a box is equipped with a heavy lever and a delicate lever, and can either emit a loud electric bell sound or a very gentle sound. The typical assumption in these circumstances is that the delicate lever is the cause of the gentle sound and the heavy lever is the cause of the loud sound. Similarly, if you are shown two small bottles of clear fluid, one with a pink cap and one with a blue cap, and you are also shown a flask of water that is tinged pink, then the typical assumption is that the pink colouring was caused by fluid from the bottle with the pink cap. Of course, if you are then shown that a drop of fluid from the bottle with the blue cap turns the water pink, and a drop of fluid from the bottle with the pink cap has no effect on the colour of the water, then this covariation information is likely to change your causal attribution. Similarly, if you are shown that a drop of fluid from the bottle with the blue cap has no effect on water colour, and then five seconds later that a drop from the bottle with the pink cap immediately turns the water blue, this temporal contiguity information is likely to change your causal attribution as well.

KEY TERM

Similarity of cause and effect
All things being equal causes and effects should be similar in nature; e.g., mechanical effects should have mechanical causes.

Shultz and Ravinsky (1977) used a number of physical reasoning problems of this type to see whether, in the absence of information about covariation or temporal contiguity, young children would make causal inferences on the basis of the similarity of cause and effect. They were also interested in whether young children would abandon the similarity principle when it conflicted with temporal and covariance information. Shultz and Ravinsky tested children aged six, eight, ten and 12 years with a variety of physical problems, presenting a variety of covariation and temporal contiguity information in addition to similarity information. In the *absence* of information about temporal contiguity or covariation, all of the children used similarity information to make their causal attributions. When information about *covariation* conflicted with similarity information, then the older children (ten- and 12-year-olds) abandoned the use of similarity information. The younger children appeared confused about which principle to apply, and did not make consistent attributions. A similar pattern was found when information about *temporal contiguity* conflicted with similarity information, although in this case only the six-year-olds showed the confused pattern of responding. Shultz and Ravinsky concluded that similarity of causes and their effects is a potent principle of causal inference for children at all ages, but that the abandonment of this principle in situations of conflict occurs at an earlier age for conflicting temporal information than for conflicting covariation information.

Does this mean that developmentally, temporal information is recognised to have causal importance prior to covariation information? As Shultz and Ravinsky point out, this would fit Mendelson and Shultz's (1975) data (described earlier) based on the experiment with a marble and some rubber tubing. Here, too, children preferred to attribute causality to a temporally contiguous but inconsistent event rather than to a temporally non-contiguous but consistent event. Interestingly, however, it has been argued that information about temporal order provides an important clue to causal *structure*, while information about covariation provides important information about causal *strength* (see Lagnado, Waldmann, Hagmayer & Sloman, 2007). Historically, experiments on causal reasoning have been more concerned with causal strength than with causal structure. This is true of most of the experiments carried out by Shultz and his colleagues. However, causal structure may be considered prior with respect to human cognition, because the structural causal relations that hold between variables are more important than their strength. It is first important to know whether a given causal relation exists at all (does smoking cause lung cancer?), and only then to know about the strength of the relation (is there a dose–response relationship?).

In any event, given the data discussed so far, it seems that children's judgements in any particular causal reasoning paradigm will depend on their background knowledge about what is *causally relevant* to the experimental set-up. Shultz himself argued for a version of this position 30 years ago, pointing out that an effect is most likely to be attributed to a cause that seems capable of directing the appropriate sort of transmission (Shultz, Fisher, Pratt & Rulf, 1986). In other words, contextual knowledge governs children's interpretation of statistical patterns of covariation. Other researchers stress the key role for children of the identification of the causal agent/s, which may precede the effort to clarify potential causal mechanism/s (see Koslowski & Masnick, 2010).

PROBABILISTIC CAUSAL INFERENCE AND CAUSAL EXPLANATION

More recently, interest has grown in whether the development of causal reasoning can be described by causal Bayes nets (Gopnik et al., 2004). Causal Bayes nets derive from Bayes theorem, a mathematical theory that has been very productive in the fields of brain imaging and machine learning. For example, neural activity is probabilistic in nature, in that the same stimulus usually evokes slightly different neural responses each time that it is experienced. The structure of the perceptual environment is also probabilistic, as recognised by the statistical learning experiments described in earlier chapters. Mathematical theories of probability, of which Bayes theory is one, therefore play a central role in understanding perception in terms of neural encoding. Accordingly, probabilistic analyses are also relevant to cognitive development, as, for example, when automatic learning of the amplitude modulation hierarchy in speech yields phonological units (Leong & Goswami, 2015; see Chapter 6). Probabilistic approaches can also be applied to other forms of learning, such as causal learning. Bayes theorem is a probability theory that in effect enables the computation of the likelihood of a given state of affairs given prior knowledge about the probabilistically likely state of affairs, prior knowledge that is based on all the factors that typically contribute to that state of affairs, and the actual current state of affairs. As an example, we can think of visual perception. When looking at a visual scene, the eye does not simply record the light intensity at each pixel, like a camera would. Instead, the eye infers the light intensity at each pixel by automatically also taking into account the light intensity at the surrounding pixels and adding in computations based on the light intensity of the same scene in the immediate past. Bayes theorem represents a mathematical solution for capturing the nature of these automatic perceptual inferences (Aitchison & Lengyel, 2017).

The same probabilistic approach can be taken to drawing causal inferences. Formally, causal Bayes nets enable algorithms for the induction of causal structures from observed covariation data by representing both straightforward relations (A, B and C all co-vary) and conditional relations (A co-varies with C only in the presence of B). The requirement for B to be present means that either A causes B, which in turn causes C (A → B → C); or that B causes both A and C (A ← B → C); or that C causes B, which causes A (A ← B ← C; see Lagnado et al., 2007). Causal Bayes nets can represent all of these possible causal structures on the basis of the statistical information regarding how A, B and C all co-vary. Information about temporal order or real-world knowledge can also come into play (for example, if A precedes B in time, then A → B → C is the only possible causal structure).

Causal inferences and causal Bayes nets

KEY TERM

Causal Bayes nets Statistical learning algorithms that can represent the causal structure underlying covariation data by representing the statistical relations present in a body of evidence.

The introduction of Bayesian probability into studies of causal reasoning by children was pioneered by Gopnik and her colleagues (Gopnik et al., 2004, for review). They carried out a series of studies investigating whether young children can use covariation data to induce causal structure in the same ways as machine learning algorithms like causal Bayes nets. In their early studies, Gopnik and her colleagues focused on whether young children are able to 'screen off' spurious associations between variables, associations that may be present in the data but that have no causal value. Their idea was that if young children reason in accordance with the causal graphical models generated by Bayes theorem, then they should only make causal inferences about genuine relationships. Accordingly, even very young children might be able to draw correct causal inferences from patterns of evidence without necessarily having an explicit understanding of the evidence being utilised, and also without having to observe large amounts of data. If so, this would provide

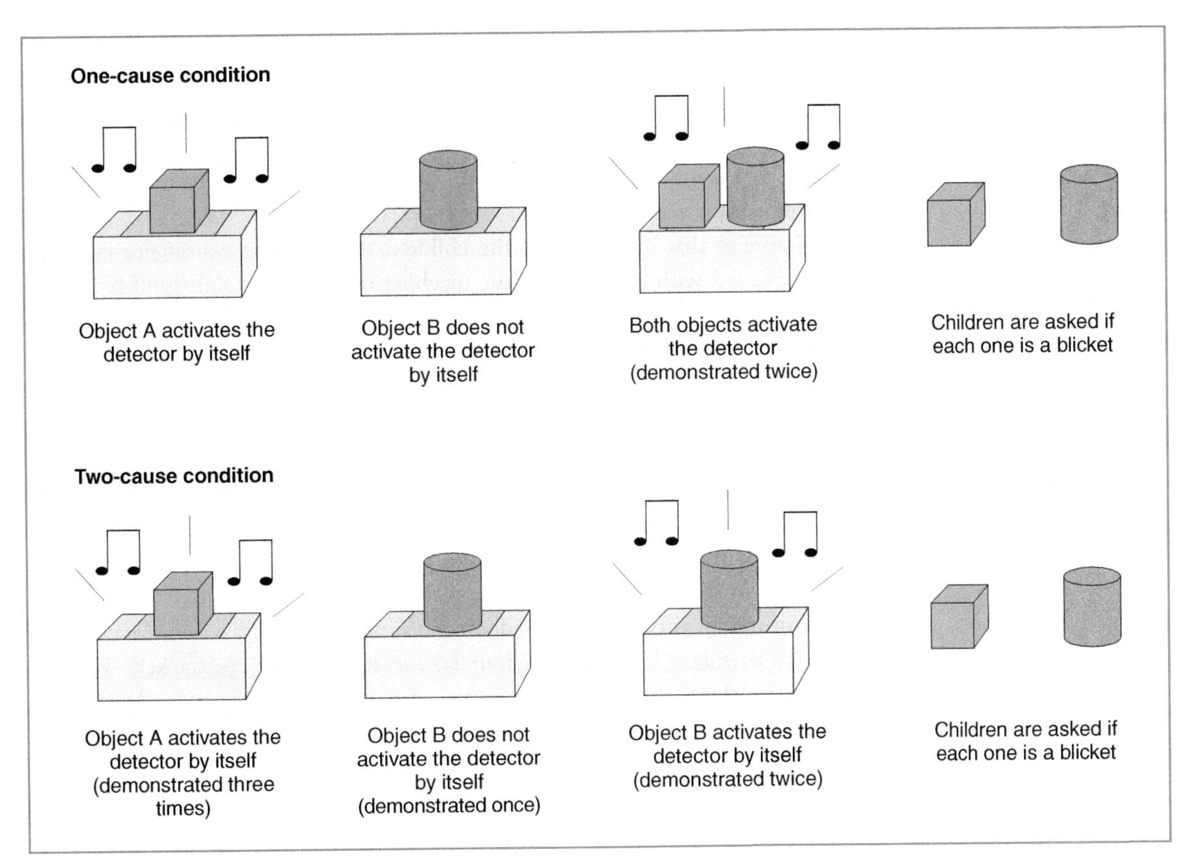

FIGURE 7.3
The procedure used in Gopnik et al. (2001), Experiment 1. From Gopnik et al. (2001). Copyright © American Psychological Association. Reproduced with permission.

evidence that causal inference is an automatic neural/perceptual computation given certain environmental experiences.

In a paradigm that has since been used extensively, Gopnik, Sobel, Schulz and Glymour (2001) examined whether children aged from two to four years could use *screening off* to work out which objects (called 'blickets') had the causal power to make a novel machine called a 'blicket detector' work. The children had to work out which objects were blickets by observing what happened when the objects were placed on the machine. The blicket detector was a box that lit up and played music when a blicket was placed on it (see Figure 7.3). Gopnik et al. (2001) also investigated whether the children could design appropriate interventions.

During the experiments, the children were told "blickets make the machine go", and were asked to "find out which things were blickets". In fact, the blicket detector was controlled by a hidden experimenter, who activated it according to various sets of covariation relations which were observed by the children. Following training to ensure that the children understood the relationship between an object setting off the machine and being labelled a blicket, children experienced both 'one-cause' and 'two-cause' tasks. In the *one-cause* task, the children saw two blocks, A and B. Block A activated the machine, and block B did not. When both blocks were placed on the machine together, it lit up and played music (the dual block event was shown twice). From observing this set of causal relations, we should infer that A is a blicket while B is not. In the *two-cause* task, the children again saw two blocks A and B, but this time each block was placed on the machine by itself on three occasions. Block A activated the machine on all three occasions, and block B on two out of three occasions. In this case, we should infer that both A and B are blickets. The presence of each block on its own increases the likelihood that the machine will be activated.

Note also that in both tasks, the children see the same contingencies. Block A is associated with activation of the machine on three occasions, and block B is associated with activation of the machine on two occasions. However, in the one-cause task, the fact that B is associated with the activation of the blicket detector on two out of three occasions is spurious with respect to causality. The requirement that block A is also present 'screens off' the possibility of a causal relation between block B and the activation of the machine. Using this simple paradigm, Gopnik et al. (2001) were able to demonstrate that children as young as two years behaved in accordance with screening off. Block A 'screened off' block B as a potential cause of the effect in the one-cause condition, even for the two-year-olds. Indeed, Sobel and Kirkham (2006) were able to extend this method to eight-month-olds infants. Using an anticipatory looking paradigm and a spatio-temporal series of events, they showed that infant looking behaviour (as measured by an eye tracker) was consistent with the automatic 'screening off' of non-causal information. Accordingly, the brain appears to compute causal inferences automatically on the basis of certain types of perceptual covariation data.

Gopnik et al. (2001) also used the 'blicket' paradigm to investigate whether the children would intervene appropriately to make the machine stop (Figure 7.4). Gopnik and her colleagues reasoned that if the children really believed that blickets had the causal power to make the machine go, then they should be able to infer that removing the blicket would make the machine stop. This was tested by showing a new group of children the following pattern of dependent probabilities. First, block B was placed on the machine, and nothing happened. After a few seconds, block A was also placed on the machine, and it began to work. After a while, the children were asked "Can you make it stop?" The majority of children removed just block A. In a second scenario, the machine was shown being activated by block A, which was then removed. Next, it was shown being activated by block B when placed alone, and then block A was added next to block B. In this scenario, the majority of children removed both blocks in order to make the machine stop. Gopnik and her colleagues argued that this showed that the children were using screening off information to make genuinely causal judgements. Schulz and Gopnik (2004) went on to show that young children also used screening off to make causal judgements in the domains of biology (which flowers make Monkey sneeze?) and psychology (which animals make Bunny scared?). They argued that children's causal inferences were consistent with the relationships between causality and probabilistic dependence represented by causal Bayes nets across domains. Even very young children appear to have the implicit ability to draw these powerful causal inferences from patterns of evidence about dependent and independent probabilities. Given that eight-month-old infants also show evidence of screening off, it is likely that causal information is computed automatically as part of how the brain encodes probabilistic information.

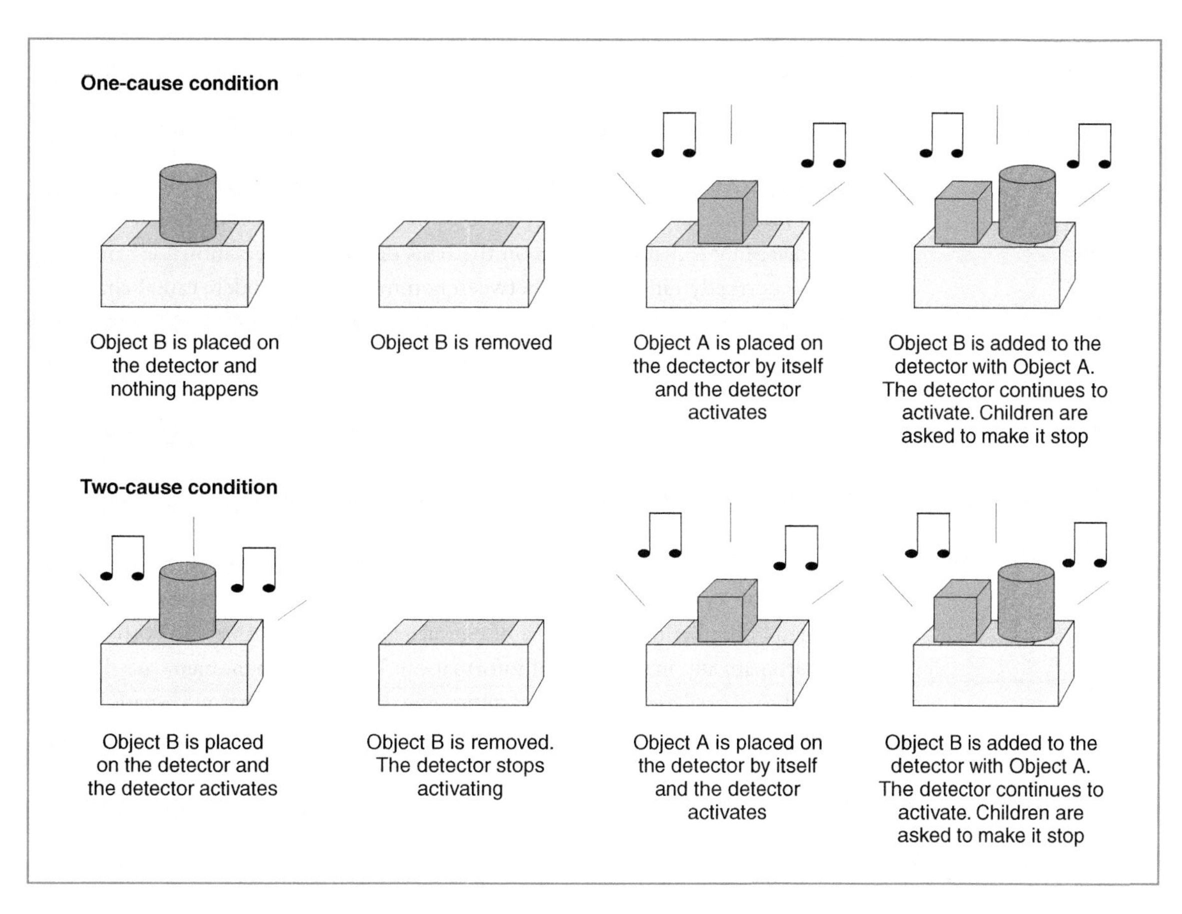

FIGURE 7.4
The procedure used in Gopnik et al. (2001), Experiment 3. From Gopnik et al. (2001). Copyright © American Psychological Association. Reproduced with permission.

Indeed, in adult neuroscience it is now recognised that experience of sensory environments leads *automatically* to the neural development of internal models of those environments, which means that perception and learning must be treated jointly (Fiser, Berkes, Orbán & Lengyel, 2010). Regarding child development, in this book we have been thinking of these internal models as being created automatically on the basis of many kinds of information, thereby forming the basis for cognitive representations in foundational domains.

Intriguingly, other species behave in similar ways to infants and children in causal inference paradigms, providing further evidence that neural learning of causal structure is automatic. For example, rats also develop accurate causal graphical models of the causal relations among events from observational data (Blaisdell, Sawa, Leising & Waldmann, 2006). Blaisdell et al. (2006) were interested in whether rats understand the relation between observations and interventions in causal reasoning as predicted by causal Bayes nets. They therefore devised experimental situations analogous to the 'screening off' experiments described above, which involved predicting the arrival of food on the basis of cues like tones and lights. For example, if an animal learns from observation that a light coming on causes a tone to sound and that the light also causes the arrival of food, then the animal should expect that the tone alone should also lead to the arrival of food, on the basis of covariation data. However, if the animal also experiences that pressing a lever causes the tone to sound, then there is no reason to expect food, as the light is absent. Blaisdell et al. (2006) trained rats with a variety of such causal relations, including whether they themselves pressed the lever to generate the tone (intervention) or whether they merely observed presentations of the tone without the involvement of the lever. They reported that rats made the correct causal inferences as predicted by causal Bayes nets on the basis of purely observational learning. The rats also correctly differentiated between common-cause models, causal chains and direct causal links. These data cannot be explained in terms of associative learning. Accordingly, automatic probabilistic learning and the automatic causal inferences that this learning supports appear to be a consequence of the neural encoding mechanisms used by the mammalian brain. By automatically encoding temporal structural information, the brain encodes causal information as well. In principle, this is the same neural explanation that we discussed in earlier chapters, concerning the imposition of causal structure on perceptual experiences *at the point of encoding* (Fekete & Edelman, 2011). As Fekete and Edelman pointed out, via the temporal/spatial sequence of neural responses, phenomenological experiences can be recorded in the brain by *trajectories* of neural activation over time and cortical space, that automatically include causal information. The same phenomena are discussed in traditional analyses of causal learning, which point out that information about temporal order provides important clues to underlying causal *structure* (Lagnado et al. 2007).

Explanations and causal reasoning

Nevertheless, showing that causal inferences are an automatic product of the way that the brain encodes probabilistic environmental information leaves open the question of how children become able to generate novel explicit causal inferences for themselves. Explicit causal reasoning requires the ability to *reflect* on your knowledge. Being asked to generate explanations for causal phenomena appears to be one important way that children enrich and make explicit their causal knowledge (Legare & Lombrozo, 2014). The developmental power of explanation-based learning has also been discussed in earlier chapters, for example in Baillargeon's work on infants' physical reasoning (see Chapter 2). In the causal realm, however, children appear to benefit from explicitly being instructed to verbalise their knowledge about a situation and to generate explanations for experienced events.

For example, Walker, Lombrozo, Williams, Rafferty and Gopnik (2017) measured the effects of explicitly being asked to generate explanations on causal reasoning by using a variant of the 'blicket' paradigm. Children aged five years were shown a machine that played music when certain Lego bricks were stuck onto the top and sides, experiencing certain patterns of covariation. For example, a blue Lego brick stuck to the top of the machine with a red brick stuck to the side might make the machine play music in 100% of observations, while a yellow Lego brick on top of the machine with a white brick stuck to the side might only make it play music in 75% of observations. For any single observation experienced by the child, the selected Lego brick pairings were *probabilistically* either causally inert or causally active. The causal rule that best explained the most outcomes observed by the children was a '100% colour' hypothesis, that certain brick colours were causal rather than inert. The important contrast in the experiment was between children in the Explanation condition and children in the Control condition.

In the Explanation condition, children were asked to explain the outcomes that they were observing on each trial ("Why did/didn't this block make my machine play music?"). In the Control condition, the children were asked to report the outcomes that they were observing on each trial ("What happened to my machine when I put this block onto it?"). Children were then asked to make causal predictions about novel blocks of Lego ("Which will make my machine play music?"). Walker et al. (2017) found that children who had to explain rather than report the outcomes were significantly more likely to favour the 100% colour hypothesis when evaluating novel pairs of bricks. This was also the hypothesis consistent with the greatest number of cases that they had observed. Walker et al. hence argued that explanation made children more sensitive to causal evidence.

Using a different paradigm based on a mechanical toy, Legare and Lombrozo (2014) investigated the effects of explanation on children's causal learning about an interlocking gear mechanism (see Figure 7.5). Children aged from three to six years were shown the toy, which had a visible and colourful interlocking gear mechanism involving five gears, and participated in either an Explanation condition or an Observation condition. All children were first introduced to a simple gear mechanism, shown how to put it together, and asked to assemble it themselves. They were then introduced to the novel machine with five gear wheels, the first of which could be turned using a crank. The first gear wheel turned the second gear wheel and so on, resulting in the fifth gear wheel operating a fan. Children were first shown the machine being operated by the experimenter, and then observed the machine in a static condition. During this period, they were either told "Let's look at this!" (Observation condition), or "Can you tell me how this works?" (Explanation condition). Causal learning was assessed by removing the machine from view, taking out the second gear, and then asking the child to choose a novel part from a selection of five parts to make it work again. The correct choice was the

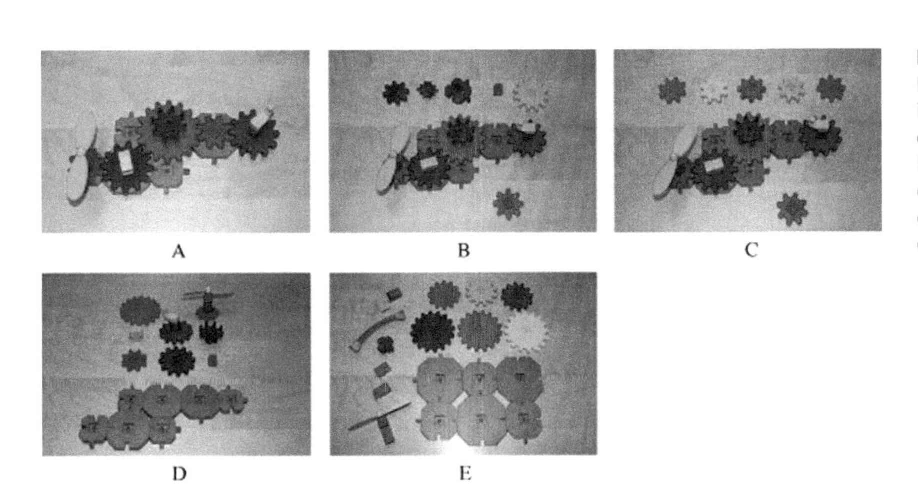

FIGURE 7.5 Examples of the different interlocking gear wheel configurations used by Legare and Lombrozo (2014) to investigate the effects of explanation on children's causal learning.

gear wheel of the same size as the missing part, which in the novel choice task was a different colour. The experimenter then removed more parts of the machine out of view of the child, and the children were asked "Can you put my machine back together the way it was before and make it work?" They were given ten minutes to recreate the machine. Children were scored for their ability to select the correct missing part, and for their overall ability to reassemble the machine. Legare and Lombrozo (2014) found that children who had been asked for explanations were significantly more likely to show causal learning, at all ages. They concluded that explanation promotes causal learning and generalisation.

Legare and her colleagues have also investigated whether young children can revise their causal explanations on the basis of new evidence (Legare, Schult, Impola & Souza, 2016). Legare et al. suggested that explanation may be particularly beneficial when it is harnessed to understand inconsistency in causal outcomes. To investigate this, they showed children aged from three to six years video clips that imparted information about preferences such as liking apples. For example, the children might see a video clip of two female actors, one of whom then stated that she liked apples, while the other stated that she did not like apples. In a second video clip (Confirmation trial), the same actors might appear at a table with a plate containing apples. At the same time, the first actor would reach for an apple while the second actor would cross her arms across her chest. The child was then asked to explain these two actions. A new clip (Test trial) was then shown with the same actors and more apples, but this time both actors reached for an apple. The child was asked "Why did that happen?" In a final clip (New Evidence trial), the scene concluded with the first actor moving the apple plate towards herself, revealing a second hidden plate containing cookies, which the second actor moved towards. The child was again asked "Why did that happen?"

Legare et al. predicted that for the Test and New Evidence trials, the children would first choose to explain the actions of the second actor, who behaved inconsistently. This was exactly what they found. In fact 92% of the children first explained the behaviour of the second actor in the Test trial, and 57% of the children first explained the behaviour of the second actor in the New Evidence trial. Legare and her colleagues concluded that children choose to explain observations that have the greatest potential to teach them something new. Accordingly, while causal understanding in children indeed benefits from explanation-based learning, children also appear to have intuitive knowledge about when generating explanations can be helpful. Once again, we see that the child is an *active* learner.

THE UNDERSTANDING OF CAUSAL CHAINS

The gear mechanism experiments discussed above also involve reasoning about *causal chains*. For example, for the novel five-gear machine used by Legare and Lombrozo (2014), each gear wheel in turn activates the next gear wheel (since the spokes of the gears are interlocking). So even though only the first gear wheel is actively turned by the human actor, an understanding of causal chains is an implicit aspect of understanding the gearing mechanism. Historically, the understanding of causal chains has been a topic in the development of causal reasoning in its own right. Most of the experiments discussed so far, such as the blicket experiments, have involved two-term causal chains (A → B). In a three-term causal chain (A → B → C), the reasoning task is more complex, as the presence of a *mediate cause* requires a transitive inference as well as an understanding of physical causal contingencies (see Shultz, Pardo & Altmann, 1982). Understanding the causal structure of a three-term causal chain is crucial for making accurate causal inferences. If an event A causes an event B to occur, which in turn results in an event C, then there is no direct causal link between A and C. For example, a tennis ball (A) can be rolled so that it strikes a golf ball (B), which in turn strikes a light plastic ball (C), dislodging it from its resting position. The golf ball is a *mediate* cause of the covariation between A and C (and is analogous to the middle term in a transitive inference problem, see Chapter 11). Although event C is caused by event A in this causal chain, the research question of interest was whether the child understands that this causal relationship only holds true when B functions as a causal mediator.

The understanding of mediate transmission

Shultz et al. (1982) gave three- and five-year-old children simple causal chains of the type described above in two conditions. In one condition, the mediate causal event (B) was effective, and in the other it was not. For example, to administer the tennis ball problem, Shultz et al. devised an apparatus in which balls of different sizes could roll along converging lanes (see Figure 7.6). The first pair of lanes was wide enough for tennis balls, the second pair for golf balls, and the lanes then converged onto a single lane for a light plastic ball. The light plastic ball was positioned at the point of convergence, so that following impact from the golf ball, it would roll to the end of the apparatus. On each side of the apparatus an arch was created that separated the tennis ball lane from the golf ball lane. These arches were too narrow for the golf ball to pass through. In order for the golf ball to act as a mediate cause, it therefore had to be on the far side of an arch. In order to create the two conditions, on one side of the apparatus the golf ball was on the near side of the arch (at Y' – ineffective mediate cause), and on the other it was on the far side of the arch (at Y – effective mediate cause).

The children's job was to choose which of the lanes to roll the tennis ball along in order to dislodge the light plastic ball from its position at the point of convergence of the lanes. If the child chose lane Y', then the plastic ball would not be dislodged, whereas if the child chose lane Y then the desired outcome could occur. The sides of the ineffective and effective mediate causes were varied at random over ten trials. Shultz et al. found that the children were able to select the correct lane for the tennis ball on the majority of trials. The correct lane was chosen on 69% of trials by the three-year-olds and on 86% of trials by the five-year-olds. Most of the errors were on the first trial, and also on the third trial for the three-year-olds. From the fourth trial onwards, children of both ages were consistently correct in choosing the lane with the effective mediate cause.

KEY TERMS

Mediate transmission
A three-term causal chain, where A impacts B, which in turn impacts C, with B being the causal mediator.

Logical search
Effective search based on an understanding of causal chains of events and causal necessity.

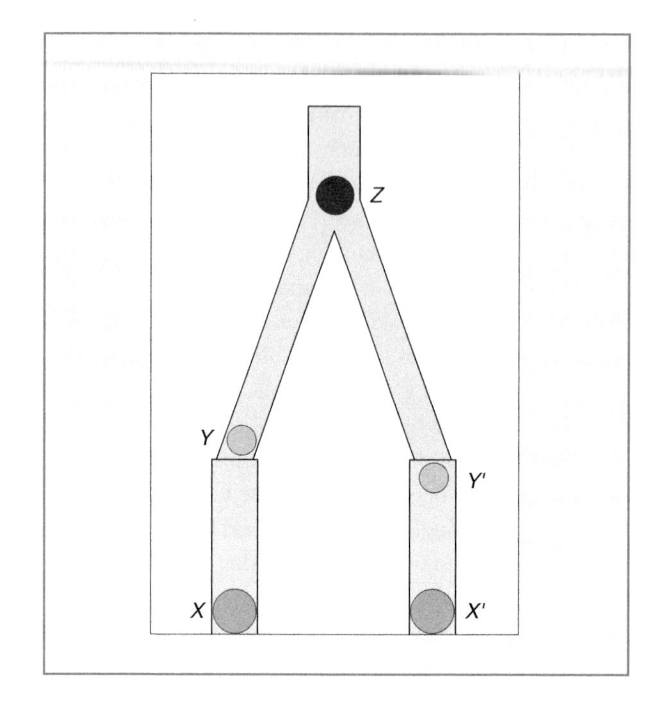

FIGURE 7.6
The runway apparatus used to study children's understanding of mediate causal transmission by Shultz et al. (1982). From Shultz et al. (1982). Copyright © The British Psychological Society. Reproduced with permission.

Children's understanding of three-term causal chains has also been investigated by Baillargeon and R. Gelman (1980, described in Bullock, Gelman & Baillargeon, 1982). They designed a 'Fred-the-Rabbit' apparatus, in which the final step in the causal chain consisted of a rabbit (Fred) falling into his bed (a mat at the end of the apparatus). Fred was first presented to the children standing on a platform above his bed. The mediate cause for getting Fred into his bed was a series of wooden blocks, which were arranged in a row in front of the platform like a series of dominoes (see Figure 7.7). Each block in turn could fall onto the block in front (in a 'domino effect'), thereby causing the final block to fall onto a lever that pushed Fred off his platform and into his bed. The initial cause in the chain was a rod positioned in a post, which could be pushed through the post to activate the first block in the series. The children's task was to explain how to get Fred into his bed.

The children were first shown the entire apparatus during pre-test demonstration trials. The mid-portion of the apparatus (the blocks) was then covered, leaving only the rod in its post and Fred on his platform visible to the child. The rod and Fred were separated by a distance of around one metre. The children (four- and five-year-olds) correctly predicted that Fred could be got into his bed by pushing the rod through the post. The experimenter then introduced two new rods, a short one that was of insufficient length to reach the first block, and a long one that was of sufficient length. Both rods then failed to get Fred into his bed (the longer rod was prevented from working by a trick). The children were asked to explain why each rod had failed to get Fred into his bed.

Baillargeon and Gelman found that the children were able to offer causally coherent explanations in both cases, distinguishing between relevant (short rod) and irrelevant (long rod) modifications. For the short rod they said that the rod was too short to reach the first block, and for the long rod they said that the experimenter must have done something to disrupt the mediating event, such as taken some of the blocks away. Baillargeon, Gelman and Meck (1981, reported in Bullock et al., 1982) then extended the Fred-the-Rabbit task to three- and four-year-olds. The children were asked to predict whether Fred would fall into his bed following a variety of modifications to either the initial or the mediate event. Modifications to the initial event included substituting a soft, flexible rod for the wooden rod, and substituting a rod with a stopper on the end that could not pass through the post. Modifications to the mediate event included the experimenter moving Fred's platform away from the final block in the series, and the experimenter moving the platform to one side of the blocks. Children's predictions were highly accurate whether the initial or the mediate event was changed (81% and 78% correct, respectively, for the three-year-olds, 87% and 85% correct, respectively, for the four-year-olds).

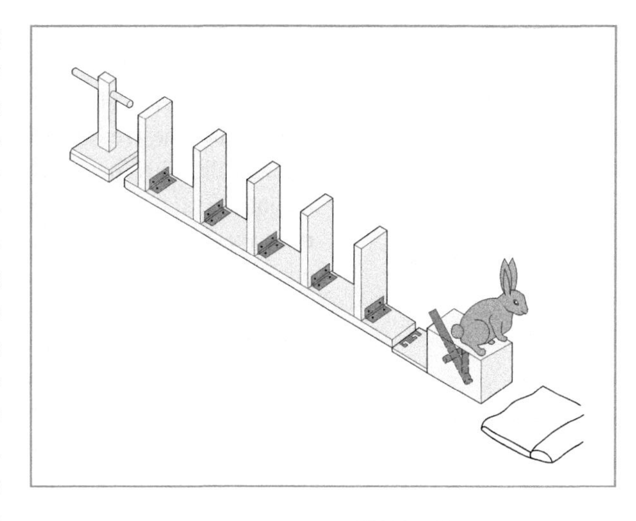

FIGURE 7.7 The 'Fred-the-Rabbit' apparatus used by Bullock et al. (1982). Copyright © 1982 Elsevier. Reproduced with permission.

From the research discussed above, it seems that even very young children can use information about three-term causal chains to reason about event sequences. This holds true whether causal reasoning is measured via a prediction task (Baillargeon et al., 1981) or via a problem-solving task (Shultz et al., 1982). As performance in prediction tasks is often inferior to performance in problem-solving tasks (e.g., Goswami & Brown, 1990), Baillargeon et al.'s data provides particularly strong evidence that children understand mediate transmission by three years of age. More recently, another important developmental variable has been suggested that supports children's understanding of mediate transmission. This is *embodiment*.

Consider the five-wheel gear system discussed earlier (Legare & Lombrozo, 2014). One way of analysing children's casual learning with this system would be to consider how long it would take them to reassemble the gear system *themselves* when given the opportunity (this analysis was not reported in Legare and Lombrozo's paper). Children who could *predict* how to reassemble the system before beginning the task would in principle have a better causal understanding than children who had to use trial and error. Boncoddo, Dixon and Kelley (2010) noted this point, and argued that gear systems are an interesting task for studying causal learning, as the turning direction of the gear wheels alternates. If one wheel is turning clockwise, and its teeth interlock with a second gear wheel, that second gear wheel will turn anti-clockwise. A child who understands these causal relationships should thus be able to predict which way the final wheel in a multiple-gear system will turn. Boncoddo et al. (2010) suggested that an intermediate strategy aiding accurate prediction could be 'force tracing', namely physically tracing the forces across the system by manually checking each gear. Boncoddo et al. were interested in whether embodiment, here measured by the amount of force tracing carried out by individual children, would be related to how early the alternation principle was discovered.

To find out, Boncoddo and her colleagues used different simple gear systems based on at least four interlocking wheels, and studied preschool children aged three to five years. The children were first introduced to two real gear wheels, and shown how the teeth interlocked by holding one in a vertical position and one horizontally. This meant that when the first wheel made the second wheel turn, the information relevant to the alternation principle was not visible. Children were then asked to solve a sequence of 40 simple gear problems depicted on a computer screen as a 'race' game. The game involved getting your train to a station as fast as possible. Each gear system, which varied in complexity from two to seven gears, had a final gear wheel that caused fuel to fall into the engine of the child's train. If the final wheel turned clockwise, then the engine had to be at one side of the final gear wheel to catch the fuel, and if it turned anti-clockwise, then the engine had to be at the other side of the final gear wheel to catch the fuel. The child responded in each trial by moving their engine to the correct place to catch the fuel. Although this task was quite unfamiliar in terms of mediate causal transmission, all the preschoolers were above chance in the task, with three-year-olds as effective as five-year-olds. Force tracing was higher in the early trials, with alternation behaviour more likely in the later trials. The critical prediction, that the number of force tracing actions would predict the growth of the alternation strategy, was supported. Boncoddo et al. (2010) concluded that new causal understanding can emerge from action. Accordingly, embodiment plays a role in the development of children's causal reasoning.

The understanding of logical search

A different way of measuring children's understanding of causal chains is to use search tasks. For example, imagine that you are on a visit to the zoo and that you want to photograph the chimpanzees, but you find that you have lost your camera. As the chimpanzees are about the eighth group of animals that you have visited, one strategy for finding the camera is to try to remember the last animals that you photographed. If you clearly remember taking a photo of the lions, but you don't recall taking a photo of the elephants, then you probably lost your camera somewhere after the lion enclosure but before you reached the elephants. If young children can use this kind of causal logic when searching for objects, then this must entail some understanding of causal chains and causal necessity.

In a playground version of the logical search task described above, Wellman, Somerville and Haake (1979) took children aged three, four and five years around eight different locations in a playground (see Figure 7.8). Each location was visited in turn. Upon arrival in each new location, the experimenter and the child played a distinctive game, such as jumping in the sandbox or hopping in tyres. At location 3, the experimenter took a photograph of the child doing the long jump as this made a good 'action shot'. At location 7, the experimenter was about to take another photograph when he discovered that his camera was missing. The children's job was to help him to find it. Wellman et al. were interested in whether the children would limit their searching behaviour to the critical area (between locations 3 and 7), or would search all of the areas that had been visited in turn. In a control

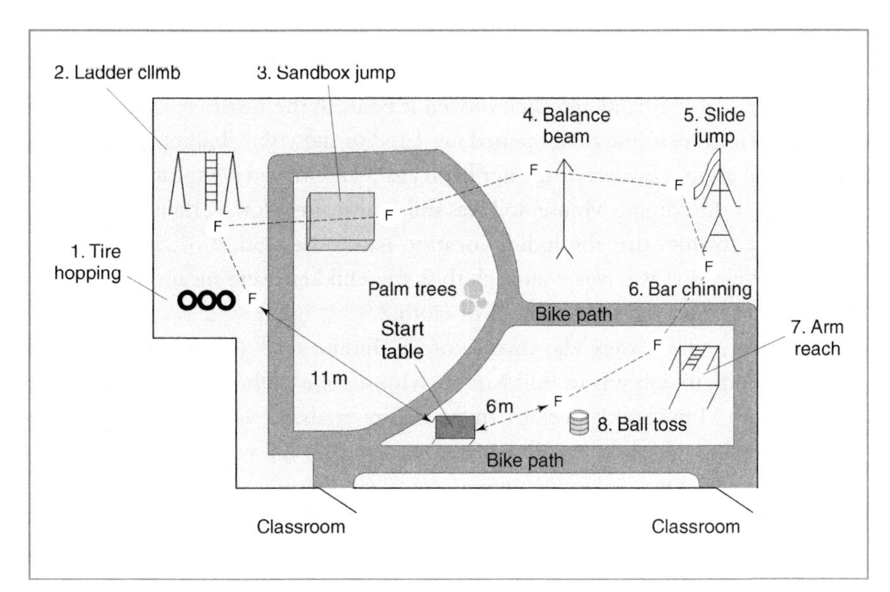

FIGURE 7.8 Schematic depiction of the playground used in the logical search study devised by Wellman et al. (1979), showing the eight locations and their associated games. From Wellman et al. (1979). Copyright © 1979 American Psychological Association. Reprinted with permission

condition, the experimenter discovered the loss of a calculator that had been in his bag throughout the experiment. The loss was discovered at location 8, and so in this control condition searching at each location in turn was an appropriate search strategy.

The important measure was the number of searches that were in the critical area (between locations 3 and 7) in both conditions. Wellman et al. found that most of the children concentrated their searches in the critical area in the camera condition, but not in the calculator condition. There were no marked age differences. The three-year-olds seemed as capable of logical searching behaviour as the five-year-olds. However, closer inspection of the data revealed that half of the searches in the critical target area were actually searches at location 3, the location where the camera had last been seen. Because of this, it is not clear whether the children in Wellman et al.'s study understood that each of the locations between 3 and 7 was *equally likely* to contain the missing camera.

This point was made by Somerville and Capuani-Shumaker (1984), who set out to investigate more directly the question of whether young children understand the causal implications of a sequence of events. Again using a search task as their critical measure, they devised a hiding and finding task in which two locations were at any one time equally likely to contain a hidden toy. This toy was a small Minnie Mouse doll, which could be concealed at one of four possible locations by the experimenter. However, in some of the hiding and finding trials it was more logical to go *forwards* and search in the next two locations from where the Minnie Mouse doll had last been seen, and in others it was more logical to go *backwards* and search in the previous two locations. Somerville and Capuani-Shumaker were interested in whether young children would recognise that some search sequences were more logical than others.

The experimental set-up consisted of a dark tablecloth with the four possible hiding locations marked by smaller stiff white cloths. Each cloth was pulled up into a peak so that it was unclear whether it concealed a Minnie Mouse doll or not. In a given *hiding* trial, the experimenter showed the children the Minnie Mouse doll in her hand, closed her hand and then moved it beneath the first two cloths, pausing beneath each in turn. She then opened her hand to show the children whether the doll was still present or not, closed her hand again and moved it beneath the second two cloths. If the Minnie Mouse doll was still *present* after cloth 2, then the children were meant to infer that the hiding location had to be cloth 3 or cloth 4. If the Minnie Mouse doll was *absent* after cloth 2, the children were meant to infer that the hiding location had to be cloth 1 or cloth 2.

The *finding* tasks were the inverse of the hiding tasks (see Figure 7.9). This time the children's job was to find Minnie Mouse's *sister*, who always liked to hide together with Minnie under a cloth. In the finding trials, the adult's hand was always empty to begin with, and then after passing beneath cloth 2 was either shown to be still empty or to now contain Minnie Mouse. The question was whether the children could infer that Minnie's sister must be hiding under cloth 3 or cloth 4 when Minnie Mouse was still absent after cloth 2, but must be hiding under cloth 1 or cloth 2 when Minnie Mouse was present after cloth 2. Three- and four-year-olds were tested.

Somerville and Capuani-Shumaker scored the first location at which the children searched. They found that the children were able to narrow down the potentially correct locations to two out of the possible four, as searching behaviour was above chance at both age levels. Interestingly, first searches were also significantly more likely to be correct when the children had to reason from the continued *presence* of Minnie after cloth 2 in the hiding task, and from the continued *absence* of Minnie after cloth 2 in the finding task. This suggests that the causal implications of

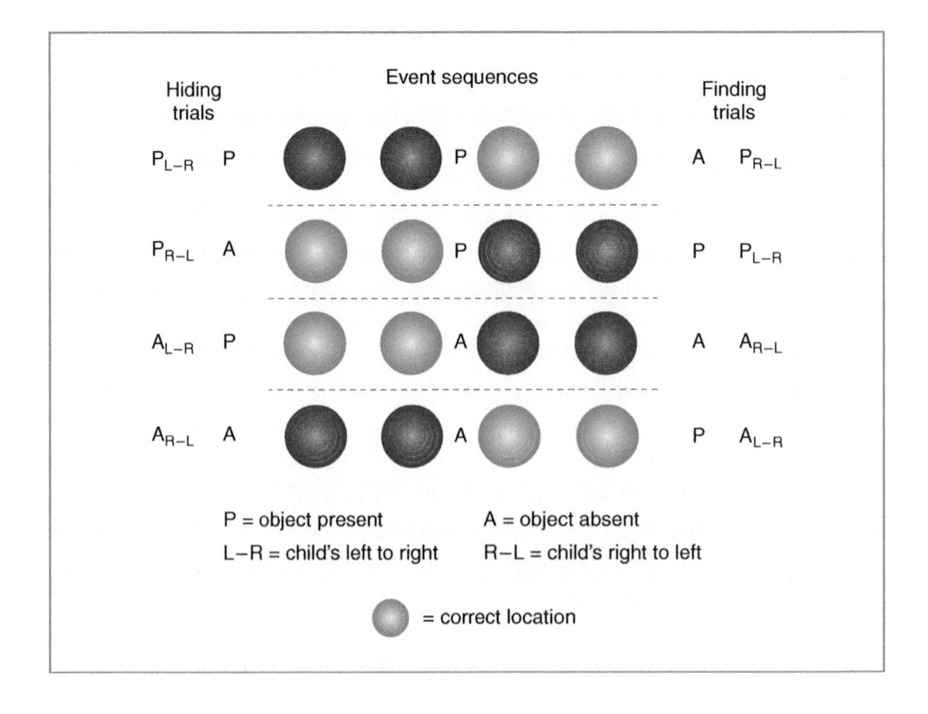

FIGURE 7.9
Schematic depiction of the sequences of events in the hiding and finding tasks devised by Somerville and Capuani-Shumaker (1984). Copyright © The British Psychological Society. Reproduced with permission.

the event sequences were easier to understand when the cause and the effect were temporally more contiguous. This reflects one of the principles of causal reasoning discussed earlier. However, when the experimenters scored the children's second searches, they found that they were by no means always correct. This was surprising given that an incorrect first search in the target area left only one plausible alternative hiding or finding location. Subsequent experiments replicating and extending Somerville and Capuani-Shumaker's basic design have shown that, although even two-year-olds respond logically on their first searches (Haake & Somerville, 1985), children younger than four appear to search by considering only one possibility at a time (Sophian & Somerville, 1988). Having searched in the critical area on their first search, they sometimes search outside the critical area on their second search. It appears that the causal implications of the hiding and finding event sequences studied by Somerville and her colleagues are not fully understood before the age of four years.

SCIENTIFIC REASONING

KEY TERM

Scientific method
Testing hypotheses in a systematic way and controlling for potential confounding variables.

The research discussed so far has shown that young children's causal reasoning is usually in accord with accurate representations of causal structure. However, even though the different causal principles appear to be available to young children, difficulties appear to arise when they have to *rule out* potential causal variables as the cause of a particular effect. This involves an understanding of the 'scientific method'. When young children are asked to determine the causes of a particular phenomenon, they may fail to test a hypothesis in a systematic way, omitting to control for confounding variables. They may also fail to seek evidence that could disconfirm their hypotheses, and they may accept causes that account for only part of the available data (Sodian, Zaitchek & Carey, 1991). In short, younger children appear to have little explicit understanding of the components of scientific reasoning. Scientific reasoning is complex, as it usually involves the understanding of multiple causal variables.

Co-ordinating theories and evidence

Most of the evidence supporting the position that young children are poor scientific reasoners has come from the work of Kuhn and her colleagues. They have conducted extensive studies of children's understanding of hypotheses and evidence (e.g., Kuhn, Amsel & O'Loughlin, 1988). Kuhn's studies have suggested that children have little understanding of how hypotheses are supported or contradicted by causal evidence until around 11 or 12 years of age (e.g., Kuhn, 1989). Accordingly, Kuhn has argued that younger children are incapable of 'scientific thinking' – the kind of thinking that requires the co-ordination and differentiation of theories and evidence, and the evaluation of hypotheses via evidence and experimentation (see also Klahr, Fay & Dunbar, 1993). Young children do not seem to know what kind of causal evidence does or could support a particular hypothesis, or what kind of evidence does or would contradict a particular theory. However, in many of Kuhn's studies, children's pre-existing background knowledge made it difficult for them to consider the covariation evidence being presented in purely statistical terms. This background or contextual knowledge may have interfered with their ability to demonstrate scientific reasoning.

For example, Kuhn et al. (1988) reported an experiment in which participants were asked to evaluate evidence about the covariation of the various foods that children ate at a hypothetical boarding school, and their susceptibility to colds. The participants, who were aged 11 and 14 years, were given the covariation information about the different foods pictorially (see Figure 7.10). For example, in the figure, apples and French fries co-vary perfectly with colds, and Special K and Coca-Cola do not. The children were then asked questions like "Does the kind of drink the children have make any difference in whether they get lots of colds or very few colds?" Only the older children showed an ability to evaluate the covariation evidence effectively, although their performance was far from perfect. The spontaneous evidence-based responses given by the 11-year-olds constituted 30% of responses, and of the 14-year-olds, 50% of responses. Adults performed at the same level as the 14-year-olds.

One reason for these performance patterns could be the background knowledge that participants brought with them to the experiment. We are often told (e.g., by advertisers of orange juice) that what we choose to drink does affect our susceptibility to colds. For example, children may have learned that drinks high in vitamin C, such as orange juice, will actually protect them against getting colds. In studies like Kuhn's, children are only credited with true scientific reasoning if they can *override* such background information and focus purely on the covariation information being presented during the experiment. The children did not do this, but real scientists may not do this either. In science, information about mechanism can guide the assessment of covariation, while systematic and unexpected covariations can also guide the discovery of new mechanisms (see Koslowski & Masnick, 2010). For example, car colour may be systematically correlated with driving speed, but pre-existing background information about mechanism makes us unlikely to conclude that red paint causes cars to go faster than blue paint. We already know that paint colour is not a mechanism for speed. However, if we learn that cautious drivers tend to choose blue cars, then a possible mechanism for the correlation becomes evident. Blue cars may go slower because they are more likely to be driven by cautious drivers.

Kuhn et al. (1988) also reported that children showed a strong tendency to make incorrect inferences about causality based on 'inclusion errors'. Inclusion errors involve the attribution of causal status to variables that only co-vary with the outcome on a single occasion. In our example, the children would accept a single instance of a food co-varying with colds as evidence that the food was a cause of susceptibility to colds (e.g., granola in Figure 7.10). Such incorrect single-instance inclusion inferences were made on 47% of occasions by the 11-year-olds, and on 65% of occasions by the 14-year-olds. Kuhn points out that this is an error-prone strategy for inferring causal relations, because even though inclusion inferences may on occasion be correct, they may also be false.

Kuhn et al.'s investigations (e.g., Kuhn, Garcia-Mila, Zohar & Andersen, 1995) have suggested that an important source of the persistence of inclusion errors

FIGURE 7.10
Four examples of the pictorial covariation evidence concerning foods eaten and susceptibility to colds (Kuhn et al., 1988).

is the prior theories that children hold about the causal status of the variables being investigated. These prior beliefs influence the selection of instances that are attended to, and which instances are relied upon when justifying conclusions about causality. Work with adults has shown similar effects. Even eminent scientists in their laboratories are more likely to attend to data that is consistent with their prior theories (Fugelsang, Stein, Green & Dunbar, 2004). Inconsistent data is not treated as 'real' until repeated observations of unexpected relationships force theory revision. There is a 'confirmation bias' in human reasoning – a tendency to seek out causal evidence that is consistent with one's prior beliefs. This is a major source of inferential error in fields as disparate as science, economics and the law, and is not restricted to young children. Kuhn et al.'s observations about the influence of prior theories on reasoning also fits with the importance of background knowledge and context in scientific reasoning. Both children and adults may be concerned with arriving at a *plausible* causal explanation of a given scientific phenomenon, and in certain cases this may justify reliance on a single instance. Koslowski (1996) has argued that the aim of scientific reasoning is to assess plausibility rather than possibility or certainty. Scientific inquiry can seldom conclude that a given explanation is definitely correct, as new evidence may always emerge in the future to show that it is wrong.

Indeed, prior knowledge can be shown to affect the hypotheses that children propose at ages as young as three years. Schulz, Bonawitz Baraff and Griffiths (2007) gave preschool children aged from three to five years statistical evidence about events and effects via storybooks that focused on the activities of a story character. The character experienced a certain outcome in the story each day for the seven days of the week. For example, in one storybook the outcome of 'Bambi has itchy spots on his legs' occurred each day when Bambi got excited and ran among plants called cat-tails. When he ran in other environments such as cedar trees, Bambi didn't get itchy spots. This example provided 'within-domain' evidence, which it was assumed would be consistent with children's prior knowledge about what typically causes itchy spots (having something rub on your skin). In another storybook, a Bunny got a tummy ache each time he thought about having to participate in 'Show-and-tell' at his preschool, which made him feel scared. Bunny also ate different foods each time he got a tummy ache, but these foods varied across the different days of the stories. This example provided 'cross-domain' evidence, as it was assumed that children's prior knowledge about what typically causes tummy aches is a food that has been eaten. After hearing the stories, the children were asked questions like "Why does Bambi have itchy spots? Is it because of running through the garden or running through the cat-tails?" and "Why does Bunny have a tummy ache? Is it because of feeling scared or eating the sandwich?"

Schulz and her colleagues reported that the older children (3.5 year olds and 4–5 year olds) were able to use the covariation evidence appropriately in the within-domain condition, choosing the causal factor (here, running in the cat-tails) at levels well above chance. For example, the 3.5 year olds selected the correct cause on 94% of trials. Control children in a baseline condition, who were asked the question without hearing the evidence, were at chance. For the cross-domain condition, the baseline control group overwhelmingly chose the non-causal factor (eating a specific food gives you a tummy ache). The oldest children in the cross-domain condition who had received evidence (the four- to five-year-olds) showed some effects of using covariation, as they were equally likely to choose the causal factor (feeling scared) as the non-causal factor (the specific food) after hearing the storybook evidence. However, the 3.5 year olds in the cross-domain condition stuck to their prior knowledge and still chose the food as the cause of Bunny's tummy ache, with only 12% choosing 'feeling scared' as the cause of the tummy aches. The youngest children, the three-year-olds, were not swayed by the evidence in either condition. For both the within-domain and the cross-domain conditions, the youngest children relied on domain-appropriate causes. Schulz et al. (2007) concluded that young children were more likely to learn about a novel cause (running in the cat-tails) when the cause was consistent with their prior theories. Accordingly, the 'confirmation bias' in human reasoning is already present in preschoolers.

Testing hypotheses

How do young children perform in reasoning tasks that are not set in contexts (like 'catching a cold' or 'having a tummy ache') that are already the subject of strongly held beliefs? This question was investigated by Sodian et al. (1991) in a clever paradigm based on mice liking to eat cheese. In order to see whether young children have any insights into the relationship between hypotheses and evidence in a novel context, Sodian et al. asked children to choose between a conclusive and an inconclusive test of a hypothesis. The test concerned how two brothers could decide whether they had a small or a large mouse in their house. Sodian et al. devised a simple paradigm involving a single test that was sufficient to draw a causal conclusion.

In Sodian et al.'s experiment, six- and eight-year-old children were told a story about two brothers who knew that they had a mouse in their house, even though they had not actually seen it (because it only came out at night). One brother believed that it was 'a big daddy mouse' and the other believed that it was 'a little baby mouse'. The problem was how they could decide who was right. To test their hypotheses about the size of the mouse, the brothers were planning to put some cheese into a box for the mouse to eat. Two boxes were available, one with a large opening that could take either the large or the small mouse, and the other with a small opening that would only allow the small mouse to enter. The children had to decide which box the brothers should use in order to determine the size of the mouse. The majority of children in both age groups realised that the box with the small opening was required. As one child remarked "They should take the house with the small opening, and if the food is gone, this tells them that it is a small mouse, and if it's still there it is a big mouse" (Sodian et al., 1991, p. 758). Sodian et al. concluded that even quite young children understand the goal of testing a hypothesis, and can distinguish between conclusive and inconclusive tests of that hypothesis in simplified circumstances.

A complementary method for examining children's understanding of hypothesis testing is to use fake evidence. Ruffman, Perner, Olson and Doherty (1993) devised a 'Fake Evidence' task to investigate whether even younger children can work out how a pattern of evidence relates to a hypothesis when only a single cause is involved. The children's task was to work out which kind of food was more likely to lead to tooth loss, green food or red food (the 'food' consisted of bits of coloured paper). Four- and five-year-olds were shown consecutive pictures of ten boys in the act of eating food. Five of them were eating green food and had healthy teeth, and five of them were eating red food and had teeth missing. The children were asked "Which type of food makes kids' teeth fall out?" All of them answered that it was red food (showing use of the covariation principle).

The children were then shown a picture of ten boys' heads, five with missing teeth and five with intact teeth. Directly in front of each boy's mouth was a piece of red or green food, depicting the covariation information. The experimenter then 'faked' the evidence, rearranging the food so that the opposite pattern of tooth loss was suggested (that green food caused tooth loss). A doll, Sally, was introduced, who didn't know that the evidence had been faked. The children were asked "When Sally sees things the way they are now, which food will she say makes kids' teeth fall out?" The majority of the five-year-olds correctly said that Sally would arrive at a mistaken hypothesis. In a second experiment in which the faked covariation evidence was not perfect (so that the *pattern* was in favour of a particular hypothesis), five-year-olds were unsuccessful, but six-year-olds were successful (four-year-olds were not tested). Ruffman et al. concluded that, by the age of around six years, children understand how simple covariation information forms the basis for a hypothesis.

In fact, the best conclusion is probably that both children and adults are poor at scientific reasoning if scientific reasoning is defined in terms of giving evidence *priority* over background knowledge and context. When children have to make causal inferences in situations involving many potential causal variables, they experience difficulty, even though their basic causal intuitions are sound. The same is true of adults, as Kuhn et al. (1995) pointed out. Many adults also perform poorly in fully-fledged scientific reasoning tasks – for example, tasks that require them to examine a database and draw conclusions. Conducting scientific investigations into the relations between variables in real-world situations is simply not an easy task when many variables are present, because human beings find it difficult to ignore their pre-existing knowledge and to keep multiple variables in mind at once (Lagnado et al., 2007). Although the essential principles for making valid causal inferences are available early in childhood, the ability to draw multivariable causal inferences remains difficult even *beyond* childhood.

MULTI-VARIABLE CAUSAL INFERENCES

Many situations encountered in real life do require some ability to draw multi-variable causal inferences. One example is the need to *integrate* information about different causes, which is characteristic of a number of aspects of causal reasoning in everyday life. In fact, we seldom have to reason about causes and their effects in isolation. Instead, we frequently have to reason about more than one cause at a time. Even everyday problems require us to take into account many causal factors and their effects, and some of the causal factors relevant to a particular problem may interact with each other. Causal reasoning is thus *usually* multi-dimensional.

To take a trivial example, imagine that you are trying to decide whether you have enough time to go to the post office during your lunch hour. You need to consider not only how far the post office is, how long your lunch hour is and how fast you can walk, but also whether it is raining (this could affect speed and time), whether sufficient cashiers will be available to prevent long queues (this could affect time in the post office itself) and whether there are any potential hold-ups en route. In other words, you will need to consider many causal factors and how they may interact with each other before deciding whether it is actually worth trying to go to the post office.

The integration of knowledge about two dimensions

The question of when young children become able to interrelate information about different causal relations has been investigated using a variety of paradigms. We will begin by considering experiments that investigate children's ability to interrelate information about two causal dimensions, and we will then consider experiments that investigate children's ability to interrelate information about three causal dimensions. One of the best-known paradigms for investigating children's ability to interrelate information about two causal dimensions is the balance scale task.

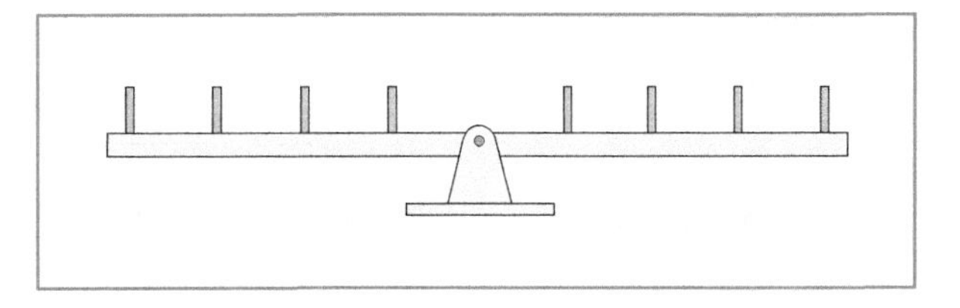

FIGURE 7.11
A balance scale apparatus.

The balance scale task measures children's ability to interrelate information about weight and distance. A typical apparatus consists of two arms of equal length that extend from a central fulcrum, like a see-saw (see Figure 7.11). Each arm can have weights attached to it at different distances from the fulcrum. The child's task is to predict which side of the balance scale will go down when different combinations of weights are placed at different distances from the centre. In order to judge this correctly, children must take *both* the relative number of weights and their relative distance from the fulcrum into account, and then combine these variables *multiplicatively*. For example, if three weights are placed on one arm of the balance scale 20 cm from the fulcrum, and six weights are placed on the other arm 10 cm from the fulcrum, the scale will balance. However, if both groups of weights are placed 10 cm from the fulcrum, then the side with six weights on it will go down.

KEY TERM

Balance scale task
Children are tested for their ability to take into account both weight and distance from the fulcrum in determining whether a balance scale will go down, up or stay level.

Siegler (1978) used a balance scale to assess the different *rules* that children use to interrelate information about weight and distance during development. His method was to ask children to make judgements about which arm of the balance scale would go down in a choice format that held one variable (weight or distance) constant while varying the other. For example, in *distance* problems, the same number of weights were placed on each arm, but at different distances from the fulcrum. In *weight* problems, different numbers of weights were placed on each arm, but at the same distance from the fulcrum. In *conflict-weight* problems, there were more weights on one arm, but the fewer weights on the other arm were at a greater distance from the fulcrum (e.g., two weights 8 cm from the fulcrum versus three weights 6 cm from the fulcrum: weight wins), and so on. Girls aged five, nine, 13 and 17 were tested.

Siegler's results led him to postulate that the development of physical understanding in the balance scale task proceeded through four different rules. Only the final rule involved information integration (see Figure 7.12). The first three rules depended on considering the dimensions of weight and distance *separately*, without trying to integrate them. Children who used rule 1 always said that the arm with the most weights would go down. Children who used rule 2 took into account distance information, but only when the two arms had equal weights. In all other cases, they ignored distance and made judgements on the basis of relative weight. Children who used rule 3 showed a developmental progression to considering distance information as well as weight information, but only when the two variables did not conflict. When one side had a greater weight and the other side had a greater distance, as in the *conflict-weight* problem given earlier, performance was at chance. Only children who used rule 4 showed an ability to integrate weight and distance information multiplicatively.

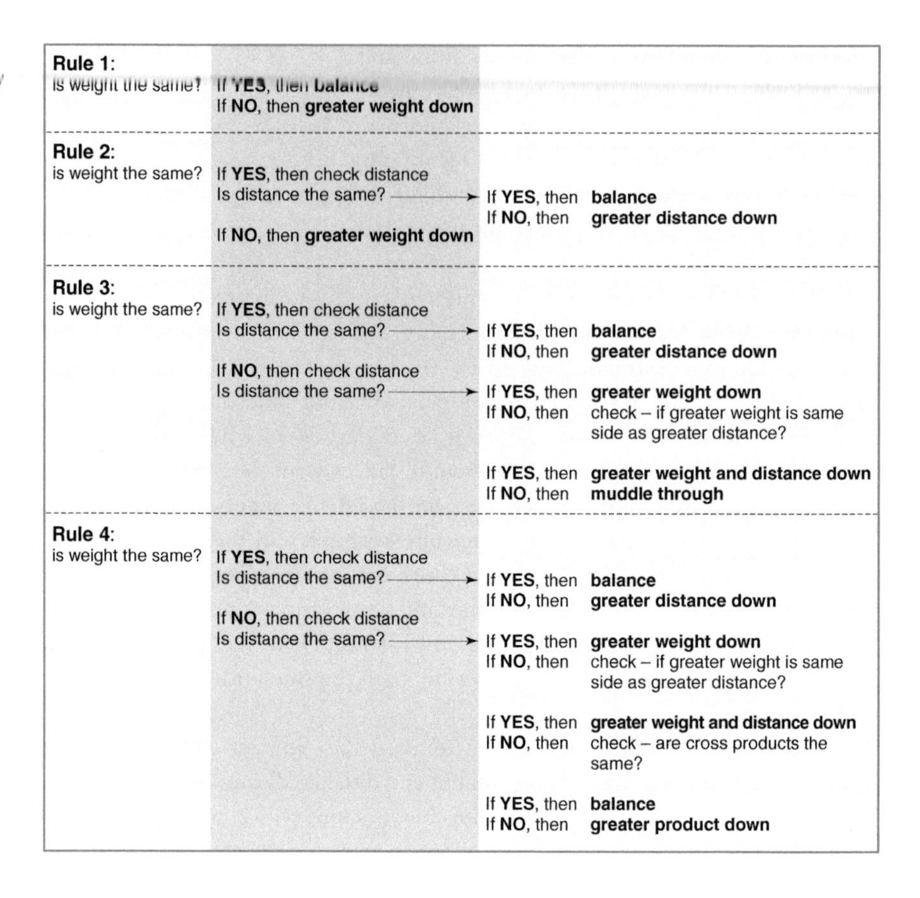

FIGURE 7.12
The four rules proposed by Siegler (1978) to explain the development of physical understanding in the balance scale task.

However, Wilkening and Anderson (1991) argued that Siegler's task could *underestimate* younger children's ability to apply integration rules because of a problem with 'false positive' responses. They pointed out that younger children might be using a simpler integration rule of *adding* weight and distance information rather than a more sophisticated rule of multiplying weight and distance. Such children would be scored by Siegler as following a non-integration rule (one of rules 1–3, and therefore a 'false positive'), whereas in fact they were following a *simpler form* of an integrative rule. To test their idea, Wilkening and Anderson asked children to *adjust* the position of a fixed set of weights on one arm of the balance scale in order to balance a set of varying weights on the other. The adjustment task thus required an *active* response from the child rather than a verbal judgement.

In the adjustment task, children were required to adjust either weight or distance. For example, if the fixed set was three weights placed 12 cm from the fulcrum and the variable set was one, two, three or four weights placed at 6, 12, 18 or 24 cm from the fulcrum, then the children had to adjust the distance of the fixed set in order to balance the scale. To balance one weight at 6 cm, the children would have to move their fixed set to 2 cm from the fulcrum; to balance one weight at 12 cm, the children would have to move their fixed set to 4 cm from the fulcrum; and so on. This was the *distance* adjustment task. An analogous *weight* adjustment task was also used, in which a fixed set of two weights at 24 cm had to be adjusted by adding more weights to balance either one, two, three or four weights placed at 8, 16, 24 or 32 cm from the fulcrum. Using the adjustment methodology, Wilkening and Anderson found that nine-year-olds, 12-year-olds and adults all used multiplicative integration rules to combine information about relative weight and distance. The youngest children tested (six-year-olds) tended to focus on either weight or distance, without trying to integrate the two. Nevertheless, Wilkening and Anderson argued that their data showed that children's causal understanding in the balance scale task was seriously underestimated by Siegler's methodology. More recently, Amsel, Goodman, Savoie and Clark (1996) tested children aged from five to 12 years in the balance scale task, and reported that children tended to recognise the importance of number of weights at about five to six years of age. By nine years of age, the majority of children tested also identified distance from the fulcrum as an important causal variable.

Another way of considering the balance scale methods used by Siegler (verbal prediction) versus Wilkening and Anderson (physically adjusting the weights) is that only the latter method enables children to benefit from embodied causal knowledge. If the brain is automatically encoding phenomenological experiences in terms of temporal structural information, and thereby encoding causal knowledge, for situations involving weight this causal knowledge will be embodied. Until the child is required explicitly to explain this embodied knowledge, it will remain embodied, and therefore may not influence verbal performance on causal reasoning tasks like the balance scale task. This possibility was explored in a study using a balance scale task reported by Pine and Messer (2000). Their task required children to balance a number of wooden beams, some of which were unevenly weighted (by having a block at one end), on a fulcrum. Pine and Messer recruited a large sample of 140 children aged from five to nine years, and first pre-tested them for their pre-existing knowledge of the principles of balance in this task. All the children who either could not explain the principles of balance or who possessed a naïve theory of balance (that all things have to balance in the centre) then experienced one of two interventions. In one intervention, the children watched as the experimenter modelled the correct solution to the balance scale task (Observe Only condition). In the second intervention, the children both watched the modelling and were asked to explain what they were seeing (Observe and Explain condition). In each condition, the children were then given the opportunity to balance the beams themselves. Pine and Messer reported that children improved in both conditions, but that significantly more children (70%) improved in the Observe and Explain condition compared to the Observe Only condition (50%). Again, we see that explanation-based learning is a powerful mechanism for causal learning. Pine and Messer asked children to explain a modelled solution that was more advanced than their own pre-existing solutions, and this led to greater learning gains than Observation Only.

The relationship between causal learning in the balance scale task and embodiment was then explored in an extension of this study by Pine, Lufkin and Messer (2004). They had noted that children in their balance scale task tended to gesture with their hands when explaining the principles of balance. Some gestures were used by virtually all the children tested, such as a closed fist to indicate a heavier weight. Accordingly, Pine et al. (2004) looked retrospectively at the videos of the children from the original Pine and Messer (2000) study, and coded their gestures at pre-test. They were interested in whether the levels of gesture displayed would relate to improvement following the interventions. In order to decide which gestures were relevant, Pine and her colleagues used the gestures produced by the 41 children who had shown a good understanding of the principles of balance at pre-test, and who did not subsequently participate in the intervention. They found that three types of gesture were regularly produced by these successful reasoners, a gesture for *weight*, a gesture for *distance* and a gesture for *middle* (that beams with equally distributed weight will balance in the centre). Pine et al. then coded the gestures of the other 99 children in their balance scale task at pre-test, those children who had subsequently experienced the interventions. Pine et al. were particularly interested in children whose gestures were concordant with their verbalisations versus children whose gestures were not concordant with their verbalisations. As the latter children's embodied knowledge about balance differed from their verbalisations, Pine et al. expected that discordant children would make greater improvements following intervention.

This was exactly what they found. Of the 99 children, 36 were classified as discordant. Of these 36 discordant children, 78% improved in the balance beam task whichever intervention they received, compared to 50% of the concordant children. In fact, 18 of the discordant children had received the Observe Only intervention, while 18 had received the Observe and Explain intervention. Post hoc analysis showed that both interventions were equally effective for discordant children, with 72% improvement for Observe Only and 83% for Observe and Explain, a non-significant difference. For the concordant children, however, receiving the Observe and Explain intervention was significantly more effective than receiving the Observe Only intervention. For concordant children, 62% improved with the Observe and Explain intervention compared to 41% for Observe Only, a significant difference. Pine et al. (2004) interpreted their findings in terms of gesture–speech mismatch, arguing that gestures can be an important index of when children are ready to learn. The idea that children's gestures show when they are on the verge of new cognitive insights was pioneered by Susan Goldin-Meadow, in the context of Piagetian logical reasoning tasks (Ping & Goldin-Meadow, 2008). However, Pine et al.'s data are also relevant to Boncoddo et al. (2010)'s conclusion that embodiment plays a role in children's causal learning. New causal understanding can emerge from action.

Integrating information about the causal effects of forces

Another type of physical information that must be integrated in the real world is information about forces. Consider a tug-of-war. This is a simple force problem. Two teams are pulling on the opposite ends of a rope. Both teams hope to move the centre of the rope beyond a certain pre-agreed point. If one team consists of 20 men and the other of ten men, then most people would predict that the team of 20 men should win the tug-of-war. The reason is that this team should exert the

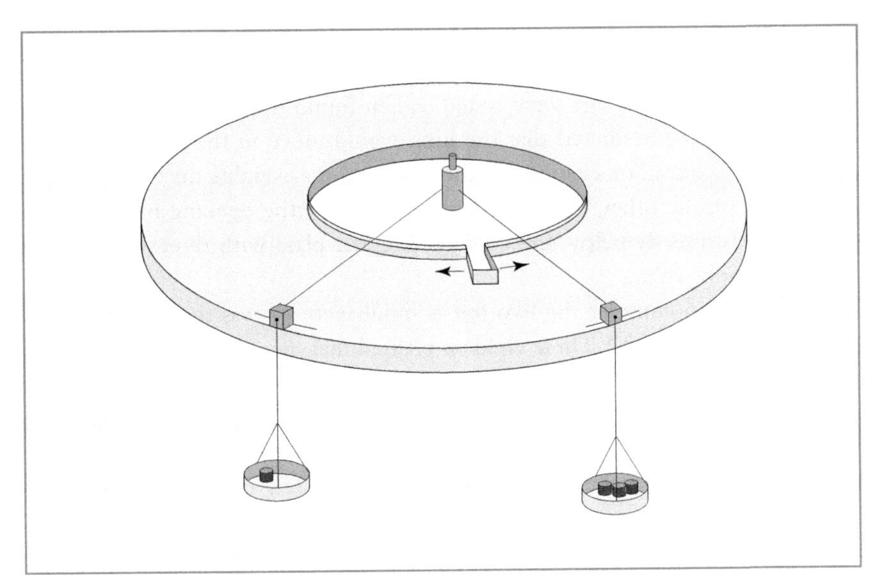

FIGURE 7.13
The force table used by Pauen (1996b).

stronger force. However, if the team of ten men all weigh 20 stones or more, and the team of 20 men all weigh eight stones or less, then the prediction might go the other way. Perhaps the combined force of ten strong men will be greater than the combined force of 20 weaker men. Alternatively, if the teams are *equal* in number and in strength, then the centre of the rope may not move at all. The forces may cancel each other out.

Pauen (1996b) gave a version of this force problem to young children. As well as the special case of two forces acting at 180° to each other (the tug-of-war), she used problems in which two forces acted at 45°, 75° and 105° to each other. The forces, which were represented by weights, were in the ratios 1:2, 1:3 and 1:6 respectively. In order to solve these problems correctly, the children had to combine two force vectors. The problem was presented using a special apparatus called a force table (see Figure 7.13).

The force table consisted of an object that was fixed at the centre of a round platform. Two forces acted on this object, both represented by plates of weights. The plates of weights hung from cords attached to the central object at either 45°, 75° or 105° to each other. The children's job was to work out the trajectory of the object once it was released from its fixed position. Although the central object was never actually released, the children had to move a barrier surrounding the platform until an opening in the barrier was in exactly the right position to catch the object. Their predictions were measured and scored in terms of whether the opening in the barrier was positioned closer to one or the other plate of weights, or was equidistant from both.

The force table problem was presented to the children in the context of a story about a king (central object) who had got tired of skating on a frozen lake (the platform) and who wanted to be pulled into his royal bed on the shore (a box behind the opening in the barrier). The children were asked to turn the barrier so that the royal bed would be in the right place for the king to slide into it. Different combinations of weights were used, and children aged six, seven, eight and nine years of age were tested. Pauen found that most of the younger children (80–85%) predicted that the king would move in the direction of the stronger force only. For example, if there were three weights on one plate and one weight on the other, these children would move the opening in the barrier so that it was directly below the cord holding the plate with three weights (the 'one-force-only' rule).

An ability to consider the two forces simultaneously was shown by some of the nine-year-olds (45%). These children realised that the correct location for the opening in the barrier was near the cord holding the heavier weights, but not exactly below it. However, although they showed this insight on some trials, they reverted to the one-force-only rule in other trials. Pauen thus judged them to be in a transitional stage regarding the integration rule. Very few children (5–10% across all groups) showed pure integration rule behaviour. Such behaviour required placing the opening in the barrier between the bisector of the angle and the stronger force. Such integration rule responses were shown by the majority of the adults tested (63%), however.

In the special cases when the forces were at 180° to each other (analogous to the tug-of-war situation described earlier), the majority of children at all ages tested gave the correct answer to the force problem. Pauen thus decided to change the situational context of the force problem, to see whether it would be easier for children to use the integration rule if the forces were represented by men pulling on ropes rather than by weights sitting on plates. In a second experiment using the force table, she replaced the plates of weights with teams of toy people pulling on ropes (see Figure 7.14). The children were told that two groups of cowboys were trying to pull a barrel to the shore in order to take it to their camp.

In this replication, fewer of the younger children used the incorrect one-force-only rule (40–50% instead of 80–85%). However, correct integration solutions did not increase. Furthermore, the solution of the special case (180°) problems actually *decreased* when Pauen used the cowboys context with the younger children. Pauen speculated that this may have been because the children who received the plates of weights applied a balance scale analogy to the force integration problem. A balance scale analogy gives rise to one-force-only solutions, which are correct in the case of the 180° problems, but not in the cases of the 45°, 75° or 105° problems. To investigate this idea further, Pauen and Wilkening (1996) gave nine-year-old children a training session with a balance scale prior to giving them the force table problem. They compared three groups, a group of children trained with a traditional balance scale, who thus learned to apply the one-force-only rule, a group of children trained with a modified balance scale

FIGURE 7.14
The force table used in Pauen's (1996b) cowboys, paradigm.

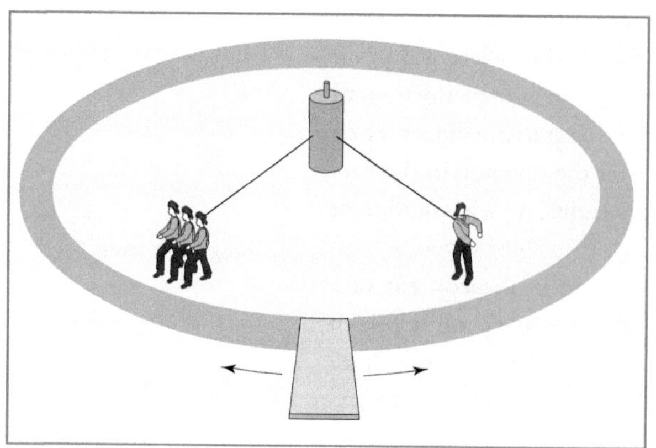

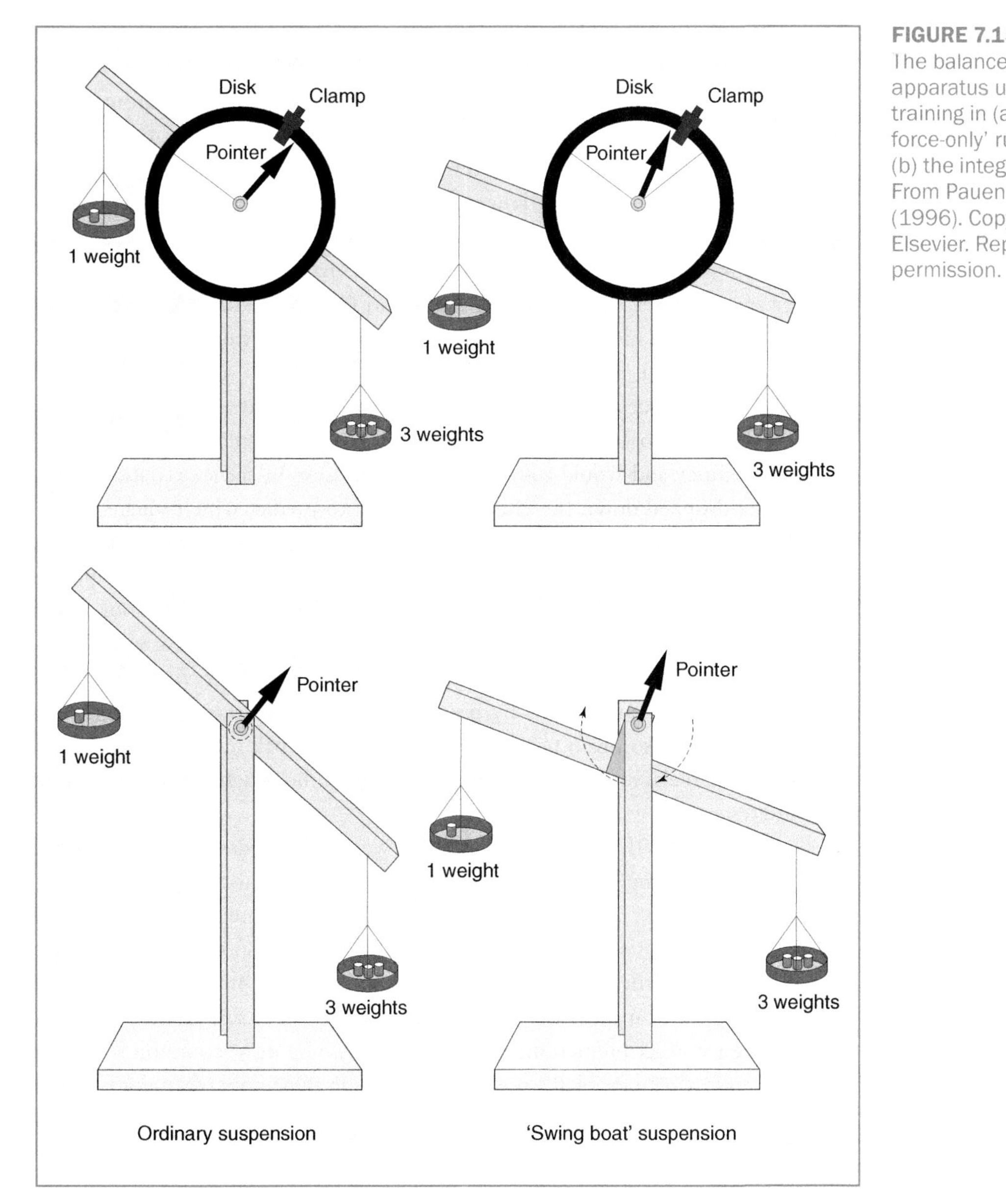

FIGURE 7.15
The balance scale apparatus used to provide training in (a) the 'one-force-only' rule versus (b) the integration rule. From Pauen and Wilkening (1996). Copyright © 1996 Elsevier. Reproduced with permission.

(shown in Figure 7.15) that had its centre of gravity below the axis of rotation (a 'swing boat' suspension), and a no-training control group. The modified balance scale provided training in the integration rule, as the swing boat suspension meant that even though the beam rotated towards the stronger force, the degree of deflection depended on the size of *both* forces. For each group, the children had to predict the location of a pointer attached to the centre of the beam for different ratios of weights. Pauen and Wilkening found that children who were trained with the traditional balance scale showed a greater tendency to use the one-force-only rule than the control group children, while the children who were trained with the modified balance scale showed a greater tendency to use the integration rule than the control group children. Hence children's responses to the force table problem varied systematically with the solution provided by the analogical model. These results suggest that the children use spontaneous analogies in their reasoning about physics, just as we have seen them do in their reasoning about biology (children's use of the personification analogy was discussed in Chapter 5) and in language acquisition. Analogies seem to play an important role in children's everyday reasoning.

Children's understanding of force dynamics has also been investigated at younger ages. Göksun, George, Hirsh-Pasek and Golinkoff (2013) devised a context for giving force integration problems to children aged from 3.5 to 5.5 years. They created a cloth game board of large 30 cm × 30 cm squares (like a giant checkerboard) and put a house in the middle. The house had a long chimney that ran down from the roof and exited at the front door. Balls could be dropped down this long chimney and would roll out of the front door. In *one-force* trials, children saw the ball dropped down the chimney, and had to predict which square on the board the ball would land on after exiting the front door. In *two-force* trials, children were shown how a hairdryer could create 'wind' that would act as an opposing force to dropping the ball, for example by blowing towards the front door of the house or across the front door of the house. Again, the children's task was to predict where the ball would come to rest when it was dropped down the chimney. Children made their predictions by physically placing a marker on their chosen square. They were also given prior experience with the ball in each condition, so that they could make reasonable predictions based on the weight of the ball. Adults were also tested.

Göksun et al. (2013) found that even the youngest children were as accurate as the adults in making causal predictions in the one-force problems. For the two-force problems, some children at all ages could successfully integrate the two forces and make accurate causal predictions. Correct integration responses were provided around 50% of the time by 3.5 year olds, 60% of the time by 4.5 year olds and 75% of the time by 5.5 year olds. The 5.5 year olds did not differ significantly from the adults in their correct predictions about two forces. The most common source of error in the two-force problems was to focus on one force only (dropping), similar to Pauen's (1996b) findings with older children. Nevertheless, Göksun et al.'s study suggests that even three-year-olds can focus on more than one force at a time, and can make causally accurate predictions that involve force integration when an age-appropriate context is provided.

The integration of knowledge about three dimensions

Children's ability to interrelate information about three different dimensions has been studied by Wilkening and his colleagues, who have examined children's ability to interrelate information about time, distance and velocity in a range of ingenious paradigms (e.g., Wilkening, 1981, 1982; Wilkening & Cacchione, 2010). Time, distance and velocity information is crucial for a decision such as whether to go to the post office in one's lunch hour (discussed earlier). The critical information is 'how long is my lunch hour?' (time), 'how far is the post office?' (distance) and 'how fast can I go?' (velocity). These variables are related by simple physical laws. For example, velocity is equivalent to distance divided by time, and distance is equivalent to time multiplied by velocity. In order to see whether children reason according to these physical laws when integrating information about time, distance and velocity, Wilkening (1981) devised a task involving a turtle, a guinea pig and a cat.

In Wilkening's task, children were shown a model of a footbridge with a turtle, a cat and a guinea pig fleeing along it. The animals were all fleeing at their own different speeds, and were running from a fierce barking dog, who was shown at the left side of the apparatus. The children's task was to judge how far each animal could run in a certain period of time. The time period was either two, five or eight seconds, and was represented in terms of the amount of time that the dog barked. The children made their judgements by moving each animal to the correct location on the footbridge after the dog had stopped barking. This version of the task required the children to integrate information about time and velocity in order to judge distance.

Wilkening also devised versions of the task that required the integration of distance and velocity, and the integration of distance and time. He argued that if the children could integrate these different sources of information successfully, then they should use multiplicative rules to make their judgements. For example, in the barking dog task described above, they should use the rule 'distance equals time multiplied by velocity'. When five- and ten-year-old children were given the barking dog task, Wilkening found that both age groups indeed used a multiplying rule, as did a control group of adults. However, in the other versions of the task the younger children did not always use the correct integration rules. For example, when asked to use information about distance and velocity to make judgements about time (judging how long the dog must have barked when shown the point the turtle, the cat or the guinea pig had reached on the footbridge), the youngest children used a subtraction rule. This suggested that they knew that the dog had barked for different amounts of time for each animal, but that they could not estimate these differences proportionally.

Even the adults used the wrong integration rule in some versions of the task, however. When asked to use information about distance and time to make judgements about velocity (deciding which animal would have been able to reach a certain point on the footbridge in the time that the dog barked), adults used a subtracting rule, just like the younger children in the time judgement task. From Wilkening's perspective, however, the fact that children and adults did not always select the correct rules was not critical. The important finding was that even five-year-olds attempted to apply algebraic rules when trying to reason about different physical dimensions. This meant that they had some conceptual understanding of the separate variables involved, even though they (and the adults) occasionally selected the wrong algebraic rules to integrate these variables. Wilkening argued that his results implied an implicit understanding of dimensional interrelations – a naïve or 'intuitive' physics. Children were adopting a practical approach to the psychological integration of separate variables in which the procedures chosen did not *violate* the physical rules, but simplified them.

More recently, Wilkening has also studied the integration of information about time, distance and velocity by using a task based on toy cars racing around a track (Wilkening & Martin, 2004). The aim was to provide a concrete test for integration problems of the kind 'If the drivers of two cars travel the same distance, one with a constant speed of 100 km/hour over the whole distance, and one with a constant speed of 75 km/hour over the first half of the distance, how fast does the second driver have to be during the second half of his journey to arrive at the same time as the first driver?' The correct answer is multiplicative – 150 km/hour (2 × 75). To provide a concrete version of this integration problem, Wilkening and Martin (2004) showed children aged six and ten years a race track, with two parallel tracks for race cars and a long tunnel that covered both tracks halfway from the race start point. The children were shown the race cars going round the tracks, one always at the same speed over the whole track, the other at a slower speed that could be varied. The children had to work out how fast the slower car should travel once it had entered the tunnel in order to finish the race at the same time as the first car.

Children either gave their responses on a quasi-numerical graphical rating scale, or by actually pushing the second car, in which case their chosen speed was recorded by a photometer. Wilkening and Martin found that in the judgement condition using the rating scale, children simply added the speed differences (this would give the answer 125 km/hour in the example above). In fact, Wilkening and Martin found that adults made the same additive error when given the race car task. However, in the action condition, when physically pushing the second race car, the ten-year-olds were able to integrate the information about time, distance and velocity, and gave accurate responses. Another way of thinking about this is that when reasoning is *embodied*, then children can reason accurately about causal information. Detailed study of the role of embodiment in the development of children's causal reasoning seems likely to yield rich rewards.

BIASES AND MISCONCEPTIONS IN CAUSAL REASONING

So far in this book, the idea of a naïve or intuitive physics has been given a central role in explaining the development of causal reasoning. Essentially, it has been argued that intuitive physics is rooted in the perception of objects and events (see also Wilkening & Cacchione, 2010). Children's perception of the world around them yields rich information about the structure and action of physical systems, and in general physical structure is a reliable guide to the causal structure of physical events. However, this is not always the case. Intuitive or naïve physics can also yield misleading models of the physical causal structure of the world. This was documented for example in pioneering work by McCloskey and his colleagues (e.g., McCloskey, 1983).

Intuitive physics and misconceptions about projectile motion

KEY TERM

Impetus theory
The theory that every motion must have a cause.

Our intuitions about projectile motion provide a good example of how our intuitions about physical causation are not always correct. In fact, when reasoning about projectile motion, most children and adults employ a pre-Newtonian, medieval theory of motion, called the impetus theory (e.g., Viennot, 1979). According to this theory, each motion must have a cause. Our everyday experience makes this a very plausible assumption. If an object is inanimate, it cannot be set into motion without a physical cause being involved. If a ball is thrown by a person, it will move upwards in a trajectory determined by the initial throw, and will fall to earth when gravity takes over. However, if a ball drops from a moving train, no impetus appears to be involved – the ball has fallen passively from the train. In this latter case, according to an impetus theory of motion, the ball will fall downwards in a straight line. If we were standing by a railway and saw a ball fall from a passing train, this would indeed appear to be the case. The train would continue moving forwards, while the ball would be perceived as falling straight down (see Wilkening & Cacchione, 2010, for a fuller analysis).

McCloskey and his colleagues showed that, when asked to predict the trajectories of objects, adults as well as children seem to follow a 'straight-down' rule. They seem to believe that if an object is dropped by a walking person, then it will fall downwards in a straight line. In fact, it will fall forwards in a parabolic arc (e.g., Kaiser et al., 1985). By Newtonian physics, the moving carrier of the object imparts a force, just as if the object was pushed from a table top. With respect to the frame of reference of the ground, the object will fall forwards in both cases. Similarly, when asked to predict the motion of a ball ejected at speed from a curved tube shaped like the letter 'C', adults and children judge that the ball will continue to move in a curvilinear arc, whereas in fact it follows a straight line with respect to the horizontal plane (e.g., Kaiser, McCloskey & Profitt, 1986; see Figure 7.16). Medieval physical theories, such as the medieval theory of impetus, depend on perceptual

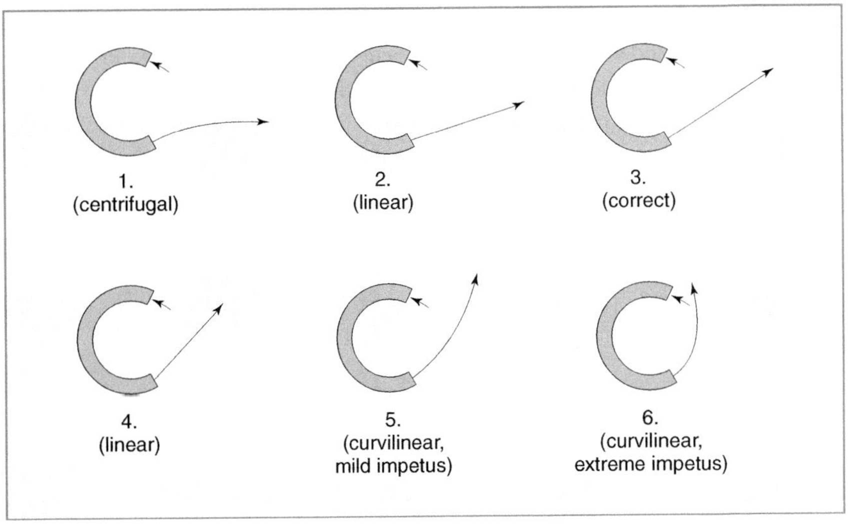

1. (centrifugal)
2. (linear)
3. (correct)
4. (linear)
5. (curvilinear, mild impetus)
6. (curvilinear, extreme impetus)

FIGURE 7.16
Schematic depiction of six alternative trajectories for the curved tube problem (Kaiser et al., 1986).

experience, and in the case of impetus theory, children are unlikely to receive many perceptual experiences that contradict the 'straight-down' rule. Because of this, Newtonian physics require direct instruction.

Gravity errors

Other research confirms that intuitive but misleading concepts of mechanics can be very difficult to dislodge. For example, the 'straight-down' rule noted above is applied not only to projectile motion, but also in tasks that test children's understanding of the effects of gravity. One example is the 'tubes task' invented by Hood (1995). In the tubes task, young children (two- to four-year-olds) are asked to find a ball that is dropped into one of three tubes. The tubes are all opaque and they can be interwoven to form a visuo-spatial maze. When the tubes are interwoven, a reliance on the 'straight-down' rule will lead the child to search in the *wrong* location for the ball. The tubes task can be administered at different levels of difficulty by increasing the number of tubes and the complexity of their inter-twining (see Figure 7.17).

Hood found that children who erred in the tubes task consistently searched in the location directly beneath the point at which the ball was dropped. He termed this a 'gravity error'. Gravity errors were the most frequent kind of error at every difficulty level tested, although the ability to solve the tubes task at different levels was related to age (more older children passed the easier levels) and sex (boys showed superior performance to girls). All erroneous search behaviour seemed to be predominantly determined by the straight-down trajectory, irrespective of the trajectory of the tube.

In later experiments using transparent tubes instead of opaque tubes, Hood showed that even the youngest children tested (two-year-olds) were able to search successfully for the ball when they could see its trajectory, and that this success occurred at all difficulty levels. Surprisingly, however, he found no evidence for transfer. Children who searched successfully with the transparent tubes were immediately given problems at the same level of difficulty with the opaque tubes, and promptly failed them. Even extensive training on the tubes task with a single tube failed to dislodge the prepotent gravity error in a group of two-year-old children. Hood concluded that his task documented a growing understanding of the operation of tubes, and of how they constrain the movements of invisibly falling objects. This explanation suggests that the gravity error is also dependent on intuitive physics. Children appear to believe that all objects fall straight down. Hence the

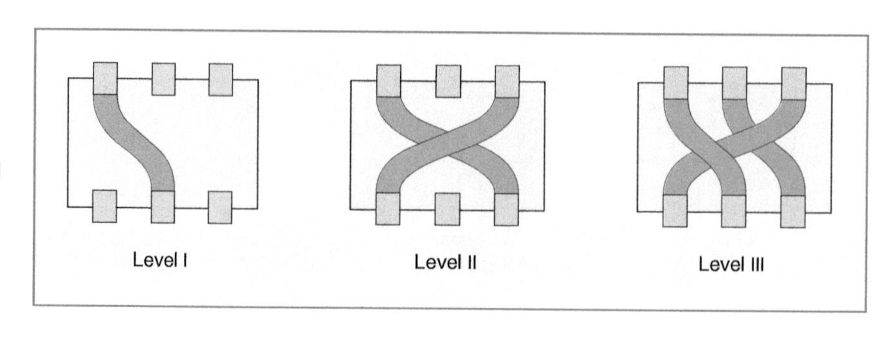

FIGURE 7.17 Schematic depiction of different levels of difficulty in the 'tubes' task. From Hood (1995). Copyright © 1995 Elsevier. Reproduced with permission.

gravity error might well recur in children who have become quite sophisticated in the tubes task if they were presented with a different apparatus or if task demands were varied.

Hood, Wilson and Dyson (2006) gave children aged 5.5 years who had passed the tubes task a modified version of the task involving two balls. Using a three-tube apparatus, in which no (opaque) tube connected to the container directly below it, the experimenter dropped both a red and a green ball simultaneously down two randomly determined tubes and asked the child to find either the red ball or the green ball (also determined randomly). Successful search required the child to monitor two potential targets. They had to assess the spatial layout of the tubes, work out which container went with which tube, and remember which tube the red versus the green ball was dropped down. Hood et al. reported that success decreased significantly in comparison to the one-ball condition, and that many search errors were gravity errors. Hood et al. argued that the increased attentional demands of the 'two ball' paradigm decreased the children's ability to inhibit the gravity error. Children need to actively suppress their naïve physical theories in order to respond correctly, even children who can demonstrate correct understanding in easier versions of particular tasks. So *inhibition* is related to accurate responding.

Reprising embodiment

It is notable that most of multi-variable causal reasoning tasks discussed so far have measured knowledge via action. For example, Shultz, Pardo and Altmann (1982) asked children to roll tennis balls to measure their understanding of mediate transmission, Wilkening and Anderson (1991) asked children to move weights on a balance beam to measure their understanding of the integration of weight and distance information, and Wilkening (1982) asked children to move toy animals along a bridge to measure their understanding of the integration of three physical variables (time, distance and velocity). One interesting possibility is that intuitive physical knowledge as measured by action may be *distinct from* intuitive physical knowledge as measured by judgement tasks. Judgement tasks require *reflection* on one's physical knowledge. For example, action-based knowledge about projectile motion may be fairly sophisticated, whereas explicit judgements about the same motions may be naïve. Action-based knowledge and verbalisable knowledge may also follow different developmental paths (see Chapter 8, for a discussion of the different developmental paths followed by implicit/procedural and explicit/declarative knowledge). The possibility that intuitive physics is best measured directly, via action, was investigated by Krist, Fieberg and Wilkening (1993), in a task using projectile motion.

The task used by Krist et al. was a throwing task. Six-year-olds, ten-year-olds and adults were asked to throw a tennis ball from different heights so that it would hit a target on the floor. The participants were asked to use a horizontal motion when throwing the tennis ball, which was somewhat awkward to produce as it involved sliding one's arm along a horizontal 'throwing' board before releasing the ball. This meant that the speed with which the ball was released was critical if the ball were to reach its target. In addition, the height of the throwing board was varied during the experiment (a lower board requires a faster throw), and so variable throwing speeds were required. As well as requiring the participants to throw the ball themselves, Krist et al. asked them to make judgements about how fast the ball should be released in order for it to reach the target when the throwing board was raised or lowered. These judgements were either required prior to or following the action phase of the experiment.

Krist et al. found that the functional relation between speed and height was present in the *action* data at all age levels tested, whereas it was only present in the *judgement* data for the adult participants. Whereas the adults judged that a lower throwing table would require a faster speed of release, the ten-year-olds did not differentiate between the speeds required for the different height levels, and some of the six-year-olds showed an *inverse* pattern, judging that greater heights would require greater speeds. In contrast, mean speeds of actual throwing represented the physical law very closely, across all age groups. Thus whereas performance in the judgement condition reflected a strong age trend towards integrating height and distance in an appropriate way, performance in the action condition showed an intuitive understanding of this relationship at all ages. These data imply that intuitive physical knowledge is present in perceptual-motor knowledge and skills. This embodied knowledge then provides a basis for the development of physical mechanical concepts. The research discussed earlier in this chapter suggested that being asked to provide *explanations* of physical phenomena may be an important mechanism supporting successful reasoning (Legare & Lombrozo, 2014). Being asked to reflect on your knowledge, for example by being asked to generate explanations for physical events, has developmental benefits for causal reasoning, improving children's successful causal predictions or judgements.

Cognitive neuroscience may also provide new insights into the development of physical causal reasoning. To date, however, only college students have been studied. Nevertheless, these studies provide interesting data relevant to child development (Dunbar, Fugelsang & Stein, 2007). Dunbar and his colleagues produced neuroimaging data that suggested that when we learn particular scientific concepts, such as the Newtonian theory of motion, they do not *replace* naïve knowledge. Rather, when faced with a novel reasoning problem, both the Newtonian theory of motion and the (medieval, naïve) impetus theory of motion are activated, and then the naïve theory is inhibited. In other words, rather than undergoing conceptual change, the brain appears to maintain *both* theories. The ability to inhibit the wrong theory then determines successful reasoning. In their experiment, Dunbar et al. (2007) showed two groups of college students – physics students and non-physics students – different videos of two balls falling while they lay in an fMRI scanner. The students were told to press one computer key if the balls were falling as they would expect in a frictionless environment, and another computer key if the balls violated their expectations. The key manipulation was whether a large and small ball fell at the same speed (Newtonian physics), or whether the larger ball fell faster than the smaller ball (naïve physics). Dunbar et al. found that the physics students correctly chose the Newtonian motion videos as correct depictions of falling, whereas the non-physics students erroneously chose the impetus theory videos (where the larger ball fell more quickly). However, when giving their correct answers, the physics students showed increased activation of the anterior cingulate, a structure typically activated by tasks requiring inhibition. Dunbar et al. (2007) argued that the physics students were inhibiting their naïve knowledge in order to respond correctly, and therefore they had not undergone conceptual change.

More recently, a similar result has been reported in college students who were asked to make judgements about the viability of electrical circuits while in an fMRI scanner (Masson, Potvin, Riopel & Brault Foisy, 2014). Only some of the electrical circuits showed physically possible outcomes regarding light bulbs being illuminated. Masson et al. reported that only physics students correctly judged which circuits these were. Again, there was extra activity for these expert students in brain regions such as anterior cingulate cortex. This neural activity was absent for the novice students, who were also making erroneous judgements. Masson et al. (2014) also concluded that their data suggested the inhibition of naïve knowledge during successful physical reasoning rather than conceptual change. Of course, as both of these fMRI studies are correlational, other interpretations of these group differences in neural activation patterns are also possible. Nevertheless, this neural imaging work is consistent with behavioural studies using reaction times, which have also suggested that generating the correct Newtonian answer to a problem about motion requires active inhibition of an incorrect answer based on impetus principles (Kozhevnikov & Hegarty, 2001). Importantly, both the behavioural and the neuroimaging data are not consistent with the notion that scientific knowledge undergoes radical restructuring with development. Accordingly, the idea that conceptual change is driven by new information that cannot be accommodated within existing conceptual frameworks (Carey, 1985; see Chapter 5) may be wrong.

These preliminary neuroimaging data suggest instead that naïve theories are retained, and then more comprehensive theories are added to one's store of knowledge, with the naïve theories requiring inhibition in order for reasoning to proceed correctly. Indeed, diSessa (2014) has made a similar point about the absence of conceptual change in the development of scientific reasoning, in his case on the basis of cognitive research. diSessa argues that conceptual change per se does not happen, as naïve knowledge is not a coherent theory. Rather, it is a series of quasi-independent elements. For example, regarding the impetus theory of motion discussed earlier, naïve physicists assume two forces, the force imparted by the hand throwing the ball, and the force of gravity. But to an expert physicist, the hand imparts *momentum*, not a second force. DiSessa points out that expert physicists reason that throwing a ball into the air only involves one force, gravity. Gravity acts on the speed of the ball, pulling it downwards until the ball reaches zero speed (at the peak of the toss), after which the ball falls to the ground. Hence to deal with misconceptions that arise from naïve physics, diSessa (2014) suggests that effective Newtonian instruction should choose the most productive aspects of naïve theories, and then refine them. For example, the recognition that gravity is the only force once the peak of the toss has been reached can be used as a basis for explaining momentum and acceleration. DiSessa (2014) discusses a range of evidence that suggests that such an instructional approach is more effective than trying to change children's prior intuitions. Children's naïve intuitions are not completely wrong – only elements of them are wrong.

SUMMARY

Causal reasoning is fundamental to cognitive development. It is particularly important for learning about the empirical relations in the world, that is for learning how the world is. Causal reasoning enables children to learn about likely causal agents and causal mechanisms in all of the foundational domains of naïve biology, naïve psychology and naïve physics. For example, children need to understand agency in terms of animate agents with intentions, and also to understand agency in terms of physical causal agents and mechanisms. They also need to develop explanations for why an agent has a certain effect. In the psychological domain this comprises theory of mind, and in the physical domain it concerns plausible causal mechanisms. Children develop these understandings and explanations by notising associations, tracking statistical dependencies (conditional probabilities), making analogies, connecting causes and effects and producing explanations. In the developed world, many physical mechanisms have to be taken on trust. For example, most people do not understand the causal mechanisms behind electricity, or how simple devices such as zips function. Nevertheless, they can still operate electrical appliances and zips. Background or collateral knowledge enables them to make sensible choices (e.g., that experts have built certain artefacts like irons and Hoovers because electricity works, see Koslowski & Masnick, 2010). Casual reasoning and explanation-based learning continue to be important throughout the lifespan.

One perspective discussed in this chapter was that causal reasoning could only offer impressive cognitive benefits if it followed recognised causal principles. A variety of evidence has shown that children do indeed reason in accordance with causal principles, for example the principle of causal priority and the principle of covariation. Further, they do so from surprisingly early in development (e.g., Shultz & Kestenbaum, 1985). Similarly, children seem to have some understanding of the relationship between hypotheses and evidence, although this understanding is less impressive early in development (Sodian et al., 1991; Somerville & Capuani-Shumaker, 1984). Even teenagers and adults have difficulties in reasoning about multiple causal variables, and in integrating causal information on a variety of dimensions (Kuhn et al., 1995; Wilkening, 1982). Nevertheless, overall the evidence suggests that causal reasoning shows impressive continuities from childhood to adulthood. Causal inferences that come easily to adults also come easily to children, and causal inferences that are difficult for adults are also difficult for children. Both children and adults also hold some intuitive *misconceptions* about physical phenomena, not all of which decline with age (Kaiser et al., 1985).

The use of causal Bayes nets to describe children's causal inferences has been an important development in the field, and offers one useful way of integrating behavioural developmental research with cognitive neuroscience. Bayes theorem has been very productive in the fields of brain imaging and machine learning, since neural activity is probabilistic in nature. The demonstration that children's causal behaviour can be described in Bayesian terms supports the view taken in this chapter, that causal inferences can result from automatic neural/perceptual computations that encode environmental experiences. Explicit observation of observed effects is then required in order to generalise such knowledge. The possibility that much of children's causal knowledge is initially *embodied* offers further opportunities for devising cognitive neuroscience studies. Studying the mechanisms that underpin children's causal computations, and studying how targeted instruction or self-explanation affects these mechanisms, could change our understanding of causal reasoning by children and its developmental trajectory.

CHAPTER 8

CONTENTS

The development of memory

8

Memory is a remarkable facility. The ability consciously to retrieve autobiographical happenings from the past – "to travel back in time in [our] own minds" (Tulving, 2002, p. 2) – was once thought to be unique to the human species. More recently, it has been shown that some animals (e.g., scrub jays) can display some behaviours suggestive of retrieving autobiographical events (e.g., they can recall where they hid food in order to conceal it from other birds, see Clayton, Griffiths, Emery & Dickinson, 2001). The retrieval of events and experiences from one's past (sometimes referred to as 'what, where, when') is usually called 'episodic memory'. Episodic memory is usually contrasted with semantic memory, our generic, factual knowledge about the world, such as knowledge of concepts and language (discussed in Chapters 5 and 6). Both episodic memory and semantic memory are forms of explicit or *declarative* memory, memories that can be brought consciously and deliberately to mind. Explicit and declarative memories are typically contrasted with implicit and *procedural* memories. These are unconscious memories, indexed by changes in performance without the involvement of conscious memory content. Examples include skill learning, habit formation, associative learning (e.g., classical conditioning) and habituation. All species appear to show implicit and procedural memory.

Cognitive psychology assumes that memory is a modular system. Semantic memory, recognition memory, working memory, implicit memory, episodic memory and procedural memory are all considered to be distinctive in various ways. Experimentally, these different types of memory are thus usually considered independently of each other. More recently, this 'modular' approach to memory has been adopted by cognitive neuroscience. Brain imaging research suggests that activating different types of memory is associated with activation of partly different brain structures. For example, episodic memory tasks are associated with activation of the hippocampus and the medial temporal lobe, whereas skill learning is associated with activation of the striatum (basal ganglia). Comparable neuroimaging studies of memory are now being carried out with children, and representative studies will be discussed in this chapter. Meanwhile behavioural studies of memory development are now so numerous that there is sufficient material for a dedicated handbook of memory development (Bauer & Fivush, 2013). Accordingly, this chapter comprises an edited highlight of available developmental data.

An important factor to bear in mind regarding memory research is that children (and adults) do not record events that occur in their lives into their memories verbatim. Even though it may feel as though you can remember 'exactly what

It is known that infants can encode relevant information, store it and later retrieve it, as evidenced by their ability to recall how they should play with a certain toy.

happened' when you went to visit your friend, your friend is bound to have a somewhat different recollection of events to you. As originally demonstrated by Bartlett (1932), children and adults *construct* memories, and the process of construction depends on prior knowledge and personal interpretation. It also depends on how much sense the memoriser can make of the *temporal structure* of their experiences. Very young children, for example, may not structure their experience in memorable ways, particularly if they do not understand particular experiences (e.g., being born, someone dying, being sexually abused), or if they do not have a clear temporal framework. Very young children are also still acquiring language, and language itself is important for memory. For example, language helps in rehearsing one's own experiences or in recounting them to someone else, and these verbal narratives help to order memories temporally and establish them more firmly. As with the development of theory of mind, the development of memory clearly cannot be isolated from the development of other cognitive processes. Remembering is embedded in larger social and cognitive activities. Thus the knowledge structures and learning skills that young children bring to their experiences are likely to be a critical factor in explaining memory development.

Despite this embedding of memory in other aspects of cognitive development, many studies of memory have used tasks that were purposely disembedded from larger social/cognitive activities, in order to provide a 'pure' measure of the memory system of interest. An unintended result has been that the applicability of many research findings are limited. "Students of human memory... [ignored] almost everything that people ordinarily remember. Their research did not deal with places or stories or friends or life experiences, but with lists of syllables and words... [leading] to a preference for meaningless materials and unnatural learning tasks" (Neisser, 1987). Wherever possible in this chapter, I will focus on studies of memory development that use less artificial memory situations and have greater 'ecological validity'. Finally, it is increasingly recognised that an important factor for memory development is a child's 'metaknowledge' of memory processes and contents. Memory can be improved if a child is aware of explicit aspects of memorising, such as the need for mnemonic strategies, and if a child can self-monitor their efficiency in memorising and keep track of the sources of different kinds of knowledge. These metacognitive aspects of memory will be discussed in Chapter 9, where the role of 'meta-knowing' will be considered with respect to both memory and learning.

KEY TERM

Recall
Actively recalling or bringing to mind previous experiences.

EARLY MEMORY DEVELOPMENT

Some aspects of early memory development have been discussed already, for example habituation, recognition memory and memory for causal event sequences (see Chapters 1 and 2). Even very young infants show good evidence of memory in such paradigms. Although habituation and recognition are implicit forms of memory, studies of deferred imitation or of elicited imitation are generally accepted as measures of the development of declarative memory (Bauer, 2006). In Chapter 1, we saw that researchers such as Rovee-Collier and Mandler have successfully used deferred imitation tasks with infants as young as six months, using novel paradigms (such as putting a mitten on a puppet) that clearly demonstrated that the infants were bringing a past event to mind. Bauer (2006) has used both deferred and elicited imitation to track the early development of declarative memory. She has documented important changes in the reliability with which recall can be observed, and in the temporal extent of memory, in very young children.

Early memory for temporally ordered events: The importance of goals

In her studies, Bauer usually requires young children to reproduce an ordered sequence of actions. Her argument is that reproduction of the *ordering* is a critical measure of explicit recall. The temporal ordering information must be encoded during the presentation of the event sequence, and subsequently recalled from a *representation* of the event, as the event itself has gone. For example, Bauer and Shore (1987) modelled 'having a bath' to young children aged from 17 to 23 months by demonstrating giving a teddy bear a bath. The sequence of events was that the teddy bear's T-shirt was removed, he was put in a toy tub, and he was washed and then dried by the experimenter (in pretend mode!). The teddy bear was then handed to the child, who was asked "Can you give the dirty bear a bath?" The children proved quite capable of reproducing the modelled event sequence when tested for immediate recall. They also remembered the correct sequence of events six weeks later, when they returned to the laboratory and were simply handed the teddy without prior modelling (delayed recall). These data suggest that very young children's representations of events are temporally ordered. Their event memory is not composed of a series of disorganised snapshots of individual components of the event. Instead, like adults, their representations display temporal ordering and are arranged around a goal.

Event sequences like 'having a bath' are very familiar to young children, however, and this familiarity may aid temporally ordered recall. An important question is thus whether the experimenters would have found similar effects if the temporally ordered events had been novel instead of familiar. To find out, Bauer and Shore invented a novel causal event sequence called 'building a rattle'. During this event the experimenter modelled putting a plastic ball into a stacking cup, covering it with a slightly smaller stacking cup and shaking the 'rattle' near to her ear. The children again showed both immediate and delayed recall for the elements of the sequence, and for their temporal order. Bauer and her colleagues argued that even very young children were sensitive to the causal relations underlying event sequences from their very first experience of them, and that this early causal sensitivity meant that very young children's representations displayed goal-oriented temporal ordering, just like adults.

However, in these early studies the temporal ordering was an intrinsic part of achieving the goal, such as the 'building a rattle' paradigm. In real life, maintaining the temporal order in which events are experienced may not always be critical to successfully achieving our goals. To study the importance of organising experienced events around a goal for effective episodic memory, Loucks, Mutschler and Meltzoff (2016) manipulated the order in which novel events were experienced experimentally. They tested slightly older children than Bauer and her colleagues (three-year-olds), also using deferred imitation as the measure of episodic memory. Children were shown a set of novel actions called 'zavving', with novel objects. 'Zavving' was demonstrated on day 1, when children watched an experimenter perform a sequence of three actions, namely putting a bead on a square block, pushing a stick through a hole in the block, and then moving the bead to a flap sticking out of the block. These were actions 1, 2 and 3. These were described in neutral terms "First, we do this. Then we do this, and then we do this. There, I did it! I zavved! That's called zavving!" Zavving was then demonstrated again using a new bead and a new square block with a flap. Finally, the child was given a turn with another new bead and new square block. A control group of children watched three different actions (rotate a rectangular block, rub it with fabric, jump a bottle cap over the block). These three actions were not given a goal-oriented description like 'zavving', but otherwise were also repeated with novel materials, after which the child could imitate with further new materials. All children performed at 100% for deferred imitation on day 1.

The children then visited the laboratory again on day 2, entering a different room with different materials, which enabled the critical declarative memory test. Children in both groups now watched the same events, one being 'zavving', although this was not labelled on day 2, and a control event. We can describe the three actions for 'zavving' as 1, 2 and 3, and the three actions for the control event as A, B and C. On day 2, children either saw a grouped version of each event, such as 1,2,3,A,B,C, or A,B,C,1,2,3, or they saw an interleaved version. The interleaved version mixed consecutive actions from each event, for example 1,2,A,B,3,C. The children then experienced a delay (they did some colouring). Remember that the experimental group had experienced 'zavving' on day 1, and the control group had not. Loucks et al. (2016) argued that if the experimental group had organised their memory of 'zavving' around a goal, then they should organise their actions on day 2 accordingly, even if they had watched the interleaved version of events. They should imitate 1,2,A,B,3,C as 1,2,3,A,B,C. This was exactly what they found. Children in the experimental group organised their imitation according to goals, reproducing the two events as 1,2,3 and A,B,C as frequently for the grouped demonstration as for the interleaved demonstration. Children in the control group produced haphazard and idiosyncratic actions in the interleaved condition. Loucks and his colleagues argued that information about goals bounds sub-actions together, enabling better memory. Goal organisation trumps veridical temporal ordering of action sequences. These data are consistent with the view that human memory is constructive rather than veridical.

We saw in Chapter 2 that infants are especially interested in the goal-directed actions of agents. The goal structure of human actions is typically determined by cause–effect relations (e.g., to make pasta for supper, we first have to boil some water). To test the idea that causal relations play a special role in organising the temporal order of events for young children, Bauer and Mandler (1989) carried out a study that used two novel-causal event sequences and two familiar sequences. Their subjects were younger children aged either 16 or 20 months. The novel-causal event sequences were 'build a rattle' and 'make the frog jump'. 'Make the frog jump' involved building a see-saw by putting a wooden board onto a wedge-shaped block, putting a toy frog at one end of the board and making him 'jump' by hitting the other end of the board. The familiar event sequences were 'give teddy a bath' and 'clean the table'. For the 'clean the table' event, a wastebasket, paper towel and empty spray bottle were used. The experimenter mimed spraying the table, wiping it with the towel and then throwing the towel away. In order to separate temporal from causal information, Bauer and Mandler also included 'novel arbitrary' event sequences. An example of a novel event whose temporal ordering was arbitrary was the 'train ride' event sequence. For this event, two toy train cars were linked together. A toy driver was then put into one of the cars, and a piece of track was produced for the train to sit on. Although the events were modelled in this order, there was no causal necessity in this ordering, and so these components could be reproduced in any order without affecting the final event.

Bauer and Mandler found that recall for the temporal order of events was indeed significantly lower for the novel arbitrary events than for the novel causal events, even though recall levels for the former were still significant. Furthermore, they found that when an irrelevant component was inserted into each kind of novel event sequence (such as attaching a sticker either to one of the cups making up the rattle or to the toy train driver), then this irrelevant component was far more likely to be displaced in the causal event sequences. Attaching the sticker to the cup was frequently displaced to another position in the 'building a rattle' sequence, or was even left out entirely. In contrast, attaching the sticker to the train driver was treated no differently to any of the other components in the 'train ride' sequence. This finding suggests that causally related pairs of elements enjoy a privileged organisational status. Bauer and Mandler concluded that causal relations were an important organising principle both for constructing event memories and for aiding recall. The importance of causal relations in structuring episodic memory fits nicely with the research discussed in Chapter 5, in which we saw the importance of causal information for developing and organising semantic memory (e.g., Pauen, 1996a).

Bauer and her colleagues have also used temporal ordering tasks to explore age-related aspects of encoding and forgetting. Bauer, Wenner, Dropnik and Wewerka (2000) studied the development of explicit memory in 360 children aged from 1–3 years. The children were studied from the ages of either 13, 16 or 20 months, for a period of a year. Memory for six event sequences was tested over time. The event sequences were similar to those used in prior work, and included building a rattle and making a gong. When recall was tested, the props were provided as reminder cues, and if the props alone were insufficient to generate recall, then verbal prompts were used (e.g., "You can use this stuff to make a gong. Show me how you make a gong"). Bauer et al. found that whereas almost 80% of 13-month-olds could retain temporally ordered memories for around a month, over 80%

TABLE 8.1 Percentage of 13-, 16- and 20-month-olds showing evidence of temporally ordered recall memory

Delay interval	Age at experience of to-be-remembered event sequences		
	20 months	**16 months**	**13 months**
1 month	100*	94*	78*
3 months	100*	94*	67
6 months	83*	72*	39
9 months	78*	50	44
12 months	67*	61	36

Data are from Bauer et al. (2000). An asterisk indicates that the number of children exhibiting the pattern of ordered recall (i.e., higher level of performance on previously experienced than on new event sequences) was reliably greater than chance. Because determination of chance levels is affected both by the number of observations and by the number of tied observations, identical values will not necessarily yield identical outcomes (e.g., 13-month-old three-month delay and 20-month-old 12-month delay).

of 20-month-olds retained such memories for at least six months. In fact, almost 70% of 20-month-olds retained these memories for at least a year. Developmental differences in retention are shown in Table 8.1. In related work, Carver and Bauer (1999, 2001) showed that even younger infants also retained temporally ordered memories. They reported that about 50% of nine-month-olds could retain temporally ordered event sequences for a month, but not for three months. Infants aged ten months could retain temporally ordered event sequences for three months. Hence there are developmental changes in both the reliability of explicit recall and in its temporal extent.

Cognitive neuroscience studies of episodic memory in infants

More recently, Bauer has used some of her explicit memory tasks as a basis for EEG studies with infants. The approach is to use ERPs such as the Nc (also used as a neural marker of infant categorisation, see Chapter 5) as an index of encoding. For example, using the deferred imitation task with nine-month-old infants, Bauer, Wiebe, Carver, Waters and Nelson (2003) recorded ERPs as indices of recognition memory, delayed recognition memory and recall one month later. When the infants first visited the laboratory, they watched three two-step event sequences, for example putting a toy car into an apparatus and then operating a rod to turn on a light. Each sequence was demonstrated twice, without the infant being allowed to imitate it. This experience was repeated on two more visits within the same week (the average gap between visits was 1.5 days). This was the *learning phase* of the experiment. Recognition memory for the learned sequences was then tested at the end of the third visit. The infants were shown pictures of the steps and the end state of one familiar sequence and of one novel sequence. The electrophysiological recordings focused on activity between 260 and 870 ms following the appearance of each picture, termed a middle latency or Nc component. The EEG was also recorded to the same pictures in a fourth visit one week later, to measure delayed recognition.

Bauer et al. (2003) reported that recordings taken on the third visit suggested that all the infants encoded the events. The Nc component showed a significantly greater negativity for the pictures of the novel sequence than for the pictures of the familiar sequence, for all the babies. However, when tested for their recall of the event sequences a month later (using behavioural re-enactment), the babies fell into two groups. One group of infants showed recall of at least one of the original three event sequences (46% of babies), as indicated by their accurate manipulation of the props used in the original events. The other group of babies did not (54%). Exploration of the Nc component at delayed recognition for these two groups showed that the infants who showed no behavioural evidence of recall after one month also showed no increased negativity to the familiar sequence after one week. Bauer et al. argued that this was suggestive of a failure of *consolidation* of the memories. The EEG at recall after one month also showed group differences. The babies who recalled at least one event sequence showed a significantly shorter latency to peak amplitude for novel sequences. However, they showed a longer latency to peak amplitude for familiar sequences. Bauer et al. argued accordingly that remembering was indexed by longer latency to peak amplitude, because of the reintegration processes associated with long-term recognition. Bauer (2006) has subsequently used EEG to demonstrate developmental differences consistent with her idea about differences in memory emerging at the point of encoding. For example, she has reported that infants aged ten months show significantly larger ERP amplitudes compared to infants aged nine months in the immediate recognition task used by Bauer et al. (2003), and she has argued that this is indicative of differential encoding. From her electrophysiological data, Bauer has concluded that encoding and consolidation rather than retrieval are the significant sources of developmental change in early explicit memory. For example, the medial temporal lobe is related to successful consolidation of declarative memories by adults, and this structure is still developing in the infant years.

Another approach that has been used successfully with infants to index episodic memory is pupillometry. For adults, pupil dilation is larger to previously seen items than to novel items, providing an index of episodic recognition memory. Indeed, pupil dilation in adults is reliably different to memories that are explicitly recollected versus simply familiar, the so-called *remember/know distinction*. Accordingly, Hellmer, Söderlund and Gredebäck (2016) set out to investigate pupil dilation in infants aged four and seven months in an episodic memory paradigm. Infants were shown coloured pictures of unfamiliar toys and of office and household items. During the encoding phase, infants were shown the pictures in one category, for example toys, for five seconds each. During the test phase, the infants saw each familiar picture paired with a novel picture from a second category, for example household items. There was then a second test phase, in which the familiar pictures were shown a second time, paired with a novel picture from the third category, such as office items. Adults were also tested.

Hellmer et al. reported that both the adults and the seven-month-old infants showed greater pupil dilation to the familiar items than to the new items, although for the adults this was the case in the first test phase and for infants it occurred in the second test phase. The four-month-olds did not show reliable differences in pupil dilation, a null finding that is difficult to interpret. Hellmer et al. argued that the seven-month-old infants required more exposure to form stable memory traces than adults, hence only showed the pupil dilation effect on the second test phase. Nevertheless, they argued that episodic memory processes were remarkably similar in the infants and the adults. It would be interesting to use pupillometry as an index of remembering in some of the paradigms used by Bauer and her colleagues, comparing pupil dilation data with ERPs. Currently there is no independent measure in her studies to support Bauer's view that larger ERP amplitudes are indicative of better encoding.

Finally, a topic of current interest in adult cognitive neuroscience is the role of sleep in memory consolidation. Sleep has been shown to affect both the quality and the quantity of adult declarative memories. Sleep appears particularly important for memory consolidation, particularly slow wave sleep (0.5–4 Hz). Accordingly, Seehagen, Konrad, Herbert and Schneider (2015) decided to investigate whether sleep would enhance memory consolidation in infants aged six and 12 months, using a deferred imitation paradigm. In their study, infants watched a puppet scenario in which a mitten was placed on the puppet's right hand (a task developed by Rovee-Collier and her colleagues, see Chapter 1). Imitation was tested after a four-hour delay, during which half of the infants took a nap averaging 106 minutes. Seehagen et al. reported that the infants who had slept performed significantly better in the deferred imitation test, at both ages. Indeed, infants who had not slept did not show evidence of declarative memory in the deferred imitation test, again at both ages. However, given the length of the nap taken by those infants allowed to sleep, it could be argued that the 'no nap' infants were simply too tired to perform well. To check this explanation, Seehagen and her colleagues repeated the experiment, but this time the deferred imitation test was given a day later. Again, half of the infants in each age group were allowed to sleep soon after learning, and half were kept awake for at least four hours after learning. All infants then went home and had a full night of sleep. The deferred imitation data were comparable to the first experiment. Only infants who had slept soon after learning showed declarative memory on day two as measured by the deferred imitation task. Seehagen et al. concluded that sleeping soon after learning acted to safeguard memory traces from interference from subsequent new information, enabling memory consolidation. Seehagen et al. suggested that as the infant brain likely has limited storage capacity, consolidation during sleep may have to occur more frequently if episodic memories are to be stored efficiently. Notice that on this view, individual differences may depend more on consolidation than on encoding.

An EEG study carried out by Friedrich and her colleagues supports the idea that sleep is important for consolidation (Friedrich, Wilhelm, Born & Friederici, 2015). Slow wave sleep in adults is thought to drive the reactivation of recently encoded declarative memories in the hippocampus, with sleep spindles (synchronised oscillatory activity of 10–15 Hz) in thalamocortical loops inducing integration of new memories with related information in the neocortex. To investigate sleep spindles and slow wave sleep in infants, Friedrich and her colleagues used a word learning task while recording EEG. Infants aged nine to 16 months first received a training session in which repeated picture-word pairings of unknown objects were presented, either in a *consistent word–object* pairing condition, or in a *consistent category* pairing condition, or in an *inconsistent pairing* condition. In the first condition, infants could learn specific word meanings, while in the second condition they could learn general word meanings (category formation). The inconsistent pairing condition was a baseline condition and did not support new learning. ERPs indicated learning of the word–object pairs during training, but no learning of the novel categories. Infants then either took a nap or went for a walk in the park with their parents, remaining awake. Retention was tested on average 84 minutes after learning. The ERPs measured at retention testing showed that those infants who had taken a nap showed memory for both the word–object pairings and the novel categories. This appears to indicate that sleep had enabled memory integration, so that the general word meanings were now also available for recall. The infants who had not taken a nap showed no memories at retention for either specific word–object pairings nor general word meanings. For the infants who had slept, sleep spindle activity was correlated to the measures of retention. EEG power in the 10–15 Hz range was significantly correlated with the ERP measure of category formation, the infant N400. Slow wave activity was not related to any of the ERP measures; however, as the infants spent most of their time asleep in slow wave sleep, this could have been a ceiling effect. Infant age did not make a difference. The best correlate of category formation was high sleep spindle activity. Accordingly, Friedrich et al. (2015) concluded that infant sleep enables both the consolidation and the reorganisation of recently encoded memories. This interesting study shows the importance of studying underlying neural mechanisms when seeking to understand cognitive development. The finding that sleep spindle activity may operate to support memory consolidation in the same way in the infant brain as in the adult brain is particularly important given the many speculations that age differences in memory are explained by the key neural structures activated by declarative memory tasks only coming on-line later in development (Schacter & Moscovitch, 1984). So far, the data from EEG measurements, pupillometry and sleep all converge on the conclusion that most of the mechanisms supporting episodic memory are on-line from early in development, before structures such as the medial temporal lobe are fully myelinated and developed.

Infantile amnesia – a real phenomenon?

The encoding and retrieval of specifically *autobiographical* memories has not been studied systematically in very young children. One reason is that it is difficult to devise robust methodologies. A second reason is that adults report surprisingly few autobiographical memories of the period of their lives before the age of around three years. This later absence of early episodic memories is often referred to as *infantile amnesia*.

KEY TERM

Infantile amnesia
The apparent inability to recall autobiographic memories before the age of around three years.

Infantile amnesia is surprising given that the first three years of life are an active period for conceptual development, which is a form of declarative memory (semantic memory), and also for language development, which involves conscious recall of verbal knowledge. The early years also see the rapid growth of causal and psychological reasoning, which depend on stable and efficient memory for events and their outcomes (see Chapters 3, 4 and 7). Further, as discussed above, when measured in terms of the retention of temporally ordered and goal-driven events it is clear that episodic memories are developing in infants. One possibility, as argued by, for example, Bauer (2015), is that there is developmental continuity of personal memory. Infantile amnesia is then the product of two complementary effects. First, as the encoding and consolidation of memories becomes more robust with age, the quality of the traces that are stored improves. These traces include more features that make memories distinctively relevant to the self, for example better elaborated and more tightly integrated features. Second, these increases in the quality of autobiographical memories are accompanied by decreased forgetting. Bauer (2015) reviews evidence that the rate of forgetting is much higher in children below the age of around six years than in later childhood and adulthood. So memory traces become less vulnerable to loss at the same time as they are encoded more robustly. From a *developmental continuity* perspective, therefore, infants may be laying down autobiographical memories even if we have not yet devised ways to measure them.

Nevertheless, the general absence of autobiographical memories before the age of about three years has been repeatedly documented and observed (see Howe & Courage, 1993; Courage & Howe, 2004). Even when we feel convinced that we can recall events from our own infancy, these events often turn out to have happened to someone else. For example, the memory researcher David Bjorklund reports a vivid memory of having croup (bronchitis) as an infant. "My crib was covered by a sheet, but I remember looking past the bars into the living room. I can hear the whir of the vaporiser, feel the constriction in my chest, and smell the Vicks Vaporub. To this day the smell of Vicks makes my chest tighten." However, when he reminded his mother about this memory, it turned out that he had never had croup. She told him "You were such a healthy baby… That was your brother, Dick. You were about 3 years old then" (Bjorklund & Bjorklund, 1992).

The interesting question is why we can access so few memories of the earliest period of our lives. One of the first explanations for infantile amnesia came from Freud (1938), who argued that early amnesia was caused by the repression of the emotionally traumatic events of early childhood. For Freud, the problem was not encoding, consolidation or storage, as early memories were assumed to be intact. Rather, repression was used to keep these memories from invading consciousness. Although the idea of repression can account for the active rejection of emotionally troublesome material from consciousness, it does not explain why memories for pleasant events are also later inaccessible. Another possibility is that early memories are coded in terms of physical action or pure sensation. Early memories are thus irretrievable, as they are stored in a different format to later memories, which depend on linguistically based encoding and storage. The finding that females tend to have earlier memories than males appears to be consistent with this explanation, as language development is usually more advanced in girls than in boys. According to this idea, early memories survive intact, but the context in which these memories were laid down is so discrepant from the one in which we seek to retrieve them (during later childhood or adulthood) that it is rarely possible to make contact with the relevant memory traces. This idea was tested in an experiment reported by Simcock and Hayne (2002).

Freud (1938) argued that infantile amnesia was caused by the repression of the emotionally traumatic events of early childhood.

Simcock and Hayne were interested in whether children could have verbal access to memories that were acquired when the children were pre-verbal. To find out, they visited children aged 27, 33 and 39 months at home, and gave them a very memorable experience. This was to play with a Magic Shrinking Machine. Children could put toys into the machine, turn some handles, and the toys would magically shrink to a much smaller size. Language skills were assessed during the visit, and of the 23 words specifically associated with the target events (e.g., the labels for the toys), the 27-month-olds knew on average 16, the 33-month-olds knew on average 19 and the 39-month-olds knew on average 20. The children were then revisited after either six months or a year, and their memory for the Magic Shrinking Machine was explored. Memory was tested both by eliciting verbal recall, and non-verbally, for example by showing the children the machine and pictures of various toys, and asking them to choose the toys that had gone into the machine.

Simcock and Hayne (2002) reported that the children in general showed good memory of the Magic Shrinking Machine when tested with the non-verbal measures, but poor memory when tested using verbal recall, at both delays. Further, even though more relevant words had typically been acquired during the delay interval (for example, the children who had been 27 months at the time of the encoding experience now had around 21 of the 23 relevant words in their productive vocabularies rather than 16), the children never showed verbal recall for these aspects of the procedure, even though they showed non-verbal recall. Simcock and Hayne concluded that the idea that it is impossible to make contact with early memories is wrong, as the children that they tested still had non-verbal access to these memories. However, they speculated that the inability to translate early, pre-verbal experiences into language prevented these experiences from becoming part of autobiographical memory

In contrast, Bauer (2015) argues that the absence of verbal recall by young children is not evidence that memory traces are being stored in ways that cannot be translated into language. She points out that the absence of evidence is not evidence of absence (i.e., that negative results cannot be interpreted causally in developmental psychology). In support of this view, Bauer and Larkina (2013) reported a prospective study of autobiographical memory in which children were asked to recall events from their recent past at ages three and four years, and were then tested again six years later. Bauer and Larkina (2013) found that many more autobiographical features were included when the children were aged nine and ten years, when they also had better verbal skills. Indeed, Jack, Simcock and Hayne (2012) reported similar data in a prospective study using the Magical Shrinking Machine. They revisited children who had experienced the machine when aged 27–51 months of age six years later, when the children were aged from eight to ten years. Some of the children were able to provide a full verbal account of their prior experience with the machine, even children who had only been two years old for their first experience. Jack et al. (2012) concluded that Simcock and Hayne's (2002) account of the role of language in infantile amnesia was too strong. Children can have verbal access to memories that were encoded before their language skills were fully formed.

Another proposal has been that infantile amnesia disappears when the child gains a cognitive sense of self, argued to occur at around two years (Howe & Courage, 1993; Courage & Howe, 2004). The cognitive self is thought to provide a new organiser of information, thus facilitating the personalisation of memory for events as specific events that happened 'to me'. However, as discussed in Chapter 3, the cognitive self may emerge earlier than two years. Another proposal has been that the memory systems that support the formation of autobiographical memories are late-developing because the brain structures that underlie these systems are not functional at birth (e.g., Schacter & Moscovitch, 1984). One speculation was that the structures essential for the formation of conscious memories only began to function properly at around 2–3 years of age. However, such arguments depended largely on drawing parallels between infant humans and infant monkeys. More recent data suggest that the maturation of the neural structures that are crucial for autobiographical memory, namely the medial temporal structures and the frontal lobes, does not map in any neat way onto infantile amnesia. In particular, while the medial temporal structures seem able to support the formation of explicit memories by the end of the first year, the frontal lobes do not mature until the early twenties. Hence the maturation of neural structures per se seems unlikely to provide a satisfactory account of infantile amnesia.

Finally, the development of knowledge structures may be important in explaining infantile amnesia. Fivush and Hamond (1990) argued that infantile amnesia may be due to a combination of the absence of distinctive memory cues and the fact that young children have yet to learn a *framework* for recounting and storing events. Because young children are in the process of trying to understand the world around them, they focus on what is similar about events, namely routines. The *routine* aspects of novel events do not make good retrieval cues for future recall. Similarly, because young children do not possess their own frameworks for constructing memories, early memories are fragmented, also making them more difficult to recall.

This more socio-linguistic account explains childhood amnesia as a natural byproduct of the development of the constructive process of memory itself (Nelson & Fivush, 2004). Fivush and Hamond's argument is appealing, because it places infantile amnesia firmly within the context of memory development in general. According to this argument, the lack of early memories is not the result of basic structural changes in the memory system with development. Instead, it is a result of the absence of abstract knowledge structures for describing the temporal and causal sequences of events. I discuss the development of these abstract knowledge structures below. By adulthood, the routine events that young children prefer to recall have merged into *scripts* (or generic knowledge structures) about specific events like 'what happens when we go to a restaurant'. Childhood amnesia is therefore due to a combination of script formation and the forgetting of novel events (see also Nelson, 1993; Nelson & Fivush, 2004).

Understanding symbolic representation as an aid to memory

Children use a number of symbolic systems in addition to language, such as making gestures (pointing to things) and engaging in symbolic play.

Abstract knowledge structures such as scripts for describing the temporal and causal sequences of events depend in part on language development, but language is not our only symbolic system. Words stand for or represent concepts and events in the everyday world, and of course we use them as symbols to encode our experiences. However, we also use other symbols to encode and communicate our experiences, such as pictorial ones. These symbols also represent or stand for objects or events, and include drawings, photographs and sculptures. All of these symbols bring to mind something other than themselves. Children, too, use a number of symbolic systems in addition to language, for example in communication. Young children make gestures, they point to things and they engage in symbolic (pretend) play. They also use culturally determined symbols. These include symbols such as maps and models. The use of many of these forms of symbolic coding enables children to represent information in memory in a form that will be accessible later on.

Symbolic understanding *itself* develops, and this development is another factor in explaining why older children have better memories than younger children. One of the most intriguing sets of experiments investigating the development of symbolic understanding comes from work on young children's understanding of models (e.g., DeLoache, 1987, 1989, 1991). The basic paradigm used in DeLoache's model studies is always the same. A 2.5- or three-year-old child is shown a scale model of a room, containing various pieces of furniture such as a couch, a dresser, a chair and some pillows (see Figure 8.1). The child is then introduced to two central characters, the stuffed toy animals Little Snoopy and Big Snoopy, who both like hiding. The scale model is introduced as Little Snoopy's room, and an adjacent room, which contains the same furniture as the model in the same spatial layout, is introduced as Big Snoopy's room. The child is told "Look, their rooms are just alike. They both have all the same things in their rooms!" Each correspondence is demonstrated "Look – this is Big Snoopy's big couch, and this is Little Snoopy's little couch. They're just the same."

Following this 'orientation phase', the child watches as the experimenter hides one of the Snoopy toys in the appropriate room. For example, Little Snoopy might be hidden under the little couch in the model room. The child is then asked to find Big Snoopy in the real room. The child is told "Remember, Big Snoopy is hiding in the same place as Little Snoopy". Three-year-old children go straight to the big couch and find Big Snoopy. Two-and-a-half-year-old children do not. They search around the big room at random, even though a memory post-test shows that they can remember perfectly well where Little Snoopy is hiding. DeLoache argues that the problem for the younger children is that they do not understand the *correspondence* between the model room and the real room. They do not seem to appreciate that they have a basis for knowing where to search for Big Snoopy.

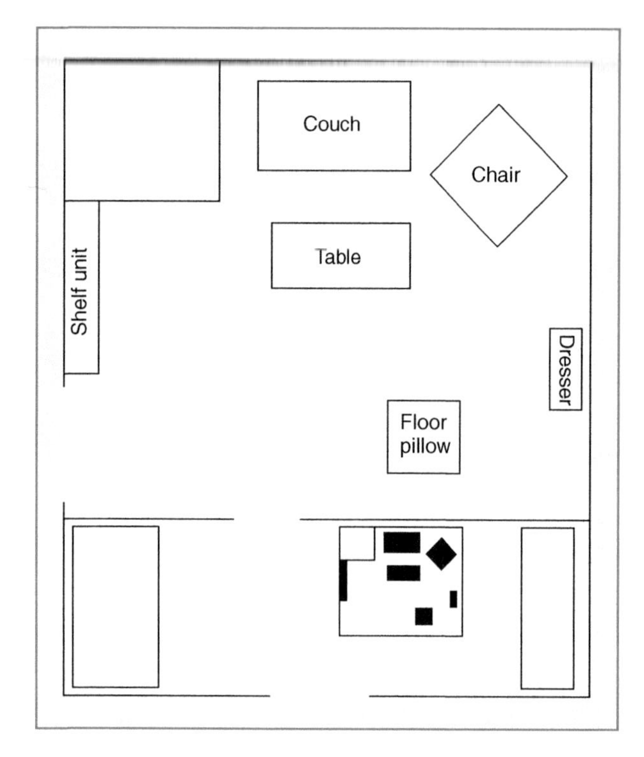

FIGURE 8.1
Diagram of the experimental room used in DeLoache's (1989) scale model studies, with the scale model shown below (the darkened areas in the model correspond to the labelled items of furniture in the room). From DeLoache (1989). Copyright © 1989 Elsevier. Reproduced with permission.

The most compelling reason for believing that the younger children's problem lies in their lack of awareness of the correspondence between the model and the room comes from DeLoache's 'magical shrinking room' studies. In these studies, 2.5-year-old children were persuaded that the experimenters had built a 'shrinking machine' that could shrink a doll and a room (DeLoache, Miller, Rosengren & Bryant, 1993). They were then shown where Big Snoopy was hiding in the big room, and asked to find Little Snoopy in the model room. As the children believed that the model was the shrunken big room, there was no representational relationship between the model and the room to confuse them, and indeed the children were very successful at searching for Little Snoopy in this task. DeLoache has also shown that younger children can find Big Snoopy when they are shown Little Snoopy's hiding place in a picture, which implies that they do understand the representational relation between the *picture* and the room (DeLoache, 1991).

Furthermore, experience with the picture task transfers to the model task. Experience with a symbolic medium that is understood (or partially understood), as pictures are, seems to facilitate the use of an unfamiliar symbolic medium (the model). More recently, Marzolf and DeLoache (1994) have shown that experience with a model–room relation can help 2.5-year-old children to appreciate a map–room relation. DeLoache argues that early experience with symbolic relations helps the development of symbolic sensitivity, which is a basic readiness to recognise that one object or event can stand for another. Thus in order to use these other symbols in memory and in learning, children may first have to learn what it is to *represent* something.

DeLoache's work shows clearly that symbol–referent relations are not always transparent to young children. As she finds experience to be important in her studies, the recent ubiquity of tablet computers for very young children offering games using such relations may already be changing this developmental picture. This is an empirical question. Nevertheless, the ability to represent the higher-order relation between symbol and referent (to adopt a 'representational stance') appears to develop fairly rapidly during the first years of life (see Chapters 2 and 3). As DeLoache has argued, the ability to *map* similarities between symbol and referent is an important component of the developing understanding of symbol–referent relations, and frequently depends on the ability to map *relational* similarities. The importance of relational mappings or analogies in cognitive development has been noted at a number of other points in this book. Given the centrality of symbol use in human cognition, it is interesting to note that DeLoache gives the ability to make relational mappings a key role in the development of early symbolic understanding.

Relational mappings can be drawn by infants in the first year of life, as discussed in Chapter 2 (Chen et al., 1997). The links between early symbolic understanding and memory development in general are not well-understood, however.

THE DEVELOPMENT OF RECOGNITION MEMORY

KEY TERM

Recognition memory
Recognising that something is familiar and has been experienced before.

Recognition memory is the ability to recognise that something is familiar and has been experienced before. It is usually considered to be a form of implicit memory. We have already seen that infants have good visual recognition memories (see Chapter 1). Most of the habituation studies of memory in infancy discussed in Chapters 1 and 2 also concerned recognition memory in various forms. Experimental paradigms that use conditioned responses are also based on recognition, for example motor paradigms (such as Rovee-Collier's kick-to-work-a-mobile paradigm) and auditory paradigms (such as DeCasper and Fifer's suck-to-hear-your-mother paradigm).

Recognition memory seems to be fairly ubiquitous in animals as well as in humans, and so this early-developing memory system is far from unique. For example, pigeons can 'remember' 320 pictures for 700 days when tested in a recognition memory paradigm (Vaughan & Greene, 1984). Given its ubiquity, the status of recognition memory as a *cognitive* skill has been questioned. For example, Fagan has argued that recognition memory may actually be a measure of *processing* rather than a measure of cognitive ability per se (see Fagan, 1992). Studies of cognitive development might thus expect to find little development in recognition memory with age. This is in fact the case.

The traditional way of examining recognition memory in young children has been to show them a series of pictures, and then to measure the number of pictures that they recognise as familiar after a certain period of time. In a classic study of this type, Brown and Scott (1971) showed children aged from three to five years a series of 100 pictures drawn from four familiar categories: people, animals, outdoor scenes/objects and household scenes/objects. Forty-four of the pictures recurred and 12 were seen only once. The pictures that recurred were seen after a lag of either zero, five, ten, 25 or 50 items. The children's task was to say "yes" if they had seen a picture before, and "no" if the picture was novel.

Brown and Scott found that the children showed accurate recognition memory on 98% of trials. There was also little difference in recognition accuracy depending on the lag between the items. In fact, the children were equally accurate for lags of zero and 50 pictures, showing 100% recognition accuracy for each. Accuracy levels for lags of five and 25 pictures were around 95%, and for a lag of ten pictures, 98%. These remarkable levels of performance fell slightly on a long-term retention test which was given after one, two, seven or 28 days. In the long-term retention test, the children were shown the 12 pictures that had been seen only once, 24 of the 44 pictures that had been seen twice, and 36 new pictures. For intervals of up to seven days, recognition memory levels were above 94% for pictures that had been seen twice. The level was somewhat lower for pictures that had been seen only once, falling from 84% after one day to around 70% after seven days. After 28 days, recognition accuracy for pictures that had been seen twice was 78%, and for pictures that had been seen only once, 56%. In a subsequent study, Brown and Scott showed that the superior memory for items seen twice was due to *both* the extra exposure to the items and to the need to make a judgement in the recognition task given in the first phase of the study. The previous need to make a "yes" judgement in itself seemed to act as a retrieval cue for the twice-seen items.

The excellent levels of recognition memory found in young children suggest that there is little for the developmental psychologist of memory to study here. However, interest in children's memory for what is familiar has been revived by the study of the development of implicit memory.

Implicit memory

KEY TERM

Implicit memory
Memory for information that the individual is not consciously aware of having.

Implicit memory is 'memory without awareness'. In implicit memory tasks, children and adults behave in ways that demonstrate that they have memory for information that they are not consciously aware of having. Although most of us would measure our memories in terms of what we can *recall* rather than in terms of what we can *recognise*, the possibility that previous experiences can facilitate performance on a particular memory task even though the subject has no conscious recollection of these previous experiences is a very intriguing one. Implicit memory has also been called 'unintentional memory', or 'perceptual learning'.

Perceptual learning tasks

One of the first studies of implicit memory in children was carried out by Carroll, Byrne and Kirsner (1985). They measured 'perceptual learning' in five-, seven- and ten-year-old children using a picture recognition task. In the first phase of the experiment, the children were shown some pictures and either had to say whether each picture contained a cross (crosses had been drawn at random on 33% of the pictures), or to say whether the picture was of something portable. The 'cross detection' task was intended to induce 'shallow processing' of the pictures at a perceptual level only, and the 'portability detection' task was intended to induce 'deep processing' at the level of meaning.

Memory for the previously experienced pictures was then studied in an unexpected recognition task. In this task, the children were asked to name a mixture of the pictures that they had already seen along with some new ones. Implicit memory was measured by the difference in the children's reaction times to name the old versus the new pictures. Half of the children received this *implicit* memory task, and the other half were asked to say whether the old and the new pictures were familiar or not. The latter was the measure of *explicit* memory.

Carroll et al. predicted that implicit memory for the pictures would not vary with depth of encoding, whereas explicit memory would. In other words, deep processing should lead to better explicit memory for the previously experienced pictures than shallow processing, whereas implicit memory levels should be identical for both processing manipulations. This was essentially what they found.

Carroll et al. concluded that perceptual learning (implicit memory) does not develop with age.

Fragment completion tasks

KEY TERM

Fragment completion task Where the individual is shown fragments of words or pictures and asked to name or recall the whole object.

Another way of measuring whether implicit memory develops or not is to use a fragment completion task based on either words or pictures. For example, Naito (1990) devised a word-fragment completion task to measure implicit memory in children aged five, eight and 11. The children were given some of the letters in a target word, and were asked to complete each fragment into the first meaningful word that came to mind. Although Naito used words written in Japanese characters, her task was equivalent to presenting a fragment like CH---Y for the target word CHERRY. This is the example given in Naito's paper, and actually the fragment CH---Y could also be CHEERY or CHUNKY. However, each Japanese fragment was chosen to have only *one* legitimate completion.

Prior to receiving the fragment completion task, the children were given two other tasks based on 67% of the target words. For half of these words, the children were asked to make a *category* judgement in a forced choice task ("Is this a kind of? – fruit/clothes"), intended to induce 'deep' processing. For the other half of the words, they were asked to judge whether the target word contained a certain letter ('shallow' processing). Naito then measured whether more word fragments were completed correctly for the 32 previously experienced target words than for the 16 novel items in each case. She found that the 'old' items were completed correctly significantly more frequently than the 'new' items at all ages, and that implicit memory did not vary with depth of processing (deep versus shallow). She also found that implicit memory levels were invariant across age group (even though a group of adults were also included in the study). In a related experiment in which children were asked to recall the target words explicitly, Naito found a strong improvement in recall with age and an effect of depth of processing. Taken together, her results suggest that implicit memory does not develop, but that explicit memory does. Naito argued that her results showed that the two types of memory were developmentally dissociable.

In *picture*-fragment completion tasks, the child is shown an increasing number of fragments of a picture of a familiar object, such as a saucepan or a telephone, until the object is recognised (see Figure 8.2). If the complete object has been presented in a prior task, such as a picture naming task, then implicit learning should result in faster recognition for fragments of previously experienced objects than for fragments of completely novel objects.

Russo, Nichelli, Gibertoni and Cornia (1995) used a picture completion paradigm of this type to measure implicit memory in four- and six-year-old children. The children were first shown a series of 12 pictures for three seconds each, and were required to name each in turn. After a ten-minute break spent playing with blocks, the children were shown the fragmented versions of the familiar pictures along with fragmented versions of 12 new pictures, in random order. For each set of fragments, they were asked to say as quickly as possible what they thought the fragments were a picture of. The number of fragments that were presented

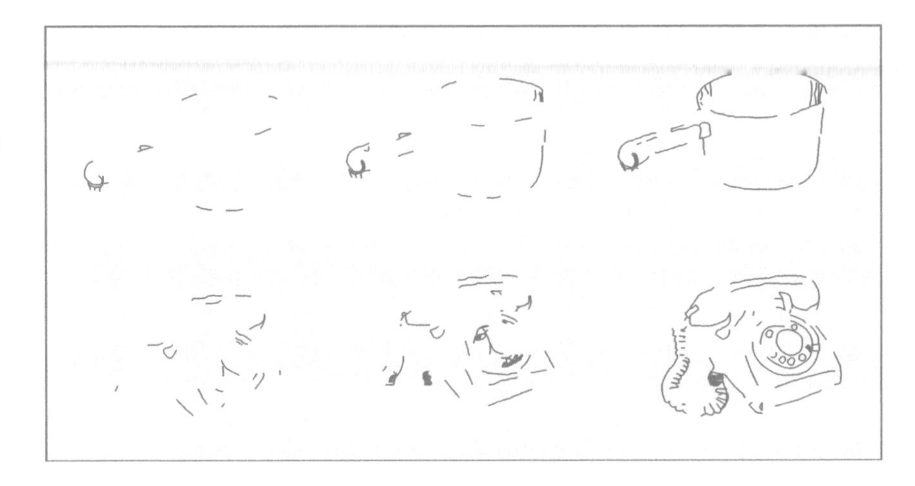

FIGURE 8.2
Examples of the fragmented pictures used by Russo et al. (1995). Copyright © 1995 Elsevier. Reproduced with permission.

were increased until the child recognised the picture. Performance in this implicit memory condition was contrasted with performance in an explicit version of the task, which was presented without time constraints. In the explicit memory task, the children were asked to use the fragments as cues to try and recall the pictures presented during the naming phase of the experiment. Russo et al. found that children of both ages recognised the familiar pictures from fewer fragments than the novel pictures, showing implicit memory. A group of young adults who were given the same picture completion task performed at similar levels to the children. Significant age differences were found in the explicit memory task, however, with the six-year-olds showing better recall than the four-year-olds. Russo et al. concluded that implicit memory as measured by fragment completion tasks is equivalent in children and in adults, and that the memory processes supporting implicit memory are fully developed by four years of age.

Perez, Peynircioglu and Blaxton (1998) also failed to find age differences in an implicit picture fragment completion task carried out with children aged four and eight years and with university students. They added an implicit *conceptual* memory task to their study. In this task, black-and-white line drawings were presented for study, and the children were then given category labels (e.g., clothing, animals) and asked to produce the first exemplars that happened to come to mind. In an explicit version of the task, the children were given the same labels and asked to recall the pictures from those categories. Age differences were found for the explicit task, but not for the implicit task. When asked effortfully to recall the original pictures, the four-year-olds recalled 33% of the pictures, the eight-year-olds 58% and the college students 76%. When simply asked to produce the first exemplars that came to mind, all age groups produced the names of around 45% of the previously studied pictures (a kind of priming effect).

A study carried out by Bullock Drummey and Newcombe (1995) suggested that even three-year-olds may have fully-fledged implicit memory processes. Their measure of implicit memory was the recognition of blurred pictures after long delays. Bullock Drummey and Newcombe showed three-year-olds, five-year-olds and adults blurred versions of pictures that they had seen three months previously in a reading book. They found that all groups showed comparable levels of implicit memory for the pictures. However, the adults had better *explicit* memory of the pictures than the children.

By contrast, Cycowicz, Friedman, Snodgrass and Rothstein (2000) have used the picture fragment completion task to argue that there *are* some developments in implicit memory with age. Cycowicz et al. make the reasonable point that studies of explicit memory in children reveal that processes like encoding and storage show age-related improvements (see Bauer, 2006). Hence some, albeit minor, age-related improvements might be expected in implicit memory tasks. Cycowicz et al. gave children aged five, nine and 14 years and college students a picture fragment completion task in which images were presented on a computer to allow easy presentation of the next level of fragmentation. All participants performed extremely well in this implicit task, correctly identifying over 90% of previously seen pictures. However, the degree of savings was lower for the youngest age group. The younger children needed more information for identifying the familiar fragmented pictures than the older children. In the explicit memory task, when the children had to recall the pictures, the usual age-related effects were found. Cycowicz et al. (2000) argued that there were developmental trends for both implicit and explicit tasks. However, they accepted that the implicit and explicit memory systems might develop at different rates.

Memory for faces

A third measure that has been used to study implicit memory in children is memory for faces. Faces have the advantage of being salient and important stimuli that are not dependent on verbal recall. For example, if the same face is presented to adult subjects on two occasions, the reaction time to recognise the face as familiar on the second occasion is dramatically reduced. This is known as a 'priming' effect. Ellis, Ellis and Hosie (1993) investigated whether young children would also show 'priming' effects for faces.

In their experiment, children aged five, eight and 11 years were shown pictures of both their classmates and unfamiliar children, and were asked to judge whether the children were smiling or not (half were smiling) and whether the picture was of a boy or of a girl. Following this 'priming' stage, the pictures of the children's classmates were presented for a second time, mixed in with previously unseen pictures of other classmates and with pictures of other unfamiliar children. On this second showing, the children were asked to judge whether the children depicted were familiar. Ellis et al. found that children of all ages were quicker to make judgements about the familiarity of the classmates that they had just seen in the priming phase of the experiment than about the familiarity of their non-primed classmates. The amount of implicit memory as measured by the proportional differences in primed and non-primed reaction times was the same for the five- and eight-year-olds, and was slightly *less* for the 11 year olds, again suggesting that the memory processes supporting implicit memory are well-developed early in childhood.

A different way to measure memory for faces is to study children's implicit memory of their classmates over time. Newcombe and Fox (1994) showed a group of ten-year-old children slides of three- and four-year-old children who had been their classmates in preschool. These slides were intermixed with slides of other children from the same preschool who had attended the school five years later. In order to see whether the children had implicit memories of their familiar classmates, galvanic skin response measures were recorded. The children were then shown the slides again, and were asked to say whether the depicted children were familiar and how much they liked them. The liking measure was included to see whether the children would show a preference for their previous classmates, even if they could not remember them.

Newcombe and Fox found that the children showed recognition of their former classmates according to *both* the implicit and the explicit measures. Overall recognition rates were fairly low (26% on the implicit measure and 21% on the explicit measure), but there were large individual differences in recognition rates. When the experimenters divided the children into two groups, a 'high explicit recognition' group and a 'low explicit recognition' group, they found that performance on the *implicit* recognition measure was equivalent in both groups. This is a very interesting result, as it implies that implicit memory for preschool experiences can be maintained even when explicit memories are lacking. Newcombe and Fox's data thus support Naito's (1990) suggestion that implicit and explicit memories may be developmentally dissociable.

Cognitive neuroscience studies of implicit memory in children

Cognitive neuroimaging studies are also beginning to explore the basis of implicit memory in children. Most studies have focused on perceptual learning of faces. As noted in Chapter 3, it has been argued that there is a specialised area for the perceptual learning of faces in the fusiform gyrus, particularly in the right hemisphere (the 'fusiform face area' or FFA), and that infants show brain activity in the FFA as early as two months (Tzourio-Mazoyer et al., 2002). However, neuroimaging studies with adults show that car experts show specific activation in this area for cars but not birds, whereas bird experts show specific activation for birds but not cars (Bukach, Gauthier & Tarr, 2006). Accordingly, activity in this neural region may actually be associated with the acquisition of *expertise*. Chess experts (adults) show specificity in the right fusiform gyrus for the spatial layout of chess boards (Righi & Tarr, 2004). The electrophysiological signature for face recognition (the N170, also discussed in Chapter 3) is also reliably present in paradigms testing other areas of expertise than face processing. For example, fingerprint experts show an N170 effect for fingerprints (Bukach et al., 2006).

Regarding studies with children, a fascinating early fMRI study of an 11-year-old boy with autism showed robust fusiform selectivity in response to Digimon cartoon characters, in which he was obsessively interested, but not to human faces (Grelotti et al., 2005). A control autistic boy who was not interested in Digimon did not show fusiform selectivity to either the cartoon characters or to human faces. A typically-developing boy interested in Pokémon showed fusiform selectivity to faces only. This pattern of results was interpreted as being due to expertise.

The autistic boys presumably showed no face selectivity because their lack of mentalising abilities (see Chapter 4) meant that they were not interested in faces. The typically-developing boy, by contrast, had developed face expertise.

More recent fMRI work is consistent with this interpretation. Adolescents with autism exhibit *hypoactivation* in the fusiform face area when shown pictures of unfamiliar faces in comparison to typically-developing adolescents (Whyte, Behrmann, Minshew, Garcia & Scherf, 2016). Whyte and her colleagues measured BOLD (blood-oxygen-level dependent) responses in the fusiform gyrus of adolescents with and without autism in response to viewing pictures of unfamiliar and inverted human faces and animal faces (cat and dog faces). The hypoactivation in autism in the right FFA was only observed for the human faces, not for the animal faces. There were no structural brain differences in the fusiform gyrus between the two groups. Since hypoactivation was specific to human faces in autism, Whyte et al. concluded that human faces were not afforded the same attentional and motivational resources by individuals with autism, and therefore that these teenagers had not developed expertise for face processing.

Also using fMRI, Cantlon and her colleagues compared activation to faces in the fusiform gyrus for typically-developing four-year-olds and adults (Cantlon, Pinel, Dehaene & Pelphrey, 2011). Cantlon and her colleagues were interested in whether increasing expertise was associated with increased activation to specific categories such as faces, or decreased activation to unrelated categories. Accordingly, they studied brain activation in response to four categories of stimuli: pictures of faces, letters, numbers and shoes. Shoes were chosen as they are familiar to young children, are of similar visual complexity pictorially to faces, but do not typically activate the FFA. Scrambled versions of each stimulus picture (jumbled versions of the constituent pixels) were used as control stimuli. The pictures had either red, blue or green borders, and children had to press a button if the border was green. This served to maintain attention during the task. The children also received extensive practice before the fMRI scanning session.

Brain activation in the fusiform gyrus during the task was then compared for the different categories of stimuli, and contrasted with activation to the jumbled control pictures in each case. Cantlon et al. (2011) reported face selectivity in the FFA for both the children and the adults. Activation was greater in this region than for shoes, letters and numbers, both for the four-year-olds and for the adults, and showed the classic right lateralisation. The degree of activation was comparable for children and adults. Cantlon and her colleagues also found an unexpected early specialisation for symbols (letters and numbers) for the children (all pre-readers) in the left lateral fusiform gyrus. The fact that digits and letters are frequent in the everyday worlds of children growing up in Western cultures (e.g., fridge magnets, alphabet blocks) could explain this early specialisation. A few days later, the children came into the laboratory and received a recognition memory test for all of the pictures seen in the scanner. Cantlon et al. (2011) reported a significant correlation between recognising previously-seen faces and *decreases* in BOLD activation to shoes in the FFA. Accordingly, they suggested that expertise may relate more to *reduced* neural responses to non-preferred stimuli in specific brain regions such as the FFA, rather than to increased activation to preferred stimuli (here, faces). Longitudinal

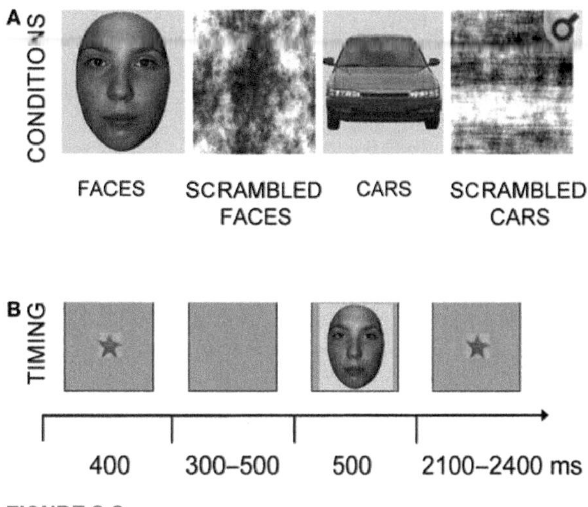

FIGURE 8.3
Examples of the stimuli used by Kuefner et al. (2010) to investigate the neural response to faces versus cars in children of different ages. In each case the veridical stimuli were compared to scrambled versions of the same pictures.

brain imaging data and training studies are required to test this idea properly.

Meanwhile, Cantlon et al.'s finding that the FFA is already showing adult-like responses to faces by four years of age is consistent with an EEG study of the face N170 in groups of children aged from four to 17 years and adults, which also reported no developmental changes with increasing age (Kuefner, de Heering, Jacques, Palmero-Soler & Rossion, 2010). Kuefner and her colleagues recorded the N170 to pictures of faces, scrambled pictures of faces, pictures of cars and scrambled pictures of cars using EEG (see Figure 8.3). To isolate responses specific to faces versus cars, the researchers then subtracted the brain response to the scrambled stimuli from the brain response to the veridical stimuli, for each category. This enabled them to control for the neural response to low-level visual features specific to each picture, such as luminance and visual contrast. To maintain attention, participants were asked to judge whether each picture was an object (a face or a car) or a 'texture' (the scrambled pictures).

Kuefner et al. (2010) reported that the target N170 had the expected occipito-temporal topography, with a right hemisphere advantage. Furthermore, once brain responses to low-level visual aspects of pictures had been removed by the subtraction method, the N170 to faces was remarkably similar across all the ages tested. Interestingly, however, a similar finding was reported for cars. When cars were the stimuli, the N170 was again 'adult-like' for all the different age groups of children tested. Kuefner and her colleagues concluded that the electrophysiological signature of implicit face processing is adult-like from four years onwards. Accordingly, available developmental neuroimaging data are supportive of conclusions drawn from behavioural studies regarding implicit memory. There appears to be little development in the neural mechanisms supporting perceptual learning, at least from the age of around four years.

THE DEVELOPMENT OF EPISODIC MEMORY

Implicit memory is usually contrasted with episodic memory, a memory system that involves conscious awareness. As mentioned earlier, episodic or declarative memory usually refers to memory for episodes or events in one's life, involving explicit recall of these episodes and events. In adults, episodic memory tends to be organised around 'schemas', or scripts, for routine events. Each script is a generic or abstract knowledge structure that represents the temporal and causal sequences of events in very specific contexts. For example, adults have a 'restaurant schema' for representing the usual sequence of events when eating in a restaurant, and a 'laundry' schema for representing the usual sequence of events when doing one's laundry. In order to study the development of episodic memory, therefore, we need to study the development of scripts and schemas. An obvious approach is to ask children about very familiar events and routines, to see whether they respond with script-like information.

Scripts or schemas are generic knowledge structures that represent the temporal and causal sequences of events in a specific context. For example, an adult will have a 'laundry' schema for the usual sequence of events when doing laundry.

The development of scripts for organising episodic memory

Asking young children questions about familiar routines was exactly the method selected by Nelson and her colleagues in their pioneering developmental work on scripts. They examined the episodic memories of three- to five-year-old children for events like going grocery shopping, attending birthday parties and baking cookies. The children were simply asked to tell the experimenters 'what happens' during such events. A series of ordered prompts was then used as necessary to prompt elaboration: "I know you know a lot about grocery shopping. Can you tell me what happens when you go grocery shopping? ... Can you tell me anything else about grocery shopping? ... What's the first thing that happens? What happens next?"

Nelson found that the youngest children gave ordered and conventionalised reports of what typically occurred during these events. For example, here is a five-year-old telling the experimenters about going grocery shopping:

> *Um, we get a cart, uh, and we look for some onions and plums and cookies and tomato sauce, onions and all that kind of stuff, and when we're finished we go to the paying booth, and um, then we, um, then the lady puts all our food in a bag, then we put it in the cart, walk out to our car, put the bags in our trunk, then leave. (Nelson, 1986, p. ix).*

Research such as this shows that episodic memory is organised around general event representations from a very early age. Nelson (1993) argues that the basic ways of structuring, representing and interpreting reality are consistent from early childhood into adulthood.

Because of this, as she points out, scripts for routine events may play a very salient role in memory development. Nelson (1988) has suggested that younger children *concentrate* on remembering routines, as routine events such as going to the babysitter are what makes the world a predictable place. The importance of this predictability means that routine events are focused on at the expense of novel and unusual events, which are forgotten.

The relationship between scripts and novel events

However, more recent studies have shown that younger children can also remember novel and unusual events over long periods of time. In a study by Fivush and

Hamond (1990), a four-year-old recalled that, when he was 2.5, "I fed my fish too much food and then it died and my mum dumped him in the toilet." Another four year-old told the experimenters that when he was 2.5 "Mummy gave me Jonathan's milk and I threw up" (this child was lactose-intolerant). Both of these events were genuine memories. These novel events had obviously made a big impression on the children concerned, as they could reproduce them accurately 18 months later! Fivush and Hamond agree with Nelson's idea that young children focus largely on routines, but suggest that children's understanding of routine events *also* helps them to understand novel events, events that differ from how the world usually works.

Fivush and Hamond's suggestion that the development of scripts enables the development of memories for novel events is at first sight inconsistent with other evidence showing that young children have a tendency to *include* novel events in their scripts, however. Whereas older children can separate novel events from the routine, tagging them separately in memory as atypical, younger children display a tendency to blur the routine with the unusual. For example, Farrar and Goodman (1990) compared four- and seven-year-old children's ability to recall novel and repeated ('script') events. In their study, the children visited the laboratory five times during a two-week period in order to play 'animal games'. These games included making bunny and frog puppets jump fences and having bears and squirrels hide from each other. Each game took place at a special table, and the games occurred in the same order on each visit. However, during one visit a novel event was inserted into the familiar routine (the event was two new puppets crawling under a bridge).

A week later, the children were interviewed about their experiences using both free recall techniques and specific questions such as "What happens when you play at this table with the puppets?" The younger children frequently reported that the novel event had occurred during the script visits as well as during the single deviational visit. They appeared unable to differentiate between a typical 'animal game' visit and the novel event that had occurred only once. The older children did not report that the novel event had occurred during both the script visits and the deviational visit. They were more likely to have formed separate and distinct memories for the two types of visit, tagging the novel event as separate and as a departure from the typical script.

Farrar and Goodman suggested that younger children relied on their general event memory when recalling events, and that this general memory had absorbed information from *both* the script visits and the novel visit. They concluded that the ability to establish separate memories of unusual episodes may still be developing at age four. However, it is also possible that both groups of researchers are correct. Younger children's tendency to merge novel events with their general event memories may depend on the *salience* of the novel event *to the child*. Highly salient novel events such as those documented by Fivush and Hamond (which may be frequently 'refreshed' in family contexts) may thus be accorded a special status in younger children's memories, while less salient events such as the deviation from the game played in Goodman's laboratory may be merged with their scripts. Certainly, Fivush and her research colleagues have demonstrated good narrative recall in three- and four-year-olds for distinctive events that were highly salient for the child. For example, here is a child aged 46 months recalling an Easter egg hunt for a researcher:

I find the basket... I won the Golden Egg... in the tree. I found... candy inside different eggs. They were green, pink, yellow, orange and umm blue. And we found candy inside. Jellybeans, suckers and tootsies rolls and and... different colour jellybeans... And yum yum yum. And we ate cupcakes with M&M sprinkles and maybe had drinks of lemonade.

Parental interaction style and the development of episodic memories

There is also growing evidence that the ways in which parents interact with their children influences the development of event memories. Parents tend to ask young children fairly specific questions about shared past events, such as "Where did we go yesterday?", "Who did we see?" and "Who was there with us?" (Hudson, 1990). Repeated experience of such questions may help young children to organise events into the correct temporal and causal order, and to learn which aspects of events are the most important to recall. If this is so, then parents who ask more of these specific questions should have children with better memories. This seems to be the case. For example, in a longitudinal study of mother–child conversations about the past, Reese, Haden and Fivush (1993) observed mother–child dyads talking about the past when the children were aged 40, 46, 58 and 70 months. The mothers were asked to talk about singular events from the past, like a special visit to a baseball game or a trip to Florida. They were asked to avoid routine events like birthday parties or Christmas, which could invoke a familiar script.

Reese et al. found that there were two distinct maternal narrative styles, which were related to the ways in which the children became able to recall their own past experiences. Some mothers consistently elaborated on the information that their child recalled and then evaluated it. Other mothers tended to switch topics and to provide less narrative structure, and seldom used elaboration and evaluation. For example, an elaborative mother who was helping her child to remember a trip to the theatre included questions like "Where were our seats?" and "What was the stage set up like?" A non-elaborative mother who was helping her child to recall a trip to Florida asked the same question repeatedly ("What kinds of animals did you see? And what else? And what else?"). The children of the elaborative mothers tended to remember more material at 58 and 70 months. Reese et al. suggested that maternal elaborativeness was a key factor in children's developing memory abilities. For example, maternal elaborativeness might be expected to lead to more organised and detailed memories, and might facilitate children's developing understanding of time. This conversational style also allows opportunities for mothers and their children to agree and disagree about the past. This negotiation could help the child to understand that the self has a unique perspective on the past.

KEY TERMS

Maternal elaborativeness
Elaborating on the child's recalled information, thought to be a key factor in helping to develop children's memory abilities.

Autobiographical memory
Remembering experiences from one's own personal history.

The construction of personal histories

Researchers such as Fivush were among the first to point out that talking about past experiences with one's parents and family enables the construction of a personal history. This implies a role for social construction in the development of autobiographical memory. If parents focus on particular events as important or self-defining when reminiscing with young children, these events may take on salient roles in the child's autobiographical self-narrative. Alternatively, the simple opportunity to talk about the past with one's parents may enhance retention. Experiences from early childhood that are frequently rehearsed may be more likely to be recalled later in life. These hypotheses about autobiographical memory are difficult to study, as they require prospective longitudinal studies. However, at least two research groups have collected relevant data. The first group conducted the longitudinal study reported by Reese et al. (1993, discussed above). This research group was able to investigate the effects of family rehearsal on the construction of personal histories, because the repeated contacts with the families allowed an estimate of rehearsal frequency.

In order to explore the construction of personal histories, Fivush and Schwarzmueller (1998) interviewed some of the children who had participated in the earlier study when they were aged eight years. Each child was asked about four to six events that had occurred and been discussed in earlier phases of the study, when the children were aged 40, 46, 58 or 70 months. Particularly distinctive events were chosen (e.g., going to Disney World, having chicken pox). The experimenter began by saying "Today I'd like to ask you about some things you did a long time ago, and see what you remember about them". Fivush and Schwarzmueller reported that the children recalled most of the events (78%) that they were asked about, including events that had occurred before 40 months of age. However, about 80% of the information that they provided about these remembered events was new. Copies of the interview transcripts were thus mailed to the children's mothers for checking. The mothers confirmed that almost all the extra information provided was accurate.

Highly distinctive events are thus extremely well-remembered by children. Fivush and Schwarzmueller concluded that much more information had been encoded and retained by the children than they had verbally reported at the time that the events were experienced. To assess the role of family reminiscence in remembering, Fivush and Schwarzmueller scored the likely frequency of rehearsal of the different events by the families, using information gathered during prior study visits. This enabled them to produce estimates of the amount of family rehearsal of the events both at the time that the event occurred, and subsequently up to the current interview at age eight. Surprisingly, no relationship was found between the amount of rehearsal across the retention interval and the amount of information that the child recalled. Fivush and Schwarzmueller concluded that it was the experience of verbalising the events at the time that they occurred that was critical for long-term retention. For verbal recall at least, the ability to give a narrative account of an experience at the time of experiencing it may be important for constructing a personal past. Fivush and Schwarzmueller suggested that this was because language enables children to construct extended, temporally organised representations of experienced events that are narratively coherent. Adult-guided reminiscing may help the child to learn more sophisticated forms of narrative organisation. These events which are organised in more narratively coherent ways may then become the first autobiographical memories that will be carried for a lifetime.

Studies from the second research group, based on the cohort recruited by Bauer et al. (2000), reached similar conclusions. For example, Abbema and Bauer (2005) followed up the children who had visited their laboratory five times between the ages of one and three years when they were aged either seven, eight or nine years. At the three-year visit, each child had talked with their parents about six relatively unique events from the recent past. Abbema and Bauer were able to explore how much the children recalled about these events after a gap of four, five or six years. Four of the six events were selected for discussion with each child. Abbema and Bauer found less recall of the distant events than Fivush and Schwarzmueller (1998), with the seven-year-olds recalling around 60% of the events and the older children around 35%. However, the events were generally less distinctive than those investigated by Fivush and Schwarzmueller (e.g., short-term outings; Abbema and Bauer found 100% recall for very distinctive events such as moving house). For the events that were recalled, the children provided as much information as they had when aged three years. In general, they provided more detailed narratives of the same events than when aged three, supporting the findings of Fivush and Schwarzmueller. Abbema and Bauer were not able to estimate the amount of rehearsal that had occurred in the interim, but they did compare participants' memory for distant events with their memory for more recent ones (broadly, events occurring within the previous nine months). They found that although memory for these more recent events was more detailed, there was little substantive difference between the ability to recall the recent versus distant past. They concluded that as long as the initial representation of an experienced event is strong enough, it can be retained over time.

A growing tendency to describe events from the perspective of the self was also evident. Abbema and Bauer noted that the function of autobiographical memory changes with age. Whereas younger children use discussion of the past to cement their understanding of their family and their role within it, older children talk about the autobiographical past to cement relationships with peers. By discussing our past with others, we 'share' ourselves with our friends, and deepen our social relationships. The idea that autobiographical memory serves largely social and cultural functions is a central theme in the paper by Nelson and Fivush (2004). They point out that the human ability to create a shared past allows each individual to enter a community or culture in which individuals "share a perspective on the kinds of events that make a life and shape a self" (p. 506). Nelson and Fivush point to important cultural differences in how shared reminiscing provides children with information about how to be a 'self' in their culture. For example, the self-definition and self-story of the individual is seen as more important in modern Western cultures than in Asian cultures.

Cognitive neuroscience studies of episodic memory in childhood

As noted earlier, cognitive neuroscience studies using episodic and declarative memory tasks with adults suggest that areas of the frontal lobes (particularly prefrontal cortex) and medial temporal lobe structures such as the hippocampus and dentate gyrus are routinely activated by such tasks. Interestingly, adults who have damage to the medial temporal lobe system have intact short-term memory (e.g., preserved digit span), but have difficulties in acquiring new long-term memories. Memories from before the neural damage are generally retained well. This has been interpreted as showing that activity in the medial temporal lobes is associated with the *consolidation* of memories rather than in their long-term storage. Long-term storage is thought to depend more on the association cortices, with prefrontal structures particularly important in retrieval (Tulving, 2002). The structure of the prefrontal cortex is still developing in adolescence, and hippocampal volume also increases during childhood. The prefrontal cortex is also typically activated by working memory tasks (Munakata, 2004). Munakata (2004) points out that computational modelling suggests that memory will be fractionated, as there is a computational trade-off between fast learning and slow learning. A system that specialises in learning rapidly is not suited to learning gradually, and vice versa. Hence memories that depend on rapid learning (e.g., a personal experience) should be sub-served by a different system to memories that depend on incremental learning (e.g., the underlying structure of the environment, as instantiated by conceptual learning).

Animal studies involving brain lesions suggest that the hippocampus plays a key role in consolidating memories and in recollection. At one time, therefore, it was thought that information had to be processed by the hippocampus in order to enter long-term memory. Accordingly, episodic memory was thought to require an intact and fully functional hippocampus. An interesting study of declarative memory in three children who had suffered hippocampal damage either at birth or in early life reported by Vargha-Khadem et al. (1997) calls this account into question. Vargha-Khadem and her colleagues studied Beth, who sustained hippocampal damage during a difficult birth; Jon, who suffered hippocampal damage during afebrile convulsions when aged four years; and Kate, who received toxic doses of an asthma drug at age nine which caused seizures that left her profoundly amnesic. All three children were referred to the research team as adolescents, because their parents

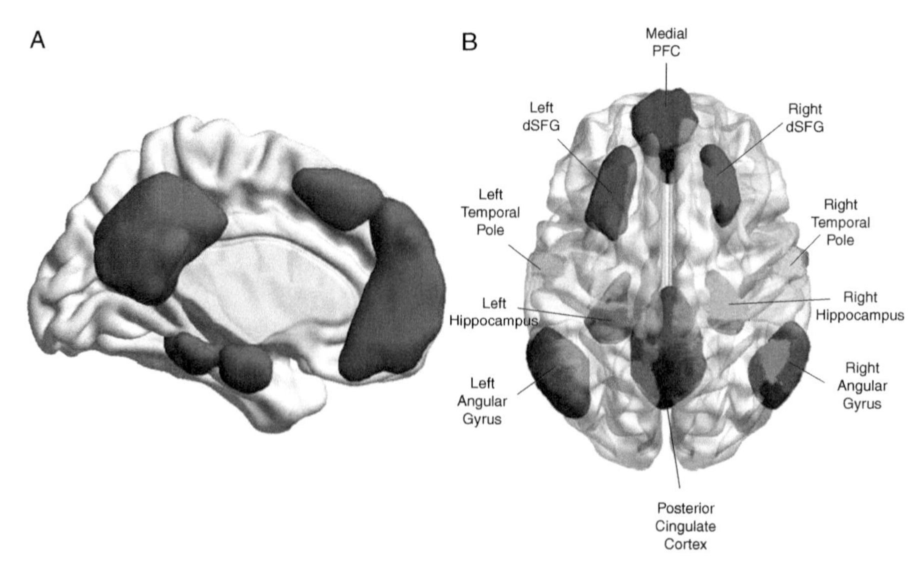

Schematic depiction of the neural regions likely to comprise the default mode network. This is the network of spontaneous intrinsic brain activity that can be observed when the brain is not thought to be engaged in any task. PFC = prefrontal cortex, dSFG = dorsal superior frontal gyrus.

complained that they were unable to remember the events of daily life. All three children were found to have lower than average development in terms of verbal and spatial intelligence, but to have memory quotients 16–20 points below their verbal IQs. Such a large discrepancy is indicative of amnesia in adults.

When given a variety of standardised tests, all three children performed within normal levels on tests of immediate memory, but not on tests of delayed recall. All three children were also unable to remember the events of their daily lives, including conversations, visitors and holidays, and had problems in remembering familiar environments, dates and times (spatial and temporal deficits). Structural imaging of their brains revealed abnormally small hippocampi (bilaterally), with severely compromised function of the remaining tissue. Amazingly, however, this extensive damage had not interfered much with their intellectual development. All three children were in mainstream schooling, could read and spell at the level expected given their IQs, and had acquired the factual knowledge about the world that is typically called semantic memory. Thus, while early hippocampal damage had had a devastating effect on the development of the children's autobiographical or episodic memory, it had not had a commensurate effect on the development of their factual knowledge, vocabulary or comprehension skills. Vargha-Khadem et al. (1997) suggested that these children's neural damage made it unlikely that episodic memory is the only gateway to semantic memory. The children appeared to have a specific difficulty in forming context-rich episodic memories, but not in forming context-free semantic memories.

KEY TERM

Structural imaging Neuroimaging to measure the structures in the brain and structural connections between different parts of the brain.

Studies measuring the functional connectivity of the hippocampus and the development of episodic memory in young children support this conclusion. Riggins, Geng, Blankenship and Redcay (2016) studied four- and six-year-old children, using fMRI to measure the 'resting state' or 'default mode' network (see also Chapter 1) and assess functional connectivity between the hippocampus and different cortical regions. For adults, variability in hippocampal connectivity is typically related to variability in performance on episodic memory tasks involving explicit recollection. An episodic memory task suited to the children was utilised, in which the participants were introduced to a series of novel toys in an experimental room, and imitated specific actions such as drumming on the toy. The children then moved on to a second experimental room, where they met other novel toys and again imitated specific actions. After an hour's delay, the episodic memory test was given. Children were shown the experienced toys mixed with previously unseen toys and asked (a) which toys were familiar and (b) which action and which location went with the toy. The children all performed well on the episodic memory test. Their performance was then correlated with their resting state data.

Riggins et al. (2016) reported that hippocampal resting state functional connectivity was similar in the four-year-olds and the six-year-olds. However, different relationships between functional connectivity and episodic memory performance were found for the younger versus the older children. For the four-year-olds, better episodic memory was related to more connectivity within middle temporal gyrus and frontal areas, and less connectivity within hippocampal areas. For the six-year-olds, better episodic memory was related to less connectivity within these frontal and temporal areas, and more connectivity within hippocampal areas, a more adult-like pattern. Riggins et al. concluded that the hippocampal memory network becomes more tightly integrated with development. However, these neural data do not reveal whether behavioural improvements cause neural restructuring or vice versa. Given the complex interactions that are likely taking place within the developing brain, converging data using other episodic memory tasks are required to support this conclusion.

Sastre, Wendelken, Lee, Bunge and Ghetti (2016) provided some converging data. They used fMRI with older children aged 8–9 years and 10–11 years to measure BOLD responses in different brain regions while an episodic memory task was being performed in the scanner. The episodic memory task depended on viewing scenes such as a city park or a farm, and deciding whether an animal or object presented subsequently 'belonged' in the scene. For example, a line drawing of a cow would belong in the farm scene but not in the city park scene. Participants subsequently saw the line drawings again, mixed in with novel line drawings, and had to identify (a) whether a drawing was familiar, and (b) which scene it had been in. A 'don't know' response was also available. The BOLD response was then compared for trials in which participants both correctly judged familiarity and recalled the relevant scene, correctly judged familiarity only, or were incorrect. Region-of-interest analyses were carried out, focusing on the hippocampus. Sastre et al. reported that the younger children did not show differential activation of the hippocampus for successful versus unsuccessful remembering. The older children, however, showed reliably greater hippocampal activation in successful memory trials. This was qualified by performance level – poor responders in the episodic memory task showed no differential activation of the hippocampus for successful versus unsuccessful remembering, just like the younger children. Successful older children showed similar hippocampal activation to adults. Sastre et al. (2016) concluded that the function of the hippocampus in supporting episodic memory changes during late childhood, with more specialisation as children get older. This mirrors the conclusions reached by Riggins et al. (2016), but shifts them along the age axis (from 4–6 years to 8–11 years). As these fMRI studies are correlational and also cross-sectional, they are suggestive of associations only. Longitudinal imaging of the same children across development, using a range of episodic memory tasks, is required to shed real light on the potential role of the hippocampus in the development of episodic memory.

THE DEVELOPMENT OF EYE-WITNESS MEMORY

A special kind of episodic memory is memory for events that may not have appeared significant at the time that they were experienced. This is eye-witness memory. Studies of eye-witness memory in adults have shown that adults have remarkably poor memories for the specifics of events that they have seen. For example, adults who have witnessed a car accident in a video can be misled into 'remembering' false details such as a broken headlight simply by the experimenter asking them leading questions like "Did you see the broken headlight?" (Loftus & Zanni, 1975). If the eye-witness memory of adults is faulty, then it has been frequently assumed that the eye-witness memory of *children* must be even worse. This is an interesting research question in its own right, but it is also an important legal issue (Ceci & Friedman, 2000). As more and more children are being called as witnesses in investigations concerning physical and sexual abuse, the status of their testimony has become of paramount importance (Goodman, Jones & McLeod, 2017). In such cases, it is critical to know whether the abuse really occurred, or whether 'memories' of abuse have been created as a result of repeated suggestive questioning by adults.

The accuracy of children's eye-witness testimony

Imagine that an experimenter comes to your school, takes you off to a quiet room to do a puzzle with two friends, and leaves you on your own. While you are working on the puzzle, a strange man comes into the room and messes around, dropping a pencil and fumbling with objects. He claims to be looking for the headmaster. He then steals a handbag and walks out. How much do you remember about these events? Ochsner and Zaragoza (1988, cited in Goodman & Aman, 1990) showed that six-year-old children remembered quite a lot. The children produced more accurate statements about the other events that they had witnessed in the room and fewer incorrect statements than a group of control children who had experienced the same events except that the man had left the room without stealing the bag. The experimental group were also less suggestible, for example being less willing to select suggested misleading alternative events during a forced-choice test. This study suggests that the eye-witness testimony of young children may be no less accurate than that of adults.

The role of leading questions

Other studies, however, have found that although young children's memory for centrally important events is equivalent to that of adults, younger children are more suggestible than adults. For example, Cassel, Roebers and Bjorklund (1996) reported a study in which six- and eight-year-old children and adults watched a video about the theft of a bike. A week later the subjects were asked to recall the events in the video, and were asked a series of increasingly suggestive questions. Cassel et al. found that children and adults showed equivalent levels of recall for items central to the event (e.g., "Whose bike was it?"). However, when Cassel et al. compared the effect of repeated suggestive questioning on the six- and eight-year-old children and the adults, they found a greater incidence of false memories in the children. Interestingly, they also found that *unbiased* leading questions such as "Did the bicycle belong to (a) the mother, (b) the boy, or (c) the girl?" were as likely to produce false memories as biased (mis)leading questions such as "The mother owned the bike, didn't she?" This is an important result, as it suggests that the mechanisms that result in false memories may be *general* mechanisms to do with the way that the developing memory system functions, rather than *specific* mechanisms related to false memories of negative events.

As Cassel et al.'s study relied on watching a video, we can hypothesise that leading questions may have an even greater effect on the recall of younger children when they actually experience an event themselves. Goodman and her colleagues devised a paradigm based on a visit to a trailer (caravan) to investigate this question (Rudy & Goodman, 1991, see also Goodman, Rudy, Bottoms & Aman, 1990). In the 'trailer experiment', children aged 4 and 7 years were taken out of their classrooms to a dilapidated old trailer, chosen to be a memorable location. The children went in pairs, and once inside the trailer they played games with a strange man. One child in each pair had the important task of watching (the bystander), and the other child (the participant) played games. This enabled the experimenters to see whether the children would show similar levels of suggestibility when they were participants or bystanders at real events. The games included 'Simon Says', dressing up, having your photograph taken, and tickling. During the 'Simon Says' game, the children had to perform various actions including touching the experimenter's knees. These games were chosen because child sexual abuse cases frequently involve reports of being photographed, of 'tickling' and of other touching.

The children were later interviewed about what had taken place in the trailer. The interview began with the interviewer asking the child to tell him or her about everything that had happened in the trailer. The interview then continued with misleading questions like "He took your clothes off, didn't he?", "How many times did he spank you?" and "He had a beard and a moustache, right?" In fact, the dressing up game did not involve removing the children's clothes, no child was spanked and the man was clean-shaven. Goodman et al. found that children of both ages recalled largely *correct* information about the games that they had played in the trailer. Neither participants nor bystanders invented information. The children were also largely accurate in their responses to specific questions about abuse, like "How many times did he spank you?" and "Did he put anything in your mouth?" The seven-year-olds answered 93% of the abuse questions correctly, and the four-year-olds answered 83% correctly.

A related study with three- and five-year-olds showed that false reports of abuse did not increase when anatomically detailed dolls were provided to enable the children to *show* as well as tell what had happened (Goodman & Aman, 1990). In this study, however, the younger children *were* more susceptible to leading questions than the older children. Under the influence of misleading questions about abuse, the three-year-olds tended to make errors of *commission* (that is, they agreed to things that had not happened) 20% of the time. Embroidery of these events was rare. In fact, the majority of commission errors occurred with the leading question "Did he kiss you?", to which the children simply nodded their heads. The five-year-olds only made commission errors on two out of 120 occasions, and both of these errors were on the question "Did he kiss you?" The younger children made very few errors on the potentially more worrying misleading abuse questions such as "He took your clothes off, didn't he?" and "How many times did he spank you?"

The striking thing about Goodman's findings in these very 'ecologically valid' studies is that, in the main, the children did *not* invent false reports of abuse. Children were also fairly resistant to misleading questioning by the adult, and this resistance remained robust in the face of anatomically detailed dolls, a factor that might have been expected to encourage invention. Goodman and her colleagues now study actual child victims of abuse, as they work with professionals conducting child maltreatment investigations in a large American city. As part of these investigations, 189 children aged from three to 17 years were taken away from their families for a five-day period, during which extensive hospital-based examinations were carried out (see Eisen, Qin, Goodman & Davis, 2002). The researchers were able to interview the children about aspects of these examinations, including an ano-genital examination and a psychological consultation. This enabled them to assess the children's responses to misleading questions, and to explore potential relations between suggestibility, intellectual ability and clinical ratings of psychopathology (global adaptive functioning). The children were divided into three broad age bands, 3–5 years, 6–10 years and 11–17 years.

In all, the professional team estimated that 101 of the 189 children had been abused, 43 had been neglected and 40 had not been abused (the other five children's status could not be determined). The majority of the children were African-American (77%) and from low socio-economic status families, thus making them rather different to the middle-class non-abused children usually studied in eye-witness memory experiments. Overall, Eisen et al. did not find that memory or suggestibility differed in maladjusted children compared to typically-developing children. Abuse status was not related to the children's eye-witness memory performance. The preschoolers (3–5-year-olds) were more suggestible, but even they made relatively few errors in response to misleading questions about abuse (16% errors). Contrary to prediction, no relationships were found between the stress experienced by the children during the different examinations and memory performance or suggestibility. In general, more accurate memory was shown by the older children, and by the more intelligent children, while less accurate memory was shown by children rated as having poor global adaptive functioning. Thus age, IQ and overall psychopathology rather than abuse status was linked to children's eye-witness memory performance. These general conclusions have held across the last decade of children's eye-witness memory research (Goodman et al., 2017). The eye-witness testimony of young children is in general highly accurate. However, children provide more detailed memories when the questioner has taken time to build rapport and uses open-ended questions.

Converging evidence from ecologically valid memory paradigms has been contributed by Ornstein and his colleagues. In order to examine how much children actually remember about salient, personally experienced events, Ornstein, Gordon and Larus (1992) investigated three- and six-year-old children's memories of a visit to the doctor for a physical examination. Each physical examination lasted about 45 minutes, and included weighing and measuring the child, checking hearing and vision, drawing blood, checking genitalia and listening to the heart and lungs. Ornstein et al. argued that such visits shared a number of features with instances of sexual abuse. These included physical contact with the child's body by an adult and emotional arousal due to injections and other procedures. Memory for the events in the physical examination was measured immediately after the examination was over, and after intervals of one and three weeks.

Ornstein et al. measured the children's memories by first asking them open-ended questions such as "Tell me what happened during your check up". More detailed questions were then asked, such as "Did the doctor check any parts of your face?" and "Did he/she check your eyes?" Misleading questions were also asked, involving features of the physical examination that had not been included in an individual child's check-up. Ornstein et al. found that children in both age groups showed good recall of the physical examination immediately after it was over, recalling 82% (three-year-olds) and 92% (six-year-olds) of the features respectively. Both groups showed some forgetting of these features after three weeks had passed, but recall was still highly accurate, being around 71% in the three-year-olds. Responses to misleading questions were also largely accurate. Children in both age groups were able correctly to reject misleading features most of the time, with correct denials on 60% of misleading questions for the three-year-olds after a three-week delay, and on 65% of misleading questions for the six-year-olds. Intrusions ('remembering' features that had not in fact occurred) were also at similar levels in the two groups after the three-week delay, being 26% for the three-year-olds and 32% for the six-year-olds. Ornstein et al. concluded that young children's recall of a personally experienced event was surprisingly good.

Although review studies of young children's eye-witness testimony also find that levels of suggestibility are higher in younger children (see Lamb, Orbach, Warren, Esplin & Hershkowitz, 2007, for a review), the effects of misleading questions are usually to increase inaccurate acquiescence (errors of commission, see Ceci & Bruck, 1995). Overall, the levels of suggestibility found in different studies appear to vary with factors such as the emotional tone of the interview itself, the child's desire to please the interviewer, characteristics of both interviewer and child, and whether the child is a participant in the action or not, among others. Almost all studies find *some* age differences in suggestibility. However, it is also worth noting that leading questions are more likely to result in new disclosures than neutral questions are (Gilstrap & Ceci, 2005). Hence leading questions cannot be dismissed as overall deleterious in eye-witness memory investigations with young children; indeed asking specific questions (rather than misleading questions) can aid younger children's recall (Goodman et al., 2017).

Links between the development of episodic memory and the development of eye-witness memory

In general, the developmental patterns for episodic memory and eye-witness memory are highly similar (Ceci & Bruck, 1993, 1995). Older children can generally provide more detailed and narratively coherent memories. Ceci and Bruck also suggested that the greater susceptibility of younger children to repeated questioning by adults might be related to the distinction between scripts and personal histories. As we have seen, as children develop, their autobiographical memories are increasingly described from the perspective of the self. Ceci and Bruck (1993) suggested that the over-dependency of younger children on scripted knowledge could mean that suggestions made by the experimenter get included into the children's script for an event, and are thereafter reported as having actually taken place. Although this fits with Farrar and Goodman's (1990) finding that, when a novel event occurs in a standard setting, younger children tend to incorporate it into their script rather than tagging it separately, in general this suggestion has not been borne out by research (Gilstrap & Ceci, 2005). The suggestibility of younger children seems to reduce their report accuracy rather than change their memories. Because younger children sometimes agree with misleading questions, their reports contain more errors. As children get older, they seem to become less susceptible to leading questions and get better at providing narrative detail. In general, the amount of information and the accuracy of the information that children report in a memory interview increases with age, mirroring developments in episodic memory skills (Peterson, 2012).

A different way of looking at the link between children's knowledge of routine, script-like information and their eye-witness recall is to investigate whether children who have *more* episodic knowledge about a certain class of events are *less* likely to demonstrate susceptibility effects. According to this hypothesis, the possession of prior knowledge about a class of events should result in the formation of more stable memories, and these more stable memories should be less susceptible to the influence of leading questions. This hypothesis can be examined by studying the role of knowledge in children's memories, and the work of Ornstein and his colleagues provides a good example of such research.

Clubb, Nida, Merritt and Ornstein (1993) looked at whether children's memories of what happens when you visit the doctor were linked to their knowledge and understanding of what happens during routine paediatric examinations. The children, who were five-year-olds, were interviewed about their knowledge of physical examinations using open-ended questions like "Tell me what happens when you go to the doctor". They were then asked a series of yes-no probe questions like "Does the doctor check your heart?" Clubb et al. found that the majority of the children remembered highly salient features such as having an injection (64%), having the doctor listen to your heart (64%) and having your mouth checked (55%). Few children remembered features such as having a wrist check (5%). These percentages were then taken as an index of knowledge. A different group of five-year-olds provided the eye-witness memory scores. This group of children had been interviewed previously about a real visit to the doctor as part of an earlier study. Clubb et al. checked the percentages of these children who had spontaneously recalled the same features (injection, heart check, mouth check, etc.) either immediately or one, three or six weeks following their real examination. These 'eye-witness memory' numbers were then correlated with the corresponding knowledge scores obtained from the first group of children.

The researchers found that the correlations between knowledge and memory were highly significant at each delay interval. From this finding, they argued that variability in knowledge in a given domain is associated with corresponding variability in recall. However, the significant correlations obtained by Clubb et al. do not tell us about the *direction* of the relationship between knowledge and memory. It could be that variability in recall determines variability in knowledge, rather than vice versa. Further, it should be remembered that normative episodic memory research shows that children get better at providing more narrative detail over time, not because they are fabricating new details, but because their linguistic skills are improving as they get older (Bauer & Larkina, 2013). Eisen et al. (2002) reported that in their study of maltreated children, overall the children with better event memories were also the children who provided more detail in their reports of abuse experiences. Girls also tended to provide more detailed disclosures than boys. This gender difference could reflect the role of verbal ability in constructing coherent narratives of one's personal history. We also saw in the autobiographical memory section of this chapter that females tend to have earlier memories than males.

THE DEVELOPMENT OF WORKING MEMORY

Both episodic memory and eye-witness memory are aspects of long-term recall. We also have a memory system for short-term recall, which is called *working* memory. Working memory is a limited capacity system or 'workspace' used to maintain information over brief periods of time. This temporary storage allows the information to be processed for use in other cognitive tasks, such as reasoning, comprehension and learning. The information that is being maintained in working memory may either be new information, or it may be information that has been retrieved from the long-term system. There are a range of theoretical accounts of working memory, but an influential model was proposed by Baddeley and Hitch (1974). In their model, working memory has three sub-components: the central executive, the visuo-spatial sketchpad and the phonological loop (see Figure 8.4). The central executive is an attentional control system, a regulatory device that co-ordinates the different working memory activities and allocates resources. The visuo-spatial sketchpad provides domain-specific storage of visual and spatial information, enabling visuo-spatial processing, including of verbal information that is being stored as an image. The phonological loop is a temporary phonological store that maintains and processes verbal and acoustic information, the former in the form of speech. The phonological loop can be conceptualised as a kind of tape-loop lasting 1–2 seconds. Decay in the phonological store is fairly rapid, and so this verbal information may need to be *refreshed* or *rehearsed* by sub-vocal articulation. Working memory performance improves substantially over childhood, and is driven by many factors including increases in storage capacity, attention, and changes in rehearsal strategies.

KEY TERMS

Working memory
A limited capacity 'workspace' system that maintains information over a short period of time. Classically conceptualised via three sub-components: central executive, visuo-spatial sketchpad and phonological loop.

Central executive
Component of working memory. It is a regulatory device that co-ordinates the different working memory activities and allocates resources.

Visuo-spatial sketchpad
Component of working memory that stores information in visual and spatial formats, rather than verbally.

Phonological loop
Component of working memory that stores information acoustically, for example by using speech sounds for coding verbal material.

Developmental psychologists have mainly been concerned with the development of the two 'slave systems' of the central executive: the visuo-spatial sketchpad and the phonological loop. One influential idea has been that children initially rely on visual codes in short-term memory, and then switch to phonological codes at the age of around five years (e.g., Conrad, 1971). The age of this switch has been seen as potentially very important, as it is similar to the age at which Piaget proposed that a fundamental shift occurred in children's logical reasoning abilities (see Chapter 11). In fact, a number of information-processing theories of cognitive development, sometimes called 'neo-Piagetian' theories, were loosely based on this temporal

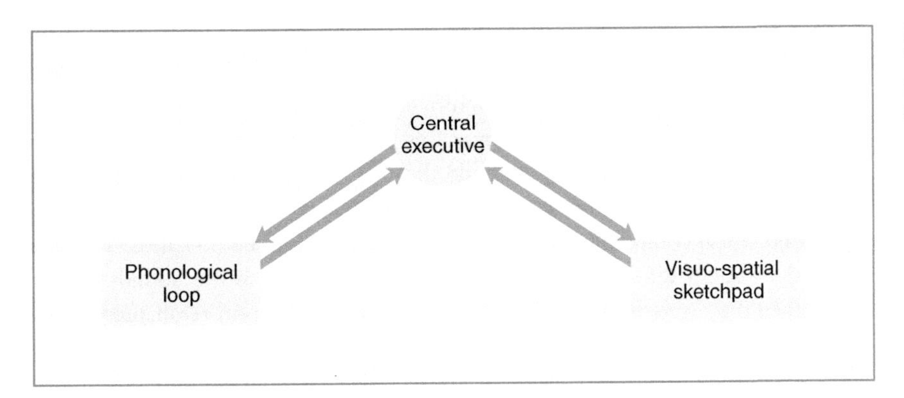

FIGURE 8.4
The model of working memory. Based on Baddeley and Hitch (1974).

co-occurrence, for example see Case (1992) and Halford (1993). As working memory has a central role in reasoning, comprehension and learning, theories of this kind argued that the development of working memory must be important for cognitive development in general. However, it is equally possible that the development of reasoning, comprehension and learning *in themselves* lead to improvements and developments in working memory. A detailed discussion is beyond the scope of this book. It is notable, however, that the advent of cognitive neuroscience has in general led to a decline in such theorising. Developmental improvements in working memory are typically now considered in terms of the maturation of relevant neural structures and age-related changes in neural organisation.

The visuo-spatial sketchpad

Most of the evidence for the idea that children rely on visual memory codes prior to around five years of age was indirect. It depended largely on showing that younger children were not susceptible to effects that are related to the use of speech sounds for coding material in working memory. As the presence of these effects is usually taken as evidence for the operation of the phonological loop (see below), the absence of these speech-based effects was taken to imply that working memory in young children relied on the visuo-spatial sketchpad.

The classic study in this tradition was performed by Conrad (1971). He gave children aged 3–11 years a series of pictures to remember. The pictures had names that either sounded similar (rat, cat, mat, hat, bat, man, bag, tap), or that sounded different (girl, bus, train, spoon, fish, horse, clock, hand). The children first learned to play a 'matching game' with the pictures. In the matching game, one complete set of pictures was presented face-up in front of the child, and then two or three pictures from a second, duplicate, set were presented for matching. After the children had grasped the idea of matching, the experimental trials began. The experimenter set out the eight cards in the full set (using either the 'sounds similar' pictures or the 'sounds different' pictures), and then concealed them from view. A subset of the duplicate pictures were then presented for matching. The experimenter named each card in the duplicate set, turned the cards face-down, and then re-exposed the full set. The children had to match the face-down cards to the correct pictures.

As adults find names that sound similar more difficult to remember over short periods of time than names that sound different, Conrad expected that the children would find the 'sounds similar' picture cards more difficult to match than the 'sounds different' picture cards. However, this 'phonological confusability' effect in adults arises because adults tend to code the picture names verbally and then to retain them in the phonological loop using rehearsal. Conrad's argument was that if a phonological confusability effect did not occur in children (i.e., if the children found the 'sounds different' cards as difficult to remember as the 'sounds similar' cards), then they were using a different memory code to support recall, presumably a pictorial one.

Conrad's results showed that only the younger children showed a 'no difference' pattern between the two picture sets (the three- to five-year-olds). The memory spans of this age group for the phonologically confusable and non-confusable pictures (measured by the number of pictures correctly recalled) were equivalent. Children aged six years and above showed longer memory spans for the 'sounds different' picture cards than for the 'sounds similar' picture cards, suggesting that they were using rehearsal strategies as a basis for recall. The possibility that the youngest subjects were also rehearsing but were idiosyncratically renaming the pictures prior to recall (e.g., 'cat' as 'pussy' or 'Tibby', thereby effectively converting the 'sounds similar' set into a 'sounds different' set) was ruled out by the children's spontaneous naming behaviour. The youngest children tended to speak aloud as they performed the task, and made comments like "cat goes with cat" or "cat here". This suggested that young, non-rehearsing children use some form of visual storage to remember visually presented materials.

If short-term storage in younger children is visually based, then visually similar objects should be easily confused in short-term memory, just as phonologically similar names are confused when short-term storage is phonological. This prediction is easily tested by using pictures of objects that look like each other in a memory span task, and then seeing whether visually similar objects are more difficult to remember than visually dissimilar objects. Hitch, Halliday, Schaafstal and Schraagen (1988) devised a picture confusion memory task of this type. Their visually similar set of pictures consisted of pictures of a nail, bat, key, spade, comb, saw, fork and pen (see Figure 8.5). Their visually dissimilar set of control pictures consisted of pictures of a doll, bath, glove, spoon, belt, cake, leaf and pig. An additional set of visually dissimilar pictures that had long names was also used in the task. This set comprised an elephant, kangaroo, aeroplane, banana, piano, policeman, butterfly and umbrella. Hitch et al. then compared five- and ten-year-old children's memory for these pictures of familiar objects.

Hitch et al.'s memory task was similar to Conrad's, except that no matching was required. Instead, the experimenters presented each picture face-up, and then turned it over, telling the child that they would have to repeat the names of the pictures in the order in which they were shown. The five-year-olds were given sequences of three pictures, and the ten-year-olds were given sequences of five pictures. Hitch et al. argued that, if the children were using rehearsal to remember the order of the pictures, then they should find the pictures with long names more difficult to recall than the visually similar pictures and the control pictures. On the other hand, if the children were using visual memory strategies, then they should find the visually similar pictures the most difficult set to recall. Hitch et al. found that, for the ten-year-olds, the pictures with long names were the most difficult to recall. For the five-year-olds, the visually similar pictures were the most difficult to recall, although there was a small effect of word length. Hitch et al. concluded that the tendency to use visual working memory becomes less pervasive as memory development proceeds.

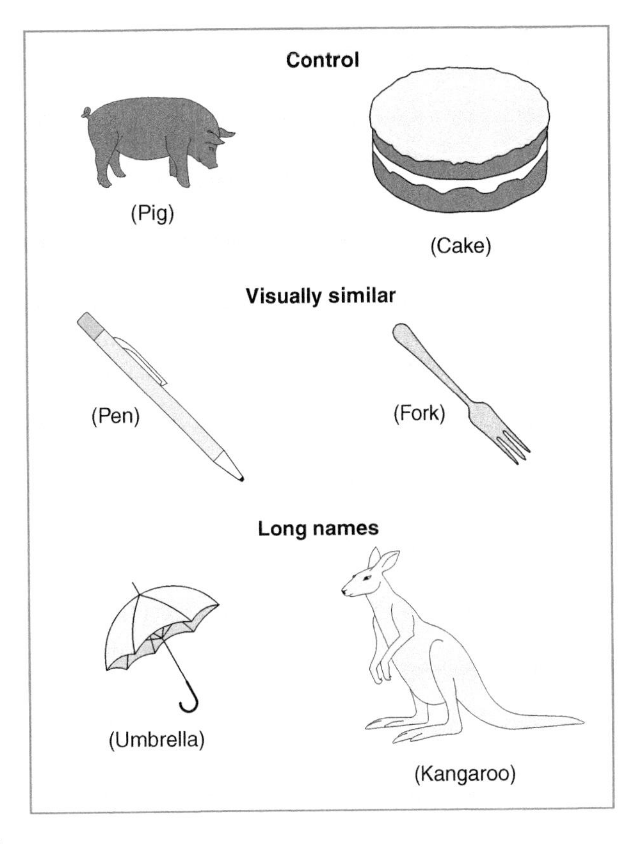

FIGURE 8.5 Examples of stimuli from each condition in Hitch et al.'s (1988) picture memory task. Copyright © The Psychonomic Society. Reproduced with permission.

All of these experiments, however, have studied the retention of *visually presented* items. Rather than showing that children rely on visual memory codes prior to around five years of age, it may thus be that younger children tend to rely on visual codes in working memory when they are given visual information to remember. Given the domain-specificity of the two slave systems, visual versus verbal, experimenters may be documenting a tendency in younger children to attempt to retain information in the modality in which it is presented, rather than a tendency to rely on visual memory codes. Older children, who typically are also learning to read, may translate visually presented material into a speech code. The visual working memory effects observed by Conrad and by Hitch et al. may thus be due to children's failure to select a particular mnemonic strategy, rather than to an early reliance on visuo-spatial memory codes. We will discuss this possibility further in the next section. Such 'production deficiencies' in strategy use (a production deficiency means that a child has a strategy available but does not think of using it) are more common in younger children, as will be discussed when we consider strategies for reasoning.

Meanwhile, it is interesting to note that deaf children continue to rely heavily on visuo-spatial codes in memory, even for material that hearing children code linguistically. O'Connor and Hermelin (1973) devised a spatial span task to measure short-term recall in the deaf. In this task, three digits were presented successively on a screen, appearing in three different windows in a horizontal visual array. The left-right order did not always correspond to the temporal-sequential order of presentation. For example, the first digit to appear might be the one in the middle window. O'Connor and Hermelin found that whereas the hearing children tended to recall the digits in their *temporal* order of appearance, the deaf children recalled them in the *spatial* order of their (left-right) appearance. This suggests that the hearing children were rehearsing the digits verbally in order to remember them, while the deaf children were representing the digits as visual images.

KEY TERM

Mnemonic strategies Strategies to aid storage and recall of memories such as rehearsal and organisation.

The phonological loop

Activity in the phonological loop is usually measured by the presence of effects that are related to the use of speech sounds for coding material. As noted, it is more difficult to remember words that sound similar over a short period of time (bat, cat, hat, rat, tap, mat) than words that do not. This is called the 'phonological confusability' or 'phonological similarity' effect. Similarly, long words like 'bicycle, umbrella, banana, elephant' take longer to rehearse than short words like 'egg, pig, car, boy', and so more short than long words can be retained in working memory. This is called the 'word length' effect.

The number of items that can be retained in the phonological loop over a short period of time is used to provide a measure of an individual's 'memory span'. Memory span gives a measure of working memory capacity, and increases with age. As span length differs with different types of material, however, such as long versus short words (and with background knowledge, see Schneider & Bjorklund, 1998), most measures of memory span are based on the retention of items like digits, which are assumed to be equally familiar to all subjects. However, number words vary in length in different languages. This can lead to different estimates of memory span in children of the same age who speak different languages. Chinese children have much longer digit spans than English children, because the Chinese number words are much shorter than the English number words (e.g., Chen & Stevenson, 1988). Welsh children have shorter digit spans than American children, as the Welsh number words are longer than the English number words. As memory span is usually one of the things that is measured in IQ tests, at one time it was wrongly thought that Welsh children had lower IQs than American children. Upon closer investigation, it was found that the IQ difference was an artefact of systematically lower Welsh scores on the digit span component of the test (Ellis & Hennelley, 1980).

Another important component of working memory capacity is speech rate or articulation rate, which also affects memory span. Children who articulate slowly tend to have shorter memory spans than children who can articulate more quickly, presumably because it takes them longer to rehearse individual items. Speech rate also increases with age. Because of this, it has been proposed that the development of memory span with age is entirely accounted for by developmental increases in speech rate. As older children speak more quickly than younger children, they can rehearse more information during the 1–2 seconds available in the phonological loop, and can thus remember more items than the younger children, giving them longer memory spans.

As required by this proposal, speech rate and memory span are highly correlated. This connection was established in a series of studies by Hulme and his colleagues, using a word repetition task (e.g., Hulme, Thomson, Muir & Lawrence, 1984; Hulme & Tordoff, 1989). For example, in Hulme et al. (1984), children aged four, seven and ten years and adults were given a pair of words, such as 'apple, tiger', to repeat as quickly as they could. The number of words produced per second provided the measure of speech rate. The results showed that speech rate was linearly related to memory span: increases in memory span were always accompanied by increases in speech rate across age. Furthermore, the relation between recall and speech rate was constant across different word lengths. When the children's memory spans for long (e.g., helicopter, kangaroo), short (e.g., egg, bus) and medium-length (e.g., rocket, monkey) words was compared, Hulme et al. found that the relationship between recall and speech rate was constant across age. This shows that, at any age, subjects can recall as much as they can say in a fixed time interval (about 1.5 seconds). Hulme et al. argued that individuals with higher speech rates could rehearse information more quickly, and could thus remember it better.

When Hitch, Halliday, Dodd and Littler (1989) replicated Hulme et al.'s work using pictures instead of words, however, the results were rather different. With *visual* presentation of the words to be remembered, only the ten-year-olds showed a correlation between speech rate and memory span, as only this group appeared to spontaneously rehearse the visual inputs. With auditory presentation, all age groups showed the correlation. Thus speech rate *does* govern the number of items that can be retained in working memory, but only when the items to be remembered are presented in speech form. This is, of course, in keeping with the findings of Conrad (1971) and Hitch et al. (1988) that we discussed earlier. Younger children appear to prefer to maintain visually presented information by using a visuo-spatial code, whereas older children spontaneously translate visual inputs into a speech code.

Henry and Millar (1993) explained this developmental pattern by proposing that rehearsal develops out of naming behaviour. They pointed out that younger children are frequently called upon to use naming to translate visual or tactile material into a verbal form, particularly as they enter school and rely increasingly on verbal strategies for learning and retaining information. Henry and Millar suggested that children's increasing speed and facility with naming leads to the discovery of rehearsal, and that the development of rehearsal probably explains the development of memory span after the age of about seven years. The development of memory span prior to this point depends both on naming and on the child's familiarity with the items to be remembered. Items that are highly familiar in long-term (semantic) memory are easier to store and to retrieve. Highly familiar items also have well-specified phonological representations in the mental lexicon, and as the speech output system is used in memory span tasks for both rehearsal and recall, words with better-specified representations will require less processing, can be articulated faster, and are more easily reconstructed when memory traces deteriorate (a process termed 'redintegration', see also Roodenrys, Hulme & Brown, 1993). According to this view, although speech rate is related to the development of memory span, speech rate is also determined by the quality of a child's phonological representations.

Henry and Millar's proposal was an interesting one, as it suggested that the *development* of working memory should be intimately related to the development of long-term or semantic memory. According to their argument, the key variable that affects the capacity of working memory, speech rate, is dependent on the development of well-specified phonological representations in semantic memory, which is linked to conceptual and linguistic development. Items that have well-specified phonological representations in long-term memory are easier to redintegrate, leading to enhanced working memory. In general, these items reside in what linguists call 'dense phonological neighbourhoods'. Phonological neighbours are words that sound similar to each other, and words with many neighbours tend to have high-probability phonotactics (see Chapter 6). There is now quite a lot of evidence that the quality of the representations of items in long-term storage affects their retention in short-term tasks. For example, Thomson, Richardson and Goswami (2005) compared serial recall for words from dense versus sparse phonological neighbourhoods in children aged seven and nine years. If long-term memory and working memory are linked, then children should show better retention of words residing in dense phonological neighbourhoods. This should be the case even when overall item frequency is matched, as was the case for the words used in this study. Thomson et al. reported that serial recall for quadruples of words like 'bone, pail, king, gum' (dense phonological neighbourhoods) was indeed significantly superior to serial recall for quadruples of words like 'wipe, bird, hook, leg' (sparse phonological neighbourhoods). Words with better-specified phonological representations appeared to enjoy better redintegration. This supports Henry and Millar's (1993) proposal that the development of long-term memory and the development of short-term memory are intimately linked.

Data such as these imply that the conventional conceptual distinction between short-term and long-term memory may be only partially applicable, at least as far as the *development* of memory is concerned. Although children do use the phonological loop to maintain information over short periods of time, just like adults, its utilisation appears to be gradual, depending on task features (e.g., verbal versus pictorial input), the age of the child and the quality of long-term representations. This pattern is reminiscent of the findings discussed earlier regarding the development of episodic memory. As the quality of a child's encoding improves, so does the quality of recall. For episodic memory, the quality of encoding refers to factors like the narrative coherence of the event representation. For working memory, the quality of encoding refers to factors like the specificity of the phonological representation. Nevertheless, in both cases, older children have more adequate representations. This suggests that Baddeley and Hitch's (1974) original proposal that working memory was a 'workspace' with two slave systems that was *distinct from* long-term memory needs to be modified, at least as the memory system is developing.

Cognitive neuroscience studies of working memory development

The study of working memory is one of the most active areas of adult cognitive neuroscience, and this is now being reflected in neuroimaging studies with children. The overwhelming majority of brain imaging studies with adults use fMRI, which has led to quite precise information about the brain regions that are active during working memory tasks. Frontal, parietal and cingulate regions are reliably activated when adults perform working memory tasks in the MRI scanner, and frontal and parietal grey matter volume, along with temporal and parietal connections of the corpus callosum, are associated with working memory capacity. These brain-behaviour correlations tell us little about the brain mechanisms that support working memory in adults. There is a smaller literature on the role of neuroelectric oscillations in encoding and retrieving items during short time periods in adults, implicating theta band responding (Sauseng, Griesmayr, Freunberger & Klimesch, 2010), and also alpha band responding. This literature is more mechanistic in focus (for example, regarding how oscillations may enable the short-term maintenance of just-experienced sensory information). In adults, event-related theta oscillations are most prominent in frontal regions during working memory tasks, whereas event-related alpha oscillations are most prominent in parietal regions, showing convergence with BOLD findings in fMRI studies. To date, normative child studies of working memory have largely focused on fMRI, with the role of oscillations only studied occasionally, and typically in children with learning difficulties. As the fMRI literature with children is now quite large, I will only give selected examples of typical studies.

Verbal working memory has been studied in children aged from nine to 15 years in a 'one-back' task, with fMRI scanning utilised to identify developmental changes in neural activation (Vogan, Morgan, Powell, Smith & Taylor, 2016). In the working memory task, children viewed letters arranged into a global form that was also a letter, such as a global form 'A' made up of the smaller letters A and P. They were taught that two of the smaller letters were irrelevant and should be ignored (these letters were O and P). The children had to identify the relevant smaller letters (overall eight relevant smaller letters were used, A, B, E, H, K, M, N and T). They then also had to decide whether the smaller letters matched those that had been used in the most-recently viewed previous global form (the 'one back' form). After practising the task, children performed the task in the scanner. Memory load was increased by using more small letters to make up the global forms. Vogan et al. were interested in which brain regions would show increased blood flow as memory load increased, and whether these regions would differ between children and adults, who were also tested.

Vogan et al. (2016) reported that children exhibited task-related activity in broadly similar brain regions to adults, with increasing activation in frontal and parietal regions as task demands increased. However, adults showed greater activity than children in brain areas thought (from adult studies) to enhance working memory performance, such as middle and inferior frontal gyrus. Although both children and adults were matched for performance in lower memory load conditions, for the highest memory loads the adults were significantly more accurate than the children. Accordingly, it is difficult to disentangle cause and effect regarding the differential brain activation observed. The simple conclusion is that verbal working memory tasks performed by children activate the same neural structures as verbal working memory tasks performed by adults. What these neural structures do, however, remains unclear.

An early neuroimaging study of visuo-spatial working memory was carried out with children aged from nine to 18 years by Klingberg, Forssberg and Westerberg (2002). They used a working memory task based on a 4 × 4 grid. Three red circles were presented sequentially in different portions of the grid, and then a final circle appeared. The children had to decide whether this final circle was in a location that had been occupied before. Klingberg et al. reported age-related bilateral activity in the superior frontal sulcus and intraparietal cortex while the children performed the task. Changes in working memory capacity (measured outside the scanner) were also correlated with increased BOLD activity in these two regions (both left-lateralised). A strong feature of this study was that the working memory task used inside the scanner was performed as well by the younger children as by the older children. The core finding, that age was correlated with increased brain activation during this task rather than with the amount of cortex activated, suggests that there might be little change in the neural substrates activated by performing visuo-spatial working memory tasks with age.

A more recent neuroimaging study of visuo-spatial working memory reached a similar conclusion. Wendelken, Baym, Gazzaley and Bunge (2011) studied children aged from eight to 13 years of age. They used a face/scene working memory task, in which the goal was to remember either faces or scenes while viewing pictures of both. Memory was tested using probe pictures which had either been recently viewed or not. A passive viewing control condition was also used, in which the same scenes were viewed without an instruction to remember either the faces or the scenes, and adults were also tested. Faces versus scenes were chosen because face processing is typically associated with increased neural activity in the fusiform face area (FFA), while scene processing is typically associated with increased neural activity in the parahippocampal place area (PPA).

Wendelken et al. (2011) found that the children activated broadly similar brain areas to the adults during task performance. In particular, when instructed to attend to scenes, the children showed greater activation in the PPA bilaterally. Activation of the PPA was also greater for the older children, who were also more successful in the task. Indeed, the degree of PPA activation was correlated with working memory performance. The FFA did not show the expected selective activation in the 'attend face' task. However, this was also the case for the adults, who were tested in the same paradigm. Again, the simplest conclusion is that visuo-spatial working memory in children activates the same neural structures as visuo-spatial working memory in adults. Once more, the brain-behaviour correlations are in themselves not informative with respect to the neural mechanisms that support working memory.

Indeed, Bathelt, Gathercole, Johnson and Astle (2017) made the important point that using individual working memory tasks in the scanner may not reveal the neural structures whose activation is related to working memory per se, as task-specific demands are also likely to change with age. For example, proficiency in a stimulus domain (such as knowledge of letters or familiarity with scenes) may cause age-related changes in neural activation that are *independent* of the working memory demands that the researchers assume that they are measuring. To remedy this, Bathelt et al. gave children aged from seven to 12 years a range of working memory tasks such as digit recall and a dot matrix (visuo-spatial) task, and then used principal component analysis to derive the three working memory factors (verbal, visuo-spatial, central executive). These three factors accounted for 92% of the variance in the data and were stable across age, suggestive of a good fit between working memory tasks and the cognitive model of working memory proposed by Baddeley and Hitch (1974). Structural MRI scans were also taken, and cortical thickness and white matter integrity (a measure of the myelination of axons) were correlated with working memory capacity as measured by the working memory task battery.

Bathelt et al. (2017) reported that myelination increased with age, reflecting the maturation of structural connections and growing integration within brain systems. Cortical thickness was not related to age, but as the youngest children studied were already relatively old, this was not unexpected. Working memory capacity was associated with white matter integrity rather than with cortical thickness. Younger children's working memory capacity was more closely associated with large white matter connections, while older children's working memory capacity was associated with cortical thickness in the left posterior temporal lobe. Bathelt et al. (2017) concluded that younger children were relying on a more distributed system of brain regions to support working memory, while older children showed more adult-like activation of specialised and localised structures. As the data are structural rather than functional, however, whether these structural-behavioural correlations reflect differential functioning within these regions during working memory task performance cannot be determined.

Regarding the possible role of oscillatory mechanisms in working memory, Wang, Tseng, Liu and Tsai (2017a) reported a study measuring neuroelectric oscillations during a visuo-spatial working memory task in children with developmental co-ordination disorder (DCD). Children aged on average ten years with DCD were compared with typically-developing children in a child-friendly task in which children viewed pictures of ladybirds sitting on grey patches (leaves). In a given trial, the children either saw two ladybirds on two patches, and had to judge whether they were sitting in the same spot, or saw the two patches one after the other (delayed condition), and then had to make the same judgement. Only the delayed condition was expected to tax visuo-spatial working memory. Children were required to make yes/no judgements by pressing different computer buttons, and were given feedback on whether they were correct or not. EEG was also recorded. Wang et al. (2017a) reported that the children with DCD were significantly less accurate and significantly slower to respond than the typically-developing children in the delayed condition only, when there was a working memory demand. They then inspected the oscillatory data. A significant increase in theta band activity over the frontal midline was found for the non-delayed condition, which was similar for both groups of children. During the delayed condition, significant increases in frontal theta activity were coupled with significant suppression of alpha band activity (~10 Hz) over parietal areas, again for both groups of children. However, the increase in theta band activity was less for the DCD children, and the suppression of alpha band activity was also reduced for these children. Brain-behaviour correlations did not reach significance. Once more, this study does not tell us much about the neural mechanisms that support working memory. The simplest conclusion is that brain activity during working memory task performance was similar in both groups, albeit this time for groups of more-able and less-able children, rather than for children versus adults.

In a study employing MEG and typically-developing six-year-olds, Sato et al. (2018) measured alpha and theta band oscillatory activity during a visual working memory task. Children were shown trials in which two coloured squares were presented onscreen, and then disappeared for 1,000 ms, after which two more coloured squares were presented. The children had to press one button if the colours were the same, and another button if the colours were different. On half of the occasions, the colours were the same, and on half they were different. Children were tested for two runs of seven minutes while MEG recordings were taken. Sato et al. reported that alpha band activity, but not theta band activity, discriminated between correct and incorrect trials, across the whole brain. Time frequency analyses were then conducted for correct trials only, to see where alpha power was focused. The data showed that alpha power in the left parietal lobe was enhanced relative to baseline at the beginning of the retention period, and then returned to baseline as the test stimulus was presented. Accordingly, Sato et al. concluded that alpha band synchrony was an important marker of working memory development. The convergence with the data reported by Wang et al. (2017a) regarding the alpha band (albeit for enhancement rather than suppression) and parietal loci is encouraging. Nevertheless, these studies are still correlational, and do not actually show *how* alpha oscillations may support working memory, for example during encoding.

As this brief selection of working memory studies indicates, despite a lot of research activity, developmental cognitive neuroscience has yet to contribute any novel evidence concerning the developmental mechanisms that support working memory. In particular, it is absolutely crucial that studies that are aimed at documenting differences between groups of different ages (or groups of different abilities) equate performance on the working memory measures being used in the scanner prior to measuring the brain. This is because if performance differs by age or by ability, the brain imaging measures chosen for inspection will inevitably differ as well. Accordingly, these neural differences cannot in themselves tell us anything about mechanisms of development. By contrast, if task performance is *equated* between groups, and it is then found that different neural structures are active when younger versus older children are scanned using fMRI, or that different neural mechanisms are implicated when oscillatory responses are measured, this could be informative with respect to development. Even then, however, only the first step has been taken in understanding potential neural causes. As in behavioural studies, to go beyond observation of age-related differences, intervention studies are required to begin to disentangle cause and effect. For example, if alpha oscillations in parietal cortex are mechanistically important for the development of working memory, then using a technique such as TMS (transcranial magnetic stimulation) to temporarily 'silence' alpha band activity should impair children's performance in visual working memory tasks. In Chapter 10, we will see that neuroimaging studies equating in-scanner performance are beginning to appear in the domains of reading and mathematics, primarily in studies exploring the mechanisms that may lead to developmental dyslexia and developmental dyscalculia.

THE DEVELOPMENT OF STRATEGIES FOR REMEMBERING

As adults, we often employ special strategies to help us to remember important information, especially when such information must be retained for relatively short periods of time. These strategies include rehearsing a phone number that we have just been given, or organising information about what we need to remember to buy at the supermarket. Henry and Millar (1993) suggested that children may discover the strategy of *rehearsal* via their increasing use of naming, for example because of activities encountered in school. The emergence of strategies such as rehearsal and *organisation* for remembering can also be studied in their own right. In fact, traditional answers to the question of 'what is memory development the development of?' (see DeMarie & Ferron, 2003) include *strategies* and *capacity*. Memory capacity is usually measured by memory span tasks, as described above.

Regarding memory strategies, younger children are typically surprisingly confident about their mnemonic abilities. They do not seem to expect that they will need to use mnemonic strategies to improve their recall. For example, in a study conducted by Yussen and Levy (1975), half of the four-year-olds tested predicted that they would remember all ten items in a standard memory span task in which they were presented with ten unrelated items to recall. In actual fact, they remembered about three! This experience did not change the children's confidence in their abilities, however. Very few of them changed their predictions about their memory capacities. Instead, they said things like "If you gave me a different list like that, I could do it!"

The emergent use of mnemonic strategies

Nevertheless, when given a specific 'ecologically valid' memory task to perform, even very young children appear to have some realisation that they will need to use mnemonic strategies to aid their memories. This was demonstrated in a study by Wellman, Ritter and Flavell (1975), who told a group of three-year-olds a story about a toy dog. Among the props used in the story were four identical plastic cups. At one point in the procedure, the experimenter put the toy dog under one of these cups, explaining that the dog would go into the doghouse while the experimenter left the room to find more things. The child was asked to remember which cup the dog was under while the experimenter was away.

During the 40 seconds that the experimenter was out of the room, most children used a variety of strategies to help them to remember which cup was the dog house. They looked at and touched the cup hiding the dog significantly more often than the other cups, they looked at the target cup and nodded to themselves "yes", looked at the other cups and shook their heads "no", they rested their hand on the target cup, and so on. Wellman et al. also found that recall for the dog's location was more successful in the children who used these strategies than in the children who didn't. An attempt to use this procedure with two-year-olds was thwarted by the restlessness of the children, who wouldn't keep still during the experimenter's absence.

Better success with two-year-olds has been reported by DeLoache, Cassidy and Brown (1985). They used a hiding game to investigate children's strategic memory for spatial locations. The children, who were aged from 18 to 24 months, watched the experimenter hide a favourite toy (e.g., Big Bird) in a natural location (e.g., under a pillow in the child's home). The children were told that Big Bird was going to hide, and that they should remember where he was hiding as they would need to find him later when the bell rang. A timer was then set for four minutes, during which time the child took part in a number of distraction activities with other toys, organised by the experimenter. The children frequently checked on Big Bird's hiding location during this distraction period, for example pointing at the pillow, saying "Big Bird!", and peeping underneath it. In a control condition in which Big Bird was put on top of the pillow, similar strategies were not observed. DeLoache et al. argued that this showed that the children's self-reminding behaviours were indeed strategic, as they were adopted as a function of the memory demands of the task.

Somerville, Wellman and Cultice (1983) also succeeded in measuring strategic recall in two-year-olds by using a highly motivating task. The task was the need to remember to buy candy at the store at a particular time specified by their mother (e.g., tomorrow morning). Memory for two events was compared, getting candy at a specified future time, and removing the washing from the washing machine at a specified future time (e.g., when Daddy gets home). Children aged from two to four years all showed better memory for the highly motivating candy event, indicating an ability to plan ahead and to keep a particular event in mind. At short delays (five minutes), even the two-year-olds achieved a level of 80% unprompted remindings for the highly motivating event. On the low motivation task (getting the washing out of the machine), overall success with unprompted reminding was much lower, falling to 26% over long delays. Somerville et al. argued that even very young children were capable of adopting a deliberate 'set' to remember at an early age.

Evidence for the strategic use of rehearsal

The spontaneous mnemonic strategies observed in these experiments were fairly task-specific, however. One mnemonic strategy that is widely used by adults when they want to remember some information over a short period of time is rehearsal. Saying things to ourselves over and over again can make it easier to remember them. We saw earlier that the spontaneous use of rehearsal may not emerge until children are at school, and are having to rely increasingly on verbal strategies for learning and retaining information. This idea of a 'production deficiency' in younger children's use of rehearsal is supported by the findings of a classic study of children's rehearsal, carried out by Flavell, Beach and Chinsky (1966).

Flavell et al. asked children aged five, seven and ten years to remember a set of pictures over a short delay of about 15 seconds while wearing a space helmet. The space helmet had a visor that concealed the children's eyes, but left their mouths visible. The short delay period began when the experimenter said "Visor down!", and any spontaneous rehearsal during this delay was measured by a trained lip-reader. While the visor was up, the children were shown up to seven pictures (apple, comb, moon, owl, pipe, flowers, American flag), and were required to remember between two and five of them while the visor was down. Flavell et al. found that only 10% of the five-year-olds used a rehearsal strategy, whereas 60% of the seven-year-olds and 85% of the ten-year-olds did so. There was also some evidence from the seven-year-olds' data that the children who rehearsed more recalled more pictures. Flavell and his colleagues argued that the majority of the younger children failed to rehearse because of a production deficiency. The younger children did not realise that they needed to use strategies such as rehearsal to help them to remember. As we saw earlier, younger children are also less likely to convert visually presented information into a verbal code.

Although Flavell et al.'s study did find some evidence for a relationship between the use of rehearsal and the accuracy of recall, later work has shown that the strategic rehearsal of seven-year-old children tends to be piecemeal and not particularly helpful to short-term memory. They tend to rehearse just the currently presented item, or the current item with very few other items. However, small amounts of training can lead to rapid improvement in the strategic use of rehearsal, with accompanying improvements in recall.

For example, Naus, Ornstein and Aivano (1977) gave eight- and 11-year-old children a list of words to remember. Some of the children were told to practise the words aloud as they normally would do to themselves, and others were told to practise the words aloud by saying the word that had just been presented in a given trial along with two other words. The children who were trained how to rehearse remembered significantly more items at test than the children who were told to practise the words as they normally would to themselves. Naus et al. argued that it was the content of rather than the activity of rehearsal that improved memory in list-learning tasks. The quality of rehearsal was more important than its frequency.

The fact that children can be trained to use rehearsal leaves open the question of whether younger children have the strategy of rehearsal available to them, but simply do not think of using it (a 'production deficiency'). One way of finding out whether younger children's lack of rehearsal is due to a 'production deficiency' is to offer them an *incentive* to use rehearsal. Kunzinger and Witryol (1984) devised an incentive-based technique that used financial rewards for studying the spontaneous use of rehearsal in seven-year-olds. They told the children that recall of some words on a list would win them ten cents, whereas recall of others would only win them one cent. If children can rehearse without being trained in efficient strategy usage, then they should be more likely to rehearse 'ten-cent words' than 'one-cent words'.

Kunzinger and Witryol indeed found that the children in their study allocated more rehearsal to the 'ten-cent words' than to the 'one-cent words'. In fact, they were six times as likely to rehearse the former as the latter at the beginning of the list. In a control condition in which every word was worth five cents, the children rehearsed less overall. Kunzinger and Witryol argued that this was because the children in the experimental condition generalised the use of rehearsal from the 'ten-cent words' to the 'one- cent words'. The extra rehearsal allocated to the 'ten-cent words' also resulted in better recall. These words were recalled significantly more often than the 'one-cent words' or the 'five-cent words'. Seven-year-olds can clearly be induced to use rehearsal, and when they do so, their memories improve.

Attractive incentives can also be used to induce memory strategies in four-year-olds, although without apparent improvement in their memories. O'Sullivan (1993) showed four-year-old children 15 different toys (doll, horse, ball, aeroplane, etc.) and told them that they would win a prize if they could remember all the toys in a recall test. Two prizes were on offer, a pencil and a box of crayons. The crayons were universally judged to be the more appealing prize. In the experiment, the toys were presented in a bag. The experimenter took them out of the bag, allowed the children to study them for three minutes, and then put them back into the bag. The children then spent 25 seconds drawing Xs on a sheet of paper to eliminate short-term memory effects, after which the recall phase of the experiment began.

O'Sullivan found that the children who were playing to win the box of crayons showed more visual examination of the toys than the children who were playing to win the pencil, and also spent less time in 'off-task' behaviour. Spontaneous use of rehearsal was not observed. However, although the possibility of winning the better prize elicited significantly more efforts to remember than the possibility of winning the poorer prize, this did not translate into superior recall for the 'crayons' group. Recall performance was in fact equivalent across the two incentive groups, and averaged eight items. It should be mentioned that all the children received both prizes at the end of the study!

Many other studies of the development of rehearsal in young children report broadly similar findings to the studies discussed here (see Schneider & Pressley, 2013; Schneider & Bjorklund, 1998, for reviews). Younger children seem disinclined rather than unable to rehearse. Rehearsal first appears when it is encouraged by training or by task-specific factors such as remembering a list of words, or by the use of verbal rather than visual presentation of the items to be remembered. Rehearsal is only used strategically somewhat later in development. The early lack of spontaneous rehearsal can thus be seen as a production deficiency rather than a competence deficiency.

Evidence for the strategic use of organisation by semantic category

Organisational mnemonic strategies, such as sorting required grocery items into related groups and using this clustering to aid recall, show a similar developmental pattern to rehearsal. Early strategic use is largely task-driven, and depends on the items to be recalled. Later strategic use is child-driven, and occurs independently of the materials to be remembered.

For example, Schneider (1986) told seven- and ten-year-old children that they would be shown a set of 24 pictures, and that they should try to do anything that would help them to remember the items in the set. The sets of pictures were either defined as having 'high category relatedness' (e.g., dog, cat, horse, cow, pig, mouse), or 'low category relatedness' (e.g., goat, deer, hippopotamus, buffalo, monkey, lamb). In addition, the sets either had high inter-item associativity according to word association norms (e.g., chair, table, bed, sofa, desk, lamp), or low inter-item associativity (e.g., refrigerator, stool, bookcase, rocking chair, stove, bench). The children were given two minutes to sort the pictures, and an additional two minutes to study them.

Schneider found that only 10% of the seven-year-olds spontaneously grouped the pictures according to their category relationships, whereas about 60% of the ten-year-olds did so. In addition, the younger children, but not the older children, were less likely to group together the items that had low inter-item associativity. Whereas the ten-year-olds used categories like 'furniture' to group the pictures of the stove, bench etc., the seven-year-olds did not. Schneider argued that the use of organisational strategies in younger children depended on the degree to which the items were associated. For the seven-year-olds, high associativity *in itself* led to the use of clustering, in a largely involuntary way. In contrast, the ten-year-olds used clustering as a deliberate strategy. The older children were apparently becoming aware of the value of organisational strategies as a mnemonic. In support of his argument, Schneider found that approximately half of the ten-year-olds in his study showed systematic and strategic behaviour that facilitated recall, and were aware of the value of organisational strategies.

Similar results were reported in a memory task devised by Bjorklund and Bjorklund (1985), in which six-, eight- and ten-year-old children were asked to recall the names of their current classmates. All of the children found this an easy task, and appeared to be behaving strategically, organising their recall in terms of grouping cues such as the seating arrangements in their classroom, the children's reading groups, or boys versus girls. However, when the experimenters asked the children how they went about remembering their classmates' names, the children were unable to outline particular strategies, suggesting that their use of clustering was involuntary. Bjorklund and Bjorklund tested this hypothesis by asking the children to use specific retrieval strategies, such as remembering all of the boys first and then all of the girls. They found that recall in this strategic condition was equivalent to recall in the free condition, when no instructions were given. Bjorklund and Bjorklund concluded that semantic associations between highly associated items can be activated with little effort, resulting in retrieval that appears to be organised and strategic when in fact it is simply a byproduct of high associativity. High associativity thus *automatically* guides the structure of recall.

However, evidence for more strategic use of semantic associativity was found in a study of younger children reported by Schneider and Sodian (1988). They asked four- and six-year-old children to hide ten pictures of people (doctor, farmer, policeman) in ten wooden houses, and then to retrieve them by matching each picture with its 'twin' (a duplicate picture provided by the experimenter). The wooden houses all had roofs that could be opened and shut like boxes, and they also had magnetic stickers on the front doors to which a picture cue could be attached. The available picture cues were either functionally related to the people who were hiding (syringe, tractor, police car) or not (key, flower, lamp).

Schneider and Sodian first asked the children to perform the hiding task without the picture cues, and measured the time taken for hiding and retrieval. They then attached a picture cue to each house, and asked the children to perform the hiding and retrieval tasks a second time. Systematic use of the semantic associations between the picture cues and the targets should lead to slower hiding times on the second trial, but more accurate retrieval. This was exactly what happened. Children of both age groups spent longer hiding the people pictures on the second trial, and also remembered more locations on the second trial. Furthermore, the six-year-olds remembered significantly more locations on the second trial than the four-year-olds, indicating that they benefited more from the semantic associations in the cue pictures. Indeed, they were more likely to hide people at semantically appropriate locations than the younger children, hiding on average 3.63 of the possible five people appropriately, compared to two out of five for the four-year-olds.

Schneider and Sodian also reported that the older children showed more understanding of the use of retrieval cues. When asked specific questions like "Which of the games I just played with you was easier?", the six-year-olds displayed greater conscious knowledge of the strategic use of the semantic cues than the four-year-olds. However, even the four-year-olds showed some knowledge of cuing as a memory aid. Furthermore, a relationship between conscious awareness of the usefulness of the cues and successful memory performance was found in both groups. Schneider and Sodian argued that the idea that preschoolers can only react automatically to highly associated cues without any awareness of their value may be misguided. Even four-year-olds appear to have some understanding of the utility of cognitive cuing. The fact that this study used a simpler and more meaningful task from the child's point of view may also be important in eradicating the apparent production deficiency in children's spontaneous use of organisation as a mnemonic strategy.

The development of multiple strategies

Research has also investigated whether young children use more than one strategy for remembering at a time, and whether the use of multiple strategies can benefit children's recall (e.g., Coyle & Bjorklund, 1997; DeMarie, Miller, Ferron & Cunningham, 2004). In general, such research finds that children who use more strategies do seem to recall more information. For example, DeMarie and Ferron (2003) gave children aged from five to ten years three different memory tasks expected to yield different mnemonic strategies. These were a picture recall task, a selective recall task and a picture sorting task. In the first task, children were given lists of 18 items to remember. The items were spoken by the experimenter, who placed a card depicting each item on the table as she spoke. The children were given two minutes to arrange the cards in any way that they liked to help them to remember the items. The cards were then concealed, and the children were asked to recall the items in any order. In the selective recall task, an apparatus with 12 doors was used. Each door had a picture on it, of either a cage or a house. Doors with cages on them concealed animals, and doors with houses on them concealed household items. The children were asked to remember where one type of item was hidden, and were given time to open the doors and check what was behind. Recall was measured by showing children pictures of the hidden items, and asking them to point to their location. In the sorting task, children were given apparatus stands which could hold cards in various positions, and were asked to place either 12 (age 5–8) or 16 (age 9–10) cards on the stand for later memory. Recall for the cards was then tested. The strategies recorded for each task included organising and grouping cards by semantic category (task 1), selectively opening relevant doors (task 2), placing cards by semantic category (task 3), rehearsal, self-testing during learning, and semantic clustering in recall.

DeMarie and Ferron (2003) reported that even the youngest children used more than one strategy for remembering. Older children used more strategies overall than younger children. The number of strategies used across tasks was significantly but modestly correlated, for younger and for older children. Finally, strategy use was strongly correlated with successful recall. Memory capacity was also measured in this study (using span tasks), and memory capacity did not predict successful recall. DeMarie and Ferron concluded that strategy use by young children was an important factor in memory development.

In a longitudinal study of 102 kindergarten children (six-year-olds), Schneider, Kron, Hunnerkopf and Krajewski (2004) also investigated the role of both strategies and capacity in memory development. The children were seen on three occasions separated by six-month intervals. Schneider et al. assessed the children's memories via an item retention task. In this task, 20 familiar items were presented for recall (the items were from five semantic categories, namely furniture, tools, fruit, clothes and body parts). The children were told that they could do whatever they wanted for a three-minute period to aid recall of these items. They were then scored for using strategies like sorting the pictures during study, clustering at recall and rehearsal. Memory capacity was measured by a span measure.

Schneider et al. reported that both strategy use and memory capacity increased over time. Strategy use was related to better recall, and children who used more strategies did particularly well. Schneider et al. were most interested in individual differences, however. They thus explored whether these general age-related increases were due to the gradual development of strategy use within individual children, or to the sudden discovery of beneficial strategies by particular children. The latter turned out to be the case. Some children (N = 57) were non-strategic at the first two measurement points, and their recall was very poor. Others (N = 9) were consistently strategic, and showed consistently high levels of recall. Another group of 28 children discovered the organisational strategy at the second measurement point, and 21 of these showed a significant increase in recall performance. The seven children who did not were termed 'utilisation-deficient'. For these children, use of the correct strategy did not enhance recall. Surprisingly, the same was found to be true six months later. This sub-group of children were still showing no recall benefit from using sorting strategies. Schneider et al. reported that memory capacity was also lower in this sub-group of children. They concluded that memory development is characterised by a rapid transition from non-strategic to strategic behaviour for most children. Memory development does not consist of a gradual increase in strategy use. Children who were consistently strategic used strategies in more than one test of memory, and the advantages of multiple strategy use were already evident in kindergarten children.

The novice–expert distinction

KEY TERM

Novice–expert distinction
The level of prior knowledge or expertise has a critical impact on the *encoding*, *storage* and *retrieval* of information, as experts organise their knowledge in different ways than novices.

The other traditional answers to the question of 'what is memory development the development of?' are *knowledge* and *metamemory*. Metamemory, or children's understanding about how we remember, is considered in the next chapter. Knowledge is considered here. Knowledge can also be conceptualised as expertise. It is now widely accepted that the knowledge base itself plays an important role in memory efficiency, and that experts organise their knowledge in different ways to novices. The level of prior knowledge has a critical impact on both the *encoding* and the *storage* of incoming information. Expertise also affects the efficiency of recall. One of the most interesting approaches to studying the influence of prior knowledge on cognitive development has been to contrast the performance of *novices* in a domain, who have little prior knowledge, with that of *experts*, who have a lot of prior knowledge. As such comparisons are usually confounded with age (novices are usually younger than experts), the most developmentally informative contrasts are those in which the experts are *younger than* the novices. Such contrasts are usually only possible in quite circumscribed domains, for example chess playing, soccer playing and expertise in physics.

As Brown and DeLoache (1978) once called young children 'universal novices', it may come as a surprise to find that young children occasionally display more expertise in circumscribed domains than older children and adults. This can occur because experts and novices in a particular domain are distinguished by differences in *experience* as well as by differences in age. If you are motivated enough, it is possible to gain a lot of experience in a domain at a relatively young age. For example, some children know an amazing amount about dinosaurs, because they find this domain so interesting that they become veritable experts in dinosaur classification and behaviour. Another such domain is chess. Some young chess players display a remarkable level of expertise, and regularly beat their adult opponents.

Children occasionally display more expertise in circumscribed domains than adults. Some young chess players, for example, are able to regularly beat their adult opponents.

If differences in experience distinguish novices and experts, then we can predict that differences in expertise should be correlated with differences in the structure and organisation of domain memory. In a pioneering study of this question, Chi (1978) examined the factors that distinguished the memories of chess experts and chess novices. Her group of experts were children aged from six to ten years, and her group of novices were graduate students who could all play chess. Chi measured the memory of both groups for 'middle game' chess positions, which involved on average 22 chess pieces. The chess players were allowed to study the chess board for ten seconds, and were then expected to recreate the middle game position from memory. Chi found that the children positioned 9.3 chess pieces accurately on the first trial, compared to 5.9 chess pieces for the adults. She then measured how long it took both groups to learn the *entire* middle game position. The children took on average 5.6 trials, and the adults 8.4 trials. Expert versus novice performance was significantly different in each case.

Although it seems plausible that the children were chess experts because they could remember more about the chess board, it is also possible that they could remember more about the chess board because they were experts. Chi argued that her data supported the latter possibility. She proposed that the child experts could see meaningful patterns in the arrays of chess pieces that were not apparent to the less-skilled adults. Schneider and his colleagues tested this interpretation in a replication of Chi's study which included additional control tasks (Schneider, Gruber, Gold & Opwis, 1993b; see Figure 8.6). Schneider et al. found that recall for random as well as meaningful chess positions was better in experts than in novices. In the control task, in which the position of geometrically shaped wooden pieces had to be reconstructed on a board that did not resemble a chess board, the effect of expertise was eliminated. These findings suggest that expertise involves both *qualitative* differences in the way that knowledge is represented and *quantitative* differences in the amount of knowledge available. The latter would in this case include knowledge about the geometrical pattern of the chess board and the form and colour of chess pieces.

The data on children's dinosaur expertise also suggests that experts structure their knowledge in qualitatively different ways from novices. Chi, Hutchinson and Robin (1988) found that experts organised their knowledge about dinosaurs in more integrated and locally coherent ways than novices. Their knowledge was more coherent at a global level, representing superordinate information such as 'meat eater' versus 'plant eater', and also at a sub-structural level, representing information about shared attributes such as 'has sharp teeth' or 'has a duckbill'. Chi et al. used a picture-sorting task to establish these differences. They found that seven-year-old dinosaur experts sorted pictures of dinosaurs on the basis of several related attributes and concepts ("He had webbed feet, so that he could swim, and his nose was shaped like a duck's bill, which gave him his name"). Seven-year-old novices sorted the same pictures on the basis of the depicted features ("He has sharp fingers, sharp toes, a big tail"). Chi and her colleagues pointed out that it was easier to understand the attributes of a dinosaur (sharp teeth, webbed feet, etc.) if one knew how they were related in a causal or correlated structure. The importance of causal relations and relational mappings for conceptual development and knowledge acquisition should already be familiar from previous chapters.

Expertise may also be *more* important for memory performance than general cognitive ability, at least when a particular domain like chess or soccer is the object of study. In a study of 'soccer experts' carried out in Germany, Schneider and his colleagues found that grade 3 soccer experts (boys and girls) recalled more information about a story concerning a soccer game than grade 7 soccer novices,

FIGURE 8.6
Examples of the chess task and of the control task (below). From Schneider et al. (1993b). Copyright © 1993 Elsevier. Reproduced with permission.

and that expertise was a stronger predictor of performance than general cognitive ability, regardless of age (Schneider, Korkel & Weinert, 1989). Similar effects have been demonstrated in adults by other researchers. For example, extensive experience of attending horse racing meetings and calculating the odds is a better predictor of who wins money at the races than IQ (Ceci & Liker, 1986). Findings such as these suggest that the old saying 'practice makes perfect' does capture something important about the development of expertise. In a longitudinal study of talented young German tennis players (the sample included Boris Becker and Steffi Graf), Schneider, Boes & Rieder (1993a) found that the amount and level of practice was an important predictor of the rankings of the players five years later. So was the level of achievement motivation. Perhaps unsurprisingly, talent per se was only one important predictor of later tennis excellence.

Overall, studies of novices and experts show that expertise plays a crucial role in the organisation of memory. The idea that knowledge enrichment leads to the reorganisation of memory has some obvious parallels with Fekete's (2010) metaphor about wine tasting by novices versus connoisseurs, discussed in Chapter 2. Fekete argued that via repeated learning experiences, or acquired expertise, the representation of the environment in the brain is irrevocably transformed, so that the connoisseur finds intricacies of flavour that are not available to the novice, even though they are tasting the same wine. Neurally, Fekete speculated that the multidimensional distribution of neural activity related to a domain of expertise could be differently organised in the expert versus the novice. However, in Chapter 7 we saw that neuroimaging experiments exploring scientific understanding, for example about electrical circuits, suggested that early misconceptions held by novices were still present in the reasoning systems of experts, but were inhibited. Neuroimaging data investigating the novice–expert distinction is not yet available. Meanwhile, it is comforting to learn that anyone can become an expert in certain domains if they are motivated enough to learn about them. For the brain, learning and memory are two sides of the same coin. Memory development thus depends on the depth of the knowledge base as well as on the use of explicit strategies such as rehearsal. High levels of expertise can even compensate for low levels of general intelligence in some memory tasks, such as recall tasks. The storage components of memory thus play a clear role in individual differences.

SUMMARY

This survey of some of the different types of memory that can be measured in children has shown that, while some memory systems develop with age, others do not. Little developmental change was found in recognition memory and in implicit memory, while large developmental changes were found to occur in episodic memory, autobiographical memory and working memory. It was argued that these developmental changes in episodic memory may in turn explain some of the developments seen in other memory systems, such as the decrease in suggestibility found in eye-witness memory and the decline in 'infantile amnesia'. It was noted that the increase in working memory capacity (memory span) seen with age may also result from the development of semantic memory, as the availability of a long-term memory representation of items seems to facilitate short-term memory performance. The *development* of short-term and long-term storage systems may thus be intimately linked, even though when measured in adults, the two systems appear to be distinct (Baddeley & Hitch, 1974).

Cognitive neuroimaging studies of episodic and working memory have multiplied in the last decade. However, the majority of these studies are not really developmental, in that they typically only document which neural areas are most active during certain memory tasks at certain ages. They do not explore the underlying neural mechanisms that enable the development of memory, nor document how these mechanisms change (or not) using longitudinal designs. Current neuroimaging studies provide convergence with adult data concerning the localisation of memory, for example with regard to the neural structures that are activated by different memory tasks, and have also recently provided some convergence regarding mechanisms, with both theta and alpha band oscillations related to working memory performance. Future cognitive neuroimaging studies of children that are more focused on underlying mechanisms seem likely to contribute unique data to understanding human memory. In principle, neuroimaging studies with young children could throw light on debates concerning implicit memory, on competence versus production deficiencies and on infantile amnesia. Meanwhile, structural fMRI studies can be relevant for assessing theories of memory. For example, the study of children with very small hippocampi suggested that semantic memory can develop normally without an intact hippocampus, whereas episodic memory cannot (Vargha-Khadem et al., 1997). Regarding functional fMRI, in earlier chapters the potential of multi-voxel pattern analysis (MVPA) for documenting the development of semantic memory was noted. Techniques such as MVPA could also potentially reveal the neural changes associated with the novice–expert distinction. MVPA could also help to map the growth and content of episodic memory and its possible restructuring with age, and may even throw light on infant and child amnesia.

CHAPTER 9

CONTENTS

Metacognition, reasoning and executive function

9

Metacognition is knowledge about cognition. Flavell (1979) was the first to propose that developing an awareness of one's own cognitive functioning might be important for cognitive development. Flavell defined metacognition as any cognitive activity or knowledge that takes as its cognitive object an aspect of cognitive activity. By this definition, metacognition encompasses factors such as knowing about your own information-processing skills, monitoring your own cognitive performance, regulating your own cognitive strategies in order to enhance your performance, knowing about the demands made by different kinds of cognitive tasks, monitoring the sources of your knowledge and developing a theory of mind. Some aspects of metacognition are research domains in their own right. These include 'metamemory' (knowledge about memory) and 'executive function' (the monitoring and self-regulation of thought and action, the ability to plan behaviour and to inhibit inappropriate responses). Clearly, the link suggested by Flavell between metacognition and cognitive performance is analogous to the links discussed in Chapter 4 between metarepresentation and social cognition ('theory of mind'). In Chapter 4, we saw that to develop metarepresentations, children need to develop the ability to take a representation itself as an object of cognition. In this chapter we will consider evidence that to develop metacognition, children need to develop the ability to take cognition itself as the object of cognition.

Whereas research exploring metarepresentational development has emphasised the critical role of communicative activities involving sharing objects with others ('shared intentionality'), metacognitive development has been studied more in terms of the important mechanisms operating within individual minds (such as metamemory and executive function). The focus in metacognition research is on developing reflective awareness of one's *own* cognition rather than of the cognitions of others. As summarised neatly by Schneider and Lockl (2002), metacognition research is concerned with what the child knows about his or her own mind. Theory of mind research is concerned with what the child knows about somebody else's mind. Nevertheless, the two areas of research are connected in important ways. This point has been made strongly by Kuhn (1999, 2000), who terms this general research field "meta-knowing". In an influential essay on the origins of metacognition, Wellman (1985) also emphasised the overlaps between metacognition and theory of mind, suggesting that metacognition consisted of a "large, multi-faceted theory of mind" (p. 29). In this chapter, I will focus on the development

KEY TERMS

Metacognition
Knowledge about cognition, including metamemory (knowledge about memory) and knowing the demands made by different cognitive tasks.

Executive function
The monitoring and self-regulation of thought and action, the ability to plan behaviour and to inhibit inappropriate responses.

of metamemory and executive function, and examine the effects of metacognitive development on reasoning and cognitive performance. I will also review a selection of studies from the growing neuroimaging literature on these topics.

METAMEMORY

Metamemory is knowledge *about* memory. Traditionally, the development of metamemory was defined as the development of the ability to monitor and regulate one's own memory behaviour (e.g., Brown, Bransford, Ferrara & Campione, 1983). This was expected to enhance performance. As children came to understand more about how their memories worked, it was expected that they should become more sensitive to the fact that certain memory tasks will benefit from particular strategies, and they should also become more aware of their own strengths and weaknesses in remembering certain types of information (Flavell & Wellman, 1977). Similarly, as metamemory develops, children should actively begin to use mnemonic strategies to improve their encoding and retrieval of information in memory. So as metamemory improves, production deficiencies in memory tasks (see Chapter 8) should decline.

There are a number of different types of metamemory knowledge that a child can acquire. These include knowledge of oneself as a memoriser, knowledge of the present contents of one's memory, and knowledge of task demands. Wellman (1978) argued that in addition to acquiring knowledge about these different metamemory 'variables', the child needs to realise that intentional memory behaviour is required ('sensitivity'). This kind of knowledge is more procedural in nature, requiring recognition that memory activity is necessary, and has been termed 'meta-strategic knowing', or knowing 'how' (Kuhn, 1999, 2000). Wellman and his colleagues researched the metamemory variables, which they defined in terms of three different areas: knowledge about tasks ("does this task require me to remember a lot of information, or a small amount?"), knowledge about persons (the child's mnemonic self-concept), and knowledge about strategies (such as the benefits of rehearsal for memory performance). All variables were expected to make overlapping contributions to metamemory. Schneider and Lockl (2002) outlined further aspects of metamemory highlighted by subsequent research. These included 'conditional metacognitive knowledge' (Paris & Oka, 1986), the ability to justify or explain your memory actions; and monitoring and self-regulation (Brown et al., 1983), encompassing the ability to select and implement memory strategies, to monitor their usefulness, and to modify them when necessary. Her analysis of metacognition led Brown to the view of the 'competent information processor', a child with an efficient 'executive' that regulated cognitive behaviours. This perspective highlights similarities between metamemory and executive function. Pressley, Borkowski and Schneider (1987) discussed the concept of the 'Good Information Processor', the child whose motivational orientation and general world knowledge enabled the *automatic* use of efficient learning strategies and procedures. Schneider and Lockl (2002) produced a taxonomy of metacognition summarising the field of behavioural constructs, which is reproduced in Figure 9.1.

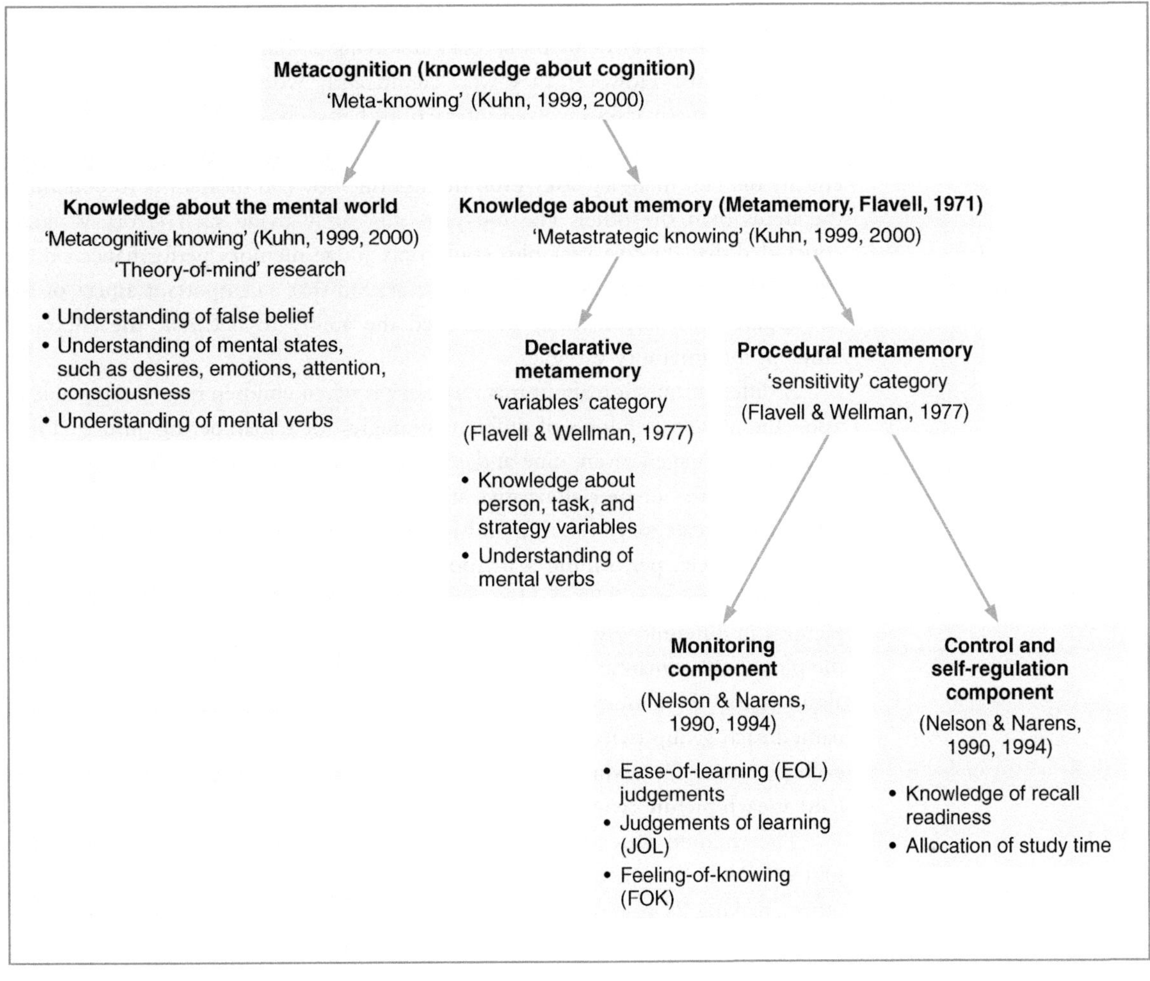

FIGURE 9.1
Schneider and Lockl's taxonomy of metacognition. From Schneider and Lockl (2002). Copyright © Cambridge University Press. Reproduced with permission.

Metamemory variables: Tasks, persons and strategies

The different aspects of metacognition outlined in Figure 9.1 have been studied largely independently of one another. For example, some research focused on the development of knowledge about metamemory 'variables'. Wellman (1978) gave five- and ten-year-old children pictures of different memory situations to assess as either easy or difficult. An example of an easy situation was a picture of a boy who had to remember three items, whereas an example of a difficult situation was a picture of a boy who had to remember 18 items. In addition, some of the situations measured children's understanding that two different variables could *interact* to determine memory difficulty. For example, a picture of a boy with 18 items to remember and a long walk during which to remember them (long recall time) was intended to be assessed as easier than a picture of a boy with 18 items to remember and a short walk during which to remember them (short recall time).

Wellman found that the five-year-olds were as good as the ten-year-olds in judging the difficulty of memory tasks involving a single variable (such as three versus 18 items). However, they were significantly worse than the ten-year-olds when the memory tasks involved interactions between variables. For example, the five-year-olds tended to predict that both boys with 18 items to remember had equally difficult memory tasks, even though one boy had more time to remember the items than the other. The ten-year-olds rarely made such errors. Wellman concluded that the five-year-olds could only judge memory performance on the basis of one of the relevant variables. He argued that an important aspect of the development of metamemory concerned the ability to *interrelate* the effects of different metamemory variables.

A different question about metamemory is when children first become able to assess the *relative* usefulness of different strategies for remembering. Justice (1985) asked children aged seven, nine and 11 years to make judgements about the relative effectiveness of four alternative strategic behaviours: rehearsing, categorising (by semantic category), looking and naming. The children first watched a video of a child, 'Lee', performing a memory task requiring the recall of a set of 12 categorisable pictures. The children were told that Lee would try to remember the pictures in different ways. For example, to demonstrate *categorisation*, Lee grouped the pictures by semantic category (e.g., apple, pear, banana) and named each group aloud twice. To demonstrate *rehearsal*, Lee grouped the pictures at random and named each group twice (e.g., truck, apple, hand). For *naming*, Lee simply named each picture twice without rearranging them spatially, and for *looking*, Lee stared hard at each picture twice without rearranging them spatially.

The children were then asked to make a series of judgements about which strategy would 'help Lee remember best'. Each strategy was paired with all of the others, making 24 paired comparison judgements in all. The seven-year-olds made no distinction between rehearsal and categorisation as the better strategy, but the nine- and 11-year-olds judged categorisation to be more effective than rehearsal (which was the case in Lee's scenario). Justice suggested that some metacognitive awareness of the usefulness of categorisation was present by at least nine years of age.

Justice and her colleagues also explored the contribution made by understanding how a strategy works to the efficacy of strategy use. Justice, Baker-Ward, Gupta and Jannings (1997) showed children aged four, six and eight years video scenarios of other students trying to remember material in a variety of circumstances. For example, the video might be of a child trying to remember some pictures, and using verbal labelling of the pictures to help him/her. Alternatively, the video might be of a child trying to remember ten things that had happened on vacation by looking at a photo album, in order to tell his/her teacher the following day. The child either labelled the events in order to aid recall ("tennis match", "museum", "shark") or looked silently at the pictures. The children viewing the videos were asked to comment on how the students shown were trying to remember, and how successful they were. Overall, four combinations of strategy and outcome were shown via these scenarios: labelling/high recall, labelling/low recall, no labelling/high recall, and no labelling/low recall. Prior to seeing the video scenarios, the children themselves received a picture recall task based on sets of ten pictures. The children were told that they could do anything they liked to help them to remember, and any strategies that they used were noted by the experimenters. The children were also asked what they did to try and remember, and why these strategies might work ("How does – work to help you remember?"). The aim in each case was to examine the relations between metacognitive understanding and recall.

When their own memories were tested (recalling ten pictures), the children participating in the study showed varying use of the labelling strategy, and varying success in remembering. In general, older children recalled more pictures than younger children, and labelling behaviour (which did not differ with age) was related to recall for the older children only. Analyses of the children's understanding of their own strategic behaviour showed that an important distinction was whether explanations given were mentalistic ("it helped get them in my mind") or non-mentalistic ("don't know"). Most children classified as non-mentalistic did not name labelling as a memory strategy (even if they had shown labelling behaviour) and continually replied "don't know" when asked how their behaviour had worked to help recall. Justice et al. (1997) found that recall was higher for the children who gave mentalistic explanations, irrespective of age. Thus labelling was a more effective strategy for children who *understood* the behaviour. The metamemory judgements made by the children of the students shown memorising in the video scenarios yielded similar findings. Children who gave mentalistic explanations of their own strategy use were more likely to demonstrate awareness of the causal relation between strategy use and performance as shown in the videos. Younger participants were in general unlikely to show awareness of this causal relation, but did show an understanding of the relationship between effort and success of recall. The data collected by Justice and her colleagues suggest that older children can make more subtle judgements about the effectiveness of various memory strategies.

Self-monitoring and self-regulation of memory

Another aspect of metamemory is 'self-monitoring', which is the ability to keep track of where you are with respect to your memory goals. A related aspect is 'self-regulation', which is the ability to plan, direct and evaluate your own memory behaviour. Both these aspects of metamemory involve executive function, and were initially studied together. More recently, the development of self-monitoring abilities has been studied in terms of ease-of-learning judgements, feeling-of-knowing and judgements-of-learning (see Figure 9.1). The development of self-regulation has been studied in terms of the different aspects of executive function (e.g., planning and directing behaviour). Clearly, adequate self-monitoring is necessary if self-regulation is to be successful.

An example of the integrated approach to studying children's ability to self-monitor and self-regulate their strategic memorial behaviour is a study by Dufresne and Kobasigawa (1989). They examined whether six-, eight-, ten- and 12-year-old children could distribute their study time efficiently between easy and hard material. In their experiment, children were given two sets of booklets of paired-associate items to study. One set of booklets contained 'easy' paired-associates, such as *dog-cat*,

KEY TERMS

Self-monitoring
An aspect of metamemory referring to the ability to keep track of where you are with respect to your memory goals.

Self-regulation
Self-regulation of memory is the ability to plan, direct and evaluate one's own memory behaviour.

Judgements-of-learning
Assessing one's learning both immediately after studying a list of items and after a delay of a few minutes.

bat-ball and *shoe-sock*. The other set of booklets contained 'hard' paired-associates, such as *book-frog*, *skate-baby* and *dress house*. The younger children were given fewer sets of booklets, to equate task difficulty across age. The children's goal was to remember all of the pairs perfectly, and they were allowed to study the booklets until they were sure that they could remember all of the 'partners'. The way in which the children allocated their study time was measured by videotaping the study period, and then timing the portions of each period spent in studying either the easy or the hard booklets.

Dufresne and Kobasigawa found that the six- and eight-year-old children did not differentiate between the easy and the hard booklets, allocating an equivalent amount of study time to each. In contrast, the ten- and 12-year-old children spent significantly longer studying the hard booklets, with the 12-year-olds spending a longer time on the hard pairs than any other group. This suggested that older children were better at self-regulating their memory behaviour. Dufresne and Kobasigawa then divided the children in each age group into those who spent 'more time on hard pairs' versus those who spent 'more time on easy pairs'. This showed that the eight-year-olds, too, had some ability to allocate study time. More eight-year-olds spent 'more time on hard pairs' than 'more time on easy pairs', and this pattern was even stronger for the ten- and 12-year-olds. Only the youngest group contained more children who spent 'more time on easy problems' than 'more time on hard problems', even though a separate test showed that the six-year-olds could differentiate between the easy and the hard booklets. Dufresne and Kobasigawa argued that the younger children lacked the *metamemorial* knowledge necessary to enable them to allocate more study time to the hard pairs. Although they were able to monitor problem difficulty, they did not use this knowledge to regulate their study time accordingly. The conclusion was that while self-monitoring might be available to younger children, younger children do not necessarily use the monitored information to improve their subsequent memory performance.

On the other hand, even some animals have been shown to be able to self-monitor their memories in implicit paradigms. For example, Hampton (2001) trained rhesus monkeys in a memory paradigm in which the monkeys viewed pictures on a computer screen, and then gained rewards if they recognised a previously experienced picture in a delayed memory test. The delayed memory test presented four images on any given trial, only one of which had been seen previously. To test self-monitoring, Hampton (2001) added a choice phase, in which monkeys were allowed to decide whether or not to take the recognition test on some trials. Accurate remembering was compared for freely chosen test trials versus forced test trials, in which monkeys were not given the option to refuse the memory test. Hampton found that the monkeys showed significantly better memory in the freely-chosen tests than in the forced tests. He argued that rhesus monkeys were able selectively to decline memory tests when they knew that they had forgotten.

Liu and her colleagues used a version of this task with four-year-olds (Liu, Su, Xu & Pei, 2018). Children were shown 216 photographs, half of which were 'easy' to remember as they depicted simple and concrete entities like dogs, and half of which were 'difficult' as they were kaleidoscope pictures. In each trial, children were shown four pictures sequentially, numbered 1 to 4. They were told to remember the number-picture pairings and were allowed to study the pictures for as long as they wished. Memory test trials followed each learning trial, in which the children were shown a number (e.g., '2'), and allowed to decide whether to go on with the memory test. For some trials, this choice was not offered, and the children were simply given the memory test. Liu et al. reported that children aged 4–4.5 years did not show better performance in the free choice trials compared to the forced choice trials, but that children aged 4.5–5 years in age did show better memory in the freely chosen tests. Accordingly, when an implicit measure of metamemory is used, older four-year-olds are able to monitor the contents of their own memories and also to use this information to improve their performance.

Dufresne and Kobasigawa's (1989) study, which tested recall for 'easy' and 'difficult' paired-associate items, found that older children were better at self-regulating their memory behaviour than younger children.

Ease-of-learning judgements

The use of implicit metacognitive measures suggests that asking young children for explicit predictions about their own memory performance might underestimate metamemory competence. In Chapter 8, we saw in Yussen and Levy's (1975) study that half of the four-year-olds tested predicted that they would remember all ten items in a standard memory span task. In fact, they recalled around three items. This aspect of self-monitoring is now termed *ease-of-learning judgements*, and can be studied via paradigms in which participants are asked to predict their own ability to remember lists or texts. Participants' actual ability to remember these materials is then tested, enabling precise measurement of the accuracy of performance prediction at different ages. The classic finding is that younger children are always worse in ease-of-learning judgement paradigms (e.g., Schneider, Borkowski, Kurtz & Kerwin, 1986). However, whether this is due to a difficulty in self-monitoring is unclear. One possibility is that younger children are more optimistic about their abilities, and that motivational factors like wishful thinking explain their unrealistic predictions. This possibility was investigated by Visé and Schneider (2000, discussed in Schneider & Lockl, 2002).

KEY TERM

Ease-of-learning judgements
Predicting one's own ability to remember items, lists or texts.

In their study, children aged four, six and nine years were asked to predict their own performance in either motor tasks (ball throwing and jumping) or memory tasks (memory span tasks or a hide-and-seek task). The children were asked to make these predictions in two conditions. In a 'wish' condition, they were asked to tell the experimenter the performance that they wished to achieve in the next trial. In an 'expectation' condition, they were asked to predict the performance that they expected to achieve in the next trial. After completing the different tasks, the children were also asked to assess their performance. This post-task assessment showed that children of all ages could monitor their performance accurately, in both the motor tasks and the memory tasks. However, in general the children did not differentiate between their wishes and their expectations. Only the nine-year-olds showed a differentiation, and only for the jumping task. Visé and Schneider concluded that overestimation in younger children was due to their belief in the causal efficacy of effort. Wishful thinking occurs in part because children believe that if they try harder, then they will be able to perform as they desire. This belief in the efficacy of effort declines as children get older, and so ease-of-learning judgements are a better index of self-monitoring behaviour in older than in younger schoolchildren.

Again, more recently implicit measures have been developed, which show that younger children do have some awareness of what they know, but that they overestimate their capabilities when asked for explicit judgements. Kim and her colleagues gave three- and four-year-old children a learning task in which toys were placed inside a box by an experimenter, either while the box was hidden behind a screen or while the box was visible to the child (Kim, Paulus, Sodian & Proust, 2016). In fact there were three conditions in the experiment: Full Knowledge, in which the child was shown the toy and watched it being placed in the box; Partial Knowledge, in which the child was shown two toys and told "I'm going to put one of these toys inside the box" and then the hiding event occurred behind the screen; and Ignorance, in which the children were not shown any toys and the hiding event took place behind the screen. Kim and her colleagues devised an implicit measure of the child's awareness of her own knowledge states, by asking "Max wants to know what's inside the box. Can you help him? If you do not want to tell him that's okay, I can tell him." This implicit measure was contrasted with an explicit measure, which was to ask the child "Do you know what's in the box, or do you not know?"

Kim et al. (2016) found that children's decision about whether to inform Max varied linearly with their level of knowledge, being highest for the Full Knowledge condition, intermediate for the Partial Knowledge condition and lowest for the Ignorance condition. This was the case for both the three-year-olds and the four-year-olds, with no age effect. For the explicit judgements, however, the four-year-olds were more accurate in assessing their own knowledge than the three-year-olds. Both groups tended to overestimate their knowledge in the Partial Knowledge condition, performing more accurately in the Full Knowledge and Ignorant conditions. Accordingly, Kim et al. (2016) concluded that young children are implicitly aware of their own knowledge.

Judgements-of-learning

Judgement-of-learning tasks provide another indication of children's ability to monitor their own memory performance. In judgement-of-learning tasks, which usually rely on paired-associate learning of pictures, participants are asked to assess their learning (a) immediately after studying a list of items, and (b) after a delay of a few minutes. In adults, the accuracy of judgements-of-learning is always better in the delayed condition, and judgements-of-learning on a trial-by-trial basis (e.g., asking participants for a judgement as they finish viewing each pair of pictures) are particularly inaccurate. Schneider, Visé, Lockl and Nelson (2000b) studied judgements-of-learning in children aged six, eight and ten years, using a paired associate learning task. In order to equate levels of recall across age, the six-year-olds were asked to recall lists of eight items (e.g., fork-mouse), the eight-year-olds were asked to recall lists of ten items and the ten-year-olds were asked to recall lists of 12 items. The children were told to try hard to learn the items, and were told that their job would be to recall the second picture in each pair when prompted with the first. In the immediate memory condition, children were asked for judgements-of-learning both immediately after studying each pair of pictures, and after finishing the list. Following ten minutes of unrelated activities, they were given the paired associate recall test. In the delayed memory condition, the children were asked for judgements-of-learning for each item about two minutes after studying the set of pairs of pictures, and were then asked for a judgement-of-learning for the whole list. They also received the paired associate recall test following ten minutes of unrelated activities.

Schneider et al. (2000b) found that the accuracy of judgements-of-learning was much higher for children in the delayed condition than for children in the immediate condition, at all ages. This shows that even kindergarten children display accurate self-monitoring as indexed by judgements-of-learning, as long as judgements are delayed (mirroring effects with adults). Differences in the accuracy of judgements-of-learning for individual trials versus for the entire list were also found, again across ages. Children tended to be over-confident in their trial-by-trial judgements-of-learning, and less over-confident when asked to make judgements-of-learning for the entire list. However, whereas the judgements-of-learning of the youngest children tended to overestimate actual performance, the judgements-of-learning by the eight- and ten-year-olds accurately reflected subsequent performance in the paired associate learning task. Schneider et al. argued that developmental effects in judgements-of-learning were negligible. Both children and adults overestimate their performance in immediate judgements-of-learning, and both children and adults are fairly accurate when judgements-of-learning are delayed. Self-monitoring is hence quite proficient even in young children, although age effects in self-monitoring may emerge if task difficulty is increased. Schneider et al. suggested that developments in self-regulation rather than in self-monitoring might explain developments in metamemory in children. Certainly, this conclusion would fit the results reported by Dufresne and Kobasigawa (1989), where self-monitoring was accurate in the younger children but self-regulation was not.

Feeling-of-knowing

Studies of feeling-of-knowing in young children in general converge with judgements-of-learning data, suggesting no significant developmental trends with age. There do not seem to be strong improvements in feeling-of-knowing accuracy as children get older. Rather, it seems that feeling-of-knowing can be dissociated from knowing per se. Feeling-of-knowing is based on the amount of information generated at *retrieval*, whether this information is correct or incorrect. In adults, feeling-of-knowing judgements are similar for correct information (information recalled that is accurate) and for errors of commission (information recalled that is inaccurate, leading to recall errors). Feeling-of-knowing judgements are much lower for errors of omission (information that is omitted, leading to recall errors). Similar effects appear to hold in children.

Lockl and Schneider (2002) investigated feeling-of-knowing in four groups of children aged respectively seven, eight, nine and ten years. The children were asked to give verbal definitions of words on the German version of the Peabody Picture Vocabulary Test (PPVT). The children were encouraged to continue until 30 words had been defined incorrectly (error of commission) or not defined (error of omission). Children were then trained on making absolute and relative feeling-of-knowing judgements. For absolute judgements, they were given words that had been defined either correctly or incorrectly, and were asked to rate how confident they were in selecting the correct picture for that word. For relative judgements, they were given a choice of two words, and asked to select the word for which it would be easier to choose the correct picture. Some pairs of words had both been defined correctly, some incorrectly, and for some one had been an error of commission and one an error of omission. Overall, for all ages, errors of omission were more frequent (64% of errors). The expectation was that stronger feeling-of-knowing judgements should be given to words that had previously shown errors of commission (incorrect definitions).

Lockl and Schneider then gave the children a second session in which absolute and relative feeling-of-knowing judgements were required for ten easy words and ten hard words. They found no age effects for either absolute or relative feeling-of-knowing judgements. Feeling-of-knowing judgements were relatively low for all of the children in the study, but there was no increase in accuracy with age. As feeling-of-knowing judgements are also relatively low in adults, Lockl and Schneider argued that even young children possess some knowledge about their mental system when making predictions about future performance.

Indeed, more recent studies suggest that children as young as 2.5 years have implicit access to whether they know something or not, even when they cannot verbalise this knowledge explicitly. Geurten and Bastin (2018) developed a memory paradigm that could be matched in difficulty for children aged 2.5 and 3.5 years, based on forced-choice recognition judgements for previously experienced pictures. In their study, children were shown pictures of familiar objects like flowers (see Figure 9.2). The pictures were presented as fragments, so that they were only partially identifiable. Prior ratings of these picture-fragments by other children aged 2.5 to 3.5 years was used to grade difficulty, so that task difficulty could be matched for the two age groups participating in the experiment. After viewing each picture-fragment, the children were given a forced choice between two complete pictures, and asked which one they had just seen. For example, the picture of a flower might be paired with a picture of a pinwheel in similar colours. Following their choice, children were asked if they were 'really sure' that they were correct, or 'not so sure' that they were correct, via a pictorial choice of a child who looked puzzled versus a child who looked confident. This comprised the explicit judgement of feeling-of-knowing. As an implicit measure of feeling-of-knowing, children were given the opportunity to ask for a cue to help them. The cue was another picture related either perceptually or conceptually to the correct answer, such as a picture of grass for the flower. If a child said that they wanted a cue to help them, this was coded as an indication that the child was unsure about her previous answer. Finally, children were given the opportunity to

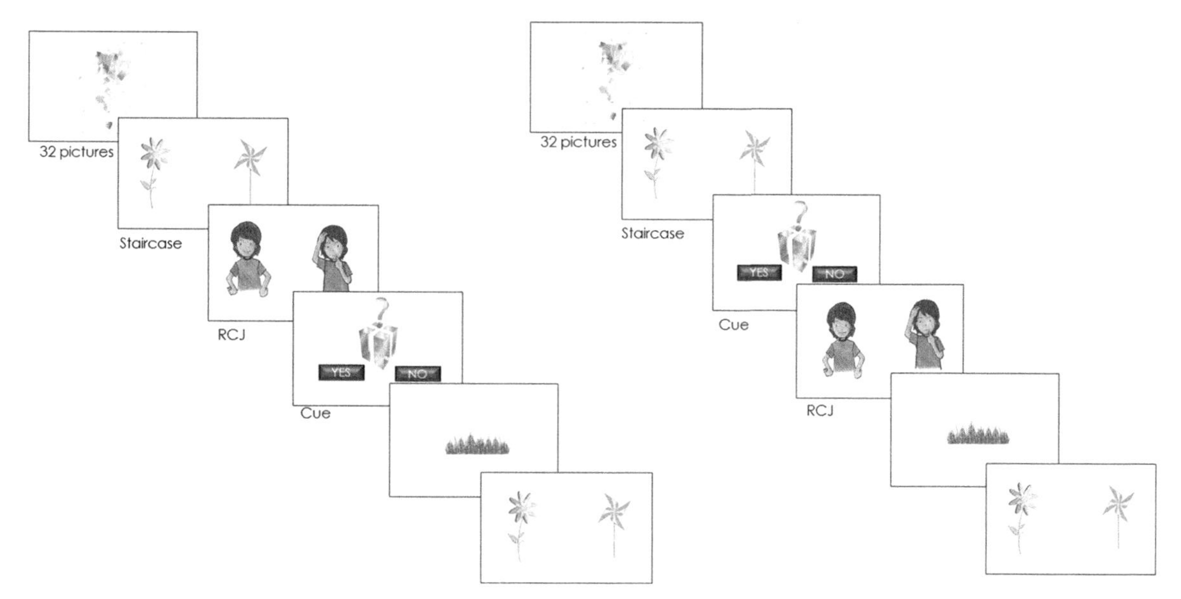

FIGURE 9.2 Examples of the picture fragments and forced choice solutions used by Geurten and Bastin (2018) to measure feeling-of-knowing judgements in children as young as 2.5 years. RCJ = retrospective confidence judgements.

change their original forced-choice judgement. However, over 90% of the time, children stayed with their original answer.

Geurten and Bastin (2018) found that response accuracy was the same for the 2.5 and 3.5 year old participants, confirming that task difficulty was equated for the younger and older children. For explicit judgements of knowing, both the younger and older children were more likely to select 'really sure' when their forced choice response had been accurate, although confidence in their correct judgements was significantly higher in the older children (76% versus 67%). For implicit judgements of knowing, however, there was no age effect. Children were more likely to ask for a cue after they had made an incorrect initial forced choice response than a correct initial forced choice response, at both ages. Geurten and Bastin then created an index of metacognitive knowing by computing a sensitivity measure based on signal detection theory for both explicit and implicit judgements. The sensitivity measure in each case took into account hits (the proportion of 'really sure' judgements when picture identification was correct) and misses (the proportion of 'really sure' judgements when picture identification was incorrect). The sensitivity index for explicit judgements did not differ from chance, but the sensitivity index for implicit judgements was significantly above chance, and this was the case for both younger and older children. Geurten and Bastin (2018) concluded that implicit metacognitive knowledge can guide task performance in children as young as 2.5 years of age.

Studies of *self-monitoring* in young children using measures of feeling-of-knowing, judgements-of-learning and judgements of ease-of-learning broadly suggest that self-monitoring by children is fairly accurate. Self-monitoring also seems present from surprisingly early in development, as indicated by more recent studies using implicit measures of the ability to monitor one's own cognitive performance. However, self-monitoring needs to result in strategy change in order to improve cognitive efficiency, and it has been suggested that children's ability to link self-regulation skills to information gained from self monitoring does show important changes with development (Schneider & Lockl, 2002). Self-regulation will be discussed in detail below, when we review studies of executive function.

Source monitoring

KEY TERM

Source monitoring
Attributing accurately the source or origins of one's memory, knowledge and beliefs.

Another aspect of metamemory is being able to attribute accurately the origins of one's memories, knowledge and beliefs. This is very important for cognitive performance, as inaccurate attribution of one's sources of knowledge can be a feature of mental illnesses such as schizophrenia (Simons, Garrison & Johnson, 2017). The ability to make accurate decisions about sources develops markedly between the ages of four and six years. For example, Bullock Drummey and Newcombe (2002) explored source memory in children aged four, six and eight years by teaching them some new facts. An example of a new fact was "What animal cannot make any sounds?" "A giraffe cannot make any sounds". Ten new facts of this kind were taught to the children, five new facts by an experimenter and five new facts by a puppet. A week later, the children were seen again, and the ten facts learned previously were presented along with five novel facts of equal difficulty, and five facts expected to be already part of the child's knowledge (learned outside the experiment). For each fact, the children were asked how they knew that. If they could not answer, they were given a forced-choice recognition test asking them whether they knew this because of their parents, a teacher, the puppet or the experimenter. Bullock Drummey and Newcombe (2002) reported that four-year-olds correctly recalled 24% of the sources of the facts, compared to 47% and 40% for the six- and eight-year-olds. The younger children were also much more likely to completely forget that they learned new facts as part of the experiment. The four-year-olds falsely attributed the sources of their knowledge to their parents, the TV or a teacher 60% of the time. In contrast, the majority of errors made by the older children were to forget whether it was the puppet or the experimenter who had told them the new fact. Bullock Drummey and Newcombe argued that the younger children were displaying 'source amnesia', which is also found in patients with frontal lobe dysfunction.

Ruffman, Rustin, Garnham and Parkin (2001) found higher levels of source monitoring in eight-year-old children in a paradigm using shorter delays. In their study, children aged six, eight and ten years were shown a video about a dog named Mick. They then listened to an audiotape that repeated some of the events seen in the video, and also added some new ones. The children were then asked immediately about what had happened (e.g., "What do you think? Did the newspaper boy walk down the path to the house?"). They were also asked source questions about their knowledge ("What do you think? Did the newspaper boy walk down the path to the house in only the tape, only the video, in both or in neither?"). Ruffman et al. reported that sources were identified correctly on 41% of occasions by the six-year-olds, 62% of occasions by the eight-year-olds and 70% of occasions by the ten-year-olds. They pointed out that the difficulties shown by younger children in attributing the sources of their memories had implications for eye-witness testimony (see Chapter 8). Children who cannot recall the sources of their memories may be more vulnerable to suggestibility.

In an experiment using more naturalistic events, Sluzenski, Newcombe and Ottinger (2004) asked four-, six- and eight-year-olds about the sources of their memories following either a delay of a week or immediately. The naturalistic events were planting a seed, making a pudding, unpacking a picnic basket and decorating a birthday invitation. Each event took around five minutes. The children performed two of the events with the experimenter, and imagined performing the other two events while participating in a script narrated by audiotape (e.g., "Imagine that you are going to plant a watermelon seed. Imagine that directly in front of you… there is a brown pot to plant the seed in, a bag of dirt, a pair of grey gloves…"). When the children returned a week later, they were asked questions about the events such as "Did you actually plant a seed, or did you just imagine planting a seed?" The four-year-olds showed accurate recall of sources for 66% of the events, compared to almost 100% accuracy for the two older groups of children. In a second experiment when source monitoring was tested immediately, over 90% of the four-year-olds could recall whether an event had been real or imagined.

It seems likely that source monitoring develops between the ages of four and eight years, and shows variation depending on the nature of the material to be remembered and its salience to the child. Younger children have more difficulty in keeping track of the sources of their memories than older children. Recent studies in adult cognitive neuroscience have linked source monitoring to activity in the medial prefrontal cortex, particularly the paracingulate sulcus (Simons et al., 2017). A number of developmental scientists have also suggested that developments in source monitoring are related to developments in executive function (e.g., Ruffman et al., 2001; Bullock Drummey & Newcombe, 2002). For example, it may be that the ability to identify accurately the sources of one's memories requires the ability to inhibit knowledge about inaccurate sources.

Metamemory and memory efficiency

Finally, an important question about metamemory considered here is whether children with better metamemories perform better in different memory tasks. After all, if the development of the ability to monitor and regulate one's own memory behaviour plays an important role in the development of memory per se, then children who are better at self-monitoring and at self-regulation should also remember more. Meta-analyses of the relationship between measures of metamemory and memory behaviour suggest that the relation is indeed a significant one. In a meta-analysis of some of the key empirical studies of this connection (27 studies and 2,231 participants), Schneider (1985) reported an overall correlation of 0.41 between metamemory and memory performance. Schneider and Pressley (1989) reported an identical correlation in an even larger meta-analysis (of 60 publications and 7,079 participants).

Individual studies suggest the same conclusion. As we saw in Chapter 8, memory-metamemory relationships can be demonstrated even in preschool children (Schneider & Sodian, 1988). In Schneider and Sodian's hide-and-seek task involving hiding people in wooden houses (the doors of which could carry a semantic reminder cue, such as a picture of a tractor for a house containing a farmer), a significant relationship between conscious awareness of the usefulness of the semantic cues and successful memory performance was found in both four-year-olds and six-year-olds. Similarly, Kurtz and Weinert (1989) showed that older German children who scored highly on a general test of cognitive ability had more metacognitive knowledge (e.g., about the usefulness of clustering by semantic category for recall) than children who scored at average levels. The high-ability children also recalled more words in a list memory task in which the words could be clustered according to categories (e.g., emotions). The relationship between metamemory and memory performance also seems to be bidirectional (see Schneider & Bjorklund, 1998). Metamemory influences behaviour, which in turn leads to improved metamemory. For example, a child's experience of the benefits of using a particular strategy, such as clustering words by semantic category, improves their memory performance and adds to their task-specific metamemory as well. A large-scale study of over 600 nine- and ten-year-old children showed that individual differences in metamemory explained a large proportion of the variance in recall (Schneider, Schlagmuller & Visé, 1998). Hence as meta-knowledge about memory develops, memory performance is indeed enhanced.

Cognitive neuroimaging studies of metamemory

There are a large number of neuroimaging studies relevant to the development of metamemory in children, typically using either EEG or fMRI. This imaging literature is also notable because typically-developing children are those most frequently studied. However, despite gathering a wealth of data, to date the field has not gone beyond reporting brain–behaviour correlations. Accordingly, at the time of writing nothing specific can be concluded about the neural mechanisms that may contribute to metamemory development. Therefore, I will only give a few examples of such studies.

EEG studies appear to confirm the conclusions reached by behavioural researchers, that self-monitoring by children is fairly accurate even in young children and does not show significant development. In research with adults, an EEG marker has been discovered that appears to relate to performance monitoring. This is an ERP labelled 'error related negativity' or ERN. In adults, the ERN typically occurs during speeded response tasks when the adult commits an error. The ERN peaks rapidly following the error, between 50 and 100 ms after making the error, and is stronger over frontal and central electrode sites. While the ERN is a robust and reliable brain response, clearly visible in the majority of studies, the mechanisms underpinning the ERN are not understood. The ERN may reflect the participant evaluating their action, or may reflect monitoring of ongoing performance, or may also reflect another form of cognitive control (the ERN is thought to originate in the anterior cingulate cortex, which is typically activated when attention must be allocated). However, the fact that the ERN can be evoked reliably makes it amenable to developmental studies of performance monitoring.

There are quite a few studies of the ERN in children, and reviews of the literature suggest that the amplitude of the ERN increases with age (Tamnes, Walhovd, Torstveit, Sells & Fjell, 2013). Another EEG marker of performance monitoring, the error positivity or Pe, typically follows the ERN after 200–400 ms. This ERP is also not well-understood, but can be reliably elicited and is thought to be independent of the ERN. The Pe also seems to increase in amplitude with age. For example, in a study assessing both the ERN and the Pe in a large sample of 326 children aged from five to seven years, Torpey, Hajcak, Kim, Kujawa & Klein (2012) recorded EEG while children performed a simple go/no-go paradigm. The children watched a screen on which a green triangle would appear in each trial, pointing either upwards, downwards, or to the left or right. The children were told to press a button as fast as they could when the triangles pointed upwards only. After a series of practice trials with feedback, the experimental trials were administered. Children were told that they could win $5 if they earned enough points by pressing the button quickly and accurately. In fact, all children got the $5 at the end of the experiment.

Torpey and her colleagues reported that the ERN and the Pe were reliably elicited by this paradigm in these relatively young children. The amplitude of both responses was also significantly related to age, with older children showing larger amplitudes for both ERPs. The older children were also faster and more accurate in the task than the younger children. Torpey et al. (2012) suggested that both the ERN and the Pe were neural markers of more efficient performance monitoring. However, whether amplitudes are greater in older children because older children are faster and more accurate in the task, or whether greater amplitudes index better performance monitoring, cannot be established from these correlational data.

Grammer and her colleagues were able to show both an ERN and a Pe in children as young as three years, using a different go/no-go task based on helping a zookeeper to capture escaped animals (Grammer, Carrasco, Gehring & Morrison, 2014). Children aged three to seven years (N = 95) played a computer game in which images of animals appeared consecutively onscreen. The child's job was to catch all the escaped animals by pressing a button except for the orangutans, who were helping the zookeeper. Accordingly, children had to inhibit pressing the button each time an orangutan was pictured. EEG was recorded over eight blocks of 40 trials each. Grammer et al. were able to record an ERN on the incorrect response trials for even their youngest participants. All children also showed a reliable Pe. Brain–behaviour correlations were observed for the size of the change in ERN amplitude and accuracy, and for the size of the change in Pe amplitude and accuracy. However, over this age range of 3–7 years, the amplitude of the ERN did not increase with age, even though older children were overall faster and more accurate in performing the go/no-go.

Indeed, it is worth noting that the ERN can even be measured in 12-month-old infants (Goupil & Kouider, 2016). Goupil and Kouider reported that the amplitude of the ERN was greater for trials in which infants made incorrect responses in a looking task. These data suggest that the ERN is an automatic brain response related to error processing. Although these neural data are consistent with behavioural data showing little developmental change in error monitoring, the fact that such young infants show an ERN makes it unlikely that this ERP is an index of *conscious awareness* of having made an error.

By contrast, behavioural data regarding source monitoring by children did suggest important developmental changes between four and eight years, although with some variability depending on the nature of the material to be remembered and its salience to the child. Source monitoring in children has also been studied using EEG. For example, Rajan and Bell (2015) taught children aged six and eight years some novel facts from two different sources, using the Bullock Drummey and Newcombe (2002) paradigm discussed earlier. The children were taught novel facts like "What animal cannot make any sounds?" "A giraffe cannot make any sounds". However, rather than a puppet being one source of the novel information and an experimenter being the other source, two different experimenters served as the sources. Following learning of the new facts, EEG recordings were taken during both fact recall and source recall. The eight-year-olds were better than the six-year-olds for fact recall (86% correct versus 61% correct), but not for source recall (both groups recalled about 60% of sources correctly). Theta band power at frontal, temporal and parietal sites was found to be higher in both groups during both kinds of recall compared to baseline, but otherwise no age differences were present. The finding that theta power did not vary with the better performance of the eight-year-olds for fact recall makes this finding rather uninformative. Further, increases in theta power compared to baseline were not tested in a control paradigm (perhaps theta power would have increased in a paradigm that wasn't measuring memory as well). Accordingly, it is difficult to interpret these brain-behaviour correlations in terms of children's memory development.

As a final example of EEG studies in the area of metamemory, Riggins, Rollins and Graham (2013) measured source memory in children aged five to six years while recording EEG. They used a paradigm based on novel toys, in which the children experienced 30 toys in one distinctive room manned by a distinctive doll and 30 more toys in a second distinctive room manned by a second distinctive doll. In each case, the child was asked to imitate an action associated with the toy. As will be recalled, the same paradigm was used by Riggins et al. (2016) to measure the neural networks active during episodic memory in an fMRI study discussed in Chapter 8. The children participating in Riggins et al. (2013) returned to the lab a week later, and were shown pictures of the previously experienced toys along with 30 pictures of new toys. EEG was recorded while the children passively viewed the pictures. Following the EEG recordings, children were presented with each item and were asked to judge whether it was novel, and were then asked to judge which room it was from (source memory).

To explore ERP markers of source monitoring, Riggins et al. computed two ERPs, the Nc and the positive slow wave (PSW) used in infant categorisation tasks (such as Quinn et al., 2006; see Chapter 5). They then compared the amplitude of the Nc and PSW in source-correct trials, source-incorrect trials and correct rejection trials. Riggins and her colleagues reported that the amplitude of both the Nc and the PSW was reliably different in each condition. Further, greater amplitude of the PSW over the left temporal lobe was related to better source memory performance by the children. Again, however, this brain–behaviour correlation does not help us to understand mechanisms underpinning the development of source memory in children. For example, greater amplitude may be related to stronger memory traces rather than to better source monitoring.

Regarding fMRI, fewer studies have focused on performance monitoring per se. However, a study utilising a longitudinal design (rather than the more typical cross-sectional fMRI design) with children aged 8–9 years and 10–12 years was reported by Fandakova et al. (2018). Fandakova and her colleagues studied children's ability to monitor their own retrieval failures using the 'scene' task discussed in a cross-sectional episodic memory study reviewed in Chapter 8 (Sastre et al., 2016). In the task, children were asked to memorise pictures of items on one of three scene backgrounds, a farm, a park or a cityscape. Participants were subsequently required both to decide if presented items were novel or familiar, and to indicate which scene the familiar items belonged to. In Fandakova et al.'s study, however, the participants were also asked to indicate if they were not sure which scene was correct. This enabled the experimenters to index a 'feeling-of-knowing', by categorising source memory as correct, incorrect or unsure. The task was given on two occasions approximately 18 months apart, and fMRI was recorded. The accuracy of source memory was found to improve with age, both when the different age groups were compared at time 1 and, importantly, when the children were compared to their younger selves. Fandakova et al. then inspected the fMRI data relevant to recollection failure, by using the 'not sure' responses compared to the incorrect responses to focus on uncertainty monitoring. As will be recalled, behavioural studies on 'feeling-of-knowing' judgements typically do not find age-related improvements, with even adults being relatively inaccurate.

The fMRI analyses showed that neural activity in the anterior insula (an area of the anterior frontal cortex) was enhanced for the uncertainty monitoring contrast, but for the older children only. Fandakova et al. argued that activation in the anterior insula was thus related to the decision to report uncertainty. Greater neural activity was also observed in the posterior parietal cortex correlated with uncertainty monitoring, but here activation did not vary with age. When the 8–9-year-old children were tested again 18 months later, their activation profiles in anterior prefrontal cortex were similar to the older children at test 1. The degree of signal change in activity in anterior prefrontal cortex over time was also significantly correlated with improved source memory. The longitudinal study design shows that this particular correlation between brain and behaviour does relate to developmental improvements in source monitoring, even though it does not throw any light on the neural mechanisms involved. Fandakova et al. concluded that changes in activation in anterior frontal cortex structures have important functional implications for memory development. Nevertheless, the nature of these functional implications remains to be discovered. Future developmental neuroimaging studies, perhaps including oscillatory analyses (see Chapter 8), may enable more specific understanding of the neural mechanisms that underpin individual differences in metamemory development between children.

METACOGNITION AND EXECUTIVE FUNCTION

KEY TERM

Inhibitory control
The ability to inhibit responses to irrelevant stimuli while pursuing a cognitively represented goal.

Historically, research on metacognition has taken a number of largely unrelated forms. As will be recalled from the last chapter, the third component of working memory hypothesised by Baddeley and Hitch (1974), namely the central executive (see Chapter 8), was proposed to play a central role in cognition via planning and monitoring cognitive activity. The central executive was thus responsible for the top-down modulation of cognitive processes. However, the central executive is not the same as metacognition, as Baddeley and Hitch limited its role to monitoring the sub-processes of working memory (the phonological loop and the visuo-spatial sketchpad). A second aspect of the study of metacognition concerned the development of 'executive function', thought to be dependent on the maturation of frontal cortex. Executive function was also thought to involve the modulation of cognitive processes in a top-down manner, but defined very broadly. This aspect of metacognitive research developed out of studies with brain-injured patients with damage to frontal cortex, who exhibited poor strategic control over behaviour (Milner, 1964). Similarities between the perseverative behaviour of these patients and of infants making search errors like the A-not-B error were noted in Chapter 2. A third aspect of metacognition research concerned inhibition behaviours in young children. Young children are typically poor at inhibiting inappropriate behaviours. Research on the possible effects of poor inhibitory control on cognitive development, and on the importance of individual differences in inhibitory control more generally, became widespread. These different strands of research have now merged into the field of executive function and metacognition. An organising construct in this field is the assumption that younger children have inadequate strategic control over their mental processes (poor executive function), and that as they gain meta-knowledge about their mental processes (metacognition, or the ability to reflect on their own cognition), strategic control improves. In general, the field is concerned with the development of conscious control over thought, emotion and action. Today most theories of executive function are based on three core factors, inhibition, working memory and cognitive flexibility (Diamond, 2013). These three factors reflect the research origins of the field.

Patients with frontal cortex damage and executive function in children

One reason for the unification of these historically separate aspects of research on metacognition was the development of cognitive neuroscience as a field. The frontal cortex turned out to be selectively activated in studies of working memory, in studies of strategic control over behaviour and in studies of the inhibition of inappropriate behaviours. The frontal cortex takes a long time to develop, with important structural changes still taking place in adolescence and young adulthood. An obvious question therefore was whether young children, who are not very good at strategic control nor at inhibiting inappropriate behaviours, would show the same kind of 'executive deficits' exhibited by patients with damage to frontal cortex. Accordingly, different behavioural measures were developed to explore this question in young children.

A typical 'executive error' seen in adults with frontal cortex damage is perseverative card sorting. As discussed earlier (regarding Diamond's analysis of A-not-B search errors in infants, see Chapter 2), if a frontal patient has been sorting a pack of cards according to a particular rule (e.g., colour) and the sorting rule is changed (e.g., to shape), then the patient finds it very difficult to change his sorting rule, and continues to sort the cards according to colour. However, at the same time as making these consistent sorting errors, the patient tells the experimenter "this is wrong, and this is wrong..." (Diamond, 1990). It is as though the patient's behaviour is under the control of their previous action (they are unable to inhibit the 'prepotent' tendency to search by the old rule), rather than under the control of their conscious intent. Clinical measures of card sorting behaviour, such as the neuropsychological test called the Wisconsin Card Sorting Test, typically require the patients to participate in many sorting trials, with a shift in sorting principle after each block of ten trials. Patients with frontal lesions are found to make significantly more sorting errors and to achieve significantly fewer shifts than control participants in the Wisconsin Card Sorting Task. Pennington (1994) has argued that 'executive errors' occur when behaviour is controlled by salient features of the environment, including prior actions, rather than by an appropriate rule held in mind. Executive errors are focused on cognitive inflexibility and perseveration, and will be discussed in detail below.

The role of inhibition in metacognitive development was first raised by Dempster (1991). Dempster argued that intelligence could not be understood without reference to inhibitory processes in the frontal cortex, and that individual differences in inhibitory processes could provide an index of the efficiency of frontal cortex in different individuals. He suggested that the critical aspect of many 'frontal' tasks like card sorting was that these tasks required the suppression of task-*irrelevant* information for effective performance. For example, frontal patients need to suppress the 'colour rule' when asked to begin sorting the cards on the basis of the 'shape rule'. Impaired inhibitory functioning could thus also be an explanation for children's difficulties in the strategic control of their behaviour. Behavioural research on inhibitory control by children suggested a particular role for representational conflict rather than simple response inhibition (e.g., Hughes, 1998; Carlson & Moses, 2001). Inhibitory control is also discussed in more detail below.

The role of working memory in metacognition was first raised by information-processing theories of cognitive development, although these theories did not address metacognition directly. Rather, these theories explored the general premise that children became capable of more sophisticated kinds of information-processing with development because of changes in working memory, and that this change in the processing components of working memory was the major factor in explaining developments in children's cognition. The dominant theories agreed that the amount of processing *capacity* that was available to the child changed with age, and that capacity increases caused developments in children's cognition. These so-called 'neo-Piagetian' theories of cognitive development were effectively theories about the role of working memory in children's improved cognition with age.

Neo-Piagetian theories invoked the brain at a descriptive level, typically assuming that cognitive development was driven by increases in the 'capacity' of the brain. These theories have largely fallen out of favour with the advent of developmental cognitive neuroscience. For example, Pascual-Leone (1970) argued that neural processing space increased with age, and that as processing space ('central computing space') increased, so did the cognitive abilities of the child. A related notion was advanced by Case (1985), whose model included a *trade-off* between 'short-term storage space' (a retention component) and 'operating space' (a processing component). Overall capacity ('executive processing space') was not thought to change with development, but the amount of available capacity was thought to increase as processing became more efficient (for example, via practice) and consequently took up less 'space'. A third neo-Piagetian model of cognitive development was proposed by Halford (1993), whose model centred on the capacity of 'active' or 'primary' memory. Halford defined primary memory as the memory system that held any information that was currently being processed, and he argued that the capacity of this system increased with development.

The common element central to these neo-Piagetian theories is the notion of the 'processing capacity' of the brain. Neo-Piagetian theories suggested that the size of available processing capacity placed an upper limit on children's cognitive performance, and that specifiable biological factors (as yet unknown) regulated the gradual shift in this upper limit with age. Cognitive development was explained by older children having more processing capacity than younger children, and qualitative improvements in cognitive performance were predicted with increasing age. These qualitative improvements were generally expected to arise from the use of more sophisticated information-processing strategies, which became available once increased processing capacity was present. For example, Halford (1993) argued that increases in processing capacity in working memory enabled children to use relational mappings of greater complexity. Case (1992) argued that the efficiency of working memory enabled the growth of reflection, which then made possible other cognitive advances (Carlson, 2003). As we saw in Chapter 8, however, neuroimaging studies of working memory capacity show that increases in capacity (measured outside the scanner) are correlated with increased neural activation levels in certain structures rather than with increasing amounts of activated cortex (e.g., Klingberg et al., 2002). Indeed, for younger children changes in working memory capacity are correlated with white matter integrity rather than with cortical thickness (see Chapter 8; Bathelt et al., 2017). Accordingly, the metaphor of unspecified biological factors enabling increased 'capacity' in the brain, which then acts as a driver for cognitive development, has not been supported by current neuroimaging studies.

KEY TERM

Dimensional Change Card Sort (DCCS) task
Card sorting task that tests the ability to sort cards by different rules or dimensions. A key feature is how children perform immediately after a rule or dimension has been switched.

Cognitive flexibility and executive function: Behavioural data

In a large and systematic programme of research, Zelazo, Frye and their colleagues have investigated the development of executive function via the Dimensional Change Card Sort (DCCS) task. Using this task, they have shown that three- to four-year-old children can experience considerable difficulty in rule shifting tasks, just like frontal patients (see Zelazo & Muller, 2010, for an overview). In a now-classic study, Frye, Zelazo and Palfai (1995) introduced the 'shape' and 'colour' sorting games. They asked children aged three, four and five years to sort a set of cards into two trays on the basis of either shape or colour. Each card depicted a single shape, a red triangle, a blue triangle, a red circle or a blue circle. To play the 'colour game', the children were told that all the red ones went into one tray, and all the blue ones went into the other tray. The instructions were quite explicit: "We don't put any red ones in that box. No way! We put all the red ones over here, and only blue ones go over there. This is the colour game." After training in *both* games (colour and shape), the children were tested for their ability to shift their sorting strategy (e.g., from colour *to* shape) over three sets of five trials. On the first set of five trials, the children were given test cards to sort according to colour. Having sorted this set, they were then told "Okay, now we're going to play a different game, the 'shape game'. You have to pay attention." Five new test cards were then presented to sort by shape. Finally, five consecutive switching trials were administered, during which the children had to sort according to a new rule (shape or colour) on each trial. Again, the instructions were quite explicit: "Okay, now we're going to switch again and play a different game, the colour game. You have to pay attention."

Frye et al. found that the three- and four-year-olds in their study experienced great difficulty in shifting their sorting strategy on the second set of five trials, despite the explicit instructions from the experimenter. They typically sorted the cards correctly for the first five trials, and then continued to sort by colour (or shape) for the second five trials, perseveratively using the wrong rule. Performance on the final set of five trials, which required consecutive switching, was at chance level. In contrast, the older children (five-year-olds) were able to switch sorting rule in the second set of five trials, and were also able to switch sorting rule on a trial-by-trial basis during the last set of five trials. They did not show the 'executive failures' characteristic of the younger children.

In order to make sure that the difficulties of the three- and four-year-olds were not due to the use of abstract and perhaps unfamiliar geometric shapes, Frye et al. carried out a similar card sorting experiment using pictures of red and blue boats and red and blue rabbits. The children were again required to sort the cards first according to one rule, then according to a second rule, and finally to alternate between the two rules on the last set of trials. Essentially the same results as with the geometric shapes were found, although the four-year-olds proved to be better at switching their sorting rule with the more familiar dimensions of boats and rabbits. Frye et al. also checked that the children's difficulties were not due to the use of the dimensions of shape and colour per se, by devising card sorting tasks based on rules about number and size. Again, essentially the same results were found, with three- and four-year-olds showing difficulties when required to switch their sorting rule.

On the basis of these findings, Frye et al. (1995) argued that children become able to make a judgement on one dimension while ignoring another between three and five years of age. Frye et al.'s results can also be explained by arguing that children cannot inhibit a prepotent tendency to use the pre-switch rule at will, even though the post-switch rule is known to them. In order to test this 'executive failure' interpretation of Frye et al.'s results, we need evidence that the post-switch rules are indeed available to the younger children.

This evidence has been provided in a replication of the original card sorting study using shape and colour rules with pictures of flowers and cars (Zelazo, Frye & Rapus, 1996). In this replication, Zelazo et al. found that 89% of the three-year-olds who failed to use the new rule on the post-switch trials could verbally report the new rule. For example, when asked "Where do the cars go in the shape game? Where do the flowers go?" the children would point to the correct boxes, and then immediately sort according to the colour rule when told "Play the shape game. Here's a flower. Where does it go?" Similar results were found even when the children received only one pre-switch trial. These findings suggested that three-year-olds fail to use post-switch rules despite knowing these rules and despite having insufficient experience with pre-switch rules to build up a habit or perseverative tendency to sort on one dimension only. In a later study in which the children watched puppets sorting the cards, very similar results were found (Jacques, Zelazo, Kirkham & Semcesen, 1999). The crucial point about making the puppets do the card sort is that the children no longer need to inhibit a prepotent motor response. They can simply judge whether the puppets are playing the game correctly. Jacques et al. (1999) found that the three- to four-year-olds that they tested consistently judged the puppets to be playing correctly when they perseverated on the pre-switch rule. When the puppet sorted correctly, the children judged him to be wrong. Jacques et al. argued that the children clearly had difficulty in formulating what actions should be done when the rule was switched. These data support the view that the important factor underlying children's behaviour is *representational conflict* rather than response inhibition.

Kirkham, Cruess and Diamond (2003) argued that the reason that three-year-olds had such difficulty with the DCCS task was that they were cognitively rather rigid. Having focused their attention on one aspect of the cards in order to sort them, they experienced great difficulty in redirecting their attention to focus on what was relevant once the rules had changed. Kirkham et al. noted that even adults have difficulty in rule-switching tasks, but their difficulty is shown by elongated reaction times rather than by sorting errors (Monsell & Driver, 2000). Kirkham et al. then argued that children might be helped to refocus their attention if they were asked to relabel the cards before sorting them. In their experiment, three-year-olds received the standard colour and shape sorting tasks, but prior to sorting each card post-switch, the child was asked "What colour is this one?" or "What shape is this one?" Hence the child rather than the experimenter labelled the relevant dimension post-switch. This led to 78% of the three-year-olds tested performing correctly after the rule had been switched, compared to 42% in the standard procedure. Kirkham et al. also devised a way of making it more difficult for older children (four-year-olds) to exercise inhibitory control over their sorting. This was done by leaving the cards face-up in the sorting boxes, so that the previously-correct dimension was very salient. In this 'face-up' condition, only 57% of four-year-olds sorted the cards correctly post-switch, compared to 92% in the standard condition. Kirkham et al. (2003) argued that the critical feature of the DCCS was to inhibit a mindset that was no longer relevant. Accordingly, younger children have difficulty in shifting their attentional focus flexibly between *conflicting* representations. Note that this explanation shares features with some theoretical explanations regarding social-cognitive development, as discussed in Chapter 4. For example, Wimmer and Perner (1983) originally argued that younger children might not yet be able simultaneously to hold two representations in mind when analysing developmental changes in performance in false belief tasks.

In contrast, Zelazo and colleagues prefer to explain the patterns of performance observed in the DCCS as arising from an inability to represent *if–then* rules (e.g., Zelazo & Frye, 1997). Older children are thought to be able to handle *if–then* rules of increased complexity. Complexity depends on the number of levels of embedding in a rule system. For example, if two rules apply to the same stimulus, as in the DCCS, then a new degree of embedding is required. During the first sort, the rules are simple. If the child is sorting by colour, the rules are "if red, place there" and "if blue, place there". During the second sort, however, the new post-switch rules "if rabbit, place there", and "if boat, place there" must be used. Zelazo, Muller, Frye and Marcovitch (2003) argued that only older children could reflect on the incompatibility of the two sets of rules and generate a higher-order rule "If we're playing by colour, then... If we're playing by shape, then...". Age-related changes in managing *if–then* complexity were thought to depend on changes in reflection (metacognition). Despite the differences in explanation that have been offered for DCCS performance, there is nevertheless widespread agreement that the task provides a good measure of executive function in young children.

Zelazo and his colleagues have also argued for a distinction between 'cool' and 'hot' executive function (Zelazo & Muller, 2010). 'Cool' executive function refers to 'purely cognitive' tasks, whereas 'hot' executive function involves making decisions about events that have emotionally significant consequences. In adults, affective decision-making is typically studied using gambling tasks. For example, participants are told that they can win money by choosing cards from particular decks. Typically, cards from disadvantageous decks provide higher initial gains, but are associated with much larger losses over the course of the game. Cards from advantageous decks provide lower gains but also much lower losses, and are hence more rewarding overall. In a classic study, Bechara, Damasio, Damasio and Anderson (1994) showed that adult participants typically prefer the disadvantageous decks at the outset of the game, but switch to preferring the advantageous decks as the game continues. Adult patients with lesions to the orbitofrontal cortex keep selecting cards from the disadvantageous decks. These and other studies with adults suggest a link between the orbitofrontal cortex and affective decision-making. Accordingly, Kerr and Zelazo (2004) set out to study 'hot' executive function in preschoolers.

To do so, they invented a gambling task suitable for children aged three and four years. This task was also based on decks of cards, but the cards depicted either happy or sad faces (see Figure 9.3). The children won sweets (M&Ms) for each happy face, and lost sweets for each sad face. Only two decks of cards were used. Cards in the disadvantageous deck depicted two happy faces and either zero, four, five or six sad faces. Cards in the advantageous deck depicted one happy face and either zero or one sad faces. Over the game (50 trials), it was hence more advantageous to select cards from the advantageous deck. The children were taught how

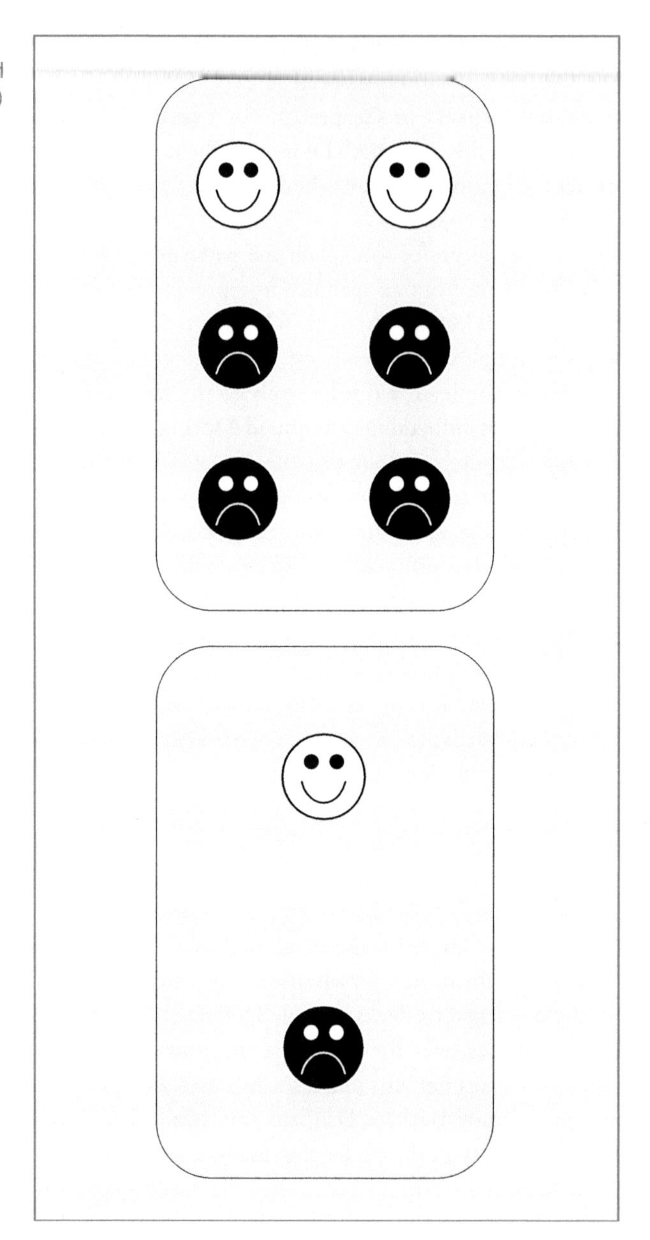

FIGURE 9.3
Examples of the cards used in Kerr and Zelazo's (2004) gambling task.

to play the game, and quickly learned the correspondence between M&Ms and happy versus sad faces. They were then scored for the number of times out of 50 that they chose cards from the advantageous versus disadvantageous deck. Kerr and Zelazo found that the three-year-olds made disadvantageous choices significantly more often than would be expected by chance. In contrast, four-year-olds made advantageous choices significantly more often than would be expected by chance. At the individual level, however, a few three-year-olds did learn to choose cards from the advantageous deck. Kerr and Zelazo argued that 'hot' executive function develops in a similar way to 'cool' executive function, with important development between the ages of three and four years. Gambling appears to be a 'conflict'

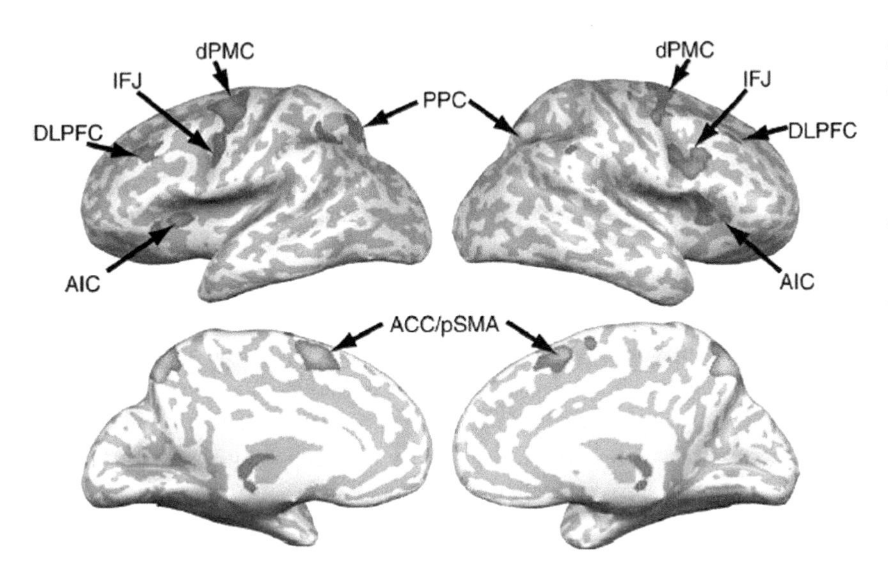

Schematic depiction of the neural regions likely to comprise the cognitive control network. These neural regions are co-activated in adults by a range of executive function tasks, and are densely interconnected by white matter fibre tracts. AIC = anterior insula cortex, DLPFC = dorsolateral prefrontal cortex, IFJ = inferior frontal junction, dPMC = dorsal pre-motor cortex, PPC = posterior parietal cortex, ACC = anterior cingulate cortex, pSMA = pre-supplementary motor area.

measure of executive function, as the gains and losses in the two decks of cards conflict. Children must learn to respond in a way that conflicts with the salient response (here, choosing the deck offering big wins) in order to maximise their gain across the experiment. Again, the key developmentally appears to be managing *conflicting* representations.

Cognitive neuroscience studies of cognitive flexibility and executive function

Despite the early patient data suggesting that activation of the prefrontal cortex plays an important role in enabling flexible rule use, more recent cognitive neuroimaging studies with adults have revealed that the prefrontal cortex does not operate in isolation during tasks such as the DCCS. Rather, it functions as part of a larger 'cognitive control' network in the adult brain that also involves anterior cingulate cortex, dorsal premotor cortex and superior parietal cortex, among other structures. These neural regions are co-activated in adults by a range of executive function tasks, and are densely interconnected by white matter fibre tracts. Age-related differences in DCCS tasks may thus be related to growing *functional connectivity* of the cognitive control network rather than to the efficient activation of individual structures per se.

Ezekiel, Bosma and Morton (2013) investigated this question by comparing functional connectivity using fMRI in relatively old children (12-year-olds) as they performed the DCCS with functional connectivity in adults. Although the children were relatively old, this meant that they could be matched for in-scanner performance with the adults. Accordingly, differences in neural activation would not be a result of age-related performance differences. The in scanner task was a simple DCCS based on shape versus colour (red or blue trucks or rabbits). A spatial independent components analysis (ICA) approach to the BOLD data was used to select cortical networks that were active at different time points during task performance.

This is a stronger approach experimentally than pre-selecting regions of interest for analysis such as prefrontal cortex, as ICA is a statistical procedure that blindly decomposes observed data into underlying components. The ICA identified 20 independent components in the imaging data. For example, there was a left fronto-parietal component, a right fronto-parietal component and a visual component. Further analyses showed that the right fronto-parietal component was significantly more active during the theoretically important switch trials. This was the case for both children and adults. However, functional connectivity with other brain regions differed between children and adults. For adults, different regions of the prefrontal cortex were tightly connected to parietal, anterior cingulate and ventral tegmental regions. For children, there were greater connections to the anterior insula, fusiform gyrus and temporal gyrus. These differing patterns of functional connectivity were interpreted to show that age-related differences in rule switching are associated with differing connectivity in the brain rather than with specific neural structures per se. However, whether differing connectivity drives improvements in behavioural performance or whether improvements in behavioural performance drive connectivity cannot be ascertained from this study.

Wendelken and his colleagues were able to use fMRI with younger children, in a study that used a rule switch task with children as young as eight years (Wendelken, Munakata, Baym, Souza & Bunge, 2012). The children were shown pictures of the Disney fish character Nemo, who was either coloured red or blue, and either faced towards the left or the right. The rule types were presented on a trial-by-trial basis inside the scanner, via the words 'colour' and 'direction', and participants responded with a button press. Rule switching was required when a 'direction' trial followed a 'colour' trial or vice versa. The children proved to be both less accurate and slower than the adults, both in following the correct rule and when switching rules. Nevertheless, the brain activation during both rule type and switch trials was very similar in the children and adults, with maximal activation in lateral prefrontal cortex, posterior parietal cortex and pre-supplementary motor cortex. Wendelken et al. reported one group difference in the whole brain analyses, with children showing greater activation in both left superior temporal gyrus and right middle temporal gyrus when switching rules compared to adults. Wendelken et al. (2012) argued that this could indicate different temporal dynamics related to different functional integration between brain regions for task switching in children versus adults. As the children were also worse at switching rules than the adults, however, this differential activation is difficult to interpret. It may reflect the poorer performance by younger children rather than the temporal dynamics of functional integration per se.

EEG has also been used to study the development of cognitive flexibility, including studies of 'hot' versus 'cold' executive functions (EF). For example, Carlson and her colleagues used a computerised gambling task called the Hungry Donkey to measure hot EF in eight-year-old children, and recorded their EEG (Carlson, Zayas & Guthormsen, 2009). In the Hungry Donkey task, the children were trying to win apples to feed their donkey. The children could win and lose apples by clicking on each of four doors shown on the screen, which over time were either more or less advantageous regarding winnings and losses. For example, door A was disadvantageous over time, yielding frequent small losses, door B was also disadvantageous over time, yielding infrequent large losses, door C was advantageous over time, but also yielded frequent very small losses, and door D was advantageous over time, but also yielded infrequent small losses. After each trial, the number of apples lost or won was indicated by apple icons appearing or disappearing on the screen from the child's overall pile of apple winnings. The DCCS was also administered, but without recording EEG.

Carlson and her colleagues were interested in two ERPs, the P300, which occurs after stimulus presentation and has been considered an index of the allocation of attentional resources, and the stimulus-preceding negativity (SPN), which despite its label also occurs just after a stimulus has been experienced but before feedback occurs. In adult studies, the SPN is thought to reflect the emotional salience of an anticipated stimulus (i.e., it is larger prior to receiving negative feedback than positive feedback). The P300 for the children was compared for net win trials and net loss trials. The SPN was compared for door A and C selections (high frequency experience of punishment) versus door B and D selections (low frequency experience of punishment). Carlson and her colleagues reported that the children learned the valence of the different doors over the course of the experiment. Regarding ERPs, punishment elicited a greater P300 than reward, as expected. Unexpectedly, girls exhibited greater P300 effects than boys, primarily as they had larger P300 amplitudes in response to punishment (losing apples). Overall, children of both sexes strongly preferred the infrequent punishment doors (D and B), even though door B led overall to them losing apples. A form of SPN was also measurable during the second half of the experiment. Children showed significantly greater negativity when they approached disadvantageous doors. Neither SPN amplitude nor P300 amplitude was correlated with behavioural performance in the DCCS however. Nevertheless, Carlson et al. (2009) have demonstrated that ERPs that are correlated with *anticipated* wins and losses can be measured in eight-year-olds.

In a follow-up study, Carlson and her colleagues then examined whether these ERPs could be predictive of later EF (Harms, Zayas, Meltzoff & Carlson, 2014). The children were invited back to the laboratory when aged 12, and were given both the Hungry Donkey task again along with the DCCS, and also a flanker task (see Figure 9.4) as an additional measure of cold EF, plus a standardised behavioural measure of thrill and adventure seeking. EEG was recorded for the Hungry Donkey task only. The ERP findings from age eight were replicated for the hot EF task at age 12, with children again showing a larger P300 to loss trials than to win trials, and a larger SPN after large losses. However, these ERPs when measured at age eight did not predict cool EF performance at age 12. Magnitude of the P300 at age eight was only a predictor of later thrill and sensation seeking, with children with high amplitudes at age eight self-reporting significantly less thrill and sensation seeking. Meanwhile, magnitude of the SPN was only a predictor of general academic competence (as rated by teachers) at age 12. These correlations are difficult to interpret. For example, Harms et al. (2014) proposed that greater P300 amplitudes indicated the devotion of more attentional resources, and hence that greater attention to punishment leads to a risk-averse developmental trajectory.

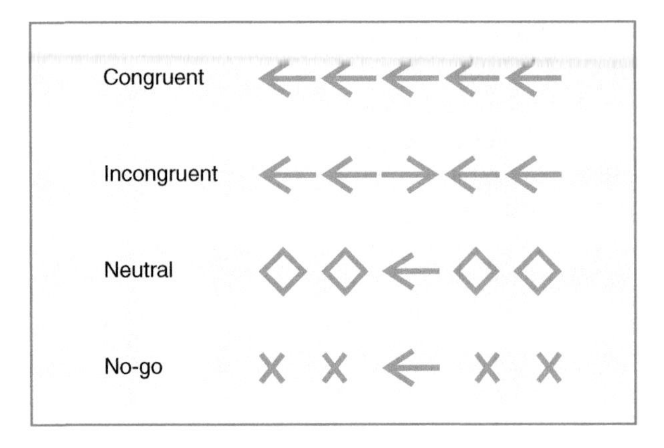

FIGURE 9.4
An example of a flanker task, from Bunge et al. (2002). Copyright © 2002 Elsevier. Reproduced with permission.

But until we understand how and why the brain generates a P300 response, this interpretation must remain speculative. Similarly, Harms et al. suggest that a greater SPN to loss may indicate better ability to learn from mistakes and hence a better academic trajectory. This is also speculative, as children's ability to learn from mistakes has not been tested experimentally. As with the fMRI studies of cognitive flexibility, researchers have discovered some significant brain–behaviour correlations, and these could prove important guidance when designing future neuroimaging studies. Currently, however, we are left unable to conclude anything specific regarding neural mechanisms related to the development of cognitive flexibility in children.

Finally, an important study by Buss and Spencer (2017) using the DCCS employed fNIRS to measure brain activity. As will be recalled, fNIRS measures changes in haemodynamic response in the brain over relatively long time windows. Buss and Spencer were able to show that children's behavioural performance was likely to be critical to the brain–behaviour relationships observed in cross-sectional imaging studies. Buss and Spencer used both 'easy' and 'hard' versions of a colour/shape DCCS with children aged three and four years. Brain activity was measured while the children played computerised versions of the DCCS, and the haemodynamic response was compared for switch trials in the easy versus hard tasks. Whereas some children performed both the easy and hard versions of the task well, other children were only able to perform well in the easy version. Of particular interest was frontal cortex activity in the children who were good at the easy version of the DCCS, but poor at the harder version. Buss and Spencer reported that frontal activation was weak in three-year-olds who were poor at the hard version of the DCCS, who perseverated in sorting by the previously-correct rule on switch trials. However, these same three-year-olds showed strong activation of frontal cortex on switch trials when performing accurately on the easy DCCS task. These data are important, as they suggest that neural activity during EF tasks does not depend on the developmental state of a child's frontal cortex. Rather, neural activity in frontal cortex depends on whether the child can successfully perform the EF task.

Carlson and her colleagues have also begun to work with younger children. As noted in Chapter 8, in studies with adults theta and alpha oscillations are thought to play a mechanistic role in working memory, although the nature of this role is not yet understood. Perone, Palanisamy and Carlson (2018) decided to measure resting state power in different EEG frequency bands in young children, and to correlate these measures with performance in a standardised computerised EF measure based on the DCCS, the Minnesota Executive Function Scale (MEFS). The MEFS presents virtual cards on a computer screen, and is suitable for ages two years upwards. As well as measuring the flexibility of children's rule switching, the computerised administration enables response time to be measured as well. The MEFS was administered to groups of children aged three years, four years, five years and nine years. Resting state EEG was measured with both eyes open and eyes closed, and relative power in the theta, alpha, beta and gamma bands was computed for each child across frontal and parietal regions of interest. Relative power was chosen as absolute power differs between individuals depending on factors such as the thickness of the skull.

Perone et al. reported that theta power decreased with age, both when eyes were open and closed. Alpha power increased with age, but for the eyes closed condition only. Beta and gamma power showed more complex patterns, differing by site (frontal versus parietal) and condition (eyes open versus closed) and by age. The key research question was whether the neural measures would be related to differences in MEFS performance. Total MEFS score controlling for age was positively related to beta power and negatively related to theta power. The ratio of theta power to beta power was also computed, as in adults this measure has been related to executive control. The theta/beta ratio declined with age and was significantly related to total MEFS score when age was controlled, with higher MEFS scores when the ratio was smaller (the same effect is found in adults, a high theta/beta ratio is associated with poor regulation of attention). Perone et al. (2018) have thus demonstrated another brain–behaviour correlation for EF tasks performed by children that is also found in adults. While this is encouraging, further work is required in order to understand the neural mechanisms that may underlie this association.

Inhibitory control, planning and executive function: Behavioural studies

Inhibitory control can be defined as the ability to inhibit responses to irrelevant stimuli while pursuing a cognitively represented goal (e.g., Carlson & Moses, 2001). Carlson and Moses (2001) identified two types of tasks that have been used to measure inhibitory control in young children. One type of task requires children to delay gratification of a desire, for example by suppressing a prepotent response such as peeking at a gift. The second type of task requires children to respond in a way that conflicts with a more salient response, thought to relate to resisting impulsive actions. For example, in the 'day/night' task, children are shown cards depicting either the sun or the moon. When they see a picture of the sun, they have to say "night". When they see a picture of the moon, they have to say "day". Performance in both of these types of task improves with age during the period from approximately three to seven years.

For example, Kochanska and her colleagues investigated children's ability to delay gratification of a desire in a longitudinal study of children aged on average 33 months when first seen, and 46 months when seen for the second time (Kochanska, Murray, Jacques, Koenig & Vandegeest, 1996). These experimenters distinguished between 'passive' inhibition (shyness, anxiety, fearfulness) and 'active' inhibition or effortful inhibitory control. To measure the latter in young children, they designed situations in which children were tempted to violate particular standards of behaviour. For example, the children were required to (a) hold a sweet (M&M) on their tongues for up to 30 seconds before eating it, (b) wait for the experimenter to ring a bell before retrieving an M&M from under a glass cup

Kochanska et al. (1996) reported that older children had better inhibitory control than younger children in terms of their ability to delay gratification.

(with the experimenter at one point lifting the bell but not ringing it), and (c) sit wearing a blind fold while the experimenter noisily wrapped a gift for the child, who was instructed not to peek. Scores on these tasks were highly correlated and were thus standardised and aggregated into one 'Home Gift' score. Other tasks were also used, for example speaking in a whisper, and moving a toy turtle slowly along a toy path. The tasks together formed an 'inhibition control' battery. Kochanska et al. reported that their battery of tasks had good internal consistencies at both measurement points. The task battery correlated with maternal ratings of impulse control, and was developmentally sensitive. Kochanska et al. suggested that the tasks tapped a stable temperament variable in young children, namely inhibitory control. They also reported that girls outperformed boys at both toddler and preschool ages. Older children had better inhibitory control than younger children.

Inhibitory control in conflict tasks was studied by Diamond and Taylor (1996) using both Luria's tapping task and a day/night task. The aim of the study was to compare children's performance on two different measures of holding two pieces of information in mind while inhibiting a strong response tendency. In the tapping task, the experimenter had a wooden dowel, which she used to tap sharply on the table before handing it to the child. If the experimenter tapped once with her dowel, the child had to tap twice. If the experimenter tapped twice with her dowel, the child had to tap once. One hundred and sixty children aged from three to seven years were tested. In the day/night task, on which 93 children were tested, the children were shown a black card with stars and were asked to say "day", and a white card with a sun to which they were asked to say "night". There were 16 experimental trials in each task.

Diamond and Taylor found age-related improvements in the tapping task, with group performance at around 65% correct for the 3.5-year-olds and approaching ceiling for the seven-year-olds. A similar pattern was found for the day/night task, although this task was more difficult than the tapping task for most age groups. Performance in both tasks decreased in later trials, although particularly so for the day/night task (e.g., 85% of four-year-olds tapped correctly on the first four trials compared to 67% on the last four trials, whereas 90% of four-year-olds responded correctly on the first four day/night trials compared to 54% on the last four trials). Diamond and Taylor concluded that the ability to exercise inhibitory control over one's behaviour developed between the ages of three and six years, with ceiling performance in seven-year-olds. They argued that most errors that they observed were suggestive of inhibitory failures (e.g., tapping many times instead of just once or twice), or of forgetting a rule or being unable to switch rules (e.g., always tapping once). They suggested that the growth of inhibitory control may be related to developments in frontal cortex.

A landmark study by Hughes (1998) created a battery of executive function tasks and gave them to 50 typically-developing young children aged 3–4 years. Hughes set out to devise tasks that could tap the three principal factors thought to underpin executive function: inhibitory control, working memory and attentional flexibility. To study inhibitory control, Hughes developed two tasks, a version of a hand game devised for adult frontal patients by Luria (Luria, Pribram & Homskaya, 1964), and a detour reaching box based on the Windows task devised by Russell and his colleagues (1991, discussed in Chapter 4). In the *hand game*, children learned to make a pointing action and to make a fist. To play the game, both the child and the adult began with their hands behind their backs. They then showed their hands, and the children had to either make the same handshape as the experimenter (*imitate*), or make the handshape opposite to that exhibited by the experimenter (*conflict*). All children received both the *imitate* and *conflict* conditions, in counterbalanced orders, and were scored as successful if they achieved a run of six consecutively correct trials. In the *detour reaching* task, the children were presented with a metal box that had a large window in the front, revealing a platform containing a marble. The box had a yellow light and a green light. If the child reached directly through the window to get a marble, an infrared beam was activated and the marble fell through a trap door on the platform. However, if the yellow light was on, the child could turn a knob to break the invisible beam, and could then retrieve the marble. Children first learned this contingency, and became successful at retrieving the marble. The apparatus was then altered so that the green light was on. Now the child had to press a button which broke the beam in order to retrieve the marble. The child was encouraged to get the marble, and the number of trials needed to achieve a criterion of three successes in a row was measured (up to 15).

In order to measure attentional flexibility, Hughes used a colour/shape set shifting task of the kind devised by Zelazo and colleagues, and also a magnets task about making patterns. In the magnets task, the child was shown a colour pattern comprising 18 circles on a card, in the sequence blue-blue-red, blue-blue-red, and so on. They were then given red and blue magnets and asked to reproduce the same pattern on a steel rule. Scoring depended on the number of correct blue-blue-red sequences made. The working memory tasks comprised a visual search task and an auditory sequencing task. In the visual search task, the child was given eight distinctive pots on a tray, and was told to place a raisin in each pot. The tray was covered with a scarf and spun around, and the child was then allowed to choose a pot and keep the raisin. The (now empty) pot was then replaced, the pots were scrambled, the tray was spun again, and the child could choose another pot. The goal was to keep selecting baited pots, to gain all the raisins, and the children were reminded on each trial to choose a pot that they hadn't yet looked in. The number of trials required to find all the raisins was scored. In the auditory sequencing task, the children were shown a book of nine pictures, each of which made a noise. After the children had checked each picture in this 'noisy book', the experimenter named some of the pictures in a given order (either two items, three items or four items). The child had to use the book to recreate the sequence of items given as a verbal list, by pressing the named pictures in turn. This generated a memory span measure for each child.

When performance on the different measures of executive function was inspected, it turned out that the three- and four-year-olds were at ceiling on the visual search task (finding the pots with raisins). However, effects of age (three versus four years) were found for four tasks, detour reaching, Luria's hand game, auditory memory span and the magnets task. In order to see whether distinct aspects of executive function could be distinguished, Hughes carried out a factor analysis controlling for verbal ability and age. This analysis suggested that the different tasks chosen did indeed load most strongly onto the expected factors of inhibitory control (detour reaching and hand game), attentional flexibility (set shifting and pattern making) and working memory (auditory span). However, when non-verbal ability was controlled as well, then the auditory working memory task loaded onto the same factor as the hand game task, while the detour reaching task had a factor of its own. Nevertheless, Hughes's data suggest that executive function can be measured in preschoolers, and that there are age-related changes in executive function in the preschool period. She concluded that executive function was a multi-faceted construct in young children. These conclusions have been supported by more recent research. As reviewed comprehensively by Diamond (2013), a range of tasks are now routinely used to measure inhibitory control in young children, including those described above. In general, children show development in these tasks between the ages of around three to nine years, and in general there is developmental continuity – performance at time 1 on a given task is generally significantly related to performance on that task at time 2. Individual differences between children in early inhibitory control is also predictive of many later outcomes. For example, one study has followed 1,000 children for 32 years. The authors reported that early measures of inhibitory control were predictive of a range of later outcomes, even once social class and intelligence had been accounted for (Moffitt et al., 2011). The outcomes were both negative, such as the likelihood of making risky choices as teenagers (e.g., smoking and drinking) and the likelihood of later criminality, and positive, such as the likelihood of being employed and the likelihood of having good physical and mental health

Another important aspect of executive function is *planning*. Carlson and Moses (2001) gave a different battery of executive function tasks to 107 three- and four-year-olds, and included a motor sequencing task intended to measure general executive planning ability. The executive function tasks were measures of inhibitory control, which were categorised as either delay or conflict tasks, comprising ten tasks in all. The ten tasks used are shown in Table 9.1. The conflict tasks comprised the day/night, grass/snow, bear/dragon, spatial conflict and card sort tasks. The delay tasks were the Pinball task ("don't release the plunger until the experimenter says 'go'"), Kochanska's gift delay and whisper measures described earlier, the tower building task and a picture-matching task requiring the child to select a match from six extremely similar pictures (KRISP task). The motor sequencing measure of planning was based on a musical keyboard with four differently coloured keys. The children had to play each key in turn using their index finger. Following practice, they had to play consecutive sequences as fast as they could until the experimenter shouted "Stop!" Children were scored for the number of sequences that didn't skip a key or involve touching the same key twice.

Carlson and Moses reported that the children's scores on the ten different tests of inhibitory control were moderately intercorrelated and appeared to tap a common underlying construct. The ten measures were thus standardised and

TABLE 9.1 Prepotent responses and correct responses on the inhibitory control tasks

Inhibitory control task	Prepotent response	Correct response
Day/night	Say "day" for the sun and say "night" for the moon	Say the opposite of what the picture shows
Grass/snow	Point to green for "grass", point to white for "snow"	Point to the colour that is opposite to its associate
Spatial conflict	Press the button on the same side as the picture	Press the button that matches the picture, irrespective of location
Card sort	Sort by a previously successful dimension	Sort by a new dimension
Bear/dragon	Follow the commands of both animals	Do what the bear says, but not what the dragon says
Pinball	Release the plunger immediately	Wait for a "Go!" signal
Gift delay	Peek while E wraps gift	Wait without peeking
Tower building	Place all the blocks oneself	Give E turns placing blocks
KRISP	Point to a similar picture right away	Wait to examine all pictures before choosing exact match
Whisper	Call out the names of familiar characters	Whisper the names

E, experimenter; KRISP, Kansas Reflection-Impulsivity Scale for Preschoolers.

KEY TERM

fNIRS (functional near-infrared spectroscopy) Allows neuroimaging of brain activation by tracking blood flow via changes in haemoglobin.

averaged to form a composite inhibitory control battery. Scores on this composite inhibitory control battery were found to be correlated with age, gender and verbal ability, and also with parental scores of inhibitory control and with performance on the motor planning task. However, one possible drawback to the motor planning task devised by Carlson and Moses (2001) is that there is no action to be planned other than playing the keyboard. Hence the planning is not really *cognitive* in nature.

In order to gain developmental information about the ability to plan more cognitively challenging tasks, Carlson, Moses and Claxton (2004b) carried out a second study of inhibitory control and planning, using three novel planning tasks. Forty-nine children aged three and four years were asked to complete the Bear/Dragon, Whisper and Gift Delay tasks described previously, along with a Tower of Hanoi planning task (Simon, 1975), a truck loading task (Fagot & Gauvain, 1997) and a kitten delivery task (Fabricius, 1988). In classic Tower of Hanoi tasks (see Figure 9.5), disks have to be transferred across wooden pegs following a set of rules. Carlson et al. presented this as a "monkey jumping" game, with a Daddy monkey (large disc), a Boy monkey (medium disk) and a Baby Sister monkey (small disc). The pegs were described as trees, and the children were told that bigger monkeys could not sit on top of smaller monkeys in case they "smushed" them. Only one monkey could jump at a time, but smaller monkeys could sit on top of bigger monkeys. Various end states (e.g., all monkeys on the end peg) then had to be created by the children. In the truck loading task, the children had to pretend to

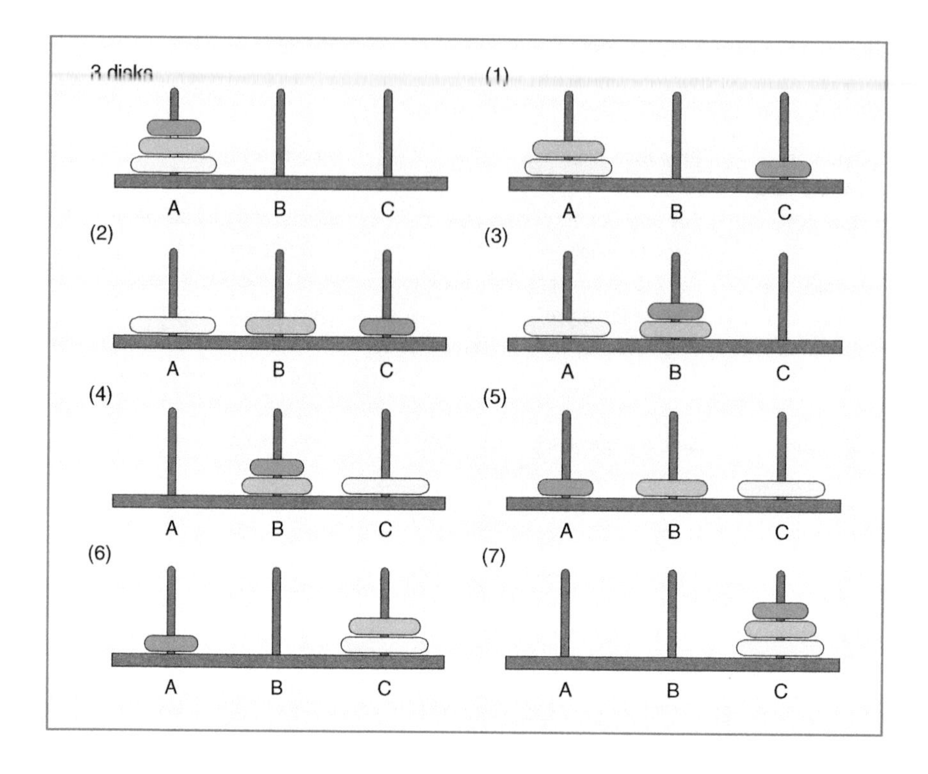

FIGURE 9.5
The Tower of Hanoi task.

be postmen (mail carriers) delivering party invitations from a toy mail truck. Each invitation was coloured differently, and had to be delivered to an appropriately coloured house set along a one-way street. The child's job was to put the invitations in the truck in the correct order to ensure the shortest journey. Finally, in the kitten delivery game, the children had to plan to minimise the distance covered in retrieving kittens from buckets placed around the room so that they could be delivered safely back to their mother. All three planning tasks were intended to require *if–if–then* reasoning as outlined by Frye, Zelazo and colleagues (see earlier discussion). For example, *if* invitations can only be delivered from the top of the stack, and *if* the pink house is last, *then* the pink invitation should be loaded first.

Carlson et al. (2004b) reported that both three- and four-year olds performed at about the same level on the Tower of Hanoi and kitten delivery tasks, but there were developmental improvements in the truck loading task. No task showed floor effects. Performance on the truck loading and Tower of Hanoi tasks was significantly related, but kitten delivery was not related to these tasks. When relations between the planning and the inhibitory control tasks were explored, no significant correlations survived controls for age and verbal ability. Carlson et al. (2004b) concluded that planning and inhibitory control were largely independent constructs in executive function, and that both developed dramatically in the preschool years.

Since the last edition of this book, there have been many more studies of the development of executive function, which has enabled refinement of these different constructs. A comprehensive model was developed by Diamond (2013) and is reproduced in Figure 9.6. As can be seen, in this model planning is now seen as a higher-level executive function, along with reasoning and problem-solving. The

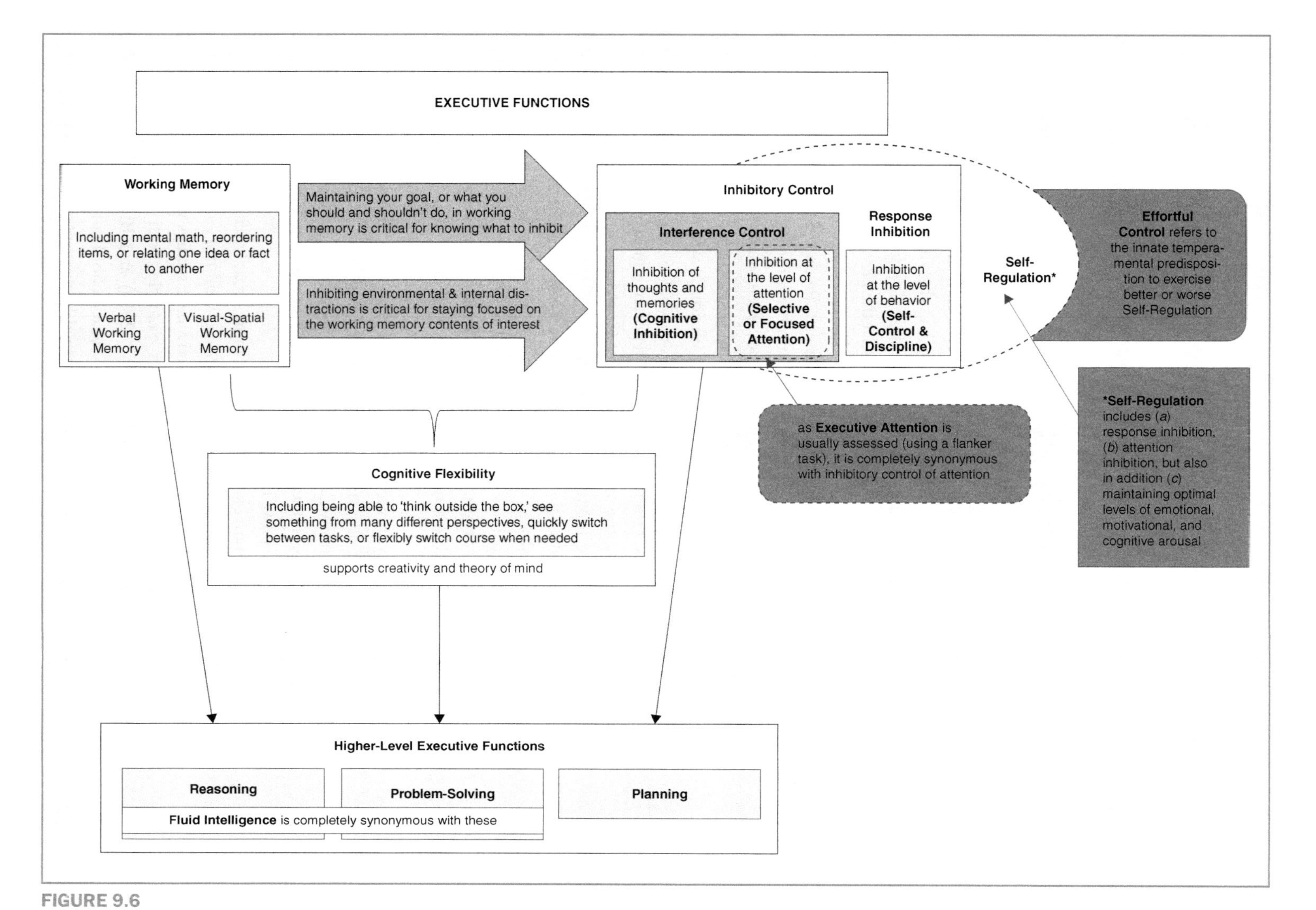

FIGURE 9.6
Depiction of a theoretical framework for understanding executive function in children, developed by Diamond (2013).

core executive functions are cognitive flexibility, inhibitory control and working memory. Given that individual differences in executive function affect so many important outcomes, the past years have also seen a growth in studies investigating whether executive functions can be trained in young children. In particular, a range of computerised games and apps to train executive function have been developed, in the hope of improving developmental outcomes for all children.

Training executive function

The first studies to investigate training executive function focused on working memory. Klingberg and his colleagues devised a computerised gaming interface to train working memory, now available as the CogMed© app. The interface trains different components of working memory such as the visuo-spatial and verbal slave systems via interactive games such as hitting targets under timed conditions. It also uses algorithms to increase difficulty level, so that the player is always being extended cognitively and performing tasks that are neither too easy nor too difficult. Despite ongoing controversy in the literature (Shipstead, Redick & Engle, 2012), studies utilising CogMed© with children have found significant and specific training benefits. For example, in a representative study, Holmes, Gathercole and Dunning (2009) trained ten-year-old children who had poor working memories with two versions of CogMed©, the standard adaptive version in which the difficulty level of the games was adapted individually, and a non-adaptive version in which training focused on easy tasks. Holmes et al. (2009) reported significant training effects for the children receiving the adaptive version of CogMed© only, for both verbal and visuo-spatial working memory. Furthermore, significant training effects were still measurable six months after the study. However, training effects were limited to working memory tasks, and did not extend to improving academic skills such as reading, nor lead to gains in IQ. In a related study, Gathercole and her colleagues found that CogMed© training was more beneficial in improving working memory skills in children with attention deficit and hyperactivity disorders (ADHD) than medication. Holmes et al. (2009) reported that adaptive CogMed© training led to greater gains than hyperidol (the most frequent medication offered for ADHD), although medication also significantly improved children's working memory performance. The gains from CogMed© were still significant at a six-month follow-up. Again, the gains observed were specific to working memory. Verbal and non-verbal IQ measures did not show improvement.

KEY TERMS

Inductive reasoning Generalising from specific examples, making an inference from a particular premise, and reasoning by analogy.

Deductive reasoning Reasoning problems that have only one logically valid answer, which include syllogistic reasoning.

Programmes to train executive function more broadly have also been developed by psychologists, typically focused on improving inhibitory control. Again, there is a degree of controversy regarding their effectiveness; studies with younger children tend to show stronger training effects. For example, both Diamond (Diamond, Barnett, Thomas & Munro, 2007; Diamond, 2013) and Blair and Raver (2014) have reported beneficial effects from an early years school curriculum developed to foster EF in young children from disadvantaged backgrounds, 'Tools of the Mind' (Bodrova & Leong, 2007). Tools of the Mind trains self-regulation and inhibitory control via 40 activities designed to promote EF, such as pretend play activities, telling oneself out loud what one should be doing, other game-like aids to facilitate memory and attention, and teacher-led scaffolding of learning activities that start from the child's intrinsic interests. There is a strong focus on shared co-operative activities between children. Blair and Raver (2014) conducted a randomised controlled trial with 759 children in kindergarten in the USA, using the Tools of the Mind curriculum. They reported positive training effects for selected EF measures for the whole cohort (e.g., working memory, response time on the DCCS), and stronger training effects on a wider range of EF measures for children attending schools in areas of high poverty. When followed up in first grade, children receiving the Tools of the Mind intervention in kindergarten also showed better vocabulary and reading development. Blair and Raver argued that improving self-regulation in young children helps them to focus on the information provided at school and hence to make academic progress.

KEY TERM

Analogical reasoning
The ability to reason from a familiar or known problem to a novel one by identifying relational correspondences between the two.

A meta-analysis of executive function training studies that either used a form of 'mixed attention' training such as the DCCS or a form of working memory training such as CogMed© supported the finding that younger children tend to benefit more from EF training (Wass, Scerif & Johnson, 2012). Wass and his colleagues found that in general, training studies targeting younger children were more likely to find significant training effects. However, although significant, the training effects were relatively weak. Wass et al. (2012) also pointed out the importance of investigating other interventions thought to boost self-regulation, such as mindfulness meditation. Diamond (2013) has also considered the possible benefits of mindfulness, yoga and other physical systems such as Taekwondo for promoting EF. She concluded that current intervention studies based on such approaches did not enable strong conclusions. The most common problem with such interventions is no active control group; that is, a comparison group of children who receive a different intervention. As Diamond noted, future EF training studies should include control groups who also receive an intervention, should set activities at the correct level for an individual via adaptive methods, and should include longitudinal follow-up tests to assess maintenance of gains. It is also notable that programmes such as Tools of the Mind are based on ways of teaching young children that build on key theories in child psychology, for example Vygotsky's theory. Vygotsky's ideas about the importance of the zone of proximal development and the role of pretend play in developing self-regulation are discussed in Chapter 11. Rather than comprising a set of training materials that the child completes in a specified period of time, therefore, it may be that truly effective EF training programmes are those that also include the *teachers* of young children. Such programmes would train teachers in how to promote EF skills via everyday classroom activities and interactions.

Cognitive neuroscience studies of inhibitory control and EF training

As noted earlier, classical analyses of executive function based on adult patient data suggested that the frontal cortex might play a critical role in inhibitory control. As there is significant development of the frontal cortex during childhood and adolescence, a simple hypothesis has been that the growth of frontal cortex enables the developments in executive function observed during childhood. Accordingly, early brain imaging studies of EF sought brain-behaviour correlations with frontal cortex activity. Such studies indeed consistently reported significant correlations between frontal cortex activity (typically measured via the BOLD response in fMRI studies) and performance in executive function tasks.

For example, Bunge and her colleagues gave a measure of response inhibition (a go/no-go task) and a measure of interference suppression (a flanker task) to children aged from 8–12 years and to adults, in a study measuring brain activity via fMRI (Bunge, Dudukovic, Thomason, Vaidya, & Gabrieli, 2002). Both tasks were based on simple visual displays involving arrows and geometrical shapes (see Figure 9.4). In the flanker task, participants were asked to press a button on the left when the central arrow was pointing to the left, and a button on the right when the central arrow was pointing to the right. The central arrow was flanked by irrelevant shapes (e.g., diamonds; neutral trials). Interference was introduced by making the flanking shapes arrows that were pointing in the wrong direction (incongruent trials). Alternatively, the flankers could be arrows pointing in the correct direction (congruent trials). In no-go trials, the arrow was flanked by crosses. When crosses appeared, participants were instructed not to press any button. Behaviourally, the children were poorer than the adults in the incongruent and the no-go trials, although accuracy levels for the children were over 90%.

The brain imaging data showed that interference suppression in the children was associated with left prefrontal activity. In particular, the left ventrolateral prefrontal cortex was more active in children when suppressing interference, whereas right ventrolateral prefrontal cortex was more active in adults. When those children best at suppressing interference were studied separately (i.e., by selecting children with equivalent behavioural performance to adults), more extensive activation of left ventrolateral prefrontal cortex was found for these children, rather than an expertise-related change to right ventrolateral prefrontal activity. For response inhibition (the go/no-go task), similar results were found. The children still failed to activate right ventrolateral prefrontal cortex, even when only those children whose behavioural performance was similar to adults were selected. However, during response inhibition children who performed well did activate a subset of the neural regions activated by the adults.

A complementary study by Durston and her colleagues used fMRI to study inhibitory control processes in children aged eight years and adults (Durston et al., 2002). They used a child-friendly go/no-go paradigm based on Pokémon characters (see Figure 9.7). Children were instructed to press a button in response to any character except Meowth ("catch all the Pokémon except for Meowth"). The number of 'go' trials preceding a no-go trial was varied (1, 3 or 5 go trials could precede a no-go trial). It was expected that children would find it more difficult to inhibit responding when more 'go' trials were used. Behaviourally, the frequency of no-go errors did indeed increase, being 8% for one preceding 'go' trial, 12.5% for three trials and 14.5% for five trials. In this study, the children and adult participants activated the same neural regions during response inhibition. Activity was greater in children, and was focused bilaterally in ventrolateral prefrontal cortex, right dorsolateral prefrontal cortex and the right parietal lobe. In adults, the degree of activation in the ventral prefrontal regions was correlated with the number of preceding

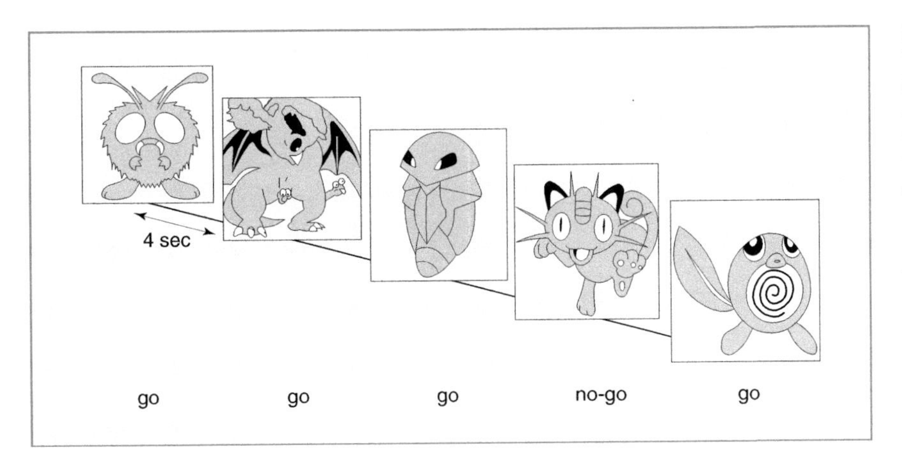

FIGURE 9.7
The no-go trial preceded by three go trials (other characters from the Pokémon series). The instruction to participants was to "catch all the Pokémon except for Meowth" by pressing the thumb button on a button box. From Durston et al. (2002). Reproduced with permission from Blackwell Publishing.

'go' trials, whereas for children it was high in all conditions. Clearly, this study differed from that of Bunge et al. (2002) in terms of the presence of adult-like right ventrolateral prefrontal activity, although both studies found that increased brain activation was correlated with better performance. Task differences and/or performance differences might provide an explanation. These two studies are typical of a number of fMRI studies with children showing that frontal activity is related to performance in EF tasks. However, as the data in such studies are correlational, it could be that improvements in behavioural abilities cause the changes in neural activation that are observed (as in the study by Buss & Spencer, 2017, discussed earlier). In order to conclude that changes in frontal cortex or frontal connectivity drive cognitive improvements, longitudinal studies are required.

Such studies have now begun to appear in the literature. Further, as in the neuroimaging literature on cognitive flexibility, it is increasingly recognised that the frontal cortex does not operate in isolation during inhibitory control, but as part of a wider network. In a longitudinal fMRI study that included children as young as two years, Long, Benischek, Dewey and Lebel (2017) investigated functional connectivity and changes with age in 44 children aged between two and five years at their first scan. Subsequent scans were taken at six-monthly intervals, although not all children contributed all scans, hence 21 children contributed longitudinal data. Children watched videos that they chose themselves during scanning. Analyses of functional connectivity showed that both local connectivity and global connectivity increased with age, in frontal, parietal and cingulate areas. In particular, the fronto-parietal network typically activated by executive function tasks in older participants showed age-related increases in connectivity from two to six years. Long et al. concluded that the development of functional connectivity in early childhood is likely to underlie major advances in cognitive abilities. However, despite the longitudinal aspect, it is important to remember that these data are still correlational. Furthermore, functional connectivity was measured while children watched a video, not while they performed an EF task.

A different neural approach to understanding individual differences in executive function in young children is to study structural brain differences between children of the same age. This approach was taken by Borst et al. (2014), who were

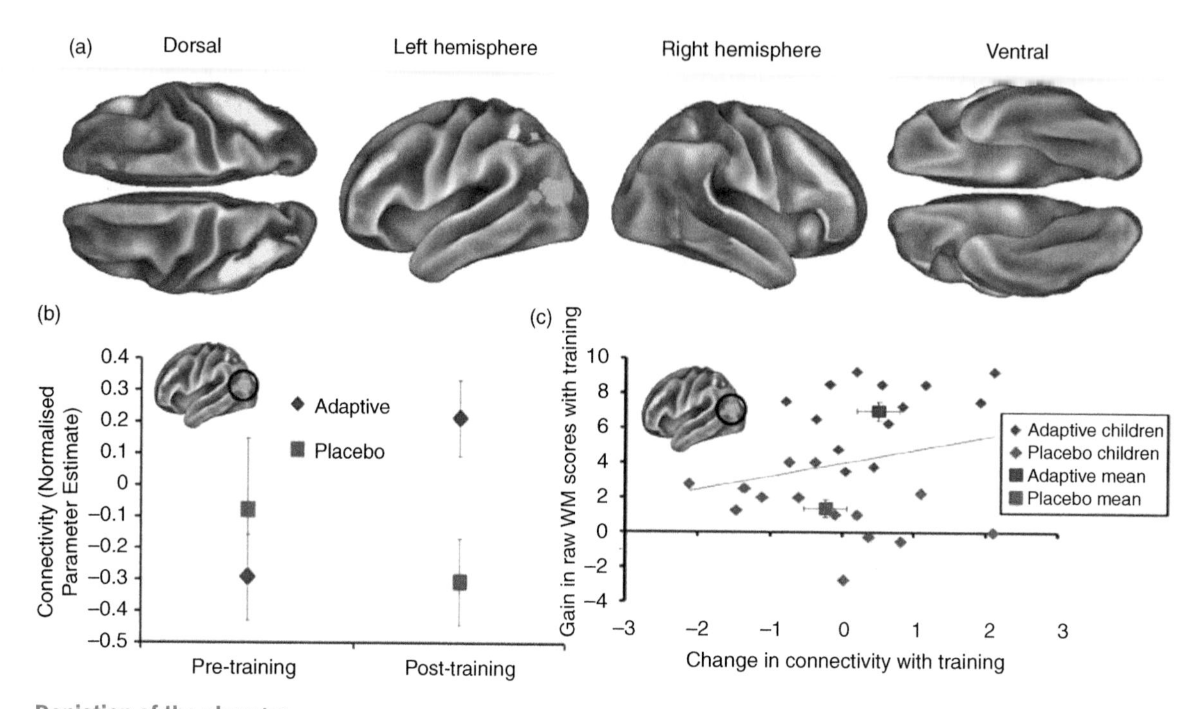

Depiction of the changes in connectivity following CogMed© training. The top panel shows the main locations of changes in connectivity, and the bottom panels show a linear relationship between these changes and the degree of training benefits.

KEY TERM

Stroop task
Participants are asked to name the ink colour and not the word for colour names; e.g., responding "green" to the word "red" written in green ink.

interested in the possible role of the anterior cingulate cortex in inhibitory control. As reviewed above, the anterior cingulate is a core structure in the 'cognitive control' network in the adult brain that also involves prefrontal cortex, dorsal premotor cortex and superior parietal cortex, and the ERN is thought to arise from anterior cingulate activity. Borst and his colleagues were interested in sulcal folding in the anterior cingulate, because asymmetric folding (i.e., different folding in the left versus right hemisphere) is associated with better inhibitory control in adults who have schizophrenia. Sulcal folding is determined in utero, and is not affected by maturation after birth or by learning. Accordingly, Borst et al. measured sulcal folding in the anterior cingulate bilaterally in children aged five years, and followed up the same children aged nine years. At both time points, the children performed a child-friendly colour Stroop task as a measure of inhibitory control, and digit span tasks as a measure of working memory.

Borst et al. used anatomical MRI scans to assess sulcal folding at age five. Less than half of the children in the study had asymmetric sulcal patterns (N = 7), the rest (N = 11) had symmetric sulcal folding. The groups differing in sulcal asymmetry did not differ in social class, non-verbal IQ (Ravens matrices) nor gender, nor in the working memory measures, but they did differ in inhibitory control. Inhibitory control as measured by the Stroop task was significantly better for the children with asymmetrical folding of the anterior cingulate, and this was the case both at age five and at age nine. The factors explaining this structure–function relationship are unclear, however. Borst et al. (2014) suggested that asymmetry in sulcal folding could constrain the development of white matter connectivity to other prefrontal structures important for the efficiency of the 'cognitive control' network. The fact that this structural characteristic of the brain is established in utero is intriguing, as it

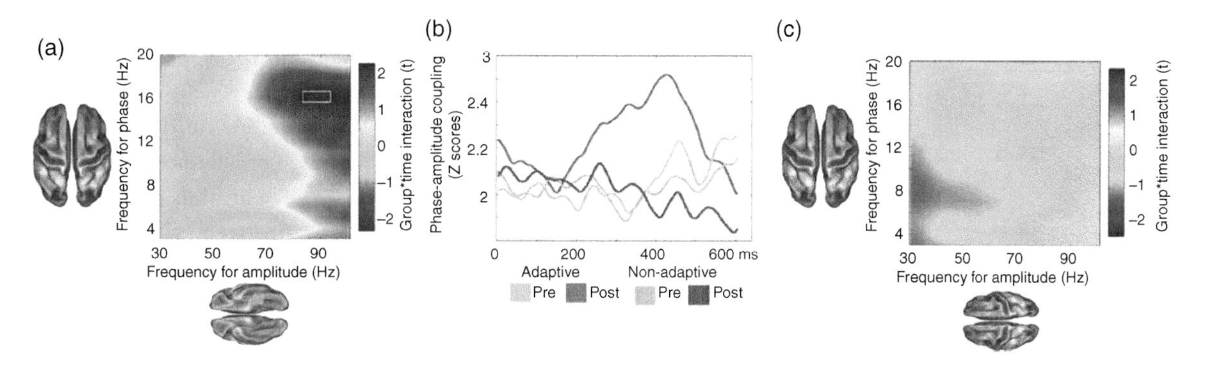

Depiction of the changes in phase-amplitude coupling (beta-gamma) following adaptive CogMed© training. Children who had received adaptive training showed a significant increase in beta-gamma (~16–90 Hz) phase-amplitude coupling between the dorsal fronto-parietal network and inferior temporal cortex while performing the spatial working memory task. Panel A shows the interaction between training group and time, with the white box denoting significant coupling. Panel B shows the time course of the coupling shown in A for the two groups by training condition. Panel C shows a control analysis based on the dorsal fronto-parietal network only, where no enhanced coupling is apparent.

may suggest that developmental disorders that involve impaired inhibitory control are determined at least in part by pre-determined structural characteristics of the brain. However, these structural differences could also be determined epigenetically, by features of the intra-uterine environment.

Another way to address the cause–effect question in brain imaging studies is to use training designs. Studies using MEG are also exploring functional connectivity as related to EF, but are contrasting connectivity in children who have either benefited from training in EF or have experienced a control intervention. If differences in connectivity are found between these two groups following training, then it is more likely that the changes in connectivity are specifically related to improvements in executive function. The best neural studies currently involve working memory training using CogMed©. For example, Astle and his colleagues reported MEG connectivity analyses based on scans of children aged eight to 11 years who had either undertaken training for 4–6 weeks on the adaptive version of CogMed© or on a non-adaptive placebo version of CogMed© (Astle, Barnes, Baker, Colclough & Woolrich, 2015). Astle measured connectivity at rest, and so the children were simply instructed to lie still with eyes closed while the MEG images were acquired. MEG scans were taken both before the CogMed© training and after training had been completed. The key question was whether changes in working memory capacity after training with CogMed© would be related to changes in intrinsic brain connectivity.

Astle et al. (2015) reported that the adaptive training group indeed showed enhanced functional connectivity at rest, indexed by greater beta band oscillatory activity (13–20 Hz). Enhanced connectivity was localised to functional networks that are typically active during EF tasks, namely right fronto-parietal networks linked to left lateral occipital cortex. This focus on the beta band is interesting given the finding discussed earlier relating children's total MEFS score (a broad EF measure) to beta power at rest in children aged 3–9 years (Perone et al., 2018). Subsequent beta-band connectivity analyses took account of individual gains in working memory to capture variable training effects across children. The fronto-parietal network both showed greater connectivity within itself when training benefits were stronger, and greater connectivity with parietal cortex. In essence, beta band activity in relevant networks that was already present before training was enhanced by CogMed© training, and enhancement showed a linear relationship with the degree of training benefits. The fact that greater connectivity was measured at rest excludes explanations based on differential task performance in the scanner.

As noted earlier in this chapter, neuroelectric oscillations are thought to be important mechanistically for working memory, although exactly how is not well understood. In order to explore how the increased co-ordination of beta band oscillations in the fronto-parietal network might support better working memory, the same children also performed a spatial working memory task while in the MEG scanner, both before and after CogMed© training (Barnes, Nobre, Woolrich, Baker & Astle, 2016). Phase-amplitude coupling between the phase of maximal responding of beta band oscillations (called preferred phase) and the amplitude of gamma band responses (power) was then explored. Children who had received adaptive training showed a significant increase in phase-amplitude coupling in the fronto-parietal network of interest while performing the spatial working memory task. Phase-amplitude coupling relations in the brain are thought to index top-down modulation of local cell assemblies, so for example the phase of the slower oscillation can help to control the power of the faster oscillatory response, enabling the faster response to occur at more optimal time points with respect to the task in hand. The enhanced phase-amplitude coupling observed by Barnes et al. was also significantly related to the children's individual gains in resting state connectivity. One interpretation of Barnes et al.'s data is that the beta oscillation generated by the fronto-parietal network is important for co-ordinating ongoing neural gamma activity related to task performance, with stronger phase-amplitude coupling following CogMed© training related to better task performance via better top-down neural control of local gamma activity. If this were the case, then naturally occurring individual differences between children in visuo-spatial working memory capacity should also be linearly related to beta-gamma phase amplitude coupling. This could be explored in future studies.

Executive function and theory of mind

As noted at the beginning of this chapter, it seems very likely that there will be important developmental connections between developing reflective awareness of one's *own* cognition (metacognition) and developing reflective awareness of the cognitions of *others* (theory of mind). This question about developmental connections was previously a very active area of research (e.g., Frye et al., 1995; Russell, 1996; Hughes, 1998; Perner & Lang, 2000). Many studies have now shown that there are significant correlations between performance on executive function tests and performance on theory of mind tests such as the false belief task, even when age and intelligence are controlled. In particular, measures of inhibitory control and of working memory show strong associations with false belief tasks. Theoretically, this was believed to reflect the need to suppress irrelevant perspectives when keeping track of false beliefs (inhibitory control) and to keep multiple perspectives in mind (working memory, see Carlson, Mandell & Williams, 2004a), although later work has shown that inhibitory control and working memory also show associations with non-false belief TOM tasks (for three- and four-year-olds, see Carlson,

Claxton & Moses, 2015). Perhaps surprisingly, planning does not seem to be related to performance in theory of mind tasks.

For example, in her landmark study, Hughes (1998) gave her three- and four-year-old participants six theory of mind tasks. These were two false belief tasks, two tasks requiring an explanation of why a doll held a false belief, and two measures of deception. The false belief tasks involved (a) false location, which was a version of the Maxi and his chocolate task discussed in Chapter 4, and (b) false contents, which was a version of the Smarties task discussed briefly in Chapter 3. Hughes found that all the theory of mind measures were correlated with her measures of inhibitory control and working memory, whereas the attentional flexibility measure (based on set shifting) was only correlated with deception. Once again, when verbal ability and non-verbal ability were controlled, inhibitory control and attentional flexibility were significantly related to deception only, and relations with working memory became non-significant. Davis and Pratt (1995) used forward and backward digit span to measure working memory in 54 children aged 3–5 years, and also administered two false belief tasks (the Smarties task and an appearance-reality task using a sponge/rock), two false photograph tasks, and a measure of vocabulary. They found that the backward digit span task was a unique predictor of children's ability to pass the false belief and false photograph tasks, even when age and vocabulary were controlled. Forward digit span did not show the same relationships. Data such as these have led to backward digit span being considered a measure of executive function rather than of memory per se.

Similar findings were reported by Carlson and Moses (2001) and Carlson et al. (2004b). Carlson and Moses included four theory of mind tasks in their study of 107 preschoolers, two false belief tasks (false location and false contents), an appearance-reality task and a deception task. Children's performance on the inhibitory control battery (based on ten tasks) was significantly correlated with their performance on the theory of mind battery, and this relationship remained significant after controls for age, verbal ability and gender. No measure of non-verbal ability was taken for this sample. The planning measure (motor sequencing) was not related to performance on the theory of mind battery after controls for age and verbal ability. Carlson et al. (2004b) included more measures of planning in their battery of executive function tasks. In their study, the theory of mind tasks were false belief tasks (false location and false contents) and appearance-reality tasks. When the relative contributions to theory of mind made by the inhibitory control measures versus the planning measures were explored, it was found that only inhibitory control was significantly related to theory of mind once age and verbal ability were controlled. Significant relations between planning and theory of mind did not survive in these more stringent analyses. Multiple regression analyses showed that whereas inhibitory control still explained significant unique variance in theory of mind performance when verbal ability and planning were entered into the equations first, planning did not explain significant unique variance in theory of mind performance when verbal ability and inhibitory control were entered into the equations first. Carlson et al. concluded that individual differences in the inhibitory control aspects of executive function rather than in the planning aspects were related to theory of mind performance. More recent studies by Carlson and her colleagues suggest that both inhibitory control and working memory play important roles in explaining these connections (Carlson et al., 2015).

THE DEVELOPMENT OF REASONING

The study of reasoning by children as distinct from executive function has focused on inductive and deductive reasoning. The literature shows that inductive and deductive reasoning are available early in development, and function in highly similar ways in children and in adults. Despite early views that reasoning was age-dependent and content-independent (see Brown, 1990), it has become clear that children do not gradually become efficient all-purpose reasoning machines. Children do not acquire and apply general reasoning strategies irrespective of the domain in which they are reasoning (and neither do adults, see Goswami, 2010). Rather, inductive and deductive reasoning show remarkable continuity across the lifespan, much like self-monitoring and implicit memory. Inductive reasoning and deductive reasoning are influenced by similar factors and are subject to similar heuristics and biases in both children and adults. Further, in the few studies that have explored links between metacognition and reasoning, the efficiency of children's reasoning is usually markedly improved by encouraging children to reflect on their reasoning processes (e.g., Brown, 1978). The same is true of adults (Gick & Holyoak, 1980).

Traditionally, the development of reasoning and problem-solving was thought to involve the acquisition of logical rules. Therefore, development was studied by seeing whether children could acquire isolated logical rules in completely unfamiliar situations. These studies were largely of deductive reasoning, as deductive reasoning problems can be solved without (or despite) real-world knowledge. In a deductive reasoning problem, there is only one logically valid answer. For example, if a child is given the two premises "All cats bark" and "Rex is a cat" (see Dias & Harris, 1988), there is only one logical deduction. Rex is a cat, all cats bark, therefore Rex must also bark. The logical deduction is counterfactual (it goes against children's real-world knowledge about the facts associated with cats), but counterfactual deductions are still logically valid.

In contrast, there is no logical justification of induction (Hume, 1748). Inductive inferences may not be logically valid, but they are still very useful in human reasoning. Generalising on the basis of a known example, making an inductive inference from a particular premise or drawing an analogy are all examples of inductive reasoning. Conceptual development involves reasoning on the basis of known examples, and as we saw in Chapter 5 this kind of induction is taken for granted in young children's reasoning. Even babies can make inductive inferences and create perceptual and conceptual prototypes (see also Chapter 2). The most important constraint on inductive reasoning is similarity. A typical inductive reasoning problem might take the form "Humans have spleens. Dogs have spleens. Do rabbits have spleens?" (see Carey, 1985). In the absence of any knowledge about spleens, it is impossible to know whether rabbits have spleens. However, as two other mammals apparently have spleens, it seems likely that rabbits might have spleens too. If the problem had been phrased as follows: "Dogs have spleens. Bees have spleens. Do humans have spleens?", then the induction becomes less intuitively compelling. Humans are reasonably similar to dogs, but they are not similar to bees. Perhaps humans don't have spleens? As we will see, beyond categorisation the development of inductive reasoning has been studied most comprehensively in terms of the ability to reason by analogy. The ability to make inductive generalisations from examples has been accepted as early-developing, but has been argued to be distinct from analogy. This is because inductive generalisations can be made on the basis of similarities in appearance ('surface similarity'), whereas the core to reasoning by analogy is relational or structural similarity. The hallmark of analogical reasoning has been seen as the ability to apply the *relational similarity constraint* (Goswami, 1992); that is, to constrain inductive inferences on the basis of relational similarity.

Reasoning by analogy

Analogical induction plays an important role in the history of science. An early example of reasoning by analogy was Archimedes's insight into the value of using water displacement to quantify mass when comparing different substances. According to the story (see Goswami, 1992), Archimedes had been asked to calculate whether base metal had been substituted for gold in an ornate and intricately designed crown that had been commissioned by his king. Archimedes knew the weight per volume of pure gold, but the crown was so ornate that he could not measure its volume. Unable to reach a solution, he went home and had a bath. According to the legend, he then cried "Eureka, I've got it". When he stepped into his bath, he had noticed that his body displaced a certain volume of water. By making an analogy between his body and the crown, the mathematical solution to calculating the gold in the crown became available: immerse the crown in water, and see whether the volume of water that was displaced was equivalent to that displaced by pure gold. This kind of 'insight' is a classical example of inductive reasoning.

Analogies are used whenever we recall familiar past situations in order to deal with novel ones, whether metacognitively (reflectively) or not. When reasoning by analogy "we face a situation, we recall a similar situation, we match them up, we reason, and we learn" (Winston, 1980, p. 689). To solve a new problem by using an analogy, we need to find the *correspondences* between the previously encountered problem and the novel one. This enables us to 'match up' the two situations. We then need to transfer knowledge from the familiar problem to the novel one. The identification of these correspondences usually requires *relational* reasoning. The solution to one problem can usually be applied to a different problem if similar sets of relations link different sets of objects in the two problems.

This point can be illustrated by thinking about some of the analogies that have led to new discoveries in science. One of the most famous was Kekule's (1865) new theory about the molecular structure of benzene, which he discovered on the basis of an analogy to a visual image that he had of a snake biting its own tail (Figure 9.8, see Holyoak & Thagard, 1995, for other examples). The carbon atoms in benzene are arranged in a ring, which shares visual similarity with a snake biting its tail, even though the objects in the analogy bear no resemblance to each other at all. The similarity is

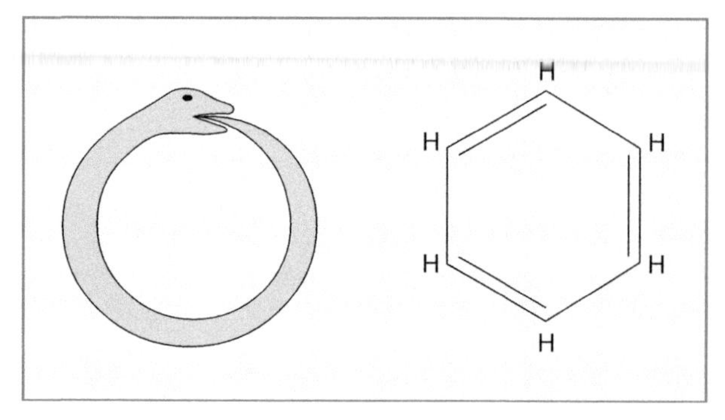

FIGURE 9.8
The visual analogy between a snake biting its own tail and the molecular structure of benzene. From Holyoak and Thagard (1995). © 1995 Massachusetts Institute of Technology, by permission of MIT Press.

purely relational – in this case, arrangement of an object or objects in a ring. Sometimes, however, there are similarities in the objects *as well as* in the relations in an analogy. An example is the invention of Velcro, which was developed in 1948 after George de Mestral noticed that burdock burrs stuck to his dog's coat because they were covered with tiny hooks (see Holyoak & Thagard, 1995). Velcro shares 'surface' similarity (similarity of appearance) as well as relational similarity (capacity to stick tight via hooks) with burdock burrs.

Most research on the development of analogical reasoning has examined whether children can recognise relational similarities between previously encountered problems and novel ones (i.e., identify correspondences), and whether they can use relational reasoning to solve analogies (the question of transfer). Related questions have been how early children are able to make relational mappings, and whether children can map relational similarities in the *absence* of surface similarities (e.g., Gentner, 1989). I will focus here on the first two of these questions, as the related questions have been covered to some extent in earlier chapters. For example, research discussed in Chapter 2 showed that even babies can make rudimentary analogies, suggesting that relational mapping abilities are present from very early in life. Similarly, research discussed in Chapter 5 showed that surface similarities between problems are not necessary for young children to use analogies. The 'personification' analogy that children use to help them to develop conceptual knowledge about biological kinds (making analogies from people to dogs and plants) is a good example. More detailed discussion of these points is available in Goswami (1992, 1996, 2001). Further, both of these processes are improved by metacognition. Identifying correspondences and transferring relations are facilitated by encouraging children to reflect on what they are doing – a metacognitive strategy.

The use of relational reasoning in childhood

The question of whether children have the cognitive ability to make relational mappings has been largely investigated through studies using *item* analogies. Item analogies provide a pure measure of relational reasoning. In an item analogy, the relation between two items *A* and *B* must be mapped to a third item *C* in order to complete the analogy with an appropriate *D* term. For example, to complete the item analogy 'bird is to nest as dog is to?' (*bird:nest::dog:?*), children must map the relation *lives in* that links *bird* to *nest* to the item *dog* in order to reason that *doghouse* is the correct solution to the analogy.

Item analogies can be given to quite young children, as long as they are set in familiar domains (see Goswami, 1991, 1992). For example, a four-year-old can be given the analogy *bird:nest::dog:doghouse* by presenting the task in the form of a game about constructing sequences of pictures (e.g., Goswami & Brown, 1990). Here is four-year-old Lucas trying to predict which picture he needs in order to complete the picture sequence *bird:nest::dog:?*, depicted in Figure 9.9.

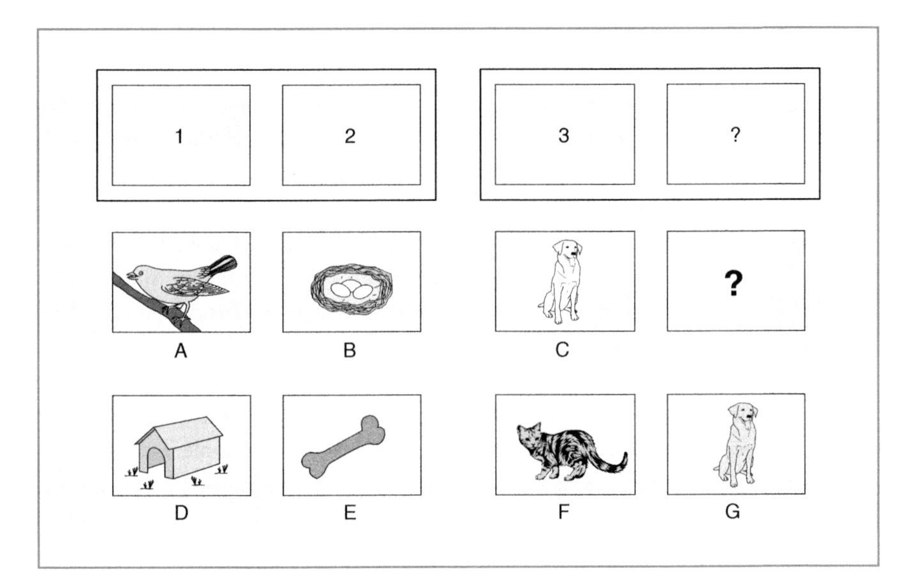

FIGURE 9.9
The game board (top row), analogy terms (middle row) and correct answer and distractors (bottom row) used for the analogy *bird:nest::dog:doghouse* by Goswami and Brown (1990). Copyright © 1990 Elsevier. Reproduced with permission.

> *Bird lays eggs in her nest [the nest in the B-term picture contained three eggs] – dog – dogs lay babies, and the babies are – umm – and the name of the babies is puppy!*

Lucas used the relation 'type of offspring' to solve the analogy. The solution that the experimenters had intended, however, was 'doghouse', as they had linked the A and B terms by the alternative relation 'lives in'. Following his verbal prediction, Lucas was shown the available completion pictures for the analogy, which did not in fact include a picture of a puppy. Instead, they depicted a *doghouse*, a *bone*, another *dog* and a *cat*. Lucas was not interested in these pictures, as he was quite certain that his answer was correct:

> *I don't have to look [at the distractor pictures] – the name of the baby is puppy!*

In the end, when Lucas was persuaded to look at the different solution pictures designed by the experimenters, he decided that the picture of the *doghouse* was the correct response. This shows the strength and flexibility of young children's analogical reasoning skills. Of course, *puppy* was an equally correct solution to the analogy given the A and B terms in the picture sequence, and Lucas's defence of his solution suggests that he fully understood the relational mapping constraint that determined the correct solution. Nevertheless, he could use this constraint flexibly when faced with alternative solutions to work out another correct answer.

Analogical reasoning as measured by the A:B::C:D item analogy format is thus available by at least age four (Goswami & Brown, 1990). If the same format is used to explore analogies based on causal relations, such as *cutting* or *melting* (e.g., *apple:cut apple::playdoh:cut playdoh; chocolate:melted chocolate::snowman:melted snowman*), then even three-year-old children succeed in the item analogy task (Goswami & Brown, 1989). It is difficult to use the item analogy format to demonstrate analogical competence in children younger than three, however, because of the abstract nature of the task. With children younger than three, it is necessary to devise ingenious *problem analogies* in order to show analogical reasoning at work.

In problem analogies, a young child is faced with a problem that they need to solve. Let us call this problem B. The use of an analogy from a previously experienced problem, problem A, offers a solution. The measure of analogical reasoning is whether the children can use the solution from problem A to solve problem B. We have already discussed one such problem analogy task, the 'reaching-for-a-toy' task devised to study analogical problem-solving in infants and toddlers (e.g., Brown, 1990; Chen et al., 1997). This problem analogy format has also been extended to two-year-olds.

Singer-Freeman (2005) devised a series of analogies for two-year-olds using real objects and models. Her analogies were based on the simple causal relations of *stretching*, *fixing*, *opening*, *rolling*, *breaking* and *attaching*. For example, a child might watch the experimenter *stretching* a loose rubber band between two plexiglass poles in order to make a 'bridge' that she could roll an orange across ("Look what I'm going to do, I'm going to use this stuff to roll the orange! Stretch it out, put it on – wow, that's how I roll the orange!"). Following an opportunity to roll the orange by themselves, the children were given a transfer problem involving a loose piece of elastic, a toy bird and a model with a tree at one end and a rock at the other. They were asked "Can you use this stuff to help the bird fly?" The intended solution was to stretch the elastic from the tree to the rock, and to 'fly' the bird along it. In a third analogy problem, the children were asked to "give the doll a ride" by stretching some ribbon between two towers of different heights that were fixed to a base board. Children in a control condition were simply asked to "help the bird fly" and "give the doll a ride" without first seeing the base analogy of rolling the orange.

Singer-Freeman found that whereas only 6% of the children in the control condition thought of the *stretching* solution to the transfer problem, 28% of 30-month-olds in the analogy condition did so, and this figure rose to 48% following hints to use an analogy ("You know what? To help the bird fly, we have to change this", said while pointing to the elastic). When the same hint was given to the children in the control condition, only 14% thought of the *stretching* solution. Although these performance levels may appear modest, they are comparable to the spontaneous levels of analogical transfer found in adults. Problem analogy studies conducted with adults typically find spontaneous transfer levels of around 30%, at least in unfamiliar problem scenarios (e.g., Gick & Holyoak, 1980). More recent work with young children (four-year-olds) also reports around 30% successful transfer when no metacognitive support is offered to help the children to notice the analogy (Tunteler & Resing, 2002).

Metacognition and reasoning by analogy: 'Learning to learn'

Of course, in problem analogy tasks children have to *notice* the analogy as well as to perform the relational mapping correctly. Notising relational similarities or correspondences between previously encountered problems and novel ones may require the child to reflect upon their knowledge; in other words, to exercise some metacognitive control. The ability to notice or recognise relational similarities between problems can also be investigated via problem analogy tasks. The factors that influence whether young children notice relational similarities between problems have been most extensively examined by Brown and her colleagues (e.g., Brown & Kane, 1988; Brown, Kane & Echols, 1986; Brown, Kane & Long, 1989). For example, Brown et al. (1986) gave four- and five-year-old children the 'Genie' problem invented by Holyoak, Junn and Billman (1984) to try to solve, and then demonstrated the solution. The children's ability to notice the correspondences between the Genie problem and a series of analogous problems was then measured in a variety of different conditions, some of which encouraged metacognitive reflection.

In the 'Genie' problem, a genie is about to move from one location to another. He needs to take some precious jewels with him, and his problem is how to move them from the old location to the new location without damaging them in any way. His solution is to roll his magic carpet into a tube, and then to roll the jewels through this tube. The children in Brown et al.'s study were shown the Genie problem via a toy scenario with toy props. They then enacted the solution with the experimenter, rolling up a piece of paper that represented the magic carpet, and rolling the jewels through the paper tube. In order to help the children to extract the *goal structure* of the problem, they were asked a series of questions to make them reflect on key aspects of the problem-solving process including "Who has a problem?", "What did the genie need to do?" and "How does he solve his problem?" The children were then shown another problem intended to be analogous to the Genie's, which also involved toy props. This was the 'Easter Bunny' problem. An Easter Bunny needed to deliver a lot of eggs to children in time for Easter, but had left things a bit late. A friend had offered to help him, but the friend was on the other side of a river, and so the eggs had to be transported across the river to this friend without getting wet. The idea was that the Easter Bunny could use an analogous solution to the genie by rolling his blanket (a piece of paper) into a tube and rolling the eggs across the river through this tube.

Brown et al. found that 70% of the children in the reflective questioning group noticed this analogy spontaneously. However, only 20% of children in a control group noticed the analogy by themselves. This control group had also experienced the Genie's problem, but had not been questioned about the goal structure of the story. Hence they had not received the metacognitive manipulation. Brown's conclusion from this and a series of similar studies was that children found it easy to recognise relational similarities between previously encountered problems and novel ones as long as they had represented the relational structure of the previously encountered problem in memory. Questioning by the experimenter facilitated this representational process, as it encouraged the children to reflect upon and represent the important relations that enabled the character to achieve his goal. Brown then investigated the effects of metacognitive support on children's analogical reasoning by investigating the effects of experiencing a *series* of analogies, and of being *taught* to look for analogies during problem-solving ('learning-to-learn'; e.g., Brown & Kane, 1988; Brown et al., 1989).

In Brown's metacognitive studies, two novel paradigms were devised, labelled the A-B-A-C paradigm and the A1-A2/B1-B2/C1-C2 paradigm. In the first paradigm, children were introduced to problem A and were asked to solve it by themselves (which they were typically unable to do). They were then given an easier problem B, which was actually similar in terms of relational correspondences to problem A. Following successful solution of problem B, problem A was re-administered, along with gentle hints about its similarity to problem B. A novel problem C was then given, to measure spontaneous analogical transfer. In the A-B-A-C paradigm, the solution to be transferred was always the same. In the A1-A2/B1-B2/C1-C2 paradigm, the children were transferring three *different* solutions, one each for problem pair A, B and C. The solution to problem A1 was analogous to the solution to problem A2, the solution to problem B1 was analogous to the solution to problem B2, and the solution to problem C1 was analogous to the solution to problem C2. In this paradigm, Brown argued that the children were forming a 'learning set' to look for analogies. In each case, the metacognitive component was implicit rather than explicit – the children had to figure out the similarities by responding to the hints from the experimenter.

For example, in Brown et al.'s (1989) investigation of the A-B-A-C paradigm, Holyoak et al.'s (1984) Genie paradigm formed the core of the procedure. Children aged seven were given the Genie problem to solve by themselves with toy props including a real toy carpet, and when they failed, they were given the Easter Bunny problem described above. The experimenter helped them to solve the Easter Bunny's problem before re-presenting the Genie problem. The children were told that problem B would help them, as the two problems were 'just the same'. Finally, a novel problem C was administered (this was about a farmer who had to transfer ripe cherries across a fallen tree without damaging them). Brown and her colleagues were most interested in performance with problem C. Success with problem C would indicate that the children had extracted 'meta-knowledge' about using analogies. Children's performance was compared to that of control children who had simply received the same problems in the A-B-A-C order without any of the help or hints given to the experimental group.

The results showed that 98% of the children in the metacognitive group solved problem C by rolling up paper, whereas only 38% of children in the control group generated the rolling solution. The children were 'learning-to-learn', learning to use analogy even though they were never instructed explicitly in how the problems were alike. Similar 'learning-to-learn' effects were demonstrated in the A1-A2/B1-B2/C1-C2 paradigm with even younger children. Here children aged three, four and five years learned to transfer different solutions (stacking objects, pulling objects, swinging over obstacles) between problem pairs. At the same time, as they progressed through the problem sequence, they extracted an abstract notion of the usefulness of problem-solving by analogy. When performance on the final problem, C2, was assessed, 85% of three-year-olds, 95% of four-year-olds and 100% of five-year-olds were successful in problem-solving by analogy (Brown et al., 1989).

Inhibition and reasoning by analogy

The role of metacognition in reasoning by analogy has been explored by Richland and her colleagues, using a picture-based 'scene analogy' item task (Richland,

Morrison & Holyoak, 2006). The item analogies in this task involved relations captured by natural language verbs such as 'kiss', 'chase' and 'feed', which are understood by very young children. The scene analogy task enabled children's success in reasoning by analogy to be compared when two relations were depicted compared to when only one relation was depicted. For example, an analogy could be *cat chases mouse::boy chases girl*, or *dog chases cat chases mouse::mum chases boy chases girl*. The latter type of analogy involving two relations was expected to require more working memory capacity, which was expected to be present in older but not younger children. Inhibitory skills were tested in each kind of analogy by the presence of object distractors. For example, for the 'boy chases girl' analogy, a cat was present in the scene as a distractor. Older children were expected to be better at inhibiting responses to distractors than younger children.

Richland et al. (2006) found that all children tested (3–4-year-olds, 6–7-year-olds, 9–11-year-olds and 13–14-year-olds) solved more single-relation analogies than two-relation analogies. Richland and her colleagues argued that this was because the two-relation analogies placed greater demands on working memory. Furthermore, only the 13–to 14-year-olds showed no effect of distraction. Richland et al. suggested that maturational changes in the availability of inhibitory control (i.e., the ability to inhibit erroneously selecting the cat picture) explained these developmental effects. Accordingly, differential success rates by age in analogies where relations are known could depend on general executive factors such as the ability to hold and integrate relations in working memory and the ability to inhibit competing irrelevant distractors.

The ability to hold and integrate information in working memory and the ability to inhibit competing irrelevant distracting information would be expected to affect other kinds of reasoning as well as reasoning by analogy, as we discussed when considering scientific reasoning (Chapter 7), and as we will also see below for deductive reasoning. Meanwhile, Brown's work shows that even very young children can extract meta-knowledge that facilitates inductive reasoning following hints, as long as they have rich conceptual representations of the domain being studied, and as long as they are interested in the subject matter. The early age at which analogies appear suggest that they provide a powerful logical tool for explaining and learning about the world. Analogies also contribute to both the acquisition and the restructuring or reorganisation of knowledge. As children's knowledge about the world becomes richer, the structure of their knowledge becomes deeper, and more complex relationships are represented, enabling deeper or more complex analogies. This means that, as children learn more about the world, the type of analogies that they make will change. While metacognitive variables are likely to affect analogy performance, the key factor affecting *competence* thus remains conceptual understanding.

Indeed, in certain circumstances younger children's lack of richly structured conceptual knowledge can enable *better* relational reasoning than shown by older children. For example, using a version of the 'blicket' paradigm discussed in Chapter 7, Walker and Gopnik (2014) showed toddlers and three-year-olds the same relational information. This was that pairs of perceptually similar blocks made the blicket machine play music, but pairs of blocks that were not perceptually similar did not activate the machine. Other toddlers were shown that pairs of perceptually different blocks activated the blicket machine, but that pairs of perceptually similar blocks did not. The toddlers (aged 18–30 months) quickly learned to select either pairs of similar blocks to make the machine play music, or pairs of dissimilar blocks. The older children did not, and neither did adults given this version of the blicket problem. Control conditions showed that the older children and adults tended to assume that individual objects caused effects, and so did not notice the higher-order relation linking the pairs of blocks (same or different, see Gopnik, Griffiths & Lucas, 2015). The toddlers, unencumbered by this prior knowledge, noticed the relational similarity constraint.

Deductive logic and deductive reasoning

Another early-developing mode of logical reasoning is deductive reasoning. Problems that can be solved by deductive reasoning have only one right answer. The problem-solver deduces this answer on the basis of the logical combination of the premises presented in the problem. For example, we use deductive logic to solve *syllogisms*, which are problems like the following:

All cats bark.
Rex is a cat.
Does Rex bark?

Given these premises, the only possible answer is that yes, Rex does bark. Although the premises in this example are obviously contrary to fact, as in the real world cats cannot bark, the plausibility or potential truth of the premises does not matter as far as the logical deduction is concerned. When children are given syllogisms to solve, the test of deductive reasoning is not whether the premises are counterfactual or not, but whether the child can draw the correct deductive inference. The critical test is whether the children can recognise that the premises, whatever they may be, *logically imply* the conclusions.

Syllogistic reasoning

Experimental research has shown that even quite young children can make deductive inferences about counterfactual premises. The problem about whether Rex barks or not was posed to five- and six-year-olds in an experiment by Dias and Harris (1988; see Table 9.2). In addition to 'contrary facts' problems such as whether Rex barks, Dias and Harris also gave children a selection of 'known facts' problems ("All cats miaow. Rex is a cat. Does Rex miaow?"), and 'unknown facts' problems ("All hyenas laugh. Rex is a hyena. Does Rex laugh?"). One group of children in their experiment were given the reasoning problems in a 'play' mode. In this condition, toy cats, dogs and hyenas were presented, and the experimenter made them miaow, bark and laugh. A second group were simply told the premises in the problems without any toys or demonstrations, and asked to judge the conclusion.

Dias and Harris found that the children in the 'play' group performed at or close to ceiling on all the different problem types. They were able to reason

TABLE 9.2 Examples of the counterfactual syllogisms used by Dias and Harris (1988)

"Yes" answers	**"No" answers**
(What noise do cats make?)	(Where do fishes live?)
All cats bark.	All fishes live in trees.
Rex is a cat.	Tot is a fish.
Does Rex bark?	Does Tot live in water?
(What are books made of?)	(What colour is milk?)
All books are made of grass	All milk is black.
Andrew is looking at a book	Jane is drinking some milk.
Is it made of grass?	Is her drink white?
(What colour is snow?)	(What colour is blood?)
All snow is black.	All blood is blue.
Tom touches some snow.	Sue has blood on her hand.
Is it black?	Is it red?
(How do birds move?)	(What is the temperature of ice?)
All birds swim.	All ice is hot.
Pepi is a bird.	Ann has some ice.
Does Pepi swim?	Is it cold?

deductively whether the problems were contrary to fact, used known facts, or used unknown facts. In contrast, the children in the verbal group only showed high levels of responding in the 'known facts' problems ("All cats miaow. Rex is a cat. Does Rex miaow?"). These problems could have been solved by using real-world knowledge rather than by using deductive logic.

In a follow-up experiment using only the 'contrary facts' problems, Dias and Harris tried presenting the premises verbally to *both* groups of children. Their aim was to rule out the possibility that the presence of the toy animals was acting as a memory prompt for the children in the 'play' group. This time, the children in the 'play' group were told that they should pretend that the experimenter was on another planet, and that everything on that planet was different. For example, the experimenter would say "All cats bark. On that planet I saw that all cats bark", using a 'make-believe' intonation, and would then verbally present the syllogism. In these 'make-believe' conditions, the levels of reasoning shown by the 'play' group remained close to ceiling. In a later study (Dias & Harris, 1990), children as young as four were found to be capable of syllogistic reasoning. This effect was robust whether the premises were presented as referring to another planet, were presented using a make-believe intonation or were presented using visual imagery. Dias and Harris concluded that young children were capable of deductive reasoning, even about counterfactual premises, as long as logical problems were presented in the context of play.

Leevers and Harris (2000) went on to show that the play context was not critical to children's capacity for deductive logic, however. Leevers and Harris gave four-year-old children counterfactual syllogisms similar to those used by Dias and Harris, but the children in their study were simply told to *think* about the problems (e.g., "I want you to *think about* it. I want you to close your eyes and make a picture in your head… so the x that you are *thinking about*, in the picture in your head…"). Examples of the problems used by Leevers and Harris included "All snow is black. Len is a snowman made of snow. Is Len white?" and "All ladybirds have stripes on their backs. Daisy is a ladybird. Is Daisy spotty?" The four-year-olds showed good syllogistic reasoning, and this transferred to new counterfactual problems given 2–3 weeks later without the mental imagery instructions. Leevers and Harris argued that their manipulations may have improved counterfactual reasoning because they encouraged the children to process the premises mentally instead of dismissing them as absurd. This idea was supported by the types of justifications given by the successful children. These were largely theoretical in nature. For example, one four-year-old girl commented "All ladybirds have stripes on their back. But they don't" before reasoning that Daisy's ladybird was not spotty. Syllogistic reasoning thus appears to be present by at least four years of age.

Conditional reasoning and the selection task

KEY TERM

Wason selection task Participants have to reason about conditional states of affairs of the 'if *p* then *q*' variety.

Another widely-used measure of deductive reasoning is the *selection* task, developed by Wason (1966). In the selection task, the subject is told about a certain (conditional) state of affairs 'if *p* then *q*'. For example, the subject might be told "If a letter is sealed, then it has a 50p stamp on it". The task for the subject is to decide on the minimum number of pieces of evidence that are needed to validate the rule. The pieces of evidence available are usually *p* (e.g., a sealed letter, shown face-down), *q* (e.g., a letter with a 50p stamp, shown face-up), *not-p* (e.g., an unsealed letter, shown face-down), and *not-q* (e.g., a letter with a 40p stamp, shown face-up, see Figure 9.10). Most adult participants can solve the selection task when the problem is presented in a familiar context such as sorting letters in the post office (Johnson-Laird, Legrenzi & Sonino-Legrenzi, 1972). The correct answer is that the minimum pieces of evidence required are *p* and *not-q*. This answer is given by the majority of adult participants in tasks using familiar contexts, but in more formal versions of the same task performance can be as low as 10% correct. A typical formal version of the selection task is "if there is a vowel on one side of a card, there is an even number on the other side" (see Wason & Johnson-Laird, 1972).

Cheng and Holyoak (1985) have argued that this huge discrepancy in adults' ability to use deductive logic in the selection task depends on whether the selection task taps into familiar knowledge structures called *pragmatic reasoning schemas*. Pragmatic reasoning schemas describe permission scenarios in real life. For example, adults frequently encounter permission rules such as "If you want this letter to arrive tomorrow, it must go first class", "If you are 18, you can drink alcohol in a pub" and "If you are 17, you can legally drive a car". Children probably encounter even more permission rules than adults. These may include "If it is 9 o'clock, you must be in bed", "If you are wearing your school blazer, you must wear your school cap" and "If the whistle has gone, you are not allowed to stay in the playground". Following Cheng and Holyoak's logic, we can predict that children should also show successful deductive reasoning in the selection task if it taps into a familiar type of permission schema.

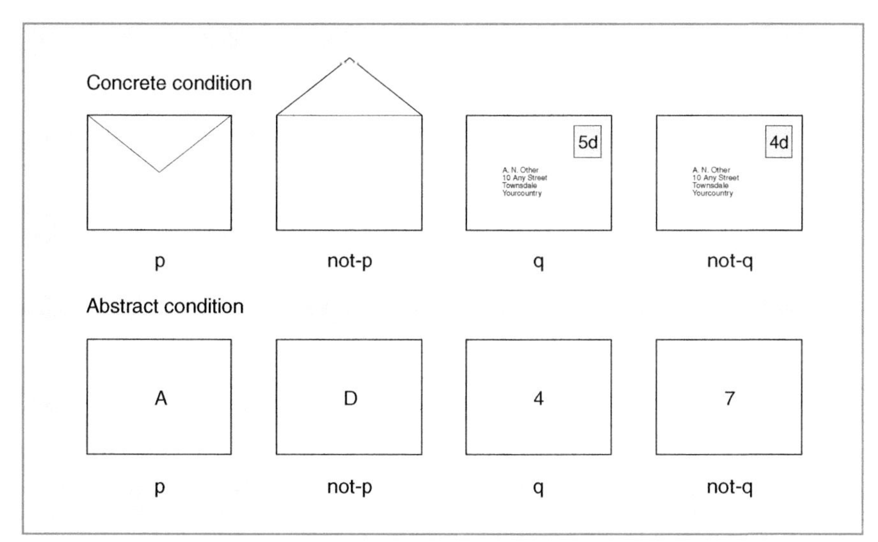

FIGURE 9.10
Two versions of the selection task, using concrete and abstract stimuli. From Johnson-Laird and Wason (1977). Copyright © Cambridge University Press. Reproduced with permission.

This idea was tested by Light, Blaye, Gilly and Girotto (1989), who devised permission rules that would be interpretable to six- and seven-year-old children. They used two rules: "In this town, the police have made a rule which says that all the lorries must be outside of the centre" and "In this game, all the mushrooms must be outside of the centre of the board". The first rule, which was designed to have underlying pragmatic force, was demonstrated by showing the children a game board with a brown centre and a white surround (see Figure 9.11). Pictures of lorries and cars were shown inside and outside the centre area. The second rule, which was designed to be arbitrary, was demonstrated using the same board but with pictures of flowers and mushrooms inside and outside the centre area instead of lorries and cars. Two lorries (or mushrooms) and one car (or flower) were always shown in the brown centre of the board, and one lorry (or mushroom) and three cars (or flowers) were always shown in the white surround.

The children's first job was to rearrange the pictures on the game board so that they obeyed the rule (e.g., moving the two offending lorries out of the town centre). The experimenter then tested the children's understanding of the rule by carrying out a potential violation. The violation was moving a picture of a car or a flower outside the centre and asking if that disobeyed the rule (this was permissible). Next, the children themselves were asked to move a picture so that it *did* disobey the rule. Finally, a version of the selection task was given. The children were shown the game board with two pictures on it, both upside-down, one in the brown area and one in the white area. The children were asked (a) which picture they would need to turn over to check whether the rule had been disobeyed, (b) whether the picture that they had turned over disobeyed the rule, and (c) whether the other picture disobeyed the rule. Light et al. found that 45% of the six-year-olds and 77% of the seven-year-olds succeeded in answering these three components of the selection task correctly in the lorry condition. However, only 5% of the six-year-olds and 23% of the seven-year-olds succeeded in the mushroom condition.

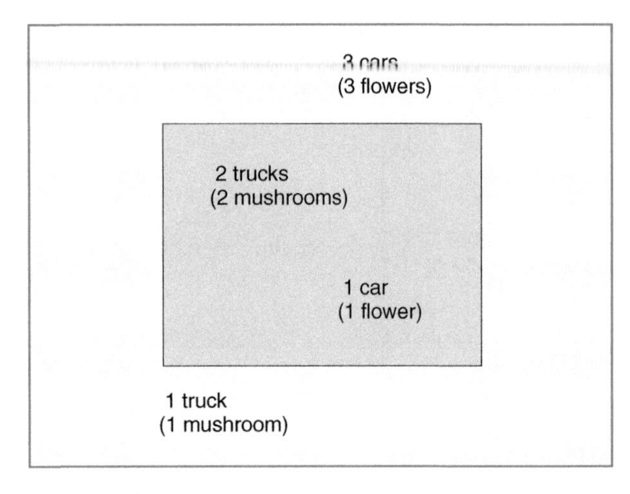

FIGURE 9.11
Schematic depiction of the game board used to test children's understanding of permission rules by Light et al. (1989). Copyright © 1989 Elsevier. Reproduced with permission.

These results suggest that six- and seven-year-old children, like adults, can use deductive logic in the selection task as long as an appropriate permission schema has been activated. Furthermore, Light et al. showed that activation of a permission schema in a *pragmatic* context could transfer to an *abstract* context. Some of the children who were successful in the lorry task were then given an abstract version of the selection task involving squares and triangles ("all the triangles must be in the centre"). Fifty-nine per cent of the successful seven-year-olds and 30% of the successful six-year-olds showed transfer from the lorry task to the triangles. Light et al. argued that this showed that the children understood the logic behind their correct choices in the lorry version of the selection task. The children were using their grasp of the pragmatics of permission and inhibition to help them to solve the task, and this understanding was then transferred to the triangles.

More recently, Harris and Nunez (1996) have shown that even three- and four-year-olds are sensitive to the pragmatics of permission and inhibition in the selection task. Harris and Nunez used a variety of story formats to present different permission rules, and then asked the children to select the correct picture from a set of four which depicted a breach of the rule. For example, in one study using familiar permission rules, the children were told "This is a story about Sally. One day Sally wants to play outside. Her Mum says that if she plays outside, she must put her coat on." The children were then shown four pictures: a picture of Sally outside with her coat on; a picture of Sally outside without her coat on; a picture of Sally inside with her coat on; and a picture of Sally inside without her coat on (see Figure 9.12). The children were asked to select the picture in which the protagonist was breaking the permission rule ("Show me the picture of where Sally is being naughty and not doing what her mum told her"). This required them to select the picture that depicted the combination of *p* and *not-q*.

The majority of children in both age groups chose the picture of Sally outside without her coat on. Similar results were found when the children were asked about novel rules ("Carol's Mum says if she does some painting she should put her helmet on"). Obviously, Harris and Nunez's paradigm is less demanding than the traditional selection task paradigm, as they required their participants to select the picture that depicted the combination of *p* and *not-q* rather than to make an independent identification of *p* and *not-q*. Nevertheless, their conclusion that three- and four-year-old children are quite capable of identifying breaches of a permission rule seems a convincing one. Furthermore, they argued that children's grasp of permission rules was not restricted to *familiar* rules, as the link between condition and action could be quite arbitrary. Even so, the actions and conditions used by Harris and Nunez in their unfamiliar condition (painting, wearing a cycling helmet) were in themselves familiar to the children. The links were not quite as arbitrary as in the formal versions of the selection task used with adults ('if there is a vowel on one side of a card, there is an even number on the other side').

Metacognition and conditional and deductive reasoning

Markovits (e.g., Markovits & Barrouillet, 2002; Markovits, 2000) has argued for a role for metacognition in successful conditional reasoning by children. He has suggested important developmental links between conditional reasoning, working memory capacity and inhibition. In general, Markovits accepts that younger children can solve conditional reasoning tasks. For example, Markovits and Barrouillet (2002) proposed that children did have an understanding of *if–then* propositions, that this understanding was inherently relational, and that rich linguistic and pragmatic experiences supported this understanding. They argued that conditional reasoning depended on relational mappings and was based on relational structures (in a similar fashion to reasoning by analogy).

FIGURE 9.12
The set of choice pictures used to test children's understanding of the permission rule "If Sally wants to play outside, she must put her coat on". From Harris and Nunez (1996). Reproduced with permission from Blackwell Publishing.

For example, they proposed that when children were given a simple *if–then* statement (e.g., "If it rains, then the street will be wet"), they constructed a mental representation of the major premise based on prior knowledge (e.g., their prior experience of rain causing streets and fields to get wet). This mental representation represented relational structure, in terms of elements (streets, fields) and relations between elements (wetting by rain) in the environment. However, the mental representation was held in working memory, which has limited capacity. This capacity was also affected by the cognitive demands made by retrieving the familiar relational structure and the cognitive demands made by inhibiting irrelevant aspects of this structure. Because retrieval processes were assumed to be less efficient for younger children than for older children, because younger children were assumed in general to have less relevant knowledge than older children, and because younger children were less good at inhibiting inappropriate information (as in the case of counterfactuals, for example), conditional reasoning overall was accordingly less efficient in younger children compared to older ones.

More recently, Markovits (2017) has argued for the importance of other factors in children's conditional reasoning aside from pure logical ability. For example, Markovits (2017) has shown that children find it easier to reason conditionally about premises that draw on category knowledge, such as "If a plant is a cactus, then the plant will have thorns. This plant is a cactus. Does it have thorns?" than about premises that draw on causal knowledge, such as "If it rains, then the street will be wet. It has rained. Is the street wet?" Another factor that influences children's success is the number of potential alternative causes that are known from life experience. For example, for a statement like "If it rains, then the street will be wet. The street is wet. Did it rain?", there are rather few alternative potential causes of the street being wet. Rain is the most likely cause, although it is possible that a different cause such as a fire hose in fact made the street wet. However, for an apparently similar *if–then* statement like "If a rock is thrown at a window, then the window will break", prior knowledge tells us that there are more potential causes for broken windows than rocks, even though these other causes are not mentioned in the premises. We know from prior experience that windows can be broken for many reasons. So when asked "If a rock is thrown at a window, then the window will break. The window is broken. Was a rock thrown?", children find it more difficult to make a valid deductive inference.

Markovits's observations are similar in kind to the points made by Koslowski and Masnick (2010) concerning the role of background knowledge in children's successful scientific reasoning (discussed in Chapter 7). Accordingly, the fact that research on both conditional and syllogistic reasoning has also uncovered the important developmental role played by the retrieval of background knowledge is not particularly surprising. The developmental effects of the need to inhibit inappropriate or irrelevant information in order to make a successful deduction is also unsurprising. Inhibition of intuitive (but false) physical knowledge was also shown to be important for successful scientific reasoning, for example when reasoning about projectile motion.

Indeed, Stanovich and West (2000) have suggested that there might be a curvilinear age trend in the success of deductive reasoning. Over a wide age span, we might first expect age-related increases in successful reasoning, as children get better at inhibiting competing information, but then a decline in success at older ages (65-plus) as inhibitory processes become less efficient. Studies have provided evidence for this curvilinear trend, although with relatively old cohorts (average age from 12 to 66 years, De Neys & Van Gelder, 2009). However, when young children lack relevant background knowledge and hence have no competing information to inhibit, this perspective also predicts that they should *out-perform* adults in deductive logic. A critical test of belief inhibition in deductive reasoning tasks is provided by the 'conflict syllogism', when belief and logic conflict (as in "All mammals can walk. Whales are mammals. Therefore, whales can walk"). In these 'conflict' syllogisms, the conclusion is logically valid, but it is difficult to draw because it is unbelievable (Kokis, Macpherson, Toplak, West, & Stanovich, 2002).

Houde and his colleagues have devised a method of using 'conflict' syllogisms to investigate the role of inhibition in children's deductive reasoning. They used *negative priming* (Moutier, Plagne-Cayeux, Melot, & Houde, 2006). Moutier et al. devised 'conflict' syllogisms in which the children had to inhibit 'unbelievable = invalid' reasoning strategies, and reason purely on the basis of the premises. For example, for a conflict syllogism such as "All elephants are hay eaters. All hay eaters are light. All elephants are light", the child must inhibit real-world knowledge about elephants being heavy in order correctly to judge the syllogism as valid. In Moutier et al.'s experiment, these 'conflict' syllogisms were followed immediately by 'non-conflict' syllogisms, in which the deduction was both believable and valid. In these non-conflict syllogisms, the opposite deduction was required (so for the non-conflict syllogism, children had to confirm that all elephants are heavy). The experimenters argued that a role for inhibitory control in syllogistic reasoning would be revealed by a significant drop in reasoning success on the non-conflict syllogisms when preceded by the conflict syllogisms (a negative priming effect). Moutier et al. (2006) tested children aged eight to ten years, and reported a strong negative priming effect. They concluded that the inhibitory and logical components of syllogistic reasoning were to some extent dissociated. Overall, we can conclude that when children are reasoning about pragmatically plausible content, then their deductive reasoning can be as efficient as that of adults. Difficulties in deductive reasoning do not usually appear to be determined by the intrinsic logical structure of a particular task. Rather, difficulties are determined by the problem content or the mode of presentation, by the availability of relevant background knowledge that interferes with reasoning, and by the efficiency with which such irrelevant background knowledge can be inhibited. When children lack relevant background knowledge, they may actually reason *more* efficiently than older children and adults.

Cognitive neuroimaging studies of deductive and relational reasoning

We saw in earlier sections of this chapter that fMRI studies of the development of cognitive flexibility and response inhibition have moved beyond a simple view that the frontal cortex is critical to developmental changes, to recognising that studying changes in connectivity in an extensive fronto-parietal *network* is likely to be a more fruitful approach. Cognitive neuroimaging studies of reasoning are following a similar pattern, although studies with children focused on specific tasks such as logical syllogisms, conditional reasoning or reasoning by analogy are still sparse in this literature. Meta-analyses of studies with adults suggest a range of differential activation within a broad fronto-parietal network for different types of logical syllogisms and conditional reasoning problems (Prado, Chadha & Booth, 2011). Studies with adults that increase the relational complexity of task content, for example by varying the format of the Wason selection task discussed above, tend to report increased activity in this fronto-parietal system with increased problem complexity (Cocchi et al., 2014). Meanwhile, studies of reasoning by analogy in adults have shown that multiple regions in frontal cortex are activated when analogy tasks are solved in the fMRI scanner (Cho et al., 2010).

In an fMRI study with 27 children aged from eight to 13 years, Prado and his colleagues focused on conditional reasoning, and specifically on whether children could correctly reject conclusions that do not follow logically (Schwartz, Epinat-Duclos, Léone & Prado, 2017). Children were given problems such as "If a baby is hungry, then she will start crying. The baby starts crying. Is she hungry?" As noted above, Markovits (2017) has shown that children's performance in these conditional forms depends in part on how many potential alternative causes are known from life experience. Children are more likely correctly to reject the offered consequence ("She is hungry", which is not logically necessary given the form of the *if–then* statement) if they know that babies also cry for other reasons, such as being cold. Accordingly, as well as measuring deductive validity, Schwartz et al. (2017) also measured children's likelihood judgements on a sliding five-point scale from "not sure at all" to "very sure" (that the baby was hungry). Both forms of problem were solved while fMRI measurements were recorded.

Schwartz et al. were interested in whether similar brain regions would be active during valid logical deductions and likelihood estimation. The data showed that *different* brain regions were most active during the two kinds of trials. Correct judgements about logical validity were related to activation of left-lateralised frontal regions, whereas differences in likelihood ratings were related to activation of right-lateralised frontal regions and bilateral parietal regions. Schwartz et al. (2017) concluded that the brain mechanisms underlying each type of judgement were qualitatively different. However, nothing specific about mechanisms can be concluded from these data. The spatial dissociation is interesting, but it reflects the neural activity correlated with task performance rather than providing mechanistic neural information.

For reasoning by analogy, there are also a few studies with children. Bunge and her colleagues reported a neuroimaging study of the item analogy task (*Dress is to closet as milk carton is to?*) in 138 children aged from six to 18 years (Whitaker et al., 2017). The children had to solve the picture-based analogies by choosing from four possible solution pictures (e.g., *cow, clock, tennis racquet, fridge*) while fMRI data were collected. In this example, the *cow* is a semantic associate of milk, the *clock* was shaped like the milk carton in the picture, the *tennis racquet* was unrelated and the *fridge* was the correct answer. Accuracy increased with age, from around 50% in the six-year-olds to 100% in the 18-year-olds. Analyses focused on the brain regions for which activation was related to individual differences in analogical reasoning after correcting for age. Increased activation in left anterior inferior prefrontal cortex was specifically related to performance in the analogy task in these correlational analyses. Also notable was the finding that all participants activated a common network of frontal, parietal and occipital regions, as might be expected for a visually delivered reasoning task. Hence reasoning by analogy activates largely similar brain regions in children and adults.

Bunge and her colleagues have also explored structural connectivity in the Matrix Reasoning task from the Wechsler Intelligence Scale for Children, which is a form of visual analogy task. They used MRI to compare structural connectivity to functional connectivity as a correlate of reasoning ability, focusing on connections between the rostrolateral prefrontal cortex and parietal cortex (Wendelken et al., 2017). Functional connectivity was also measured by neural activation in a visual analogy task and a source memory task (from Sastre et al., 2016; see Chapter 8). Cross-sectional analyses showed that matrix reasoning ability was most strongly related to *functional* connectivity between right rostrolateral prefrontal cortex and the inferior parietal lobe for teenagers, but to fronto-parietal *structural* connectivity for younger children aged up to 11 years. The data were interpreted to show that structural connections precede functional connections. Many participants also contributed longitudinal data, typically via measurements separated by at least a year. When the same children were compared at time 1 and time 2, it was structural connectivity that predicted future improvements in reasoning, and not functional connectivity. Functional connectivity did not predict future changes in structural connectivity. Wendelken et al. (2017) concluded that the integrity of white matter connections between parietal and frontal cortex are important for developmental changes in reasoning ability. However, whether improved relational reasoning increased white matter integrity or vice versa is unclear, as both factors increased with age. Further research, ideally involving training designs, is required in order to understand the nature of the computations afforded by these structural connections that might make a difference to children's reasoning.

SUMMARY

Metacognition, or developing an awareness of one's own cognitive functioning, is important for many aspects of cognitive development. The study of metacognition focuses on reflective awareness of one's *own* cognition, contrasting with the development of reflective awareness of the cognitions of others, which is theory of mind. Children with good metacognitive skills are 'good information processors'. They can use metacognitive strategies to improve their memories, for example by adopting efficient strategies for remembering. They can monitor their own performance, keeping track of where they are with respect to their memory goals, and they can evaluate their memory behaviour. They can also self-regulate their own memory behaviour, for example by planning and directing their own activities. Self-monitoring behaviours appear to develop relatively early. For example, there is little change with age in ease-of-learning judgements, judgements-of-learning and feeling-of-knowing. There is also little change with age in neural markers of error monitoring, such as the error-related negativity (ERN). Self-regulation appears to undergo more protracted development. The development of self-regulation is intimately related to the development of executive function.

Executive function refers to the control aspects of gaining awareness of one's mental processes. Strategic control over one's mental processes is thought to improve as metacognition develops. Executive function involves cognitive flexibility, for example in shifting attentional focus between conflicting representations. It also involves the ability to exercise conscious control over one's thoughts, actions and emotions. Inhibitory control can be measured both by delay tasks (e.g., delaying gratification of a desire) and by conflict tasks (e.g., responding in a way that conflicts with a more salient response, as in the day/night task). Planning is another important aspect of executive function, and appears to be relatively independent of other EF skills. The development of working memory is another. Performance in behavioural executive function tasks is strongly correlated with performance in behavioural theory of mind tasks, as might be expected.

Developments in executive function were classically related to developments in the frontal cortex, which is not fully formed until early adulthood. However, most neuroimaging studies cannot tell us whether structural developments in the brain enable the development of executive function, or vice versa. Logically, any changes documented in the frontal cortex could reflect cognitive-behavioural developments rather than cause them. Indeed, an important study by Buss and Spencer (2017) with three-year-olds showed little frontal activity in a difficult version of the DCCS task that measures cognitive flexibility, but extensive frontal activity in the same children in an easy DCCS task. Ways in which neuroimaging might be used to disentangle cause and effect in cognitive development include longitudinal studies and training studies (see Chapter 11). Very few such studies are currently available. It is also possible to measure connectivity of the brain at rest, and then to measure whether differential task performance at different ages is associated with different patterns of connectivity.

In contrast to executive function, inductive and deductive reasoning show remarkable similarities in children and adults. The same factors seem to govern successful reasoning across age, and similar levels of spontaneous performance are found in both children and adults when unfamiliar versions of the different experimental paradigms are administered. Inductive and deductive reasoning in both children and adults is affected by background knowledge, and younger children are more affected when irrelevant background information must be inhibited. On the other hand, younger children can be more successful in reasoning when they lack relevant background knowledge, as shown in the relational reasoning form of the 'blicket' task used by Walker and Gopnik (2014) and in the conflict syllogisms discussed in this chapter. In Chapter 8 we saw that similar effects in children and adults have been observed in some studies of memory, and in Chapter 7 we saw that similar effects in children and adults have been observed in scientific reasoning. For example, there is remarkable continuity from childhood to adulthood in certain kinds of memory, such as implicit memory and recognition memory, and in a neural correlate of performance monitoring, the ERN. We also saw that the effects on particular memory systems of using unfamiliar stimuli and contexts were highly similar in children and in adults. This is probably a general phenomenon that applies equally to reasoning. Attempts to design 'pure' measures of reasoning, in which the reasoning process is measured *independently* of the context in which it is required, are generally counterproductive to high levels of performance. This is true for both children and adults. When familiar, 'child-centred' materials are used, then even very young children show efficient reasoning. They can reason by analogy, they can solve problems requiring deductive logic, and they can display successful conditional reasoning. However, metacognitive awareness and reflection improves performance for all these kinds of reasoning. Knowledge about your own reasoning skills, and actively monitoring your own performance, enable you to behave strategically in order to improve efficiency. Consequently, 'meta-knowing' is important for children's performance in all kinds of reasoning tasks.

CHAPTER 10

CONTENTS

Reading and mathematical development

10

Children's acquisition of reading and mathematics depends in part on their ability to learn symbolic systems. Symbolic systems are human cultural inventions for representing information, systems that are detached from the information itself. During the course of human development, a variety of different symbolic systems have been invented. The most ubiquitous symbolic systems today are orthographic systems (for example the alphabet, Chinese characters, Sanskrit characters) and the number system (for example Arabic numerals). Other symbolic systems include art, musical notation and maps. Once acquired, these symbolic systems become instrumental in shaping children's cognitive development. Vygotsky (see Chapter 11) called symbolic systems 'sign systems', and pointed out that sign systems have transformed human culture and society. In order to participate fully in their cultural and social environments, therefore, children need to acquire these sign systems. Symbolic systems have enabled human cognition to develop beyond the constraints of biology. As well as responding to direct stimulation from the environment, humans can organise their cognitive behaviour by creating and using symbols that can be responded to psychologically – such as print and Arabic digits. The acquisition of both print and Arabic number requires social transmission (e.g., by parents and teachers), but also requires some cognitive prerequisites on the part of the child. This will be the main focus of this chapter – the cognitive developmental origins of literacy and numeracy. Detailed treatment of their further development is beyond the scope of this book, as is detailed treatment of atypical development (dyslexia and dyscalculia). I will only discuss briefly what I consider to be the most promising explanations for developmental dyslexia and developmental dyscalculia.

Since the last edition of this book, the cognitive neuroscience of reading development and developmental dyslexia has expanded enormously, so much so that handbooks devoted solely to this topic are now available (Eden & Flowers, in press). Accordingly, I will only give a flavour of the available research in this book. The cognitive neuroscience of mathematics has similarly expanded, and has played an important role in developmental theorising. I will also assess this research in brief. In the last edition of this book, I reviewed the claim that the human brain has dedicated neural circuits for recognising numerosity. The claim was that number knowledge was 'innate', with "two distinct core systems of numerical representations… [which] therefore do not emerge through individual learning or cultural transmission" (Feigenson, Dehaene & Spelke, 2004, p. 307). More recent evidence suggests that this theoretical claim was an over-interpretation

KEY TERM

Sign systems
Human-invented symbolic systems for representing information, including orthographic (e.g., alphabetic), number (e.g., Arabic) and musical notation systems.

of available data. Reading development is widely accepted to require individual learning and cultural transmission, and mathematical development turns out to be broadly similar. The perspective I will adopt here is that there are sensory processes and physiological structures underlying the cognitive acquisition of reading and number. Other species may share some of these sensory processes and physiological structures, suggestive of evolutionary antecedents. However, such demonstrations do not mean that these sensory processes and physiological structures in themselves constitute cognitive systems. For example, children with dyslexia have prosodic difficulties with speech rhythm, and some animals are able to recognise some prosodic parameters of spoken language (e.g., domestic dogs show sensitivity to the emotional content of language, which is prosodically determined). Apes produce calls with pure tonal notes, repetition, rhythm and phrasing, reminiscent of the rhythmic sounds produced during canonical babbling by babies at around eight months (ape 'singing', see Masataka, 2007). Meanwhile, recordings from cells in the monkey brain appear to show the existence of 'number neurons' tuned to specific quantities (Nieder, 2016). Perceptual systems for tracking and individuating small numbers of objects also seem to be shared with other species (Ansari, 2008). However, apes and dogs cannot read and monkeys cannot do calculus.

READING DEVELOPMENT

Different cultures have invented different symbol systems for representing spoken language. For example, China and Japan use character-based scripts, while most European languages use the alphabet. These different types of printed symbol share one core feature, which is that they are a visual code for spoken language. Reading is essentially the cognitive process of understanding speech written down. Meaning is communicated via print. Skilled readers access meaning directly from this visual code, but phonology does not become irrelevant. In fact, phonological activation appears mandatory during reading, even by highly skilled readers (see Ziegler & Goswami, 2005). Where languages differ is in the units of sound that are represented by print. This has been termed a difference in psycholinguistic 'grain size'. Japanese characters (called *Kana*) represent individual syllables. Chinese characters (called *Kanji*) represent morphemes. The alphabet represents phonemes, although it does so with more transparency in some languages than in others. For example, Italian, Greek and Spanish are all highly consistent in their spelling-sound correspondences: one letter makes only one sound. English, Danish and French are markedly less consistent: one letter can make multiple sounds. Consider the letter A in English, which makes different sounds in the highly familiar words *man*, *make*, *car* and *walk*. These two factors of phonological complexity and orthographic transparency lead to cross-cultural differences in reading acquisition by children.

KEY TERM

Phonological awareness Ability to detect and manipulate the component sounds that comprise words.

Phonological awareness and learning to read

As we saw in Chapter 6, an important part of language acquisition is phonological development. Children need to learn the sounds and combinations of sounds that are permissible in a particular language, so that their brains can develop phonological representations of the sound structure of individual words. As might be expected, there are individual differences in the quality of the phonological representations developed by children. These individual differences turn out to be very important for reading development, and affect children's 'phonological awareness'. Phonological awareness refers to the child's ability to detect and manipulate the component sounds that comprise words, at different linguistic grain sizes.

In Chapter 6, phonological development was discussed largely in terms of syllables and phonemes. We saw that the prosodic cues that are exaggerated in infant-directed speech – namely changes in pitch, duration and stress – carried important information about word boundaries and about how sounds were ordered into words. We also saw that the physical changes where languages placed the boundaries determining phonemes were not random. Rather, general auditory perceptual abilities seemed to influence where these 'basic cuts' were made. In addition to awareness of syllables and phonemes, an important level of phonological awareness for literacy acquisition is awareness of 'onset-rime' units. *Onsets* and *rimes* represent a grain size intermediate between syllables and phonemes. While the primary phonological processing unit across the world's languages is the syllable, each syllable comprising a word can be decomposed into onsets, rimes and phonemes in a hierarchical fashion. This is shown in Figure 10.1. The onset-rime division of the syllable depends on dividing at the vowel. For English, words like *sing*, *sting* and *spring* all share the same rime, the sound made by the letter string 'ing'. The onset of *sing* is /s/, the onset of *sting* is /st/ and the onset of *spring* is /spr/. These onsets comprise one, two and three phonemes respectively. However, in many languages in the world, syllable structure is simple. Syllables are CV (consonant–vowel) units. Hence for many languages, onsets, rimes and phonemes are equivalent. Each onset and each rime in the syllable is also a single phoneme.

The development of phonological awareness appears to follow a very similar developmental sequence across languages. Children first gain awareness of syllables, then of onset-rime units and finally of phonemes. Despite the fact that infants can distinguish different phonemes (as syllable onsets, see Chapter 6), the ability to

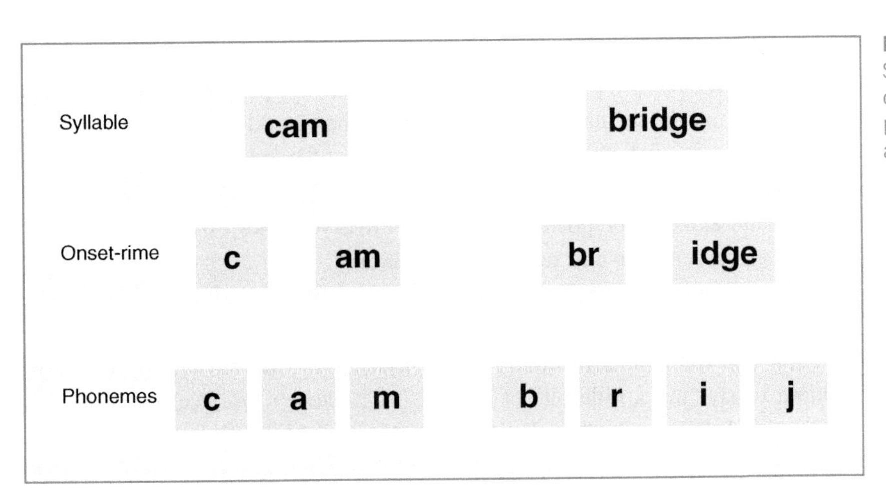

FIGURE 10.1 Schematic depiction of the hierarchical psycholinguistic structure of a bisyllabic word.

KEY TERM

Metalinguistic awareness
Ability to reflect on one's knowledge of the sound structure of words.

reflect on one's knowledge of the sound structure of words develops gradually, particularly at the phonemic level. Hence phonological awareness is also called *metalinguistic* awareness, highlighting the fact that the child needs to reflect on and become consciously aware of knowledge that is already present in the mental lexicon. While metalinguistic awareness of syllables and onset-rimes appears to develop prior to learning to read, direct tuition is usually required in order for phonemic awareness to develop. The intimate relationship between becoming a reader and becoming phonemically aware means that phonemic awareness develops at a faster rate for children who are learning to read transparent orthographies (e.g., Italian and Greek children) compared to children who are learning to read opaque orthographies (e.g., English and French children).

KEY TERM

Syllabic awareness
Being able to recognise the number of syllables in words.

Becoming aware of syllables

In a pioneering study, Liberman, Shankweiler, Fischer and Carter (1974) devised a tapping task to measure syllable awareness in pre-reading children. They gave American children aged from four to six years a small stick, and asked them to tap once for words that had one syllable (*dog*), twice for words that had two syllables (*dinner*) and three times for words that had three syllables (*president*). A criterion of six correct responses in a row was required in order for children to be accorded syllabic awareness. This criterion was passed by 46% of the four-year-olds, 48% of the five-year-olds and 90% of the six-year-olds. The four- and five-year-olds were pre-readers, and the six-year-olds had been learning to read for about a year.

Another measure used to assess syllabic awareness in young children is the counting task. Devised by the Russian psychologist Elkonin (1963), in the counting task children are given plastic counters and are asked to use them to represent the number of syllables in words of increasing length. Treiman and Baron (1981) gave a syllable counting task to five-year-old pre-readers. For example, if the experimenter said "butter", the child had to set out two counters. Treiman and Baron (1981) also reported good syllable awareness in these pre-readers. Treiman and Zukowski (1996) devised a same-different task to assess syllable-level skills. In this task, the children were introduced to a puppet who felt happy when he heard two words that had some of the same sounds in them. The children had to listen to pairs of words like 'hammer, hammock' and 'compete, repeat', repeat them, and then decide whether the puppet would like them. The first pair of words shares the first syllable, and the second pair of words shares the second syllable, so the puppet should like both of these pairs. In contrast, pairs of words like 'delight, unique' and 'plastic, heavy' do not share sounds, so the puppet should dislike these pairs. Treiman and Zukowski reported that 100% of five-year-olds, 90% of six-year-olds and 100% of seven-year-olds succeeded in this task.

These different tasks suggest that syllable awareness has developed in young children before they learn to read. Similar data has been found in deletion tasks (deleting a syllable from a word, e.g., *party* → *part*, see Bruce, 1964); tasks requiring children to say "just a little bit" of a word (e.g., *Peter* → *Pete*, see Fox & Routh, 1975); and blending tasks (blending together two syllables to form a word, e.g., *sis* + *ter* makes *sister*, Anthony, Lonigan, Driscoll, Phillips & Burgess, 2003). These tasks have been given to children as young as three years of age, who perform above chance at the syllable level. Clearly, there are some task-dependent factors, as the success levels seem to vary with the different tasks. For example, same-different judgement tasks appear to be easier than tapping tasks. However, without giving a range of tasks to the *same children*, it is difficult to be sure about the relative cognitive demands made by tasks like tapping, counting, blending and same-different judgement. The key finding is that at the syllable level, children as young as three years perform at above-chance levels in *all* of these tasks.

Similar results are found in other languages. For example, Cossu, Shankweiler, Liberman, Tola and Katz (1988) gave the *tapping* task to Italian pre-readers aged four and five years, and to schoolchildren being taught to read, aged 7–8 years. The children were asked to tap once for each syllable in words like 'gatto', 'melone' and 'termometro'. Syllable awareness was shown by 67% of the four-year-olds, 80% of the five-year-olds and 100% of the school-age sample. Durgunoglu and Oney (1999) gave the tapping task to Turkish kindergartners. Performance was 94% correct. Hoien, Lundberg, Stanovich and Bjaalid (1995) gave the syllable *counting* task to 128 Norwegian preschoolers aged on average six years 11 months. For the syllable task, the children had to make pencil marks for each syllable in a word (e.g., 'telephone' = 3 marks). The children performed at 83% correct. Counting tasks were also given to German preschoolers by Wimmer, Landerl, Linortner and Hummer (1991) and to French kindergarteners by Demont and Gombert (1996). The German preschoolers performed at 81% correct in the syllable counting task and the French children performed at 69% correct. Durgunoglu, Nagy and Hancin-Bhatt (1993) used a blending task to assess phonological awareness at the syllable level in Spanish-speaking children living in the USA. The children were asked to blend pairs of syllables into words (e.g., 'do-ce'). Performance was 85% correct. Hence Italian children, Norwegian children, French children, Turkish children, Spanish children and German children, just like English-speaking children, seem to have good syllable awareness prior to receiving literacy teaching.

"Hey, diddle, diddle! The cat and the fiddle, the cow jumped over the moon." The nursery rhyme may be helpful in establishing onset-rime awareness in preschoolchildren.

Becoming aware of onsets and rimes

KEY TERM

Onset-rime awareness Dividing the syllable at the vowel, as in s–ing, str–ing.

As the nursery rhyme is an intimate part of an English-speaking childhood, it might be expected that onset-rime awareness is also well-developed in young children prior to schooling. Popular nursery rhymes have strong rhythms that emphasise syllabification via metrical patterning ("HUMP-ty DUMP-ty sat on a wall"), and many nursery rhymes contrast rhyming words in ways that distinguish the onset from the rime (e.g., 'Twinkle Twinkle Little Star' rhymes 'star' with 'are'; 'Incy Wincy Spider' rhymes 'spout' with 'out'). The most widely used measure of onset-rime awareness in preschool children is the oddity task devised by Bradley and Bryant (1978). In this task, children are asked to select the 'odd word out' of a group of three words on the basis of either the initial sound, the medial sound or the final sound (e.g., bus, bun, *rug*; *pin*, bun, gun; top, *doll*, hop). The initial sound task can be solved on the basis of the different onset, and the medial and final sound tasks can be solved on the basis of the different rime. Bradley and Bryant (1983) gave the oddity task to around 400 preschool English children aged four and five years. The children scored above chance levels in both the onset and rime versions of the task, with average performance being 56% correct with onsets and 71% correct with rimes.

Another task used to measure onset-rime awareness is the same-different judgement task. For example, Treiman and Zukowski (1996) asked their children whether the puppet would like word pairs like 'plank, plea' (shared onset), and 'spit, wit' (shared rime). Fifty-six per cent of the five-year-olds, 74% of the six-year-olds, and 100% of the seven-year-olds, made accurate judgements. Blending and segmentation have also been used to study onset-rime awareness, with similar results. For example, Anthony et al. (2003) asked children to help a puppet to blend units like 'h' and 'at' into *hat*. Children can also be asked to complete nursery rhymes, enabling younger children to be tested. Bryant, Bradley, Maclean and Crossland (1989) asked three-year-olds to complete familiar nursery rhymes such as "Jack and Jill went up the? [hill]". Five nursery rhymes were used for testing, and only one of the 64 three-year-olds in the study knew none of these nursery rhymes. On average, the children knew about half of the nursery rhymes (they could score 1 for partially completing the rhymes and 2 for fully completing them, making a total possible score of 10, the mean score for the group was 4.5). Onset-rime awareness is clearly present in English-speaking children prior to schooling.

Again, rhyme awareness is also found prior to literacy tuition in other languages. Chukovsky (1963) collected a large corpus of Russian children's language games and poems, and noted that the children were fascinated by rhymes. For example, Tania, aged 2.5 years, made up the following poem:

Ilk, silk, tilk
I eat Kasha with milk.
Ilks, silks, tilks,
I eat Kashas with milks.

The oddity task has been given to pre-readers in a variety of languages. For example, Wimmer, Landerl and Schneider (1994) developed a version for German kindergartners, testing 138 children. They used four words in order to increase task difficulty for these older children (German children do not go to school until age six). The onset task was made up of words like 'Korn, Kopf, *Rock*, Korb', and the two rime tasks were made up of words like 'Bund, Hund, *Wand*, Mund' (middle sound different) and 'Haus, *Baum*, Maus, Laus' (end sound different). The children were tested about four months prior to beginning schooling, which meant that they were pre-readers aged on average six years. Wimmer et al. reported that performance was above chance, and that the onset task was more difficult than the rime task for these children (44% correct versus 73% correct respectively). Ho and Bryant (1997) gave Chinese three-year-olds a rhyme oddity task, and found that performance was 68% correct. Hoien et al. (1995) gave their Norwegian preschoolers a match-to-sample rhyme task, in which children had to select one picture out of three that rhymed with a target picture. Performance was 91% correct. Porpodas (1999) devised a Greek version of the oddity task. He reported that first-grade children in Greece scored 90% correct. As with syllables, therefore, children across languages seem to develop good onset-rime awareness prior to receiving literacy teaching.

Becoming aware of phonemes

The same is not true for the development of phoneme awareness. Although some children in some studies in some languages (e.g., Turkish, Czech; see Goswami & Ziegler, 2006) have been reported to develop some phoneme awareness prior to schooling, in general studies find that phoneme awareness develops as a result of direct teaching, usually the direct teaching of literacy. This is not particularly surprising, as the phoneme is not a natural speech unit. This was discussed briefly in Chapter 6. In Chapter 6, we saw that the concept of a phoneme is an abstraction from the physical stimulus (memorably described as a 'cerebral construction' by Mersad & Dehaene-Lambertz, 2016). The mechanism for learning about the abstract unit of the phoneme seems to be learning about letters. Letters are used to symbolise phonemes, even though the physical sounds corresponding (for example) to the 'P' in *pit* and *spoon* are rather different. Hence the development of phonemic awareness depends in part on the consistency with which letters symbolise phonemes. Accordingly, there is cross-language divergence in the rate of development of phonemic awareness.

> **KEY TERM**
>
> **Phonemic awareness** Being able to divide spoken words into the smallest sound elements, phonemes, that are an abstraction from the speech signal and are usually learned via learning letter–sound correspondences.

This has been shown using a variety of cognitive tasks. One task that has been used in many languages is phoneme counting. For example, Wimmer et al. (1991) gave a phoneme counting task to their German preschoolers, Demont and Gombert (1996) gave a phoneme counting task to their French kindergartners, Hoien et al. (1995) gave a phoneme counting task to their Norwegian preschoolers and Durgunoglu and Oney (1999) gave a phoneme tapping task to their Turkish kindergartners. The German children performed at 51% correct, the French children at 2% correct, the Norwegian children at 56% correct and the Turkish children at 67% correct. The Italian children studied by Cossu et al. (1988) were also given a phoneme tapping task. Criterion was reached by 13% of the four-year-olds and 27% of the five-year-olds. In contrast, 97% of the school-aged sample (who were being taught to read) reached criterion. One reason for this variability in results is that not all of the studies of kindergartners checked that the children were pre-readers, but another is that the written languages varied in transparency. In English, counting tasks at the phoneme level are generally performed rather poorly. For example, Liberman et al. (1974) reported levels of 0% correct at the phoneme level for their four-year-olds and 17% correct for their five-year-olds. By age six, when the children had been learning to read for about a year, phoneme tapping was at 70% correct. Studies of first-grade children by Tunmer and Nesdale (1985) and of second grade children by Perfetti, Beck, Bell and Hughes (1987) report success levels of 71% and 65% respectively. By the end of first grade, children learning to read transparent languages typically score at much higher levels than this (e.g., Turkish 94%; Greek 100%; German 92%; see Durgunoglu & Oney, 1999; Harris & Giannouli, 1999; Wimmer et al., 1991). In contrast, Demont and Gombert (1996) found that by the end of grade 1, French children scored 61% in phoneme counting tasks, very similar to the achievement levels shown by English children.

> **KEY TERM**
>
> **Orthographic transparency** Consistency of symbol to sound (e.g., grapheme–phoneme) correspondence.

Again, similar effects of orthographic transparency are found with other tasks. For example, using the same-different judgement task, Treiman and Zukowski (1996) reported that 25% of American five-year olds could recognise shared beginning phonemes (steak, sponge) and final phonemes (smoke, tack), compared to 39% of six-year-olds. By seven years, performance was at 100% correct. Goswami and East (2000) tested phoneme awareness in English five-year-olds using an oddity task (final phoneme shared, as in *cliff, drum, swam*), a blending task and a segmentation task. They found different performance levels depending on which task was used. The five-year-olds scored 0% correct in the phoneme segmentation task, 54% correct in the phoneme blending task and 37% correct in the oddity task. In Greek first-grade children, Porpodas (1999) found a performance level of 98% correct for phoneme deletion. Using a final phoneme deletion task (in which children had to delete the final phoneme in spoken CVC non-words), Durgunoglu and Oney (1999) reported that Turkish first grade children scored 98% correct. Ideally, rather than relying on cross-sectional comparisons, the same children should be followed longitudinally in order to track developmental relationships. Nevertheless, the general picture is clear. Phoneme awareness develops at different rates in children who are learning to speak and read different languages.

Theoretically, it has been proposed that there are at least two language-dependent reasons for these developmental differences between languages (Ziegler & Goswami, 2005, 2006). One is the phonological structure of the syllable. As noted above, for many of the world's languages, the most frequent syllable type is CV. For these languages, onset-rime segmentation of the syllable is equivalent to phonemic segmentation. An Italian child who segments early-acquired words like *Mamma* and *casa* (house) at the onset-rime level will thereby arrive at the phonemes comprising these words (e.g., /m/ /a/ /m/ /a/). Only 5% of English monosyllables follow the CV pattern (see De Cara & Goswami, 2002; examples are *go* and *see*). The most frequent syllable type in English is CVC (*cat, dog, soap*). Other phonological factors might also contribute to the relative ease or difficulty of becoming aware of phonemes. For example, languages differ in the sonority profile of their syllables. Vowels are the most sonorant sounds, followed in decreasing order by glides (e.g., /w/), liquids (e.g., /l/), nasals (e.g., /n/) and obstruents (e.g., /p/). Whereas the majority of syllables in English end with obstruents (almost 40%), the majority of syllables in French either end in liquids or have no coda at all (almost 50%).

The second important factor in explaining cross-language differences in phonemic awareness is orthographic transparency. Some languages have a 1:1 mapping between letters and sounds. For these languages, letters correspond consistently to one phoneme. Examples include Greek, German, Spanish and Italian. Other languages have a 1:many mapping between letters and sounds. A good example is English. Some letters or letter clusters can be pronounced in more than one way, for example O in *go* and *to*, EA in *speak* and *steak* and G in *magic* and *bag* (see Berndt, Reggia & Mitchum, 1987; Ziegler, Stone & Jacobs, 1997). It is much easier to become aware of phonemes if one letter consistently maps onto one and the same phoneme. It is relatively difficult to learn about phonemes if a letter can be pronounced in multiple ways (see Ziegler & Goswami, 2005, 2006, for more detailed arguments). This

TABLE 10.1 Data (% correct) collated from studies comparing phoneme counting in different languages in kindergarten or early grade 1

Language	% phonemes counted correctly
Greek[1]	98
Turkish[2]	94
Italian[3]	97
Norwegian[4]	83
German[5]	81
French[6]	73
English[7]	70
English[8]	71
English[9]	65

1 Harris & Giannouli, 1999; 2 Durgunoglu & Oney, 1999; 3 Cossu et al., 1988; 4 Hoien et al., 1995; 5 Wimmer et al., 1991; 6 Demont & Gombert, 1996; 7 Liberman et al., 1974; 8 Tunmer & Nesdale, 1985; 9 Perfetti et al., 1987 and grade 2 children.

theoretical analysis predicts that children learning to read in languages like Italian and Spanish should find it easiest to become aware of phonemes. They are learning languages with predominantly CV syllables, so that onset-rime segmentation and phonemic segmentation are equivalent, and their written language consistently represents one phoneme by one letter. Children learning to read in languages like German should have a more difficult time, because spoken syllables are complex in structure (in fact, German has the same syllable structure as English). Nevertheless, German has a 1:1 mapping from letter to sound, facilitating the process of becoming aware of phonemes. It is children who are learning to read in languages like English and French who should have the most difficult time. These children are learning languages with a complex syllable structure, and an inconsistent orthography. In fact, cross-language comparisons indeed suggest that it takes children longer to learn about phonemes in languages like English and French compared to languages like Italian and Spanish. This is shown in Table 10.1.

Longitudinal connections between phonological awareness and reading

Despite cross-language variability in the rate at which children acquire phoneme awareness, preschool differences in phonological sensitivity appear to predict reading and spelling development across languages. The strongest research designs measure phonological awareness prior to school entry, and then explore whether individual differences in phonological awareness predict children's performance in standardised tests of reading and spelling two or three years later. Other cognitive variables that might cause a longitudinal relationship, such as individual differences in intelligence or in memory, are also measured, and are then controlled in the longitudinal analyses. This enables the researchers to measure whether there is a *specific* connection between phonological awareness and progress in literacy.

In one of the first longitudinal studies to demonstrate a specific connection, Bradley and Bryant (1983) followed up the 400 preschool children to whom they had administered the oddity task at ages four and five years. At follow-up, the children were aged on average eight and nine years. The children were given standardised tests of reading, spelling and reading comprehension, and their performance was adjusted for age and IQ. Bradley and Bryant found high correlations between performance on the oddity task at ages four and five and reading and spelling performance three years later. When the effects of IQ and memory were controlled in multiple regression equations, the oddity task accounted for up to 10% of unique variance in reading. In a study with three-year-olds reported by Bryant et al. (1989), a significant relationship between nursery rhyme knowledge at age three and success in reading and spelling at ages five and six was found, again controlling for factors such as social background and IQ. Similar results with English-speaking samples have been reported by a number of other research groups. For example, Baker, Fernandez-Fein, Scher & Williams (1998) measured nursery rhyme knowledge in 39 kindergarten children, and reported that it was the strongest predictor of word attack and word identification skills measured in grade 2. Rhyme knowledge at time 1 accounted for 36% and 48% of unique variance in reading at time 2 respectively. The second strongest predictor of reading at time 2 was letter knowledge, which accounted for an additional 11% and 18% of the variance respectively.

Similar longitudinal connections between phonological awareness and reading have been found in other languages. In a landmark study, Lundberg, Olofsson and Wall (1980) gave 143 Swedish children a range of phonological awareness tests in kindergarten. The tests used included syllable blending, syllable segmentation, rhyme production, phoneme blending, phoneme segmentation and phoneme reversal. When Lundberg et al. examined the predictive relationships between these tests and reading attainment in second grade, both the rhyme test and the phoneme tests were found to be significant predictors of reading almost two years later. In a German replication of Bradley and Bryant's study, Wimmer et al. (1994) followed up the 183 German kindergartners who had received the oddity task at age six (in kindergarten) one year later (at grade 1), and again when they were almost ten years old. Wimmer et al. reported that performance in the oddity task was only minimally related to reading and spelling progress in German children when they were 7–8 years old (the same age as the English children in Bradley and Bryant's study). However, at the three-year follow-up, when the children were aged on average nine years nine months, rime awareness (although not onset awareness) was significantly related to both reading and spelling development. In the study with Norwegian preschoolers by Hoien et al. (1995) mentioned earlier, the children were given measures of syllable, onset-rime and phoneme awareness. When reading was tested in first grade, it was found that syllable, rhyme and phoneme awareness all made independent contributions to variance in reading. In the study with Chinese preschoolers that was mentioned earlier, Ho and Bryant (1997) gave the rime oddity task to 100 Chinese children at age three years, and measured their progress in reading and spelling two years later. Phonological awareness was found to be a significant predictor of reading even after other factors such as age,

IQ and mother's educational level had been controlled. Individual differences in phonological sensitivity hence predict individual differences in reading attainment across languages, even for children who are learning to read non-alphabetic scripts. Because the children are so young, the measures of phonological sensitivity used in these longitudinal studies are usually syllable, onset and rime measures. Although different tasks measure phonological awareness at different 'grain sizes', the general assumption is that syllable, onset-rime and phoneme measures are all tapping the same cognitive construct, namely phonological awareness (Anthony et al., 2003).

Training children's phonological skills: The impact on reading

As always in developmental psychology, however, the existence of a robust longitudinal association between factor A (e.g., phonological awareness) and outcome B (e.g., literacy) does not in itself show a causal relationship. In order to test whether a developmental relationship is a causal one, an intervention study is required. In the area of reading and spelling, there are now many such intervention studies, because improving children's literacy is an important goal for governments in many cultures. However, not every training study has followed a stringent research design. When carrying out studies aimed at isolating causal variables in developmental psychology, it is crucial to have the correct control groups (Goswami, 2003). In the case of phonology and reading, it is important that some groups of children receive an intervention designed to improve their performance that does not manipulate phonological sensitivity. When exploring causality, it is not sufficient to compare a group of children who receive a phonological intervention to a group of children who do not. Although the children in such 'unseen' control groups are receiving training in reading from their classroom teachers, they are not receiving the individualised attention of an eager researcher who is piloting a special new training programme. Effects due simply to participating in an intervention are called 'Hawthorne' effects. They arise not from cognitive changes due to targeted training, but from the generalised motivational and self-esteem effects of participating in something extra and unusual. This aspect of receiving an intervention, as well as the cognitive aspects of the intervention itself, must be controlled if a training study is to provide unambiguous information about cognitive variables.

One training study that did follow a stringent research design was that of Bradley and Bryant (1983), mentioned earlier. As part of their study, Bradley and Bryant selected the 65 children in their cohort of 400 who had performed most poorly in the oddity task when they were aged four and five years. These 65 children were divided into four matched groups, three of which were provided with two years of specialised training. One group, the phonological awareness group, were given training in grouping words on the basis of sound. This was done using a picture sorting task and various word games. For example, the children were taught to put the pictures of a *hat*, a *rat*, a *mat* and a *bat* together when grouping by rhyme (see Figure 10.2). During the course of the two years, the children were taught to group words by a variety of grain sizes, namely onset, rime, vowel phonemes and coda phonemes. A control group spent the same amount of time with the same

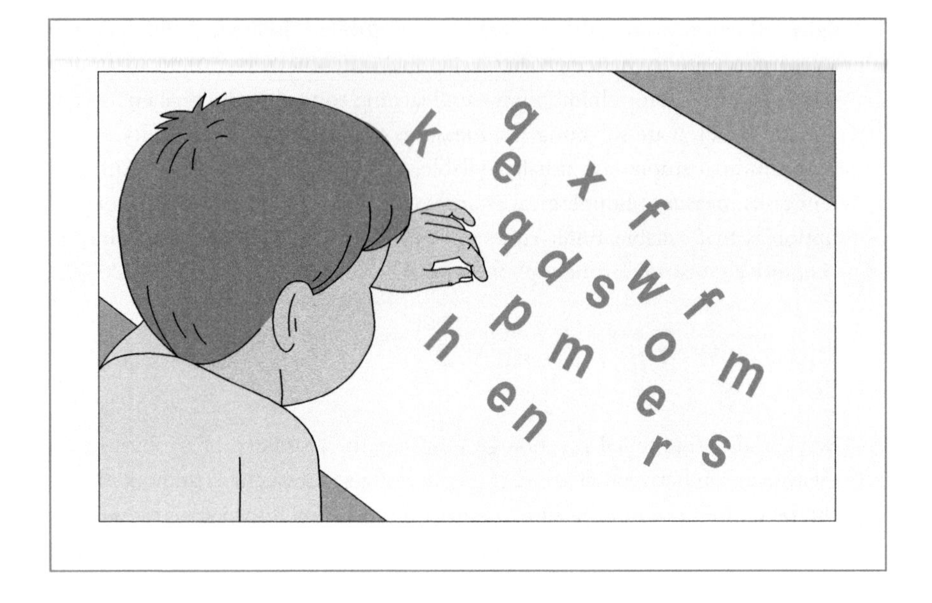

FIGURE 10.2
Bradley and Bryant (1983) used plastic letters to teach children to group words on the basis of shared sounds and shared spellings.

researchers and the same games, but learned to group words by semantic category. For example, in the picture card game they might group the cards into 'farmyard animals'. A second experimental group also did the phonology picture and word games, but in addition spent the second year of the study matching plastic letters to shared phonology. For example, they learned to use plastic letters to represent the rime 'at' in words like *hat*, *rat* and *mat*. The fourth group of children were an unseen control group. Following the intervention, the children who had received training in phonological awareness alone were four months ahead in reading and spelling compared to the children in the semantic control group (a non-significant difference). The children in the group who had received both phonological awareness and plastic letters training were a significant eight months further on in reading than the children in the semantic control group, and 12 months further on in spelling, even after adjusting post-test scores for age and IQ. Compared to the children who had spent the intervening period in the unseen control group, the phonology + plastic letters group were an astonishing 24 months further on in spelling and 12 months in reading. Clearly, combining phonological awareness training with instruction in orthography–phonology relations yields the largest benefits.

Again, training studies in other languages have reported similar beneficial effects (e.g., Lundberg, Frost & Petersen, 1988, Danish; Schneider, Kuespert, Roth, Visé & Marx, 1997, German; Schneider, Roth & Ennemoser, 2000a, German). For example, Schneider et al. (2000a) developed a six-month metalinguistic training programme covering syllables, rhymes and phonemes for young German children. They then identified 138 German children as being 'at risk' for dyslexia while in kindergarten on the basis of rhyme production, rhyme matching and syllable segmentation skills that were significantly poorer than those of German control kindergartners. Schneider et al. developed an innovative research design in which *all* of the children designated 'at risk' received training. One group received metalinguistic training alone, a second group received letter–sound training alone, and a third group received combined phonological + letter–sound training. Progress in reading and spelling for these three experimental groups was then compared to that of children from the same kindergartens who had never been at risk for reading difficulties. As might be expected from the results reported by Bradley and Bryant (1983), the German children receiving the phonological + letter–sound training programme had the best outcome. A year into first grade, this group showed comparable attainment in literacy to those children who had never been at risk and who had received no training. The other two training groups were still significantly impaired in literacy attainment in comparison to those children who had never been at risk. The importance of phonology was demonstrated by the finding that the 'at risk' group who had received letter–sound training alone, without metalinguistic training, either performed at comparable levels in later reading and spelling progress to the 'metalinguistic training alone' group, or performed at lower levels than this group. Clearly, training either phonological awareness alone or training letter–sound recoding alone is insufficient for effective progress in literacy. Training one set of skills without the other will not prevent literacy difficulties. The cognitive profiles of children with developmental dyslexia in different languages will be discussed later in this chapter.

Learning to decode in different languages

So far, we have seen that the two factors of phonological complexity and orthographic transparency lead to cross-cultural differences in the emergence of phoneme awareness in children learning to read different languages, but that phonological awareness at all grain sizes (syllable, onset-rime, phoneme) is a significant predictor of reading acquisition across the world's languages. As might be expected, the factors of phonological complexity and orthographic transparency also lead to cross-cultural differences in the rate at which children learn to decode printed symbols into sound. Children who are learning to read spoken languages with simple (CV) syllables and consistent letter–sound correspondences acquire decoding skills most rapidly (for example, Italian and Spanish children). Cross-language differences in phonological complexity and orthographic transparency also have effects on the brain. Both the grain size of lexical representations and the reading strategies that children develop for decoding show systematic differences across orthographies (Ziegler & Goswami, 2005).

A detailed analysis of these differences is beyond the scope of this book (see Ziegler & Goswami, 2005, 2006, for detail), nevertheless in general there are more developmental similarities than differences across languages. We will focus on the similarities here. The similarities arise from the learning problem facing a child who is attempting to learn the orthographic system that their culture uses as a visual code for spoken language. Beginning readers across languages are faced with three problems: availability, consistency and the granularity of symbol-to-sound mappings. The *availability* problem reflects the fact that not all phonological units are accessible prior to reading. As we have seen, phonemes in particular may be inaccessible to pre-readers, and the speed with which phonemic awareness develops varies with orthographic consistency. The grain sizes that are most accessible prior

KEY TERM

Consistency problem
The fact that the alphabet represents phonemes with more transparency in some languages than in others.

to reading (syllables and onset-rimes) hence may not correspond to the visual symbols used to represent phonology. We can predict that a Japanese child, who is learning visual symbols that represent syllables (an early-developing grain size), will be at an advantage compared to an English child, who is learning letters that represent phonemes (a later-developing grain size, at least for languages where onset-rime units are not equivalent to phonemes).

The *consistency* problem refers to the fact that the alphabet represents phonemes with more transparency in some languages than in others. As discussed earlier, Italian, Greek and Spanish are all highly consistent in their spelling–sound correspondences. For these languages, one letter makes only one sound for reading. English, Danish and French are markedly less consistent in their spelling–sound correspondences, as one letter can make multiple sounds in reading. Sound–spelling consistency can vary across languages as well. While most alphabetic languages are inconsistent for sound–spelling correspondence, with one sound corresponding to more than one letter (a 1:many relationship), some languages are consistent for spelling as well as for reading (e.g., Italian, Serbo-Croatian). In terms of initial acquisition, 'feed-forward inconsistency' (from spelling to sound) appears to be most influential in slowing development (see Ziegler & Goswami, 2005). English has an unusually high degree of feed-forward inconsistency, and this appears to cause problems for beginning readers in English, who have to decode words that sound different yet share spelling patterns, like *though*, *cough*, *through* and *bough*.

KEY TERM

Granularity problem There are more orthographic units to learn when access to the phonological system is based on bigger grain sizes.

Finally, the *granularity* problem refers to the fact that there are many more orthographic units to learn when access to the phonological system is based on bigger grain sizes as opposed to smaller grain sizes. That is, there are more words than there are syllables, there are more syllables than there are rimes, there are more rimes than there are graphemes and there are more graphemes than there are letters (graphemes are alphabetic units that make a single sound, for example the phoneme /f/ can be represented by the grapheme 'ph'). It seems likely that reading proficiency in a particular language will depend on the resolution of all three of these problems, a resolution which will of necessity vary by orthography (Ziegler & Goswami, 2005, 2006, for detail). For example, children learning to read English must develop multiple strategies in parallel in order to become successful readers. They need to develop whole word recognition strategies in order to read words like *cough* and *yacht*, they need to develop rhyme analogy strategies in order to read irregular words like *light*, *night* and *fight*, and they need to develop grapheme-phoneme recoding strategies in order to read regular words like *cat*, *pen* and *big*.

For many of the world's languages, children only need to develop grapheme-phoneme recoding strategies in order to become highly skilled readers. Languages like German, Italian, Turkish and Finnish can be read very successfully in letter-by-letter fashion. Indeed, experiments across orthographies suggest that children learning to read these languages do rely on grapheme-phoneme recoding, and that this strategy develops to an efficient level within the first months of learning to read (e.g., Cossu, Gugliotta, & Marshall, 1995; Durgunoglu & Oney, 1999; Wimmer, 1996). There are various cognitive 'hallmarks' that suggest a reliance on grapheme–phoneme recoding. One is a length effect: children who are applying grapheme–phoneme correspondences should take longer to read words with more letters/

phonemes. Children learning to read consistent orthographies like Greek show reliable length effects compared to children learning to read English (e.g., Goswami, Porpodas & Wheelwright, 1997). Another is skilled non-word reading: children applying grapheme–phoneme correspondences should be as efficient at reading letter strings that do not correspond to real words (e.g., *grall*, *tegwump*) as they are at reading letter strings that do correspond to real words (*ball*, *wigwam*). Young readers of consistent orthographies like German are usually much better at reading non-words like *grall* compared to English children (Frith, Wimmer & Landerl, 1998). Of course, there is more than one way of reading a non-word. A non-word like *grall* can either be read by applying grapheme–phoneme correspondences, or can be read by analogy to a familiar real word like *ball*. German children show no difference in reading accuracy for non-words that can be read by analogy compared to non-words that cannot be read by analogy. English children do show a difference. For example, when given non-words that could be read by analogy to real English words (e.g., *dake* [cake], *murn* [burn]), English children were more accurate than when given phonologically matched non-words that did not have orthographic analogies (e.g., *daik*, *mirn*; see Goswami, Ziegler, Dalton & Schneider, 2003). German children showed no difference, and also read the non-words very efficiently when the two types of non-word (large grain size, small grain size) were mixed together into one list. English children, in contrast, showed a strategy switching cost with the mixed list (they were apparently alternating between using rime analogies and using grapheme–phoneme recoding).

When the efficiency of grapheme–phoneme recoding strategies is compared in children who are learning to read different languages during the first year of reading instruction, there is a clear advantage for children who are learning to read consistent alphabetic orthographies. In the largest cross-language study carried out to date, scientists in 14 of the countries comprising the European Union in 2000 measured simple word and non-word reading in first grade children attending schools using 'phonics' based (grapheme–phoneme based) instructional programmes (Seymour, Aro & Erskine, 2003). The word and non-word items were matched for difficulty across the languages, and length of tuition in reading was also equated, although the ages of the children differed. The age differences were unavoidable, as (for example) the children in England and Scotland began school at age five, whereas the Scandinavian children began school at age seven. The data are shown in Table 10.2. As can be seen, the efficiency of grapheme–phoneme recoding approached ceiling level during the first year of teaching for most of the European languages. Children learning to read languages like Finnish, German, Spanish and Greek were decoding both words and non-words with accuracy levels above 90%, In contrast, children learning to read French (79% correct), Danish (71% correct) and Portuguese (73% correct) were not as advanced, reflecting the reduced orthographic consistency of these languages. The children learning to read in English showed the slowest rates of acquisition, reading 34% of the simple words correctly and 29% of the simple non-words. When followed up a year later, these children were achieving levels of 76% for real words and 63% for non-words, still short of the early efficiency shown by the Finnish and German children. This is not surprising, of course. The English children were learning a symbol system with inconsistent correspondences at the

TABLE 10.2 Data (% correct) adapted from the COST A8 study of grapheme–phoneme recoding skills for monosyllables in 14 European languages (Seymour et al., 2003)

Language	Familiar real words	Non-words
Greek	98	97
Finnish	98	98
German	98	98
Austrian German	97	97
Italian	95	92
Spanish	95	93
Swedish	95	91
Dutch	95	90
Icelandic	94	91
Norwegian	92	93
French	79	88
Portuguese	73	76
Danish	71	63
Scottish English	34	41

phoneme level, and they also had to match these symbols to phonemes that were embedded in complex syllables. These two factors of phonological complexity and orthographic transparency make the learning problem more challenging for a young reader of English.

Indeed, more recent studies suggest that the inconsistent English orthography coupled with its complex syllable structure makes English an 'outlier' language regarding reading acquisition. It is more difficult to acquire reading in English than in any other language, and accordingly the generality of cognitive factors for reading based on English-language studies (which have dominated the experimental literature) can be questioned. However, studies of reading acquisition in other languages are to date generally supportive of the language-universal approach adopted by psycholinguistic grain size theory. For example, Ziegler and his colleagues tested 1,265 children learning to read Finnish, Hungarian, Dutch, Portuguese and French in their second year of schooling, measuring phonological awareness, phonological memory, vocabulary, rapid object naming and non-verbal IQ as predictors of reading (Ziegler et al., 2010). Phonological awareness was the main factor associated with reading performance in each language, although its effects were stronger for the less transparent orthographies. Ziegler et al. concluded that the predictors of reading performance are relatively universal across languages, particularly regarding alphabetic orthographies.

Another factor that has been of interest more recently regarding individual differences in reading acquisition across languages is morphological awareness. Morphological units (such as *ness* in *darkness*) are larger orthographic units that repeat in different words (*illness*, *happiness*, *sadness*) and are likely to become important for recoding print to sound once longer words are being encountered (single syllable words are always mono-morphemic). Morphological units usually have consistent spelling across different words. Accordingly, awareness of morphological units may also assist the achievement of efficient phonological decoding. Interestingly, languages with more inconsistent orthographies, like English, tend to be morphologically simple in comparison to languages like Finnish, which has a very consistent orthography but complex morphology. Hence it may follow that children who are learning to read consistent orthographies may rely relatively *more* on morphological units as they become more skilled in reading. Alternatively, it is possible that readers of less consistent orthographies might rely more on morphological units, as these units offer another form of large-unit consistency (like rime analogies) that make the process of phonological recoding to sound more accurate. In his recent work, Ziegler and his colleagues have been comparing children's use of morphological units in reading across languages. For example, Mousikou et al. (in press) compared third-grade children's use of morphological units by

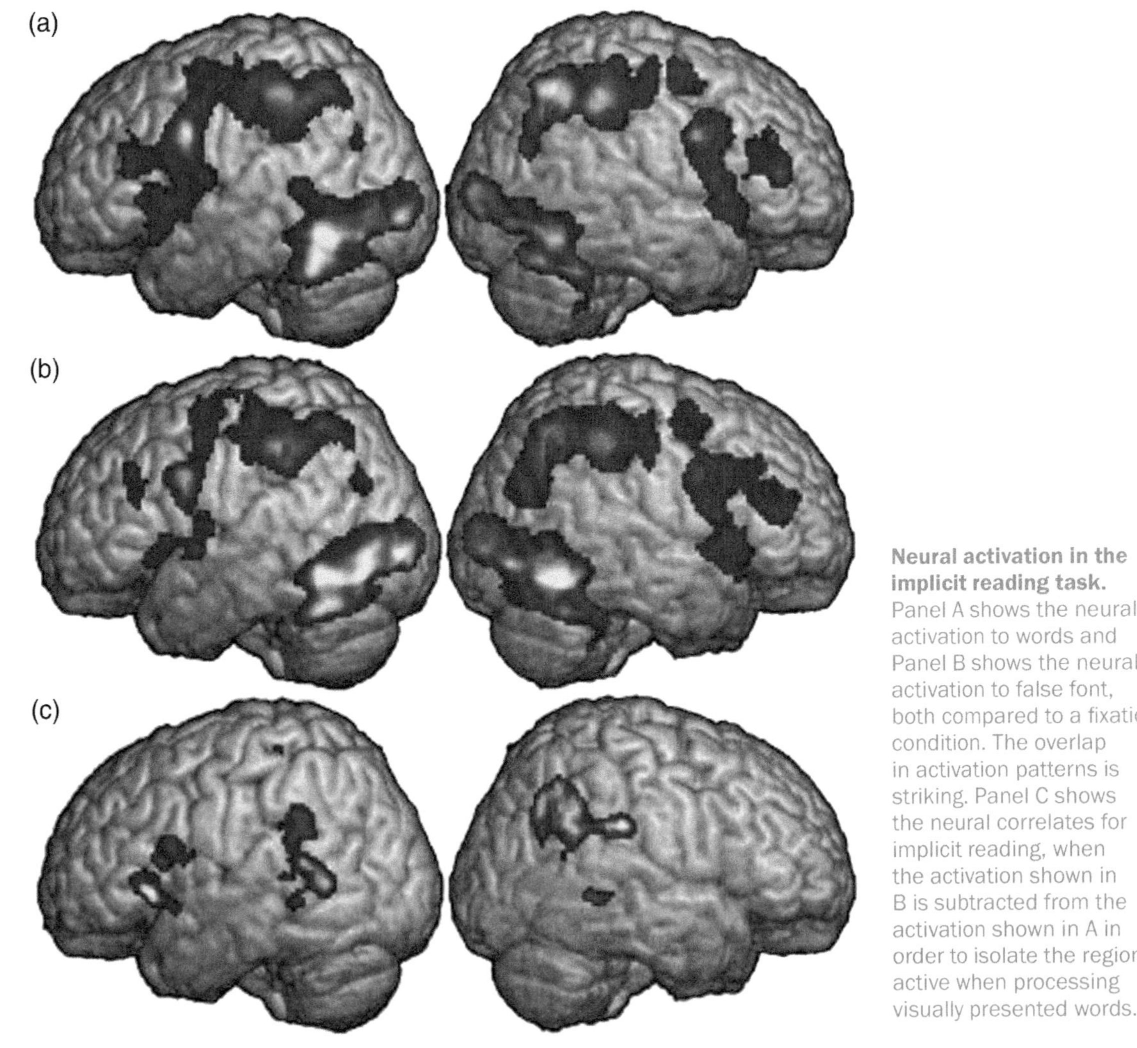

Neural activation in the implicit reading task. Panel A shows the neural activation to words and Panel B shows the neural activation to false font, both compared to a fixation condition. The overlap in activation patterns is striking. Panel C shows the neural correlates for implicit reading, when the activation shown in B is subtracted from the activation shown in A in order to isolate the regions active when processing visually presented words.

comparing children learning to read English, French, German and Italian. Both speed and accuracy data were collected. Mousikou et al. reported that all the children tested showed sensitivity to morphological units. However, gains in reading speed and accuracy were greatest for the children learning the most inconsistent orthography, namely English. The French, German and Italian children did not differ, even though French is also a relatively inconsistent orthography. Mousikou et al. concluded that children rely on morphological units most when they are learning orthographies characterised by very low spelling-to-sound consistency. Indeed, some authors suggest that explicit morphological instruction could help to facilitate reading acquisition in English (Bowers & Bowers, 2017). Training studies testing this idea have yet to be conducted.

Cognitive neuroscience of reading acquisition

Most neuroimaging studies of reading development are with English-speaking children, although the field is currently expanding rapidly. The growing number of studies in other orthographies are important, as the neural networks that are developed by a brain learning to read the 'outlier' English orthography might be somewhat different to the neural networks that are developed by a brain learning to read a more transparent orthography. However, given the central role of phonological awareness in learning to read in all orthographies so far studied, most core features of developing a neural system for reading seem likely to be relatively universal. Indeed, studies of adult readers across languages show that the neural networks that develop for reading involve very similar loci across languages. The 'reading network', or the network of brain regions typically active during silent adult reading, appear to comprise left-lateralised frontal, temporoparietal and occipitotemporal regions.

A representative neuroimaging study of normative development using fMRI was reported by Turkeltaub and his colleagues (Turkeltaub, Gareau, Flowers, Zeffiro & Eden, 2003) for English. Turkeltaub et al. measured neural activation in children and young adults aged from seven years to 22 years while they were performing a 'false font' task (see Figure 10.3). As will be recalled, when making comparisons in brain activation across different ages and interpreting them developmentally, it is important to equate performance on the in-scanner task. The false font task was chosen because the seven-year-olds could perform it as well as the adults. The task is based on meaningless symbols which have similar visual features to letters, and is designed to mimic the demands of reading an alphabetic script. The children were asked to detect certain features such as ascenders (b, d and k are examples of letters with ascenders). Turkeltaub et al. argued that comparison of the neural activity found when detecting features like ascenders in real words with the neural activity found during the false font task yielded a measure of 'implicit reading'. The implicit reading-related activity in the college students was found to be located in the left hemisphere sites identified in other studies of adult readers, including left posterior temporal and left inferior frontal cortex and also right inferior parietal cortex. To explore developmental effects, Turkeltaub et al. then restricted the analyses to children under the age of nine years. Now

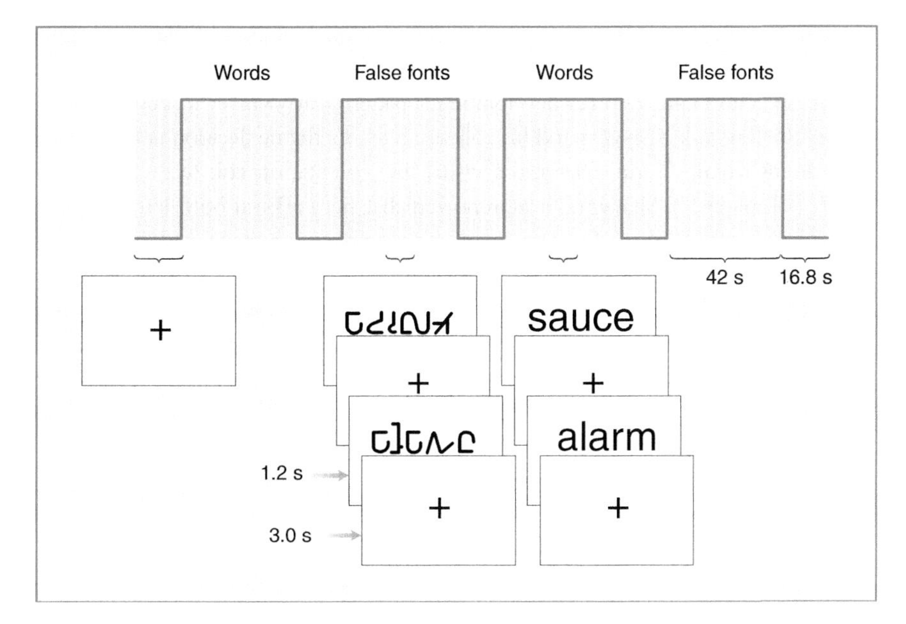

FIGURE 10.3
In the 'false font' task used by Turkeltaub et al. (2003) subjects were instructed to press a button held in their right hand if the stimulus contained an ascender or 'tall letter' or a button held in their left hand if it did not. In the examples given here, 'sauce' would be given a "no" response, as would the first false font string shown. The word 'alarm' would be given a "yes", as would the second false font string. From Turkeltaub et al. (2003). Copyright © 2003 Macmillan Magazines Limited. Reproduced with permission.

the main neural region activated by the false font task was left posterior superior temporal cortex (traditionally considered to be an area activated by phonological processing, and possibly active during grapheme–phoneme translation). As reading developed, activity in left temporal and frontal areas increased, while activity in right posterior areas declined. This pattern suggested that reading-related activity in the brain becomes more left-lateralised with development, perhaps as decoding becomes more efficient.

Turkeltaub et al. (2003) also explored the neural activation associated with the three core measures of phonological processing used as predictors of individual differences in reading acquisition, namely phonological awareness, phonological short-term memory and rapid automatised naming. Although neural activity in all three tasks was highly interrelated, Turkeltaub et al. were able to explore potential functional differences by calculating partial correlations between activated brain regions and each of the three measures while controlling for the effects of the other two measures. They reported that the three different measures did seem to correlate with distinct patterns of brain activity. Phonological awareness tasks appeared to be associated with activity in a network of areas in left posterior superior temporal cortex and inferior frontal gyrus. The degree of activity in this region varied with the level of children's phonological skills, a parametric relationship. Activity in the inferior frontal gyrus also increased with reading ability. Phonological short-term memory (measured by digit span) appeared to activate left intraparietal sulcus. This is a region typically activated in working memory studies with adults (see Chapter 8). Digit span tasks also activated the middle frontal gyri (bilaterally) and the right superior temporal sulcus. Rapid automatised naming appeared to be associated with activity in a spatially different, bilateral network, including the right posterior superior temporal gyrus, right middle temporal gyrus and left ventral inferior frontal gyrus.

Converging data were reported in an fMRI study of 119 typically-developing readers aged from seven years to 17 years by Shaywitz and colleagues, also conducted in English, who found a similar developmental pattern (Shaywitz et al., 2007). Instead of the false font task, Shaywitz and his colleagues used a rhyme decision task (e.g., "do these items rhyme?": *leat, kete*) and a visual line orientation task (e.g., "Do [\\V] and [\\V] match?"). Shaywitz et al. reported that networks in both left and right superior and middle frontal regions were more active in younger readers, with the amount of activation declining as reading developed. In contrast, activity in the left anterior lateral occipitotemporal region increased as reading developed. Hence both Turkeltaub et al. (2003) and Shaywitz et al. (2007) reported decreased right hemisphere activation as reading developed, albeit in different neural networks. The difference in the behavioural tasks used (e.g., false font versus rhyme judgement) may explain some of these differences. However, the left occipitotemporal cortex has been a focus of more recent studies, as this neural region is typically active during object and face recognition, and appears to become specialised over time for print. Dehaene and his colleagues have labelled this region the 'Visual Word Form Area' or VWFA. The VWFA is active whenever printed words are shown to the brain (even if the words are only shown to the left visual field, which means that they first activate visual areas in the right hemisphere, see Cohen, Atkinson & Chaput, 2004). The VWFA becomes increasingly activated by tasks involving print as children get older and become better readers (Pugh, 2006). Indeed, Pugh and others have suggested that the amount of activity in the VWFA is the best fMRI correlate that we have of reading expertise. However, it is important to note that the VWFA is also activated by nonsense words. In fact, the VWFA shows greater activity the more 'word-like' the nonsense words become. Given that nonsense words that contain large fragments of real words elicit greater brain activity, activation in the VWFA is probably most accurately described as orthographically related, with activation both when whole words are viewed and when fragments of familiar words, such as orthographic rimes, are seen.

Studies by Dehaene and his colleagues have explored the development of the VWFA in some detail (Dehaene, Cohen, Morais & Kolinsky, 2015). The location of the VWFA in left ventral occipitotemporal cortex is similar for a range of orthographies with different visual characteristics, for example Hebrew, English, French and Chinese. The VWFA area is adjacent to the fusiform face area, and as literacy increases, the activity of the VWFA to letters increases while responsivity to faces decreases, with specialised face processing shifting more to the right hemisphere. Other researchers have used fMRI to track the early development of specialised letter knowledge in young children in fine detail. For example, James (2010) gave letter writing practice to young preschool children who were just beginning to learn to read. The children learned to pick out target letters from four alternatives (including reversed letters), to recognise the letters in storybooks, and also to form and write the letters. Significant improvement in writing the letters was found following the training. fMRI was recorded as children passively viewed both the letters and control visual stimuli, both before and after training. James (2010) reported that prior to receiving writing training, presentation of the letters either caused no activation in the left fusiform gyrus (the area that becomes the VWFA), or modest activation. Following training, visual presentation of the letters led to strong selective activation of the left fusiform gyrus, and also led to activity in ventral premotor areas. Control children received matched training focused on visual recognition of letters without writing them. These children did not show activity to the learned letters in either neural area. The activation of *motor* areas during a visual task (letter recognition) following training suggests that learning to write letters created multi-sensory mental representations for familiar letters for these young children. In more recent work, James and Engelhardt (2012) have reported that learning to write letters by hand appears to be preferable for supporting literacy development in young children. Pre-reading five-year-old children were given training in either hand-drawing letters and shapes, or in copying letters and shapes by typing on a keyboard. The children then viewed both familiar and novel letters and shapes passively during fMRI scanning. Only children who had self-drawn the letters showed greater activation of the left fusiform gyrus to familiar letters. James and Engelhardt made the interesting observation that when children self-draw letters, they produce quite variable outputs. Developmentally, this may help to create abstract knowledge of letters, such that A is the same letter even when the visual appearance of a particular letter A is variable.

Studies using EEG have been less dominant in the cognitive neuroscience of reading literature, even though the superior time course of EEG makes it very suited to tracking developmental changes in reading expertise. A typical paradigm has been presenting real words and nonsense words while recording EEG, thereby keeping the visual/spatial demands associated with text processing constant, and varying only whether the target is a lexical word form or not. Such EEG studies show that the brain responds differently to real versus nonsense words within one-fifth of a second, suggesting that lexical access (contact between the visual word form and its meaning) occurs very rapidly during reading. Indeed, the speed of this differentiation has been shown to be similar for both children and adults across languages (a negative deflection 160–180 ms following stimulus onset, now called the N170, the same label used for the ERP marking face-specific processing, see Csépe, Szücs & Honbolygó, 2003; Sauseng, Bergmann & Wimmer, 2004). EEG data thus show that the time course of visual word recognition is very rapid even for children. Indeed, the N170 can be used to track the early development of reading expertise by measuring when neural activity in the VWFA first becomes specialised for letter strings that are real words.

This approach was pioneered by Maurer, Brandeis and their colleagues, who studied the development of the N170 in Swiss children learning to read in German (Maurer, Brem, Bucher & Brandeis, 2005; Maurer et al., 2007). Maurer et al. followed the children from the very beginning of learning to read, using a repetition detection task. The children saw printed symbol strings and had to detect repetition of either real words or of meaningless strings, yielding a measure of word-specific neural processing. Brain activity was recorded in kindergarten, before the children had received any instruction in reading, and again in the second grade. Maurer et al. reported that before any reading instruction had commenced, the children did not show an N170 to printed words, despite having considerable knowledge about individual letters in the words. After approximately 1.5 years of reading instruction, the children did show a reliable N170 to words, described by the authors as evidence for a 'coarse tuning' to print. Perhaps unsurprisingly, children at family risk for dyslexia who were also included in the study had yet to develop an N170 to words. In more recent studies, Maurer and his colleagues have shown that the emergence of the N170 is related more to reading fluency and vocabulary size (semantic knowledge) rather than to phonological recoding skills (Eberhard-Moscicka, Jost, Raith & Maurer, 2015). Accordingly, the emergence of the N170 appears to reflect a mixture of the brain's increasing perceptual specialisation for print, along with the increasing influence of top-down influences from semantic knowledge on reading efficiency. Nevertheless, although both fMRI and EEG studies have produced reliable neural correlates of growing expertise in reading, studies to date have not added any new knowledge regarding the neural mechanisms involved in becoming a good reader.

Developmental dyslexia

KEY TERM

Developmental dyslexia Describes children who have unexpected difficulties in learning to read and to spell despite adequate educational instruction, good vocabularies and normal intelligence.

Developmental dyslexia, or unexpected difficulties in learning to read and to spell despite adequate educational instruction and normal intelligence has been found in all languages in the world so far studied (Goswami, 2015). The core cognitive problem in developmental dyslexia is thought to be a phonological one. Despite apparently normal language acquisition, sensitive behavioural tasks show that children with dyslexia have not developed well-specified phonological representations of the *sound structure* of the individual words in their mental lexicons (Snowling, 2000). Three types of behavioural task reveal phonological difficulties across languages. These are phonological awareness tasks (e.g., the tapping task, the oddity task), phonological short-term memory tasks (such as digit span, see Chapter 8), and 'rapid automatised naming' or RAN tasks – tasks requiring children to name familiar items like colours, pictures or digits as fast as they can. Children diagnosed with developmental dyslexia find it difficult to perform these phonological tasks in languages as diverse as Chinese (e.g., Ho, Law & Ng, 2000), Japanese (e.g., Kobayashi, Kato, Haynes, Macaruso & Hook, 2003), German (e.g., Wimmer, 1993) and English (e.g., Bradley & Bryant, 1978).

The two critical factors governing reading acquisition of phonological complexity and orthographic transparency also lead to cross-cultural differences in the *manifestation* of developmental dyslexia. For example, a large-scale cross-language study of the predictors of dyslexia in six European languages involving over 1,000 children with dyslexia showed that orthographic complexity increased the strength of some predictors of dyslexia, particularly phonological awareness and RAN (Landerl et al., 2013). Landerl et al. reported that phoneme deletion and RAN were the strongest predictors of dyslexia in all six languages being compared (Finnish, Hungarian, German, Dutch, French and English), with phonological STM playing a comparatively minor role. Dyslexia in the more transparent orthographies was typically manifest by very slow reading and inaccurate spelling. As learning to read in itself improves the specificity of phonological representations, it is also important to compare children with dyslexia to younger children who are reading at the same level as they are (a 'reading level' or 'RL match' design). Matching for reading level goes some way to equating the experience of the brain with print. If the phonological skills of the dyslexic children are inferior even to those of these *younger* children, then it can be assumed that any cognitive deficits found (such as phonological impairments) are fundamental to dyslexia rather than simply being a consequence of poorer reading experience.

A short survey of available studies confirms that the phonological deficit has been established in developmental dyslexia using RL match designs in many languages (Goswami, 2015, for an overview). For example, Bradley and Bryant (1978) gave English ten-year-old children with developmental dyslexia the oddity task described earlier, and compared their performance with that of seven-year-old typically-developing readers. The children with dyslexia were significantly worse than the younger controls in each version of the oddity task. For onsets, they scored 54% correct compared to 89% for the younger children, for the middle sound different (rime) task they scored 75% correct compared to 94% correct for the younger RL controls and for the final sound different (rime) task they scored 81% correct compared to 97% correct for the seven-year-olds. In a study of German children with developmental dyslexia, Landerl, Wimmer and Frith (1997) gave an onset-rime Spoonerism task to 11-year-old German dyslexics, 11-year-old German chronological age (CA) controls and eight-year-old German RL controls. The Spoonerism task required the children to substitute onsets between two words (e.g., *boat-fish* to *foat-bish*). When performance was scored in terms of correct Spoonerisms (i.e., *foat-bish*), the German dyslexic group made significantly more errors than both their RL and CA controls (German dyslexic = 37% correct, RL controls = 57% correct, CA controls = 68% correct).

Despite this shared 'phonological deficit' across languages, children with dyslexia who are learning to read transparent orthographies develop better decoding skills than children with dyslexia who are learning to read opaque languages. Similarly, children with dyslexia who learn to read spoken languages with a simple (CV) syllable structure become more competent than children with dyslexia who learn to read spoken languages with a complex syllable structure. This developmental pattern means that reading levels can be very accurate in older children with developmental dyslexia who are learning to read consistent orthographies. An accuracy deficit in reading is often only visible earlier in the developmental process. On the other hand, reading remains extremely slow and effortful even in these consistent orthographies. This speed impairment means that the children are functionally dyslexic, even if decoding is relatively accurate. For example, recoding a sentence to sound can take so long that the beginning of the sentence is lost from short-term memory, meaning that reading comprehension is severely affected. In addition, spelling skills remain poor in children with developmental dyslexia who are learning consistent orthographies. In fact, developmental dyslexia is more usually diagnosed on the basis of poor spelling in these languages. The persistent difficulties with spelling likely reflect the 1:many correspondence between sound and spelling in most alphabetic orthographies (discussed earlier).

Illustrative studies can be described for the consistent orthographies German (by Wimmer, 1993, 1996) and Greek (by Porpodas, 1999). Porpodas (1999) selected a sample of 16 Greek first-graders with literacy difficulties out of an initial cohort of 564 children. The dyslexic children were at least two standard deviations below the other children in spelling accuracy, and at least one standard deviation below the other children in decoding time. Porpodas then compared these children to a sample of 16 CA-matched control Greek children from the same cohort. All children were given a set of 24 two- and three-syllable non-words to try and read. The non-words were created by changing the initial and middle letters of real words. Porpodas reported that the Greek dyslexic children read 93% of these non-words correctly, compared to 97% for the CA controls. Although this was a significant difference, even the struggling Greek readers were clearly very accurate. They took twice as long as the typically-developing children to recode the non-words to sound, however. When given a Greek version of the oddity task, the poor readers scored 51% correct, and the controls scored 89% correct. Wimmer (1996) reported on the reading skills of a group of German children in grade 1 who later became dyslexic. The children were given simple non-words like *Mana* (*Mama*) and *Aufo* (*Auto* – car) to read. He found that seven out of 12 children who later became dyslexic read less than 60% of these non-words accurately, compared to an average performance of 96% correct for beginning readers who did not subsequently become dyslexic. In a phoneme reversal task given at the same time (e.g., reverse 'ob' to 'bo'), the to-be-dyslexic children scored on average 22% correct, compared to an average of 69% correct for the control children. In fact, 42% of the dyslexic children could not attempt the phonological task at all. Wimmer (1993) reported on the reading and phonological skills of the dyslexic group when they were ten-year-olds. In a timed non-word reading task based on 'Italian' type non-words with open syllables (*ketu*, *heleki*, *tarulo*), the children with dyslexia scored as well as RL controls, reading on average 92% of items correctly. They also performed as well as younger RL controls in a phoneme awareness task requiring them to substitute one vowel phoneme for another (e.g., "Mama ist krank" to "Mimi ist krink"). Both groups succeeded on 86% of trials. Clearly, after three years of reading instruction, the German dyslexics had developed accurate phonological recoding skills. They had also used print–sound relationships to improve the quality of their phonological representations, as they were now performing at comparable levels in phonological awareness tasks to RL controls. However, they were still significantly poorer than CA controls, who scored 95% correct in the vowel substitution task.

A carer may help to compensate to some extent for a child's genetic predisposition to dyslexia through the use of phonological improvement tools, such as language games and nursery rhymes.

Dyslexia is also heritable. When children are born into families where one or both parents have dyslexia, between 35% and 65% are later diagnosed with dyslexia themselves (Fisher & DeFries, 2002). Most studies exploring heritability have been family and twin studies, particularly in English (e.g., Gayan et al., 1999; Gayan & Olson, 2001). Of course, the idea of 'genes for dyslexia' does not make sense, as reading is a culturally determined activity. While a child's genes can determine their eye colour, a child's genes cannot determine their reading level. Children at genetic risk for dyslexia are those with a genetic risk of experiencing phonological difficulties. Levels of association reported so far in behavioural and molecular genetics are not strong enough to translate into reliable predictors of risk for a single child (Fisher & Francks, 2006). Nevertheless, research exploiting our growing understanding of the neural encoding of speech reveals that phonological difficulties in children with dyslexia are reliably associated with sensory differences, and these sensory differences could be inherited. One plausible sensory candidate is perceiving amplitude envelope rise times (the sensory cues that enable the automatic phase realignment of oscillating cell networks in auditory cortex with patterns of amplitude modulation in the speech signal, see Chapter 6). During the last decade, impaired rise time discrimination has been documented for children with dyslexia who are learning to read English, French, Spanish, Chinese, Dutch, Finnish and Hungarian (Goswami, 2015, for review). Impairments in rise time discrimination are related to phonological awareness in these languages, using language-appropriate measures (for example, tone awareness in Chinese and phoneme awareness in Spanish). Furthermore, infants who are born into families where one or both parents have dyslexia already show impaired rise time discrimination by ten months of age (Kalashnikova, Goswami & Burnham, 2018a). These sensory difficulties with rise time seem likely to be related to impairments in the neural encoding of speech, which could be associated with the development of impaired phonological representations of words in the mental lexicon. This potential causal hypothesis, from auditory sensory difficulties to phonological difficulties to reading difficulties, is discussed further below.

Cognitive neuroimaging studies of developmental dyslexia

The growth in cognitive neuroimaging studies of developmental dyslexia since the last edition of this book has been remarkable, with more than 100 studies for fMRI alone. A good survey of the literature is available from Hoeft and her colleagues (Xia, Hancock & Hoeft, 2017; Black, Xia & Hoeft, 2017). Most fMRI studies have relied on comparisons with age-matched control children, however. As age-matched controls will be reading at a significantly higher level compared to the children with dyslexia, it is impossible to draw conclusions about causality from differences in brain activation in these studies. An exception is if the groups are matched for in-scanner task performance, although this is rarely done. In such cases, different activation patterns for the dyslexic brain could be informative. Nevertheless, such designs only provide the first step in establishing neural causation (see Goswami, 2015), as interventions are needed to test causal hypotheses regarding development. It is also impossible to draw causal conclusions about developmental trajectories from fMRI studies of dyslexic *adults*, yet this design is the one most frequently employed. Many fMRI studies recruit university students who have a childhood history of dyslexia, and compare their brain activity to that of other university students without dyslexia. As the dyslexic students may have developed a series of compensatory strategies over childhood, any group differences in neural activation may simply reflect these compensatory strategies. The two groups will also have a very different history of reading experience, with the students with dyslexia typically having read thousands fewer words (Goswami, 2015). Accordingly, the best fMRI data regarding potential causal factors in dyslexia come from longitudinal neuroimaging studies, which use repeated assessments of the same cohort of children over time, with scanning beginning long before the children receive any tuition in reading. I will give a flavour of the data available from such studies here, focusing largely on those that relate to the cross-language cognitive phenotype, namely the phonological deficit. I will then consider EEG studies that relate to the phonological deficit.

A structural fMRI study with a longitudinal design of a relatively small sample of 39 children was reported by Clark et al. (2014). Using a family risk design, Clark and colleagues took structural scans of children at high versus low risk of dyslexia the year before reading was taught (at age 6–7, as this was a Norwegian cohort), a year after reading tuition began (at age 8–9), and after dyslexia had been diagnosed (at age 11–12). Norwegian is a relatively transparent orthography, and reading instruction commences at age seven. Most seven-year-olds quickly attain high efficiency in recoding print to sound (reaching over 90% accuracy in reading words or pseudo-words within the first year of schooling). However, the children who were later identified as dyslexic in Clark et al.'s study showed slow and inefficient learning of letters even as preschoolers, and showed severe impairments in recoding print to sound once reading instruction began. Remarkably, however, Clark et al. did not find structural differences in any of the neural regions typically thought to comprise the 'reading network' in their pre-reading scans, that is the left-lateralised frontal, temporoparietal and occipitotemporal regions. Instead, the pre-reading neuroanatomical regions that differed significantly were in sensory areas, specifically cortical thickness in primary auditory cortex (Heschl's gyrus) and primary visual cortex. The only neural structure in which group differences were consistent over development (in that group differences were found at all three scans) was primary auditory cortex. Heschl's gyrus was still significantly thinner in the children with dyslexia at the end of the study. These structural data are consistent with a sensory basis for impaired phonological development in basic auditory processing.

White matter tractography showing the arcuate fasciculus in infants either at risk for dyslexia or not at risk. White matter integrity (fractional anisotropy) was measured in infants at family risk for dyslexia (FHD+) aged from six to 18 months and in infants with no family history of dyslexia (FHD–). The structures are already visibly different, although this could be a developmental result of differential acoustic processing of language input.

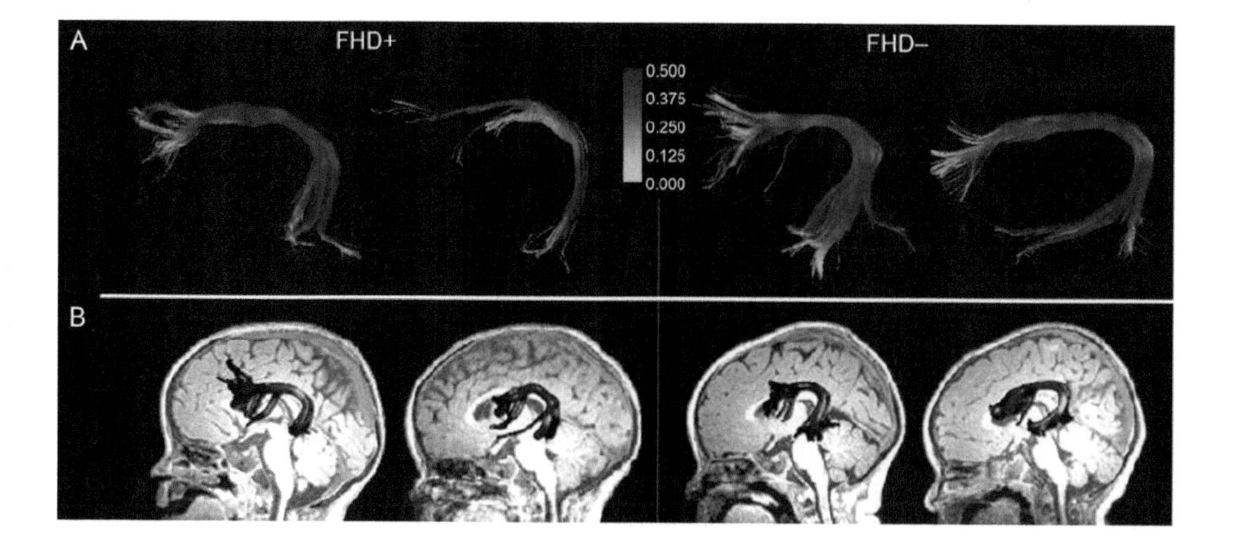

It should be noted that group differences in the left-lateralised 'reading network' were evident for these Norwegian children, but only by age 11, indicating that the 'reading network' emerges with growing expertise. Cortical thickness in parts of the reading network, such as temporoparietal cortex, was reduced in comparison to the control children by age 11. This suggests that the hypo-activation of the 'reading network' reported in many fMRI studies of dyslexia (Richlan, Kronbichler & Wimmer, 2011, for a meta-analysis), including some using an RL match design (Hoeft et al., 2006), is probably a consequence of reduced reading experience. It does not seem to be a structural difference in the dyslexic brain that predates the onset of reading tuition, as these structural differences only emerged in the 11-year-old scans.

Indeed, a later report on these Norwegian children measured connectivity between the different neural regions comprising the 'reading network' (Morken, Helland, Hugdahl & Spechtad, 2017). For the connectivity analyses, functional MRI was used, and children received a reading task in the scanner. For the youngest test points (six and eight years) children had to recognise logographs like LEGO, simple regular words like SNOP and longer more complex Norwegian words like LØRDAGSGODT. The in-scanner task was to decide whether different words fitted a category such as 'things to play with'. At age 11–12 years, a sentence reading task was added. Morten et al. reported connectivity analyses based on the two word reading conditions, which were given at all ages. Pre-determined connections were specified between regions of interest within the left-lateralised 'reading network' (namely the superior temporal gyrus, occipitotemporal cortex, the inferior parietal lobule, inferior frontal gyrus and precentral gyrus) and then functional connectivity was compared at different ages. The pattern of results suggested a delay in connectivity in the children with dyslexia. While the control group seemed to stabilise connection strengths over time, with steep decreases in connectivity from six to eight years and then less steep decreases in connectivity, the dyslexic group began with connectivity at a level well below the control group, showed a marked increase in connectivity between six and eight years, and then stabilisation from eight to 12 years. At age 12, there were no longer group-level differences in connectivity for readers of this shallow orthography. Although this is discussed as *normalisation* by the authors, the group differences in literacy abilities had become much greater over developmental time. It may thus be that for these consistently spelled Norwegian words, the in-scanner tasks were no longer sensitive to potential group differences.

A longitudinal fMRI study of children in the USA at family risk for dyslexia took a different approach to measuring connectivity, focusing on white matter pathways (Langer et al., 2017; Wang et al., 2017b). Fractional anisotropy is a measure that can be used to assess white matter integrity (i.e., degree of myelination, axon density and the calibre of axons). During development, increased myelination leads to increased fractional anisotropy, with thousands of axons entering a given white matter tract at different locations. Wang and her colleagues focused their interest on the left arcuate fasciculus, a white matter tract thought to be related to language processing that links key areas in the 'reading network' including the superior temporal gyrus, the inferior parietal lobule and inferior frontal gyrus. They used fractional anisotropy to measure the integrity of the left arcuate fasciculus in 21 at risk and 24 control children when they were pre-readers, and then looked at developmental changes in fractional anisotropy as the children learned to read.

Wang et al. reported that fractional anisotropy in the left arcuate fasciculus was significantly lower for the at-risk children when they were pre-readers. Ten of the children subsequently developed into poor readers. When the rate of development of fractional anisotropy in the left arcuate fasciculus was compared for these children, it was significantly lower for the poor readers compared to children who became good readers. Accordingly, there are some structural differences in the brain in pre-reading children at risk for developmental dyslexia within the core 'reading network', at least for this measure. Langer et al. (2017) have further reported significantly lower fractional anisotropy in the left arcuate fasciculus in 14 infants at family risk for dyslexia aged from six to 18 months, who were compared to 18 infants of a similar age range with no family history of dyslexia. Mean fractional anisotropy in the left arcuate fasciculus was also significantly related to a measure of expressive vocabulary for the infants. The data suggest a structural difference prior to reading tuition in at-risk infants for a key white matter tract linking areas of the brain active during language processing. Given early atypical auditory sensory processing in infants at family risk for dyslexia (Kalashnikova et al., 2018a), however, it is important to remember that neural encoding of language may also be atypical from birth. Logically, therefore, atypical linguistic processing by infants at family risk for dyslexia could already be affecting the development of the white matter tracts related to language processing, resulting in lower fractional anisotropy.

Finally, functional fMRI studies can be useful for understanding how atypical auditory processing affects letter learning in children with dyslexia. For example, Blau et al. (2010) asked nine-year-old Dutch children with or without a dyslexia diagnosis to listen to single speech sounds like /a/ or /b/ while in the scanner. The children either heard just the speech sounds by themselves, or heard the speech sounds in the presence of visually presented letters. The letters were either congruent (letter A for sound /a/) or incongruent (letter A for sound /e/). The children were asked to press a button whenever they heard a sound or saw a letter. Blau et al. were particularly interested in the incongruent condition, when speech sounds and letters did not match. By age nine, letter–sound integration should be fairly automatic, and so the incongruent condition was expected to discriminate between good and poor readers. Data analysis showed that while the children in each group were equally fast at detecting the stimuli, there were group differences in neural activity for both hearing speech sounds and seeing letters. The children with dyslexia showed significantly weaker activation of both the left superior temporal gyrus and the left fusiform gyrus (VWFA) when hearing speech sounds, as well as weaker activation of the fusiform gyrus bilaterally when seeing letters. When speech sounds and letters matched in the congruent condition, the control children, but not the dyslexic children, showed significant activation in the left superior temporal gyrus close to Heschl's gyrus. This activation was reduced for control children in the incongruent condition, when the letters and speech sounds did not match. The children with dyslexia did not show any change in activation for the mismatch condition. Blau et al. (2010) concluded that the neural integration of speech sounds with letters is impaired for children with dyslexia. However, these data do not show that impaired neural integration of orthographic and phonological information causes dyslexia. Rather, the finding that there is reduced activation in the 'reading network' simply for hearing speech sounds is consistent with a developmental causal model based on impaired auditory processing of language, which would in turn affect letter learning.

All the fMRI studies discussed so far are consistent with the behavioural research on developmental dyslexia, with atypical neural processing demonstrated either during reading itself, during phonological tasks, or for specific tasks like letter–sound matching. No study has yet revealed new causal factors in dyslexia. However, an attractive feature of fMRI is that it can reveal changes in neural organisation following interventions intended to ameliorate dyslexia. For example, if an intervention is targeted at improving phonological skills, then fMRI can show whether neural activity in the regions active during phonological tasks changes or normalises following the intervention. If any behavioural changes found are due to Hawthorne effects (non-specific effects related to being singled out to be in an intervention), then activation in the neural networks related to phonology per se is less likely to change. A number of fMRI studies of targeted phonological remediation have now been carried out and provide converging data, showing that neural activity in the left-lateralised 'reading network' normalises (i.e., increases) following intervention. For example, Shaywitz and colleagues carried out a training study based on three groups of children aged from six to nine years (Shaywitz et al., 2004). One group comprised children with a diagnosed reading disability who were receiving a daily targeted intervention based on phonology. A second group comprised children with a diagnosed reading disability who were receiving the typical community-based interventions on offer in their schools. The third group were non-impaired controls. The children receiving the targeted remediation improved their reading accuracy and reading fluency and also their reading comprehension. Comparing brain activation following remediation to fMRI scans taken before the intervention began, Shaywitz et al. (2004) reported that children receiving the phonological remediation showed increased activation in the left posterior temporal and inferior frontal regions associated with efficient reading in typically-developing children. The reading disabled children who were receiving community-based interventions did not show comparable gains, neither behaviourally nor neurally (in fact, these children were really an unseen control group, as they did not receive any extra attention during the study). A seen control group (for example, receiving semantic training) is required in order to strengthen this conclusion.

More recently, a similar conclusion has been reported by Olulade and his colleagues (Olulade, Napoliello & Eden, 2013). In their study, children with dyslexia were compared to RL-matched children before and after receiving a phonological intervention programme lasting eight weeks. This study also used a nice design of having the children with dyslexia act as their own controls, by also receiving a maths intervention. Following the maths intervention, no improvements in literacy or phonology were shown by the dyslexic children, but significant improvements in both literacy and phonology followed the phonological intervention. The phonological intervention also led to neural improvements in visual motion processing for the children with dyslexia. This was an unexpected and important result. Dysfunction of the visual magnocellular system for processing motion is sometimes suggested to be a cause of dyslexia, even though all available data are correlational. fMRI studies using age-matched designs typically find that the children with dyslexia show less activity in regions in visual cortex active during motion processing. Olulade et al. (2013) replicated this finding with age-matched controls, but also tested an RL control group. There was no difference in neural activity for visual motion processing between children with dyslexia and RL controls. Coupled with the finding that neural activity during motion processing *increased* for the dyslexic children following the *phonological* intervention, we can conclude that the reading problems experienced by children with dyslexia are not caused by magnocellular dysfunction. Rather, it seems likely that the reduced visual reading experience accompanying being dyslexic alters the functional organisation of the brain.

As a final example of interventions, we will consider a training study reported by Heim, Pape-Neumann, van Ermingen-Marbach, Brinkhaus and Grande (2015). This study is relevant to another visual theory of dyslexia, based on 'sluggish' attention-shifting. The core idea is that children with dyslexia cannot move their visuo-spatial attention smoothly from letter to letter, which impairs their reading (Facoetti et al., 2010; most data supporting this theory come from Italian children with dyslexia). Working with children learning to read the transparent German orthography, Heim et al. selected three groups of nine-year-olds with a diagnosis of dyslexia. They then compared the efficacy of three interventions, a phonological intervention, an attention intervention based on orienting visual attention to word fragments, and an orthographic reading intervention aimed at training sight vocabulary by rapid recognition of common spelling sequences like 'ing'. The interventions lasted for four weeks, and the children received fMRI before and after the interventions. An age-matched typically-reading control group who did not get any interventions also received fMRI at the same time points. The task in the scanner was single word reading. The children first saw a word and could read it silently while scanning took place. Then there was a break in scanning, and the child read the word aloud. All three dyslexic groups showed significant improvements in reading after the interventions, and this was matched by a significant increase in neural activation of the VWFA for all groups during the in-scanner reading task. The phonological and orthographic intervention groups also showed similar effects regarding increased activation in parts of the left-lateralised 'reading network', such as the superior temporal gyrus. The only differences in post-training neural activation were between the phonological and attention intervention groups. The attention group showed stronger activation in the left Heschl's gyrus than the phonology group, a surprising result given that Heschl's gyrus is in primary auditory cortex. This appears to suggest that the sluggish shifting of visuo-spatial attention in dyslexia is a result of children's phonological difficulties. Training children in orienting visual attention to word fragments appears to alter brain activation in primary auditory areas rather than in attention areas. Accordingly, it may be reduced practice in recoding letters to sounds that causes the slow attention shifting found in Italian dyslexic children, rather than slow attention shifting causing poor reading. Overall,

Heim et al. concluded that while specific training effects of different interventions can be demonstrated using fMRI, the changes in VWFA activation appeared to be key concerning improvements in reading performance. The specific training effects found are also useful for evaluating competing theories, and again appear to support a core role for impairments in auditory/phonological processing in the aetiology of developmental dyslexia.

Most recently, the auditory 'rise time' hypothesis regarding the phonological 'deficit' in dyslexia has received support from EEG and MEG studies. Some of these new studies are unique in providing a direct measure of the quality of children's sensory/neural 'phonological representations'. This has become possible because new mathematical approaches enable the neuronal oscillatory response to speech to be reverse-engineered, thereby going backwards from brain activation to language input. In Chapter 6, we reviewed EEG and MEG data on neural speech encoding illustrating the core role played by neuronal oscillations at different temporal rates. We saw that oscillatory rates in auditory cortex broadly match temporal rates of amplitude modulations nested in the amplitude envelope of speech (delta, theta, beta/low gamma). We saw that amplitude envelope rise times trigger automatic phase-resetting of these neuronal populations by the brain, so that oscillating networks can match their peaks and troughs in excitability with the peaks and troughs in amplitude modulations that are hierarchically nested at different temporal rates in the speech signal. We also saw that oscillatory entrainment plays a key mechanistic role in the development of phonological awareness. For example, a neural network oscillating at ~2 Hz (in the oscillatory delta band) will align itself (via phase locking) to the modulation pattern of the stressed syllables in the speech signal, while another network oscillating at ~5 Hz (in the oscillatory theta band) will align itself (via phase locking) to the modulation pattern of the syllables. Given that amplitude rise times are important mechanistically for the automatic parsing of phonological units from continuous speech, and given that children with developmental dyslexia across languages show impaired discrimination of amplitude envelope rise times, we have a potential neural mechanism for explaining why phonological development is impaired in dyslexia. The automatic neural phase resetting process, which depends on rise time discrimination, may be impaired in children with dyslexia. If this were the case, then encoding of the speech signal would be less accurate in the dyslexic brain, as the brain response would be 'out of time' with respect to the speech input, providing a less faithful representation of the sensory input.

At the time of writing, there are only a few studies exploring this hypothesis. In the first such study, Power and his colleagues gave children with developmental dyslexia aged 13 years a rhythmic syllable listening task while recording EEG (Power, Mead, Barnes & Goswami, 2013). The children heard the syllable 'ba' repeated every 500 ms (2 Hz), but occasionally the syllable was out of time, and children had to press a button when this occurred. Rhythmic violations were calibrated individually, so that performance accuracy was 79.4% for each child. This ensured that dyslexic and age-matched control children were performing at the same level while EEG was recorded. Power et al. (2013) reported that while both groups of children showed neural entrainment in the delta band, the children with dyslexia showed a different preferred phase. Preferred phase is the point in time when most neurons are discharging their electrical potentials and brain excitation is maximised. The brains of the children with dyslexia were showing their peak neural response at a less informative point in the stimulus, which would impair the quality of phonological information. Indeed, Power et al. noted that the consistent timing difference in preferred delta phase shown by the dyslexic children would have cascading consequences for the optimal encoding of faster-rate information, such as phonetic information, via the oscillatory hierarchy.

In a follow-up study, Power, Colling, Mead, Barnes and Goswami (2016) tested the same dyslexic children a year later with a sentence listening task, using degraded speech to make the task more difficult. Children had to report the sentences that they heard while EEG was recorded. Both CA and RL controls were also tested. Power et al. then used the electrical brain responses to recreate the speech envelopes of the sentences that the children had listened to. To equate the groups for behavioural performance, similar analyses were also run for sentences that all groups had reported accurately. This reverse-engineering technique allows the amplitude envelope of the speech signal to be reconstructed from the responses of the neuronal populations that encode it. By going from the brain response back to the speech, a direct measure of encoding accuracy can be obtained. Using this envelope reconstruction measure, Power et al. found that the accuracy of speech encoding in the delta band (relevant to representing speech prosody, see Chapter 6) was significantly poorer for the children with dyslexia compared to *both* RL controls and CA controls. Accordingly, the brains of the dyslexic children were encoding a significantly less accurate representation of the speech envelopes that they were hearing regarding one key band of amplitude modulations, the relatively slow modulations centred on ~2 Hz. This is an interesting result, as here cognitive neuroscience enables a *direct measure* of the quality of children's phonological representations, thereby going beyond behavioural data. As might be expected given the core phonological deficit in developmental dyslexia, the children with dyslexia were encoding a poorer-quality representation of the speech signal. Individual differences in encoding accuracy in the whole sample were also related to individual differences in a prosodic awareness task, a brain–behaviour correlation. Power et al. (2016) suggested that children with dyslexia have difficulty in recovering prosodic structure from the speech signal. As delta-band amplitude modulations sit at the top of the modulation hierarchy discussed in Chapter 6, these difficulties would have negative consequences for all other levels of phonological representation.

A study using MEG with Spanish children with developmental dyslexia found very similar results to Power et al. (2016), but using natural speech rather than degraded speech (Molinaro, Lizarazu, Lallier, Bourguignon & Carreiras, 2016). Although the study relied on a CA-match design, it is interesting because the MEG method enables precise spatial localisation of group differences. Molinaro et al. gave ten-year-old children with and without developmental dyslexia a sentence listening task in the MEG scanner, with questions asked about the meaning of each sentence to ensure children's attention. They then measured coherence (a phase locking measure) between the brain response at different frequencies and the speech, with significant coherence assumed when the brain response to the speech was significantly greater than in a resting baseline. The data showed significantly greater coherence (stronger oscillatory synchronisation) for the control children compared to the dyslexic children in the delta band response. This effect was significant in the right and left auditory cortex, and also in the right and left superior temporal gyrus, and in the right middle temporal region and the left inferior frontal regions. Connectivity analyses showed that reduced neural synchrony originated in right auditory cortex, and then moved across to the left inferior gyrus, suggesting that the sensory processing of auditory information was the primary impairment for these children. Correlational analyses showed a brain–behaviour relationship for performance in a phoneme deletion task, with greater connectivity related to better phoneme awareness. Despite the absence of an RL matched group, Molinaro et al. (2016) were also able to test Spanish adults with dyslexia in the same paradigm. These adult dyslexics had similar levels of word reading proficiency to the Spanish control children, but had significantly poorer phonological skills. The Spanish dyslexic adults showed the same neural response pattern as the Spanish dyslexic children, with impaired oscillatory synchronisation to speech in the delta band, originating in right auditory cortex. The delta band data suggest that in Spanish, a syllable-timed language (English is stress-timed), children with dyslexia also have difficulty in recovering prosodic structure from the speech signal

Most recently, it has been discovered using stimulus reconstruction in EEG that low frequency oscillatory responses (adding across delta and theta band responses) also represent some phonetic information directly. This means that faster neural responses are not the only mechanism for accurate phoneme perception. Impaired neural entrainment to delta-band and theta-band speech information could also directly affect the quality of children's phonetic perception. Di Liberto et al. (2018) used this new EEG reconstruction technique with a story listening task, comparing the neural encoding of phonetic aspects of speech by English-speaking children with and without dyslexia aged on average eight years. Both RL and CA control groups were utilised. The new approach to stimulus reconstruction effectively adds an estimate of phonetic detail to the reconstructed speech envelope. This enables the responses of the neuronal populations that encode the speech signal to be used to estimate how well phonetic detail is being represented on top of envelope information. Di Liberto et al. (2018) reported atypical encoding of phonetic detail by the children with dyslexia in right frontal, central and occipital cortex, compared to both RL and CA controls. The speech signal estimates were either significantly lower or higher for the children with dyslexia in these right-lateralised areas, indicating atypical neural encoding. The speech signal estimates were significantly correlated with individual differences in phonological awareness and phonological memory, brain–behaviour correlations.

The studies by Power et al. (2016) and Di Liberto et al. (2018) demonstrate the promise of methods using the responses of neuronal populations to reconstruct the speech signal for our understanding of reading development and dyslexia. This method in essence uses neural responses to recreate the sensory input. Accordingly, it has relevance to developmental questions beyond the study of reading, for example regarding the quality of the sensory information available for statistical learning. Stimulus reconstruction can also be applied *across languages* to investigate the role of degraded neural encoding of speech information by the dyslexic brain.

In theory, once the signal processing is sophisticated enough, the brain response could be 'played back' and the original speech should be heard. The quality of the phonological representations developed by different children could thus be compared. Meanwhile, the studies conducted to date suggest that neural encoding of the speech signal is indeed impaired in developmental dyslexia. Accordingly, the dyslexic brain has to use degraded phonological representations as a basis for developing phonological awareness and acquiring literacy. While this neural insight may appear unsurprising when viewed from the perspective of cognitive findings in behavioural developmental psychology, it is notable that auditory theories of dyslexia have not previously found much favour in the wider field of dyslexia research.

More importantly, precise information about the source of the phonological deficit enables better targeted remediation. The EEG data suggest that optimal remediation should be addressed to enhancing neural encoding of the speech signal from as early as possible in development, for example by an early focus on activities involving speech rhythm. The EEG data also provide information about a neural mechanism that could be targeted technologically. For example, assistive listening technology could be developed that amplifies the slow envelope information in the speech signal, the sensory information that is perceived poorly by the dyslexic brain according to EEG and MEG studies. If such listening devices could amplify the rise time information and enhance the delta-band amplitude modulations that support prosodic awareness from infancy, it is even possible that in future years a diagnosis of dyslexia would be no more inconvenient than a diagnosis of short-sightedness. In each case, technology would be able to offer a device that reduced the impact of these sensory differences on children's learning, thereby enabling more normative developmental trajectories.

MATHEMATICAL DEVELOPMENT

In the last edition of this book, I noted that cognitive neuroimaging studies of number had offered a novel explanation for a confusing and contradictory developmental literature concerning mathematical development. This was because pioneering work concerning the mental representation of mathematical knowledge by Dehaene and his research group had led to the proposal of the 'triple code' model of the representation of number in the brain. Cognitive neuroimaging work, largely with adults, had provided evidence for three numerical coding systems which activated distinct brain regions. One was a visually based code for Arabic numerals (visual number forms) which primarily activated the fusiform gyrus. The second was a linguistic system for storing 'number facts'. Tasks requiring use of the multiplication tables or other overlearned arithmetic knowledge, such as '2 + 2 = 4', typically activated left-lateralised language areas (left angular gyrus), as did other overlearned verbal sequences such as the days of the week and the months of the year. Dehaene and his colleagues also produced evidence for a novel idea, that the human brain possessed a 'number sense' system that was activated when comparing quantities of different magnitudes, located to the horizontal intraparietal sulcus. The idea of an approximate, analogue magnitude representation in the human brain

KEY TERM

Analogue magnitude representation
Idea that the brain uses an internal continuum for representing quantity that is an analogue of the external stimulus, hence is approximate and follows Weber's law.

was very appealing, and led many to assume that mathematical development in children therefore depended simply on associating symbolic number (e.g., Arabic numerals) with the relevant non-symbolic quantity in the analogue magnitude representation. The idea of an approximate analogue magnitude representation has generated considerable research on children's mathematical development in the last decade. The related idea of embodiment as a possible developmental driver of mathematical understanding has yet to receive much research attention, however, despite the research discussed in Chapter 7 showing that children have embodied knowledge of magnitudes for variables like weight and distance. One counter-example is a study reported by Giles et al. (2018), who found that the precision with which children aged 5–11 years could hit a moving target with a bat was a unique predictor of their mathematical ability. Mathematics, like reading, is a cultural invention. Mathematics is an abstract system by which humans learn to manipulate symbols that represent numbers in non-intuitive ways. These abstract manipulations enable a deeper understanding of physical principles by which the world operates, for example via wave theory or calculus. In order to organise the rest of this chapter, I will discuss the cognitive and neuroimaging data supporting the analogue magnitude claim first.

The analogue magnitude representation

An analogue representational format suggests that when the brain makes comparisons between continuous quantities, some kind of internal continuum is used. This internal continuum is an analogue of the external stimulus. For example, when judging size, weight or number, the brain might be relying on a representation of quantitative information that follows Weber's law. Weber's law is a general law of human sensory perception and is logarithmic rather than linear. For example, if I keep adding 2, then 2 + 2 is a doubling of the original quantity, or a proportional increase of 100%. However, 20 + 2 is a proportional increase of only 10%. An analogue magnitude representation would imply similar proportional increases neurally, for example in the level of neural activation or in the number of neurons activated. There should be a doubling of neural activity for 2 + 2 and an increase of 10% for 20 + 2. In the case of number, therefore, an analogue representation would mean that numbers are not stored mentally as discrete entities reflecting exact quantities, but as *approximations* of quantity. Hence as quantities got larger, the representations for these numbers would get less precise. The representation for 10 would be more precise than the representation for 100 or 1,000.

Cognitive data

If the analogue magnitude representation codes quantity in an imprecise, approximate way in the human brain, then number discrimination should be *ratio-sensitive*. Weber's law reflects the fact that our ability to make physical discriminations is ratio-sensitive. Our performance depends on the proportion by which stimuli differ on the relevant dimension. This was first noticed by psychophysicists who were interested in people's ability to make different physical comparisons between,

KEY TERM

Weber's law
Discovery that our ability to make physical discriminations is ratio-sensitive: human performance depends on the proportion by which stimuli differ on the relevant dimension.

for example, lines of different lengths and squares of different luminance. For a wide range of stimuli, the threshold of stimulus discrimination was found to increase with stimulus intensity. On Dehaene's proposal of an analogue magnitude system, the same ratio-sensitive relationship should be found for number.

Analogue coding can be tested by asking people to solve numerical tasks without using language, and measuring whether their responses are ratio-sensitive. For example, if asked to judge whether there are 12 dots in a briefly presented visual display, adults are less precise with ten or 11 dots than with four or 20 dots (Van Oeffelen & Vos, 1982). The ratio-dependence of number discrimination is found in other species, too. For example, Agrillo, Piffer, Bisazza and Butterworth (2012) tested whether fish (guppies) would behave in a ratio-sensitive fashion. Guppies are social fish and like to join other guppies, preferring large groups of other fish over smaller groups. Agrillo et al. devised a fish tank with dividers, in which two shoals of fish were visible at each end of the tank, one larger than the other. An individual guppy was then released into the middle of the tank. Agrillo et al. recorded which end of the tank the guppy swam towards. The numerical ratios of the shoals increased from 4:16 (0.25), 4:12 (0.33), 4:8 (0.5), 4:6 (0.67) to 6:8 (0.75). The individual guppies spent significantly more time swimming near the larger shoal for the ratios 0.25–0.5, but not for 0.67 and 0.75. Hence their shoal numerosity judgements were ratio-sensitive. Piffer, Miletto Petrazzini and Agrillo (2013) subsequently tested newborn guppies rather than adult fish. The guppies were trained to discriminate between arrays of dots such as seven versus 14 for a food reward when aged from four days old. By day ten, they were able to perform the task successfully for a ratio of 0.50 even when the dot displays were controlled for cumulative area and density. The controls for perceptual factors suggest that the fish were responding on the basis of number per se, rather than continuous quantity. Dehaene has pointed out that infants, too, appear to be ratio-sensitive in their discrimination of numbers. Experiments with infants have also highlighted the importance of controlling for perceptual variables.

The idea of an analogue magnitude representation that is intrinsic to the human brain has been useful for organising the infant number literature. As discussed briefly in Chapter 2, infant studies of number discrimination also utilise dot displays, but typically depend on habituation rather than training. Typically, infants are habituated to a display representing one number, and then a new number is presented. If dishabituation occurs, it is assumed that the infants can distinguish the two quantities. As we saw in Chapter 2, Cooper (1984) used habituation to show that ten-month-old infants could distinguish relations such as 'greater than' and 'less than'. Wynn (1992) used habituation to explore whether babies could add and subtract small numbers. She found that babies dishabituated when the 'wrong' answer to addition or subtraction problems involving one and two was presented. From such data, she argued that infants could compute the numerical results of simple arithmetical operations. More recent investigations have focused on possible confounds between changes in number and changes in basic perceptual variables like total surface area in the visual display. Experiments using large numbers with infants that control for perceptual variables in the displays show that infants aged six months can (for example) discriminate eight from 16 (ratio 0.50), but not eight from 12 (ratio 0.75, see Xu & Spelke, 2000). Similarly, they can discriminate 16 from 32 when perceptual variables are controlled, but not one from two (Xu et al., 2005), and they can discriminate four from eight, but not two from four (Xu, 2003). Other infant paradigms, for example change detection, have provided converging data (Libertus & Brannon, 2010).

If infants are using an analogue magnitude representation to make judgements about quantity, this would provide a simple explanation for their ratio-sensitive behaviour, which is observed with numbers larger than three or four. The ratio for the comparison of eight to 16 is 1:2 (0.5), whereas for the comparison of eight with 12 it is 2:3 (0.75), therefore closer to one. Accordingly, infants do worse with eight to 12 than with eight to 16. The ratio for 16:32 and for 4:8 is also 1:2, therefore infants perform well. The apparent *ratio dependence* of infant sensitivity to numbers larger than three or four can thus be explained by a reliance on the analogue magnitude representation. As the analogue representation codes magnitude in an imprecise, approximate way, it yields two important properties concerning infants' representations of quantity. One is that the representation of numerically close quantities is similar (e.g., four and five, nine and ten). Another is that the precision of encoding gets worse and worse for larger and larger quantities. Infants, therefore, are responding to *quantity* or magnitude rather than to number per se.

What about numbers smaller than three or four? It has been argued that small numbers depend on a different representational process, called *subitising*. Subitising yields precise representations of distinct objects (e.g., Feigenson et al., 2004), and refers to the fast enumeration of the numerosity of very small sets, typically one, two and three. Subitising is thought to be influenced more by perceptual processes, such as the object tracking system discussed in Chapter 1 based on 'object files'. The object individuation system is a perceptual system, thereby explaining infant sensitivity to perceptual variables like total surface area when judging smaller numbers. In fact, studies with adults also show a difference in behavioural performance with the numbers one, two and three compared to larger numbers. For example, in tasks requiring the identification of the number of dots in a briefly presented display, reaction times were the same for the numbers one, two and three. Above three, reaction time increases steadily with display size (e.g., Kaufman, Lord, Reese & Volkmann, 1949). The discrimination of small numbers may thus depend on fast, automatic processes that do not involve the analogue magnitude representation. Early in development, these discriminations may not reflect number per se either. This idea is discussed later, in the section on language and number.

KEY TERM

Subitising
Distinguishing between the numerosity of very small sets of numbers without counting.

Analogue magnitude representation in older children

A number of studies suggest that as children get older their analogue magnitude representation supports magnitude discrimination between increasingly smaller ratios. For example, Huntley-Fenner and Cannon (2000) asked children aged three, four and five years to compare two rows of black squares and to decide whether one row had more squares. The rows of black squares contained between one and 15 squares, and varied systematically in their ratio and their interval distance. Ratios were either 1:1, 1:2 or 2:3, and for the latter ratios the interval distance varied from

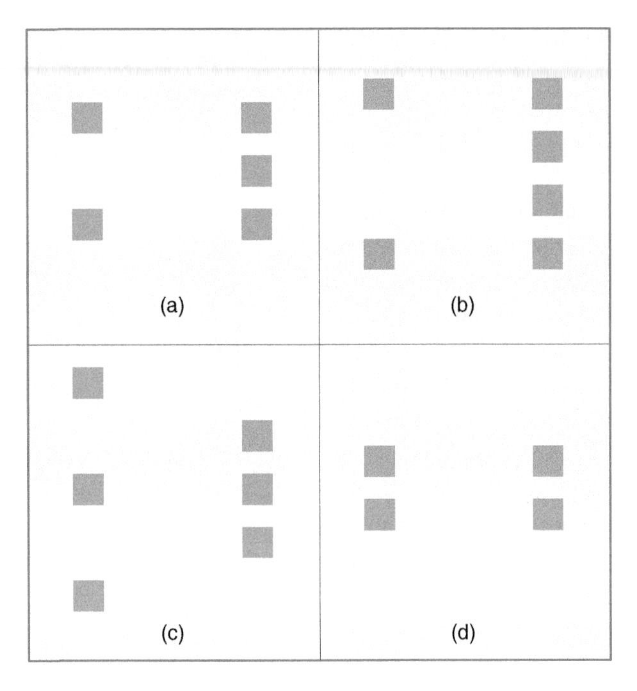

FIGURE 10.4
Examples of the stimuli used by Huntley-Fenner and Cannon (2000). From Huntley-Fenner and Cannon (2000). Reproduced with permission from Blackwell Publishing.

one to five. Hence the ratio 1:2 was tested by comparisons between one and two, two and four, three and six, four and eight, and five and ten. The ratio 2:3 was tested by comparisons between two and three, four and six, six and nine, eight and 12, and ten and 15. Examples of the stimuli are shown in Figure 10.4. The children's counting skills were also measured. Huntley-Fenner and Cannon reported that most children made their choices without counting the arrays, and that those children who did count only did so on a minority of trials (average = eight of the 19 trials). Counting did not improve accuracy: the 'counters' were correct on 70% of trials and the non-counters on 67% of trials. Consistent with the analogue magnitude representation account, the children were significantly more successful in choosing the numerically larger row for the displays with a ratio of 1:2 (81% correct) than for the displays with a ratio of 2:3 (53% correct), at all ages.

Although the data reported by Huntley-Fenner and Cannon (2000) are consistent with reliance on an analogue magnitude representation, they did not control for perceptual variables in their displays (apart from row length). Barth et al. (2005a) asked five-year-olds to compare the numerosity of two arrays of dots, a red array and a blue array. The children were asked to decide whether there were more blue dots or more red dots. To control for perceptual variables, dots could either be the same size in the two displays, or of different sizes. The overall area occupied by the arrays could either be the same or different. These displays controlled for surface area, but not for summed contour length. The magnitude comparison task was presented via an animated display, depicted in Figure 10.5. The ratios used for red versus blue dots were 5:3 and 3:5. Barth et al. reported that the five-year-olds in their study performed at above chance level in all the displays. As performance did not vary with the perceptual variables being controlled, they argued that this showed a reliance on an analogue magnitude representation for number. In a further experiment controlling for contour length, similar results were found. Barth et al. argued that their data showed a reliance on primitive approximate number representations.

In related work, Barth, La Mont, Lipton and Spelke (2005b) used similar animated displays to test magnitude comparisons by five-year-olds for displays of 10–58 dots, presented very briefly to preclude counting. This time three ratios were used, 0.57, 0.67 and 0.80. On half of the trials, dot size, total contour length, summed dot area and density were negatively correlated with number. On the other half of the trials, they were positively correlated. If children were relying on perceptual variables to make their magnitude judgements, then they should be wrong systematically in the former (negatively correlated) set of trials. Barth et al. reported that

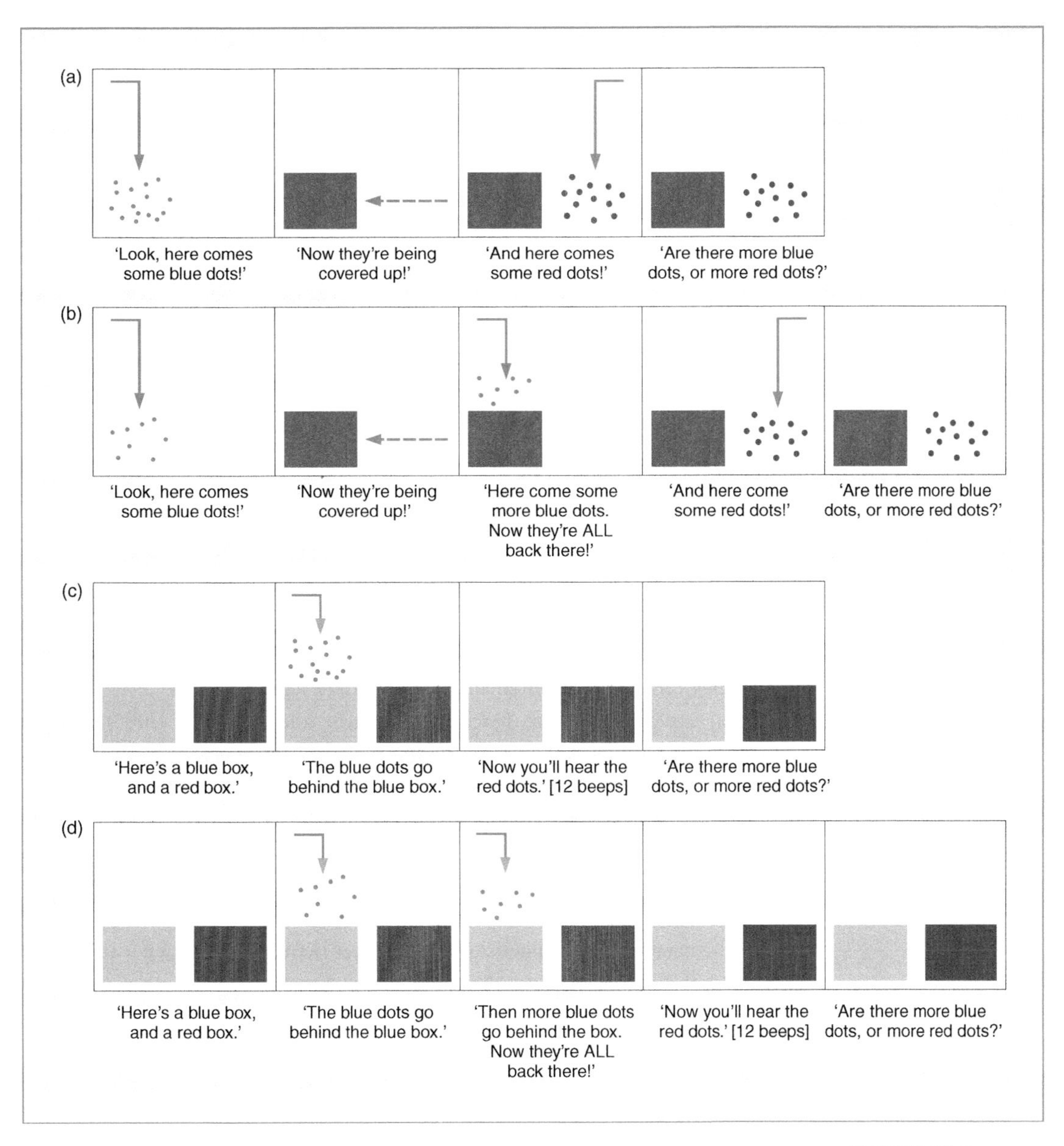

FIGURE 10.5
The magnitude comparison task used by Barth et al. (2005a). (a) Comparison of visual arrays. (b) Addition and comparison of visual arrays. (c) Comparison of visual arrays and auditory sequences. (d) Addition and comparison of visual arrays and auditory sequences. From Barth et al. (2005a). Copyright © 2005 National Academy of Sciences, USA. Reproduced with permission.

the five-year-olds performed well above chance level (67% correct), and showed a significant effect of ratio. As ratio approached one, performance declined. As ratio-dependence is the signature of the analogue magnitude representation, Barth et al. argued that children's judgements depended on this abstract and non-linguistic system. A similar ratio-dependence was found in a cross-modal version of the task, in which one set of dots was presented auditorily ("Now you'll hear the red dots. Are there more blue dots, or more red dots?"). Again, performance was significantly above chance, in fact children were as accurate in the cross-modal task as in the visual comparison task.

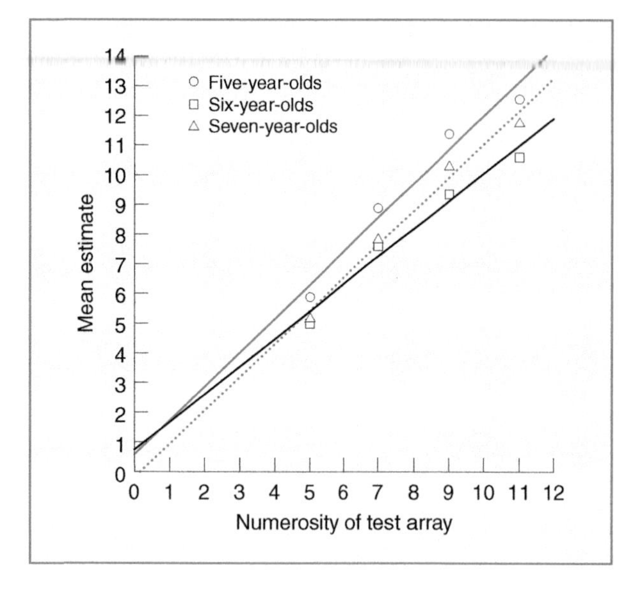

FIGURE 10.6
Five-, six- and seven-year-olds' mean estimates graphed as a function of the display numerosity. Participants' estimates increased proportionally to the presented numerosity. From Huntley-Fenner (2001). Copyright © 2001 Elsevier. Reproduced with permission.

Using a number estimation task, Huntley-Fenner (2001) found further evidence consistent with reliance on the analogue magnitude representation, this time in children aged from five to seven years. The children were shown an array of either five, seven, nine or 11 black squares on a computer screen, for an extremely brief presentation (250 ms). The display was then masked, and the children were asked to point out the numerosity of the array using a number line from one to 20. Overall the children received 40 trials for each numerosity (160 trials in all). Huntley-Fenner reported that there was no improvement over trial number in the task, suggesting that perceptual learning of the task was not occurring. There was, however, a systematic change in the distribution of responses proportional to target quantity, which is another signature of the analogue magnitude representation. The standard deviations of the children's estimates increased proportionally to the means of their estimates. This is shown in Figure 10.6. Huntley-Fenner concluded that when children were asked to estimate number, they relied upon their analogue magnitude representation.

Similar conclusions were reached by Jordan and Brannon (2006) using a numerical similarity task. Six-year-old children were first trained to match exact numbers using a touch-sensitive computer screen. For example, a computer display would show a target number of dots, say two dots. The display would then disappear, and two new displays would be shown side by side, one also containing two dots, and one containing say eight dots. The child's job was to touch the display with two dots. Perceptual variables were controlled. Once the child had learned the task, the arrays were changed so that an exact match was no longer presented. Instead, the child had to choose the most similar match. For example, if the target display was four dots, the child might have to choose between two dots and eight dots. Children were not explicitly told to select the most similar number, but were told to "play the game, the same way you played before… Just make your best guess. I'm sure you'll do a good job, you've been doing so well."

In fact, although the target displays varied in number, the choices were always between two versus eight and three versus 12. This enabled the researchers to calculate whether the children judged smaller targets as nearer to the small value in each case, and larger targets as nearer to the large value. If similarity judgements are made according to an analogue representation of the dimension being tested, then the probability that children choose the larger value (e.g., eight or 12) as the target value increases should itself increase. This follows Weber's law. In addition, the target value for which children are equally likely to choose two or eight (or three or 12) should be at the geometric mean (the square root of the product of those two numbers). This is the point of subjective similarity on an analogue magnitude representation. If the representation of number is linear, then the point of subjective similarity should be the arithmetic mean (e.g., five or 7.5). Jordan and Brannon

(2006) reported that the probability of choosing eight or 12 indeed increased with the numerosity of the target, and that the point of subjective similarity was indeed the geometric mean. This pattern confirmed the predictions made by Weber's law. Jordan and Brannon argued that their data suggested that children were relying on an analogue magnitude representation to make numerical similarity judgements. Remarkably, Jordan and Brannon also demonstrated that the psychophysical functions for the six-year-old children were identical to those of rhesus monkeys trained in the same task. This is consistent with Dehaene's (1997) suggestion of an evolutionarily grounded approximate magnitude representation found in animals and in man. More recent studies broadly confirm these conclusions (Halberda & Feigenson, 2008). Children become better able to discriminate sets with ratios closer to one as they get older (Siegler, 2016).

Cognitive neuroscience of number processing

The idea of an evolutionarily driven 'number sense' that is common across species has been highly appealing to researchers, with the specific neural claim of a supramodal representation for number in intraparietal cortex receiving much research attention (Dehaene et al., 1998). A number of neuroimaging studies suggested that the intraparietal sulcus was reliably activated by tasks requiring knowledge of numerical quantities and their relations. For example, Dehaene, Spelke, Pinel, Stanescu and Tsivkin (1999) compared brain activation using fMRI for two arithmetic tasks in adults, one involving exact addition (e.g., 4 + 5 = 9) and one involving approximate addition (e.g., 4 + 5 = 8). They reported that during exact calculation participants showed greatest relative activation in a left-lateralised area in the inferior frontal lobe, traditionally regarded as a language area. During approximate calculation, participants showed greatest relative activation in a bilateral parietal area involved in visuo-spatial processing. Dehaene et al. also used EEG to track the precise time course of brain activation. They found that the ERPs to exact versus approximate trial blocks were already different by 400 ms, before the possible answers to the additions were displayed. Dehaene et al. argued that their data supported the idea that exact calculation relies on knowledge of 'number facts' or verbal associations stored in the language areas of the brain. Approximate calculation activates visuo-spatial parietal networks, which thus may support a language-independent representation of quantity, the analogue magnitude representation.

Another marker of analogue magnitude coding is the *distance* effect. If adults have to decide under speeded conditions whether an Arabic numeral is larger or smaller than five, they are slower and less accurate with numbers close to five (like four and six) than with numbers distant from five (like one or nine, Moyer & Landauer, 1967). In a related fMRI experiment using the distance effect (Pinel, Dehaene, Riviere & Le Bihan, 2001), adults were given a number comparison task involving two-digit numbers from 30 to 99, and were asked to decide whether a given number was smaller than or larger than 65. The numbers used were classified as either close to 65 (60–64, 66–69), intermediate from 65 (50–59, 70–79) or distant from 65 (30–49, 80–99). In a second condition, the numbers were spoken rather than presented visually. Both conditions yielded a behavioural distance effect, with faster reaction times for more distant numbers. Numerical distance also had an effect on the degree of brain activation in the parietal cortex, for both conditions. There was greater activity for smaller distances. Pinel et al. argued that as brain activation decreased quasi-monotonically with increasing numerical distance, this was consistent with a semantic representation for number in the left and right inferior parietal areas, based on analogue magnitude. Dehaene, Piazza, Pinel and Cohen (2003) also carried out a meta-analysis of different fMRI studies with adults that involved number processing. By comparing activations across studies and tasks, they argued that the core of the analogue magnitude representation was the bilateral horizontal segment of the intraparietal sulcus. This area was active whenever adults had to access the meanings of the quantities that numbers represented, or the proximity relations of numbers. More recent reviews of adult fMRI studies have supported the notion that activation in the intraparietal sulcus is present whenever the analogue magnitude representation is required (Ansari, 2008). Indeed, Dehaene and his colleagues have recently provided evidence that *blind* expert mathematicians activate the same intraparietal, inferior temporal and dorsal prefrontal regions of the brain as sighted expert mathematicians while evaluating complex mathematical functions (Amalric, Denghien & Dehaene, 2018).

An important question for cognitive development is whether the same is true for children. Cantlon, Brannon, Carter and Pelphrey (2006) used fMRI to explore the neural correlates of the analogue magnitude representation in four-year-old children. They devised a task suitable for both children and adults, and compared neural activity in both groups. The task involved viewing a series of visual displays of 16 circles (see Figure 10.7). The circles were coloured blue, and could vary in size, density, visual surface area and spatial arrangement. There were two types of deviant trial. Occasionally, participants would see a display of 32 circles instead of 16 circles. This was a number deviant trial. At other times, they would see a display of 16 squares or 16 triangles instead of circles. This was a shape deviant trial. Deviants and standards were carefully constructed so that perceptual variables such as cumulative surface area and density overlapped (in fact, the standard for some participants was 32 instead of 16, and number deviants could be in the ratio of 1:2 or 2:1, so number deviants could be 16 or 64). Cantlon et al. reported that number deviants led to increased activation in the right intraparietal sulcus for children across conditions, and to increased bilateral activation in the same area for adults. Shape deviants led to increased activation in visual areas such as the fusiform gyrus for both groups. Cantlon et al. concluded that the intraparietal sulcus is activated for non-symbolic numerical processing early in development, before formal schooling has begun.

Regarding EEG studies, Temple and Posner (1998) used EEG to explore the neural basis of the distance effect in five-year-old children, comparing their brain activation to that of adults tested previously. As will be recalled, the distance effect is one of the hallmarks of the analogue magnitude representation. Temple and Posner asked the children to make judgements concerning whether numbers presented on a computer screen were larger than or smaller than five. The numbers (one, four, six and nine) were either represented by Arabic digits or by groups of dots. The children had to press a response key as rapidly as possible to make their judgement.

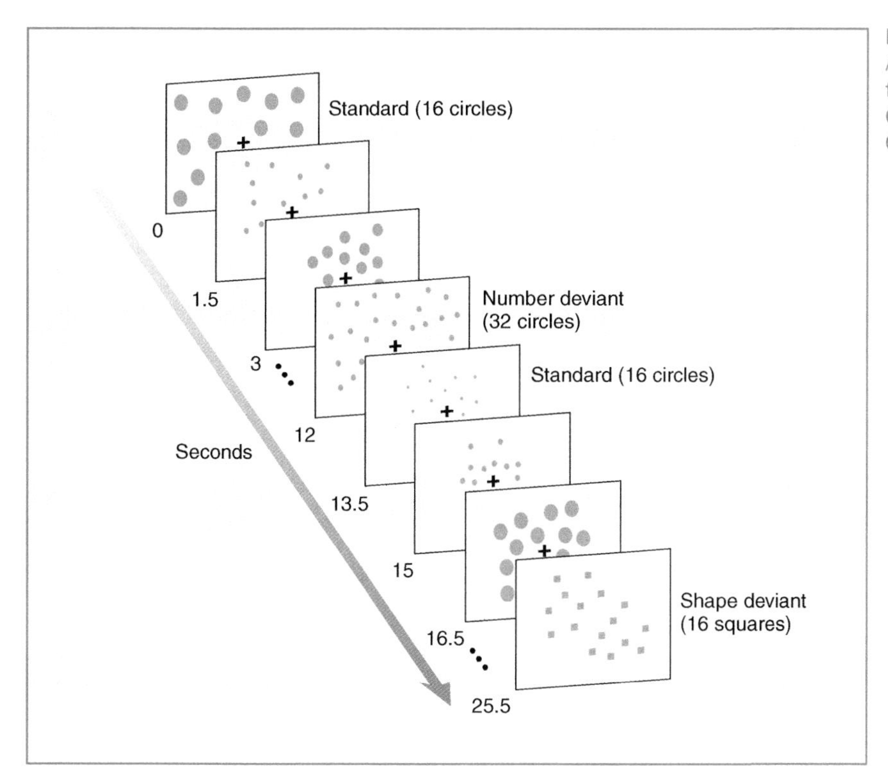

FIGURE 10.7
A schematic depiction of the visual displays used by Cantlon et al. (2006). From Cantlon et al. (2006).

The distance effect is shown when reaction times (RTs) to smaller distances (e.g., deciding that six is larger than five) are longer than RTs to larger distances (e.g., deciding that nine is larger than five). The task is shown in Figure 10.8. Temple and Posner reported that the components of the EEG waveform affected by distance were remarkably similar in the children and the adults. Neural distance effects also occurred at very similar time points in both groups (approximately 200 ms after stimulus onset), even though the children showed much longer response times than the adults. Most children did not actually press the response key until over 1.5 seconds after stimulus onset, compared to around half a second for adults. The distance effect was also centred on the same parietal electrodes. Temple and Posner argued that the same parietal cerebral circuit underpinned abstract magnitude comparisons by children and by adults.

EEG has also been used to compare the object individuation system thought to underlie subitising and the analogue magnitude representation in the same children. Hyde, Simon, Berteletti and Mou (2016) tested a large group of 100 children aged 3–4 years and used two tasks while recording EEG, a task in which children viewed arrays of 1–4 dots to tap the object individuation system, and a version of the task used by Cantlon et al. (2006) to tap the analogue magnitude system. Here children viewed a stream of sequentially presented novel arrays of dots numbering between eight and 32. Controls for perceptual variables were employed and Hyde et al. expected to record two ERPs found in work with adults, a P2 over posterior sites for the analogue magnitude representation, and an N1 over posterior sites for the object individuation system. As Hyde et al. were testing young children, the N1 was

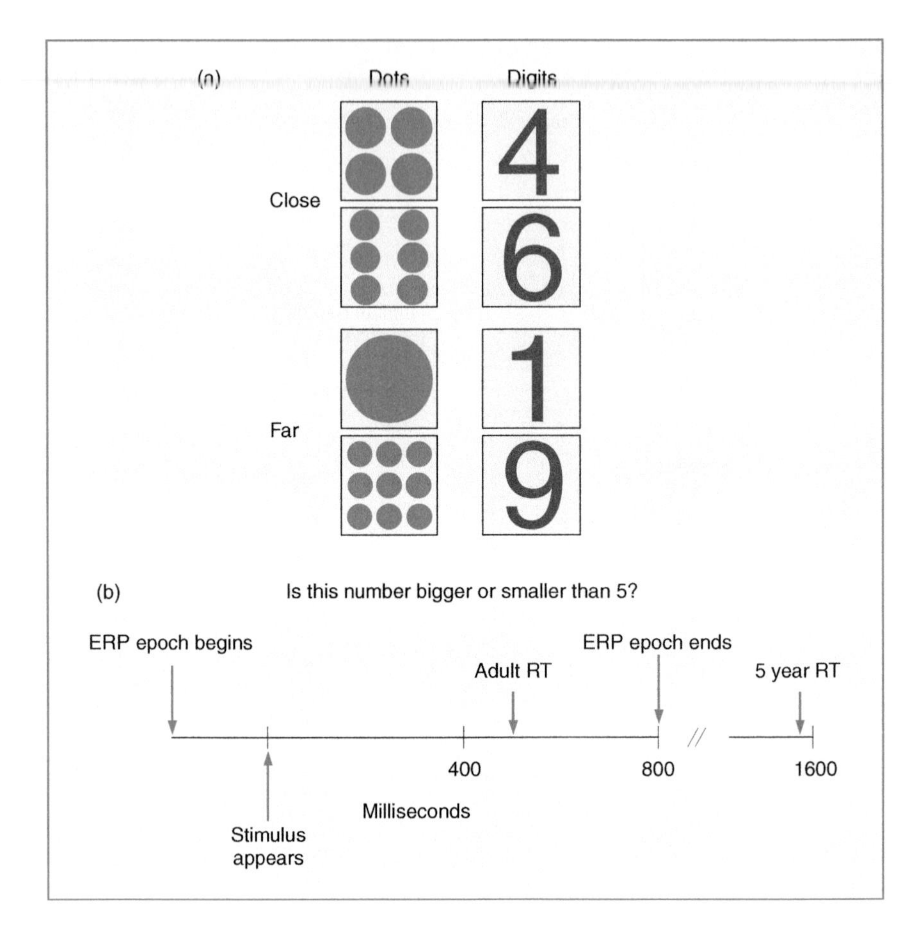

FIGURE 10.8 Temple and Posner asked children to make judgements concerning whether numbers or dots presented on a computer screen were larger than or smaller than five. From Temple and Posner (1998). Copyright © 1998 National Academy of Sciences, USA. Reproduced with permission.

expected to occur in a time window of 235–275 ms, while the P2 was expected around 350–450 ms. Hyde et al. reported distinct ERPs for object individuation versus analogue magnitude activation in their young participants. However, only the N1 response for object individuation was correlated with children's counting proficiency. The development of counting is considered in detail in the following section.

Finally, as noted earlier, single cell recording data from monkeys has suggested that some neurons in the brain specifically encode numerosity (Nieder, 2016). These 'number neurons' are predominantly in the intraparietal sulcus and lateral prefrontal cortex, and show tuning curves, in that they respond most strongly to a particular number (say, six), but also respond to a lesser extent to adjacent numbers (say, five or seven). The idea that the brain may contain 'number neurons' has generated a lot of interest among cognitive researchers, particularly as the fact that these neurons have tuning curves would yield the symbolic distance effect. For numbers that are close, such as five and six, the tuning curves (the degree of neural activation) will overlap more than for numbers that are distant, such as one and six. In monkeys the 'number neurons' also respond to numerosity when presented temporally rather than visually (e.g., as a sequence of six tones). This suggests that 'number neurons' are responsive to number supra-modally. Finally, the peaks of

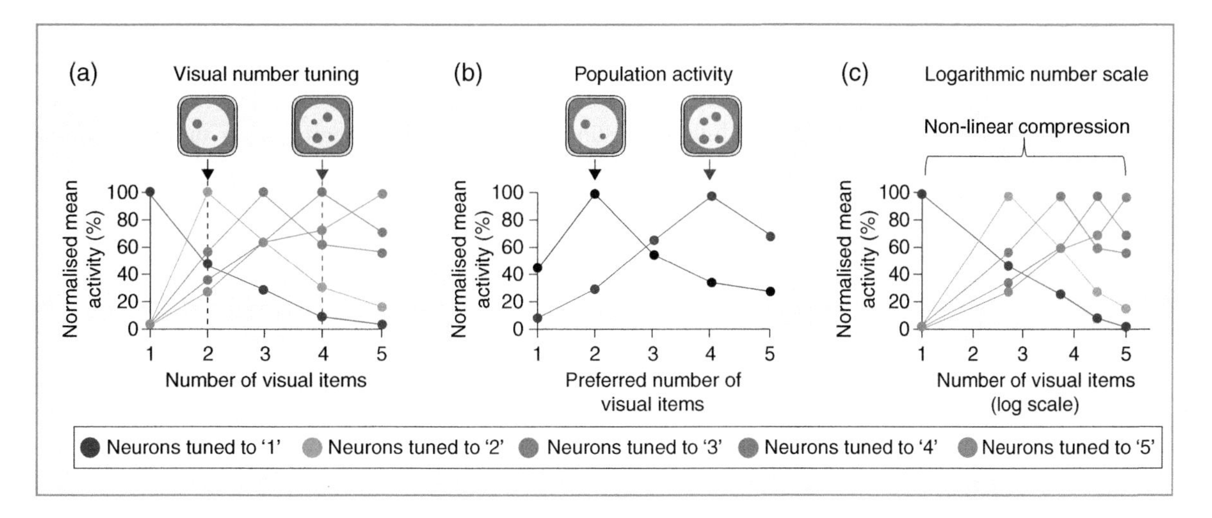

FIGURE 10.9
Schematic depiction of the response of number neurons tuned to different quantities, showing how such neural coding could yield a mental number line. Adapted from Nieder (2016).

the tuning curves seem to follow a logarithmic rather than a linear scale, and logarithmic coding fits Weber's law. Data such as these have led Nieder (2016) to claim that 'number neurons' provide an abstract representation of number in the brain. To date, it has not been possible to verify these findings in the human brain, as single cell recordings in humans can only occur as part of brain surgery.

It is also important to realise that Nieder's data show many responsive neurons for a given number, not just one single neuron, and so it is the *relative* response of the entire population that encodes a particular number. These populations of 'number neurons' in the monkey brain also represent information such as 'greater than' or 'less than', shown by experiments in which monkeys consistently have to select the greater or the lesser quantity. Finally, population coding enables a mental 'number line', in which the ordinal relationship between numbers is represented spatially, with small numbers to the left and large numbers to the right. By averaging neural activation across the whole population of number neurons to different numbers, overlapping tuning curves are generated which preserve spatially the ordinal relations between numbers. The array of tuning curves can thus be 'read out' to give the position on the number line (this is shown schematically in Figure 10.9).

Interestingly, experiments with baby chicks show that they behave as though they have such a spatially ordered mental number line. Rugani, Vallortigara, Priftis & Regolin (2015) taught three-day old chicks to go round a panel with dots on it to get a food reward. They then presented the chicks with two panels, one to the left and one to the right. Each panel had the same number of dots. If the panel had fewer dots than the training panel (e.g., two dots rather than five dots, or eight dots rather than 20 dots), the chicks chose the left panel 70% of the time. If the panel had more dots than the training panel (e.g., eight dots rather than five dots, or 32 dots rather than 20 dots), the chicks chose the right panel 70% of the time. This was also true when perceptual controls were included. Nevertheless, neural data on monkey 'number neurons' has only touched the surface of what the brain may be doing. Most recently, Nieder has found that there is also temporal coding of number in the monkey brain. The temporal pattern of the firing of neurons in a given interval also provides information about number. Hence just as we saw in the case of phonology, the *phase relations* between firing patterns at different temporal rates (such as theta and gamma) may also be crucial for representing information. This could provide a mechanistic link to working memory. Working memory plays an important role in dyscalculia, a selective difficulty in mathematics in children, as we will see below. First, we consider cognitive factors in the development of numerical understanding by children.

Counting

KEY TERMS

Cardinality
Understanding that numbers refer to distinctive individual quantities.

Ordinality
Understanding that each number has a fixed place among other numerical entities that is dependent on increasing magnitude.

Despite the existence of a neural analogue magnitude representation for encoding quantity in the human brain, a second representational system for small numbers which is precise and distinct, and possibly also populations of 'number neurons' that enable computation of a mental number line, the role of individual learning and culture in mathematical development is still profound. Even if these neural representational systems are accepted to be part of our physiology, with clear evolutionary antecedents, there is a large difference between selecting three crackers rather than two, or distinguishing eight dots from 16 dots, and having a symbolic number system. Indeed, there is heated debate concerning whether the analogue magnitude system, our 'number sense', plays any role in children's acquisition of a symbolic number system (as discussed below). Developmentally, acquisition of a symbolic number system probably begins with counting. Children learn to count when they are relatively young, but it is not clear that they understand what they are doing when they are counting (e.g., Piaget, 1952). Counting appears to be learned first as a linguistic routine. Gradually, children learn the principles underlying counting (Gelman & Gallistel, 1978). For example, they come to understand the principle of *cardinality*, which is the understanding that all sets with the same number are qualitatively equivalent. They also come to understand the principle of *ordinality*, which is the understanding that numbers come in an ordered scale of magnitude. The count sequence represents this ordered scale of magnitude, and a number label, such as 'five', represents the fact that five horses is the equivalent amount to five biscuits. The language of the count sequence captures number meaning in terms of both a distinctive individual quantity and an ordinal entity with a fixed place among other numerical entities. Hence the count sequence enables the precise interpretation of any number in non-analogue fashion, including placing it on the mental number line.

The development of counting

Counting by young children has been studied extensively. By the age of around three years, most children can recite the number words 'one' through to 'five' while pointing to one object at a time (e.g., Gelman & Gallistel, 1978; Fuson, 1988). From their studies of counting, Gelman and Gallistel argued that children as young as two years used number words in systematic ways. These ways suggested some appreciation of the principles underlying counting, namely cardinality, ordinality and one-to-one correspondence. One-to-one correspondence is important for counting, as in order to count accurately it is necessary to count each member of a set once and only once. Each label must be used for one unique object. Gelman and Gallistel (1978) did not claim a full understanding of cardinality nor ordinality from children's early counting behaviour, however. The count principle relevant to cardinality was suggested to be the recognition that the last number counted represents the value of the set (this is termed the 'cardinal word principle' by Wynn, 1990). The count principle relevant to ordinality was the need to count in a stable order, using the same oral sequence each time. These principles do not fully capture the understanding of symbolic number. As pointed out by Bryant and Nunes (2002), cardinality is also about the relations between sets of numbers, and ordinality is also about an ordered scale of magnitude, and these aspects of numerical understanding are not captured by behaviours such as counting in a stable order. Nevertheless, children's early counting is suggestive of some insight into what number words mean. For example, even very young children do not give one object when asked for two, three or four (Wynn, 1990).

One of the first systematic studies of counting was reported by Saxe (1977). He gave children aged three, four and seven years (who could all count up to nine on a pre-test) a range of counting tasks in order to trace underlying development. For example, the children were shown an array of beads (up to nine) and were asked to "put out just the same number"; they were shown some drawn circles and asked to "draw just the same number" themselves; they were asked to give a puppet on the table "just the same number [of model animals] to eat" as another puppet underneath the table; and they were shown two linear arrays (e.g., nine toy horses, 11 toy pigs) and asked "Are there just the same number of pigs and horses, or does one have more?" Children who did not count spontaneously in these tasks were prompted to do so (e.g., the experimenter asked "Would counting help?"). Saxe reported that the accuracy of counting rose sharply between ages three and four years. He also observed a developmental shift from 'pre-quantitative' counting (where counting was not used in order to produce the same number), and 'quantitative' counting (where it was). For example, a pre-quantitative child might count 14 toy ducks and then count six toy fish, and say that there are same number "because I counted". A quantitative child would count the toy ducks and fish accurately using one-to-one correspondence.

By the age of around three years, most children can recite the number words 'one' through to 'five' while pointing to one object at a time.

Saxe then carried out a longitudinal study, following a group of three-year-old children for 18 months, using the same counting tasks. As in the cross-sectional study, a change from pre-quantitative to quantitative counting was found with age. Saxe suggested that progression from pre-quantitative to quantitative counting was related to the accuracy of counting, because both were regulated by the 'same cognitive development'. Saxe's suggestion was that this cognitive development was the acquisition of one-to-one correspondence. Once the 'logic' underlying counting was understood, children began to construct quantitative counting strategies, and counting accuracy improved. However, Saxe speculated that a different developmental process might apply to the counting of small numbers. Indeed, work on children's counting behaviour with small numbers has not made much reference to one-to-one correspondence.

The best-known research on counting with small numbers is that of Wynn (1990, 1992). Wynn (1990) gave children aged 2.5 years, three years and 3.5 years a Give-a-Number task based on toy dinosaurs. The dinosaurs were presented in a pile on the table, and the children were asked to give a puppet a certain number of dinosaurs. For example, they were asked "Could you give Big Bird two dinosaurs to play with, just give him two and put them here… can you get two dinosaurs for Big Bird?". Children were asked to give the puppet either one, two, three, five or six toys. They were then asked to check that they had given the correct number (to prompt counting behaviour), and were also prompted to fix things if counting revealed the incorrect number ("But Big Bird wanted two. How can we make it so there's two?"). The children tested appeared to respond in one of two ways, either by systematically counting the toys or by grabbing a handful for Big Bird. In fact, only four of the oldest children qualified as 'Counters'. The 'Grabbers' tended to respond accurately if Big Bird wanted one dinosaur, showed a trend towards systematicity for two dinosaurs, but otherwise did not even approximate the number asked for. Wynn concluded that the Grabbers did not understand Gelman and Gallistel's (1978) cardinal word principle (that is, they did not understand that counting determines numerosity). A follow-up study supported this conclusion. A new group of two- to three-year-olds were administered the Give-a-Number task along with a counting task and a Give Some task (e.g., "Give Big Bird some toy pigs"). Wynn found that each child succeeded up to a certain numerosity, and then failed for all higher ones. The shift from grabbing to counting appeared to occur at around 3.5 years. Wynn suggested that a major conceptual change in children's understanding of counting occurred at this time. Wynn suggested that after acquiring the number words 'one', 'two' and 'three' in sequence, children then acquire the meaning of *all* the number words in their counting range.

To explore these possibilities developmentally, Wynn (1992) carried out a seven-month longitudinal study of two- to three-year-olds, using the Give-a-Number task and two related tasks, a How Many task and a Point-to-x task. In the How Many task, children were asked to count sets of 2–6 items and were then asked "So how many are there?" In the Point-to-x task, children were shown pictures of two sets, and were asked for a specific number x (e.g., "Can you show me the three flowers?"). Children were tested for all combinations of numbers for which they appeared to have cardinal knowledge. For example, a child who could reliably give *two* in the Give-a-Number task would be tested with two versus three, three versus four, one versus three and one versus four. This checked for knowledge of the cardinal meanings of two and three, and the partial meanings of three and four. Performance in the Give-a-Number task was used to compute the highest number which children could succeed consistently in giving, which then determined developmental group (I, II, III, IV). Children in Group I could distinguish one from two, those in Group II could distinguish one, two and three, those in Group III could distinguish one, two, three and four and those in Group IV could distinguish one, two, three, four and five. Even Group I children could count accurately to numbers higher than these (e.g., in Group I most children counted accurately to five).

Wynn's longitudinal data are consistent with her claim that children learn the cardinal meanings of the smaller number words sequentially, and then simultaneously learn the cardinal meanings of the remaining number words in their counting range. However, it is important to note that she did not check this prediction empirically for numbers greater than six. Once the children knew the cardinal meaning of four, they also seemed simultaneously to understand five and six. Children who knew the cardinal meanings of all the number words used counting in both the Give-a-Number and the Point-to-x tasks, whereas those who knew the cardinal meanings of only the smaller number words did not. Even very young children understand that number words refer to specific numerosities. The youngest children tested (2.5 years) could identify two and three in the Point-to-x task when each number was paired with one, even though they could not consistently give Big Bird two items.

Wynn's assumption that after acquiring the number words 'one', 'two' and 'three' in sequence, children then acquire the meaning of *all* the number words in their counting range has been questioned, for example by Carey and her colleagues. It is possible that younger children simply regard all larger number words as referring to 'a lot'. Rather than conceiving of number words for larger numbers as corresponding to specific numerosities, perhaps they do not even conceive of large, exact numerosities at all. Because younger children can individuate objects up to around three or four using the perceptual object tracking system, researchers such as Carey argue that they gradually come to understand the number words 'one', 'two', 'three' and 'four' to mean sets of these sizes, first becoming a 'one knower', and then over a few months a 'two knower', a 'three knower' and a 'four knower' (Carey, Shusterman, Haward & Distefano, 2017). On this view, young children (typically younger than 3.5 years) understand number words larger than 'four' simply to mean 'a lot'. It is learning to count that allows children to 'bootstrap' their way to a concept of large exact numbers, and this concept is claimed not to be present prior to acquiring quantification labels such as the number words (e.g., Carey, 2004). Accordingly, children first learn the meanings of the natural language labels 'one', 'two', 'three' and probably also 'four' by mapping these number words onto the outputs of the subitising process that determines small cardinal values. Number words beyond 'four' are only mapped onto analogue magnitudes after children have constructed the count principles.

To test this idea, Carey et al. (2017) recruited three-year-olds who could solve Wynn's Give-a-Number task for set sizes up to three, and then taught them the meaning of 'four' by training them with pairs of animal pictures showing different set sizes (e.g., four horses versus ten horses). The pictures were first labelled during a demonstration of all the cards prior to the training (e.g., "This card has four horses, this card has ten horses, but this card has four horses"). The children then had to choose the cards showing four animals, without counting, as quickly as possible during the game, first for the animals experienced during training, and then for novel animals. Carey et al. found that the children quickly learned the meaning of 'four' following training, thereby becoming 'four knowers'. The critical test was whether they could also become 'ten knowers' if the same training paradigm was used for sets of ten animals. Carey et al. argued that if the analogue magnitude representation was underpinning the successful transition for these children to becoming 'four knowers', it should be just as easy for them to become 'ten knowers'. However, if the object individuation system was the key to successful performance, then three-year-olds should be at chance with 'ten', despite the fact that they could count to ten. Children indeed performed at chance levels for ten in three further experiments. Despite learning the label 'ten' in the training trials, they could not generalise this knowledge to novel sets of ten animals. As children of this age succeed in ratio-dependent contrasts using the same set sizes (e.g., they can discriminate ten from 20 or ten from 30), Carey et al. (2017) argued that the analogue magnitude system does not underlie children's understanding of the meaning of the words in the count sequence. Carey et al.'s data also question Wynn's assumption that after becoming 'four knowers', young children then acquire the meaning of *all* the number words in their counting range.

Of course, a strong mechanistic argument cannot be made from a negative result in developmental psychology. When children fail at something, we cannot draw strong conclusions. Nevertheless, Carey's idea that enrichment of the object individuation system is the basis for children's early numerical understanding of the number words from one to four is nicely illustrated by a boy aged 2.5 years (tested by Sarnecka and Gelman, 2004), who was given a rather small amount of chocolate candies. This boy protested "I like some *plenty*! I like some *too much*! I like some *lot*! I like some *eight*!" Did this child equate 'eight' with a lot, or did he expect that 'eight' was a large and exact numerosity that was more than he had been given? To find out, Sarnecka and Gelman tested a whole group of children aged from two years seven months to three years six months. The children were first given a pretest, in which they were asked to give puppets one, two, three, five or six erasers in order to sort them into the levels of understanding defined by Wynn (1992). The aim was to find children who did not yet know the meaning of 'six'. Seventeen children were identified at Levels I, II and III, and were then given the Six-Versus-A-Lot task. The idea was to see whether the children would treat 'six' and 'a lot' as synonymous.

The Six-Versus-A-Lot task proceeded as follows. The children were introduced to a game about putting pennies into bowls. They were then told:

> *I'm going to put six pennies in here [one bowl], and six pennies in here [another bowl]. All right! So this bowl has six pennies and this bowl has six pennies. Here are some more pennies [pours all remaining pennies into one of the two bowls]. Okay, now I'm going to ask you a question about six pennies. Which bowl has six pennies?*

Alternatively, they were told:

> *I'm going to put a lot of pennies in here [places six pennies in a bowl], and a lot of pennies in here [places six pennies in another bowl]. All right! So this bowl has a*

lot of pennies and this bowl has a lot of pennies. Here are some more pennies [pours all remaining pennies into one of the two bowls]. Okay, now I'm going to ask you a question about a lot of pennies. Which bowl has a lot of pennies?

Sarnecka and Gelman predicted that if children conceived of 'six' as a precise and large numerosity, then they should not choose the bowl with the most pennies in the 'six' condition, but they should be happy to choose it in the 'a lot' condition. On the other hand, if they conceived of 'six' as 'a lot', then they should also choose the fullest bowl of pennies in the 'six' condition. The data showed that even children at Level I did not conceive of 'six' and 'a lot' as synonymous. The children almost never chose the fuller bowl in the 'six' condition, but they chose it around half the time in the 'a lot' condition. Sarnecka and Gelman argued that children did not treat the quantifier 'a lot' as specific, but did treat the number word 'six' as specific. They argued that even unmapped number words (i.e., the child does not yet have the exact mapping between 'six' and sets of six items) are treated as referring to exact numerosities. In contrast to the claims made by Carey and her colleagues, on these data children at Wynn's levels of I, II and III do appear to have some notion of exact numerosity for large numbers. This interpretation is also supported by a longitudinal study of 198 preschool children reported by vanMarle et al. (2016). VanMarle and her colleagues assessed the children's object tracking and analogue magnitude systems at two time points during the kindergarten year, when the children were aged on average three years ten months and four years two months. They then investigated how well the two types of knowledge predicted children's performance in Wynn's Give-a-Number task (with sets from one to six), used as a measure of cardinal knowledge. The data showed that both systems predicted cardinal knowledge at age three years ten months, while the analogue magnitude system was the only significant predictor at four years two months. VanMarle et al. argued that both perceptual systems contributed to learning the meaning of the count words in early childhood.

Experiments such as the Six-Versus-A-Lot coins task demonstrate that acquisition of the count sequence enables children to conceive of the precise interpretation of a given number.

As can be seen from these selected studies, it is not yet agreed whether once children understand the exact numerosity of one, two, three and four, then the acquisition of the count sequence helps children to conceive of the precise interpretation of *any* number. For example, Siegler (2016) argues that *in principle* this should be the case, as children will reason by analogy. Siegler also makes the point that an easy concrete way to extend the object file system beyond 'four' is for a child to count their fingers, which enables a consistent representation of set sizes up to ten. Indeed, children frequently perform simple arithmetic by using their fingers. Note that this is another form of embodied number knowledge. Counting captures number meaning for the child, in terms of both a distinctive individual quantity and an ordinal entity with a fixed place in a sequence. Even before the child has fully mapped this sequence to the symbolic representation of number, experience with

the language of counting provides social and cultural support for cognitive development. Cultural transmission of the count sequence enables children to organise their cognitive structures for number (i.e., to organise subitising and the analogue magnitude representation into a coherent system).

However, the claim that learning to count might facilitate the development of a symbolic number system in children does not preclude the possibility that with development, language and number become independent. Indeed, we know from adult patient data that number skills can be preserved even when the comprehension and production of language has been severely compromised (e.g., by semantic dementia, see Remond-Besuchet et al., 1999). Patients can have exceptional calculation abilities even when they have lost considerable linguistic skills.

The role of language in counting

Even if the relationship between counting and children's understanding of number is not yet agreed, all researchers agree that possessing a system of number names is helpful in learning to enumerate sets and in calculation. Counting also provides a concrete representation for the ordinal aspects of number that can be learned by heart.

An interesting test of the idea that counting has an impact on children's number processing is to see whether cross-cultural differences in the set of number names have any cognitive consequences. Some languages use language labels that provide more concrete support for the ordinal aspects of number than other languages. For example, the Asian spoken numeral system specifies precisely what is represented by each number in terms of place value. Rather than using opaque labels for numbers above ten (eleven, twelve, thirteen...), these numbers are labelled ten-one, ten-two, ten-three, ten-four, and so on. Twenty is labelled two-tens, 43 is labelled four-tens-three, eighty is labelled eight-tens, and so on. There is also clear phonetic distinction of number labels. Whereas English uses the phonetically similar labels 'fourteen' and 'forty' for quite distant numbers in terms of their place value, in Chinese, Japanese and Korean the labels are the more distinctive 'ten-four' and 'four-tens'. On a linguistic analysis, therefore, the Asian spoken numeral system might assist the development of cognitive representations for number meanings and their ordinal aspects more than the English spoken numeral system.

Miura, Kim, Chang and Okamoto (1988) asked children from China, Japan, Korea and America to construct numbers from sets of wooden blocks (ten blocks and unit blocks, hence 12 requires one ten block and two units). All participants were in first grade, and were aged on average six or seven years. The children were asked to read numbers presented on cards, and then to represent the numbers using the blocks. They were then asked whether they could show each number in a different way using the blocks. The numbers tested were 11, 13, 28, 30 and 42. Miura et al. (1988) found that 91% of the American first-graders used unit blocks to represent these numbers on the first trial. Sixty-six per cent of these children then could not make a correct representation on the second trial (i.e., they could not use the ten blocks correctly). In contrast, 81% of the Chinese children, 83% of the Korean children and 72% of the Japanese children used the ten blocks when representing the numbers on the first trial. Over 90% of these children could also make an alternative correct construction on the second trial, either by using the unit blocks or by using ten blocks and more than nine unit blocks (e.g., 28 as one ten block and 18 units). Miura et al. concluded that the cognitive representation of numbers differed for the Asian and American children. By first grade, place value was an integral component of numerical representations for the Asian children, but not for the American children. These differences in cognitive representation were ascribed to language. However, there is evidence that the cognitive differences found may be more apparent than real. When children are shown how to use the ten units in practice trials before the experiment, then English-speaking children's performance becomes statistically equivalent to that of Japanese children (63% for the English children, 87% for the Japanese; see Saxton & Towse, 1998).

On the other hand, linguistic effects on the cognitive representation of symbolic number could arise earlier in development, during the acquisition of counting. Hodent, Bryant and Houdé (2005) have argued that the number words that toddlers learn could have specific effects on number understanding, contrasting French and English. Hodent et al. pointed out that in French, the number word for 'one', '*un*', is also used to distinguish singular and plural (*un* = a, *des* = some). They hypothesised that the fact that the same label is used to represent singularity in the count sequence (un, deux, trois…) and when contrasting 'one' with 'a lot' could lead to interference for young children who are attempting to judge the accuracy of simple computations, such as 1 + 1 = 3. Indeed, Houdé (1997) found that French-speaking two-year-olds accepted simple sums such as 1 + 1 = 3 "because there are lots". Hodent et al. hence contrasted the performance of French toddlers with that of English toddlers in simple computational problems analogous to those used with infants by Wynn (1992). Mickey Mouse dolls or Babar the Elephant dolls were used to present simple arithmetic problems. Hodent et al. predicted that French toddlers, but not English toddlers, should experience difficulties with sums like 1 + 2 = 4, when the starting point of the operation was singular. However, neither nation was expected to experience difficulties when the starting point of the operation was plural, as in 2 + 1 = 4. This was exactly what Hodent et al. found. The French two-year-olds were significantly worse than the English two-year-olds for the sum 1 + 2 = 4 only. By three years of age, this cross-cultural difference had disappeared. Hodent et al. argued that language played a role in the cognitive representation of small numbers during a precise window of symbolic number development.

The converse idea, that it is not necessary to have a system of number names in order to have an idea of larger numbers, has also been supported by linguistic studies. One example is studies of indigenous tribes in the Amazon, who lack counting words. In particular, the Piraha only have the number words 'one', 'two' and 'many' (Gordon, 2004). The Piraha do not even use the words 'one' and 'two' consistently. The Munduruku have number words from one to five (Pica, Lemer & Izard, 2004). The words 'one', 'two' and 'three' are used consistently, but not the words 'four' or 'five'. Above five, the Munduruku show little consistency in how they use words like 'some', 'many' or 'small quantity'. Despite this lack of counting language, both tribes did very well in non-verbal number tasks involving set sizes up to 80. For example, in one task participants were asked to point to the display of dots in a pair of displays that was more numerous. Interestingly, the performance of both tribes suggested the use of Weber's law. As the ratio of the to-be-compared quantities approached one, discrimination performance decreased. Further, accuracy in number discrimination decreased as target numbers increased. Both tribes hence appeared to be responding as would be predicted on the basis of the analogue magnitude representation.

One developmental function of acquiring the count sequence might thus be to *detach* numerosity from the approximate analogue representation. Number names enable any number to be interpreted precisely in non-analogue fashion. The experimental studies of children learning to count reviewed above support this view. The studies suggest that children start to treat number words as referring to specific, unique numerosities even before they know exactly which numerosity each word refers to (e.g., Sarnecka & Gelman, 2004). According to this analysis, counting is very important for the development of a symbolic number system. Indeed, training studies that focus on training young children's counting skills show strong effects (Jordan, Glutting, Dyson, Hassinger-Das & Irwin, 2012). Jordan and her colleagues trained low-SES kindergarten children on a range of what they termed 'number sense' activities, which had a focus on counting, including finger counting and using fingers for simple addition and subtraction. A control group received a language intervention focused on number-related terms such as 'more than'. The number intervention children showed much greater gains on math outcome measures, with large effect sizes. On the other hand, many studies have looked specifically at the analogue magnitude system (e.g., measured via the distance effect or the ratio effect) as a predictor of later mathematical performance. Here studies have typically reported a relatively weak relationship, with low power. For example, one representative meta-analysis with over 17,000 participants reported a modest correlation of 0.24 and a small effect size (Schneider et al., 2016). Indeed, studies that aim to train core aspects of the analogue magnitude system via computer games typically show little transfer to symbolic arithmetic. Szücs and Myers (2017) conducted a systematic review of such training studies and concluded that there was little evidence that training focused purely on 'number sense' had an impact on classroom performance. Accordingly, training counting skills in the early years may offer the best educational pay-off, despite differences in the way that different languages label numbers.

The development of other skills with numbers, such as addition, subtraction, multiplication and division, cannot be described in this book due to space constraints. There are other cognitive prerequisites for understanding mathematical procedures than those considered here, and a thorough review is offered by Bryant and Nunes (2002). For example, the understanding of Piagetian reversibility and inversion (see Chapter 11) is likely to be related to the understanding of additive composition – that numbers are composed of other numbers, and hence that 4 + 5 = 9, and 9 – 5 = 4. The understanding of sharing is likely to be related to the understanding of division. Even preschoolers are good at sharing toys or sweets between a group of children, but this does not mean that they understand that the more potential recipients, the smaller the bounty per child. The understanding of ratio (one-to-many correspondence) appears to be important for multiplicative and proportional reasoning. Other Piagetian logical operations such as class inclusion are also important for mathematical reasoning. These logical operations are discussed in Chapter 11. I will finish this chapter by a brief review of the literature on developmental dyscalculia. The 'number sense' research has had an important and beneficial effect on research into specific mathematical difficulties in children, and has also led to some interesting cognitive neuroimaging studies.

Developmental dyscalculia

Developmental dyscalculia can be defined as a specific difficulty in learning mathematics despite access to good teaching, which affects 3–6% of children, and which is equally common in males and females. There is much less research on developmental dyscalculia than on developmental dyslexia, and consequently the field is still evolving. In contrast to developmental dyslexia, a single cognitive 'deficit' comparable to the phonological deficit has yet to be identified for developmental dyscalculia, and there are few data concerning impaired sensory processing. Further, mathematics is a collection of various competences, and is not a well-defined single skill as in the case of reading. However, a significant impetus to research on developmental dyscalculia came from the behavioural, cross-species and neural evidence pioneered by Dehaene and his colleagues regarding a basic 'number sense' in the brain, the analogue magnitude representation (Dehaene, 2011). A seductive and intuitively appealing implication of the 'number sense' research was that children with an impaired analogue magnitude representation would have developmental dyscalculia. Accordingly, both behavioural and neuroimaging studies sought evidence consistent with this proposal.

The strongest behavioural markers of the analogue magnitude system are the sensitivity of numerical discrimination judgements to ratio (Weber's law) and the distance effect, both discussed earlier. Both of these measures have been used to study children with developmental dyscalculia. For example, Landerl and Kölle (2009) recruited a relatively large sample of 51 children with dyscalculia aged 8–10 years from an initial sample of 2,635 children aged 8–10 years attending schools in an Austrian urban district. Typically-developing control children were selected from the same sample. The children were then given a range of number tasks including speeded magnitude discrimination for visual arrays matched for perceptual variables and digit comparison to measure the distance effect. All children showed the distance effect in the digit task, and the nature of this effect (e.g., their reaction times) did not differ between the children with dyscalculia and the typically-developing controls. No significant group differences were found in the magnitude comparison task either. Landerl and Kölle concluded that children with dyscalculia did not seem to process numbers in a qualitatively different way to other children.

Mussolin, Mejias and Noël (2010b) reported a similar study focused on the distance effect with a smaller sample of 15 French children with developmental dyscalculia aged 10–11 years. The children received a range of tasks to measure the distance effect, for example digit comparison, groups of dots, random stick patterns and number words. Perceptual variables were controlled for the non-symbolic formats like the random sticks, and typically-developing control children were also tested. Mussolin et al. (2010b) reported that both the children with developmental dyscalculia and the control children showed distance effects, across all formats tested, symbolic and non-symbolic. Furthermore, overall the children with developmental dyscalculia were as accurate in making the comparisons as the control children. However, the children with developmental dyscalculia were significantly slower than controls when the numbers to be compared were close together. Mussolin et al. argued that the distance effect was therefore stronger in the children with developmental dyscalculia, indicating a deficit in the analogue magnitude system. However, without a second control group matched for mathematical level, we cannot be sure that this is the case. As the children with developmental dyscalculia were at a lower level of mathematical development than their age-matched controls, it could be that that the stronger distance effect is actually typical of that level of mathematical development. A deficit in the analogue magnitude system as the primary cognitive deficit in developmental dyscalculia is thus not supported unambiguously by these data.

Indeed, a study reported by Piazza and her colleagues suggests that the inclusion of a mathematical level matched (younger) control group could be helpful in clarifying cognitive deficits in developmental dyscalculia. Piazza et al. (2010) reported a study of magnitude discrimination in which children saw pairs of visual displays of random dots (e.g., 16 versus 32, 16 versus 12) and had to choose the larger quantity. Perceptual variables like surface area were controlled, and 25 children with developmental dyscalculia aged on average ten years were tested. Adults, five-year-olds and typically-developing ten year olds were also tested, and the sensitivity of numerical discrimination judgements to ratio (Weber's law) was checked for each group. All groups showed ratio effects in their magnitude judgements. However, when the response distributions of each group were considered in terms of the internal Weber fraction (called the w statistic), it was found that the children with developmental dyscalculia had similar Weber fractions to the five-year-olds rather than to the ten-year-olds. Nevertheless, the children with developmental dyscalculia responded as quickly in the paradigm as the ten-year-olds. Hence although Piazza et al. concluded that their data indicated a severe impairment of the analogue magnitude representation in the children with developmental dyscalculia, it is too early to draw this conclusion. For example, a deficit in the analogue magnitude system might be expected to remove ratio effects in discrimination tasks. This selection of studies suggests instead that children with developmental dyscalculia show distance and ratio effects when making judgements about magnitude, just like other children.

More recently, a literature has been developing that argues for impairments in visuo-spatial working memory as the primary cognitive impairment in developmental dyscalculia (Szücs, 2016). As noted earlier, mathematics is a collection of various competences. Mathematics as a subject is also incremental, in that developing mathematical knowledge requires the incremental acquisition of several layers of information. Children must learn and understand one set of mathematical facts in order to be able to acquire the next level of mathematical knowledge. In this way acquiring mathematics is quite different from acquiring reading, and it is

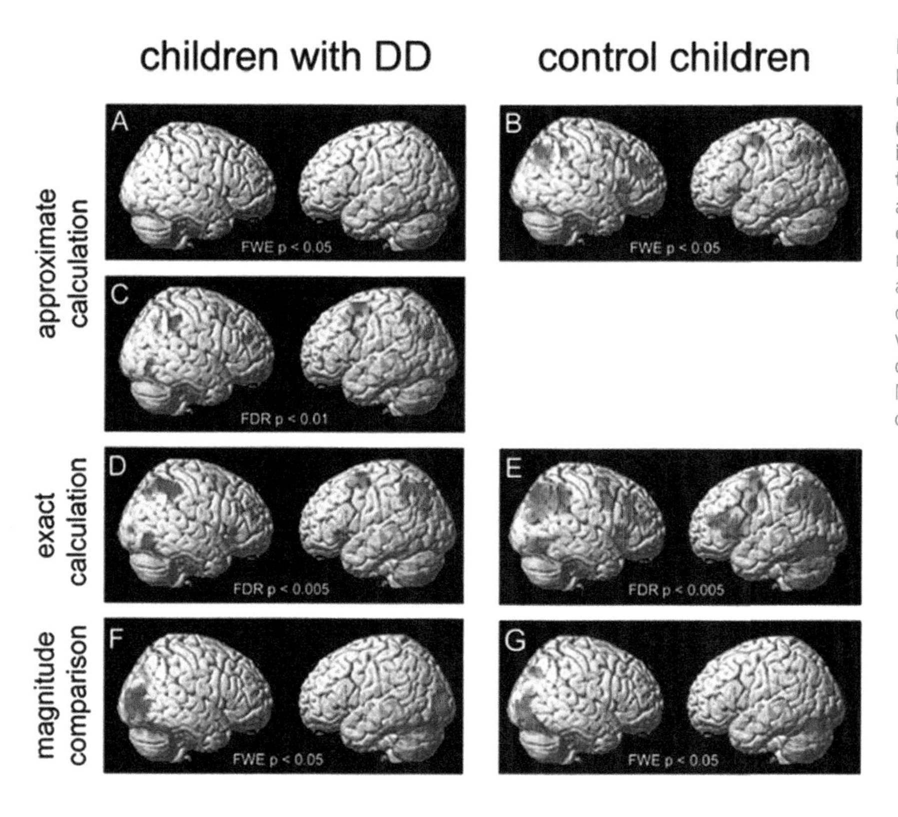

Neural activation patterns in children with developmental dyscalculia (DD) and control children in different mathematical tasks. The tasks were approximate calculation, exact calculation and magnitude comparison, and comparisons were corrected for family-wise error (FWE) or false discovery rate (FDR). No significant group differences were observed.

clear that working memory could play an important role in individual differences. Working memory tasks typically also activate the intraparietal sulcus, the putative location of the analogue magnitude representation. To test working memory against the analogue magnitude representation as an explanatory construct for developmental dyscalculia, Szücs, Devine, Soltesz, Nobes and Gabriel (2013) tested a group of 12 children with developmental dyscalculia aged 9–10 years and an age-matched control group drawn from an initial sample of over 1,000 children. The selected children received a battery of computerised tasks that included distance effect measures, subitising measures, tests of reaction time and inhibition, and Stroop (control) tasks, as well as five working memory tests and a mental rotation task. Stringent statistical comparisons including permutation testing to control for false positives and computing confidence intervals showed surprisingly few group differences. The children with developmental dyscalculia differed from controls only in visuo-spatial working memory, visuo-spatial short-term memory and inhibitory function (the go/no-go task). These differences showed large effect sizes. Regarding the analogue magnitude representation, the distance effect was present in both groups and the Weber fraction (w statistic) was equivalent in both groups, for both accuracy and speed of responding. The groups were also equivalent in the verbal working memory tasks. Szücs et al. concluded that the dominant cognitive features of developmental dyscalculia were deficient visuo-spatial working and short-term memory, and impaired inhibition. Szücs (2016) provides a comprehensive review of studies of developmental dyscalculia that reaches similar conclusions. Menon (2016) has also argued for a key role for visuo-spatial working memory in developmental dyscalculia.

Cognitive neuroscience of developmental dyscalculia

As reviewed earlier, pioneering adult fMRI studies by Dehaene and his colleagues have shown that the intraparietal sulcus is reliably activated by tasks requiring knowledge of numerical quantities and their relations, interpreted as indicating the location of the hypothesised analogue magnitude representation in the human brain. Further, the 'number neurons' identified in the monkey brain by Nieder and his colleagues are also in the intraparietal sulcus. Accordingly, an obvious prediction for developmental studies is that children with developmental dyscalculia will show atypical activation of the intraparietal sulcus. This prediction has been tested in a number of studies. Note that in order to interpret any group differences found in brain activation, the children with developmental dyscalculia should be performing at an equal level to control children on the behavioural task used in the scanner. The fact that the distance effect is reliably found both in control children and in children with developmental dyscalculia is hence an advantage for neural studies.

For example, Kucian et al. (2006) gave 18 Swiss-German children with developmental dyscalculia aged 11 years and 20 typically-developing controls matched for age a series of tasks in the fMRI scanner, including approximate and exact calculation and a magnitude comparison task to measure the distance effect (e.g., visual display of six raspberries versus four hazelnuts, children had to choose the larger number). Both groups showed activation of the right intraparietal sulcus in the magnitude comparison (distance effect) task, and similar activation in a connected network of other brain regions including the fusiform gyrus. However, there were no group differences in neural activation in any areas in the network for this task. Kucian et al. concluded that children with developmental dyscalculia activated similar neural networks for number processing as typically-developing children.

Price, Holloway, Räsänen, Vesterinen and Ansari (2007) gave a magnitude comparison task based on displays of dots to eight Finnish children aged 11 years who had been diagnosed with developmental dyscalculia while recording fMRI, and also studied eight age-matched control children. A whole-brain analysis identified three regions in which brain activity differed by group, the right intraparietal sulcus, the left fusiform gyrus and the left medial prefrontal cortex. Regarding the right intraparietal sulcus, neural activity levels were responsive to numerical distance in the control children but not in the children with developmental dyscalculia. Accordingly, Price et al. argued that there was a weakened parietal representation of numerical magnitude in developmental dyscalculia. Convergent data were reported by Mussolin et al. (2010b), who carried out fMRI with the children in their behavioural study (discussed above). The typically-developing control children showed modulation of activation in both the right and left intraparietal sulcus related to the distance effect, but the children with developmental dyscalculia did not. Rather, they showed modulation of activity in the middle frontal gyrus and the left supra-marginal gyrus related to the distance effect. Mussolin et al. also concluded that developmental dyscalculia was associated with an impaired parietal representation of magnitude.

As far as I am aware, no brain imaging studies of developmental dyscalculia to date have included a mathematical level matched group of (younger) children. This is important, as it may be that earlier in mathematical development modulation of activation in other neural regions such as the fusiform gyrus or middle frontal gyrus is related to the distance effect. In such a case, the apparently atypical activation shown by the children with developmental dyscalculia by Price et al. and Mussolin et al. would not necessarily indicate a neural impairment of the analogue magnitude representation. The children with developmental dyscalculia could be showing a normative pattern of brain activation given their level of mathematical attainment. In contrast to developmental dyslexia, there are also no longitudinal brain imaging studies at the time of writing. Accordingly, it is too early to conclude that the neural analogue magnitude representation is impaired in children with developmental dyscalculia, particularly given the small number of participants in all studies to date.

A further issue is that working memory tasks are also reliably associated with activation in the parietal lobe. For his review of the role of working memory in mathematical development and developmental dyscalculia, Menon (2016) conducted a meta-analysis of fMRI studies and provided a brain map of the regions activated by numerical tasks, arithmetical tasks, working memory tasks and visuo-spatial tasks. This is reproduced as Figure 10.10. As can be seen, the neural regions active during these different tasks are highly similar, and parietal activation is common to all of the different tasks. Accordingly, disentangling any specific role in developmental dyscalculia for an impaired neural analogue magnitude representation in the intraparietal sulcus from atypical activation due to visuo-spatial working memory impairments may not be possible using fMRI as the key imaging tool.

SUMMARY

Acquisition of the two dominant symbol systems in our culture, the alphabet and Arabic number, depend on different cognitive prerequisites. For the alphabet and other writing systems, the key is phonological development. Phonological development is a natural part of language acquisition. Recent research on neural speech encoding has revealed a novel set of acoustic statistics, the amplitude modulation phase hierarchy, that yield different phonological units like syllables, rhymes and phonemes. There are individual differences in the quality of the phonological representations for word forms developed by children, and these individual differences appear to be related to auditory processing, particularly of amplitude envelope rise times. Individual differences in both phonology and rise time discrimination predict individual differences in the acquisition of literacy across languages. Children with better phonological awareness and better rise time discrimination acquire reading more easily. Children who experience unusual difficulties in developing

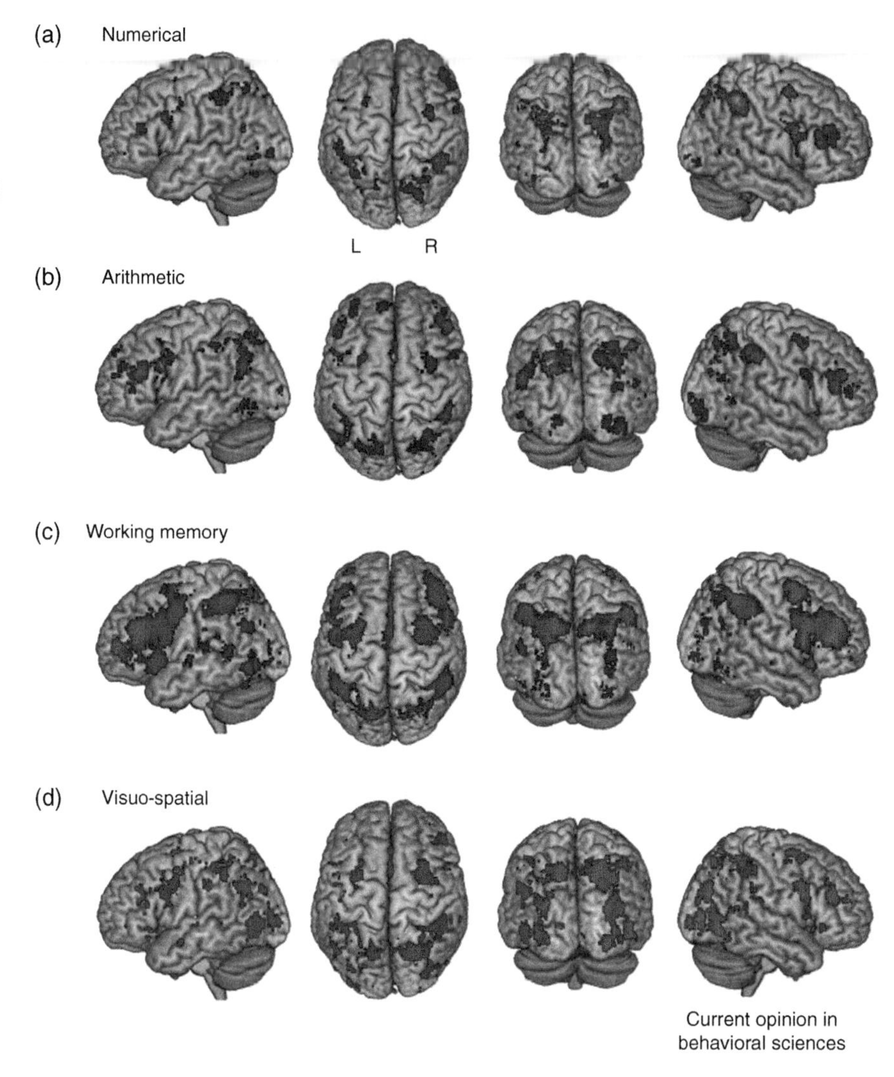

FIGURE 10.10 Patterns of fronto-parietal activations elicited by different cognitive tasks as summarised by Menon (2016). The substantial overlap between the neural networks activated by number tasks, arithmetic tasks, working memory tasks and visuo-spatial tasks is clearly apparent.

well-specified phonological representations are at risk for developmental dyslexia.

Across languages, there is an apparently universal developmental sequence for acquiring phonological awareness of syllables, onsets and rimes (larger grain sizes). Children learning very different spoken languages (e.g., English versus Chinese) develop syllable and onset-rime awareness prior to schooling. This could reflect automatic extraction of these phonological units via statistical learning of the acoustic statistics comprising the amplitude modulation phase hierarchy. For phonological awareness of phonemes, there is cross-language divergence. Phoneme awareness typically develops *in response* to learning to read, and it develops more rapidly in children who are learning to read transparent alphabetic orthographies. These children also learn to decode faster than children who are learning to read less consistent spelling systems. Hence for reading, the nature of the symbol system itself affects cognitive development. For typically-developing children, however, these cognitive differences are transient in nature. Most children learn to read to an efficient level whether they are learning a consistent alphabetic orthography, an inconsistent alphabetic orthography or a non-alphabetic orthography. The neural systems activated during skilled reading show more similarity than divergence across orthographies (Paulesu et al., 2001). Indeed, atypical neural encoding of the speech signal, characterised by sub-optimal synchronisation between oscillatory neural networks and slow temporal information (delta band speech envelope information), has been found for dyslexic children learning both inconsistent (English) and consistent (Spanish) orthographies.

Acquisition of the number system is supported in part by an analogue magnitude representation that codes quantity via an internal continuum, and in part by automatic perceptual processes (referred to as 'subitising') that code the numerosity of very small sets (one, two and three; the object tracking system). The neural location of the analogue magnitude representation appears similar in children and adults, and this approximate number system appears to function in similar ways in children and adults. The acquisition of number as a symbolic system depends critically on language. The acquisition of the count sequence, which is first learned as a verbal routine, appears to be very important for developing an understanding of symbolic number. The language of the count sequence seems to help children to organise their physiological/cognitive structures for number. The small numbers yielded by subitising, and the larger approximate quantities that map onto the analogue magnitude representation, can be coded precisely in non-analogue fashion by using the language of the count sequence. The count sequence thus provides a linguistic structure for two of the central principles of a symbolic number system – the cardinal principle that numbers mean a distinctive individual quantity, and the ordinal principle that each number has a fixed place among other numerical entities that is dependent on increasing magnitude.

Research on the analogue magnitude system has also fostered a new literature on the cognitive neuroscience of developmental dyscalculia, a specific difficulty in learning mathematics that affects between 3–6% of children. Unlike developmental dyslexia, developmental dyscalculia affects boys and girls equally, however its neural underpinnings are not yet well-understood. In particular, a sensory/neural system that could yield impaired information and thereby impair symbolic learning has yet to be agreed upon. Children with developmental dyscalculia typically show the hallmarks of the analogue magnitude representation (such as the distance effect and the ratio effect), although they may be slower in certain tasks. They also appear to have intact subitising processes, although again speed decrements can be identified. Interpretation of the functional impact of slower performance is difficult, as both cognitive and neural studies including a mathematical level matched control group (comparable to the reading level matched control group that has been important regarding causal research in developmental dyslexia) are currently absent from the literature. Cognitively, the best-agreed impairment in developmental dyscalculia is in visuo-spatial working memory. Studies of embodied knowledge of magnitude, and whether such embodied knowledge is impaired in developmental dyscalculia, are not yet in the literature but could offer an interesting novel avenue for the field.

The development of symbolic systems for reading and mathematics also provides a classic example of Vygotsky's idea that symbol systems transform human cognitive development. Cultural transmission of the count sequence to young children plays an important role in organising the physiological/cognitive structures upon which numbers build. Similarly, cultural transmission of the alphabet (and other spelling systems) transforms the representation of spoken language. The physiological/cognitive structures for spoken language, and particularly for phonology, underpin the acquisition of literacy, but these structures are also transformed by literacy. As Frith (1998) observed, learning the alphabet is like catching a virus. "This virus infects all speech processing, as now whole word sounds are automatically broken up into sound constituents. Language is never the same again" (p. 1051). For both reading and number, symbol systems enable cognition to develop beyond the constraints of biology. According to the data reviewed here, neither numerical nor orthographic representations are innate. Cognitive development is required. Learning both reading and mathematics is supported by pre-existing neural representations for coding visual and auditory information. However, both reading and mathematics depend also on the child learning culturally determined symbolic systems, which shape cognitive development in particular ways.

CHAPTER 11

CONTENTS

Theories of cognitive development

11

Theories of cognitive development are explanatory systems that account for the data regarding 'what develops'. The most comprehensive theory of cognitive development has been Piaget's theory of logical development. Piaget was by training a biologist, and in his theory the primary causal mechanism for building knowledge was the adaptation to and refinement of existing cognitive 'schemes' by the environment. Knowledge was constructed by the child as a consequence of his or her active experiences with the external world. An alternative explanatory account of cognitive development was proposed by Vygotsky, a Russian psychologist trained in philology. His theory had a more cultural focus, recognising the key role of social interaction in cognitive development, and the important role of adults in mediating cultural knowledge for children and in supporting them in acquiring it. More recently, as cognitive neuroscience has assumed a more prominent role in studies of cognition, connectionist modelling has provided an explanatory framework for thinking about cognitive development. 'Neuroconstructivism' provides an example of a theoretical framework that includes insights from both neuroscience and connectionist modelling. Neuroconstructivism uses concepts such as 'enbrainment' to describe biological aspects of cognitive development, for example changes in the physical structure of the brain as a consequence of learning. However, neuroconstructivism operates at a purely descriptive level. By contrast, connectionism has generated working models of how complex cognition (e.g., conceptual development; see Chapter 5) can arise from the learning activity of simple on/off nodes in a connected network. Originally these computer models were intended to be analogous to networks of neurons, which are either generating action potentials ('on') or recovering from firing ('off'). Hence connectionism represented the first attempts to explain cognitive development in terms of simple (neural) networks that learn complex structure from 'input'. More recently, machine learning and deep neural nets have provided more sophisticated models of how the brain may build cognitive systems from 'the input'. In my view, these models go beyond neuroconstructivism in terms of their potential theoretical contributions.

For example, it has become clear that the human brain uses both rate coding (the number of action potentials) and temporal coding (the timing of action potentials with respect to each other), as well as more complex algorithms such as nested temporal hierarchies of activation (for example, the phase of one temporal rate of firing [peak versus trough] may govern the number of action potentials [power] at a faster rate of firing) and temporal synchronisation across neural networks in different parts of the brain during learning. 'Deep learning' models are powerful enough computationally to use similar learning algorithms, and consequently are proving influential in adult cognitive neuroscience. In deep learning, sophisticated machine learning algorithms are trained with large input databases (for example, of real clusters of medical symptoms) and then acquire expertise that can exceed that of human operators (for example, in medical diagnosis, see LeCun, Bengio & Hinton, 2015). One limitation of early connectionist models was that they could only learn the 'input' that was programmed into them by the experimenter. This input did not necessarily reflect the input used by infant and child learners. Deep learning networks can typically deal with real world environmental input, such as a huge database of 2D or 3D visual scenes, and give efficient output (e.g., 'this is a woman throwing a Frisbee in the park'). Currently it is quite difficult to 'reverse-engineer' deep learning networks, to find out *how* they acquire their high levels of expertise. Nevertheless, given the increasing importance of insights from cognitive neuroscience to adult cognition, the future of cognitive developmental theorising seems likely to follow a hybrid route. Contemporary theories of cognitive development must acknowledge the importance of reflective knowledge construction (Piaget) and of the child's social and cultural world (Vygotsky). At the same time, contemporary theories must be consistent with biological constraints on how the brain actually learns. These different explanatory frameworks will be discussed in turn, and where possible evaluated in terms of the experiments that have been discussed in earlier chapters.

PIAGET'S THEORY

Piaget's fundamental interest was in the origins of knowledge, and he applied the principles of biology to the study of the development of knowledge in children. One of his central assumptions was that knowledge structures (or schemes) adapt themselves to their environments. Piaget suggested that the cognitive system would naturally seek equilibrium, and that cognitive development was caused by two processes. These were 'accommodation' and 'assimilation'. Accommodation was the process of adapting cognitive schemes for viewing the world (general concepts) to fit reality. Assimilation was the complementary process of interpreting experience (individual instances of general concepts) in terms of current cognitive schemes. As every cognitive equilibrium can be only partial, every existing equilibrium must evolve towards a higher form of equilibrium – towards a more adequate form of knowing. This process of knowledge evolution was thought to drive cognitive development. When one cognitive scheme became inadequate for making sense of the world, it was replaced by another.

KEY TERMS

Accommodation
Adapting and changing cognitive schemes or general concepts for viewing the world to better fit reality.

Assimilation
Interpreting new experiences in terms of current cognitive schemes.

Sensory-motor period
First stage in Piaget's theory, covering 0–2 years: cognition is based on the infant's physical interactions with the world.

Stages in cognitive development

Piaget is usually characterised as proposing a stage model of cognitive development, as his observations led him to propose that a major overhaul of current cognitive schemes occurred four times between infancy and adulthood. Hence he proposed four major cognitive stages in logical development, corresponding to four successive forms of knowing. During each of these stages, children were hypothesised to think and reason in a different way. Each stage was thought to require fundamental cognitive restructuring on the part of the child. However, Piaget recognised that the acquisition of each new way of thinking would not necessarily be synchronous across all the different domains of thought (see Chapman, 1988). Instead, he argued that the chronology of the stages might be extremely variable, and that such variability might also occur within a given stage.

Piaget's stages, and their approximate ages of occurrence, were:

1. The sensory-motor period: 0–2 years.
2. The period of pre-operations: 2–7 years.
3. The period of concrete operations: 7–11 years.
4. The period of formal operations: 11–12 years on.

Sensory-motor cognition was based on the infant's physical interactions with the world. The onset of representational thought came as pre-operational thinking developed, involving the internalisation of action on the mental plane. When the results of such internalisations (called compositions) became mentally reversible, concrete operational cognition developed, marking the beginning of truly mental operations. Finally, during formal operational cognition, certain concrete operations became linked together, marking the onset of scientific thought. The ages of attainment that Piaget gave for the different cognitive stages were only approximations. For example, he suggested that the concrete operations of class inclusion and seriation (see below) might pursue slightly different developmental courses, the former being related to linguistic development and the latter to perceptual development. Nevertheless, the key idea was of an internal programming of cognitive change, based on states of 'disequilibrium' between the child's mental states and the external world.

The sensory-motor stage

One of Piaget's major claims, a claim that has received support from data from cognitive neuroscience, was that thought develops from action. In his view, a 'logic of action' existed prior to, and in addition to, the logic of thought. In Piaget's terms, a practical logic of relations and classes in terms of sensory-motor action was the precursor of the representational logic of relations and classes that emerged at the concrete operational stage. Piaget pointed out that babies are born with many means of interacting with their environment. Their sensory systems are functioning at birth, as we saw in Chapter 1. Babies also have a range of motor responses that are ready for use, such as sucking and grasping. Piaget argued that the presence of these reflexes meant that babies were born with the potential to know everything about their worlds, even though at birth they knew almost nothing. These basic abilities allowed infants to gain knowledge of the world and to build up hypotheses about it. Piaget's baby was conceptualised as busily interpreting and reinterpreting perceptual information in the light of his or her hypotheses, hypotheses that were drawn from sensory-motor experiences in the everyday world.

Piaget identified six sub-stages of development within the sensory-motor stage. The first was the modification of reflexes. For example, the baby could learn to modify its sucking reflex in order to fit the contours of its mother's nipple (accommodation). At the same time, the baby assimilated the sucking response to an increasing range of objects, and gradually became able to distinguish between objects that would satisfy hunger and objects that would not. The second stage was called primary circular reactions. A circular reaction is a repetitive behaviour. Babies seem to enjoy engaging in repetitive behaviours, and another important insight of Piaget's was that this repetition might have cognitive value. The first repetitive behaviours were concerned with the self, and thus Piaget labelled them 'primary'. Primary circular reactions involved the recreation of sensory experiences. A good example is thumb-sucking.

The third sensory-motor stage was called secondary circular reactions. Secondary repetitive behaviours involve the outside world. For example, a baby might seek to recreate interesting events in its environment, such as dropping an object. The circular reaction would be to repeatedly drop the object without getting bored – a behaviour that also requires repetitive behaviour on the part of the baby's caretaker, who has to repeatedly pick the object up! The fourth stage was called the co-ordination of circular reactions. At this point, the baby became able to co-ordinate a series of behaviours in order to attain a goal. Piaget called this goal-oriented behaviour 'means–ends' behaviour. An example is pulling on a blanket so that a desired toy at the edge of the blanket moves to a location within the baby's reach.

The fifth sensory-motor stage was called tertiary circular reactions. By this stage, babies' ability to recreate events in the outside world had become more sophisticated. Stage 5 infants could conduct different trial-and-error explorations in order to determine the results of certain actions. For example, a baby may

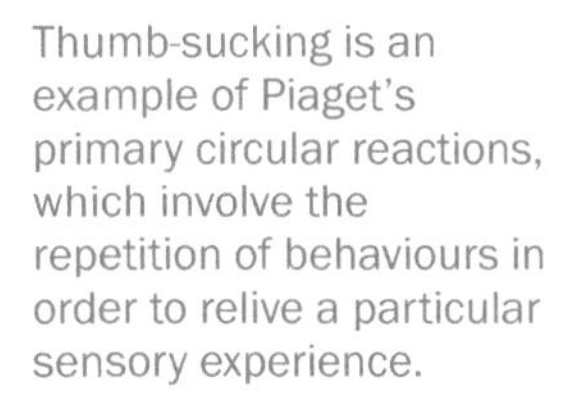

Thumb-sucking is an example of Piaget's primary circular reactions, which involve the repetition of behaviours in order to relive a particular sensory experience.

repeatedly drop an object in a variety of ways in order to examine the different trajectories that the object will take. The focus of interest for the infant is the variation of these trajectories, rather than the repetition of the action of dropping (as in secondary circular reactions). Such actions can be viewed as hypothesis-testing behaviour, and Piaget argued that tertiary circular reactions led to the discovery of the spatial and causal relations between the objects involved.

The final stage of sensory-motor cognition was called the interiorisation of schemes. At this point, the baby became able to anticipate the consequences of certain actions, and thus to work out the sequence of actions required to attain a desired goal prior to performing the actions themselves. This anticipation occurred via mental combination of the actions and their consequences, without the need for trial-and-error exploration. The interiorisation of schemes hence marked the *cognitive* representation of actions and their consequences. These representations were detached from immediate action and were liberated from direct perception – they were fully symbolic. According to Piaget, stage 6 of sensory-motor cognition marked the beginning of conceptual thought.

Probably the most famous example of how Piaget thought that sensory-motor information led to the development of conceptual thought was his analysis of object permanence. Piaget measured the emerging conception of the permanence of objects by studying the development of babies' *searching behaviour*. As we saw in Chapter 2, infants begin to search for partially hidden objects at the age of around 4–5 months. However, an object must be partially visible for the infant to try and retrieve it. By around nine months, the infant becomes able to search for fully hidden objects. Now the A-not-B error appears. Search behaviour is apparently determined by previously successful actions. By around 12 months, the A-not-B error disappears. As long as the infants see the object being moved to a new hiding location, they can retrieve it over multiple hidings. Piaget acknowledged that this behaviour indicated an understanding of the object itself and of its relation to other objects, but noted that search difficulties remained when displacement was 'invisible' (i.e., not witnessed by the infant). For Piaget, a full conception of objects was only present once invisible displacements were solved, at around 18 months. Now the infant could find objects wherever they were hidden. For Piaget, this marked the attainment of a cognitive representation of the object, detached from motor action and from sensory perception. More direct measures of mental representations, for example by using neuroimaging, suggest that relying on search behaviour to index cognitive representations was misguided (see Chapter 2).

Evaluation

Clearly, Piaget was wrong about how slowly early *cognitive* representations develop. However, if we allow much younger babies the ability to develop cognitive representations, even for objects that are out of view, then Piaget's description of sensory-motor cognition is actually remarkably consistent with the current data on infant development discussed in Chapters 1 and 2. The idea that sensory-motor responses are a primary source of information for infants must be correct, and the idea that sensory-motor behaviours play an important role in knowledge acquisition fits with statistical learning (which is primarily sensory), learning by imitation (primarily motor), learning by analogy, and explanation-based learning. Piaget argued that sensory-motor responses were foundational because of the 'logic of action'. Sensory-motor behaviours *became* thought. Previously, I argued that this was unlikely. However, the neuroimaging data now available suggest that this idea is very plausible. Similar ideas are now proposed by the adult literature on 'embodied cognition'. The recognition of the distributed nature of mental representations also supports the view that sensory-motor representations are intrinsic to thought. For example, in Chapter 5 we saw that sensory-motor experiences associated with different concepts like 'cup' or 'bicycle' are retained as part of the concept of a cup or bicycle. Cognition appears to be distributed, partly embodied, and not amodal. Piaget's view that the development of the object concept took 18 months to achieve was incorrect. However, his focus on the importance of action and his recognition that the repetition and recreation of sensory-motor experience were an important means of learning were fundamental insights into infant cognition.

KEY TERM

Explanation-based learning
Machine learning which depends on construing causal explanations for phenomena on the basis of single training examples.

Piaget's ideas about the interiorisation of action are also interesting in the light of more recent infant data. Piaget argued that sensory-motor behaviours became representational via interiorisation, and that this interiorisation occurred via 'motor analogies'. For example, he had noticed that his own children imitated certain spatial relations that they had observed in the physical world with their own bodies. They imitated the opening and closing of a match-box by opening and closing their hands and mouths. Piaget suggested that this behaviour showed that the infants were trying to understand the mechanism of the match-box through a motor analogy, reproducing a kinaesthetic image of opening and closing. This is reminiscent of the 'like me' analogy discussed in Chapter 3. Although that analogy was discussed in relation to the development of social cognition, Piaget is using a similar explanatory framework involving motor imitation for understanding the physical world. An outstanding empirical question is whether action provides a separate and autonomous source of knowledge from representational or reflective understanding. Given the massive interconnectivity in the brain, it seems unlikely that we develop separate and autonomous sources of knowledge about anything in the world. However, this is an empirical question, and was previously hotly debated in adult cognition (e.g., Caramazza & Mahon, 2003). As noted in earlier chapters, sources of knowledge that are intimately linked during development may achieve a degree of independence in the developed system (e.g., language and number; see Chapter 10). Hence empirical longitudinal data are required to answer this question.

In Piaget's theory, analogies were also thought to play a role in the generalisation of sensory-motor schemes to new objects. In fact, Piaget argued that analogical transfer was rapid once a new physical concept had been understood. Commenting on the acquisition of the 'pull' schema (the ability to use string-like objects as a 'means for bringing'), Piaget wrote:

> *Let us note that once the new schema is acquired, it is applied from the outset to analogous situations. The behavior pattern of the string is without any difficulty applied to*

the watch chain. Thus, at each acquisition we fall back on the application of familiar means to new situations. (Piaget, 1952, p. 297, cited by Brown, 1990)

As discussed in Chapter 2, recent experiments on analogy have supported Piaget's view. Simple relational mappings are available to infants as young as three months (see Greco, Hayne & Rovee-Collier, 1990) and are readily transferred to new objects (e.g., to novel mobiles). Analogy is also used by theorists to explain children's cognitive development in many other domains, for example syntax (Tomasello, 2008, see Chapter 6) and knowledge about large numbers (Siegler, 2016, see Chapter 10).

The pre-operational and concrete operational stages

In order to document cognitive changes beyond stage 6 of sensory-motor cognition, Piaget investigated children's understanding of the properties of concrete objects and the relations between those objects. The set of logical concepts that described classes of objects and their relations were called the 'concrete operations'. Children were thought to develop a full understanding of the properties and relations of concrete objects rather gradually, between the ages of around two and seven years. The key concrete operations were children's understanding of transitive relations between objects of different lengths or heights ('transitivity'), children's understanding of classes of objects and their part-whole relations ('class inclusion'), and children's understanding of addition, subtraction and equivalence ('conservation'). Along with related logical concepts such as seriation, these operations of transitivity, class inclusion and conservation have been the main focus of later research.

Piaget's explanation for these logical developments was that during pre-operational cognition, the schemes or concepts of sensory-motor thought were redeveloped in the mental realm. For example, the interiorisation of operations such as addition and subtraction on concrete objects led to the representation of the formal properties of whole numbers, such as that $2 + 2 = 4$ simultaneously implies that $4 - 2 = 2$. The recognition of the 'reversibility' of this operation was thought to be a critical feature of concrete operational cognition. The functional units of thought were not isolated concepts and judgements, but an integrated system of reversible and interdependent structures:

The child can grasp a certain operation only if he is capable, at the same time, of correlating operations by modifying them in different, well-determined ways – for instance, by inverting them … the operations always represent reversible structures which depend on a total system. (Piaget, 1952, p. 252)

The main characteristics of pre-operational thought were that it was egocentric, in that the child perceived and interpreted the world in terms of the self, that it displayed centration, in that the child tended to fix on one aspect of a situation or object and ignore other aspects, and that it displayed a lack of reversibility, in that the child was unable mentally to reverse a series of events or steps of reasoning. The pre-operational child was thus seen as pre-logical, having a subjective and

KEY TERMS

Seriation
Understanding that objects can form an ordered series in term of their physical attributes.

Pre-operational
Second stage in Piaget's theory covering 2–7 years: the child is developing some understanding of the properties of concrete objects and the relations between these objects, but is hampered by egocentricity, centration and lack of reversibility.

Egocentric thought
Perception and interpretation of the world in terms of the self (e.g., inability to 'put oneself in another's shoes').

Centration
Tendency to focus on one aspect of a situation or object and ignore other equally relevant aspects.

Reversibility of thought
Inability mentally to reverse a series of events or steps of reasoning.

self-centred grasp of the world. As with sensory-motor cognition, Piaget's insights clearly capture aspects of young children's behaviour in certain situations. The acquisition of the concrete operations was thought to be marked by the gradual waning of egocentricity. Children became able to 'decentre' or to consider multiple aspects of a situation simultaneously, and they became able to grasp 'reversibility' or the ability to understand that any operation on an object simultaneously implied its inverse. Piaget argued that the organisation of the concrete operations could be described in terms of mathematical groupings, for example $A + A' = B$ (class inclusion), $A > B$, $B > C$, therefore $A > C$ (transitivity), and $A = B$, A transforms to A', therefore $A' = B$ (conservation). Piaget's idea was that cognitive structures developed to mirror mathematical logic. The psychological reality of the logical structures representing truly logical thought and their reversibility could be represented formally. Most subsequent research on pre-operational and concrete operational thought has neglected Piaget's focus on mathematical groupings, however. Instead, it has focused on whether logical concepts like conservation, transitivity and class inclusion are present at an earlier age than Piaget supposed. Rather than assuming that children who fail a Piagetian logical task are incompetent, such research explores the possibility that children's knowledge becomes obscured by misleading aspects of the tasks. For example, the language that Piaget used to ask children about class inclusion has been shown to be misleading. Other researchers investigate whether younger children's errors across the concrete operations can be accounted for by other developmental factors. For example, younger children, with their poorer executive skills, may be unable to inhibit responses triggered by misleading aspects of some of Piaget's paradigms (e.g., Houdé, 2000; Houdé & Borst, 2015).

Transitivity

KEY TERM

Transitivity
Transitive relations hold between any entities that can be organised into an ordinal series; e.g., John is taller than Paul, Paul is taller than Peter, so John is taller than Peter.

One key logical concept identified by Piaget was transivity (the transitive inference). Transitive relations hold between any entities that can be organised into an ordinal series and are fundamental to basic mathematical concepts such as measuring. Piaget and Inhelder (1956) proposed that a typical transitive inference problem of the form "If Tom is bigger than Mark, and Mark is bigger than John, who is bigger, Tom or John?" could not be solved until concrete operational reasoning had been acquired. Children's ability to make these kinds of transitive inferences before the age of 6–7 years has been hotly debated. Much of the debate focused on an experiment carried out by Bryant and Trabasso (1971), which showed that even four-year-olds could make transitive inferences as long as they were trained to remember the premises comprising the inferential problem. This study suggested that the logical ability to make a transitive inference was present in quite young children, and that sensory-motor concepts did not have to be redeveloped in the mental realm.

Bryant and Trabasso used a five-term series ($A > B > C > D > E$) in order to provide a true test of inferential ability for four-, five- and six-year-old children. In a five-term series, two components will be both 'larger' and 'smaller'. These components are B and D ($B > C$, $A > B$; $D > E$, $C > D$). In a three-term series, in contrast, A is always large and C is always small (A > B > C). If children are trained to remember the premises A > B and B > C in a three-term series, then in theory they could work out the relationship between A and C without using a transitive inference, by remembering that A is large and that C is small. However, when children are trained to remember the premises in a five-term series, a memory strategy will not work for an inference about the relationship between B and D. B and D have been both large and small.

Bryant and Trabasso's (1971) paradigm was based on coloured wooden rods. Children were shown five rods (red, white, yellow, blue and green), each of which was a different length (three, four, five, six and seven inches long). The rods were always presented in pairs. For example, the child might learn that blue was larger than red, and that red was larger than green. The rods were presented via a container box that had holes of different depths bored into it, so that only one inch of each rod was seen in any training trial. During this training phase of the experiment, the child was asked which rod was taller (or shorter). After making a choice, the two rods were removed from the container and shown to the child, providing direct visual feedback about their relative length. During the testing phase of the experiment, the children were asked to make a comparison without feedback regarding the true length of the rods. As well as the four direct comparisons on which they had been trained (A > B, B > C, C > D, D > E), the children were asked about the six possible inferential comparisons in a random order (A ? C, A ? D, A ? E, B ? D, B ? E, C ? E). For the critical B ? D comparison, 78% of the four-year-olds, 88% of the five-year-olds and 92% of the six-year-olds were successful. Bryant and Trabasso concluded that children at all three age levels were able to make genuine transitive inferences very well.

This conclusion did not go undisputed (see Breslow, 1981, for a useful review). One problem with the rods task was that the container used during the comparisons provided children with a visual reminder of which were the 'long' rods and which were the 'short' rods. For example, for the critical B ? D comparison, rod B was necessarily near the 'long' end of the box and rod D was near the 'short' end of the box. This spatial cue might have enabled the children to solve the comparison correctly by associating the respective rods with the large or the small end points of the box. Pears and Bryant (1990) thus eliminated the memory load in the transitive task completely, by using visible premises. Children were shown pairs of coloured bricks presented in little 'towers' one on top of the other. The child's task was to build a complete tower of bricks from single bricks of the appropriate colours, using the premise pairs as a guide.

Before being allowed to build the target towers, the children were asked a series of inferential questions such as "Which will be the higher in the tower that you are going to build, the yellow brick or the blue one?" Three kinds of tower had to be constructed during the experiment: four-brick towers (involving three premises), five-brick towers (involving four premises), and six-brick towers (involving five premises. To take a five-brick tower as an example, if the little towers showed red on top and blue beneath (RtB), blue on top and green beneath (BtG), green on top and yellow beneath (GtY) and yellow on top and white beneath (YtW), then the target tower was (Rt Bt GtYt W). The different kinds of problems used are shown

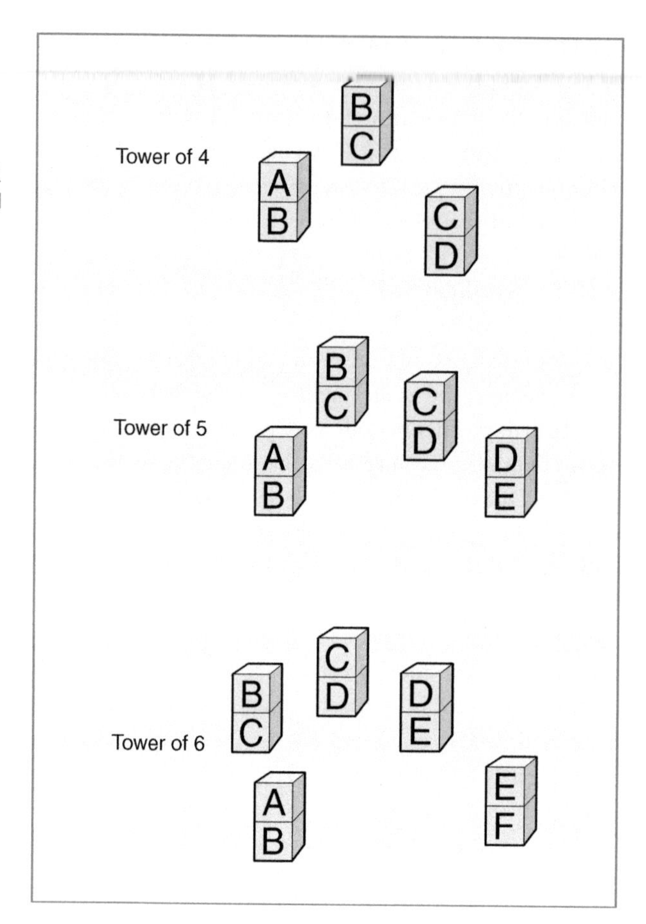

FIGURE 11.1 Examples of the premise towers used by Pears and Bryant (1990). The letters A, B, C, D, E and F denote different colours. Copyright © The British Psychological Society. Reproduced with permission.

in Figure 11.1. Pears and Bryant found that the children were significantly above chance in their performance on two-thirds of the critical inferential questions. From this finding, they argued that four-year-olds do possess the ability to make transitive inferences, at least about the continuum of space.

Conservation and invariance

KEY TERM

Conservation Understanding that simply changing the appearance of a quantity does not affect the amount that is present, as the change in appearance is reversible.

Another logical concept investigated by Piaget was conservation, which is the ability to conserve quantity across changes in appearance. This logical operation underpins the understanding of invariance, an important logical insight that in turn underpins the number system and gives stability to the physical world. Children who understand the principle of invariance understand that simply changing the appearance of a quantity does not affect the amount that is present, as the change in appearance is reversible. Changes in one dimension can be compensated for by changes in another. Piaget designed the conservation task as a measure of children's understanding of the principle of invariance.

In the conservation task, children's understanding of invariance was assessed by asking them to compare two initially identical quantities, one of which was then transformed. For example, a child could be shown two rows of five beads arranged in 1:1 correspondence, or two glasses of liquid filled to exactly the same level (Figure 11.2). An adult experimenter would then alter the appearance of one of these quantities while the child was watching. For example, the adult could pour the liquid in one of the glasses into a shorter, shallower beaker, or could spread out the beads in one of the rows so that the row looked longer. Piaget's experimental question was whether children understood that quantity remained invariant despite these changes in perceptual appearance. The answer to this question appeared to be "no". Most children below the age of around seven who were given the conservation task told the experimenter that there was now less water in the shallower beaker, or that there were more beads in the spread-out row.

The simplest way to test children's understanding of invariance, however, would have been to show them a single quantity, such as a glass of liquid or a row of pennies, and then to transform the appearance of this single quantity. As pointed out by Elkind and Schoenfeld (1972), the traditional conservation task involved a hidden transitive inference. In the traditional task, the child is shown two identical quantities, Q1 and Q2, and one is then transformed (e.g., Q1 to Q1A). The child is usually asked "Is there more, less, or the same as before?" To answer correctly,

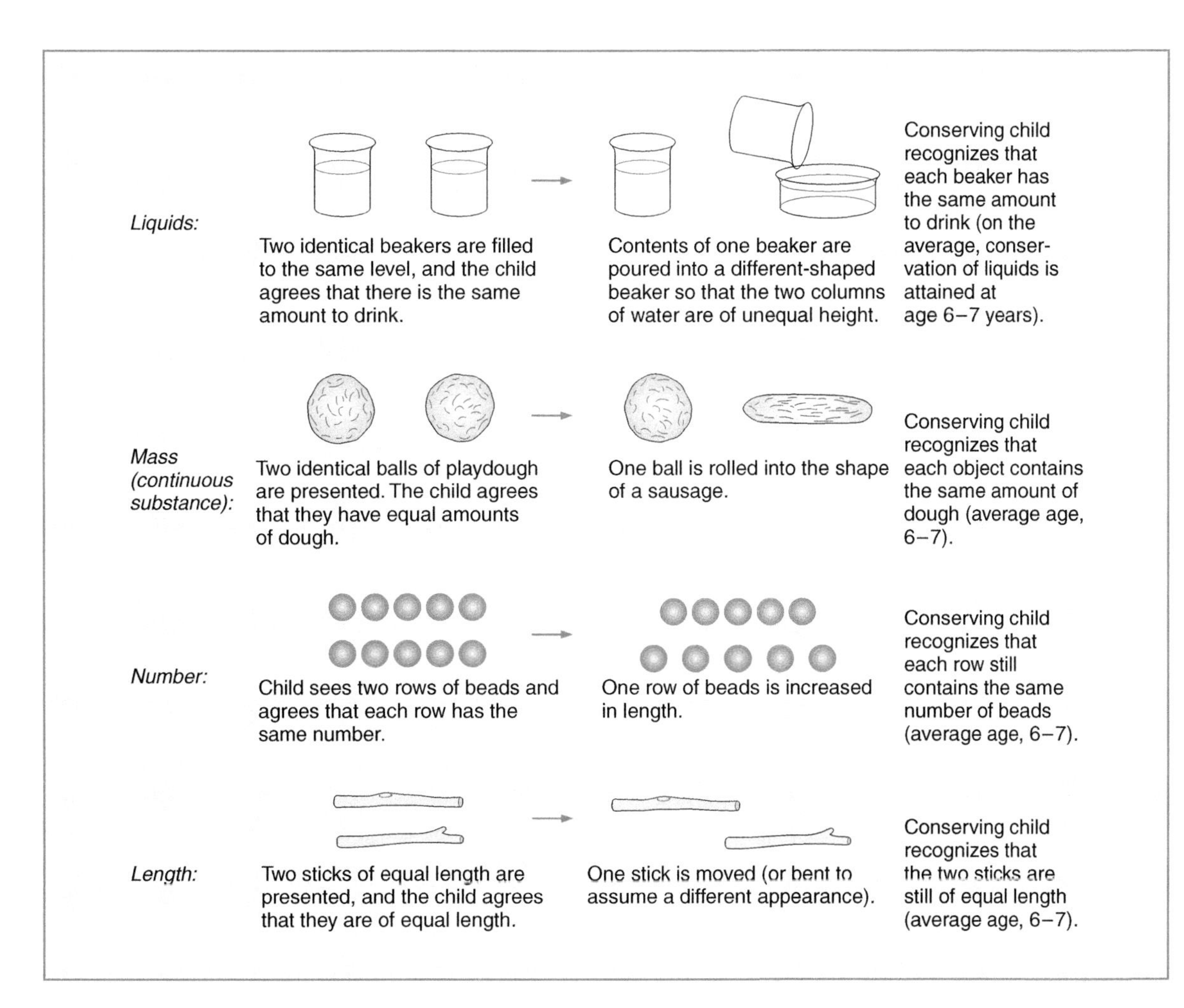

FIGURE 11.2 Examples of different versions of the conservation task. From Schaffer (1985). Adapted from *Developmental Psychology: Childhood and Adolescence*. By D. R. Schaffer. Copyright © 1996, 1993, 1989, 1985 Brooks/Cole Publishing Company, Pacific Grove, CA 93950, a division of International Thompson Publishing, Inc.

TABLE 11.1 The hidden transitive inference in the traditional conservation task

Step (A):	Q1 =	Q2
Step (B):	Q1 =	Q1A
(therefore, via transitive inference)		
Step (C):	Q1A =	Q2

Based on Elkind and Schoenfeld (1972).

the child must reason that: (1) quantity 1 = quantity 2 (Q1 = Q2); (2) quantity 1 has changed to quantity 1A, but the two are still equivalent (Q1 = Q1A); and therefore (3) that quantity 1A = quantity 2 (Q1A = Q2; see Table 11.1). The child is being asked to conserve equivalence rather than to conserve invariance. Elkind and Schoenfeld hence compared the 'conservation of identity' (one entity is transformed) with the conservation of equivalence by four- and six-year-old children. The children had little difficulty in judging the invariance of a single quantity, but the four-year-olds were poor at judging quantitative equivalence. Some understanding of the conservation of quantity is thus present by age four.

Most investigators who have used the traditional conservation task, however, have replicated Piaget's findings. Children even make conservation errors when they are reasoning about natural transformations in equivalence paradigms. For example, if children are shown two rows of toy boats floating in a pool, and then one row is allowed to drift apart so that it appears longer than the other row, the children claim that there are more boats in the longer row (Miller, 1982). Children can even fail in identity paradigms. For example, Miller (1982) showed that if children are told to watch their classmates playing a game of 'Simon Says', and Simon Says "spread out" or "bunch together", the children making the conservation judgements claim that there are more children when the group is spread out, and fewer children when the group is bunched together.

One aspect of conservation that has come under the spotlight is the pragmatics of the traditional Piagetian conservation task. The intentional structure of the experimenter's non-linguistic behaviour is at variance with the answer that is desired. The child is asked a question about two quantities ("are there more, less or the same?"), an adult carries out an apparently salient transformation, and then asks them the same question again. If someone asks you the same question twice, it usually means that you should change your answer. The experimental set-up may lead children to answer the question that they think the tester plans to ask, rather than attending to the precise question that is in fact asked. Donaldson (1978), Rose and Blank (1974) and Siegal (1991) have all argued that different aspects of the pragmatics of the conservation task may mislead children into giving non-conserving responses.

Perhaps the most compelling study for the influence of pragmatics was the ingenious 'naughty teddy' paradigm devised by McGarrigle and Donaldson (1975). In their studies, four- and five-year-old children were told that they would play a special game with the experimenter. The children were shown a cardboard box containing a teddy bear, and were told that the teddy was very naughty and was liable to escape from his box from time to time and try to "mess up the toys" and "spoil the game". The conservation materials were then brought out (e.g., two rows of counters in one-to-one correspondence). The child was asked "Are there more here or more here or are they both the same number?" All of a sudden, the naughty teddy appeared and altered the length of one of the rows by shoving the counters together. The teddy received the appropriate scolding and the children were then asked again: "Are there more here or more here or are they both the same number?" Under these conditions of an accidental transformation of the arrays, the majority of four- and five-year-old children in the experiment gave conserving responses.

However, given the findings of Miller (1982) concerning natural transformations, it must be asked whether children who experience accidental versions of the conservation task are receiving a proper test of conservation. Piaget's idea was that younger children make mistakes because they tend to focus on one perceptual dimension of the transformed quantity and ignore the compensating dimension. In exciting and accidental paradigms, the children may not actually look at the transformed arrays (see Bryant, 1982). One way to explore whether there is genuine logical development in the conservation task is to study the same children over time. Siegler (1995) has used a 'microgenetic' method to study the mastery of number conservation in five-year-olds using a task based on rows of buttons.

Siegler tried to induce cognitive change by giving the children different types of feedback in a conservation training paradigm. Three types of training were studied during four successive training sessions. One group of five-year-olds was simply told whether their answers to different problems were correct or incorrect (the 'feedback' training group). Children in a second group were asked to explain their reasoning, and were then given feedback concerning the correctness of their answers to the problems (a kind of explanation-based learning). Members of a third group received feedback about their answers, and were then asked by the experimenter "How do you think I knew that?" This last type of training required the children to explain the experimenter's reasoning. This condition was expected to encourage the children to search for the causes of the observed events (as we saw in Chapter 7, requiring children to provide explanations promotes causal understanding). This third condition indeed had the largest effect on conservation performance. Learning in this condition was found to involve two distinct realisations: that relative length did not predict which row had the greatest number of objects, and that the type of quantitatively relevant transformation did (i.e., whether buttons were added to or subtracted from the rows, or whether the appearance of the rows was altered by lengthening or shortening).

Interestingly, understanding of the importance of the type of transformation did not lead to the immediate rejection of the less advanced forms of reasoning, even when the same problem was presented several times during the experiment. Instead, children's understanding of conservation occurred gradually rather than suddenly. Siegler also found large individual differences in children's ability to benefit from having to explain the experimenter's reasoning. The children who benefited most tended to be those who had displayed greater variability of reasoning in the pre-test. Siegler concluded that children use several types of reasoning before as well as during transitional periods of cognitive development.

He suggested that an 'overlapping waves' model of logical development may be more appropriate, with certain ways of thinking being prevalent at different times. Siegler's data may also be understood in terms of the child's growing efficiency in inhibiting misleading aspects of the transformations. As will be recalled, we saw in Chapter 7 that adults who have learned Newtonian physics still activate misleading representations based on impetus theories of motion. Those adults who are better at inhibiting these misleading representations are more accurate in reasoning tasks (Dunbar et al., 2007). Cognitive neuroscience studies of children experiencing Siegler's microgenetic paradigms would be very interesting. This would be one way of testing Houdé's (2000) hypothesis about the role of inhibitory processes in logical development.

Class inclusion

KEY TERM

Class inclusion
Understanding that a set of items can be simultaneously part of a combined set and part of an embedded set (e.g., brown cows belong both to the class of 'brown cows' and to the broader class of 'cows in general').

The third major logical operation among Piaget's concrete operations was class inclusion. The logical concept of class inclusion involves understanding that a set of items can be simultaneously part of a combined set and part of an embedded set. For example, imagine a bunch of six flowers, four of which are red and two of which are white. The combined set is the six flowers, and the embedded sets are the white flowers and the red flowers. To see whether young children understand the logical concept of class inclusion, Piaget devised the class inclusion task. The child was shown a combined set, such as the flowers, and was then asked "Are there more red flowers or more flowers here?" Children younger than approximately six years of age usually responded that there were more red flowers. Piaget argued that children could only deal with the parts or with the whole separately. They could not think about the flowers in two ways simultaneously as they lacked reversibility – just as they could not simultaneously think about total length and gap size when trying to judge the quantity in a row of pennies.

Markman and Seibert (1976) argued instead that the children could be failing to make part–whole comparisons because Piaget's class inclusion question sounded a bit strange. In natural speech, we do not usually contrast a whole and a part by asking "Are there more red flowers or more flowers here?" For a part–part comparison we would ask "Are there more red flowers or more white flowers?", and for a part–whole comparison we might say "Are there more red flowers, or are there more flowers in the bunch?" The use of the term 'bunch' is a natural linguistic device for referring to a collection of objects. Collections have some degree of internal organisation and form natural units which are marked in the spoken language. Collection terms alert language users to wholes rather than parts. Piaget's class inclusion question relies on the repeated use of the class (part) term 'flower', which might make the listener assume that a part–part comparison was required.

To test this idea, Markman and Seibert contrasted five- to six-year-old children's class inclusion performance in two different versions of the class inclusion task. In one version, the standard class inclusion question was asked using Piaget's class-term format. In the second version, the class inclusion question was asked using the collection terms found in natural language. The children in the collection version of the experiment were asked about four different types of collection, a bunch of grapes, a class of children, a pile of blocks and a family of frogs. For example, the children were shown some grapes and told:

> *Here is a bunch of grapes, there are green grapes and there are purple grapes, and this is the bunch. Who would have more to eat, someone who ate the green grapes or someone who ate the bunch?*

The children in the standard version of the experiment were asked about the same stimuli (grapes, children, etc.), but were told:

> *Here are some grapes, there are green grapes and there are purple grapes. Who would have more to eat, someone who ate the green grapes or someone who ate the grapes?*

Markman and Seibert found that the children performed at a significantly higher level when the collection term was used to pose the class inclusion question (70% correct) than when the class term was used (45% correct). They argued that this showed that the psychological coherence of collections was greater than that of classes, as collections are more readily conceptualised as wholes. The data suggest that even four-year-olds have the logical concept of class inclusion. They appear illogical because the unnatural language of the standard class inclusion task induces them to make part–part comparisons.

Markman and Seibert's results were replicated by Fuson, Lyons, Pergament, Hall and Kwon (1988) with five- and six-year-olds, and also by Hodges and French (1988) with younger children (three- and four-year-olds). Fuson et al. also found that experience with the collection terms transferred to classic Piagetian class inclusion problems employing class terms. However, Dean, Chabaud and Bridges (1981) argued that collection terms could facilitate performance in the class inclusion task simply because collection terms imply large numbers. They pointed out that if children interpret collection nouns as synonymous with 'a lot', then their performance on the class inclusion question would improve simply because of the connotation of a large number of objects. Dean et al. gave five- and six-year-old children part–whole problems in which the collection term was used to describe the part rather than the whole. For example, Dean et al. told the children:

> *Here are some ants, some red ants and some brown ants. Suppose the red ants were an army, but the brown ants were not an army. Do you think there are more in the army of red ants, or more ants?*

Forty-five per cent of children responded that there were more red ants, even though they could see that there were fewer red ants than brown ants in the display. Dean et al. argued that these errors showed the influence of the 'large number' connotation of collection terms.

A different way of understanding these data is to accept that language can bias our cognitive responses. Armies are usually enormous, and perhaps the children assumed that the red ants visible on the table were only representatives of the whole army. Even adult logic can be misled by clever wording, which is a staple of the advertising industry. The real question with respect to Piaget's theory is whether the logical structures of class inclusion, transitivity and conservation are available to young children. Regarding class inclusion reasoning, the structural organisation of the family is a highly familiar example of an inclusive set (see Halford, 1993). Most young children know that a family is made up of parents and children. The combined set of the family thus has two natural embedded sets, adults and children. These embedded sets also have their own natural language labels ('parents', 'children'). Goswami and Pauen (2005) investigated whether family structure could provide a useful analogy for more traditional class inclusion problems involving piles of blocks and bunches of balloons.

The children in Goswami and Pauen's study (four- to five-year-olds) had all failed the traditional Piagetian class inclusion task, which was given as a pre-test ("Are there more red flowers or more flowers?"). They were then shown a toy family, for example a family of toy mice (two large mice as parents, three small mice as children) or a family of yo-yos (two large yo-yos as parents, three small yo-yos as children). Their job was to create analogous families (two parents and three children) from an assorted pile of toy animals (such as fluffy toy bears, ladybirds, ducks and crocodiles) or from a pile of other toys (such as toy cars, spinning tops, balls and helicopters). After the children had correctly created four families that were analogous to the mice/yo-yo families (having two parents and three children), they were given four class-inclusion problems involving toy frogs, sheep, building blocks and balloons. The class inclusion problems were posed using collection terms ('group', 'herd', 'pile', 'bunch'). A control group of children received the same class-inclusion problems using collection terms, but did not receive an analogy training session in which they learned to create families. The results showed that the children in the 'create-a-family' condition solved more of the class inclusion problems. They appeared to be using analogies to family structure.

Evaluation

This survey of some of the key experiments that have challenged Piaget's view that pre-operational children lack logical concepts like class inclusion and transitivity has been necessarily brief. However, it has echoed many of the conclusions reached in Chapter 9. 'Pure' measures of reasoning, in which logical abilities are measured independently of the context in which they are required, are difficult to devise. When familiar, 'child-centred' materials are used, and when attention is paid to linguistic and non-linguistic aspects of the experimental set-up, then the logic of the concrete operations appears to be understood by four-year-olds. Furthermore, as also seen in Chapter 9, these experiments document the facilitatory effects of accessing familiar relational or organisational structures on children's performance in reasoning tasks. For example, in deductive reasoning, Harris and Nunes (1996) and Light et al. (1989) found that the activation of an appropriate permission schema helped children to reason successfully in the selection task. In this chapter, we have seen that a family structure scenario or a tower-building scenario can reveal logical competence with class inclusion or with transitivity. In terms of children's logical development, the concrete operations appear to be similar to deductive and inductive reasoning. Children reason in similar ways to adults, but are more easily misled by interfering variables such as contextual variables, in part because they are worse at inhibiting irrelevant information.

Developmentally, therefore, sensory-motor cognition does not seem to be redeveloped in the mental realm, as Piaget argued. Instead, the crucial factors in explaining development appear to be the augmentation of sensory-motor knowledge by increasing experience, some of which is active and some of which is transmitted by language (discussed below). Logical development is more likely to depend on the ability to reflect on one's knowledge (metacognition), and efficiently to inhibit competing knowledge that is interfering with the application of logic (an executive function). With developments in metacognition and executive function, children can apply concepts like transitivity and invariance in a strategic way to new situations. Houdé (2000) offers an interesting example of this framework for explaining cognitive development that is based on inhibition. Thoughtful treatments of Piaget's theory that disagree with the arguments given in this chapter can be found in Chapman (1988) and Smith (1992, 2002).

Formal operational thought

According to Piaget's theory, cognitive change beyond the concrete operational period depended on the emerging ability to take the results of concrete operations and to generate hypotheses about their logical relationships. This 'formal operational' reasoning became available at the age of approximately 11 or 12 years. Piaget described this level of reasoning as 'operating on operations', or 'second-order' reasoning:

> *This notion of second-order operations also expresses the general characteristic of formal thought – it goes beyond the framework of transformations bearing directly on empirical reality (concrete operations) and subordinates it to a system of hypothetico-deductive operations, i.e. operations which are possible. (Inhelder & Piaget, 1958, p. 254)*

KEY TERM

Formal operational thought
The ability to take the results of concrete operations and to generate hypotheses about their logical relationships, for example applying a formal system such as propositional logic to elementary operations concerning classes of objects and their relations.

Piaget characterised formal operational reasoning in terms of the ability to apply a formal system such as propositional logic to the elementary operations concerning classes of objects and their relations. Again, the underlying idea was that cognitive structures developed to mirror mathematical logic.

Piaget described the basis of this propositional logic in terms of the combinatorial system describing all 16 possible binary relations between the entities *p*, *q*, *not-p* and *not-q* (such as the conditional rule 'if p then q', discussed in Chapter 9). He also described the subsystem of transformations that could operate on these relations (such as finding the inverse relation or the reciprocal relation). This latter analysis has been called the INRC grouping (I for identity of the relation, N for negation or inverse, R for reciprocity and C for correlation). The presence of these binary combinatorial relations, and of the INRC operations, were thought to be the hallmarks of formal operational thought. As with concrete operational thought, however, most subsequent research has neglected Piaget's focus on mathematical groupings, and has focused instead on younger children's performance in the tasks that Piaget used to demonstrate the presence of the formal operations.

Formal operational tasks

Formal operational thought was scientific thought, as Piaget believed that the attainment of the formal operations allowed the child to represent alternative hypotheses and their deductive implications. Indeed, many of Piaget's tests for the presence of formal operational structures involved scientific tasks, such as discovering the rule that determines whether material bodies will float or sink in water, discovering the rule between weight and distance that will enable a balance beam to balance, and discovering the rule that governs the oscillation of a pendulum. All of these rules were part of the combinatorial or INRC groupings. For example, the rule governing the behaviour of the balance beam is an inverse proportional relation between weight and distance (that simultaneously increasing the weight and decreasing the distance on one arm of the balance is equivalent to decreasing the weight and increasing the distance on the other arm). Discovery of such proportionality was thought to be a key feature of formal operational reasoning.

Piaget's experimental method was to allow children to manipulate the independent variables (e.g., the length of the string and the weight of the bob in the pendulum task), and then see whether they could arrive at the correct rule. Most children did not manage to discover the appropriate rules before the age of around 11 years. For example, in the pendulum task, children needed to use strings of different lengths and bobs of different weights in order to discover that the period of a pendulum is a function of its length. As children usually began the task by believing that the weight of the bob must be an important factor in determining the oscillation of the pendulum, they needed to hold the length of the string constant while experimenting with a variety of weights in order to conclude that weight alone does not affect the pendulum's period. Children younger than 11–12 years usually failed to see the necessity for holding other variables such as string length constant, and thus failed to reason according to Piaget's combinatorial system (use of the system can be expressed as follows: if p is 'increases in the period of oscillation', and q is 'decreases in the length of the string', then p implies q and vice versa). Piaget argued that children who manipulated the two variables correctly showed an awareness of this combination, and also of all the other possible combinations in the group, since the other combinations were discarded as being irrelevant to the problem at hand. More recent experiments suggest that the key determinants of scientific reasoning in these paradigms may be somewhat different, however. For example, the work of Howe on scientific reasoning has suggested that hypothesis testing only changes conceptual knowledge when children debate outcomes with their peers or teachers (e.g., Howe, Tolmie, Duchak-Tanner & Rattray, 2000). Nine-year-olds who debated their conceptual knowledge before beginning a scientific task (about shadows) and reached a consensus, and who then tested that consensual hypothesis experimentally as a group, did show conceptual changes in understanding.

The emphasis on the discovery of proportionality as a key factor in formal operational reasoning also led Piaget to include reasoning by analogy among his formal operational tasks. His logic was simple. As a full appreciation of the possible relations between objects was a concrete operational skill, the construction of relations between those relations must be a formal operational skill. Analogical reasoning was higher-order reasoning in the sense that the simple relations in an analogy ('lower-order relations') had to be linked by a relation at a higher level in order to make the analogy valid. Analogies also involved proportional reasoning, as an item analogy of the form *Rome is to Italy as Paris is to France* (*Rome:Italy::Paris:France*) was logically equivalent to a proportional expression like *3:4=15:20*.

Evaluation

Evidence discussed in earlier chapters has shown that formal operational reasoning is available prior to adolescence in a variety of paradigms. For example, in Chapter 9 we discussed a series of experiments showing that the ability to reason by analogy in the item analogy format is present in children as young as three years of age (e.g., Goswami & Brown, 1989, 1990). In Chapter 7, we reviewed a number of experiments on scientific reasoning. In the balance scale task, we saw that nine-year-olds could use the multiplicative integration rule to link weight and distance (Wilkening & Anderson, 1991). When the ability to combine information about different dimensions was tested using the barking dog and fleeing animals task (requiring the integration of time and velocity to judge distance), even five-year-olds showed an ability to use multiplicative rules (Wilkening, 1982). Wilkening (1981) noted that this finding completely contradicted Piaget's notion that time had to be derived from information about speed and distance, and that the operations for deriving time required formal operational reasoning. Chapter 7 also reviewed work on hypothesis testing by children as well as their understanding of the deductive implications of evidence (e.g., as conclusive or inconclusive; Sodian et al., 1991; Ruffman et al., 1993). We saw that children as young as four years could test hypotheses systematically in certain contexts. Both children and adults are, however, subject to the confirmation bias, and tend to reason poorly when evidence must be given *priority* over existing background knowledge. Meanwhile, the work of Howe and colleagues showed that nine-year-olds can evaluate scientific hypotheses by testing relevant variables when they are working in collaborative peer groups supported by teachers.

Again, the core factors in successful formal operational reasoning appear to be familiarity and context or circumstance, including educational context. Successful scientific reasoning can be promoted by teachers who guide students in testing hypotheses that have been reached through group discussion and consensus (Howe & Tolmie, 2003). Recall that we also saw in Chapter 7 that causal inference can be an automatic neural/perceptual computation given certain environmental conditions. Even infants will automatically 'screen off' non-causal perceptual information given particular covariation data (Sobel & Kirkham, 2006), while rats tested in food prediction scenarios make the causal inferences predicted by causal Bayes nets on the basis of purely observational learning. Data such as these suggest that any model of logical development based on *qualitatively different* kinds of reasoning becoming available at different ages will be wrong. Instead, the core factors to explore with respect to developmental changes in reasoning are most likely to be familiarity and context (including contextual support, e.g., from teachers), language abilities, prior knowledge, metacognitive skills such as reflection, and executive function skills such as inhibition.

Nevertheless, Piaget's core idea that cognitive structures mirror mathematical systems was very insightful. Similar arguments have been applied here to the learning brain. In earlier chapters, we have seen that mathematical advances in signal processing have enabled deeper understanding of the rich knowledge structures that the brain can construct on the basis of sensory experience. For example, the brain seems to record the causal structure of perceptual experiences *at the point of encoding* (Fekete & Edelman, 2011; Moors et al., 2017). Studies in adult cognitive neuroscience show that experience of sensory environments leads *automatically* to the neural development of internal models of those environments (Fiser et al., 2010; Hochstein et al., 2015). Accordingly, neural systems that learn the patterns or regularities in environmental input captured by different statistical algorithms can, in principle, acquire complex cognitive structures like language and conceptual knowledge (see Chapters 5 and 6). Although the INRC grouping chosen by Piaget may not have been the appropriate mathematical model for cognition, future advances in cognitive neuroscience seem likely to depend on advances in mathematical theories. Deep learning provides a good example. Modern mathematics has enabled computational models that use multiple processing layers simultaneously to learn from natural data in their raw form (e.g., visual scenes) and models that automatically discover representations at multiple levels of abstraction (LeCun et al., 2015). To date, these models have not been applied to problems in developmental psychology, Further, reverse-engineering the models to discover exactly what is being learned is technically challenging. Nevertheless, deep learning models offer enormous promise for testing developmental theories.

VYGOTSKY'S THEORY

The notion that developmental changes in reasoning might be better explained by cognitive factors such as familiarity, context, metacognition and language fits well with the theoretical emphasis of Vygotsky's theory of cognitive development. Vygotsky recognised the importance of language and of cultural context in cognitive development. Although he did not live long enough to generate the wealth of experimental data contributed by Piaget, his ideas about cognitive development were of comparable importance. Whereas Piaget focused on how the individual child constructed knowledge for him- or herself, Vygotsky argued that knowledge originated in socially meaningful activity and was shaped by language. Social context and culture were crucial in explaining cognitive development, and language played an essential role in the organisation of 'higher psychological functions' (Vygotsky, 1978, p. 23). Language was seen as the primary symbolic system that children could respond to psychologically. Language hence mediated cognition. Vygotsky argued that cognition developed prior to language, as demonstrated by the cognitive activities of babies. However, "the most significant moment in the course of intellectual development... occurs when speech and practical activity, two previously completely independent lines of development, converge" (p. 24). This convergence was marked by egocentric or private speech, which Vygotsky interpreted as fundamental in organising the child's cognitive activities. Egocentric speech was seen as part of goal-directed behaviour. Eventually, children were thought to internalise egocentric speech as the 'inner speech' that organised mental life.

Vygotsky conceptualised early thought as pre-linguistic, and early language as pre-intellectual, with purely social functions:

> *In the first year of the child's life (that is, during the preintellectual stage of development in speech), we find rich development in the social function of speech. The relatively complex and rich social contact of the child leads to a very early development of a 'means of contact'... babbling, behavioral displays and gestures emerge as a means of social contact. (Vygotsky, 1934, p. 88)*

Via the stage of egocentric speech, language gradually became internalised and became the means of organising thought. Vygotsky pointed out that when children were put into problem-solving situations, they:

> *not only act in attempting to achieve a goal, they also speak. As a rule, this speech arises spontaneously and continues almost without interruption throughout the experiment. It increases and is more persistent every time the situation becomes more complicated and the goal more difficult to attain. Attempts to block it... are either futile, or lead the child to 'freeze up'. (Vygotsky, 1978, p. 25)*

Vygotsky argued that language was as important as action in attaining goals, and that language and action were part of "one and the same complex psychological function" (1978, p. 25). Indeed, as argued in Chapter 6, language is an action as well as a means of communication. For Vygotsky, language enabled children to disconnect themselves from the immediate, concrete situation and to generate possibilities and plans for solving problems. Language was also thought to play a role in controlling the child's own behaviour – by speaking of their intentions, children guided their actions. Finally, language enabled children to ask adults to help them, thus acting as a problem-solving tool. In fact, Vygotsky suggested that children did not at first distinguish the roles played by the child and the helper in problem-solving, experiencing a 'syncretic whole'. When this 'social speech' became internalised, children effectively appealed to themselves for help. Language and thought were also inevitably interdependent. One example given by Vygotsky was visual perception (Vygotsky, 1978). Initially, children use language to label their visual perceptions, thus enhancing individual objects in the visual field ("ball", "car"). Soon, however, language can be used to create a 'time field' in addition to the visual–spatial field of current perception. The child can now view changes in the immediate situation from the viewpoint of past activities, and can act in the present from the viewpoint of the future. Language enabled the child's field of attention to embrace "a whole series of potential perceptual fields that form successive, dynamic structures over time" (Vygotsky, 1978, p. 36). Vygotsky speculated that animals cannot do this, and noted that the ability to combine past and present perceptual fields via language also enabled reconstructive memory:

> *Created with the help of speech, the time field for action extends both forward and backward... [creating] the conditions for the development of a single system that... encompasses two new functions: intentions and symbolic representations of purposeful action. (Vygotsky, 1978, pp. 36–7).*

Socio-cultural tools for mediating knowledge

KEY TERM

Sign systems
Cultural semiotic systems; psychological tools that enable the symbolic representation and organisation of knowledge, which include drawing pictures, writing, reading, using number systems, maps, and diagrams.

As well as language, Vygotsky identified a number of 'sign systems' or cultural semiotic systems that enabled the symbolic representation of knowledge. These included drawing pictures, writing, reading, using number systems, maps and diagrams. These symbolic systems were seen as psychological tools for organising cognitive behaviour. Cognitive development, therefore, did not just happen in the brain of the individual child. Rather, it depended on *interactions* between the child and the cultural tools available for mediating knowledge. These interactions with cultural tools first depended on interpersonal communication with adults and other teachers, and were then internalised by the child, the tools (symbols) thereby becoming *internal* mediators of cognitive processes. Sign systems had a crucial role in an individual's intellectual development, as they extended the operation of psychological processes beyond the individual. For example, memory could be extended by recording memories in writing, or by drawing a picture of a salient event, thereby fundamentally changing the memory itself via symbolic activity. Wertsch has called this idea 'mediated cognition' (e.g., Wertsch, 1985; Rowe & Wertsch, 2002). Because these symbol systems are the product of socio-cultural evolution (they are not reinvented by each individual), Vygotsky saw sign systems as social in nature. Children gained access to these psychological tools because of the social/cultural context in which they developed, and because of face-to-face communication and interaction with others. Hence the communicative and intellectual functions of sign systems were inherently related.

The interrelatedness of social and cognitive processes in the child was repeatedly emphasised in Vygotsky's writings:

> *Sign-using activity in children is neither simply invented nor passed down by adults... within a general process of development, two qualitatively different lines of development, differing in origin, can be distinguished: the elementary processes, which are of biological origin... and the higher psychological functions, of sociocultural origin... the history of child behavior is born from the interweaving of these two lines. (Vygotsky, 1978, p. 46)*

By elementary processes, Vygotsky meant basic perceptual processes such as vision, touch and hearing, and basic aspects of behaviour such as goal-directed action and social communication. Two fundamental forms of cultural behaviour were thought to emerge in infancy, psychological tool use and language. Both were mediated by adults. Psychological tools meant tools invented by human society, such as signs, symbols and concepts. The combination of psychological tool use and language in psychological activity changed psychological functioning in a profound way, enabling higher levels of cognitive activity in the child. These higher levels of cognition were thought to be specific to human beings. Only humans had mediated cognition, as animals do not create artificial signs to communicate with each other:

> *We know that social interaction such as that found in the animal world... is not mediated by speech or any other system of signs ... [this] is social interaction of only the most primitive or limited type. (Vygotsky, 1934, p. 11)*

The zone of proximal development

KEY TERM

Zone of proximal development
The difference between what children can do on their own and what they can do under adult guidance or in collaboration with more able peers.

In addition to his emphasis on the social and cultural nature of cognitive development, Vygotsky illustrated the importance of learning from others to cognitive development. He rejected the Piagetian idea that development unfolds according to its own timetable, without any influence from school learning. For Vygotsky, development could not be independent of learning, particularly during the school years. Clearly, learning began prior to schooling, and:

> *any learning a child encounters in school always has a previous history. For example, children begin to study arithmetic in school, but long beforehand they have had some experience with quantity... children have their own preschool arithmetic, which only myopic psychologists could ignore. (Vygotsky, 1978, p. 84)*

But school learning introduced something fundamentally new into the child's development. This something new was captured by Vygotsky's concept of the 'zone of proximal development'. This concept captured the insight that "what children can do with the assistance of others might be in some sense even more indicative of their mental development than what they can do alone" (Vygotsky, 1978, p. 85).

While acknowledging that it was important to measure a child's actual level of development, for example via tests of mental function, Vygotsky argued that it was also important to investigate how much further a child could go under the guidance of a teacher. He gave the example of two children who entered school aged ten years, and who could deal with standardised tasks up to the degree of difficulty typical of the eight-year-old level. These two children would have a mental age of eight years. However, suppose the experimenter then showed them different ways of dealing with some of the problems. Suppose that with assistance one child could deal with problems up to a 12-year-old's level, the other with problems up to a nine-year-old's level. Vygotsky argued that mentally these children were clearly not the same. They differed in terms of their zone of proximal development. He defined this as:

> *The distance between the actual developmental level, as determined by independent problem solving, and the level of potential development as determined through problem solving under adult guidance or in collaboration with more capable peers. (Vygotsky, 1978, p. 86)*

The zone of proximal development enabled a *prospective* rather than retrospective characterisation of cognitive development.

The concept of a zone of proximal development has had a profound impact in education. Vygotsky's insight that school learning can affect a child's cognitive development is important for how we set about teaching children in schools. For example, the zone of proximal development contradicts the notion that learning should be matched with the child's developmental level. Learning can *change* the child's developmental level, and hence it is better for teachers to discover the child's zone of proximal development and teach to that in order for instruction to bring optimal benefits. Vygotsky's view that school learning mediated cognitive development in the middle childhood years can be illustrated by his contrast between *scientific* and *spontaneous* concepts. Spontaneous concepts are defined as those generated by children on the basis of their observations and experiences. Sometimes these concepts are wrong (an example is the medieval impetus theory of motion, discussed in Chapter 7). Scientific concepts were thought to be acquired consciously and effortfully in school, and were thought to transform students' knowledge. For example, scientific concepts were thought to restructure spontaneous concepts and raise them to a higher level (an example might be restructuring the erroneous impetus theory of motion into a Newtonian theory).

Although Vygotsky himself did not produce evidence for these ideas, neo-Vygotskians did attempt to explore them empirically. As discussed in Chapter 9, syllogistic reasoning was originally considered a form of higher-level reasoning. Vygotsky's student Luria carried out an extensive study of syllogistic reasoning in peasants living in remote villages of Uzbekistan and Kirghizia (Luria, 1976). He found evidence apparently suggesting that these unschooled peasants were incapable of syllogistic reasoning, being hampered by an 'empirical bias'. For example, Uzbekistan is a hot, plains region, which does not have snow. When unschooled villagers from Uzbekistan were given syllogisms such as "In the Far North, where there is snow, all bears are white. Novaya Zemla is in the Far North. What colour are the bears there?", they did very poorly (Luria, 1976). They would give responses along the lines of "I cannot tell, you would have to ask people who had been there and seen the bears". Even when syllogisms were based on familiar information, such as the factors that affect the growth of cotton, invoking an unfamiliar setting led to logical failures. With premises such as "Cotton grows well where it is hot and dry. England is cold and damp. Can cotton grow there or not?", these villagers refused to commit themselves, although some would make pertinent observations (e.g., "If it's cold there, it won't grow. If the soil is loose and good, it will"). Neo-Vygotskians argued that such data showed the dominant role of schooling in the development of higher-order logical thought (see Karpov, 2005, p. 175).

According to Vygotsky, since learning can change a child's developmental level, it would be optimally beneficial for a child to be taught in line with his or her zone of proximal development, rather than at his or her current developmental level.

As we saw in Chapter 9, however, even four-year-olds can use syllogistic reasoning when premises are familiar, and with appropriate manipulations (e.g., a make-believe world) they can make logical deductions about counterfactuals as well. Anthropologists have also challenged this simplistic view of the effects of schooling (e.g., Cole & Scribner, 1974). Despite lacking firm empirical evidence, however, Vygotsky's idea that school learning can systematically target the zone of proximal development and transform knowledge is a very important one. Russian neo-Vygotskians (e.g., Karpov, 2005) have also stressed that *joint activity* with adults is critical to the effective use of the zone of proximal development in teaching. Verbal mediation is not enough to optimise learning. The adult must mediate the child's acquisition, mastery and internalisation of new content via *shared activity*.

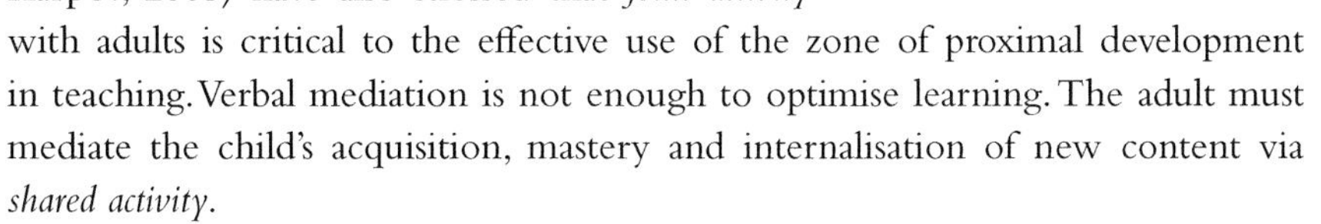

Neo-Vygotskians argue that mediation should begin with the adult explaining and modelling the procedure or material to be learned. The adult should then involve the child in joint performance of this procedure or material, thereby creating the zone of proximal development of a new mental process. The child's mastery and internalisation of the material should then be guided until the adult can begin to withdraw, eventually passing responsibility for further development to the child. Karpov (2005) argues that while some instructional programmes that have been developed to exploit Vygotsky's ideas about mediated cognition and the zone of proximal development fulfil neo-Vygotskian criteria, others do not. An example of a successful instructional programme is 'reciprocal teaching', designed to improve reading comprehension by Palincsar and Brown (1984). Reciprocal teaching begins with the teacher modelling the optimal strategies for reading comprehension to a small group of children. These strategies include summarising what has been read and predicting what will happen next. First the teacher leads the group in analysing the text in accordance with these strategies, and gradually leadership passes to the children.

Vygotsky argued that a full understanding of the zone of proximal development also entailed a re-evaluation of the role of imitation in learning. He suggested that, using imitation, children are capable of doing much more in collective activity or under the guidance of adults:

> *Learning awakens a variety of internal developmental processes that are able to operate only when the child is interacting with people in his environment and in cooperation with his peers. Once these processes are internalized, they become part of the child's independent developmental achievement... learning is not development, however properly organized learning results in mental development and sets in motion a variety of developmental processes that would be impossible apart from learning. Thus, learning is a necessary and universal aspect of the process of developing culturally organized, specifically human, psychological functions. (Vygotsky, 1978, p. 90)*

Sadly, Vygotsky died before he could carry out a systematic analysis of what the internal developmental processes activated by education were and which aspects of the educational process led to the internalisation that caused further cognitive development. However, he recognised the importance of investigating the relations between the cognitive levels achieved with adult or peer guidance and the internalisation of these levels, both for understanding cognitive development and for assessing the value of school learning:

> *If successful [this research] should reveal to the teacher how developmental processes stimulated by the course of school learning are carried through inside the head of each individual child. The relevance of this internal, subterranean developmental network of school subjects is a task of primary importance for psychological and educational analysis. (Vygotsky, 1978, p. 91)*

Surprisingly, such research still remains to be done. Neo-Vygotskians have focused instead on an approach called 'theoretical learning', which offers an interesting alternative to the constructivist learning pedagogies based on Piaget's theory. In theoretical learning, children are not required to rediscover scientific knowledge for themselves. Instead, they are taught precise definitions of scientific concepts. They then master and internalise the procedures related to these concepts by using the conceptual knowledge that they have been taught to solve subject-domain problems. This is said to lead to cognitive benefits, as children eventually adopt a general strategy of searching for a general principle or theory when faced with new problems in new domains, hence demonstrating 'formal-logical thought' (see Karpov, 2005, pp. 182–202). One reason why researchers have not tackled the question of "how developmental processes stimulated by the course of school learning are carried through inside the head of each individual child" may be the complexity of the research question. As Vygotsky noted, cognitive development was unlikely "to follow school learning the way a shadow follows the object that casts it... extensive and highly concrete research based on the concept of the zone of proximal development is necessary to resolve the issue" (Vygotsky, 1978, p. 91).

Play

Vygotsky also recognised the importance of play for child development, an aspect of his theorising that has been the focus of extensive research attention. Vygotsky argued that the world of the imagination fulfilled a crucial psychological function in development, enabling children to realise desires that could not otherwise be fulfilled (e.g., being the mother):

> *We often describe a child's development as the development of his intellectual functions, every child stands before us as a theoretician who, characterized by a higher or lower level of intellectual development, moves from one stage to another. But if we ignore a child's needs... we will never be able to understand his advance from one developmental stage to the next... it is impossible to ignore the fact that the child satisfies certain needs in play. (Vygotsky, 1978, pp. 92–3)*

Vygotsky thought that the imagination represented a specifically human form of cognitive activity, was absent in animals, and arose during the preschool years: "Like all functions of consciousness, it originally arises from action" (Vygotsky, 1978, p. 93). Vygotsky argued that the defining characteristic of play was the creation of an imaginary situation. Play was not simply symbolic, because of the role of motivation. Play also involved rules of behaviour.

Vygotsky illustrated his ideas with an example of two sisters who decided to "play" at being sisters:

> *The vital difference... is that the child in playing tries to be what she thinks a sister should be. In life the child behaves without thinking that she is her sister's sister. In the game of sisters... both are concerned with displaying their sisterhood... both acquire rules of behavior... they dress alike, talk alike... as a result of playing, the child comes to understand that sisters possess a different relationship to each other than to other people. What passes unnoticed by the child in real life becomes a rule of behavior in play. (Vygotsky, 1978, pp. 94–5)*

Accordingly, for Vygotsky, play enabled the child to act in a purely cognitive realm, rather than relying on motivational incentives supplied by external things. For infants and toddlers, *things* decide action: doors demand to be opened, staircases to be climbed, switches to be operated. Vygotsky suggested that early in development, perception and motives were interdependent: "Every perception is a stimulus to activity" (Vygotsky, 1978, p. 96). In play, *things* lost their determining force and the child acted independently of what was seen. In play, thought was separated from objects and action arose from *ideas* rather than from things. A piece of wood became a doll and a stick became a horse. Symbolic play was crucial developmentally, as the meaning of things to the child was no longer dominated by their status as real objects in the perceptual world. In play, their meanings could be detached from perceptual reality (as in the mental decoupling discussed by Leslie, 1987; see Chapter 3). Vygotsky argued that play provided a critical bridge between the perceptual/situational constraints of early childhood, and adult thought, which was totally free of situational constraints.

According to Vygotsky's analysis of play, a central developmental function was that the 'rules of the game' required children to act against their immediate impulses. This enabled the development of self-regulation:

> *At every step the child is faced with a conflict between the rules of the game and what he would do if he could suddenly act spontaneously... he achieves the maximum display of willpower when he renounces an immediate attraction in the game (such as candy, which by the rules of the game he is forbidden to eat because it represents something inedible). Ordinarily, a child experiences subordination to rules in the renunciation of something he wants, but here subordination to a rule and renunciation of action on immediate impulse are the means to maximum pleasure. Thus, the essential attribute of play is a rule that has become a desire. (Vygotsky, 1978, p. 99)*

Hence play facilitated the development of self-regulation, as the child's desire in playing was to control his or her immediate impulses. Vygotsky pointed out that play also enabled children to decouple action from meaning. In infancy and toddler-hood, actions were dominant over meaning, as children could do more than they could understand:

> *Internal and external action are inseparable: imagination, interpretation and will are the internal processes carried by external action... [in play] action retreats to second place... just as operating with the meaning of things leads to abstract thought... the ability to make conscious choices occurs when the child operates with the meaning of actions. (Vygotsky, 1978, p. 101)*

Meaning is detached from action by means of a different action. A child who wishes to ride a horse stamps the ground and imagines him- or herself riding a horse, rather than performing the actions required to ride a real horse.

Vygotsky thus regarded play as a major factor in cognitive development. He also argued that play itself created a zone of proximal development:

> *In play a child always behaves beyond his average age, above his daily behavior; in play it is as though he were a head taller than himself... play contains all developmental tendencies in a condensed form and is itself a major source of development... action in the imaginative sphere, in an imaginary situation, the creation of voluntary intentions, and the formation of real-life plans and volitional motives – all appear in play and make it the highest form of preschool development... From the point of view of development, creating an imaginary situation can be regarded as a means of developing abstract thought. The corresponding development of rules leads to actions on the basis of which the division between work and play becomes possible, a division encountered at school age as a fundamental fact. (Vygotsky, 1978, pp. 102–3)*

Vygotsky regarded play as a major factor in cognitive development and argued that play itself created a zone of proximal development where children behaved beyond their age.

Russian neo-Vygotskians have developed this idea further, arguing that teachers have an important role in creating a zone of proximal development for play. For example, Karpov (2005, p. 142) notes that some groups of children (for example, children of low socio-economic status) engage in little socio-dramatic play, and that some teachers regard socio-dramatic play as a waste of time. Adult mediation is required to initiate or extend socio-dramatic play, for example in explaining real-world situations like attending a hospital clinic and explaining the causes of adult actions in those situations. The goal is for play to become "a micro-world of active experiencing of social roles and relationships" (see Karpov, 2005, p. 140). Russian neo-Vygotskians have also argued that the motive to engage in imaginary play (i.e., to fulfil desires that cannot be fulfilled otherwise) leads in turn to the motive to study at school:

> *One of the major outcomes of sociodramatic play is that, by the end of the period of early childhood, children become dissatisfied with such a pseudo-penetration into the world of adults... the only 'real' and 'serious' role that is available for the child in an industrialized society... is the role of a student in school. (Karpov, 2005, p. 153)*

Karpov notes, however, that empirical evidence to support this idea is currently lacking.

In other areas, the Russian neo-Vygotskians have produced quite a lot of empirical evidence to support Vygotsky's major ideas, and have also augmented them. Karpov's (2005) elegant survey of a literature primarily available in Russian is extremely useful for illustrating this extension of Vygotsky's ideas. For example, regarding Vygotsky's theory about the role of play in the development of self-regulation, the Russian neo-Vygotskians noted that the child's internal desire was not the only critical factor. The child's playmates exerted an important regulatory function as well. Karpov (2005, p. 157) cites Elkonin's observation of six-year-old boys playing at being fire-fighters. One boy is the chief fire-fighter, another is the driver of the fire engine, others are fire-fighters. Play begins, the chief shouts "Fire", everyone takes a seat in the engine and the driver pretends to drive. When they reach the fire, the fire-fighters jump out to extinguish the fire. The driver jumps out too, as this is the most fun part of the game, but the others remind him that he is the driver and that he has to stay with the fire engine. So the boy returns to sit in the engine. This mutual regulation coupled with the child's own growing abilities to suppress immediate desires is said to lead to the development of self-regulation. This occurs first in the zone of proximal development provided by play, and eventually in non-play situations. Karpov (2005, p. 158) cites further relevant data from Manuilenko (1948). In this study, children aged from three to seven years were required to stand motionless for as long as they could. For example, they had to stand motionless alone in a room, alone in a room in the play context of being 'a sentry', or they had to be a sentry in a room full of their playmates. The children were able to stand still longest in the third situation. The playmates were monitoring the sentry, and this helped him to stand still for longer. Again, the zone of proximal development provided by play enabled superior performance to the actual developmental level of self-regulation (Karpov, 2005). Karpov notes that, to date, Russian neo-Vygotskians have not provided causal evidence for Vygotsky's idea that imaginary play leads to the development of self-regulation in non-play situations.

Vygotsky's ideas that imaginary pretend play promoted cognitive development via fostering what we now term metacognition and executive function have been questioned by some Western psychologists, however. For example, Lillard et al. (2013) carried out a comprehensive review of experimental studies of play, evaluating the impact of play on many different aspects of cognitive development including theory of mind, self-regulation and problem-solving. Lillard et al. concluded that there was little evidence for play as a *unique* causal agent for development, even though there was firm evidence that play can be beneficial (i.e., it is one of many routes to positive developmental outcomes). On the other hand, individual experimental studies that have included pretend play as an intervention have tended to report positive outcomes. For example, Thibodeau, Gilpin, Brown and Meyer (2016) randomly assigned 110 children aged three to five years to one of three conditions, a fantastical pretend play intervention in which the children were supported in developing fantasy situations (e.g., going to the moon) and acting them out in small groups, a non-imaginative play condition in which the children played games like 'Simon Says' with support, and a control condition comprising normal classroom activities. Prior to and after the interventions, a series of measures of executive function were administered, including an inhibitory control measure (the day/night task) and the DCCS (see Chapter 9). The children experiencing the fantasy play intervention showed significantly greater improvement on some measures (e.g., DCCS) than the children experiencing the non-imaginative play intervention, but not on other measures (e.g., inhibitory control). In another representative study, White et al. (2017) investigated whether 180 children aged 4 and 6 years would persevere longer on a dull and repetitive task (a slowed down go/no-go task) if they were pretending to be the cartoon characters Batman or Dora the Explorer. The task was introduced as "This is a very important activity and it would be helpful if you worked hard on this for as long as you could". Control conditions were thinking about one's own thoughts and feelings ("Am I working hard?") or thinking about oneself in the third person ("Is [child's name] working hard?"). The children who were pretending to be Batman or Dora the Explorer spent significantly more time on the dull repetitive task than the children in the other two conditions, at both ages. This latter study is consistent with the data reported by the Russian neo-Vygotskians.

Evaluation

Vygotsky's theory of cognitive development has probably had a bigger impact on education than on developmental psychology. This is unsurprising, as Vygotsky accorded education a key role in cognitive development, whereas Piaget did not. Yet, ironically, Russian neo-Vygotskians argue persuasively that many of these educational impacts have involved *misunderstandings* of Vygotsky's ideas, such as the idea of the zone of proximal development and the idea that all meaning is socially constructed (Karpov, 2005). Vygotsky's notion of the zone of proximal development was a sophisticated one, which recurred in other spheres of activity such as play, and which required sensitive mediation by adults. Similarly, although Vygotsky argued that cognitive development originated in socially meaningful activity, he also argued that teachers should teach children directly the knowledge that has been acquired over the course of human socio-cultural evolution. He did not argue that children should be always scaffolded and supported to discover this knowledge for themselves.

These ideas have received new impetus from investigations of 'natural pedagogy' and the 'pedagogical stance' by Gergely, Csibra and their colleagues. For example, Csibra and Gergely (2006, 2009) proposed that humans are adapted to transfer relevant cultural knowledge spontaneously to conspecifics. Younger humans are adapted to fast-learn the contents of such teaching, via a species-specific social learning system that they called 'pedagogy' (see Chapter 3). For Csibra and Gergely, 'pedagogy' is a neglected aspect of the triadic interactions with objects that infants and their caretakers engage in. One function of these triadic interactions is *epistemic*, namely to provide infants with reliable, new and relevant information about the objects concerned (Csibra & Shamsudheen, 2015). Knowledge transfer during these triadic interactions is thought to be triggered by communicative (ostensive) cues such as eye contact and the use of Parentese. Although not part of Csibra and

Gergely's argument, it is easy to imagine that this species-specific social learning system also enables cultural knowledge transfer to older children via collaborative learning in the zone of proximal development. As Gergely, Egyed and Kiraly (2007) point out, such a social learning system requires a default assumption about other agents, which is that they are trustworthy and benevolent sources of universally shared cultural knowledge. As children get older, they become more selective about who is regarded as trustworthy (Harris, 2007). Nevertheless, this social learning system clearly overlaps conceptually with key elements of Vygotsky's theory.

Another critical aspect of Vygotsky's theory for cognitive development is the idea that language and other *symbol systems* play a causal role in organising cognition. For Vygotsky, language is as important as action. In Western psychology, there has been less emphasis on the causal role of language as a symbol system, as the focus has been more on how children discover knowledge for themselves, primarily via action. This is beginning to change, however. In Chapter 7, we reviewed experiments on the importance of explanation by verbalising knowledge for children's developing causal understanding. Being asked to generate verbal explanations for causal phenomena appears to be one important way that children enrich and make explicit their causal knowledge (Legare & Lombrozo, 2014). Clearly, language as a symbol system here plays a key developmental role. Vygotsky's emphasis on the importance of interpersonal communication for learning in young children has other parallels with Western cognitive developmental work. For example, in areas such as theory of mind and social cognition (see Chapter 4), the importance of *communication* within families and with peers is a central aspect of empirical investigation (e.g., Dunn et al., 1991a). Vygotsky's emphasis on the world of the imagination and the value of socio-dramatic play, including in young children's education, is valued even by those who reject a unique causal developmental role for fantastical pretend play (Lillard et al., 2013). In their analytical review, Lillard et al. state "Despite the poor state of the evidence on pretend play's benefits... research does not advocate what is offered as the only alternative to a playful approach in educational settings, adult-centred instruction" (p. 26). They add "The hands-on, child-driven educational methods sometimes referred to as 'playful learning'... are the most positive means yet known to help young children's development" (p. 29). However, Vygotsky's ideas about how play will lead to an appetite for school are less compelling.

THEORETICAL FRAMEWORKS RELATED TO COGNITIVE NEUROSCIENCE

KEY TERM

Neuroconstructivism Explains the mechanisms of cognitive change by considering the biological constraints on the neural activation patterns that comprise mental representations.

Throughout this book, we have seen that new insights from cognitive neuroscience are requiring us to adapt traditional theoretical models of aspects of cognitive development. As anticipated in the Foreword, traditional explanatory constructs such as 'constraints on learning', 'modules' and 'domain-general' versus 'domain-specific' analyses of learning are changing in light of new understandings, particularly concerning the distributed nature of neural (mental) representations. Indeed, the machine learning approach that produces outputs closest to the human brain, deep learning, relies on distributed representations, not symbolic representations. Clearly, cognitive neuroscience coupled with advances in machine learning may in time offer an alternative theoretical framework for explaining cognitive development. Deep learning networks have architectures that can involve billions of connections between units and hundreds of millions of weights on those connections, and the child's brain is likely to have equivalent complexity. Deep learning systems can also in principle capture the role played by cognitive factors internal to the child, such as imaginative play and inner speech, in cognitive development. Indeed, current deep learning systems are already capable of representing the thoughts represented by words in a sentence. Language by its very nature appears to be an amodal symbolic system, yet deep learning models of language comprehension based on distributed representations are already capable of highly skilled performance. As emphasised by LeCun et al. (2015), this skilled performance is not based on internal symbolic representations (e.g., 'words'), but on distributed knowledge. Indeed, LeCun et al. conclude that current deep learning models are more compatible with the view that everyday reasoning involves many simultaneous analogies that each contribute plausibility to a conclusion, such as a conclusion about what a word means. Therefore, while language clearly provides a means for organising the child's inner mental life, and creating (as Vygotsky put it) a 'time field for action' extending both backwards and forwards, at the level of mechanism this explanation is simply a good metaphor.

At the time of writing, it seems plausible that in the next decade, artificial intelligence will become able to deal with inputs like language and imaginative play, inputs generated by the child and then in turn responded to psychologically by the child. This will occur not because these processes will be specified by the engineer, but because they will be extracted by a learning system that learns in the same natural learning environments that are available to children. However, to date the possible insights about cognitive representations offered by deep learning have not been applied to cognitive development. I will therefore primarily review two related but older frameworks here, 'neuroconstructivism' and 'connectionism'. Finally, I will consider the implications of the experiments discussed in this book concerning a new theoretical framework for cognitive development based on cognitive neuroscience. It should be noted that neither connectionism nor neuroconstructivism can incorporate the critical roles of social-cognitive factors on cognitive development, such as intention-reading and communication, whereas a deep learning system is *in principle* capable of incorporating such factors.

Neuroconstructivism

The important contribution of neuroconstructivism to theorising about cognitive development was to consider explicitly the biological constraints on the neural activation patterns that comprise mental representations. These biological constraints affect brain development, and hence will affect the development of the neural substrates underpinning mental representations as well as the mechanisms that contribute to cognitive change. Neuroconstructivists argued that environmental experiences are key to development, as these experiences will change the brain's 'hardware' (grey and white matter), leading to changes in the nature of cognitive representations. In turn, these changes will lead to new experiences and thus further changes to neural systems (Westermann et al., 2007, p. 75; see also Mareschal et al., 2007). By acknowledging the key role of the environment (broadly defined), neuroconstructivism avoids neural reductionism. "Algorithm and hardware change each other in development... they cannot be studied in isolation" (Westermann et al., 2007, p. 75).

Neuroconstructivism is based on the notion of biological 'constraints', but describes these as constraints on *development* rather than as constraints on learning (Table 11.2). One such constraint is the biological action of genes. Genes cannot turn themselves 'on' or 'off', but require signals to tell them to do so. These signals can originate within the cell, outside the cell, or outside the organism. Therefore, as discussed in the Foreword, genetic activity is modified by neural, behavioural and external environmental events (epigenetics). To describe cognitive development adequately, the theory argues that all of these interactions must be understood. For example, low levels of the neurotransmitter serotonin are associated with depressed cognition. Maternal and infant serotonin levels are correlated in monkeys, and also correlate with insecure attachment. Hence it is plausible that low serotonin could be associated with aberrant maternal care or with shared genes (or both). Neuroconstructivism argues that a full understanding of these interactions is required to explain the development of disorders of cognition such as depression (Gottlieb, 2007).

A second important constraint identified by neuroconstructivism is 'encellment', the fact that the development of neurons is constrained by their cellular environments. Neural activation patterns themselves constitute 'experiences' for the neural networks involved, and this experience of connection patterns (plus underlying morphology) is hypothesised to affect the ways in which progressively more complex representations are formed. A similar biological constraint is 'enbrainment', which refers to the fact that the functional properties of brain regions are constrained by their interactions with other regions, for example via

KEY TERMS

Encellment
The development of neurons is constrained by their cellular environments.

Enbrainment
The functional properties of brain regions are constrained by their co-development with other regions.

TABLE 11.2 Biological constraints on neuroconstructivism, following Westermann et al. (2007)

Constraint	Example
Genes	Genes do not encode structure deterministically; e.g., gene expression is influenced by the environment
Encellment	Development of neurons is constrained by the cellular environment
Enbrainment	The brain is not really modular, as the functional properties of individual regions are constrained by their connectivity
Embodiment	The brain is in a body, which is embedded in a physical and social environment
Ensocialment	The social and physical environment constrains the emergence of neural representations
Interactions between constraints	The interactive working of these constraints shapes the neural structures that form the basis of mental representations

feedback processes and top-down interactions. These interregional interactions will affect the development of the neural structures involved and therefore of neural representations. 'Embodiment' is an analogous constraint. Embodiment refers to the fact that the brain is in a body, and that the body will act as a 'filter' for information from the environment. The ways in which the senses function will hence constrain the development of mental representations. The body also allows the infant to manipulate the environment. "The embodiment view emphasizes that pro-activity in exploring the environment is a core aspect of cognitive development... the child... selects the experiences from which to learn" (Westermann et al., 2007, p. 78).

The final constraint on development discussed by Westermann et al. (2007) is called 'ensocialment'. Social aspects of the environment will have profound effects on social and behavioural development, for example by affecting gene expression. Westermann et al. (2007) argue that putting these constraints together (and acknowledging that they will interact with each other) constitutes a theoretical framework for cognitive development. "Put together, in the development of cognitive processing, these constraints form an interactive network shaping the neural structures that form the basis of mental representations" (p. 79). Although the proactivity of the child is stressed theoretically, the emergence of the neural representations supporting cognitive behaviour is said to be strongly constrained by the ontological history of the individual: "The events occurring at a given time constrain the range of possible adaptations available to the system in the future" (Westermann et al., 2007, p. 80). The authors note that this notion of *progressive specialisation* is shared with constructivist theories of cognitive development, such as Piaget's theory.

Neuroconstructivism has many appealing features but it does not really provide a *theory* of cognitive development. However, its focus on the biological constraints that affect the development of the neural structures and neural networks that underlie cognitive processing is very important, and this focus fits well with the viewpoint adopted in this book. These biological constraints necessarily impact cognitive theories. A good example is theories of conceptual development, as discussed in Chapter 5. In Chapter 5 I reviewed some neuroimaging studies showing the distributed nature of conceptual representations. Cognitive neuroimaging studies with adults now enable voxel-level precision regarding which neural networks are active during the processing of specific concepts. Further, we can reverse-engineer this knowledge and predict which concept has just been activated in the mind of a person on the basis of which voxels in the brain are active. I suggested that developmentally, such data mean that there is no amodal or abstract 'concept' for entities such as 'robins' or 'cups' that is held in a separate conceptual system. Rather, the 'cup' network would be activated to different degrees depending on the context in which the 'cup' concept was activated. I also suggested that such a modality-specific system would lend itself naturally to development. As infants and children gained more conceptual knowledge about the world, modality-specific knowledge would change, and so would the activity weightings between neurons or networks. Children's concepts could thus naturally differ from those of adults, depending on the sensory and social/emotional information being attended to. This kind of explanatory framework for conceptual development would arise naturally from neuroconstructivism. Nevertheless, mapping such correlations does not in itself illuminate the *mechanisms* of conceptual development.

Another appealing feature of neuroconstructivism is that developmental disorders are explained via altered constraints on brain development that thereby alter the developmental trajectory (Karmiloff-Smith, 2007). This idea also fits the viewpoint adopted in this book. For example, as discussed in Chapter 10, the core problem in developmental dyslexia is a (cognitive) phonological deficit. This cognitive deficit may develop because children with developmental dyslexia have brains that are less efficient in auditory sensory processing, for example in discriminating amplitude envelope rise times. This sensory constraint would affect the automatic extraction of the amplitude modulation phase hierarchy that is present in infant- and child-directed speech (a statistical structure), which would affect the cognitive development of phonological awareness (Goswami, 2015). Altered auditory sensory functioning would thus act as a constraint on the developmental trajectory. Indeed, in Chapter 10 I suggested that such altered sensory functioning could be expected to affect the development of neural structures, such as the arcuate fasciculus (a white matter tract) from infancy. The integrity of the arcuate fasciculus has been related to language acquisition. The altered constraint on brain development in cases of dyslexia (the atypical auditory sensory processing) would thus affect structural brain development from the get-go. Further, affected children would not develop well-specified phonological representations as part of natural language acquisition, as this aspect of statistical learning (the amplitude modulation phase hierarchy statistics) would be impaired by the different sensory information received by the brain. Thus the structure and function of the brains of children with a sensory impairment in amplitude rise time discrimination would already differ from those of other children prior to schooling and tuition in reading.

Another important theoretical implication of neuroconstructivism is that adult cognition can only be fully understood by considering cognitive development. Adult cognition is an outcome of development, and so developmental constraints must be taken seriously in formulating adult cognitive theories. This is a central message of this book. The importance of developmental constraints has also been argued in specific areas of adult cognition, such as skilled reading (see Ziegler & Goswami, 2005, for a discussion of developmental 'footprints' in the behaviour of skilled readers). Ziegler and Goswami (2005) also discussed the shortcomings of the 'dual route' model of reading, originally proposed on the basis of adult patient data, as a model of reading development.

One drawback of neuroconstructivism is that it relied heavily on analogies from sensory neuroscience. Constraints such as encellment and enbrainment have been largely derived from the behaviour of cells and neural networks in the visual cortex. For example, if axons have committed themselves to particular connections, they cannot become uncommitted. As *cognitive* neuroscience gathers pace, different constraints may become apparent. For example, for the cognitive system it is not clear that "The events occurring at a given time constrain the range of possible adaptations available to the system in the future" (Westermann et al., 2007, p. 80). This is actually unlikely to be true, as both cognitive neuroscience and machine learning (deep learning) show that cognitive knowledge is highly distributed.

Accordingly, the cognitive system can always adapt, sometimes to the detriment of the individual, as in certain mental health disorders. Further, deep learning models show that one part of the system can decide to clear the contents of another part of the system (akin to particular 'memories') if that leads to more effective long-term learning for the system as a whole. This is an automatic outcome of the learning algorithms used by the models, but it has important implications for cognitive development. In fact, axonal commitment may not even be true of the sensory systems that are traditionally used as examples in sensory neuroscience, such as visual development in kittens. The development of visual cortical connections in kittens was thought to be irreversibly constrained by rearing environment (e.g., Blakemore & Cooper, 1970), but was later shown to be more malleable than originally supposed (e.g., Rauschecker & Singer, 1981).

In the Foreword, I also discussed examples from adult cognitive neuroscience in which primary sensory processing of input (such as flashes of light or simple tones) was rapidly modulated by experience (Noesselt et al., 2007). Here the brain was shown to be responding to *abstracted dependencies* that it had learned were present in the environment, rather than encoding the exact particular sensory features of each individual environmental event. This automatic abstraction process was also discussed in the early chapters of this book in terms of prototype formation. Visual neuroscience researchers have shown that perceptual encoding mechanisms adapt automatically and rapidly to changes in the statistical patterns presented by environments, even though the conscious observer is not aware of these changes nor these adaptations (Hochstein et al., 2015). This is because a core job of the brain is accurate prediction. Learning that is relevant to cognitive development is likely to be similar. There will be top-down modulation of sensory processing by expectations built up through the experiences of the infant and young child, and these expectations will rapidly affect the neural encoding and processing of new sensory and environmental information.

The fundamental theoretical point is that there is always a role for learning in human cognition. Indeed, recent cultural innovations have required our cognitive systems to become adaptive in novel ways (e.g., in using new technologies). Pensioners can learn to use computers and mobile telephones. Obviously, biological constraints will affect how new interactive networks are formed during this new learning, and how the neural structures that form the basis of new mental representations are shaped. But a more efficient level of theoretical description may be a cognitive one. For example, individual differences in whether pensioners become efficient computer users are likely to reflect factors such as encapsulated knowledge, motivation to learn and self-belief. Describing these cognitive variables may be as useful (or more useful) for theories of cognitive development than trying to understand every aspect of encellment or enbrainment.

Finally, there are ample data showing that novel inputs late in life can alter neural structures. An example is the development of the hippocampus in London taxi drivers (see Goswami, 2004, for more examples). London taxi drivers must pass an examination to demonstrate detailed knowledge of the street map of London, called 'The Knowledge'. London taxi drivers show enlarged hippocampal formations compared to adults who do not drive taxis, and hippocampal volume is correlated with the amount of time spent as a taxi driver (Maguire et al., 2000). Although neuroconstructivism emphasises the proactive child, a comprehensive theory of cognitive development requires an analysis of whether 'the range of possible adaptations available to the system in the future' is *ever* restricted for the cognitive system. The effects of different types of learning on the brain (e.g., learning by imitation versus learning by analogy), the important roles of motivation and metacognition, the role of representations internal to the child that themselves affect cognitive development (such as language), will be greater than those captured by constraints such as 'ensocialment'. This caveat is discussed further below, with respect to connectionist models.

Connectionism

Whereas neuroconstructivism aims to capture development, connectionism is concerned with learning. Connectionist models provided early computational models of cognition. The computational architecture in these models was built from networks of simple processing units. Each unit had an output that was a simple numerical function of its inputs. This was intended as a loose representation of a neural network. Cognitive entities such as concepts or aspects of language were represented by patterns of activation across several units, just as representations are distributed in the brain. Early connectionist models were bottom-up, in that information could only flow forwards in the models (Figure 11.3). The input units received *experimenter-determined information* intended to be representative of the input of a given cognitive type, such as language. The input was thus dependent on the expertise of the programmer in the domain being modelled. This information was fed forwards to a layer of 'hidden' units, a layer of units lying between the input and output units that learned features of the input and fed the output units. A major contribution of early connectionist models was the *in-principle demonstration* that a simple network could learn the structure of the input (e.g., linguistic structure). This showed that structure could be learned in the absence of innate knowledge (e.g., about language).

Connectionist models were then developed that not only fed information forward during learning, but allowed 'back-propagation' as well. Now the hidden units could affect input units as well as output units, leading to the development of simple recurrent networks and simple interactive activation networks (see Figure 11.3). In adult cognition, such networks were used to demonstrate further important *in-principle effects*. One was that bottom-up processing could generate effects that had been assumed from behavioural data to require top-down processing. For example, phoneme restoration (i.e., recreating a 'missing' phoneme that has been obliterated by someone coughing when hearing speech) was shown to be possible in the absence of knowledge about real words if a network was trained on statistical regularities at the phoneme level (Pitt & McQueen, 1998). Hence, in principle, phoneme restoration need not be a top-down, lexical effect – although of course, this in-principle demonstration does not mean that phoneme restoration is *not* a top-down effect. Similarly, it was shown that words could be segmented from the speech stream on the basis purely of bottom-up, statistical information (Christiansen, Allen,

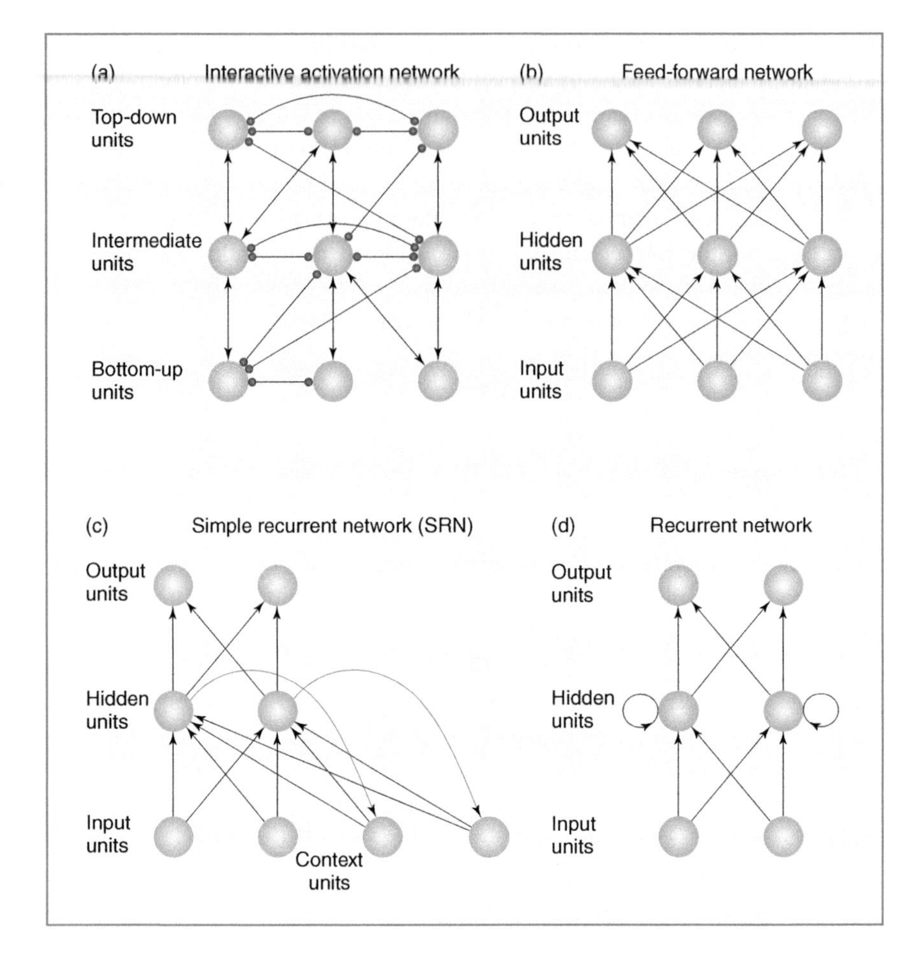

FIGURE 11.3 Network architectures used in connectionist psycholinguistics. From Christiansen and Chater (2001). Copyright © 2001 Elsevier. Reproduced with permission.

& Seidenberg, 1998). The importance of such acoustic statistical learning had an enormous impact on infant studies (see Chapter 6). Connectionist models of adult speech processing, sentence processing, language production and reading aloud converged to show that aspects of language and language processing can be successfully captured without assuming discrete symbolic representations for concepts or words (Christiansen & Chater, 2001). Prior to connectionism, most cognitive theories assumed symbolic representations, which were discrete or amodal in form and all-or-nothing in terms of functionality (they were either engaged during a cognitive task, or not). This is the metaphor of the 'algebraic' mind (see Elman, 2005).

Early connectionism was thus important for foregrounding the importance of distributed and graded representations. Distributed representations have been discussed throughout this book, but graded representations are also important. A representation can be graded in terms of (for example) the number of relevant neurons firing, their firing rates, the coherence of the firing patterns, and how 'clean' they are for signalling the appropriate information (see Munakata, 2001). These types of information are represented by different connection weights between units. Graded representations offer an alternative explanation of behavioural dissociations, for example in infant looking and search behaviour (see Munakata, 2001, for more examples). As we saw in Chapter 2, babies appear to attend to certain variables such as height at different ages in occlusion events (four months) versus containment events (seven months) versus covering events (12 months). Rather than postulating separable cognitive systems for, say, occlusion events versus containment events, it could be that a weaker (lower-grade) representation of a hidden object suffices for occlusion events but not for containment events. Alternatively, as also discussed in Chapter 2, there will by necessity be some differences in the spatio-temporal dynamics of events like containment versus covering. If dynamic experiences like these are encoded by *trajectories* of neural activation over time and cortical space, then these neural trajectories would also by their very nature be contrastive – as they would differ at the point of encoding. In principle, graded representations can also explain apparent double dissociations, when one group succeeds on task A but not task B, whereas another group succeeds on task B but not task A. Double dissociations are seen in adult cognitive neuropsychology as evidence for functional separation. Connectionism shows that rather than indexing separable systems, these double dissociations could, in principle, arise from graded representations (see Munakata, 2001).

Early connectionist models were also able to demonstrate computational trade-offs that may be relevant for explaining the development of different types of cognitive system, such as semantic memory versus episodic memory. For example, a single system can either learn rapidly, with non-overlapping representations, or slowly, using overlapping representations. A system that learns rapidly could be crucial for the development of episodic memory (Munakata & McClelland, 2003). As discussed in Chapter 8, episodic memory is memory for episodes or events in one's life, and involves explicit recall of these episodes and events, for example meeting a particular person at a particular place. Episodic memories must be encoded quickly and kept distinct from other episodic memories, such as meeting another person at the same place. In contrast, semantic memory is our generic, factual knowledge about the world, such as our knowledge of concepts and language. These kinds of memories are better learned slowly, with overlapping representations that collapse across particular instances. Munakata and McClelland (2003) argued that different aspects of the memory system might rely on different types of learning, and hence show the observed specialisations (for example, the hippocampus might specialise in fast learning while posterior cortex specialises in slow learning). This is an important in-principle idea, and as we saw in Chapter 8, this idea still governs some memory research. However, as also discussed in Chapter 8, developmental data do not necessarily fit easily with this particular principle. Vargha-Khadem and colleagues (1997) showed that children with hippocampal damage have preserved semantic memory, but grossly impaired episodic memory. These dissociations are explicable in terms of the kind of learning required for each type of memory system, but not if the hippocampus is hypothesised to specialise in fast learning.

Connectionist models of aspects of cognitive development have also been important in demonstrating *in-principle* effects regarding the nature of learning by naïve systems. For example, 'critical period' effects, when learning appears to be particularly effective during a particular developmental time window, can arise in neural networks that rely on error-driven learning (via the back-propagation algorithm;

see Figure 11.3) even though the underlying learning mechanism remains constant in these models. A 'novice' back-propagation system is very responsive to learning from its errors, whereas an 'expert' system is less responsive, hence more entrenched in its learning. This feature was used to explain why it is easier to learn a first language early in life than a second language later in life (see Johnson & Munakata, 2005). During learning of the first language, learning becomes entrenched, and so the system is less sensitive to errors when learning the second language. On the other hand, if both languages are acquired at the same time, before learning has become entrenched, both languages are learned easily.

The error-correction learning algorithm underlying entrenchment is only one possible form of learning that can be adopted by connectionist models. If a model depends on 'Hebbian' learning (units that 'fire together, wire together', so that the response that the system makes to received inputs is strengthened over time), it becomes difficult to reorganise the system when a new learning environment is experienced. This form of entrenchment is suggested to underlie the critical period effects found for learning native phonemes (see Chapter 6). For example, Munakata and McClelland (2003) suggested that this kind of entrenchment could explain why Japanese native speakers find it difficult to learn the English /l/–/r/ distinction. These two phonemes are the same (single phoneme) in Japanese, and Hebbian learning would produce a single representation for this phoneme, which would continue to be strengthened by input comprising *both* English /l/ and /r/. Hence the system would never learn to distinguish English /l/ and /r/. However, it is possible behaviourally to train Japanese adults to recognise the English /l/–/r/ distinction using auditory-visual training methods (Hardison, 2003). This suggests that entrenchment is not a necessary characteristic of linguistic critical period phenomena in humans.

Another important in-principle demonstration was that so-called 'U-shaped curves' in development, such as the apparent decline and then recovery in the accuracy of young children's use of the past tense rule (see Chapter 6), can arise from item-based learning (Rumelhart & McClelland, 1986). There was thus no need to invoke symbolic rules of grammar. The statistical structure of the input turned out to be sufficient for non-monotonic changes in performance when the frequency of different forms was taken into account. Such demonstrations led to important debates about the nature of the input in language learning, and about whether there was sufficient information in the input alone (e.g., the language tokens experienced by the child) to explain the complexity of the system that was learned (natural language). In an influential paper, Elman (2005) discussed this general issue in terms of the 'poverty of the stimulus' argument. Prior to connectionism, a popular view was that extra innate 'pre-knowledge' was required to fully explain language learning, as the actual input experienced by the child was too impoverished to explain the complexity of language development. An example of such innate pre-knowledge was the language acquisition device postulated by Chomsky (1957; see Chapter 6). Elman showed that connectionist models can learn compositional and hierarchical structure from learning conditional probabilities in the input, and from using more powerful learning algorithms such as Bayes nets (see Chapter 7 for a discussion of

Bayes nets in casual learning). Connectionist models of language acquisition hence demonstrate that, in principle, the available input is rich enough to enable a simple network that learns according to different statistical algorithms to behave like a child's brain acquiring language.

Elman (2005) identified two challenges for connectionist models of cognitive development. One was to capture social cognition, and the other was to capture the fact that infants and children are not passive learners, but are instead active learners with motivations and desires. Elman made the key point that children may not attend to 'the input' in the same way that the connectionist model does. The input as decided by the engineer of the model is not necessarily equivalent to what is taken in by the learning child. This point draws attention to a larger conceptual point concerning the value of existing connectionist models for cognitive development. These models can only model learning in terms of their inputs, and decisions about these inputs are taken *a priori* by the modeller whose decisions are necessarily limited by current behavioural data and current theories about the structure of the input. For example, we saw in Chapter 5 that Quinn and Johnson (1997, 2000) produced a successful connectionist model of conceptual development that learned the contrastive concepts of animals and furniture. From information about (for example) the different leg lengths associated with tables, chairs, dogs and cats, the model first devoted large numbers of hidden units to coding the 'global' conceptual level of animals versus furniture. Subsequently, the model devoted more and more hidden units to distinctions in attributes such as leg length that yielded the 'basic' level of categorisation. As noted in Chapter 5, however, the infant brain might be attending to different aspects of the stimuli, aspects that have not been considered by the modellers. Therefore, these aspects will not have been built into the connectionist simulation. The example given for conceptual development was that infants might devote attention to features that afford actions, such as 'being graspable', rather than to features such as 'leg length'. A model that attended to features that afford actions might not first develop a global level of representation – but this is an empirical question.

Further, we know from cognitive neuroscience studies that actions on objects are coded in part by a special neural system, the mirror neuron system. We also know that with growing expertise mirror neurons are eventually tuned to respond specifically to biological actions (Tai et al., 2004 for adults, versus Lloyd-Fox et al., 2011, for infants). As argued in Chapter 5, all of these different aspects of the input must be taken into account when developing an adequate connectionist model of how young children develop conceptual knowledge from their real-world experience. This is why the in-principle demonstrations from connectionism concerning principles of learning are currently more important for theories of cognitive development than the successes of particular models in supporting particular theories (e.g., in suggesting a 'global to basic' shift in infant categorisation; see Chapter 5). Another example of this general point about the nature of 'the input' concerns language acquisition. As discussed in Chapter 6, Kuhl et al. (2003) showed that babies did not learn language from the television. Even though the babies in her study of non-native phoneme discrimination received the same amount and quality of

(Mandarin Chinese) speech 'input' from videos as the babies who were interacting with real people, being exposed to this acoustic and visual speech 'input' without a live person present was not sufficient for learning. A real person to exchange gaze with also appeared to be important in terms of how the brain learns language.

Accordingly, although early connectionist models of development were clearly limited in a number of ways, they did show that in principle, aspects of cognitive development can be captured by applying statistical learning algorithms to complex input. Most of the limitations were related to the design decisions made by the modellers when constructing a given connectionist network. For example, the initial state of 'knowledge' of the network, the nature of its input and output representations, the patterns of connectivity that it could form, and the learning algorithms that it used all limited the potential insights regarding cognitive development. In the previous edition of this book, I concluded that the optimistic view would be that once all of these modelling parameters were understood in detail, then connectionist modelling would enable a comprehensive understanding of cognitive development. In practice, the past decade has produced advances on all these fronts.

Current deep learning models use general purpose learning algorithms to learn from natural input, for example processing a million visual images of natural scenes or processing real conversational human speech. Deep learning models use both feed-forward and feed-back architectures, they have multiple hidden layers, they use non-linear transformations between layers, and they compute both local and global statistics. Perhaps most importantly of all, their architectures can involve billions of connections between units and hundreds of millions of connection weights, just like the human brain. On some estimates, the infant brain at birth has 100 billion neurons and during infancy and early childhood, the brain is making around one million new connections per second. By these estimates, it is clear that older connectionist models cannot hope to provide comprehensive models of real-time learning, whereas deep learning models do appear to hold out this promise. Currently, deep learning is largely being applied to problems with commercial pay-offs, such as developing driverless cars and developing 'helper' robots who can take care of elderly people. However, in the future deep learning models are also likely to be applied to cognitive development, the pre-eminent natural case of 'unsupervised' learning.

As we have seen throughout this book, children discover the structure of the world by observing it and acting upon it. They build their own cognitive systems via experience and because they develop in a cultural context affording natural pedagogy. Soon, deep learning systems will be able to model this too. Reverse-engineering their sophisticated computations to discover *how* the discoveries that children make naturally are supported algorithmically may take rather longer, however. An interesting question for theories of cognitive development is whether unsupervised deep learning systems will turn out to develop internal mediators of cognitive processes just as children do, internal mediators such as inner speech, a notion of psychological 'essences', and an imagination. As Vygotsky made clear, part of the 'input' for human cognitive development is generated internally to the brain, by the child's own mind.

COGNITIVE NEUROSCIENCE AND COGNITIVE DEVELOPMENT: THE WAY FORWARD

How can the new methods offered by cognitive neuroscience and machine learning combine with the behavioural methods and important insights contributed by the traditional cognitive developmental theories of Piaget and Vygotsky? We have seen a number of exciting possibilities in this book. In my view, our field will advance most rapidly if cognitive neuroscience studies of cognitive development respect important experimental constraints related to applying brain imaging to child development. After discussing these constraints, I will summarise what we know now from cognitive neuroscience that is relevant to cognitive developmental theories. I will end by considering what we may become able to know in the future.

Experimental designs for interpreting age-related differences in brain activity

Longitudinal research designs

The first key experimental point to make is that longitudinal studies are of paramount importance in order to interpret age-related differences in brain activity. Neural structures and neural activity will always change with age, and so in order to interpret age-dependent differences in brain data, we need to be observing the *same brain* at different developmental stages. For example, we saw in Chapter 10 that in the area of developmental dyslexia, longitudinal studies are emerging which measure the brain both before children at family risk for dyslexia begin learning to read, and then at subsequent points as reading as acquired. Clark et al. (2014) reported a relevant study, in which structural MRI scans of Norwegian children at high versus low risk of dyslexia were taken the year before reading was taught (at age 6–7), a year after reading tuition began (at age 8–9) and again after dyslexia had been diagnosed (at age 11–12). Clark et al. did not find structural differences in any of the neural regions typically thought to comprise the 'reading network' in their pre-reading scans, namely left-lateralised frontal, temporoparietal and occipitotemporal regions. Instead, the pre-reading neuroanatomical regions that differed significantly in structure were in sensory areas, specifically cortical thickness in primary auditory and visual cortex. The only neural structure in which group differences were *consistent* over development (in that group differences were found at all three scan points) was primary auditory cortex. Heschl's gyrus was still significantly thinner in the children with dyslexia at the end of the study. These structural data are consistent with a sensory basis for impaired reading in impaired auditory processing, leading to impaired phonological development, as discussed in Chapter 10. In Clark et al.'s study, group differences in the left-lateralised 'reading network' did become evident at age 11, indicating that the 'reading network' emerges with growing reading expertise.

Another example of longitudinal research designs in dyslexia comes from the longitudinal MRI study of US children at family risk for dyslexia that focused on white matter pathways (Wang et al., 2017b). Wang et al. reported that white matter integrity in the left arcuate fasciculus was significantly lower for the at-risk children when they were pre-readers. Ten of the children subsequently developed into poor readers, and these ten children showed a significantly lower rate of development of white matter integrity in this pathway compared to children who subsequently became good readers. Wang et al.'s study showed that there are also some structural differences in the brain in pre-reading at risk children within the core 'reading network'. However, as noted earlier, these structural differences could still be the result of differing input to the brain prior to reading. For example, as key parameters of the speech signal are perceived differently by at-risk versus not-at-risk infants (see Kalashnikova et al., 2018a), these differences in sensory processing could cause the differences in how the neural structures develop. Logically, impaired processing of amplitude envelope rise time by the neonate would affect how the white matter pathways comprising the arcuate fasciculus develop. Longitudinal research designs incorporating both sensory and neural measures at multiple time points prior to reading tuition are required to find out. It should also be noted that both of these longitudinal studies included relatively small samples of children.

A larger sample of older children contributed data to a longitudinal study of neural changes associated with the development of reasoning by analogy reported by Wendelken et al. (2017). When the same children were compared at time 1 and time 2, it was structural connectivity that predicted future improvements in analogical reasoning, and not functional connectivity. Interestingly, functional connectivity did not predict future changes in structural connectivity. Wendelken et al. (2017) concluded that the integrity of white matter connections between parietal and frontal cortex is important for developmental changes in reasoning ability. Again, however, we are left with the question of which factors caused these observed changes in structural connectivity. In Wendelken et al.'s study, the maximal change in functional connectivity occurred at age 13 years. The maximum rate of increase in structural connectivity occurred at age seven years. Further studies exploring time-lagged relationships of this nature may help to explain what these patterns of structural and functional connectivity mean regarding the development of the fronto-parietal network that is activated by analogical reasoning tasks.

Finally, we can consider the longitudinal fMRI study of executive function reported by Long et al. (2017). Long et al. investigated functional connectivity and changes with age in 44 children aged between two and five years at their first brain scan. Subsequent scans were taken at six-monthly intervals, although as not all children contributed all scans, only 21 children contributed longitudinal data. Children watched videos that they chose themselves during scanning. Analyses of functional connectivity showed that both local connectivity and global connectivity increased with age, in frontal, parietal and cingulate areas. In particular, the fronto-parietal network typically activated by executive function tasks in older participants showed age-related increases in connectivity from two to six years. Long et al. concluded that the development of functional connectivity in early childhood is likely to underlie major advances in cognitive abilities. However, although these changes in functional connectivity are interesting, the data show correlational relationships with the development of executive function rather than a causal relationship. Indeed, it should be recalled that Buss and Spencer (2017) reported opposite results in a study of executive function (using the colour/shape DCCS) with children aged three and four years. Buss and Spencer used both 'easy' and 'hard' versions of the DCCS, and brain activity was compared for switch trials in the easy versus hard versions. Regarding frontal activity in the children who were good at the easy version of the DCCS but poor at the more difficult version, Buss and Spencer reported that frontal activation was weak in three-year-olds when measured during the difficult version of the task, but strong in these same three-year-olds when they were performing accurately in the easy DCCS task. Accordingly, these cross-sectional data suggest that neural activity in frontal cortex depends on whether the child can successfully perform the EF task being used during scanning, rather than determining a child's performance. Even for longitudinal studies, disentangling cause and effect in brain changes will require careful parametric variation of the paradigms used and of task difficulty, with training designs being the optimal choice (see below).

Equating in-scanner task performance across ages

If longitudinal work is not possible, then it is important to equate in-scanner behavioural performance across group differences in age. There are a range of options for doing this. For studies of sensory processing, experimental paradigms can be adjusted so that all participants are performing at sensory threshold (that is, at the just-noticeable-difference level for each child). For example, in their rhythmic listening study with children with dyslexia discussed in Chapter 10, Power et al. (2013) calibrated the detection of rhythmic violations individually for each child, so that performance accuracy for both dyslexic and control children was 79.4%. This enabled group differences in the EEG recordings to be interpreted unambiguously. The EEG data showed that the dyslexic children showed a different preferred phase in the delta band oscillatory brain response for the auditory and auditory-visual conditions only. The analyses suggested that the peak dyslexic brain response was 'out of time' for the children with dyslexia, hence the maximal neural response was not matched to the maximal information points in the incoming signal as well as for control children. This atypical neural response would affect the quality of automatic auditory statistical learning of the amplitude modulation phase hierarchy that is important for phonological development (see Chapter 10).

For studies using behavioural rather than sensory measures, children can either be trained so that they are performing at the same level in the task that will be used in the scanner, or a task can be selected that is *a priori* known to be as easy for younger children as for older children. For example, in their fMRI study of normative reading development, also discussed in Chapter 10, Eden and her colleagues used a 'false font' task that was performed at an equivalent level by seven-year-old children and by adults as a way of measuring 'implicit reading' processes (Turkeltaub et al., 2003). Turkeltaub et al were able to show that as reading skill developed with age, activity in left temporal and frontal areas of the brain increased, while activity in right posterior areas declined. This pattern suggested that reading-related activity in the brain became more left-lateralised with development. Although these data do not show the mechanisms involved in these changes, mapping a template of normative development is very useful when it provides a comparison for indexing atypical development of a particular cognitive skill. Turkeltaub et al. (2003) were also able to match differences in neural activation in different brain regions with performance in key behavioural measures related to reading development, such as phonological awareness. These behavioural measures were taken *outside* the scanner.

A study in which children received training prior to receiving the same in-scanner task has also been contributed by Eden and her colleagues (see Chapter 10). In a study of children with developmental dyslexia, Olulade et al. (2013) were interested in the theory that dysfunction of the magnocellular system for processing visual motion may be a sensory cause of dyslexia. They provided a phonological training intervention lasting eight weeks to children with dyslexia, and tested visual motion processing using fMRI before and after training. Olulade et al. reported that neural activity during motion processing increased for the dyslexic children following the phonological intervention. They also found no difference in neural activity for visual motion processing between children with dyslexia and reading level matched control children. The training data suggest that the reading problems experienced by children with dyslexia are not caused by neural magnocellular dysfunction. Rather, atypical motion processing appears to be associated with dyslexia because of the extended reduction in reading experience that accompanies being dyslexic.

Focusing on resting state activity

A third option is to measure resting state brain activity at different ages, and then to correlate resting state activity with performance in behavioural tasks measured outside the scanner. This method enables developmental changes in connectivity to be related to developmental changes in behavioural performance, and has been widely adopted for studies of working memory. As we saw in Chapter 8, both structural MRI and functional MRI studies have correlated age-dependent performance in episodic and working memory tasks with differences in the brain at rest. For structural brain differences, Bathelt et al. (2017) measured structural characteristics of the brain at rest in children of different ages, focusing on white matter integrity and cortical thickness. They reported that myelination increased with age, reflecting the maturation of structural connections and growing integration within brain systems. Working memory capacity assessed outside the scanner was more closely associated with large white matter connections for younger children, but with cortical thickness in the left posterior temporal lobe for older children. Older children thus showed a more adult-like activation of specialised and localised neural structures.

Regarding functional data, Riggins et al. (2016) measured the default mode network at rest in children aged four and six years, and then measured episodic memory outside the scanner with a task involving toys performing memorable actions at different locations. Both groups of children performed well on the memory task, but different correlations were found by age with the resting state network measures. For the four-year-olds, better episodic memory was related to more connectivity within the middle temporal gyrus and frontal areas, and less connectivity within hippocampal areas. For the six-year-olds, better episodic memory was related to less connectivity within these frontal and temporal areas, and more connectivity within hippocampal areas, a more adult-like pattern of activation. Again, while these neural data do not reveal the episodic memory *mechanisms* underlying developmental changes, mapping a template of normative development is very useful as a comparison for atypical development.

Increasing the study of brain mechanisms of information encoding and information processing

Neural statistical learning and the structure of neural representations

Throughout this book, I have stressed that a key consideration in applying cognitive neuroscience methods to cognitive development is to design neuroimaging studies that focus on neural *mechanisms* of encoding and processing information. For example, the mechanisms that encode the types of sensory information that appear to underpin key aspects of cognitive development is one important target, as in the case of spatio-temporal dynamic events that specify animacy or agency. A second key target is *sensory statistical learning* of visual and auditory information, emphasised in the early chapters in this book. For example, visual sensory statistical learning is the basis of prototype extraction, and we have seen that the brain automatically generates neural representations classically considered to be cognitive, such as prototypes (Chapter 5). Automatic neural visual learning mechanisms generate summary statistics of features that recur across visual scenes (Hochstein et al., 2015, although recall that the data discussed by Hochstein and his colleagues depended on experiments with colours or facial expressions). Nevertheless, in principle 'prototypical' members of different conceptual entities such birds, dogs and trees should also be automatically generated by the brain on the basis of sensory-motor experiences. Designing neuroimaging experiments to reveal the nature of such automatic visual statistical learning has already begun with adults, as discussed in the Foreword (Noesselt et al., 2007). Similar methods could be applied with infants and children, but using real-world inputs.

An important consideration for such cognitive neuroscience studies of development is that automatic statistical learning will only be as good as the sensory information upon which it is based. Hence individual differences in how children's brains process visual, auditory and tactile sensory information will lead to individual differences in the quality of the information that is extracted by these automatic statistical processes. Given that this sensory information is the foundation for developing different cognitive systems, understanding these individual differences is of critical importance. We saw in Chapter 6 that automatic auditory statistical learning generates prototypical phonemes from speech input, as documented in EEG mismatch response experiments with infants (Dehaene-Lambertz & Gliga, 2004). We also saw that the brain automatically computes conditional probabilities between speech sounds, thereby grouping phonemes or syllables into possible 'words', shown, for example, by the phase locking experiments carried out by Kabdebon et al. (2015). A newer set of auditory statistics was also discussed, the hierarchical nesting of changes in intensity (amplitude modulation) at different temporal rates in the speech amplitude envelope, the amplitude modulation phase hierarchy (Leong & Goswami, 2015). The different temporal rates of amplitude modulation were shown to be dependent in timing on each other (phase dependency). This 'amplitude modulation phase hierarchy' was then shown via modelling studies to provide sufficient information for the automatic extraction of phonological units of different sizes with over 90% efficiency by the listening child.

Using such information, it has been possible to demonstrate that individual differences in sensory processing (how well children perceive amplitude modulation information) indeed impact cognitive outcomes in the field of language development. New studies focusing on oscillatory entrainment, the neural mechanism that encodes amplitude modulation information in speech, have documented group differences in such mechanisms in children who have poor versus good phonological development. We saw in Chapter 10 that the dyslexic brain is impaired in detecting changes in amplitude envelope 'rise times', the rates of change in signal intensity that denote different rates of amplitude modulation. These rise times are one of the sensory triggers used by the brain for phase-resetting the neuroelectric oscillations that encode the speech signal (another is the timing of visual speech information). The dyslexic brain also shows atypical phase entrainment to slower rates of amplitude modulations in acoustic speech input, in the delta band (~2 Hz; Power et al., 2013). Accordingly, cognitive neuroscience research has identified a neural mechanism that may underlie the impaired development of phonological awareness by children diagnosed with developmental dyslexia. An initial sensory difference in processing an important acoustic parameter (rise time) is associated with atypical neural entrainment. Over developmental time, this would lead to impaired automatic learning of one key set of acoustic statistics, the amplitude modulations at different rates that are hierarchically nested in the speech amplitude envelope. This in turn would impair phonological development for affected children.

Developmental dyslexia research hence provides a nice example of how automatic neural statistical learning and cognitive development will only be as good as the sensory information upon which it is based. As we saw in Chapter 10, impairments in 'rise time' discrimination have been found across languages in children with developmental dyslexia and, as already noted, a recent study showed that impaired rise time discrimination can also be measured in infants at family risk for developmental dyslexia (Kalashnikova et al., 2018a). Thus it is plausible that one developmental pathway to dyslexia across languages arises from impaired auditory sensory processing of amplitude 'rise times', associated with atypical automatic neural entrainment to the amplitude modulation patterns in the speech signal, which is assumed to affect the automatic extraction of phonology, which then in turn affects the ease of learning to read (see Goswami, 2015, for a full description). Although systematic studies across languages are required to test this 'temporal sampling' framework further, the 'rise time' research provides an example in some detail of how a small difference in initial sensory processing can have a major effect on children's cognitive development (Goswami, 2011).

Furthermore, the knowledge about mechanism gained by brain imaging that we discussed in Chapter 10 offers novel avenues for remediation in dyslexia. We saw that EEG studies showed atypical neural entrainment and impaired low-frequency speech envelope representation for English children with dyslexia, while an MEG study showed weaker oscillatory synchronisation to low-frequency speech information for Spanish children with dyslexia (Power et al., 2013, 2016; Molinaro et al., 2016). Remember that children with dyslexia do develop some phonological awareness and reading skills, but with considerably more effort (less automaticity) than other children, and much more slowly. As with most developmental disorders of learning, learning is not absent, rather it is impaired and delayed, and this may have knock-on effects for other processes, perhaps for example for visuo-spatial attention. Given this information about mechanism, it should be possible to make learning easier for children at risk for developing dyslexia by adjusting the *input*. The available data suggest that we should target auditory input rather than visual input. We can do this at the sensory level by trying to amplify the acoustic aspects of the speech signal that a dyslexic brain appears to be processing less efficiently. As discussed earlier, we can engineer the speech signal to amplify the rise times and the low-frequency envelope information. Such modifications of the sensory input may enable better automatic extraction of the relevant acoustic statistics by the dyslexic brain, such as the amplitude modulation phase hierarchy. This in turn should enhance phonological awareness and thus facilitate the task of learning to read. We can also intervene at the behavioural level, by designing interventions to facilitate rhythmic synchrony across modalities (Bhide, Power & Goswami, 2013).

As this extended example drawn from the dyslexia studies shows, without a clear understanding of the *statistical structure of natural inputs* to the child's brain, it is difficult to understand developmental mechanisms in cognitive development. Cognitive neuroscience has shown us that the way in which neural representations are structured can make learning easier or more difficult for children, and the 'rise time' studies in developmental dyslexia provide one example of how multi-level developmental understanding (cognitive, sensory and neural) can be achieved by an integrated research programme that includes cognitive neuroscience. As cognitive neuroimaging studies of developmental dyscalculia increase, particularly studies including a group of younger children matched for mathematical level to the children with dyscalculia, similar progress may be made. For example, we may gain better insight into how the neural analogue magnitude system is structured in the dyscalculic brain. On current data, the analogue magnitude system does not appear to be structured differently in dyscalculia, but visuo-spatial short-term memory and working memory do appear to function less effectively. Accordingly, it could be that developmental dyscalculia arises because information that is appropriately structured cannot be held over a short period in analogue form in order to perform the mathematical task in hand. While currently speculative, this explanatory framework is open to exploration at the cognitive, sensory and neural level by an integrated research programme that includes cognitive neuroscience. The behavioural tasks required for such a programme of research have already been developed (Szűcs et al., 2013).

Another potential example of how the structure of children's representations can make learning easier or more difficult comes from the behavioural studies of working memory considered in Chapter 8. We saw that when younger children are given information to remember via pictures (*bus*, *rocket*, *kangaroo*, etc.), they do not appear spontaneously to translate the information into a verbal code. Rather, they retain it in visuo-spatial form. This is assumed to be the case on the basis of the absence of word length effects on working memory capacity for younger children (Hitch et al., 1989). Older children do spontaneously translate the information into a verbal code, hence they can use rehearsal strategies to improve their memories, and consequently they both show word length effects and better memories. As noted many times throughout this book, however, it is imprudent to make a causal claim about development on the basis of a negative result. It would be relatively simple to add cognitive neuroscience methods to test this basic theoretical framework. For example, we could use brain imaging to see whether differential neural activation of visual versus auditory cortex correlates with these assumed strategy differences for younger versus older children. It would also be interesting to use multi-voxel pattern analysis (MVPA) to measure the voxels activated by the different types of task materials used in such working memory studies. For example, MVPA could reveal the voxels activated by items such as 'bus' versus 'kangaroo' when being retained by children when they are younger versus older. We would expect different as well as overlapping voxels to be activated in Hitch et al.'s pictorial version of the working memory task. We could then study how these specific neural activations changed with strategy use. Regarding episodic memory, in Chapter 8 we concluded that organisation by semantic category, another efficient behavioural memory strategy, may be an automatic byproduct of the way in which memories are stored. Activation of highly associated items could lead to spreading activation and thus activate related items. Accordingly, memory retrieval which appears to be organised and strategic may in fact be the result of neural encoding mechanisms. Carefully designed cognitive neuroscience studies that use MVPA offer one way of testing these cognitive theories about memory development.

Increasing the study of oscillatory processes in infants and children

Another factor stressed in this book regarding neural studies of developmental mechanisms is the potential importance of understanding the role of neuroelectric oscillations. We first encountered oscillations via the mu rhythm discussed in Chapter 1, also called an alpha rhythm in some infant literature. This is a relatively slow neural rhythm or oscillation, with networks of brain cells oscillating or rhythmically alternating between activation and inhibition 6–8 times per second (6–8 Hz). We saw that many studies of infant action-effect learning used the mu rhythm as a marker of learning, because the mu rhythm reduces in amplitude (desynchronises) with experience of goal-directed actions. Although the reasons for this desynchronisation are not yet understood, mu desynchronisation was used in some clever experimental designs to explore infant cognition. For example, Southgate and Vernetti (2014) found differential neural activity as indexed by the mu rhythm for two hiding events observed by infants. In event 1, the experimenter believed that a box was empty, whereas the infant knew that it contained a ball. In event 2, the experimenter believed that the ball was in a box, whereas the infant knew that the box was empty. The six-month-old infants observing event 2 knew that the box was empty, but they also knew that the agent thought that the ball was in the box, and so they expected an action. They expected the agent to open the box to retrieve the ball, and indeed the mu rhythm, the electrophysiological signature of action-effect learning, became suppressed in these infants.

We also saw in Chapter 1 that alpha band oscillations in the adult brain are an important gatekeeper for attention. Adults are most likely to detect visual events that coincide with the oscillatory peak of an alpha band oscillation in parietal cortex, while visual events that arrive 'out of phase' (during the oscillatory trough) do not reach conscious awareness (Mathewson et al., 2009). Xie et al. (2017) showed that oscillations in the infant alpha (6–9 Hz) and theta (2–6 Hz) bands were also related to attention for infants aged six, eight, ten and 12 months. However, Xie and colleagues only measured oscillatory power and not oscillatory phase. Future studies could measure phase as well as power, to ascertain whether visual attention is also more efficient during the peaks of the alpha oscillation compared to the troughs for infants and children. The development of visual attention is also related in important ways to the development of eye movements or saccades (Amso & Scerif, 2015). In the adult cognitive neuroscience literature, oscillations are most studied in the areas of visual attention, working memory and linguistic processing.

Indeed, linguistic processing is one area in which oscillations are already being linked to cognitive development, as discussed in Chapters 6 and 10. We saw that adult cognitive neuroscience studies have shown that oscillating networks of neurons in auditory and motor cortex at certain preferred rates (delta, theta, beta and low gamma) play a mechanistic role in speech encoding via phase alignment (or phase-power relations) with the peaks and troughs in amplitude modulations in speech at matching temporal rates. For example, a cortical network oscillating at ~2 Hz (delta band) will phase-align itself with amplitude modulations at ~2 Hz, while another cortical network oscillating at ~5 Hz (theta band) will phase-align itself with amplitude modulations at ~5 Hz (Gross et al., 2013). The phase of the theta oscillation also controls the power of the gamma band response in the adult brain. We saw that the relevant neural networks *automatically* phase-reset their activity to be aligned with the speech information on the basis of amplitude rise times. Studies with children and infants are now showing phase alignment to both the speech envelope and to long-distance dependencies between syllables, providing potential models for syntactic learning in language development as well as for phonological learning (Kabdebon et al., 2015; Power, Mead, Barnes & Goswami, 2012; Power et al., 2013, 2016). In particular, the dyslexic brain seems to be out of phase regarding delta band neural entrainment, for both stress-timed (English) and syllable-timed (Spanish) languages (Power et al., 2016; Molinaro et al., 2016). If the oscillatory peak of cortical responding does not match the most informative points in the speech signal, then important speech information may be arriving during the oscillatory trough, and either failing to be encoded or being encoded with less specificity. Accordingly, this would be expected to affect the quality of children's speech-based representations and their phonological learning. More in-depth study of the oscillatory mechanisms involved in speech encoding also seem likely to increase our understanding of atypical oral language development (developmental language disorder or DLD), as well as our understanding of developmental dyslexia across languages. Such studies also have the potential to provide objective measures for deciding whether interventions are necessary in early childhood for 'late talkers' or slow readers.

Another area of cognitive development in which oscillations are being linked to children's cognitive performance is working memory. In Chapter 8, we saw that studies with adults suggest a role for neuroelectric oscillations in encoding and retrieving items during working memory tasks, particularly regarding links between theta band and alpha band responding (Sauseng et al., 2010). In adults, event-related theta oscillations are most prominent in frontal regions during working memory tasks, whereas event-related alpha oscillations are most prominent in parietal regions. In Chapter 8, we also reviewed a study by Wang et al. (2017a) measuring oscillatory activity during a visuo-spatial working memory task in children with developmental co-ordination disorder (DCD). The children were aged on average ten years, and were compared with typically-developing children in a child-friendly task using pictures of ladybirds sitting on leaves. Children had to judge whether two ladybirds were sitting in the same spot either when the leaves were presented simultaneously or one after the other (delayed presentation condition, expected to be more difficult). Wang et al. (2017a) reported that the children with DCD were significantly less accurate and significantly slower to respond than the typically-developing children in the delayed condition only. The EEG data showed a significant increase in theta band activity over the frontal midline for both groups in the non-delayed condition, and significant increases in frontal theta activity were coupled with significant suppression of alpha band activity over parietal areas for both groups of children in the delayed condition. However, the increase in theta band activity in the delayed condition was less for the DCD children, and the suppression of alpha band activity was also reduced for these children. Hence phase-amplitude coupling mechanisms appeared to be less efficient. Although this particular study does not tell us *how* theta and alpha oscillations support working memory performance in children, the child-friendly paradigm could be used in more targeted studies.

For example, looking for potentially atypical phase-power relations between different oscillatory bands may be informative. In adult cognition, phase-amplitude coupling relations in the brain are thought to index top-down modulation of local cell assemblies to improve task performance. The top-down modulation enables the slower oscillation to control the faster oscillatory response, enabling the faster response to occur at more optimal time points given the task in hand. Even measuring the phase of the oscillatory response as well as its power could be informative regarding cognitive development. As noted previously, when detecting visual targets, adults are unaware of visual stimuli that occur during the trough (least excitable phase) of a parietal alpha oscillation (Mathewson et al., 2009). Visual events that arrive 'out of phase' do not reach conscious awareness. When adults monitor two visuo-spatial locations, then a theta-rhythmic process (~4 Hz) appears to alternately sample each spatial location, with target detection benefits alternating in a 4 Hz rhythm (Landau et al., 2015). Similar mechanisms could underpin developmental differences in children's visuo-spatial working memory performance, and targeted developmental studies including cognitive neuroscience measures could help us to find out whether this is the case.

Increasing cognitive neuroscience studies of cognitive interventions

Finally, as in behavioural studies of cognitive development, the optimal research design for testing whether neural factor A causes changes in neural factor B is to use intervention studies. For example, if increased functional connectivity in a certain neural system, such as the cognitive control network, is causally connected to changes in a cognitive system such as executive function, then training children on executive function tasks should improve functional connectivity in the cognitive control network (see Chapter 9). Similarly, if atypical phase synchronisation of delta-band oscillations in auditory cortex to delta-band amplitude modulation information in the speech signal is causally related to the poor phonological development of children with developmental dyslexia (see Chapter 10), then training dyslexic children's phonological awareness should enhance the synchronisation of these delta-band oscillatory cortical networks with the delta-band amplitude modulation information in speech. Currently there are rather few intervention studies of this nature in the developmental neuroimaging literature, and indeed such studies are expensive and difficult to carry out. The exception is the area of working memory. As we saw in Chapter 9, a computerised adaptive gaming interface to train working memory developed by Klingberg and his colleagues, the CogMed© app, has been the focus of a number of training studies with children, both behavioural training studies and studies using neuroimaging. Neural studies have measured changes in functional connectivity with working memory training, and broadly suggest that successful training with CogMed© (i.e., training that brings behavioural improvements) also leads to enhanced functional connectivity.

For example, as discussed in Chapter 9, using MEG Astle and his colleagues reported greater beta band oscillatory activity (13–20 Hz) at rest in children who had received CogMed© training (Astle et al., 2015). Enhanced connectivity was localised to functional networks that are typically active during executive function tasks, namely right fronto-parietal networks linked to left lateral occipital cortex. Subsequent beta-band connectivity analyses captured variable training effects across children by taking account of *individual gains* in working memory. The fronto-parietal network both showed greater connectivity within itself when training benefits were stronger, and greater connectivity with parietal cortex. This is a nice parametric effect, with greater learning following training associated with greater changes in the brain. The beta band activity in the relevant neural networks showed a linear relationship with the degree of training benefits. As noted in Chapter 9, the fact that Astle and his colleagues measured greater connectivity at rest excludes explanations of the data based on differential task performance in the scanner.

Astle and his colleagues also carried out an MEG study of changes in phase-amplitude coupling during performance of a visuo-spatial working memory task in the scanner following training with CogMed©. The same children performed the visuo-spatial working memory task during MEG scanning both before and after CogMed© training (Barnes et al., 2016). Phase-amplitude coupling between the phase of maximal responding of beta band oscillations (preferred phase) and the amplitude of gamma band responses (power) was then explored. Children who had received adaptive training showed a significant increase in phase-amplitude coupling in the fronto-parietal network of interest during performance of the spatial working memory task. As noted above, phase-amplitude coupling relations in the brain are thought to be one neural mechanism for improving task performance. The enhanced phase-amplitude coupling observed was also significantly related to individual children's gains in resting state connectivity. As concluded in Chapter 9, one interpretation of Barnes et al.'s data is that the beta oscillation generated by the fronto-parietal network helps to co-ordinate ongoing neural gamma activity related to task performance. Accordingly, stronger phase-amplitude coupling following CogMed© training is related to better task performance via better top-down control of local gamma activity. These neural studies of the effects of working memory interventions on the brain show that, in principle, developmental cognitive neuroimaging studies can be designed in ways that do provide information about novel mechanisms and causal factors in cognitive development. More such studies, utilising equally stringent research designs, are now required.

What we know now: Structure and function in brain development related to cognitive development

Despite the relative paucity of available studies, the cognitive neuroscience studies reviewed in this book do appear to support some general conclusions about cognitive neuroscience and cognitive development. For example, it appears that the infant and child brain comprises essentially the same *structures* (neural networks, such as the default mode network) as the adult brain, and that these structures are carrying out essentially the same *functions* via the same *mechanisms*. Studies of diverse topics (discussed in Chapters 8 and 9) such as reasoning by analogy (Whitaker et al., 2017), sleep spindle activity in memory consolidation (Friedrich et al., 2015) and asymmetrical folding of the anterior cingulate in executive function performance (Borst et al., 2014) have shown activation patterns in similar brain networks for both children and adults and similar brain-behaviour correlations for both children and adults. In Chapter 6, we reviewed infant and child neuroimaging studies related to language development. We saw that regarding shared neural structures, French researchers using fMRI found that three-month-old infants listening to speech while asleep activated the same brain areas that are active in adults during speech processing (sensory areas such as Heschl's gyrus in primary auditory cortex and left hemisphere structures such as Broca's area, see

Dehaene-Lambertz et al., 2004, 2006). Regarding shared neural mechanisms for language, we saw that using EEG, German researchers have found that neuronal oscillatory entrainment to delta-band amplitude modulations in rhythmic complex sounds is present from birth (Telkemeyer et al., 2011). Accordingly, the infant brain seems to use the same structures for processing language as the adult brain, and the same mechanisms.

Other data for similarity in structure and function between adult and infant brains come from studies of face processing, mirror neurons and working memory. For example, regarding face processing, Tzourio-Mazoyer et al. (2002) showed via fMRI that two-month-old infants viewing unfamiliar female faces showed activation in the fusiform face area identified by adult studies (see Chapter 3). Meanwhile, Farroni et al. (2002) using EEG showed that four-month-old infants displayed an 'infant N170' when processing human faces, as adults do (also reviewed in Chapter 3). Similar neural processes in infants and adults thus appear to be activated when faces are being viewed. In Chapter 8, however, we also discussed the possibility that activation in the fusiform face area may be related to increased expertise. Car experts show more activity in this brain region when viewing cars, and bird experts show more activity in this brain region when viewing birds. Fingerprint experts show an N170 effect for fingerprints and skilled readers show an N170 effect for words (Chapter 10). Regarding mirror neurons, we saw in Chapter 4 that Iacoboni et al. (2005) suggested using neuroimaging data from adults that the mirror neuron system does not simply provide a substrate for action recognition, but also for intention coding. These ideas about action recognition and intention coding require that the mirror neuron system is activated only by biological actions (Tai et al., 2004). In an fNIRS study with infants aged five months, Lloyd-Fox and her colleagues showed selective activation to biological motion when contrasted with motion from mechanical toys (Lloyd-Fox et al., 2011). Infants who watched biological motions such as a hand opening and closing, eyes opening and closing, or a mouth opening and closing, showed differential neural activation to the three types of biological motion in widespread frontal and temporal regions. We also saw that Turati and her colleagues (Turati et al., 2013; Natale et al., 2014) have used electromyography (EMG) to measure infant motor activation and thereby document mirror-like neural motor mechanisms. EMG measures the electrical potentials generated by muscle cells, and was used to record from the mouth-opening muscles of young infants. The infants watched a video of an actress reaching for an object and then bringing it to either her mouth or to her head. The neural motor responses of the infants were reliably different in the two conditions by six months of age. However, in Lloyd-Fox et al.'s study, the activated neural areas were much larger area than the mirror neuron area per se. Also, it was only the older infants studied by Turati and her colleagues (nine-month-olds) who showed predictive differential neuromotor activity, as the female grasped the object, suggestive of activity related to her *intention* to take the dummy to her mouth.

Accordingly, it is likely that neural development related to cognitive development consists largely of enriching connections between neural structures that are already specialised for certain types of sensory information, and developing novel pathways or even novel functions via experience. This enrichment process would depend largely on the quality of the infant and child's learning environments. Nevertheless, available studies suggest that normative learning environments are quite sufficient for adaptive brain development. The brain may benefit from some enrichment, but in terms of cost-benefit relations, it is much worse for cognitive development if the child suffers an adverse set of learning environments (Goswami, 2008). Regarding the development of novel pathways and functions, reading development again provides us with empirical data. By utilising a combination of longitudinal and training research designs, some researchers are already documenting the development of such novel pathways/functions as young children learn to read.

For example, in Chapter 10 we saw that James (2010) gave letter-writing training to young preschool children who were just beginning to learn to read. Significant improvement in writing the letters was found following the training. fMRI was also recorded both before and after training, while the children viewed letters and control visual stimuli. James (2010) reported that prior to receiving writing training, presentation of the letters either caused no activation in the left fusiform gyrus or modest activation. Following training, visual presentation of the letters led to strong selective activation of the left fusiform gyrus, and *also* led to activity in ventral premotor areas. This link to motor cortex was apparently a result of the writing training, as control children who received matched training focused on visual recognition of letters without writing them did not show neural activity to the learned letters. The activation of *motor* areas during a visual task (letter recognition) following training suggests that learning to write letters created multisensory distributed representations for familiar letters for these young children. Again, as with the work of Barnes et al. (2016), James's (2010) study shows that in principle, cognitive neuroimaging studies can be designed in ways that do provide novel information about mechanisms in cognitive development.

SUMMARY

As I hope to have demonstrated in this book, while challenging, applying cognitive neuroscience methods to understand cognitive development has the potential to yield rich rewards for our field. Going forward, we need to combine our new knowledge gleaned from machine learning and cognitive neuroscience about the power of automatic neural learning algorithms to represent conceptual structure with our 'old' knowledge about the importance of language and imaginative pretend play, knowledge construction via explanation and inductive generalisation (analogies), direct teaching and being part of a community to cognitive development. In the past decade, cognitive neuroscience has helped us to understand, at least in principle, how the brain can create cognitive representations and cognitive systems from perceptual inputs using large networks of simple cells that can either be in an excitatory or an inhibitory state.

We have seen that the brain uses a complex and sophisticated combination of rate-based and temporal coding, utilising functionally connected hierarchies of phase-amplitude coupling as for example in encoding the speech signal (Chapter 6). We have reviewed evidence that causal inferences are an automatic product of the way that the brain encodes probabilistic environmental information (for example, via causal Bayes nets, Chapter 7). However, to generate novel explicit causal inferences *for themselves*, children likely need to reflect on their world knowledge using language. Generating *explanations* for causal phenomena appears to be one important way that children enrich their causal knowledge and grow their expertise. Accordingly, one plausible theoretical framework for cognitive development is that the brain uses automatic information-processing mechanisms based on statistical learning to create knowledge underpinning emergent cognitive domains such as causality, agency and animacy. The young child as an active learner then develops these emergent conceptual representations into a rich and sophisticated cognitive system via child-led mechanisms such as imitation, explanation, pretend play, analogy and experiencing both natural and school-based pedagogy. The growing evidence that all knowledge is *distributed* in the brain, even apparently symbolic knowledge such as concepts and words, means that as argued by Clark (2006) the entire cognitive system should be conceptualised as a "loose-knit, distributed representational economy" (p. 373). Some elements in the economy might conflict with other elements in the economy during cognitive development, but this is inevitable as there is no homunculus or 'single central reasoning engine' that determines cognition. Rather, there are many interacting parts of the overall reasoning machinery that can be maintained at the same time, or even wiped completely if they are not enhancing performance, as we saw in our discussion of insights from machine learning earlier. The activity of all these distributed representations operating together *is* cognition. Children may indicate more cognitive insight via gesture than they can indicate via verbal explanation, or they may be able to regulate their own thoughts and actions when playing in ways that they cannot yet manage in the absence of pretending. These behaviours are all consistent with truly distributed representations. Indeed, this perspective fits with the idea that there is no all-knowing, inner 'central executive' that governs what is 'known' and that orchestrates cognitive development. Rather, there is a "vast parallel coalition of more-or-less influential forces whose largely self-organizing unfolding makes each of us the thinking beings that we are" (Clark, 2006, p. 373).

In the last edition of this book, I proposed that the most exciting ways in which cognitive neuroimaging might contribute to understanding cognitive development would be: (1) in revealing more about developmental trajectories; (2) in revealing more about this 'vast coalition' of distributed representations and how they interact; and (3) in adding to our understanding of how sensory input gives rise to cognitive knowledge. To date, in my judgement, most progress has been made for the third of these proposals. I also argued that the truly ambitious goal for a new theory of cognitive development informed by cognitive neuroscience would be to cross disciplinary boundaries and integrate theoretical frameworks derived from fields like biology, culture, cognition, emotion, perception and action into an overarching explanatory framework (following Diamond, 2007). As Diamond argued, cognitive development is a lifelong process, and we must pay more than lip-service to the complexity of human experience. Ultimately, it is this very complexity that enables the remarkable cognitive developmental changes that are observed as babies develop into children, and as children develop into adults. All of this complexity of experience shapes the brain ('neuroplasticity'). Because cognitive neuroscience methods offer the potential of measuring this shaping directly, in my view they have the potential in the future to play a unique role in developing new explanatory frameworks for children's cognitive development.

References

Abbema, D. L.V., & Bauer, P. J. (2005). Autobiographical memory in middle childhood: Recollections of the recent and distant past. *Memory, 13*, 829–845.

Adolph, K. E., Cole, W. G., Komati, M., Garciaguirre, J. S., Badaly, D., Lingeman, J. M., Chan, G., & Sotsky, R. B. (2012). How do you learn to walk? Thousands of steps and dozens of falls per day. *Psychological Science, 23*(11), 1387–1394.

Adolph, K., & Robinson, S. R. (2013). The road to walking: What learning to walk tells us about development. In P. D. Zelazo (Ed.), *Oxford Handbook of Developmental Psychology*, pp. 403–443. Oxford: Oxford University Press.

Agrillo, C., Piffer, L., Bisazza, A., & Butterworth, B. (2012). Evidence for two numerical systems that are similar in humans and guppies. *PLOS One*, 7(2), e31923.

Aitchison, L., & Lengyel, M. (2017). With or without you: Predictive coding and Bayesian inference in the brain. *Current Opinion in Neurobiology, 46*, 219–227.

Akhtar, N., & Tomasello, M. (1997). Young children's productivity with word order and verb morphology. *Developmental Psychology, 33*, 952–965.

Amalric, M., Denghien, I., & Dehaene, S. (2018). On the role of visual experience in mathematical development: Evidence from blind mathematicians. *Developmental Cognitive Neuroscience, 30*, 314–323.

Amsel, E., Goodman, G., Savoie, D., & Clark, M. (1996). The development of reasoning about causal and noncausal influences on levers. *Child Development, 67*, 1624–1646.

Amso, D., & Scerif, G. (2015). The attentive brain: Insights from developmental cognitive neuroscience. *Nature Reviews Neuroscience, 16*, 606–619.

Amsterdam, B. (1972). Mirror self-image reactions before age two. *Developmental Psychobiology, 5*(4), 297–305.

Anderson, J. R. (1990). *Cognitive Psychology and Its Implications*, 3rd edition. Hillsdale, NJ: Erlbaum.

Ansari, D. (2008). Effects of development and enculturation on number representation in the brain. *Nature Reviews Neuroscience, 9*, 278–291.

Ansari, D., Garcia, N., Lucas, E., Harmon, K., & Dhital, B. (2005). Neural correlates of symbolic number processing in children and adults. *NeuroReport, 16*, 1769–1773.

Anthony, J. L., Lonigan, C. J., Driscoll, K., Phillips, B. M., & Burgess, S. R. (2003). Preschool phonological sensitivity: A quasi-parallel progression of word structure units and cognitive opertations. *Reading Research Quarterly, 38*, 470–487.

Arterberry, M. E., & Bornstein, M. H. (2001). Three-month-old infants' categorization of animals and vehicles based on static and dynamic attributes. *Journal of Experimental Child Psychology, 80*, 333–346.

Astington, J. W. (2001). The future of theory-of-mind research: Understanding motivational states, the role of language, and real-world consequences. *Child Development*, 72(3), 685–687.

Astle, D. E., Barnes, J. J., Baker, K., Colclough, G. L., & Woolrich, M. W. (2015). Cognitive training enhances intrinsic brain connectivity in childhood. *Journal of Neuroscience, 35*(16), 6277–6283.

Atkinson, J., & Braddick, O. (1989). Development of basic visual functions. In A. M. Slater & G. Bremner (Eds.), *Infant Development*, pp. 7–41. Hove, UK: Lawrence Erlbaum Associates.

Atsumi, T., Koda, H., & Masataka, N. (2017). Goal attribution to inanimate moving objects by Japanese macaques (Macaca fuscata). *Scientific Reports*, 7, 40033.

Baddeley, A. D., & Hitch, G. (1974). Working memory. In G. H. Bower (Ed.), The *Psychology of Learning and Motivation*, Vol. 8, pp. 47–90. London: Academic Press.

Bahrick, L. E., & Watson, J. S. (1985). Detection of intermodal proprioceptive-visual contingency as a basis of self perception in infancy. *Developmental Psychology, 21*, 963–973.

Baillargeon, R. (1986). Representing the existence and location of hidden objects: Object permanence in 6- and 8-month-old infants. *Cognition, 23*, 21–41.

Baillargeon, R. (1987a). Object permanence in 3.5- and 4.5-month-old infants. *Developmental Psychology, 23*, 655–664.

Baillargeon, R. (1987b). Young infants' reasoning about the physical and spatial properties of a hidden object. *Cognitive Development, 2*, 179–200.

Baillargeon, R. (2001). Infants' physical knowledge: Of acquired expectations and core principles. In E. Dupoux (Ed.), *Language, Brain and Cognitive Development: Essays in Honour of Jacques Mehler*, pp. 341–361. Cambridge, MA: MIT Press.

Baillargeon, R. (2002). The acquisition of physical knowledge in infancy: A summary in eight lessons. In U. Goswami (Ed.), *Blackwell Handbook of Childhood Cognitive Development*, pp. 47–83. Oxford: Blackwell.

Baillargeon, R. (2004). Infants' physical world. *Current Directions in Psychological Science, 13*, 89–94.

Baillargeon, R., & DeVos, J. (1991). Object permanence in young infants: Further evidence. *Child Development, 62*, 1227–1246.

Baillargeon, R., & DeVos, J. (1994). *Qualitative and Quantitative Reasoning about Unveiling Events in 12.5- and 13.5-month-old Infants.* Unpublished manuscript, University of Illinois.

Baillargeon, R., DeVos, J., & Graber, M. (1989). Location memory in 8-month-old infants in a non-search AB task: Further evidence. *Cognitive Development, 4*, 345–367.

Baillargeon, R., & Gelman, R. (1980). *Young Children's Understanding of Simple Causal Sequences: Predictions and Explanations.* Paper presented to the meeting of the American Psychological Society, Montreal.

Baillargeon, R., Gelman, R., & Meck, B. (1981). *Are Preschoolers Truly Indifferent to Causal Mechanisms?* Paper presented at the Biennial Meeting of the Society for Research in Child Development, Boston.

Baillargeon, R., & Graber, M. (1987). Where is the rabbit? 5.5-month-old infants' representation of the height of a hidden object. *Cognitive Development, 2,* 375–392.

Baillargeon, R., & Graber, M. (1988). Evidence of location memory in 8-month-old infants in a non-search AB task. *Developmental Psychology, 24,* 502–511.

Baillargeon, R., Graber, M., DeVos, J., & Black, J. (1990). Why do young infants fail to search for hidden objects? *Cognition, 36,* 255–284.

Baillargeon, R., Li, J., Ng, W., & Yuan, S. (2008). An account of infants' physical reasoning. In A. Woodward & A. Needham (Eds.), *Learning and the Infant Mind,* pp. 66–116. New York: Oxford University Press.

Baillargeon, R., Needham, A., & DeVos, J. (1992). The development of young infants' intuitions about support. *Early Development & Parenting, 1,* 69–78.

Baillargeon, R., Scott, R. M., & Bian, L. (2016). Psychological reasoning in infancy. *Annual Review of Psychology, 67,* 159–186.

Baillargeon, R., Spelke, E. S., & Wasserman, S. (1985). Object permanence in 5-month-old infants. *Cognition, 20,* 191–208.

Baillargeon, R., & Wang, S.-H. (2002). Event categorization in infancy. *Trends in Cognitive Sciences, 6,* 85–93.

Baker, L., Fernandez-Fein, S., Scher, D., & Williams, H. (1998). Home experiences related to the development of word recognition. In J. L. Metsala & L. C. Ehri (Eds.), *Word Recognition in Beginning Literacy.* Mahwah, NJ: Lawrence Erlbaum Associates, Inc.

Baldwin, D. A., & Markman, E. M. (1989). Establishing word–object relations: A first step. *Child Development, 60,* 381–398.

Bardi, L., Regolin, L., & Simion, F. (2011). Biological motion preference in humans at birth: Role of dynamic and configural properties. *Developmental Science, 14*(2), 353–359.

Barnes, J. J., Nobre, A. C., Woolrich, M. W., Baker, K., & Astle, D. E. (2016). Training working memory in childhood enhances coupling between frontoparietal control network and task-related regions. *Journal of Neuroscience, 36*(34), 9001–9011.

Baron-Cohen, S. (1989a). Perceptual role-taking and protodeclarative pointing in autism. *British Journal of Developmental Psychology, 7,* 113–127.

Baron-Cohen, S. (1989b). The autistic child's theory of mind: A case for specific developmental delay. *Journal of Child Psychology and Psychiatry, 30,* 285–297.

Baron-Cohen, S. (1995). *Mindblindness: An Essay on Autism and Theory of Mind.* Cambridge, MA: MIT Press.

Baron-Cohen, S., Ring, H. A., Wheelwright, S., Bullmore, E. T., Brammer, M. J., Simmons, A., & Williams, S. C. R. (1999). Social intelligence in the normal and autistic brain: An fMRI study. *European Journal of Neuroscience, 11*(6), 1891–1898.

Barr, R., Marrott, H., & Rovee-Collier, C. (2003). The role of sensory preconditioning in memory retrieval by preverbal infants. *Learning & Behaviour, 31*(2), 111–123.

Barr, R., Rovee-Collier, C., & Campanella, J. (2005). Retrieval protracts deferred imitation by 6-month-olds. *Infancy,* 7(3), 263–283.

Barsalou, L. W. (2017). What does semantic tiling of the cortex tell us about semantics? *Neuropsychologia, 105,* 18–38.

Barsalou, L. W., Simmons, W. K., Barbey, A. K., & Wilson, C. D. (2003). Grounding conceptual knowledge in modality-specific system. *Trends in Cognitive Sciences,* 7, 84–91.

Barth, H., La Mont, K., Lipton, J., Dehaene, S., Kanwisher, N., & Spelke, E. S. (2005a). Nonsymbolic arithmetic in adults and young children. *Cognition, 98,* 199–222.

Barth, H., La Mont, K., Lipton, J., & Spelke, E. S. (2005b). Abstract number and arithmetic in preschool children. *Proceedings of the National Academy of Sciences, 102*(39), 14116–14122.

Bartlett, F. C. (1932). *Remembering.* Cambridge: Cambridge University Press.

Bates, E., Camaioni, L., & Volterra, V. (1975). The acquisition of performatives prior to speech. *Merrill Palmer Quarterly, 21,* 205–226.

Bates, E., Devescovi, A., & Wulfeck, B. (2001). Psycholinguistics: A cross-language perspective. *Annual Review of Psychology, 52,* 369–398.

Bathelt, J., Gathercole, S. E., Johnson, A., & Astle, D. E. (2017). Differences in brain morphology and working memory capacity across childhood. *Developmental Science, 21*(3), e12579.

Bauer, P. J. (2006). Constructing a past in infancy: A neuro-developmental account. *Trends in Cognitive Sciences, 10,* 175–181.

Bauer, P. J. (2015). A complementary processes account of the development of childhood amnesia and a personal past. *Psychological Review, 122*(2), 204–231.

Bauer, P. J., Dow, G. A., & Hertsgaard, L. A. (1995). Effects of prototypicality on categorisation in 1- to 2-year-olds: Getting down to basic. *Cognitive Development, 10,* 43–68.

Bauer, P. J., & Fivush, R. (2013). The development of memory: Multiple levels and perspectives. In P. J. Bauer & R. Fivush (Eds.), *The Wiley Handbook on the Development of Children's Memory,* Vol. I/II, pp. 1–13. New York: John Wiley & Sons.

Bauer, P. J., & Larkina, M. (2013). The onset of childhood amnesia in childhood: A prospective investigation of the course and determinants of forgetting of early-life events. *Memory, 22*(8), 907–924.

Bauer, P. J., & Mandler, J.M. (1989). One thing follows another: Effects of temporal structure on 1- to 2-year-olds' recall of events. *Developmental Psychology, 25,* 197–206.

Bauer, P. J., & Shore, C. M. (1987). Making a memorable event: Effects of familiarity and organisation on young children's recall of action sequences. *Cognitive Development, 2*, 327–338.

Bauer, P. J., Wenner, J. A., Dropnik, P. L., & Wewerka, S. S. (2000). Parameters of remembering and forgetting in the transition from infancy to early childhood. *Monographs of the Society for Research in Child Development, 65*(4), i–vi, 1–213.

Bauer, P. J., Wiebe, S. A., Carver, L. J., Waters, J. M., & Nelson, C. A. (2003). Developments in long-term explicit memory late in the first year of life: Behavioral and electrophysiological indices. *Psychological Science, 14*, 629–635.

Bechara, A., Damasio, A. R., Damasio, H., & Anderson, S. W. (1994). Insensitivity to future consequences following damage to human prefrontal cortex. *Cognition, 50*, 7–15.

Becker, J. (1994). 'Sneak shoes', 'sworders' and 'nose beards': A cause study of lexical innovation. *First Language, 14*(2), 195–211.

Behne, T., Carpenter, M., Call, J., & Tomasello, M. (2005a). Unwilling versus unable: Infants' understanding of intentional action. *Developmental Psychology, 41*, 328–337.

Behne, T., Carpenter, M., & Tomasello, M. (2005b). One-year-olds comprehend the communicative intentions behind gestures in a hiding game. *Developmental Science, 8*, 492–499.

Benavides-Varela, S., & Gervain, J. (2017). Learning word order at birth: A NIRS study. *Developmental Cognitive Neuroscience, 25*, 198–208.

Benedict, H. (1979). Early lexical development: Comprehension and production. *Journal of Child Language, 6*, 183–200.

Bergelson, E., & Swingley, D. (2012). At 6–9 months, human infants know the meanings of many common nouns. *Proceedings of the National Academy of Sciences of the United States of America, 109*, 3253–3258.

Berko, J. (1958). The child's learning of English morphology. *Word, 14*, 150–177.

Berndt, R. S., Reggia, J. A., & Mitchum, C. C. (1987). Empirically derived probabilities for grapheme-to-phoneme correspondences in English. *Behavior Research Methods, Instruments, & Computers, 19*, 1–19.

Bertenthal, B. I., Proffitt, D. R., Spetner, N. B., & Thomas, M. A. (1985). The development of infant sensitivity to biomechanical motions. *Child Development, 56*, 531–543.

Bhide, A., Power, A. J., & Goswami, U. (2013). A rhythmic musical intervention for poor readers: A comparison of efficacy with a letter-based intervention. *Mind, Brain and Education, 7*(2), 113–123.

Bigelow, A. E., MacLean, K., & Proctor, J. (2004). The role of joint attention in the development of infants' play with objects. *Developmental Science, 7*(5), 518–526.

Bjorklund, D. F., & Bjorklund, B. R. (1985). Organisation vs. item effects of an elaborated knowledge base on children's memory. *Developmental Psychology, 21*, 1120–1131.

Bjorklund, D. F., & Bjorklund, B. R. (1992). *Looking at Children: An Introduction to Child Development*. Pacific Grove, CA: Brooks/Cole.

Black, J. M., Xia, Z., & Hoeft, F. (2017). Neurobiological bases of reading disorder part II: The importance of developmental considerations in typical and atypical reading. *Language and Linguistics Compass, 11*, e12252.

Blair, C., & Raver, C. C. (2014). Closing the achievement gap through modification of neurocognitive and neuroendocrine function: Results from a cluster randomized controlled trial of an innovative approach to the education of children in kindergarten. *PLOS One, 9*(11), e112393.

Blaisdell, A. P., Sawa, K., Leising, K. J., & Waldmann, M. R. (2006). Causal reasoning in rats. *Science, 311*, 1020–1022.

Blakemore, C., & Cooper, G. (1970). Development of the brain depends on the visual environment. *Nature, 228*, 477–478.

Blau, V., Reithler, J., van Atteveldt, N., Seitz, J., Gerretsen, P., Goebel, R., & Blomert, L. (2010). Deviant processing of letters and speech sounds as proximate cause of reading failure: A functional magnetic resonance imaging study of dyslexic children. *Brain, 133*(3), 868–879.

Bloom, L. (1973). *One Word at a Time*. Paris: Mouton.

Bodrova, E., & Leong, D. J. (2007). *Tools of the Mind: The Vygotskian Approach to Early Childhood Education*. London: Pearson.

Bogartz, R. S., Shinskey, J. L., & Speaker, C. J. (1997). Interpreting infant looking: The event set × event set design. *Developmental Psychology, 33*, 408–422.

Bonatti, L., Peña, M., Nespor, M., & Mehler, J. (2005). Linguistic constraints on statistical computations: The role of consonants and vowels in continuous speech processing. *Psychological Science, 33*, 451–459.

Boncoddo, R., Dixon, J. A., & Kelley, E. (2010). The emergence of a novel representation from action: Evidence from pre-schoolers. *Developmental Science, 13*(2), 370–377.

Borst, G., Cachia, A., Vidal, J., Simon, G., Fischer, C., Pineau, A., Poirel, N., Mangin, J.-F., & Houdé, O. (2014). Folding of the anterior cingulate cortex partially explains inhibitory control during childhood: A longitudinal study. *Developmental Cognitive Neuroscience, 9*, 126–135.

Bortfeld, H., Morgan, J., Golinkoff, R., & Rathbun, K. (2005). Mommy and me: Familiar names help launch babies into speech stream segmentation. *Psychological Science, 16*, 298–304.

Bowerman, M. (1982). Reorganizational processes in lexical and syntactic development. In E. Wanner & L. R. Gleitman (Eds.), *Language Acquisition: The State of the Art*, pp. 319–346. New York: Cambridge University Press.

Bowers, J. S., & Bowers, P. N. (2017). Beyond phonics: The case for teaching children the logic of the English spelling system. *Educational Psychologist, 52*(2), 124–141.

Bowlby, J. (1969). *Attachment: Vol. 1 of Attachment and Loss*. London: Hogarth Press; New York: Basic Books.

Bowlby J. (1971). *Attachment and Loss.* London: Routledge; Harmondsworth: Penguin Books.

Bowlby, J. (1973). *Separation: Anxiety & Anger: Vol. 2 of Attachment and Loss.* London: Hogarth Press.

Bradley, L., & Bryant, P. E. (1978). Difficulties in auditory organisation as a possible cause of reading backwardness. *Nature, 271*, 746–747.

Bradley, L., & Bryant, P. E. (1983). Categorising sounds and learning to read: A causal connection. *Nature, 301*, 419–421.

Braine, M. (1963). The ontogency of English phrase structure: The first phase. *Language, 39*, 1–13.

Braine, M. (1971). The acquisition of language in infant and child. In C. Reed (Ed.), *The Learning of Language.* New York: Appleton-Century-Croft.

Braine, M. D. (1976). Children's first word combinations. *Monograph of the Society for Research in Child Development, 41*(1), 104.

Brand, R. J., & Shallcross, W. L. (2008). Infants prefer motionese to adult-directed action. *Developmental Science, 11*(6), 853–861.

Brandone, A. C., Horwitz, S. R., Aslin, R. N., & Wellman, H. M. (2014). Infants' goal anticipation during failed and successful reaching actions. *Developmental Science, 17*(1), 23–34.

Bremner, J. G. (1988). *Infancy.* Oxford: Blackwell.

Breslow, L. (1981). Re-evaluation of the literature on the development of transitive inferences. *Psychological Bulletin, 89*, 325–351.

Bretherton, E., & Beeghly, M. (1982). Talking about internal states: The acquisition of an explicit theory of mind. *Developmental Science, 18*, 906–921.

Brooks, R., & Meltzoff, A. N. (2002). The importance of eyes: How infants interpret adult looking behavior. *Child Development, 38*, 958–966.

Brooks, R., & Meltzoff, A. N. (2005). The development of gaze following and its relation to language. *Developmental Science, 8*, 535–543.

Brown, A. L., & Scott, M. S. (1971). Recognition memory for pictures in preschool children. *Journal of Experimental Child Psychology, 11*, 401–412.

Brown, A. L. (1978). Knowing when, where and how to remember: A problem of metacognition. In R. Glaser (Ed.), *Advances in Instructional Psychology*, Vol. 1. Hillsdale, NJ: Erlbaum.

Brown, A. L. (1990). Domain-specific principles affect learning and transfer in children. *Cognitive Science, 14*, 107–133.

Brown, A. L., & DeLoache, J.S. (1978). Skills, plans and self-regulation. In R. S. Siegler (Ed.), *Children's Thinking: What Develops?* Hillsdale, NJ: Erlbaum.

Brown, A. L., Bransford, J. D., Ferrara, R. A., & Campione, J.C. (1983). Learning, remembering and understanding. In J. H. Flavell & E. M. Markman (Eds.), *Handbook of Child Psychology*, Vol. 3. New York: Wiley.

Brown, A. L., & Kane, M. J. (1988). Preschool children can learn to transfer: Learning to learn and learning by example. *Cognitive Psychology, 20*, 493–523.

Brown, A. L., Kane, M. J., & Echols, C.H. (1986). Young children's mental models determine analogical transfer across problems with a common goal structure. *Cognitive Development, 1*, 103–121.

Brown, A. L., Kane, M. J., & Long, C. (1989). Analogical transfer in young children: Analogies as tools for communication and exposition. *Applied Cognitive Psychology, 3*, 275–293.

Brown, G. D. A. (1984). A frequency count of 190,000 words in the London–Lund Corpus of English Conversation. *Behavior Research Methods, Instruments, and Computers, 16*, 502–532.

Brown, J. R., Donelan-McCall, N., & Dunn, J. (1996). Why talk about mental states? The significance of children's conversations with friends, siblings and mothers. *Child Development, 67*, 836–849.

Brown, R., & Hanlon, C. (1970). Derivational complexity and order of acquisition in child speech. In R. Hayes (Ed.), *Cognition and the Development of Language.* New York: Wiley.

Bruce, D. J. (1964). The analysis of word sounds. *British Journal of Educational Psychology, 34*, 158–170.

Brusini, P., Dehaene-Lambertz, G., Dutat, M., Goffinet, F., & Christophe, A. (2016). ERP evidence for on-line syntactic computations in 2-year-olds. *Developmental Cognitive Neuroscience, 19*, 164–173.

Bryant, P. E. (1982). *Piaget: Issues and Experiments.* Leicester: British Psychological Society.

Bryant, P. E., Bradley, L., Maclean, M., & Crossland, J. (1989). Nursery rhymes, phonological skills and reading. *Journal of Child Language, 16*, 407–428.

Bryant, P. E., & Nunes, T. (2002). Children's understanding of mathematics. In U. Goswami (Ed.), *Blackwell Handbook of Childhood Cognitive Development*, pp. 412–439. Oxford: Blackwell.

Bryant, P. E., & Trabasso, T. (1971). Transitive inferences and memory in young children. *Nature, 232*, 456–458.

Buchsbaum, D., Griffiths, T. L., Plunkett, D., Gopnik, A., & Baldwin, D. (2015). Inferring action structure and causal relationships in continuous sequences of human action. *Cognitive Psychology, 76*, 30–77.

Bukach, C. M., Gauthier, I., & Tarr, M. J. (2006). Beyond faces and modularity: The power of an expertise framework. *Trends in Cognitive Sciences, 10*, 159–166.

Bulf, H., Johnson, S. P., & Valenza, E. (2011). Visual statistical learning in the newborn infant. *Cognition, 121*, 127–132.

Bullock, M., & Gelman, R. (1979). Preschool children's assumptions about cause and effect: Temporal ordering. *Child Development, 50*, 89–96.

Bullock, M., Gelman, R., & Baillargeon, R. (1982). The development of causal reasoning. In W. J. Friedman (Ed.), *The Developmental Psychology of Time*, pp. 209–254. New York: Academic Press.

Bullock Drummey, A., & Newcombe, N. (1995). Remembering vs. knowing the past: Children's implicit and explicit memory for pictures. *Journal of Experimental Child Psychology, 59*, 549–565.

Bullock Drummey, A., & Newcombe, N. (2002). Developmental changes in source memory. *Developmental Science, 5*(4), 502–513.

Bunge, S. A., Dudukovic, N. M., Thomason, M. E., Vaidya, C. J., & Gabrieli, J. D. E. (2002). Immature frontal lobe contributions to cognitive control in children: Evidence from fMRI. *Neuron, 33*, 301–311.

Bush, G., Luu, P., & Posner, M. I. (2000). Cognitive and emotional influences in anterior cingulate cortex. *Trends in Cognitive Sciences, 4*(6), 215–222.

Bushnell, I. W. R., McCutcheon, E., Sinclair, J., & Tweedie, M. E. (1984). Infants' delayed recognition memory for colour and form. *British Journal of Developmental Psychology, 2*, 11–17.

Buss, A. T., & Spencer, J. P. (2017). Changes in frontal and posterior cortical activity underlie the early emergence of executive function. *Developmental Science, 21*(4), e12602.

Buttelmann, D., Over, H., Carpenter, M., & Tomasello, M. (2014). Eighteen-month-olds understand false beliefs in an unexpected-contents task. *Journal of Experimental Child Psychology, 119*, 120–126.

Buttelmann, F., Suhrke, J., & Buttelmann, D. (2015). What you get is what you believe: Eighteen-month-olds demonstrate belief understanding in an unexpected-identity task. *Journal of Experimental Child Psychology, 131*, 94–103.

Butterworth, G. (1977). Object disappearance and error in Piaget's stage 4 task. *Journal of Experimental Child Psychology, 23*, 391–401.

Butterworth, G. (1998). What is special about pointing in babies? In F. Simion & G. Butterworth (Eds.), *The Development of Sensory, Motor and Cognitive Capacities in Early Infancy: From Perception to Cognition*, pp. 171–190. Hove: Psychology Press.

Callaghan, T., Rochat, P., Lillard, A., Claux, M. L., Odden, H., Itakura, S., Tapanya, S., & Singh, S. (2005). Synchrony in the onset of mental-state reasoning evidence from five cultures. *Psychological Science, 16*, 378–384.

Callanan, M., & Oakes, L. M. (1992). Preschoolers' questions and parents' explanations: Casual thinking in everyday activity. *Cognitive Development, 7*, 213–233.

Cameron-Faulkner, T., Lieven, E., & Tomasello, M. (2003). A construction based analysis of child directed speech. *Cognitive Science, 27*, 843–873.

Campanella, J., & Rovee-Collier, C. (2005). Latent learning and deferred imitation at 3 months. *Infancy, 7*(3), 243–262.

Cantlon, J. F., Brannon, E. M., Carter, E. J., & Pelphrey, K. A. (2006). Functional imaging of numerical processing in adults and 4-year-old children. *PLOS Biology, 4*(5), 844–854.

Cantlon, J. F., Pinel, P., Dehaene, S., & Pelphrey, K. A. (2011). Cortical representations of symbols, objects, and faces are pruned back during early childhood. *Cerebral Cortex, 21*(1), 191–199.

Caramazza, A., & Mahon, B. Z. (2003). The organisation of conceptual knowledge: The evidence from category-specific semantic deficits. *Trends in Cognitive Sciences, 7*(8), 354–361.

Carey, S. (1978). The child as word learner. In M. Halle, J. Bresnan, & G. A. Miller (Eds.), *Linguistic Theory and Psychological Reality*. Cambridge, MA: MIT Press.

Carey, S. (1985). *Conceptual Change in Childhood*. Cambridge, MA: MIT Press.

Carey, S. (2004). On the origin of concepts. In J. Miller (Ed.), *Daedalus*, pp. 59–68. Cambridge, MA: MIT Press.

Carey, S., & Gelman, R. (1991). *The Epigenesis of Mind: Essays on Biology and Cognition*. Hillsdale, NJ: Lawrence Erlbaum.

Carey, S., Shusterman, A., Haward, P., & Distefano, R. (2017). Do analog number representations underlie the meanings of young children's verbal numerals? *Cognition, 168*, 243–255.

Carey, S., & Spelke, E. (1994). Domain-specific knowledge and conceptual change. In L. A. Hirschfeld & S. A. Gelman (Eds.), *Mapping the Mind*, pp. 169–200. New York: Cambridge University Press.

Carlson, S. M. (2003). Executive function in context: Development, measurement, theory, and experience. *Monographs of the Society for Research in Child Development, 68*(3), 138–151.

Carlson, S. M., Claxton, L. J., & Moses, L. J. (2015). The relation between executive function and theory of mind is more than skin deep. *Journal of Cognition and Development, 16*(1), 186–197.

Carlson, S. M., Mandell, D. J., & Williams, J. L. (2004a). Executive function and theory of mind: Stability and prediction from ages 2 to 3. *Developmental Psychology, 40*(6), 1105–1122.

Carlson, S. M., & Moses, L. J. (2001). Individual differences in inhibitory control and children's theory of mind. *Child Development, 72*, 1032–1053.

Carlson, S. M., Moses, L. J., & Claxton, L. (2004b). Individual differences in executive functioning and theory of mind: An investigation of inhibitory control and planning ability. *Journal of Experimental Child Psychology, 87*(4), 299–319.

Carlson, S. M., Zayas, V., & Guthormsen, A. (2009). Neural correlates of decision making on a gambling task. *Child Development, 80*(4), 1076–1096.

Carpendale, J. I. M., & Lewis. C. (2004). Constructing an understanding of mind: The development of children's social understanding within social interaction. *Behavioural and Brain Sciences, 27*, 79–151.

Carpenter, M., Akhtar, N., & Tomasello, M. (1998a). Fourteen- through 18-month-old infants differentially imitate intentional and accidental actions. *Infant Behavior & Development, 21*, 315–330.

Carpenter, M., Call, J., & Tomasello, M. (2005). Twelve- and 18-month-olds copy actions in terms of goals. *Developmental Science, 8*, F13–F20.

Carpenter, M., Nagell, K., & Tomasello, M. (1998b). Social cognition, joint attention, and communicative competence from 9 to 15 months of age. *Monographs of the Society for Research in Child Development, 63*(4), 1–43.

Carroll, M., Byrne, B., & Kirsner, K. (1985). Autobiographical memory and perceptual learning: A developmental study using picture recognition, naming latency and perceptual identification. *Memory & Cognition, 13*, 273–279.

Carvalho, A., He, A. X., Lidz, J., & Christophe, A. (2019). Prosody and function words cue the acquisition of word meanings in 18-month-old infants. *Psychological Science, 30*(3), 319–332.

Carver, L. J., & Bauer, P. J. (1999). When the event is more than the sum of its parts: Nine-month-olds' long-term ordered recall. *Memory, 7*, 147–174.

Carver, L. J., & Bauer, P. J. (2001). The dawning of a past: The emergence of long-term explicit memory in infancy. *Journal of Experimental Psychology: General, 130*, 726–745.

Case, R. (1985). *Intellectual Development: Birth to Adulthood.* New York: Academic Press.

Case, R. (1992). Neo-Piagetian theories of child development. In R. J. Sternberg & C. J. Berg (Eds.), *Intellectual Development*, pp. 161–196. Cambridge: Cambridge University Press.

Casey, B. J., Galvan, A., & Hare, T. A. (2005). Changes in cerebral functional organization during cognitive development. *Current Opinion in Neurobiology, 15*, 239–244.

Cassel, W. S., Roebers, C. E. M., & Bjorklund, D. F. (1996). Developmental patterns of eyewitness responses to repeated and increasingly suggestive questions. *Journal of Experimental Child Psychology, 61*, 116–133.

Cazden, C. (1970). The neglected situation in child language research and education. In F. Williams (Ed.), *Language and Poverty*. Chicago: Markham.

Ceci, S. J., & Bruck, M. (1993). The suggestibility of the child witness: A historical review and synthesis. *Psychological Bulletin, 113*, 403–439.

Ceci, S. J., & Bruck, M. (1995). *Jeopardy in the Courtroom.* Washington, DC: American Psychological Association.

Ceci, S. J., & Friedman, R. D. (2000). The suggestibility of children: Scientific research and legal implications. *Cornell Law Review, 86*, 34–108.

Ceci, S. J., & Liker, J. (1986). A day at the races: A study of IQ, expertise and cognitive complexity. *Journal of Experimental Psychology: General, 115*, 255–266.

Cernock, J. M., & Porter, R. H. (1985). Recognition of maternal axillary odors by infants. *Child Development, 56*, 1593–1598.

Chapman, M. (1988). *Constructive Evolution: Origins and Development of Piaget's Thought.* Cambridge: Cambridge University Press.

Chen, C., & Stevenson, H.W. (1988). Cross-linguistic differences in digit span of preschool children. *Journal of Experimental Child Psychology, 46*, 150–158.

Chen, X., Striano, T., & Rakoczy, R. (2004). Auditory-oral matching behavior in newborns. *Developmental Science, 7*, 42–47.

Chen, Z., Campbell, T., and Polley, R. (1997). From beyond to within their grasp: The rudiments of analogical problem solving in 10- and 13-month-olds. *Developmental Psychology, 33*(5), 790–801.

Cheng, P.W., & Holyoak, K. J. (1985). Pragmatic reasoning schemas. *Cognitive Psychology, 17*, 391–416.

Cheour, M., Ceponiene, R., Lehtokoski, A., Luuk, A., Allik, J., Alho, K., & Näätänen, R. (1998). Development of language-specific phoneme representations in the infant brain. *Nature Neuroscience, 1*, 351–353.

Cheung, C. H. M., Bedford, R., Johnson M. H., Charman, T., Gliga, T., & the BASIS Team (2018). Visual search performance in infants associates with later ASD diagnosis. *Developmental Cognitive Neuroscience, 29*, 4–10.

Chi, M.T. H. (1978). Knowledge structure and memory development. In R. S. Siegler (Ed.), *Children's Thinking: What Develops?*, pp. 73–96. Hillsdale, NJ: Erlbaum.

Chi, M.T. H., Hutchinson, J. E., & Robin, A. F. (1988). Knowledge-constrained inferences about new domain-related concepts: Contrasting expert and novice children. *Merrill-Palmer Quarterly, 35*(1), 27–62.

Cho, S., Moody, T. D., Fernandino, L., Mumford, J. A., Poldrack, R. A., Cannon, T. D., Knowlton, B. J., & Holyoak, K. J. (2010). Common and dissociable prefrontal loci associated with component mechanisms of analogical reasoning. *Cerebral Cortex, 20*(3), 524–533.

Chomsky, N. (1957). *Syntactic Structures.* The Hague and Paris: Mouton.

Chouinard, M., & Clark, E. (2003). Adult reformulations of child errors as negative evidence. *Journal of Child Language, 30*, 637–669.

Christiansen, M. H., Allen, J., & Seidenberg, M. (1998). Learning to segment speech using multiple cues: A connectionist model. *Language and Cognitive Processes, 12*(2/3), 221–268.

Christiansen, M. H., & Chater, N. (2001). Connectionist psycholinguistics: Capturing the empirical data. *Trends in Cognitive Sciences, 5*(2), 82–88.

Chukovsky, K. (1963). *From Two to Five.* Berkeley, CA: University of California Press.

Clark, A. (2006). Language, embodiment, and the cognitive niche. *Trends in Cognitive Sciences, 10*(8), 370–374.

Clark, E.V. (1973). Non-linguistic strategies and the acquisition of word meanings. *Cognition, 2*, 161–182.

Clark, E.V. (2003). *First Language Acquisition.* Cambridge: Cambridge University Press.

Clark, E.V. (2004). How language acquisition builds on cognitive development. *Trends in Cognitive Sciences, 8*, 472–478.

Clark, K. A., Helland, T., Specht, K., Narr, K. L., Manis, F. R., Toga, A. W., & Hugdahl, K. (2014). Neuroanatomical precursors of dyslexia identified from pre-reading through to age 11. *Brain, 137*(12), 3136–3141.

Clayton, N. S., Griffiths, D. P., Emery, N. J., & Dickinson, A. (2001). Elements of episodic-like memory in animals. *Philosophical Transactions of the Royal Society B: Biological Sciences, 356*, 1483–1491.

Clearfield, M. W., & Mix, K. S. (1999). Number versus contour length in infants' discrimination of small visual sets. *Psychological Science, 10*(5), 408–411.

Clubb, P. A., Nida, R. E., Merritt, K., & Ornstein, P. A. (1993). Visiting the doctor: Children's knowledge and memory. *Cognitive Development, 8*, 361–372.

Cocchi, L., Halford, G. S., Zalesky, A., Harding, I. H., Ramm, B. J., Cutmore, T., Shum, D. H. K., & Mattingley, J. B. (2014). Complexity in relational

processing predicts changes in functional brain network dynamics. *Cerebral Cortex, 24*(9), 2283–2296.
Cohen, L. B., Atkinson, D. J., & Chaput, H. H. (2004). *Habit X: A New Program for Obtaining and Organizing Data in Infant Perception and Cognition Studies (Version 1.0)* [Computer software]. Austin: University of Texas.
Cohen, L. B., & Caputo, N. F. (1978). *Instructing Infants to Respond to Perceptual Categories.* Paper presented at the Midwestern Psychological Association Convention, Chicago, IL.
Cole, M., & Scribner, S. (1974). *Culture and Thought.* New York: John Wiley & Sons.
Conrad, R. (1971). The chronology of the development of covert speech in children. *Developmental Psychology, 5*, 398–405.
Cooper, R. G. (1984). Early number development: Discovering number space with addition and subtraction. In C. Sophian (Ed.), *The Origins of Cognitive Skills*, pp. 157–192. Hillsdale, NJ: Lawrence Erlbaum.
Cornell, E. H. (1979). Infants' recognition memory, forgetting and savings. *Journal of Experimental Child Psychology, 28*, 359–374.
Cornell, E. H., & Bergstrom, L. I. (1983). Serial position effects in infant's recognition memory. *Memory & Cognition, 11*, 494–499.
Cossu, G., Gugliotta, M., & Marshall, J. C. (1995). Acquisition of reading and written spelling in a transparent orthography: Two non parallel processes? *Reading and Writing*, 7(1), 9–22.
Cossu, G., Shankweiler, D., Liberman, I. Y., Tola, G., & Katz, L. (1988). Awareness of phonological segments and reading ability in Italian children. *The Haskins Laboratories Status Report on Speech Research, SR-91*, 91–103.
Courage, M. L., & Howe, M. L. (2004). Advances in early memory development research: Insights about the dark side of the moon. *Developmental Review, 24*, 6–32.
Coyle, T. R., & Bjorklund, D. F. (1997). Age differences in, and consequences of, multiple- and variable-strategy use on a multitrial sort-recall task. *Developmental Psychology, 33*, 372–380.
Csépe, V., Szücs, D., & Honbolygó, F. (2003). Number word reading as challenging task in dyslexia. *International Journal of Psychophysiology, 51*, 69–83.
Csibra, G., Bíró, S., Koós, O., & Gergely, G. (2003). One-year-old infants use teleological representations of actions productively. *Cognitive Science, 27*, 111–133.
Csibra, G., & Gergely, G. (2006). Social learning and social cognition: The case for pedagogy. In Y. Munakata & M. H. Johnson (Eds.), *Processes of Change in Brain and Cognitive Development. Attention and Performance XXI*, pp. 249–274. Oxford: Oxford University Press.
Csibra, G., & Gergely, G. (2009). Natural pedagogy. *Trends in Cognitive Sciences, 13*(4), 148–153.
Csibra, G., & Shamsudheen, R. (2015). Nonverbal generics: Human infants interpret objects as symbols of object kinds. *Annual Review of Psychology, 66*, 689–710.
Cuevas, K., Raj, V., & Bell, M. A. (2012). A frequency band analysis of two-year-olds' memory processes. *International Journal of Psychophysiology, 83*(3), 315–322.
Cuevas, K., Rovee-Collier, C., & Learmouth, A. E. (2006). Infants form associations between memory representations of stimuli that are absent. *Psychological Science, 17*(6), 543–549.
Cumming, R., Wilson, A., & Goswami, U. (2015). Basic auditory processing and sensitivity to prosodic structure in children with specific language impairments: A new look at a perceptual hypothesis. *Frontiers in Psychology, 6*, 972.
Curtin, S., Mintz, T. H., & Christiansen, M. H. (2005). Stress changes the representational landscape: Evidence from word segmentation. *Cognition, 96*, 233–262.
Cutler, A., & Norris, D. (1988). The role of strong syllables in segmentation for lexical access. *Journal of Experimental Psychology: Human Perception and Performance, 14*, 113–121.
Cutting, J. E., Proffitt, D. R., & Kozlowski, L. T. (1978). A biomechanical invariant for gait perception. *Journal of Experimental Psychology: Human Perception & Performance, 4*, 357–372.
Cycowicz, Y. M., Friedman, D., Snodgrass, J. G., & Rothstein, M. (2000). A developmental trajectory in implicit memory is revealed by picture fragment completion. *Memory, 8*, 19–35.
Dale, P. S. (1980). Is early pragmatic development measurable? *Journal of Child Language, 7*, 1–12.
Das Gupta, P., & Bryant, P. E. (1989). Young children's causal inferences. *Child Development, 60*, 1138–1146.
Davis, H. L., & Pratt, C. (1995). The development of children's theory of mind: The working memory explanation. *Australian Journal of Psychology, 47*(1), 25–31.
Dean, A. L., Chabaud, S., & Bridges, E. (1981). Classes, collections and distinctive features: Alternative strategies for solving inclusion problems. *Cognitive Psychology, 13*, 84–112.
De Boysson-Bardies, B., Sagart, L., & Durand, C. (1984). Discernible differences in the babbling of infants according to target language. *Journal of Child Language, 11*, 1–15.
De Cara, B., & Goswami, U. (2002). Statistical relations among spoken words: The special status of rimes in English. *Behavior Research Methods, Instruments, and Computers, 34*, 416–423.
DeCasper, A. J., & Fifer, W. P. (1980). Of human bonding: Newborns prefer their mothers' voices. *Science, 208*(4448), 1174–1176.
DeCasper, A. J., & Spence, M. J. (1986). Prenatal maternal speech influences newborns' perception of speech sounds. *Infant Behaviour & Development, 9*, 133–150.
Dehaene, S. (1997). *The Number Sense: How the Mind Creates Mathematics.* New York: Oxford University Press.
Dehaene, S. (2011). *The Number Sense: How the Mind Creates Mathematics*, updated edition. New York: Oxford University Press.

Dehaene, S., Cohen, L., Morais, J., & Kolinsky, R. (2015). Illiterate to literate: Behavioural and cerebral changes induced by reading acquisition. *Nature Reviews Neuroscience, 16*, 234–244.

Dehaene, S., Dehaene-Lambertz, G., & Cohen, L. (1998). Abstract representations of numbers in the animal and human brain. *Trends in Neuroscience, 21*, 355–361.

Dehaene, S., Piazza, M., Pinel, P., & Cohen, L. (2003). Three paretial circuits for number processing. *Cognitive Neuropsychology, 20*, 487–506.

Dehaene, S., Spelke, E., Pinel, P., Stanescu, R., & Tsivkin, S. (1999). Sources of mathematical thinking: Behavioural and brain-imaging evidence. *Science, 284*, 970–974.

Dehaene-Lambertz, G., Dehaene, S., & Hertz-Pannier, L. (2004). Functional neuroimaging of speech perception in infants. *Science, 298*(5600), 2013–2015.

Dehaene-Lambertz, G., & Gliga, T. (2004). Common neural basis for phoneme processing in infants and adults. *Journal of Cognitive Neuroscience, 16*, 1375–1387.

Dehaene-Lambertz, G., Hertz-Pannier, L., Dubois, J., Mériaux, S., Roche, A., Sigman, M., & Dehaene, S. (2006). Functional organisation of perisylvian activation during presentation of sentences in preverbal infants. *PNAS, 103*(38), 14240–14245.

De Jong, G. (2006). Toward robust real-word inference: A new perspective on explanation-based learning. In J. Firnkranz, T. Scheffer, & M. Siliopoulou (Eds.), *ECML06, the Seventeenth European Conference on Machine Learning*, pp. 102–113. Berlin: Springer Verlag.

DeLoache, J. S. (1987). Rapid change in the symbolic functioning of very young children. *Science, 238*, 1556–1557.

DeLoache, J. S. (1989). Young children's understanding of the correspondence between a scale model and a larger space. *Cognitive Development, 4*, 121–139.

DeLoache, J. S. (1991). Symbolic functioning in very young children: Understanding of pictures and models. *Child Development, 62*, 736–752.

DeLoache, J. S., Cassidy, D. J., & Brown, A. L. (1985). Precursors of mnemonic strategies in very young children's memory. *Child Development, 56*, 125–137.

DeLoache, J. S., Miller, K. F., Rosengren, K. S., & Bryant, N. (1993). *Symbolic Development in Young Children: Honey, I Shrunk the Troll.* Paper presented at the meeting of the Psychonomic Society, Washington, DC.

DeMarie, D., & Ferron, J. (2003). Capacity, strategies, and metamemory: Tests of a three-factor model of memory development. *Journal of Experimental Child Psychology, 84*, 167–193.

DeMarie, D., Miller, P. H., Ferron, J., & Cunningham, W. (2004). Path analysis tests of theoretical models of children's memory performance. *Journal of Cognition and Development, 5*, 461–492.

Demont, E., & Gombert, J. E. (1996). Phonological awareness as a predictor of recoding skills and syntactic awareness as a predictor of comprehension skills. *British Journal of Educational Psychology, 66*, 315–332.

Dempster, F. N. (1991). Inhibitory processes: A neglected dimension of intelligence. *Intelligence, 15*, 157–173.

De Neys, W., & Van Gelder, E. (2009). Logic and belief across the lifespan: The rise and fall of belief inhibition during syllogistic reasoning. *Developmental Science, 12*(1), 123–130.

Dennett, D. 1978. *Brainstorms: Philosophical Essays on Mind and Psychology*. Cambridge, MA: Bradford Books/MIT Press.

DeVries, J. I. P., Visser, G. H. A., & Prechtl, H. F. R. (1985). The emergence of fetal behaviour II: Quantitative aspects. *Early Human Development, 12*(2), 99–120.

Diamond, A. (1985). The development of the ability to use recall to guide action, as indicated by infants' performance on A-not-B. *Child Development, 56*, 868–883.

Diamond, A. (1988). Differences between adult and infant cognition: Is the crucial variable presence or absence of language? In L. Weiskrantz (Ed.), *Thought without Language*, pp. 337–370. Oxford: Clarendon Press.

Diamond, A. (1990). Developmental time course in human infants and infant monkeys, and the neural bases of inhibitory control in reaching. *Annals of the New York Academy of Science, 608*, 637–676.

Diamond, A. (2007). Interrelated and interdependent. *Developmental Science, 10*(1), 152–158.

Diamond, A. (2013). Executive functions. *Annual Review of Psychology, 64*, 135–168.

Diamond, A., Barnett, W. S., Thomas, J., & Munro, S. (2007). Preschool program improves cognitive control. *Science, 318*(5855), 1387–1388.

Diamond, A., & Taylor, C. (1996). Development of an aspect of executive control: Development of the abilities to remember what I said and to "Do as I say, not as I do". *Developmental Psychobiology, 29*(4), 315–334.

Dias, M. G., & Harris, P. L. (1988). The effect of make-believe play on deductive reasoning. *British Journal of Developmental Psychology, 6*, 207–221.

Dias, M. G., & Harris, P. L. (1990). The influence of the imagination on reasoning by young children. *British Journal of Developmental Psychology, 8*, 305–318.

Diesendruck, G., & Peretz, S. (2013). Domain differences in the weights of perceptual and conceptual information in children's categorization. *Developmental Psychology, 49*(12), 2383–2395.

Di Giorgio, E., Lunghi, M., Simion, F., & Vallortigara, G. (2017). Visual cues of motion that trigger animacy perception at birth: The case of self-propulsion. *Developmental Science, 20*, e12394.

Di Liberto, G., Peter, V., Kalashnikova, M., Goswami, U., Burnham, D. , & Lalor, E. (2018). Atypical cortical entrainment to speech in the right hemisphere underpins phonemic deficits in dyslexia. *Neuroimage, 175*, 70–79.

diSessa, A. A. (2014). A history of conceptual change research: Threads and fault lines. In K. Sawyer (Ed.), *Cambridge Handbook of the Learning Sciences*, 2nd edition, pp. 88–108. Cambridge: Cambridge University Press.

Dodd, B. (1979). Lipreading in infancy: Attention to speech in- and out-of-synchrony. *Cognitive Psychology, 11*, 478–484.

Dodd, B., & Crosbie, S. (2010). Language and cognition: Evidence from disordered language. In U. Goswami (Ed.), *The Wiley-Blackwell Handbook of Childhood Cognitive Development*, pp. 604–625. Oxford: Blackwell.

Dodge, K. A. (2006). Translational science in action: Hostile attributional style and the development of aggressive behaviour problems. *Development and Psychopathology*, *18*, 791–814.

Doelling, K. B., Arnal, L. H., Ghitza, O., & Poeppel, D. (2014). Acoustic landmarks drive delta-theta oscillations to enable speech comprehension by facilitating perceptual parsing. *Neuroimage*, *85*, 761–768.

Dollaghan, C. A. (1994). Children's phonological neighbourhoods: Half empty or half full? *Journal of Child Language*, *21*, 257–271.

Donaldson, M. (1978). *Children's Minds*. Glasgow: William Collins.

Dufresne, A., & Kobasigawa, A. (1989). Children's spontaneous allocation of study time: Differential and sufficient aspects. *Journal of Experimental Child Psychology*, *47*, 274–296.

Dunbar, K. N., Fugelsang, J. A., & Stein, C. (2007). Do naïve theories ever go away? Using brain and behavior to understand changes in concepts. *American Psychological Association*, *8*, 193–206.

Dunn, J., Brown, J., & Beardsall, L. (1991a). Family talk about feeling states and children's later understanding of others' emotions. *Developmental Psychology*, *27*, 448–455.

Dunn, J., Brown, J., Slomkowski, C., Tesla, C., & Youngblade, L. (1991b). Young children's understanding of other people's feelings and beliefs: Individual differences and their antecedents. *Child Development*, *62*, 1352–1366.

Dunn, J., & Cutting, A.L. (1999). Understanding others, and individual differences in friendship interactions in young children. *Social Development*, *8*, 201–219.

Dunn, J., & Hughes, C. (2001). "I got some swords and you're dead!": Violent fantasy, antisocial behavior, friendship, and moral sensibility in young children. *Child Development*, *72*(2), 491–505.

Durgunoglu, A. Y., & Oney, B. (1999). A cross-linguistic comparison of phonological awareness and word recognition. *Reading & Writing*, *11*, 281–299.

Durgunoglu, A. Y., Nagy, W. E., & Hancin-Bhatt, B. J. (2002). Cross-language transfer of phonological awareness. *Journal of Educational Psychology*, *85*, 453–465.

Durston, S., Thomas, K. M., Yang, Y., Uluğ, A. M., Zimmerman, R. D., & Casey, B. J. (2002). A neural basis for the development of inhibitory control. *Developmental Science*, *5*(4), F9–F16.

Eberhard-Moscicka, A. K., Jost, L. B., Raith, M., & Maurer, U. (2015). Neurocognitive mechanisms of learning to read: Print tuning in beginning readers related to word-reading fluency and semantics but not phonology. *Developmental Science*, *18*(1), 106 118.

Eden, G., & Flowers, L. (Eds.) (in press). *Wiley-Blackwell Handbook of Cognitive Neuroscience of Developmental Dyslexia*. Oxford: Blackwell.

Edwards, J., Beckman, M. E., & Munson, B. (2015). Frequency effects in phonological acquisition. *Journal of Child Language*, *42*, 306–311.

Eimas, P. D., & Quinn, P. C. (1994). Studies on the formation of perceptually-based basic-level categories in young infants. *Child Development*, *65*, 903–917.

Eimas, P. D., Siqueland, E. R., Jusczyk, P., & Vigorito, J. (1971). Speech perception in infants. *Science*, *171*, 303–306.

Eisen, M., Qin, J. J., Goodman, G. S., & Davis, S. (2002). Memory and suggestibility in maltreated children. *Journal of Experimental Child Psychology*, *83*, 167–212.

Elkind, D., & Schoenfeld, E. (1972). Identity and equivalence conservation at two age levels. *Developmental Psychology*, *6*, 529–533.

Elkonin, D. B. (1963). The psychology of mastering the elements of reading. In B. Simon & J. Simon (Eds.), *Educational Psychology in the USSR*. Stanford, CA: Stanford University Press.

Ellis, N. C., & Hennelly, R. A. (1980). A bilingual word-length effect: Implications for intelligence testing and the relative ease of mental calculation in Welsh and English. *British Journal of Psychology*, *71*, 43–51.

Ellis, H. D., Ellis, D. M., & Hosie, J. A. (1993). Priming effects in children's face recognition. *British Journal of Psychology*, *84*, 101–110.

Elman, J. L. (2005). Connectionist models of cognitive development: Where next? *Trends in Cognitive Sciences*, *9*(3), 111–117.

Elsner, B., Jeschonek-Seidel, S., & Pauen, S. (2013). Event-related potentials for 7-month-olds' processing of animals and furniture items. *Developmental Cognitive Neuroscience*, *3*, 53–60.

Emberson, L. L., Zinszer, B. D., Raizada, R. D. S., & Aslin, R. N. (2017). Decoding the infant mind: Multivariate pattern analysis (MVPA) using fNIRS. *PLOS One*, *12*(4), e0172500.

Ezekiel, F., Bosma, R., & Morton, J. B. (2013). Dimensional change card sort performance associated with age-related differences in functional connectivity of lateral prefrontal cortex. *Developmental Cognitive Neuroscience*, *5*, 40–50.

Fabricius, W.V. (1988). The development of forward search planning in pre-schoolers. *Child Development*, *59*, 1473–1488.

Facoetti, A., Trussardi, A. N., Ruffino, M., Lorusso, M. L., Cattaneo, C., Galli, R., Molteni, M., & Zorzi, M. (2010). Multisensory spatial attention deficits are predictive of phonological decoding skills in developmental dyslexia. *Journal of Cognitive Neuroscience*, *22*, 1011–1025.

Fagan, J. F. III (1992). Intelligence: A theoretical viewpoint. *Current Directions in Psychological Science*, *1*, 82–86.

Fagot, B. I., & Gauvain, M. (1997). Mother-child problem solving: Continuity through the early childhood years. *Developmental Psychology*, *33*, 480–488.

Fandakova, Y., Bunge, S. A., Wendelken, C., Desautels, P., Hunter, L., Lee, J.K., & Ghetti, S. (2018). The importance of knowing when you don't remember: Neural signaling of retrieval failure predicts

memory improvement over time. *Cereb Cortex, 28*(1), 90–102.
Fantz, R. L. (1961). The origin of form perception. *Scientific American, 204*, 66–72.
Fantz, R. L. (1966). Pattern discrimination and selective attention as determinants of perceptual development from birth. In A. H. Kidd & J. J. Rivoire (Eds.), *Perceptual Development in Children*, pp. 143–173. New York: International Universities Press.
Farrar, M. J., & Goodman, G. S. (1990). Developmental differences in the relation between script and episodic memory: Do they exist? In R. Fivush & J. Hudson (Eds.), *Knowing and Remembering in Young Children*. New York: Cambridge University Press.
Farroni, T., Csibra, G., Simion, F., & Johnson, M. H. (2002). Eye contact detection in humans from birth. *Proceedings of the National Academy of Sciences, 99*, 9602–9605.
Fearon, R. M. P., & Roisman, G. I. (2017). Attachment theory: Progress and future directions. *Current Opinion in Psychology, 15*, 131–136.
Feigenson, L., Carey, S., & Spelke, E. S. (2002). Infants' discrimination of number vs. continuous extent. *Cognitive Psychology, 44*, 33–66.
Feigenson, L., Dehaene, S., & Spelke, E. S. (2004). Core system of number. *Trends in Cognitive Sciences, 8*(7), 307–314.
Fekete, T. (2010). Representational systems. *Minds and Machines, 20*(1), 69–101.
Fekete, T., & Edelman, S. (2011). Towards a computational theory of experience. *Consciousness and Cognition, 20*, 807–827.
Fenson, L., & Ramsay, D.S. (1981). Effects of modeling action sequences on the play of twelve-, fifteen-, and nineteen-month-old children. *Child Development, 52*, 1028–1036.
Fenson, L., Dale, P. S., Reznick, J. S., Bates, E., Thal, D., & Pethick, S. (1994). Variability in early communicative development. *Monographs of the Society for Research in Child Development, 59*(5), 1–173.
Fernald, A. (1993). Approval and disapproval: Infant responsiveness to vocal affect in familiar and unfamiliar languages. *Child Development, 64*, 657–674.
Fernald, A., & Mazzie, C. (1991). Prosody and focus in speech to infants and adults. *Developmental Psychology, 27*, 209–221.
Fiser, J., & Aslin, R. N. (2002). Statistical learning of new visual feature combinations by infants. *Proceedings of the National Academy of Sciences, 99*, 15822–15826.
Fiser, J., Berkes, P., Orbán, G., & Lengyel, M. (2010). Statistically optimal perception and learning: From behavior to neural representations. *Trends in Cognitive Sciences, 14*(3), 119–130.
Fisher, S. E., & DeFries, J. C. (2002). Developmental dyslexia: Genetic dissection of a complex cognitive trait. *Nature Reviews Neuroscience, 3*, 767–780.
Fisher, S. E., & Francks, C. (2006). Genes, cognition and dyslexia: Learning to read the genome. *Trends in Cognitive Sciences, 10*, 250–257.
Fivush, R., & Hamond, N. R. (1990). Autobiographical memory across the preschool years: Toward reconceptualising childhood amnesia. In R. Fivush & J. Hudson (Eds.), *Knowing and Remembering in Young Children*. New York: Cambridge University Press.
Fivush, R., & Schwarzmueller, A. (1998). Children remember childhood: Implications for childhood amnesia. *Applied Cognitive Psychology, 12*, 455–473.
Flanagan, S., & Goswami, U. (2018). Modelling the amplitude modulation structure of child phonology and morphology speech tasks. *Journal of the Acoustical Society of America, 143*(3), 1366–1375.
Flavell, J. H. (1979). Metacognition and cognitive monitoring: A new area of cognitive-developmental inquiry. *American Psychologist, 34*, 906–911.
Flavell, J. H., Beach, D. R., & Chinsky, J. H. (1966). Spontaneous verbal rehearsal in a memory task as a function of age. *Child Development, 37*, 283–299.
Flavell, J. H., Flavell, E. R., & Green, F. I. (1983). Development of appearance-reality distinction. *Cognitive Psychology, 15*, 95–120.
Flavell, J. H., & Wellman, H. M. (1977). Metamemory. In R. Kail & J. Hagen (Eds.), *Perspectives on the Development of Memory and Cognition*. Hillsdale, NJ: Lawrence Erlbaum Associates, Inc.
Fox, B., & Routh, D. K. (1975). Analyzing spoken language into words, syllables, and phonemes: A developmental study. *Journal of Psycholinguistic Research, 4*(4), 331–342.
Fremgen, A., & Fay, D. (1980). Overextensions in production and comprehension: A methodological clarification. *Journal of Child Language, 7*, 205–211.
Freud, S. (1938). The psychopathology of everyday life. In A. A. Brill (Ed.), *The Writings of Sigmund Freud*, pp. 317–385. New York: Modern Library.
Friedrich, M., & Friederici, A. D. (2004). N4000-like semantic incongruity effect in 19-month-olds: Processing known words in picture contexts. *Journal of Cognitive Neuroscience, 16*(8), 1465–1477.
Friedrich, M., & Friederici, A. D. (2005). Phonotactic knowledge and lexical-semantic processing in one-year olds: Brain responses to words and nonsense words in pictures contexts. *Journal of Cognitive Neuroscience, 17*(11), 1785–1802.
Friedrich, M., & Friederici, A. D. (2006). Early N400 development and later language acquisition. *Psychophysiology, 43*, 1–12.
Friedrich, M., & Friederici, A. D. (2011). Word learning in 6-month-olds: Fast encoding–weak retention. *Journal of Cognitive Neuroscience, 23*(11), 3228–3240.
Friedrich, M., & Friederici, A. D. (2017). The origins of word learning: Brain responses of 3-month-olds indicate their rapid association of objects and words. *Developmental Science, 20*(2), e12357.
Friedrich, M., Wilhelm, I., Born, J., & Friederici, A. D. (2015). Generalization of word meanings during infant sleep. *Nature Communications, 6*, 6004.
Frith, U. (1998). Editorial: Literally changing the brain. *Brain, 121*, 1051–1052.
Frith, U., & Frith, C.D. (2003). Development and neurophysiology of mentalizing. *Philosophical Transactions of the Royal Society: Biological Sciences, 358*, 459–473.

Frith, U., Wimmer, H., & Landerl, K. (1998). Differences in phonological recoding in German- and English-speaking children. *Scientific Studies of Reading, 2*, 31–54.
Frye, D., Zelazo, P. D., & Palfai, T. (1995). Theory of mind and rule-based reasoning. *Cognitive Development, 10*, 483–527.
Fugelsang, J. A., Stein, C. B., Green, A. E., & Dunbar, K. N. (2004). Theory and data interactions of the scientific mind: Evidence from the molecular and the cognitive laboratory. *Canadian Journal of Experimental Psychology, 58*, 86–95.
Fuson, K. C. (1988). *Children's Counting and Concepts of Number*. New York: Springer-Verlag.
Fuson, K. C., Lyons, B., Pergament, G., Hall, J. W., & Kwon, Y. (1988). Effects of collection terms on class inclusion and on number task. *Cognitive Psychology, 20*, 96–120.
Galazka, M. A., Bakker, M., Gredebäck, G., & Nyström, P. (2016). How social is the chaser? Neural correlates of chasing perception in 9-month-old infants. *Developmental Cognitive Neuroscience, 19*, 270–278.
Galazka, M. A., Roche, L., Nystrom, P., & Falck-Ytter, T. (2014). Human infants detect other people's interactions based on complex patterns of kinematic information. *PLOS One, 9*(11), e112432.
Gallagher, H. L., & Frith, C. D. (2003). Functional imaging of 'theory of mind'. *Trends in Cognitive Sciences, 7*, 77–83.
Gao, W., Zhu, H., Giovanello, K. S., Smith, J. K., Shen, D., Gilmore, J. H., & Lin, W. (2009). Evidence on the emergence of the brain's default network from 2-week-old to 2-year-old healthy pediatric subjects. *PNAS, 106*(16), 6790–6795.
Gayan, J., & Olson, R. K. (2001). Genetic and environmental influences on orthographic and phonological skills in children with reading disabilities. *Developmental Neuropsychology, 20*, 483–507.
Gayan, J., Smith, S. D., Cherny, S. S., Cardon, L. R., Fulker, D. W., Kimberling, W. J., Olson, R. K., Pennington, B., & DeFries, J. C. (1999). Large quantitative trait locus for specific language and reading deficits in chromosome 6p. *American Journal of Human Genetics, 64*, 157–164.
Geerdts, M. S., Van De Walle, G. A., & LoBue, V. (2016). Learning about real animals from anthropomorphic media. *Imagination, Cognition and Personality, 36*(1), 5–26.
Gelman, R. (1990). First principles organise attention to and learning about relevant data: Number and the animate-inanimate distinction as examples. *Cognitive Science, 14*, 79–106.
Gelman, R., Bullock, M., & Meck, E. (1980). Preschooler's understanding of simple object transformations. *Child Development, 51*, 691–699.
Gelman, R., & Gallistel, C. R. (1978). *The Child's Understanding of Number*. Cambridge, MA: Harvard University Press.
Gelman, S. A. (1988). The development of induction within natural kind and artifact categories. *Cognitive Psychology, 20*, 65–90.
Gelman, S. A. (1998). Categories in young children's thinking. *Young Children, 53*, 20–26.
Gelman, S. A. (2004). Psychological essentialism in children. *Trends in Cognitive Sciences, 8*, 404–409.
Gelman, S. A., & Coley, J. D. (1990). The importance of knowing a dodo is a bird: Categories and inferences in 2-year-old children. *Developmental Psychology, 26*, 796–804.
Gelman, S. A., Coley, J. D., and Gottfried, G. M. (1994). Essentialist beliefs in children: The acquisition of concepts and theories. In L. A. Hirschfeld & A. S. Gelman (Eds.), *Mapping the Mind: Domain Specificity in Cognition and Culture*, pp. 341–365. Cambridge, MA: Cambridge University Press.
Gelman, S. A., & Gottfried, G.M. (1993). *Causal Explanations of Animate and Inanimate Motion*. Unpublished manuscript.
Gelman, S. A., & Markman, E. M. (1986). Categories and induction in young children. *Cognition, 23*, 183–209.
Gelman, S. A., & Markman, E. M. (1987). Young children's inductions from natural kinds: The role of categories and appearances. *Child Development, 58*, 1532–1541.
Gelman, S. A., & O'Reilly, A. W. (1988). Children's inductive inferences within superordinate categories: The role of language and category structure. *Child Development, 59*, 876–887.
Gelman, S. A., & Kremer, K. E. (1991). Understanding natural cause: Children's explanations of how objects and their properties originate. *Child Development, 62*, 396–414.
Gelman, S. A., & Opfer, J. (2002). Development of the animate-inanimate distinction. In U. Goswami (Ed.), *Blackwell Handbook of Childhood Cognitive Development*, pp. 151–166. Oxford: Blackwell.
Gelman, S. A., & Wellman, H. M. (1991). Insides and essences: Early understandings of the non-obvious. *Cognition, 38*, 213–244.
Gentner, D. (1989). The mechanisms of analogical learning. In S. Vosniadou & A. Ortony (Eds.), *Similarity and Analogical Reasoning*, pp. 199–241. Cambridge: Cambridge University Press.
Gergely, G. (2001). The obscure object of desire: 'Nearly, but clearly not, like me': Contingency preference in normal children versus children with autism. *Bulletin of the Menninger Clinic, 65*, 411–426.
Gergely, G. (2010). Kinds of agents: The origins of understanding instrumental and communicative agency. In U. Goswami (Ed.), *The Wiley-Blackwell Handbook of Childhood Cognitive Development*, pp. 76–105. Oxford: Blackwell.
Gergely, G., Bekkering, H., & Király, I. (2002). Rational imitation in preverbal infants. *Nature, 415*, 755.
Gergely, G., & Csibra, G. (2003). Teleological reasoning in infancy: The naive theory of rational action. *Trends in Cognitive Sciences*, 7, 287–292.
Gergely, G., Egyed, K., & Kiraly, I. (2007). On pedagogy. *Developmental Science, 10*(1), 139–146.
Gergely, G., Nádasdy, Z., Csibra, G., & Bíró, S. (1995). Taking the intentional stance at 12 months of age. *Cognition, 56*, 165–193.
Geurten, M., & Bastin, C. (2018). Behaviors speak louder than explicit reports: Implicit metacognition in 2.5-year-old children. *Developmental Science, 22*(2), e12742.

Gibson, E. J., & Walk, R. D. (1960). The "visual cliff". *Scientific American, 202*, 64–71.
Gick, M. L., & Holyoak, K. J. (1980). Analogical problem solving. *Cognitive Psychology, 12*, 306–355.
Giles, O. T., Shire, K. A., Hill, L. J. B., Mushtaq, F., Waterman, A., Holt, R. J., Culmer, P. R., Williams, J. H. G., Wilkie, R. M., & Mon-Williams, M. (2018). Hitting the target: Mathematical attainment in children is related to interceptive-timing ability. *Psychological Science, 29*(8), 1334–1345.
Gillan, D. J., Premack, D., & Woodruff, G. (1981). Reasoning in the chimpanzee I: Analogical reasoning. *Animal Behaviour Processes*, 7, 1–17.
Gilmore, R. O., & Johnson, M. H. (1995). Working memory in infancy: Six-month-olds' performance on two versions of the oculomotor delayed response task. *Journal of Experimental Child Psychology, 59*, 397–418.
Gilstrap, L. L., & Ceci, S. J. (2005). Reconceptualizing children's suggestibility: Bidirectional and temporal properties. *Child Development, 76*, 40–53.
Giraud, A. L., & Poeppel, D. (2012). Cortical oscillations and speech processing: Emerging computational principles and operations. *Nature Neuroscience, 15*, 511–517.
Gleason, J. B. (1980). The acquisition of social speech and politeness formulae. In H. Giles, W. P. Robinson, & S. M. P. (Eds.), *Language: Social Psychological Perspectives*. Oxford: Pergamon.
Göksun, T., George, N. R., Hirsh-Pasek, K., & Golinkoff, R. M. (2013). Forces and motion: How young children understand causal events. *Child Development, 84*(4), 1285–1295.
Goldfield, B. A., & Reznick, J. S. (1990). Early lexical acquisition: Rate, content and the vocabulary spurt. *Journal of Child Language, 17*, 171–183.
Goldin-Meadow, S., & Wagner, S. M. (2005). How our hand helps us learn. *Trends in Cognitive Sciences, 9*(5), 234–241.
Goldstein, T. R., & Lerner, M. D. (2017). Dramatic pretend play games uniquely improve emotional control in young children. *Developmental Science, 21*(4), e12603.
Goodman, G. S., & Aman, C. (1990). Children's use of anatomically detailed dolls to recount an event. *Child Development, 61*, 1859–1871.
Goodman, G. S., Jones, O., & McLeod, C. (2017). Is there consensus about children's memory and suggestibility? *Journal of Interpersonal Violence, 32*(6), 926–939.
Goodman, G. S., Rudy, L., Bottoms, B. L., & Aman, C. (1990). Children's concerns and memory: Issues of ecological validity in the study of children's eyewitness testimony. In R. Fivush & J. Hudson (Eds.), *Knowing and Remembering in Young Children*, pp. 331–346. New York: Cambridge University Press.
Gopnik, A., Glymour, C., Sobel, D. M., Schulz, L. E., Kushnir, T., & Danks, D. (2004). A theory of causal learning in children: Causal maps and Bayes nets. *Psychological Review, 111*, 3–32.
Gopnik, A., Griffiths, T. L., & Lucas, C. G. (2015). When younger learners can be better (or at least more open-minded) than older ones. *Current Directions in Psychological Science, 24*(2), 87–92.
Gopnik, A., Sobel, D., Schulz, L., & Glymour, C. (2001). Causal learning mechanisms in very young children: Two, three, and four-year-olds infer causal relations from patterns of variation and covariation. *Developmental Psychology, 37*(5), 620–629.
Gordon, P. (2004). Numerical cognition without words: Evidence from Amazonia. *Science, 306*, 496–499.
Goswami, U. (1991). Analogical reasoning: What develops? A review of research and theory. *Child Development, 62*, 1–22.
Goswami, U. (1992). *Analogical Reasoning in Children*. London: Lawrence Erlbaum.
Goswami, U. (1996). Analogical reasoning and cognitive development. *Advances in Child Development and Behaviour*, 26, pp. 91–138.
Goswami, U. (2001). Analogical reasoning in children. In D. Gentner, K. J. Holyoak, & B. N. Kokinov (Eds.), *The Analogical Mind: Perspectives from Cognitive Science*, pp. 437–470. Cambridge, MA: MIT Press.
Goswami, U. (2003). Why theories about developmental dyslexia require developmental designs. *Trends in Cognitive Sciences*, 7, 534–540.
Goswami, U. (2004). Neuroscience and education. *British Journal of Educational Psychology, 74*, 1–14.
Goswami, U. (2008). *Foresight Mental Capital and Wellbeing Project, Learning Difficulties: Future Challenges*. London: Government Office for Science.
Goswami, U. (2009). Mind, brain, and literacy: Biomarkers as usable knowledge for education. *Mind, Brain, and Education, 3*, 176–184.
Goswami, U. (2010). Inductive and deductive reasoning. In U. Goswami (Ed.), *Blackwell-Wiley Handbook of Childhood Cognitive Development*, 2nd edition. Oxford: Blackwell.
Goswami, U. (2011). A temporal sampling framework for developmental dyslexia. *Trends in Cognitive Sciences, 15*, 3–10.
Goswami, U. (2015). Sensory theories of developmental dyslexia: Three challenges for research. *Nature Reviews Neuroscience, 16*, 43–54.
Goswami, U. (2016). Educational neuroscience: Neural structure-mapping and the promise of oscillations. *Current Opinion in Behavioural Sciences, 10*, 89–96.
Goswami, U. (2018). A neural basis for phonological awareness? An oscillatory temporal sampling perspective. *Current Directions in Psychological Science, 27*(1), 56–63.
Goswami, U., & Brown, A. (1989). Melting chocolate and melting snowmen: Analogical reasoning and causal relations. *Cognition, 35*, 69–95.
Goswami, U., & Brown, A. L. (1990). Higher-order structure and relational reasoning: Contrasting analogical and thematic relations. *Cognition, 36*, 207–226.
Goswami, U., & Bryant, P. E. (1990). *Phonological Skills and Learning to Read*. Hillsdale, NJ: Lawrence Erlbaum.
Goswami, U., & East, M. (2000). Rhyme and analogy in beginning reading: Conceptual and methodological issues. *Applied Psycholinguistics, 21*, 63–93.
Goswami, U., & Pauen, S. (2005). The effects of a "family" analogy on class inclusion reasoning by young children. *Swiss Journal of Psychology, 64*, 115–124.

Goswami, U., Porpodas, C., & Wheelwright, S. (1997). Children's orthographic representations in English and Greek. *European Journal of Psychology of Education, 12*, 273–292.

Goswami, U., & Szücs, D. (2011). Educational neuroscience: Developmental mechanisms; towards a conceptual framework. *Neuroimage, 57*, 651–658.

Goswami, U., & Ziegler, J. C. (2006). Fluency, phonology and morphology: A response to the commentaries on becoming Literate in different languages. *Developmental Science, 9*(5), 451–453.

Goswami, U., Ziegler, J., Dalton, L., & Schneider, W. (2003). Nonword reading across orthographies: How flexible is the choice of reading units? *Applied Psycholinguistics, 24*, 235–247.

Gottleib, G. (2007). Probabilistic epigenesis. *Developmental Science, 10*(1), 1–11.

Goupil, L., & Kouider, S. (2016). Behavioral and neural indices of metacognitive sensitivity in preverbal infants. *Current Biology, 26*(22), 3038–3045.

Grammer, J. K., Carrasco, M., Gehring, W. J., & Morrison, F. J. (2014). Age-related changes in error processing in young children: A school-based investigation. *Developmental Cognitive Neuroscience, 9*, 93–105.

Greco, C., Hayne, H., & Rovee-Collier, C. (1990). The roles of function, reminding and variability in categorization by 3-month-old infants. *Journal of Experimental Psychology: Learning, Memory & Cognition, 16*, 617–633.

Grelotti, D. J., Klin, A. J., Gauthier, I., Skudlarski, P., Cohen, D. J., Gore, J. C., Volkmar, F. R., & Schultz, R. T. (2005). fMRI activation of the fusiform gyrus and amygdale to cartoon characters but not to faces in boy with autism. *Neuropsychologia, 43*, 373–385.

Griffiths, T. L., & Tenenbaum, J. B. (2009). Theory-based causal induction. *Psychological Review, 116*(4), 661–716.

Gross, J., Hoogenboom, N., Thut, G., Schyns, P., Panzeri, S., Belin, P., & Garrod, S. (2013). Speech rhythms and multiplexed oscillatory sensory coding in the human brain. *PLOS Biology, 11*(12), e1001752.

Grossmann, T., Lloyd-Fox, S., & Johnson, M. H. (2013). Brain responses reveal young infants' sensitivity to when a social partner follows their gaze. *Developmental Cognitive Neuroscience, 6*, 155–161.

Haake, R. J., & Somerville, S. C. (1985). The development of logical search skills in infancy. *Developmental Psychology, 21*, 176–186.

Haith, M. M. (1998). Who put the cog in infant cognition? Is rich interpretation too costly? *Infant Behavior & Development, 21*, 167–179.

Haith, M. M., Hazan, C., & Goodman, G. S. (1988). Expectation and anticipation of dynamic visual events by 3.5-month-old babies. *Child Development, 59*, 467–479.

Halberda, J., & Feigenson, L. (2008). Developmental change in the acuity of the "number sense": The approximate number system in 3-, 4-, 5-, and 6-year-olds and adults. *Developmental Psychology, 44*(5), 1457–1465.

Halford, G. S. (1993). *Children's Understanding: The Development of Mental Models*. Hillsdale, NJ: Erlbaum.

Hamlin, J. K., Wynn, K., & Bloom, P. (2010). Three-month-olds show a negativity bias in their social evaluations. *Developmental Science, 13*(6), 923–929.

Hampton, R. R. (2001). Rhesus monkeys know when they remember. *PNAS, 98*(9), 5359–5362.

Happe, F., and Frith, U. (2006). The weak coherence account: Detail-focused cognitive style in autism spectrum disorders. *Journal of Autism and Developmental Disorders, 36*(1), 5–25.

Hardison, D. M. (2003). Acquisition of second-language speech: Effects of visual cues, context, and talker variability. *Applied Psycholinguistics, 24*(4), 495–522.

Harms, M. B., Zayas, V., Meltzoff, A. N., & Carlson, S. M. (2014). Stability of executive function and predictions to adaptive behavior from middle childhood to pre-adolescence. *Frontiers in Psychology, 5*, 331.

Harris, M., & Giannouli, V. (1999). Learning to read and spell in Greek: The importance of letter knowledge and morphological awareness. In M. Harris & G. Hatano (Eds.), *Learning to Read and Write: A Cross-Linguistic Perspective*, pp. 51–70. Cambridge: Cambridge University Press.

Harris, P. L. (2007). Trust. *Developmental Science, 10*(1), 135–138.

Harris, P. L., Brown, E., Marriott, C., Whittall, S., & Harmer, S. (1991). Monsters, ghosts, and witches: Testing the limits of the fantasy-reality distinction in young children. *British Journal of Developmental Psychology, 9*, 105–123.

Harris, P. L., Koenig, M. A., Corriveau, K. H., & Jaswal, V. K. (2018). Cognitive foundations of learning from testimony. *Annual Review of Psychology, 69*, 251–273.

Harris, P. L., & Nunez, M. (1996). Understanding of permission rules by preschool children. *Child Development, 67*, 1572–1591.

Hart, B. H., & Risley, T. R. (1995). *Meaningful Differences in the Everyday Experience of Young American Children*. Baltimore, MD: Paul H. Brookes.

Hauk, O., Johnsrude, I. S., & Pulvermuller, F. (2004). Somatotopic representation of action words in human motor and premotor cortex. *Neuron, 41*, 301–307.

Hayes, L. A., & Watson, J. S. (1981). Neonatal imitation: Fact or artifact? *Developmental Psychology, 17*, 655–660.

Heibeck, T. H., & Markman, E. M. (1987). Word learning in children: An examination of fast mapping. *Child Development, 58*, 1021–1034.

Heider, F., & Simmel, M. (1944). An experimental study of apparent behavior. *American Journal of Psychology, 57*, 243–259.

Heim, S., Pape-Neumann, J., van Ermingen-Marbach, M., Brinkhaus, M., & Grande, M. (2015). Shared vs. specific brain activation changes in dyslexia after training of phonology, attention, or reading. *Brain Structure and Function, 220*(4), 2191–2207.

Hellmer, K., Söderlund, H., & Gredebäck, G. (2016). The eye of the retriever: Developing episodic memory mechanisms in preverbal infants assessed through pupil dilation. *Developmental Science, 21*(2), e12520.

Henderson, L. M., Yoder, P. J., Yale, M. E., & McDuffie, A. (2002). Getting to the point: Electrophysiological

correlates of protodeclarative pointing. *International Journal of Developmental Neuroscience, 20*, 449–458.
Hendrickson, T. J., Mueller, B. A., Sowell, E. R., Mattson, S. N., Coles, C. D., Kable, J. A., Jones, K. L., Boys, C. J., Lim, K. O., Riley, E. P., & Wozniak, J. R. (2017). Cortical gyrification is abnormal in children with prenatal alcohol exposure. *NeuroImage: Clinical, 15*, 391–400.
Henry, L. A., & Millar, S. (1993). Why does memory span improve with age? A review of the evidence for two current hypotheses. *European Journal of Cognitive Psychology, 5*, 241–287.
Hepper, P. G. (1988). Foetal 'soap' addiction. *The Lancet* (11 June), 1347–1348.
Hepper, P. G. (1992). Fetal psychology: An embryonic science. In J. G. Nijhuis (Ed.), *Fetal Behaviour: Development and Perinatal Aspects*, pp. 129–156. Oxford: Oxford University Press.
Hepper, P. G., Wells, D. L., & Lynch, C. (2005). Prenatal thumb sucking is related to postnatal handedness. *Neuropsychologia, 43*, 313–315.
Herrmann, P. A., Waxman, S., & Medin, D. L. (2010). Anthropocentrism is not the first step in children's reasoning about the natural world. *Proceedings of the National Academy of Sciences, 107*(22), 9979–9984.
Hespos, S. J., & Baillargeon, R. (2001). Reasoning about containment events in very young infants. *Cognition, 78*, 207–245.
Hespos, S. J., & Baillargeon, R. (2006). Décalage in infants' knowledge about occlusion and containment events: Converging evidence from action tasks. *Cognition, 99*, B31–B41.
Heyes, C. (2014). False belief in infancy: A fresh look. *Developmental Science, 17*(5), 647–659.
Hitch, G. J., Halliday, S., Dodd, A., & Littler, J.E. (1989). Development of rehearsal in short-term memory: Differences between pictorial and spoken stimuli. *British Journal of Developmental Psychology, 7*, 347–362.
Hitch, G. J., Halliday, S., Schaafstal, A. M., & Schraagen, J. M. (1988). Visual working memory in young children. *Memory & Cognition, 16*, 120–132.
Ho, C. S.-H., & Bryant, P. (1997). Phonological skills are important in learning to read Chinese. *Developmental Psychology, 33*, 946–951.
Ho, C. S.-H., Law, T. P.-S., & Ng, P. M. (2000). The phonological deficit hypothesis in Chinese developmental dyslexia. *Reading & Writing, 13*, 57–79.
Hochstein, S., Pavlovskaya, M., Bonneh, Y. S., & Soroker, N. (2015). Global statistics are not neglected. *Journal of Vision, 15*(7), 1–17.
Hodent, C., Bryant, P., & Houdé, O. (2005). Language-specific effects on number computation in toddlers. *Developmental Science, 8*, 420–423.
Hodges, R. M., & French, L. A. (1988). The effect of class and collection labels on cardinality, class-inclusion and number conservation tasks. *Child Development, 59*, 1387–1396.
Hoeft, F., Hernandez, A., McMillon, G., Taylor-Hill, H., Martindale, J. L., Meyler, A., Keller, T. A., Siok, W. T., Deutsch, G. K., Just, M. A., Whitfield-Gabrieli, S., & Gabrieli, J. D. E. (2006). Neural basis of dyslexia: A comparison between dyslexic and nondyslexic children equated for reading ability. *Journal of Neuroscience, 26*(42), 10700–10708.
Hogrefe, J., Wimmer, H., & Perner, J. (1986). Ignorance versus false belief: A developmental lag in attribution of epistemic states. *Child Development, 57*, 567–582.
Hoien, T., Lundberg, L., Stanovich, K. E., & Bjaalid, I. K. (1995). Components of phonological awareness. *Reading & Writing*, 7, 171–188.
Holmes, J., Gathercole, S. E., & Dunning, D. L. (2009). Adaptive training leads to sustained enhancement of poor working memory in children. *Developmental Science, 12*(4), F9–F15.
Holyoak, K. J., Junn, E. N., & Billman, D. O. (1984). Development of analogical problem-solving skill. *Child Development, 55*, 2042–2055.
Holyoak, K. J., & Thagard, P. (1995). *Mental Leaps*. Cambridge, MA: MIT Press.
Hood, B. M. (1995). Gravity rules for 2- to 4-year-olds? *Cognitive Development, 10*, 577–598.
Hood, B. M., Wilson, A., & Dyson, S. (2006). The effect of divided attention on inhibiting the gravity error. *Developmental Science, 9*, 303–308.
Hood, L., & Bloom, L. (1979). What, when, and how about why: A longitudinal study of the early expressions of causality. *Monographs of the Society for Research in Child Development, 44*(6).
Hornik, R., Risenhoover, N., & Gunnar, M. (1987). The effects of maternal positive, neutral, and negative affective communications on infant responses to new toys. *Child Development, 58*, 937–944.
Houdé, O. (1997). The problem of deductive competence and the inhibitory control of cognition. *Current Psychology of Cognition, 16*, 108–113.
Houdé, O. (2000). Inhibition and cognitive development: Object, number, categorization and reasoning. *Cognitive Development, 15*, 63–73.
Houdé, O., & Borst, G. (2015). Evidence for an inhibitory-control theory of the reasoning brain. *Frontiers in Human Neuroscience, 9*, 148.
Howe, C. J., & Tolmie, A. (2003). Group work in primary school science: Discussion, consensus and guidance from experts. *International Journal of Educational Research, 39*, 51–72.
Howe, C. J., Tolmie, A., Duchak-Tanner, V., & Rattray, C. (2000). Hypothesis testing in science: Group consensus and the acquisition of conceptual and procedural knowledge. *Learning and Instruction, 10*, 361–391.
Howe, M. L., & Courage, M. L. (1993). On resolving the enigma of infantile autism. *Psychological Bulletin, 113*, 305–326.
Hudson, J. A. (1990). The emergence of autobiographical memory in mother-child conversation. In R. Fivush & J. Hudson (Eds.), *Knowing and Remembering in Young Children*. New York: Cambridge University Press.
Hughes, C. (1998). Executive function in preschoolers: Links with theory of mind and verbal ability. *British Journal of Developmental Psychology, 16*, 233–253.

Hughes, C. (2011). *Social Understanding, Social Lives: From Toddlerhood to the Transition to School.* Hove: Psychology Press.

Hughes, C., Devine, R. T., & Wang, Z. (2017). Does parental mind-mindedness account for cross-cultural differences in preschoolers' theory of mind? *Child Development, 89*(4), 1296–1310.

Hughes, C., & Dunn, J. (1998). Understanding mind and emotion: Longitudinal associations with mental-state talk between young friends. *Developmental Psychology, 34*, 1026–1037.

Hughes, C., & Dunn, J. (2000). Hedonism or empathy: Hard-to-manage children's moral awareness, and links with cognitive and maternal characteristics. *British Journal of Developmental Psychology, 18*, 227–245.

Hughes, C., Dunn, J., & White, A. (1998). Trick or treat? Uneven understanding of mind and emotion and executive dysfunction in hard-to-manage preschoolers. *Journal of Child Psychology & Psychiatry, 39*, 981–994.

Hulme, C., Thomson, N., Muir, C., & Lawrence, A. (1984). Speech rate and the development of short-term memory span. *Journal of Experimental Child Psychology, 38*, 241–253.

Hulme, C., & Tordoff, V. (1989). Working memory development: The effects of speech rate, word length, and acoustic similarity on serial recall. *Journal of Experimental Child Psychology, 47*, 72–87.

Hume, D. (1748). *Enquiry Concerning Human Understanding.* Sections IV–VII (paras. 20–61), pp. 25–79.

Huntley-Fenner, G. (2001). Children's understanding of number is similar to adults' and rats': Numerical estimation by 5- and 7-year-olds. *Cognition, 78*, B27–B40.

Huntley-Fenner, G., & Cannon, E. (2000). Preschoolers' magnitude comparisons are meditated by a preverbal analog mechanism. *Psychological Science, 11*(2), 147–152.

Hyde, D. C., Porter, C. L., Flom, R., & Stone, S. A. (2013). Relational congruence facilitates neural mapping of spatial and temporal magnitudes in preverbal infants. *Developmental Cognitive Neuroscience, 6*, 102–112.

Hyde, D. C., Simon, C. E., Berteletti, I., & Mou, Y. (2016). The relationship between non-verbal systems of number and counting development: A neural signatures approach. *Developmental Science, 20*(6), e12464.

Iacoboni, M., Molnar-Szakacs, I., Gallese, V., Buccino, G., Mazziotta, J. C., & Rizzolatti, G. (2005). Grasping the intentions of others with one's own mirror neuron system. *PLOS Biology, 3*, 529–535.

Inagaki, K., & Hatano, G. (1987). Young children's spontaneous personification as analogy. *Child Development, 58*, 1013–1020.

Inagaki, K., & Hatano, G. (1993). Young children's understanding of the mind-body distinction. *Child Development, 64*, 1534–1549.

Inagaki, K., & Hatano, G. (2004). Vitalistic casuality in young children's naïve biology. *Trends in Cognitive Sciences, 8*, 356–362.

Inagaki, K., & Sugiyama, K. (1988). Attributing human characteristics: Developmental changes in over- and under-attribution. *Cognitive Development, 3*, 55–70.

Inhelder, B., & Piaget, J. (1958). *The Growth of Logical Thinking from Childhood to Adolescence.* New York: Basic Books.

Jack, F., Simcock, G., & Hayne, H. (2012). Magic memories: Young children's verbal recall after a 6-year delay. *Child Development, 83*(1), 159–72.

Jackson, P. L., Meltzoff, A. N., & Decety, J. (2006). Neural circuits involved in imitation and perspective-taking. *Neuroimage, 31*, 429–439.

Jacques, S., Zelazo, P. D., Kirkham, N. Z., & Semcesen, T. K. (1999). Rule selection and rule execution in preschoolers: An error-detection approach. *Developmental Psychology, 35*, 770–780.

James, K. H. (2010). Sensori-motor experience leads to changes in visual processing in the developing brain. *Developmental Science, 13*(2), 279–288.

James, K. H., & Engelhardt, L. (2012). The effects of handwriting experience on functional brain development in pre-literate children. *Trends in Neuroscience and Education, 1*(1), 32–42.

James, K. H., & Maouene, J. (2009). Auditory verb perception recruits motor systems in the developing brain: An fMRI investigation. *Developmental Science, 12*(6), F26–F34.

Jardri, R., Houfflin-Debarge, V., Delion, P., Pruvo, J-P., Thomas, P., & Pins, D. (2012). Assessing fetal response to maternal speech using a non-invasive functional brain imaging technique. *International Journal of Developmental Neuroscience, 30*(2), 159–161.

Jenkins, J., & Astington, J. W. (1996). Cognitive factors and family structure associated with theory of mind development in young children. *Developmental Psychology, 32*, 70–78.

Johansson, G. (1973). Visual perception of biological motion and a model for its analysis. *Perception & Psychophysics, 14*, 201–211.

Johnson, C. M., Sullivan, J., Jensen, J., Buck, C., Trexel, J., & St Leger, J. (2018). Prosocial predictions by bottlenose dolphins (*Tursiops* spp.) based on motion patterns in visual stimuli. *Psychological Science, 29*(9), 1405–1413.

Johnson, M. H., & De Haan, M. (2015). *Developmental Cognitive Neuroscience: An Introduction.* Oxford: Blackwell Publishing Ltd.

Johnson, M. H., & Munakata, Y. (2005). Processes of change in brain and cognitive development. *Trends in Cognitive Sciences, 9*, 152–158.

Johnson-Laird, P. N., Legrenzi, P., & Sonino-Legrenzi, M. (1972). Reasoning and a sense of reality. *British Journal of Psychology, 63*, 395–400.

Johnson-Laird, P. N., & Wason, P. C. (1977). *Thinking: Readings in Cognitive-Science.* Cambridge: Cambridge University Press.

Jones, S. S., & Smith, L. B. (1993). The place of perception in children's concepts. *Cognitive Development, 8*, 113–139.

Jordan, K. E., & Brannon, E. M. (2006). A common representational system governed by Weber's law: Nonverbal numerical similarity judgements in 6-year-olds and rhesus macaques. *Journal of Experimental Child Psychology, 95*, 215–229.

Jordan, N. C., Glutting, J., Dyson, N., Hassinger-Das, B., & Irwin, C. (2012). Building kindergartners' number sense: A randomized controlled study. *Journal of Educational Psychology, 104*(3), 647–660.

Joseph, R. (2000). Fetal brain behaviour and cognitive development. *Developmental Review, 20*, 81–98.

Jusczyk, P.W., & Aslin, R. N. (1995). Infants' detection of the sound patterns of words in fluent speech. *Cognitive Psychology, 29*, 1–23.

Jusczyk, P.W., Houston, D. M., & Newsome, M. (1999). The beginnings of word segmentation in English-learning infants. *Cognitive Psychology, 39*, 159–207.

Justice, E. M. (1985). Categorisation as a preferred memory strategy: Developmental changes during elementary school. *Developmental Psychology, 6*, 1105–1110.

Justice, E. M., Baker-Ward, L., Gupta, S., & Jannings, L. R. (1997). Means to the goal of remembering: Developmental changes in awareness of strategy use-performance relations. *Journal of Experimental Child Psychology, 65*, 293–314.

Kabdebon, C., Pena, M., Buiatti, M., & Dehaene-Lambertz, G. (2015). Electrophysiological evidence of statistical learning of long-distance dependencies in 8-month-old preterm and full-term infants. *Brain and Language, 148*, 25–36.

Kaiser, M. K., McCloskey, M., & Profitt, D. R. (1986). Development of intuitive theories of motion: Curvilinear motion in the absence of external forces. *Developmental Psychology, 22*, 67–71.

Kaiser, M. K., Profitt, D. R., & McCloskey, M. (1985). The development of beliefs about falling objects. *Perception & Psychophysics, 38*, 533–539.

Kalashnikova, M., Goswami, U., & Burnham, D. (2018a). Mothers speak differently to infants at risk for dyslexia. *Developmental Science, 21*(1), e12487.

Kalashnikova, M., Peter, V., Di Liberto, G. M., Lalor, E. C., & Burnham, D. (2018b). Infant-directed speech facilitates 7-month-olds cortical tracking of speech. *Scientific Reports, 8*, 13745.

Kaminski, J., Call, J., & Fischer, J. (2004). Word learning in a domestic dog: Evidence for fast mapping. *Science, 304*, 1682–1683.

Kaminski, J., Schulz, L., & Tomasello, M. (2012). How dogs know when communication is intended for them. *Developmental Science, 15*(2), 222–232.

Kampis, D., Parise, E., Cisbra, G., & Kovács, A. M. (2015). Neural signatures for sustaining object representations attributed to others in preverbal human infants. *Proceedings of the Royal Society B: Biological Sciences, 282*(1819).

Karasik, L. B., Tamis-LeMonda, C. S., & Adolph, K. E. (2011). Transition from crawling to walking and infants' actions with objects and people. *Child Development, 82*(4), 1199–1209.

Karmiloff-Smith, A. (1992). *Beyond Modularity: A Developmental Perspective of Cognitive Science*. Cambridge, MA: MIT Press/Bradford Books.

Karmiloff-Smith, A. (2007). Atypical epigenesis. *Developmental Science, 10*(1), 84–88.

Karpov, Y.V. (2005). *The Neo-Vygotskian Approach to Child Development*. New York: Cambridge University Press.

Kaufman, E. L., Lord, M., Reese, T. W., & Volkmann, J. (1949). The discrimination of visual number. *American Journal of Psychology, 62*, 498–525.

Kaufman, J., Csibra, G., & Johnson M. H. (2003a). Representing occluded objects in the human infant brain. *Proceedings of the Royal Society of London B (Suppl.), 270*, S140–S143.

Kaufman, J., Csibra, G., & Johnson, M. H. (2005). Oscillatory activity in the infant brain reflects object maintenance. *PNAS, 102*(42), 15271–15274.

Kaufman, J., Mareschal, D., & Johnson, M. H. (2003b). Graspability and object processing in infants. *Infant Behavior and Development, 26*, 516–528.

Keane, M. K. (1985). *Analogical Problem-Solving*. Chichester: Ellis Horwood.

Keil, F. C. (1987). Conceptual development and category structure. In U. Neisser (Ed.), *Concepts and Conceptual Development: Ecological and Intellectual Factors in Categorisation*, pp. 175–200. Cambridge: Cambridge University Press.

Keil, F. C. (1989). *Concepts, Kinds and Cognitive Development*. Cambridge, MA: MIT Press.

Keil, F. C. (1991). The emergence of theoretical beliefs as constraints on concepts. In S. Carey & R. Gelman (Eds.), *The Epigenesis of Mind: Essays on Biology & Cognition*, pp. 237–256. Hillsdale, NJ: Lawrence Erlbaum.

Keil, F. C. (1994). The birth and nurturance of concepts of domains: The origins of concepts of living things. In L. A. Hirschfeld & S. A. Gelman (Eds.), *Mapping the Mind*, pp. 234–254. New York: Cambridge.

Keil, F. C. (2006). Explanation and understanding. *Annual Review of Psychology, 57*, 227–254.

Keil, F. C., & Batterman, N. (1984). A characteristic-to-defining shift in the development of word meaning. *Journal of Verbal Learning & Verbal Behaviour, 23*, 221–236.

Kekule, F. A. (1865). *Bulletin of the Society of Chemistry France* (Paris), *3*, 98.

Kerr, A., & Zelazo, P. D. (2004). Development of 'hot' executive function: The children's gambling task. *Brain and Cognition, 55*, 148–157.

Kim, E.Y., & Song, H. (2015). Six-month-olds actively predict others' goal-directed actions. *Cognitive Development, 33*, 1–13.

Kim, S., Paulus, M., Sodian, B., & Proust, J. (2016). Young children's sensitivity to their own ignorance in informing others. *PLOS One, 11*(3), e0152595.

Király, I., Csibra, G., & Gergely, G. (2013). Beyond rational imitation: Learning arbitrary means actions from communicative demonstrations. *Journal of Experimental Child Psychology, 116*(2), 471–486.

Kirby, L.A., Moraczewski, D., Warnell, K., Velnosky, K., & Redkay, E. (2018). Social network size relates to developmental neural sensitivity to biological motion. *Developmental Cognitive Neuroscience, 30*, 169–177.

Kirkham, N. Z., Cruess, L., & Diamond, D. (2003). Helping children apply their knowledge to their behavior on a dimension-switching task. *Developmental Science, 6,* 449–467.

Kirkham, N. Z., Slemmer, J. A., & Johnson, S. P. (2002). Visual statistical learning in infancy: Evidence for a domain general learning mechanism. *Cognition, 83,* B35–B42.

Kisilevsky, B. S., & Low, J. A. (1998). Human fetal behaviour: 100 years of study. *Developmental Review, 18,* 1–29.

Klahr, D., Fay, A. L., & Dunbar, K. (1993). Heuristics for scientific experimentation: A developmental study. *Cognitive Psychology, 24,* 111–146.

Klingberg, T., Forssberg, H., & Westerberg, H. (2002). Increased brain activity in frontal and parietal cortex underlies the development of visuospatial working memory capacity during childhood. *Journal of Cognitive Neuroscience, 14*(1), 1–10.

Kluender, K. R., Diehl, R. L., & Killeen, P. R. (1987). Japanese quail can learn phonetic categories. *Science, 237*(4819), 1195–1197.

Kobayashi, M., Kato, J., Haynes, C. W., Macaruso, P., & Hook, P. (2003). Cognitive linguistic factors in Japanese children's reading. *Japanese Journal of Learning Disabilities, 12,* 240–247.

Kochanska, G., Murray, K., Jacques, T. Y., Koenig, A. L., & Vandegeest, K. (1996). Inhibitory control in young children and its role in emerging internalization. *Child Development, 67,* 490–507.

Koenig, M. A., & Echols, C. H. (2003). Infants' understanding of false labeling events: The referential roles of words and the speakers who use them. *Cognition, 87,* 179–208.

Kokis, J., Macpherson, R., Toplak, M., West, R. F., & Stanovich, K. E. (2002). Heuristic and analytic processing: Age trends and associations with cognitive ability and cognitive styles. *Journal of Experimental Child Psychology, 83,* 26–52.

Kopera, K., Dehaene, S., & Streissguth, A. P. (1996). Impairments of number processing induced by prenatal alcohol exposure. *Neuropsychologia, 34,* 1187–1196.

Koslowski, B. (1996). *Theory and Evidence: The Development of Scientific Reasoning.* Cambridge, MA: MIT Press.

Koslowski, B., and Masnick, A. M. (2010). Causal reasoning. In U. Goswami (Ed.), *The Wiley-Blackwell Handbook of Childhood Cognitive Development,* 2nd edition. Oxford: Blackwell.

Kotovsky, L., & Baillargeon, R. (1998). The development of calibration-based reasoning about collision events in young infants. *Cognition, 67,* 311–351.

Kovács, A. M., Kühn, S., Gergely, G., Csibra, G., & Brass, M. (2014). Are all beliefs equal? Implicit belief attributions recruiting core brain regions of theory of mind. *PLOS One, 9*(9), e106558.

Kovács, A. M., Tauzin, T., Téglás, E., Gergely, G., & Csibra, G. (2014). Pointing as epistemic request. 12-month-olds point to receive new information. *Infancy, 19*(6), 543–557.

Kovács, A. M., Téglás, E., & Endress, A. D. (2010). The social sense: Susceptibility to others' beliefs in human infants and adults. *Science, 330*(6012), 1830–1834.

Kozhevnikov, M., & Hegarty, M. (2001). Impetus beliefs as default heuristic: Dissociation between explicit and implicit knowledge about motion. *Psychonomic Bulletin and Review, 8,* 439–453.

Krist, H., Fieberg, E. L., & Wilkening, F. (1993). Intuitive physics in action and judgement: The development of knowledge about projectile motion. *Journal of Experimental Psychology: Learning, Memory & Cognition, 19,* 952–966.

Kristen-Antonow, S., Sodian, B., Perst, H., & Licata, M. (2015). A longitudinal study of the emerging self from 9 months to the age of 4 years. *Frontiers in Psychology, 6*(789).

Kucian, K., Loenneker, T., Dietrich, T., Dosch, M., Martin, E., & von Aster, M. (2006). Impaired neural networks for approximate calculation in dyscalculic children: A functional MRI study. *Behavioral and Brain Functions, 2*(31).

Kuefner, D., de Heering, A., Jacques, C., Palmero-Soler, E., & Rossion, B. (2010). Early visually evoked electrophysiological responses over the human brain (P1, N170) show stable patterns of face-sensitivity from 4 years to adulthood. *Frontiers in Human Neuroscience, 3*(67).

Kuhl, P. K. (1986). Reflections on infants' perception and representation of speech. In J. Perkell & D. Klatt (Eds.), *Invariance and Variability in Speech Processes.* Norwood, NJ: Ablex.

Kuhl, P. K. (1991). Human adults and human infants show a 'perceptual magnet effect' for the prototypes of speech categories, monkeys do not. *Perception & Psychophysics, 50,* 93–107.

Kuhl, P. K. (2004). Early language acquisition: Cracking the speech code. *Nature Reviews Neuroscience, 5,* 831–843.

Kuhl, P. K. (2007). Is speech learning 'gated' by the social brain? *Developmental Science, 10*(1), 110–120.

Kuhl, P. K., Coffey-Corina, S., Padden, D., & Rivera-Gaxiola, M. (2006). The bilingual brain: A comparison of native and non-native speech perception in monolingual and bilingual infants. *The Journal of the Acoustical Society of America, 120*(5), 3135–3135.

Kuhl, P. K., & Meltzoff, A. N. (1982). The bimodal perception of speech in infancy. *Science, 218,* 1138–1141.

Kuhl, P. K., Ramírez, R. R., Bosseler, A., Lin, J. F. L., & Imada, T. (2014). Infants' brain responses to speech suggest analysis by synthesis. *Proceedings of the National Academy of Sciences, 111*(31), 11238–11245.

Kuhl, P. K., Tsao. F.-M., & Liu, H.-M. (2003). Foreign-language experience in infancy: Effects of short-term exposure and social interaction on phonetic learning. *Proceedings of the National Academy of Sciences, 100,* 9096–9101.

Kuhl, P. K., Williams, K. A., Lacerda, F., Stevens, K. N., & Lindblom, B. (1992). Linguistic experience alters

phonetic perception in infants by 6 months of age. *Science*, *255*, 606–608.
Kuhlmeier, V. A., Wynn, K., & Bloom, P. (2003). Attribution of dispositional states by 12-month-olds. *Psychological Science*, *14*, 402–408.
Kuhn, D. (1989). Children and adults as intuitive scientists. *Psychological Review*, *96*, 674–689.
Kuhn, D. (1999). Metacognitive development. In L. Balter & C. S. Tamis-LeMonda (Eds.), *Child Psychology: A Handbook of Contemporary Issues*, pp. 259–286. New York: Psychology Press.
Kuhn, D. (2000). Theory of mind, metacognition, and reasoning: A life-span perspective. In P. Mitchell & K. J. Riggs (Eds.), *Children's Reasoning and the Mind*, pp. 301–326. Hove, UK: Psychology Press/Taylor & Francis.
Kuhn, D., Amsel, E., & O'Loughlin, M. (1988). *The Development of Scientific Thinking Skills.* San Diego: Academic.
Kuhn, D., Garcia-Mila, M., Zohar, A., & Andersen, C. (1995). Strategies of knowledge acquisition. *Monographs of the Society for Research in Child Development*, *60*(4).
Kunzinger, E. L., & Witryol, S. L. (1984). The effects of differential incentives on second-grade rehearsal and free recall. *Journal of Genetic Psychology*, *144*, 19–30.
Kurtz, B. E., & Weinert, F. E. (1989). Metamemory, memory performance and causal attributions in gifted and average children. *Journal of Experimental Child Psychology*, *48*, 45–61.
Lagnado, D. A., Waldmann, M. R., Hagmayer, Y., & Sloman, S. A. (2007). Beyond covariation: Cues to causal structure. In A. Gopnik & L. Schulz (Eds.), *Causal Learning: Psychology, Philosophy, and Computation.* Oxford: Oxford University Press.
Lakoff, G. (1986). *Women, Fire and Dangerous Things: What Categories Tell Us About the Nature of Thought.* Chicago: University of Chicago Press.
Lam, C., & Kitamura, C. (2010). Maternal interactions with a hearing and hearing-impaired twin: Similarities and differences in speech input, interaction quality, and word production. *Journal of Speech, Language & Hearing Research*, *53*, 543–556.
Lamb, M. E., Orbach, Y., Warren, A. R., Esplin, P. W., & Hershkowitz, I. (2007). Enhancing performance: Factors affecting the informativeness of young witnesses. In M. P. Toglia, J. D. Read, D. F. Ross, & R. C. L. Lindsay (Eds.), *The Handbook of Eyewitness Psychology, Vol 1. Memory for Events*, pp. 429–451. Mahwah, NJ: Lawrence Erlbaum Associates Publishers.
Lamsfuss, S. (1995). *Regularity of Movement and the Animate-Inanimate Distinction.* Poster presented at the Biennial Meeting of the Society for Research in Child Development, Indianapolis, IN, March 1995.
Landau, A. N., Schreyer, H. M., van Pelt, S., & Fries, P. (2015). Distributed attention is implemented through theta-rhythmic gamma modulation. *Current Biology*, *25*(17), 2332–2337.
Landerl, K., Wimmer, H., & Frith, U. (1997). The impact of orthographic consistency on dyslexia: A German-English comparison. *Cognition*, *63*, 315–334.
Landerl, K., & Kölle, C. (2009). Typical and atypical development of basic numerical skills in elementary school. *Journal of Experimental Child Psychology*, *103*(4), 546–565.
Landerl, K. et al. (2013). Predictors of developmental dyslexia in European orthographies with varying complexity. *The Journal of Child Psychology and Psychiatry*, *54*(6), 686–694.
Langer, N., Peysakhovich, B., Zuk, J., Drottar, M., Sliva, D. D., Smith, S., Becker, B. L. C., Grant, P. E., & Gaab, N. (2017). White matter alterations in infants at risk for developmental dyslexia. *Cerebral Cortex*, *27*(2), 1027–1036.
LeCun, Y., Bengio, Y., & Hinton, G. (2015). Deep learning. *Nature*, *521*, 436–444.
Leevers, H. J., & Harris, P. L. (2000). Counterfactual syllogistic reasoning in normal four-year-olds, children with learning disabilities, and children with autism. *Journal of Experimental Child Psychology*, *76*, 64–87.
Legare, C. H. (2014). The contributions of explanation and exploration to children's scientific reasoning. *Child Development Perspectives*, *8*(2), 101–106.
Legare, C. H., & Lombrozo, T. (2014). Selective effects of explanation on learning during early childhood. *Journal of Experimental Child Psychology*, *126*, 198–212.
Legare, C. H., Schult, C. A., Impola, M., & Souza, A. L. (2016). Young children revise explanations in response to new evidence. *Cognitive Development*, *39*, 45–56.
Leong, V., & Goswami, U. (2015). Acoustic-emergent phonology in the amplitude envelope of child-directed speech. *PLOS One*, *10*(12), e0144411.
Leong, V., Kalashnikova, M., Burnham, D., & Goswami, U. (2017). The temporal modulation structure of infant-directed speech. *Open Mind*, *1*(2), 78–90.
Leong, V., Stone, M., Turner, R. E., & Goswami, U. (2014). A role for amplitude modulation phase relationships in speech rhythm perception. *Journal of the Acoustical Society of America*, *136*, 366–381.
Leslie, A. M. (1987). Pretense and representation: The origins of 'theory of mind'. *Psychological Review*, *94*, 412–426.
Leslie, A. M. (1994). ToMM, ToBY and agency: Core architecture and domain specificity. In L. A. Hirschfeld & S. A. Gelman (Eds.), *Mapping the Mind*, pp. 119–148. New York: Cambridge.
Leslie, A. M., & Keeble, S. (1987). Do six-month-old infants perceive causality? *Cognition*, *25*, 265–88.
Levitt, H. (1971). Transformed up-down methods in psychoacoustics. *Journal of the Acoustical Society of America*, *49*, 467–477.
Lewkowicz, D. J., & Hansen-Tift, A. M. (2012). Infants deploy selective attention to the mouth of a talking face when learning speech. *PNAS*, *109*(5), 1431–1436.
Liberman, I. Y., Shankweiler, D., Fischer, F. W., & Carter, B. (1974). Explicit syllable and phoneme segmentation in the young child. *Journal of Experimental Child Psychology*, *18*, 201–212.

Libertus, M. E., & Brannon, E. M. (2010). Stable individual differences in number discrimination in infancy. *Developmental Science, 13*(6), 900–906.

Light, P., Blaye, A., Gilly, M., & Girotto, V. (1989). Pragmatic schemas and logical reasoning in 6- to 8-year-old children. *Cognitive Development, 4*, 49–64.

Lillard, A. S. (2002). Pretend play and cognitive development. In U. Goswami (Ed.), *Blackwell Handbook of Childhood Cognitive Development*, pp. 188–205. Oxford: Blackwell.

Lillard, A. S., Lerner, M. D., Hopkins, E. J., Dore, R. A., Smith, E. D., & Palmquist, C. M. (2013). The impact of pretend play on children's development: A review of the evidence. *Psychological Bulletin, 139*(1), 1–34.

Liszkowski, U., Carpenter, M., Henning, A., Striano, T., & Tomasello, M. (2004). Twelve-month-old infants point to share attention. *Developmental Science, 7*, 297–307.

Liu, Y., Su, Y., Xu, G., & Pei, M. (2018). When do you know what you know? The emergence of memory monitoring. *Journal of Experimental Child Psychology, 166*, 34–48.

Lloyd-Fox, S., Blasi, A., Everdell, N., Elwell, C. E., & Johnson, M. H. (2011). Selective cortical mapping of biological motion processing in young infants. *Journal of Cognitive Neuroscience, 23*(9), 2521–2532.

Lockl, K., & Schneider, W. (2002). Developmental trends in children's feeling-of-knowing judgements. *International Journal of Behavioral Development, 26*(4), 327–333.

Loftus, E. F., & Zanni, G. (1975). Eyewitness testimony: The influence of the wording of a question. *Bulletin of Psychonomic Society, 5*, 19–31.

Lohmann, H., & Tomasello, M. (2003). The role of language in the development of false belief understanding: A training study. *Child Development, 74*, 1130–1144.

Long, X., Benischek, A., Dewey, D., & Lebel, C. (2017). Age-related functional brain changes in young children. *NeuroImage, 155*, 322–330.

Loucks, J., Mutschler, C., & Meltzoff, A. N. (2016). Children's representation and imitation of events: How goal organization influences 3-year-old children's memory for action sequences. *Cognitive Science, 41*(7), 1904–1933.

Löw, A., Bentin, S., Rockstroh, B., Silberman, Y., Gomolla, A., Cohen, R., & Elbert, T. (2003). Semantic categorization in the human brain: Spatiotemporal dynamics revealed by magnetoencephalography. *Psychological Science, 14*(4), 367–372.

Lundberg, I., Frost, J., & Petersen, O.-P. (1988). Effects of an extensive program for stimulating phonological awareness in preschool children. *Reading Research Quarterly, 23*, 263–284.

Lundberg, I., Olofsson, A., & Wall, S. (1980). Reading and spelling skills in the first school years predicted from phonemic awareness skills in kindergarten. *Scandinavian Journal of Psychology, 21*, 159–173.

Luo, Y., & Baillargeon, R. (2005). When the ordinary seems unexpected: Evidence for incremental physical knowledge in young infants. *Cognition, 95*(3), 297–328.

Luria, A. R. (1976). *Cognitive Development: Its Cultural and Social Foundations*. Cambridge, MA: Harvard University Press.

Luria, A. R., Pribram, K. H., & Homskaya, E. D. (1964). An experimental analysis of the behavioural disturbances produced by a left frontal arachnoidal endothelioma (meningioma). *Neuropsychologia, 2*, 257–280.

MacNeilage, P. F. (1998). The frame/content theory of evolution of speech production. *Behavioral and Brain Sciences, 21*, 499–511.

MacWhinney, B. (1995). *The CHILDES Project: Tools for Analyzing Talk*. Hillsdale, NJ: Erlbaum.

Maguire, E. A., Gadian, D. G., Johnsrude, I. S., Good, C. D., Ashburner, J., Frackowiak, R. S. J., & Frith, C. D. (2000). Navigation-related structural change in the hippocampi of taxi drivers. *PNAS, 97*(8), 4398–4403.

Mandel, D. R., Jusczyk, P. W., & Pisoni, D. B. (1995). Infants' recognition of the sound patterns of their own names. *Psychological Science, 6*, 314–317.

Mandler, J. M. (1990). Recall of events by preverbal children. In A. Diamond (Ed.), *The Development and Neural Bases of Higher Cognitive Functions*, pp. 485–516. New York: New York Academy of Sciences.

Mandler, J. M. (2004). *The Foundations of Mind: Origins of Conceptual Thought*. Oxford: Oxford University Press.

Mandler, J. M., & Bauer, P. J. (1988). The cradle of categorisation: Is the Basic Level basic? *Cognitive Development, 3*, 247–264.

Mandler, J. M., Bauer, P. J., & McDonough, L. (1991). Separating the sheep from the goats: Differentiating global categories. *Cognitive Psychology, 23*, 263–298.

Mandler, J. M., & McDonough, L. (1993). Concept formation in infancy. *Cognitive Development, 8*, 291–318.

Mandler, J. M., & McDonough, L. (1995). Long-term recall of event sequences in infancy. Special issue: Early memory. *Journal of Experimental Child Psychology, 59*, 457–474.

Manuilenko, Z. V. (1948). The development of voluntary behavior in preschool-age children. *Soviet Psychology, 13*(4), 65–116.

Marcus, G. F., Pinker, S., Ullman, M., Hollander, M., Rosen, T., & Xu, F. (1992). Overregularization in language acquisition. *Monographs of the Society for Research in Child Development, 57*(4), 1–178.

Mareschal, D., & Johnson, M.H. (2003). The 'what' and 'where' of object representations in infancy. *Cognition, 88*, 259–276.

Mareschal, D., Johnson, M. H., Sirois, S., Thomas, M. S. C., Spratling, M., & Westermann, G. (2007). *Neuroconstructivism: How the Brain Constructs Cognition*. Oxford: Oxford University Press.

Mareschal, D., Plunkett, K., & Harris, P. L. (1999). A computational and neuropsychological account of object-oriented behaviours in infancy. *Developmental Science, 2*, 306–317.

Marie, C., Magne, C., & Besson, M. (2011). Musicians and the metric structure of words. *Journal of Cognitive Neuroscience, 23*, 294–305.

Markman, A. B., & Dietrich, E. (2000). Extending the classical view of representation. *Trends in Cognitive Sciences, 4*(12), 470–475.

Markman, E. M., & Seibert, J. (1976). Classes and collections: Internal organization and resulting holistic properties. *Cognitive Psychology, 8*, 561–577.

Markovits, H. (2000). A mental model analysis of young children's conditional reasoning with meaningful premises. *Thinking and Reasoning, 6*(4), 335–348.

Markovits, H. (2017). In the beginning stages: Conditional reasoning with category based and causal premises in 8- to 10-year-olds. *Cognitive Development, 41*, 1–9.

Markovits, H., & Barrouillet, P. (2002). The development of conditional reasoning: A mental model account. *Developmental Review, 22*(1), 5–36.

Markson, L., & Bloom, P. (1997). Evidence against a dedicated system for word learning in children. *Nature, 385*, 813–815.

Marshall, P. J., & Meltzoff, A. N. (2011). Neural mirroring systems: Exploring the EEG μ rhythm in human infancy. *Developmental Cognitive Neuroscience, 1*(2), 110–123.

Marshall, P. J., & Meltzoff, A. N. (2014). Neural mirroring mechanisms and imitation in human infants. *Philosophical Transactions of the Royal Society B, 369*(1644).

Marshall, P. J., Saby, J. N., & Meltzoff, A. N. (2013). Imitation and the developing social brain: Infants' somatotopic EEG patterns for acts of self and other. *International Journal of Psychological Research, 6*, 22–29.

Marzolf, D. P., & DeLoache, J. S. (1994). Transfer in young children's understanding of spatial representations. *Child Development, 65*, 1–15.

Masataka, N. (2007). Music, evolution and language. *Developmental Science, 10*(1), 35–39.

Massey, C. M., & Gelman, R. (1988). Preschooler's ability to decide whether a photographed object can move itself. *Developmental Psychology, 24*, 307–317.

Masson, S., Potvin, P., Riopel, M., & Brault Foisy, L. M. (2014). Differences in brain activation between novices and experts in science during a task involving a common misconception in electricity. *Mind, Brain and Education, 8*(1), 44–55.

Mathewson, K. E., Gratton, G., Fabiani, M., Beck, D. M., & Ro, T. (2009). To see or not to see: Prestimulus alpha phase predicts visual awareness. *Journal of Neuroscience, 29*(9), 2725–2732.

Maurer, U., Brem, S., Bucher, K., & Brandeis, D. (2005). Emerging neurophysiological specialization for letter strings. *Journal of Cognitive Neuroscience, 17*(10), 1532–1552.

Maurer, U., Brem, S., Bucher, K., Kranz, F., Benz, R., Steinhausen, H. C., & Brandeis, D. (2007). Impaired tuning of a fast occipito-temporal response for print in dyslexic children learning to read. *Brain, 130*(12), 3200–3210.

May, L., Gervain, J., Carreiras, M., & Werker, J. F. (2017). The specificity of the neural response to speech at birth. *Developmental Science, 21*(3), e12564.

McCloskey, M. (1983). Intuitive physics. *Scientific American, 248*, 122–130.

McCune-Nicolich, L. (1981). Toward symbolic functioning: Structure of early pretend games and potential parallels with language. *Child Development, 52*, 785–797.

McGarrigle, J., & Donaldson, M. (1975). Conservation accidents. *Cognition, 3*, 341–350.

McGillion, M., Herbert, J. S., Pine, J., Vihman, M., DePaolis, R., Keren Portnoy, T., & Matthews, D. (2017). What paves the way to conventional language? The predictive value of babble, pointing, and socioeconomic status. *Child Development, 88*(1), 156–166.

McGurk, H., & MacDonald, J. (1976). Hearing lips and seeing voices. *Nature, 264*, 746–748.

McKenzie, B., & Over, R. (1983). Young infants fail to imitate facial and manual gestures. *Infant Behaviour & Development, 6*, 85–95.

McKenzie, B. E., & Bigelow, E. (1986). Detour behaviour in young human infants. *British Journal of Developmental Psychology, 4*, 139–148.

McKenzie, B. E., Day, R. H., & Ihsen, E. (1984). Localisation of events in space: Young infants are not always egocentric. *British Journal of Developmental Psychology, 2*, 1–10.

Medin, D. L. (1989). Concepts and conceptual structure. *American Psychologist, 44*, 1469–1481.

Medin, D. L., & Schaffer, M. M. (1978). Context theory of classification learning. *Psychological Review, 85*, 207–238.

Mehler, J., Jusczyk, P., Lambertz, G., Halsted, N., Bertoncini, J., & Amiel-Tison, C. (1988). A precursor of language acquisition in young infants. *Cognition, 29*, 143–178.

Mehler, J., Lambertz, G., Jusczyk, P. W., & Amiel-Tyson, C. (1986). Discrimination de la langue maternelle par le nouveau-né. *Comptesrendus de l'Académie des Sciences de Paris, 303*(3), 637–640.

Meins, E. (1997). *Security of Attachment and the Social Development of Cognition*. Hove, UK: Psychology Press.

Meins, E. (1999). Sensitivity, security and internal working models: Bridging the transmission gap. *Attachment & Human Development, 1*(3), 325–342.

Meins, E. (2013). *Security of Attachment and the Social Development of Cognition*. New York: Psychology Press.

Meins, E., & Fernyhough, C. (1999). Linguistic acquisitional style and mentalising development: The role of maternal mind-mindedness. *Cognitive Development, 14*, 363–380.

Meins, E., Fernyhough, C., Wainwright, R., Das Gupta, M., Fradley, E., & Tuckey, M. (2002). Maternal mind-mindedness and attachment security as predictors of theory of mind understanding. *Child Development, 73*, 1715–1726.

Meltzoff, A. N. (1985). Immediate and deferred imitation in 14- and 24-month-old infants. *Child Development, 56*, 62–72.

Meltzoff, A. N. (1988a). Infant imitation after a 1-week delay: Long-term memory for novel acts and multiple stimuli. *Developmental Psychology, 24*, 470–476.

Meltzoff, A. N. (1988b). Infant imitation and memory: Nine-month-olds in immediate and deferred tests. *Child Development, 59*, 217–225.

Meltzoff, A. N. (1988c). Imitation of televised models by infants. *Child Development, 59*, 1221–1229.
Meltzoff, A. N. (1995a). Understanding the intentions of others: Re-enactment of intended acts by 18-month-old children. *Developmental Psychology, 31*, 838–850.
Meltzoff, A. N. (1995b). What infant memory tells us about infantile amnesia: Long-term recall and deferred imitation. *Journal of Experimental Child Psychology, 59*, 497–515.
Meltzoff, A. N. (2002). Imitation as a mechanism for social cognition: Origins of empathy, theory of mind and the representation of action. In U. Goswami (Ed.), *Blackwell Handbook of Childhood Cognitive Development*, pp. 6–25. Oxford: Blackwell.
Meltzoff, A. N. (2017). Elements of a comprehensive theory of infant imitation. *Behavioral and Brain Sciences, 40*, e396.
Meltzoff, A. N., & Borton, R. W. (1979). Intermodal matching by human neonates. *Nature, 282*, 403–404.
Meltzoff, A. N., & Decety, J. (2003). What imitation tells us about social cognition: A rapprochement between developmental psychology and cognitive neuroscience. *Philosophical Transactions of the Royal Society: Biological Sciences, 358*, 491–500.
Meltzoff, A. N., & Moore, M. K. (1977). Imitation of facial and manual gestures by humamn neonates. *Science, 198*, 75–78.
Meltzoff, A. N., & Moore, M. K. (1983). Newborn infants imitate adult facial gestures. *Child Development, 54*, 702–709.
Mendelson, R., & Shultz, T. R. (1975). Covariation and temporal contiguity as principles of causal inference in young children. *Journal of Experimental Child Psychology, 22*, 408–412.
Menon, V. (2016). Working memory in children's math learning and its disruption in dyscalculia. *Current Opinion in Behavioral Sciences, 10*, 125–132.
Mercure, E., Quiroz, I., Goldberg, L., Bowden-Howl, H., Coulson, K., Gliga, T., Filippi, R., Bright, P., Johnson, M. H., & Macsweeney, M. (2018). Impact of language experience on attention to faces in infancy: Evidence from unimodal and bimodal bilingual infants. *Frontiers in Psychology, 9*, 1943.
Meristo, M., Morgan, G., Geraci, A., Iozzi, L., Hjelmquist, E., Surian, L., & Siegal, M. (2012). Belief attribution in deaf and hearing infants. *Developmental Science, 15*(5), 633–640.
Mersad, K., & Dehaene-Lambertz, G. (2016). Electrophysiological evidence of phonetic normalization across coarticulation in infants. *Developmental Science, 19*(5), 710–722.
Mervis, C. B. (1987). Child-basic object categories and early lexical development. In U. Neisser (Ed.), *Concepts and Conceptual Development: Ecological and Intellectual Factors in Categorisation*, pp. 201–233. Cambridge: Cambridge University Press.
Mervis, C. B., & Pani, J. R. (1980). Acquisition of basic object categories. *Cognitive Psychology, 12*, 496–522.
Meyer, M., Gelman, S. A., Roberts, S. O., & Leslie, S. J. (2016). My heart made me do it: Children's essentialist beliefs about heart transplants. *Cognitive Science, 41*(6), 1694–1712.
Michotte, A. (1963). *The Perception of Causality*. Andover: Methuen.
Miller, S. A. (1982). On the generalisability of conservation: A comparison of different kinds of transformation. *British Journal of Psychology, 73*, 221–230.
Mills, D., Prat, C., Stager, C., Zangl, R., Neville, H., & Werker, J. (2004). Language experience and the organization of brain activity to phonetically similar words: ERP evidence from 14- and 20-month olds. *Journal of Cognitive Neuroscience, 16*, 1452–1464.
Milner, A. D., & Goodale, M. A. (1995). *The Visual Brain in Action*. Oxford: Oxford University Press.
Milner, B. (1963). Effects of brain lesions on card sorting. *Archives of Neurology, 9*, 90–100.
Milner, B. (1964). Some effects of frontal lobectomy in man. In J. M. Warren & K. Akert (Eds.), *The Frontal Granular Cortex & Behaviour*, pp. 313–334. New York: McGraw-Hill.
Mitroff, S. R., Scholl, B. J., & Wynn, K. (2004). Divide and conquer: How object files adapt when a persisting object splits into two. *Psychological Science, 15*, 420–425.
Miura, I. T., Kim, C. C., Chang, C.-M., & Okamoto, Y. (1988). Effects of language characteristics on children's cognitive representation of number: Cross-national comparisons. *Child Development, 59*, 1445–1450.
Moffitt, T. E. (1990). Juvenile delinquency and attention deficit disorder: Boys' developmental trajectories from age 3 to age 15. *Child Development, 61*, 893–910.
Moffitt, T. E., Arseneault, L., Belsky, D., Dickson, N., Hancox, R. J., Harrington, H., Houts, R., Poulton, R., Roberts, B. W., Ross, S., Sears, M. R., Thomson, W. M., & Caspi, A. (2011). A gradient of childhood self-control predicts health, wealth, and public safety. *PNAS, 108*(7), 2693–2698.
Molinaro, N., Lizarazu, M., Lallier, M., Bourguignon, M., & Carreiras, M. (2016). Out-of-synchrony speech entrainment in developmental dyslexia. *Human Brain Mapping, 37*(8), 2767–2783.
Moll, H., & Tomasello, M. (2004). 12- and 18-month-old infants follow gaze to spaces behind barriers. *Developmental Science, 7*, F1–F9.
Monsell, S., & Driver, J. (2000). Banishing the control homunculus. In S. Monsell & J. Driver (Eds.), *Control of Cognitive Processes: Attention and Performance XVIII*, pp. 3–32. Cambridge, MA: MIT Press.
Moore, D., Benenson, J., Reznick, S.J., Peterson, M., & Kagan, J. (1987). Effect of auditory numerical information infants' looking behaviour: Contradictory evidence. *Developmental Psychology, 23*, 665–670.
Moore, C., & Corkum, V. (1994). Social understanding at the end of the first year of life. *Developmental Review, 14*, 349–372.

Moors, P., Wagemans, J., & de-Wit, L. (2017). Causal events enter awareness faster than non-causal events. *PeerJ*, *5*, e2932.

Morken, M., Helland, T., Hugdahl, K., & Spechtad, K. (2017). Reading in dyslexia across literacy development: A longitudinal study of effective connectivity. *NeuroImage*, *144*(A), 92–100.

Mousikou, B., Beyersmann, E., Ktori, M., Javourey-Drevet, L., Crepaldi, D., Ziegler, J., Grainger, J., & Schroeder, S. (in press). Orthographic consistency influences morphological processing in reading aloud: Evidence from a cross-linguistic study. *Developmental Science*.

Moutier, S., Plagne-Cayeux, S., Melot, A. M., & Houde, O. (2006). Syllogistic reasoning and belief-bias inhibition in school children: Evidence from a negative priming paradigm. *Developmental Science*, *9*(2), 166–172.

Moyer, R. S., & Landauer, T. K. (1967). Time required for judgments of numerical inequality. *Nature*, *215*, 1519–1520.

Mullaley, S. L., & Maguire, E. A. (2014). Learning to remember: The early ontogeny of episodic memory. *Developmental Cognitive Neuroscience*, *9*, 12–29.

Mumme, D., Fernald, A., & Herrera, C. (1996). Infants' responses to facial and vocal emotional signals in a social referencing paradigm. *Child Development*, *67*, 3219–3237.

Munakata, Y. (2001). Graded representations in behavioural dissociations. *Trends in Cognitive Sciences*, *5*(7), 309–315.

Munakata, Y. (2004). Computational cognitive neuroscience of early memory development. *Developmental Review*, *24*, 133–153.

Munakata, Y., & McClelland, J. L. (2003). Connectionist models of development. *Developmental Science*, *6*, 413–429.

Murphy, G. L. (1982). Cue validity and levels of categorisation. *Psychological Bulletin*, *91*, 174–177.

Mussolin, C., De Volder, A., Grandin, C., Schlögel, X., Nassogne, M. C., & Noël, M. P. (2010a). Neural correlates of symbolic number comparison in developmental dyscalculia. *Journal of Cognitive Neuroscience*, *22*(5), 860–874.

Mussolin, C., Mejias, S., & Noël, M. P. (2010b). Symbolic and nonsymbolic number comparison in children with and without dyscalculia. *Cognition*, *115*(1), 10–25.

Myers, N. A., Clifton, R. K., & Clarkson, M. G. (1987). When they were very young: Almost-threes remember two years ago. *Infant Behaviour & Development*, *10*, 128–132.

Naatanen, R., & Picton, T. W. (1987). The N1 wave of the human electric and magnetic response to sound: A review and an analysis of the component structure. *Psychophysiology*, *24*, 375–425.

Nagy, E., Pal, A., & Orvos, H. (2014). Learning to imitate individual finger movements by the human neonate. *Developmental Science*, *17*(6), 841–857.

Naito, M. (1990). Repetition priming in children and adults: Age-related differences between implicit and explicit memory. *Journal of Experimental Child Psychology*, *50*, 462–484.

Natale, E., Senna, I., Bolognini, N., Quadrelli, E., Addabo, M., Cassia, V. M., & Turati, C. (2014). Predicting others' intention involves motor resonance: EMG evidence from 6- and 9-month-old infants. *Developmental Cognitive Neuroscience*, *7*, 23–29.

Naus, M. J., Ornstein, P. A., & Aivano, S. (1977). Developmental changes in memory: The effects of processing time and rehearsal instructions. *Journal of Experimental Child Psychology*, *23*, 237–251.

Nazzi, T., Bertoncini, J., & Mehler, J. (1998). Language discrimination by newborns: Toward an understanding of the role of rhythm. *Journal of Experimental Psychology: Human Perception and Performance*, *24*, 756–766.

Needham, A., Barrett, B., & Peterman, K. (2002). A pick me up for infants' exploratory skills: Early simulated experiences reaching for objects using 'sticky' mittens enhances young infants' object exploration skills. *Infant Behavior and Development*, *25*(3), 279–295.

Neisser, U. (1987). *Concepts and Conceptual Development: Ecological and Intellectual Factors in Categorisation*. Cambridge: Cambridge University Press.

Nelson, K. (1986). *Event Knowledge: Structure and Function in Development*. Hillsdale, NJ: Erlbaum.

Nelson, K. (1988). The ontogeny of memory for real events. In U. Neisser & E. Winograd (Eds.), *Remembering Reconsidered: Ecological and Traditional Approaches to the study of Memory*, pp. 244–276. New York: Cambridge University Press.

Nelson, K. (1993). The psychological and social origins of autobiographical memory. *Psychological Science*, *4*, 7–14.

Nelson, K., & Fivush, R. (2004). The emergence of autobiographical memory: A social cultural developmental theory. *Psychological Review*, *111*, 486–511.

Nelson, T. O., & Narens, L. (1990). Metamemory: A theoretical framework and some new findings. In G. H. Bower (Ed.), *The Psychology of Learning and Motivation*, pp. 125–173. New York: Academic Press.

Nelson, T. O., & Narens, L. (1994). Why investigate metacognition? In J. Metcalfe & A. P. Shimamura (Eds.), *Metacognition: Knowing about Knowing*, pp. 1–25. Cambridge, MA: MIT Press.

Newcombe, N., & Fox, N. (1994). Infantile amnesia: Through a glass darkly. *Child Development*, *65*, 31–40.

Nicolich, L. M. (1977). Beyond sensori-motor intelligence: Assessment of symbolic maturity through analysis of pretend play. *Merrill-Palmer Quarterly*, *23*, 89–101.

Nieder, A. (2016). The neuronal code for number. *Nature Reviews Neuroscience*, *17*, 366–382.

Noesselt, T., Rieger, J. W., Schoenfeld, M. A., Kanowski, M., Hinrichs, H., Heinze, H.-J., & Driver, J. S. (2007). Audiovisual temporal correspondence modulates human multisensory superior temporal sulcus plus primary sensory cortices. *Journal of Neuroscience*, *27*(42), 11431–11441.

Noles, N. S., Scholl, B. J., & Mitroff, S. R. (2005). The persistence of object file representations. *Perception & Psychophysics*, *67*, 324–334.

Ochsner, J. E., & Zaragoza, M. S. (1988). *The Accuracy and Suggestibility of Children's Memory for Neutral and Criminal Eyewitness Events.* Paper presented at the American Psychology and Law Meetings, Miami, FL.

O'Connor, N., & Hermelin, B. (1973). Spatial or temporal organisation of short-term memory. *Quarterly Journal of Experimental Psychology*, *25*(3), 335–343.

Oller, D. K. (1980). The emergence of the sounds of speech in infancy. In G. Yeni-Komshian, J. Kavanaugh, & C. Ferguson (Eds.), *Child Phonology*, pp. 93–112. New York: Academic Press.

Oller, D. K., & Eilers, R. (1988). The role of audition in infant babbling. *Child Development*, *59*, 441–449.

Olulade, O. A., Napoliello, E. M., & Eden, G. F. (2013). Abnormal visual motion processing is not a cause of dyslexia. *Neuron*, *79*(1), 180–190.

Onishi, K. H., & Baillargeon, R. (2005). Do 15-month-old infants understand false beliefs? *Science*, *308*, 255–258.

Oostenbroek, J., Suddendorf, T., Nielsen, M., Redshaw, J., Kennedy-Costantini, S., Davis, J., Clark, S., & Slaughter, V. (2016). Comprehensive longitudinal study challenges the existence of neonatal imitation in humans. *Current Biology*, *26*(10), 1334–1338.

Ornstein, P. A., Gordon, B. N., & Larus, D. M. (1992). Children's memory for a personally-experienced event: Implications for testimony. *Applied Developmental Psychology*, *6*, 49–60.

O'Sullivan, J. T. (1993). Preschoolers' beliefs about effort, incentives and recall. *Journal of Experimental Child Psychology*, *55*, 396–414.

Oved, I., Cheung, P., & Barner, D. (2014). Concepts as representatives for essences: Evidence from use of generics. *Proceedings of the Annual Meeting of the Cognitive Science Society*, *36*, 2723–2728.

Oved, I., Nichols, S., & Barner, D. (2015). *A Learning Model for Essentialist Concepts.* 5th International Conference on Development and Learning and on Epigenetic Robotics (ICDL-EpiRob), 13–16 August 2015. Providence, RI.

Palincsar, A. S., & Brown, A. L. (1984). Reciprocal teaching of comprehension-fostering and monitoring activities. *Cognition and Instruction*, *1*, 117–175.

Paris, S. G., & Oka, E. R. (1986). Children's reading strategies, metacognition, and motivation. *Developmental Review*, *6*, 25–56.

Pascual-Leone, J. (1970). A mathematical model for the transition rule in Piaget's developmental stages. *Acta Psychologica*, *32*, 301–345.

Pauen, S. (1996a). *Wie klassifizieren Kinder Lebewesen und Artefakte? Zur Rolle der Erscheinung und Funktion von Objektteilen.* Unpublished manuscript, University of Tübingen, Germany.

Pauen, S. (1996b). Children's reasoning about the interaction of forces. *Child Development*, *67*, 2728–2742.

Pauen, S. (2002). Evidence for knowledge-based category discrimination in infancy. *Child Development*, *73*, 1016–1033.

Pauen, S., & Wilkening, F. (1996). *Children's Spontaneous Use of Analogies in Explaining Natural Phenomena.* Manuscript submitted for publication.

Pauen, S., & Wilkening, F. (1997). Children's analogical reasoning about natural phenomena. *Journal of Experimental Psychology*, *67*, 90–113.

Paulesu, E., Démonet, J. F., Fazio, F., McCrory, E., Chanoine, V., Brunswick, N., Cappa, S. F., Cossu, G., Habib, M., Frith, C. D., & Frith, U. (2001). Dyslexia: Cultural diversity and biological unity. *Science*, *291*(5511), 2165–2167.

Paulus, M., Hunnius, S., Elk, M., & Bekkering, H. (2012). How learning to shake a rattle affects 8-month-old infants' perception of the rattle's sound: Electrophysiological evidence for action-effect binding in infancy. *Developmental Cognitive Neuroscience*, *2*, 90–96.

Pears, R., & Bryant, P. (1990). Transitive inferences by young children about spatial position. *British Journal of Psychology*, *81*, 497–510.

Peccei, J. S. (2005). *Child Language: A Resource Book for Students.* London: Routledge.

Pennington, B. F. (1994). The working memory function of the prefrontal cortices: Implications for developmental and individual differences in cognition. In M. M. Haith, J. Benson, R. Roberts, & B. F. Pennington (Eds.), *The Development of Future Oriented Processes*, pp. 243–289. Chicago: University of Chicago Press.

Perez, L. A., Peynircioglu, Z. F., & Blaxton, T. A. (1998). Developmental differences in implicit and explicit memory performance. *Journal of Experimental Child Psychology*, *70*, 167–185.

Perfetti, C.A., Beck, I., Bell, L., & Hughes, C. (1987). Phonemic knowledge and learning to read are reciprocal: A longitudinal study of first grade children. *Merrill-Palmer Quarterly*, *33*, 283–319.

Perner, J., & Lang, B. (2000). What causes 3-year-olds' difficulty on the dimensional change card sorting task? *Infant & Child Development*, *11*, 93–105.

Perner, J., Ruffman, T., & Leekam, S. R. (1994). Theory of mind is contagious; you catch it from your sibs. *Child Development*, *65*, 1224–1234.

Perone, S., Palanisamy, J., & Carlson, S. M. (2018). Age-related change in brain rhythms from early to middle childhood: Links to executive function. *Developmental Science*, *21*(6), e12691.

Perris, E. E., Myers, N. A., & Clifton, R. K. (1990). Long-term memory for a single infancy experience. *Child Development*, *61*, 1796–1807.

Peterson, C. (2012). Children's autobiographical memories across the years: Forensic implications of childhood amnesia and eyewitness memory for stressful events. *Developmental Review*, *32*(3), 287–306.

Peterson, C., & Siegal, M. (1998). Changing focus on the representational mind: Concepts of false photographs, false drawings and false beliefs in deaf, autistic and normal children. *British Journal of Developmental Psychology*, *16*, 301–320.

Peterson, C., Warren, K. L., & Short, M. M. (2011). Infantile amnesia across the years: A 2-year follow-up of children's earliest memories. *Child Development, 82*(4), 1092–1105.

Petitto, L. A., & Marentette, P. F. (1991). Babbling in the manual mode: Evidence for the ontogeny of language. *Science, 251*, 1483–1496.

Petitto, L. A., Holowka, S., Sergio, L. E., Levy, B., & Ostry, D. J. (2004). Baby hands that move to the rhythm of language: Hearing babies acquiring sign language babble silently on the hands. *Cognition, 93*, 43–73.

Piaget, J. (1952). *The Child's Conception of Number.* London: Routledge Kegan Paul.

Piaget, J. (1954). *The Construction of Reality in the Child.* New York: Basic Books.

Piaget, J. (1962). *Play, Dreams, and Imitation in Childhood.* New York: Norton.

Piaget, J., & Inhelder, B. A. (1956). *The Child's Conception of Space.* London: Routledge Kegan Paul.

Piazza, M., Facoetti, A., Trussardi, A. N., Berteletti, I., Conte, S., Lucangeli, D., Dehaene, S., & Zorzi, M. (2010). Developmental trajectory of number acuity reveals a severe impairment in developmental dyscalculia. *Cognition, 116*(1), 33–41.

Pica, P., Lemer, C., & Izard, V. (2004). Exact and approximate arithmetic in an Amazonian indigene group. *Science, 306*(5695), 499–503.

Piffer, L., Miletto Petrazzini, M. E., & Agrillo, C. (2013). Large number discrimination in newborn fish. *PLOS One, 8*(4), e62466.

Pilley, J. W., & Reid, A. K. (2011). Border collie comprehends object names as verbal referents. *Behavioural Processes, 86*(2), 184–195.

Pine, K. J., Lufkin, N., & Messer, D. (2004). More gestures than answers: Children learning about balance. *Developmental Psychology, 40*(6), 1059–1067.

Pine, K. J., Messer, D. J. (2000). The effect of explaining another's actions on children's implicit theories of balance. *Cognitive and Instruction, 18*(1), 35–51.

Pinel, P., Dehaene, S., Riviere, D., & Le Bihan, D. (2001). Modulation of parietal activation by semantic distance in a number comparison task. *Neuroimage, 14*, 1013–1026.

Ping, R. M., & Goldin-Meadow, S. (2008). Hands in the air: Using ungrounded iconic gestures to teach children conservation of quantity. *Developmental Psychology, 44*(5), 1277–1287.

Pitt, M. A., & McQueen, J. M. (1998). Is compensation for coarticulation mediated by the lexicon? *Journal of Memory and Language, 39*, 347–370.

Plomin, R. (2018). *Blueprint: How DNA Makes Us Who We Are.* London: Allen Lane.

Plotnik, J. M., de Waal, F. B., & Reiss, D. (2006). Self-recognition in an Asian elephant. *PNAS, 103*(45), 17053–17057.

Poeppel, D. (2014). The neuroanatomic and neurophysiological infrastructure for speech and language. *Current Opinion in Neurobiology, 28*, 142–149

Pomiechowska, B., & Csibra, G. (2017). Motor activation during action perception depends on action interpretation. *Neuropsychologia, 105*, 84–91.

Pomiechowska, B., & Csibra, G. (2018). *Referential Understanding of Pointing Actions and its Consequences for Object Representation.* Budapest: Cognitive Development Center, Central European University.

Porpodas, C. D. (1999). Patterns of phonological and memory processing in beginning readers and spellers of Greek. *Journal of Learning Disabilities, 32*, 406–416.

Power, A. J., Colling, L. C., Mead, N., Barnes, L., & Goswami, U. (2016). Neural encoding of the speech envelope by children with developmental dyslexia. *Brain & Language, 160*, 1–10.

Power, A. J., Mead, N., Barnes, L., & Goswami, U. (2012). Neural entrainment to rhythmically-presented auditory, visual and audio-visual speech in children. *Frontiers in Psychology, 3*, 216.

Power, A. J., Mead, N., Barnes, L., & Goswami, U. (2013). Neural entrainment to rhythmic speech in children with developmental dyslexia. *Frontiers in Human Neuroscience, 7*, 777.

Prado, J., Chadha, A., & Booth, J. R. (2011). The brain network for deductive reasoning: A quantitative meta-analysis of 28 neuroimaging studies. *Journal of Cognitive Neuroscience, 23*(11), 3483–3497.

Premack, D., & Woodruff, G. (1978). Does the chimpanzee have a theory of mind? *Behavioral and Brain Sciences, 4*, 515–526.

Pressley, M., Borkowski, J. G., & Schneider, W. (1987). Good strategy users coordinate metacognition and knowledge. In R. Vasta & G. Whitehurst (Eds.), *Annals of Child Development*, Vol. 5, pp. 89–129. Greenwich, CT: JAI Press.

Price, G. R., Holloway, I., Räsänen, P., Vesterinen, M., & Ansari, D. (2007). Impaired parietal magnitude processing in developmental dyscalculia. *Current Biology, 17*(24), R1042–R1043.

Pruett, J. R. Jr et al. (2015). Accurate age classification of 6- and 12-month-old infants based on resting state functional connectivity magnetic resonance imaging data. *Developmental Cognitive Neuroscience, 12*, 123–133.

Pugh, K. (2006). A neurocognitive overview of reading acquisition and dyslexia across languages. *Developmental Science, 9*(5), 448–450.

Quinn, P. C. (1994). The categorisation of above and below spatial relations by young infants. *Child Development, 65*, 58–69.

Quinn, P. C. (2002). Category representation in infants. *Current Directions in Psychological Science, 11*, 66–70.

Quinn, P. C., Doran, M. M., Reiss, J. E., & Hoffman, J. E. (2010). Neural markers of subordinate-level categorization in 6- to 7-month-old infants. *Developmental Science, 13*(3), 499–507.

Quinn, P. C., & Eimas, P. D. (1986). On categorisation in early infancy. *Merrill-Palmer Quarterly, 32*, 331–363.

Quinn, P. C., & Johnson, M. H. (1997). The emergence of category representations in infants: A connectionist

analysis. *Journal of Experimental Child Psychology, 66,* 236–263.
Quinn, P. C., & Johnson, M. H. (2000). Global before basic category representations in connectionist networks and 2-month old infants. *Infancy, 1,* 31–46.
Quinn, P. C., Westerlund, A., & Nelson, C. A. (2006). Neural markers of categorization in 6-month-old infants. *Psychological Science, 17,* 59–66.
Rajan, V., & Bell, M. A. (2015). Developmental changes in fact and source recall: Contributions from executive function and brain electrical activity. *Developmental Cognitive Neuroscience, 12,* 1–11.
Rakoczy, H., Tomasello, M., & Striano, T. (2005). On tools and toys: How children learn to act on and pretend with 'virgin objects'. *Developmental Science, 8*(1), 57–73.
Ramirez-Esparza, N., Garcia-Sierra, A., & Kuhl, P. K. (2014). Look who's talking: Speech style and social context in language input to infants are linked to concurrent and future speech development. *Developmental Science, 17*(6), 880–891.
Rauschecker, J. P., & Singer, W. (1981). The effects of early visual experience on the cat's visual cortex and their possible explanation by Hebb synapses. *The Journal of Physiology, 310,* 215–239.
Reese, E., Haden, C. A., & Fivush, R. (1993). Mother-child conversations about the past: Relationships of style and memory over time. *Cognitive Development, 8,* 403–430.
Remond-Besuchet, C., Noel, M. P., Seron, X., Thioux, M., Brun, M., & Aspe, X. (1999). Selective preservation of exceptional arithmetical knowledge in a demented patient. *Mathematical Cognition, 5*(1), 41–63.
Repacholi, B. M., & Gopnik, A. (1997). Early reasoning about desires: Evidence from 14- and 18-month-olds. *Developmental Psychology, 33,* 12–21.
Repacholi, B. M., Meltzoff, A. W., Rowe, H., & Toub, T. S. (2014). Infant, control thyself: Infants' integration of multiple social cues to regulate their imitative behavior. *Cognitive Development, 32,* 46–57.
Reynolds, G. D., & Richards, J. E. (2008). Attention and early brain development. *Encyclopedia on Early Childhood Development,* www.child-encyclopedia.com/brain/according-experts/attention-and-early-brain-development.
Reynolds, G. D., & Romano, A. C. (2016). Development of attention systems and working memory in infancy. *Frontiers in Systems Neuroscience, 10,* 15.
Richlan, F., Kronbichler, M., & Wimmer, H. (2011). Meta-analyzing brain dysfunctions in dyslexic children and adults. *NeuroImage, 56*(3), 1735–1742.
Richland, L. E., Morrison, R. G., & Holyoak, K. J. (2006). Children's development of analogical reasoning: Insights from scene analogy problems. *Journal of Experimental Child Psychology, 94*(3), 249–273.
Riecke, L., van Opstal, J., Goebel, R., & Formisano, E. (2007). Hearing illusory sounds in noise: Sensory-perceptual transformations in primary auditory cortex. *Journal of Neuroscience, 27*(46), 12684–12689.
Rieser, J. J., Doxey, P. A., McCarrell, N. J., & Brooks, P. H. (1982). Wayfinding and toddlers' use of information from an aerial view of a maze. *Developmental Psychology, 18,* 714–720.
Riggins, T., Rollins, L., & Graham, M. (2013). Electrophysiological investigation of source memory in early childhood. *Developmental Neuropsychology, 38*(3), 180–196.
Riggins, T., Geng, F., Blankenship, S. L., & Redcay, E. (2016). Hippocampal functional connectivity and episodic memory in early childhood. *Developmental Cognitive Neuroscience, 19,* 58–69.
Righi, G., & Tarr, M. J. (2004). Are chess experts any different from face, dird, or Greeble experts? *Journal of Vision, 4,* 504a.
Rivera-Gaxiola, M., Silva-Pereyra, J., & Kuhl, P. K. (2005). Brain potentials to native and non-native speech contrasts in 7- and 11-month-old American infants. *Developmental Science, 8*(2), 162–172.
Rizzolatti, G., & Craighero, L. (2004). The mirror neuron system. *Annual Review of Neuroscience, 27,* 169–192.
Rizzolatti, G., Fogassi, L., & Gallese, V. (2001). Neurophysiological mechanisms underlying the understanding and imitation of action. *Nature Reviews Neuroscience, 2,* 661–670.
Rizzolatti, G., & Sinigaglia, C. (2010). The functional role of the parieto-frontal mirror circuit: Interpretations and misinterpretations. *Nature Reviews Neuroscience, 11,* 264–274.
Rochat, P. (2009). *Others in Mind: Social Origins of Self-Consciousness.* Cambridge: Cambridge University Press.
Rochat, P., Morgan, R., & Carpenter, M. (1997). Young infants' sensitivity to movement information specifying social causality. *Cognitive Development, 12,* 537–561.
Rodriguez, C. (2007). Object use, communication and signs: The triadic basis of early cognitive development. In J. Valsiner & A. Rosa (Eds.), *The Cambridge Handbook of Socio-cultural Psychology,* pp. 257–276. New York: Cambridge University Press.
Roodenrys, S., Hulme, C., & Brown, G. (1993). The development of short-term memory span: Separable effects of speech rate and long-term memory. *Journal of Experimental Child Psychology, 56,* 431–442.
Rosch, E. (1978). Principles of categorisation. In E. Rosch & B. B. Lloyd (Eds.), *Cognition and Categorisation.* Hillsdale, NJ: Erlbaum.
Rosch, E., & Mervis, C. B. (1975). Family resemblances: Studies in the internal structure of categories. *Cognitive Psychology, 7,* 573–605.
Rosch, E., Mervis, C. B., Gray, W. D., Johnson, M. D., & Boyes-Braem, P. (1976). Basic objects in natural categories. *Cognitive Psychology, 8,* 382–439.
Rose, S., & Blank, N. (1974). The potency of context in children's cognition: An illustration through conservation. *Child Development, 45,* 499–502.
Rose, S. A., Feldman, J. F., & Jankowski, J. J. (2001). Visual short-term memory in the first year of life: Capacity and recency effects. *Developmental Psychology, 39,* 539–549.
Rosengren, K. S., Gelman, S. A., Kalish, C. W., & McCormick, M. (1991). As time goes by: Children's

early understanding of growth in animals. *Child Development, 62*, 1302–1320.
Rovee-Collier, C. K. (1993). The capacity for long-term memory in infancy. *Current Directions in Psychological Science, 2*, 130–135.
Rovee-Collier, C. K., & Hayne, H. (1987). Reactivation of infant memory: Implications for cognitive development. *Advances in Child Development and Behaviour, 20*, 185–238.
Rovee-Collier, C. K., Schechter, A., Shyi, G.C.W., & Shields, P. (1992). Perceptual identification of contextual attributes and infant memory retrieval. *Developmental Psychology, 28*, 307–318.
Rovee-Collier, C. K., Sullivan, M. W., Enright, M., Lucas, D., & Fagen, J. W. (1980). Reactivation of infant memory. *Science, 208*, 1159–1161.
Rowe, S., & Wertsch, J. (2002). Vygotsky's model of cognitive development. In U. Goswami (Ed.), *Blackwell Handbook of Childhood Cognitive Development*. Oxford: Blackwell.
Rowland, C. F., & Pine, J. M. (2000). Subject-auxiliary inversion errors and wh-question acquisition: What children do know? *Journal of Child Language, 27*, 157–181.
Rudy, L., & Goodman, G. S. (1991). Effects of participation on children's reports: Implications for children's testimony. *Developmental Psychology, 27*, 527–538.
Rueda, M. R., Pozuelos, J. P., & Cómbita, L. M. (2015). Cognitive neuroscience of attention: From brain mechanisms to individual differences in efficiency. *AIMS Neuroscience, 2*(4), 183–202.
Ruffman, T., Perner, J., Olson, D., & Doherty, M. (1993). Reflecting on scientific thinking: Children's understanding of the hypothesis-evidence relation. *Child Development, 64*, 1617–1636.
Ruffman, T., Rustin, C., Garnham, W., & Parkin, A. J. (2001). Source monitoring and false memories in children: Relation to certainty and executive functioning. *Journal of Experimental Child Psychology, 80*, 95–111.
Rugani, R., Vallortigara, G., Priftis, K., & Regolin, L. (2015). Number-space mapping in the newborn chick resembles humans' mental number line. *Science, 347*(6221), 534–536.
Rumelhart, D. E., & Abrahamson, A. A. (1973). A model for analogical reasoning. *Cognitive Psychology, 5*, 1–28.
Rumelhart, D. E., & McClelland, J. (1986). On learning the past tenses of English verbs. In D. E. Rumelhart, J. L. McClelland, & PDP Research Group (Eds.), *Parallel Distributed Processing*, Vol. 2, pp. 216–271. Cambridge, MA: MIT Press.
Russell, J. (1996). *Agency: Its Role in Mental Development*. Hove: Psychology Press.
Russell, J. (2005). Justifying all the fuss about false belief. *Trends in Cognitive Sciences, 9*, 307–308.
Russell, J., Mauthner, N., Sharpe, S., & Tidswell, T. (1991). The 'windows task' as a measure of strategic deception in preschoolers and autistic subjects. *British Journal of Developmental Psychology, 9*, 331–349.
Russo, R., Nichelli, P., Gibertoni, M., & Cornia, C. (1995). Developmental trends in implicit and explicit memory: A picture completion study. *Journal of Experimental Child Psychology, 59*, 566–578.
Saby, J. N., Meltzoff, A. N., & Marshall, P. J. (2013). Infants' somatotopic neural responses to seeing human actions: I've got you under my skin. *PLOS One, 8*(10), e77905.
Saffran, J. R. (2001). Words in a sea of sounds: The output of infant statistical learning. *Cognition, 81*, 149–169.
Saffran, J. R., Aslin, R. A., & Newport, E. L. (1996). Statistical learning by 8-month-old infants. *Science, 274*, 1926–1928.
Sarnecka, B. W., & Gelman, S. A. (2004). Six does not just mean a lot: Preschoolers see number words as specific. *Cognition, 92*, 329–352.
Sastre III, M., Wendelken, C., Lee, J. K., Bunge, S. A., & Ghetti, S. (2016). Age- and performance-related differences in hippocampal contributions to episodic retrieval. *Developmental Cognitive Neuroscience, 19*, 42–50.
Sato, J., Mossad, S. I., Wong, S. M., Hunt, B. A. E., Dunkley, B. T., Smith, M. L., Urbain, C., & Taylor, M. J. (2018). Alpha keeps it together: Alpha oscillatory synchrony underlies working memory maintenance in young children. *Developmental Cognitive Neuroscience, 34*, 114–123.
Sauseng, P., Bergmann, J. , & Wimmer, H. (2004). When does the brain register deviances from standard word spellings? An ERP study. *Cognitive Brain Research*, 20, 529–532.
Sauseng, P., Griesmayr, B., Freunberger, R., & Klimesch, W. (2010). Control mechanisms in working memory: A possible function of EEG theta oscillations. *Neuroscience & Biobehavioral Reviews, 34*(7), 1015–1022.
Saxe, G. B. (1977). A developmental analysis of notational counting. *Child Development, 48*, 1512–1520.
Saxe, R. (2006). Uniquely human social cognition. *Current Opinion in Neurobiology, 16*(2), 235–239.
Saxton, M., & Towse, J. N. (1998). Linguistic relativity: The case of place-value in multi-digit numbers. *Journal of Experimental Child Psychology, 69*(1), 66–79.
Scaife, M., & Bruner, J. (1975). The capacity for joint visual attention in the infant. *Nature, 253*, 265–266.
Schacter, D. L., & Moscovitch, M. (1984). Infants, amnesics and dissociable memory system. In M. Moscovitch (Ed.), *Infant Memory: Its Relation to Normal and Pathological Memory in Humans and Other Animals*, pp. 173–216. New York: Plenum Press.
Schaffer, D. R. (1985). *Developmental Psychology: Childhood and Adolescence*. Pacific Grove, CA: Brooks/Cole Publishing Company.
Scheier, C., Lewkowicz, D.J., & Shimojo, S. (2003). Sound induces perceptual reorganization of an ambiguous motion display in human infants. *Developmental Sciences, 6*, 233–244.
Schleussner, E., Schneider, U., Arnscheidt, C., Kahler, C., Haueisen, J., & Seewald, H. J. (2004). Short communication: Prenatal evidence of left–right asymmetries in auditory evoked responses using fetal magnetoencephalography. *Early Human Development, 78*, 133–136.

Schneider, M., Beeres, K., Coban, L., Merz, S., Schmidt, S. S., Stricker, J., & Smedt, B. D. (2016). Associations of non-symbolic and symbolic numerical magnitude processing with mathematical competence: A meta-analysis. *Developmental Science, 20*(3), e12372.

Schneider, W. (1985). Developmental trends in the metamemory-memory behaviour relationship: An integrative review. In D. L. Forrest-Pressley, G. E. MacKinnon, & T. G. Waller (Eds.), *Cognition, Metacognition and Human Performance*, Vol. 1, pp. 57–109. Orlando, FL: Academic Press.

Schneider, W. (1986). The role of conceptual knowledge and metamemory in the development of organisational processes in memory. *Journal of Experimental Child Psychology, 42*, 218–236.

Schneider, W., & Bjorklund, D. F. (1998). Memory. In W. Damon (Ed.), *Handbook of Child Psychology*, 5th edition, pp. 467–521. New York: Wiley.

Schneider, W., Boes, K., & Rieder, H. (1993a). Performance prediction in adolescent top tennis players. In J. Beckmann, H. Strang, & E. Hahn (Eds.), *Aufmerksamkeit und Energetisierung*. Goettingen: Hogrefe.

Schneider, W., Borkowski, J. G., Kurtz, B. E., & Kerwin, K. (1986). Metamemory and motivation: A comparison of strategy use and performance in German and American children. *Journal of Cross-Cultural Psychology, 17*, 315–336.

Schneider, W., Gruber, H., Gold, A., & Opwis, K. (1993b). Chess expertise and memory for chess positions in children and adults. *Journal of Experimental Child Psychology, 56*, 328–349.

Schneider, W., Korkel, J., & Weinert, F. E. (1989). Domain-specific knowledge and memory performance: A comparison of high- and low-aptitude children. *Journal of Educational Psychology, 81*, 306–312.

Schneider, W., Kron.V., Hunnerkopf. M., & Krajewski, K. (2004). The development of young children's memory strategies: First findings from the Wurzburg Longitudinal Memory Study. *Journal of Experimental Child Psychology, 88*, 193–209.

Schneider, W., Kuespert, P., Roth, E., Visé, M., and Marx, H. (1997). Short- and long-term effects of training phonological awareness in kindergarten: Evidence from two German studies. *Journal of Experimental Child Psychology, 66*, 311–340.

Schneider, W., & Lockl, K. (2002). The development of metacognitive knowledge in children and adolescents. In T. Perfect & B. Schwartz (Eds.), *Applied Metacognition*, pp. 224–247. Cambridge: Cambridge University Press.

Schneider, W., Niklas, F., & Schmiedeler, S. (2014). Intellectual development from early childhood to early adulthood: The impact of early IQ differences on stability and change over time. *Learning and Individual Differences, 32*, 156–162.

Schneider, W., & Pressley, M. (1989). *Memory Development between 2 and 20*. New York: Springer.

Schneider, W., & Pressley, M. (2013). *Memory Development between 2 and 20*, 3rd edition. Hillsdale NJ: Erlbaum & Associates.

Schneider, W., Roth, E., & Ennemoser, M. (2000a). Training phonological skills and letter knowledge in children at-risk for dyslexia: A comparison of three kindergarten intervention programs. *Journal of Educational Psychology, 92*, 284–295.

Schneider, W., Schlagmuller, M., & Visé, M. (1998). The impact of metamemory and domain-specific knowledge on memory performance. *European Journal of Psychology of Education, 13*, 91–103.

Schneider, W., & Sodian, B. (1988). Metamemory-memory behaviour relationships in young children: Evidence from a memory-for-location task. *Journal of Experimental Child Psychology, 45*, 209–233.

Schneider, W., Visé, M., Lockl, K., & Nelson, T. O. (2000b). Developmental trends in children's memory monitoring: Evidence from a judgment-of-learning task. *Cognitive Development, 15*, 115–134.

Scholl, B. J., & Tremoulet, P. D. (2000). Perceptual causality and animacy. *Trends in Cognitive Sciences, 4*, 299–309.

Schulz, L., & Gopnik, A. (2004). Causal learning across domains. *Developmental Psychology, 40*, 162–176.

Schulz, L. E., Bonawitz Baraff, E., & Griffiths, T. L. (2007). Can being scared cause tummy aches? Naive theories, ambiguous evidence, and preschoolers' causal inferences. *Developmental Psychology, 43*(5), 1124–1139.

Schwartz, F., Epinat-Duclos, J., Léone, J., & Prado, J. (2017). The neural development of conditional reasoning in children: Different mechanisms for assessing the logical validity and likelihood of conclusions. *NeuroImage, 163*, 264–275.

Scott, R. M., & Baillargeon, R. (2013). Do infants really expect agents to act efficiently? A critical test of the rationality principle. *Psychological Science, 24*(4), 466–474.

Scott, R. M., & Baillargeon, R. (2017). Early false-belief understanding. *Trends in Cognitive Sciences, 21*(4), 237–249.

Scott, R. M., Richman, J. C., & Baillargeon, R. (2015). Infants understand deceptive intentions to implant false beliefs about identity: New evidence for early mentalistic reasoning. *Cognitive Psychology, 82*, 32–56.

Scott, S. K., & Johnsrude, I. S. (2003). The neuroanatomical and functional organization of speech perception. *Trends in Neurosciences, 26*, 100–107.

Seehagen, S., Konrad, C., Herbert, J. S., & Schneider, S. (2015). Timely sleep facilitates declarative memory consolidation in infants. *PNAS, 112*(5), 1625–1629.

Sekuler, R., Sekuler, A. B., & Lau, R. (1997). Sound alters visual motion perception. *Nature, 385*, 308.

Setoh, P., Wu, D., Baillargeon, R., & Gelman, R. (2013). Young infants have biological expectations about animals. *PNAS, 110*(40), 15937–15942.

Seymour, P. H. K., Aro, M., & Erskine, J. M. (2003). Foundation literacy acquisition in European orthographies. *British Journal of Psychology, 94*, 143–174.

Shaywitz, B. A., Shaywitz, S. E., Blachman, B. A., Pugh, K. R., Fulbright, R. K., Skudlarski, P., Mencl, W. E., Constable, R. T., Holahan, J. M., Marchione, K. E., Fletcher, J. M., Lyon, G. R., & Gore, J. C. (2004). Development of left occipitotemporal systems for skilled reading in children after a phonologically-based intervention. *Biological Psychiatry, 55*(9), 926–933.

Shaywitz, B. A., Skudlarski, P., Holahan, J. M., Marchione, K. E., Constable, R. T., Fulbright, R. K., Zelterman, D., Lacadie, C., & Shaywitz, S. E. (2007). Age-related changes in reading systems of dyslexic children. *Annals of Neurology, 61*(4), 363–370.

Shimizu, C., Norona, A., Paparella, T., Freeman, S. F. N., & Johnson, S. P. (2015). Electrophysiological evidence of heterogeneity in visual statistical learning in young children with ASD. *Developmental Science, 18*(1), 90–105.

Shimizu, Y. A., & Johnson, S. C. (2004). Infants' attribution of a goal to a morphologically unfamiliar agent. *Developmental Science, 7*, 425–430.

Shipstead, Z., Redick, T. S., & Engle, R. W. (2012). Is working memory training effective? *Psychological Bulletin, 138*(4), 628–654.

Shukla, M., White, K. S., & Aslin, R. N. (2011). Prosody guides the rapid mapping of auditory word forms onto visual objects in 6-month-old infants. *PNAS, 108*(15), 6038–6043.

Shultz, T. R. (1982). Rules of causal attribution. *Monographs of the Society for Research in Child Development, 47*(1), 194.

Shultz, T. R., Fisher, G. W., Pratt, C. C., & Rulf, S. (1986). Selection of causal rules. *Child Development, 57*, 143–152.

Shultz, T. R., & Kestenbaum, N. R. (1985). Causal reasoning in children. *Annals of Child Development, 2*, 195–249.

Shultz, T. R., & Mendelson, R. (1975). The use of covariation as a principle of causal analysis. *Child Development, 46*, 394–399.

Shultz, T. R., Pardo, S., & Altmann, E. (1982). Young children's use of transitive inference in causal chains. *British Journal of Psychology, 73*, 235–241.

Shultz, T. R., & Ravinsky, F. B. (1977). Similarity as a principle of causal interference. *Child Development, 48*, 1552–1558.

Sedlak, A. J., & Kurtz, S. T. (1981). A review of children's use of causal interference principles. *Child Development, 52*, 759–784.

Siegal, M. (1991). *Knowing Children: Experiments in Conversation and Cognition.* Hillsdale, NJ: Lawrence Erlbaum.

Siegal, M., & Beattie, K. (1991). Where to look first for children's knowledge of false beliefs. *Cognition, 38*, 1–12.

Siegler, R. S. (1978). *Children's Thinking: What Develops?* Hillsdale, NJ: Lawrence Erlbaum.

Siegler, R. S. (1995). How does change occur? A microgenetic study of number conservation. *Cognitive Psychology, 28*, 225–273.

Siegler, R. S. (2016). Magnitude knowledge: The common core of numerical development. *Developmental Science, 19*(3), 341–361.

Siegler, R. S., & Liebert, R. M. (1974). Effects of contiguity, regularity and age on children's causal inferences. *Developmental Psychology, 10*, 574–579.

Simcock, G., & Hayne, H. (2002). Breaking the barrier: Children do not translate their preverbal memories into language. *Psychological Science, 13*, 225–231.

Simion, F., Regolin, L., & Bulf, H. (2008). A predisposition for biological motion in the newborn baby. *PNAS, 105*(2), 809–813.

Simon, H. A. (1975). The functional equivalence of problem solving skills. *Cognitive Psychology, 7*, 268–288.

Simon, T. J., Hespos, S. J., & Rochat, P. (1995). Do infants understand simple arithmetic? A replication of Wynn (1992). *Cognitive Development, 10*, 253–269.

Simons, D. J., & Keil, F. C. (1995). An abstract to concrete shift in the development of biological thought: The insides story. *Cognition, 56*, 129–163.

Simons, J. S., Garrison, J. R., & Johnson, M. K. (2017). Brain mechanisms of reality monitoring. *Trends in Cognitive Science, 21*(6), 462–473.

Simpson, E. A., Miller, G. M., Ferrari, P. F., Suomi, S. J., & Paukner, A. (2016). Neonatal imitation and early social experience predict gaze following abilities in infant monkeys. *Scientific Reports, 6*, 20233.

Singer-Freeman, K. E. (2005). Analogical reasoning in 2-year-olds: The development of access and relational inference. *Cognitive Development, 20*(2), 214–234.

Sirois, S., & Mareschal, D. (2002). Models of habituation in infancy. *Trends in Cognitive Sciences, 6*, 293–298.

Slater, A. M. (1989). Visual memory and perception in early infancy. In A. M. Slater & G. Bremner (Eds.), *Infant Development*, pp. 43–71. Hove, UK: Lawrence Erlbaum.

Slater, A. M., Morison, V., & Rose, D. (1983). Perception of shape by the new-born baby. *British Journal of Developmental Psychology, 1*, 135–142.

Slaughter, V. (1998). Children's understanding of pictorial mental representations. *Child Development, 69*, 321–332.

Sluzenski, J., Newcombe, N. S., & Ottinger, W. (2004). Changes in reality monitoring and episodic memory in early childhood. *Developmental Science, 7*, 225–245.

Smiley, S., & Brown, A. L. (1979). Conceptual preferences for thematic or taxonomic relations: A nonmonotonic age trend from preschool to old age. *Journal of Experimental Child Psychology, 28*, 249–257.

Smith, L. (1992). *Jean Piaget: Critical Assessments*, 4 volumes. London: Routledge.

Smith, L. (2002). Piaget's model. In U. Goswami (Ed.), *Blackwell Handbook of Childhood Cognitive Development*, pp. 515–537. Oxford: Blackwell.

Snowling, M. J. (2000). *Dyslexia*. Oxford: Blackwell.

Sobel, D. M., & Kirkham, N. Z. (2006). Blickets and babies: The development of causal reasoning in toddlers and infants. *Developmental Psychology, 42*(6), 1103–1115.

Sodian, B., Zaitchek, D., & Carey, S. (1991). Young children's differentiation of hypothetical beliefs from evidence. *Child Development, 62*, 753–766.

Somerville, S. C., & Capuani-Shumaker, A. (1984). Logical searches of young children in hiding and finding tasks. *British Journal of Developmental Psychology, 2*, 315–328.

Somerville, S. C, Wellman, H. M., & Cultice, J. C. (1983). Young children's deliberate reminding. *Journal of Genetic Psychology, 143*, 87–96

Sommerville, J. A., Woodward, A. L., & Needham, A. (2005). Action experience alters 3-month-old infants' perception of others' actions. *Cognition, 96*, B1–B11.
Sophian, C., & Somerville, S. C. (1988). Early developments in logical reasoning: Considering alternative possibilities. *Cognitive Development, 3*, 183–222.
Sorce, J. F., Emde, R. N., Campos, J., & Klinnert, M. D. (1985). Maternal emotional signaling: Its effect on the visual cliff behavior of 1-year-olds. *Developmental Psychology, 21*, 195–200.
Soska, K. C., Adolph, K. E., Johnson, S. P. (2010). Systems in development: Motor skill acquisition facilitates three-dimensional object completion. *Developmental Psychology, 46*(1), 129–138.
Southgate, V., & Begus, K. (2013). Motor activation during the prediction of nonexecutable actions in infants. *Psychological Science, 24*(6), 828–835.
Southgate, V., Begus, K., Lloyd-Fox, S., di Gangi, V., & Hamilton, A. (2014). Goal representation in the infant brain. *Neuroimage, 85*, 294–301.
Southgate, V., & Vernetti, A. (2014). Belief-based action prediction in preverbal infants. *Cognition, 130*(1), 1–10.
Spelke, E. S. (1976). Infants' intermodal perception of events. *Cognitive Psychology, 8*, 553–560.
Spelke, E. S. (1991). Physical knowledge in infancy: Reflections on Piaget's theory. In S. Carey & R. Gelman (Eds.), *The Epigenesis of Mind: Essays on Biology and Cognition*, pp. 133–169. Hillsdale, NJ: Lawrence Erlbaum.
Spelke, E. S. (1994). Initial knowledge: Six suggestions. *Cognition, 50*, 431–445.
Spelke, E. S., Phillips, A. T., & Woodward, A. L. (1995). Infants' knowledge of object motion and human action. In D. Sperber, A. J. Premack, & D. Premack (Eds.), *Causal Cognition: A Multidisciplinary Debate*, pp. 44–78. Oxford: Clarendon Press.
Srinavasan, M., & Carey, S. (2010). The long and the short of it: On the nature and origin of functional overlap between representations of space and time. *Cognition, 116*(2), 217–241.
Stahl, A. E., & Feigenson, L. (2015). Observing the unexpected enhances infants' learning and exploration. *Science, 348*(6230), 91–94.
Stanovich, K. E., & West, R. (2000). Individual differences in reasoning: Implications for the rationality debate. *Behavioural and Brain Sciences, 23*(5), 645–665.
Starkey, P., & Cooper, R. G. (1980). Perception of number by human infants. *Science, 210*, 1033–1035.
Starkey, P., Spelke, E. S., & Gelman, R. (1983). Detection of intermodal numerical correspondences by human infants. *Science, 222*, 179–181.
Stavans, M., & Baillargeon, R. L. (2018). Four-month-old infants individuate and track simple tools following functional demonstrations. *Developmental Science, 21*(1), e12500.
Stechler, G., & Latz, E. (1966). Some observations on attention and arousal in the human infant. *Journal of the American Academy of Child Psychology, 5*, 517–525.
Striano, T., Henning, A., & Stahl, D. (2005). Sensitivity to social contingencies between 1 and 3 months of age. *Developmental Science, 8*, 509–519.
Surian, L., & Geraci, A. (2012). Where will the triangle look for it? Attributing false beliefs to a geometric shape at 17 months. *British Journal of Developmental Psychology, 30*(1), 30–44.
Sutton, J., Smith, P. K., & Swettenham, J. (1999). Social cognition and bullying: Social inadequacy or skilled manipulation? *British Journal of Developmental Psychology, 17*, 435–450.
Swingley, D. (2005). 11-month-olds' knowledge of how familiar words sound. *Developmental Science, 8*, 432–443.
Swingley, D., & Humphrey, C. (2018). Quantitative linguistic predictors of infants' learning of specific English words. *Child Development, 89*, 1247–1267.
Symons, D. K. (2004). Mental state discourse, theory of mind, and the internalization of self-other understanding. *Developmental Review, 24*, 159–188.
Szücs, D. (2016). Subtypes and comorbidity in mathematical learning disabilities: Multidimensional study of verbal and visual memory processes is key to understanding. *Progress in Brain Research, 227*, 277–304.
Szücs, D., Devine, A., Soltesz, F., Nobes, A., & Gabriel, F. (2013). Developmental dyscalculia is related to visuo-spatial memory and inhibition impairment. *Cortex, 49*(10), 2674–2688.
Szücs, D., & Goswami, U. (2013). Developmental dyscalculia: Fresh perspectives. *Trends in Neuroscience in Education, 2*(2), 33–37.
Szücs, D., & Myers, T. (2017). A critical analysis of design, facts, bias and inference in the approximate number system training literature: A systematic review. *Trends in Neuroscience and Education, 6*, 187–203.
Tai, Y. F., Scherfler, C., Brooks, D. J., Sawamoto, N., & Castiello, U. (2004). The human premotor cortex is 'mirror' only for biological actions. *Current Biology, 14*, 117–120.
Tamnes, C. K., Walhovd, K. B., Torstveit, M., Sells, V. T., & Fjell, A. M. (2013). Performance monitoring in children and adolescents: A review of developmental changes in the error-related negativity and brain maturation. *Developmental Cognitive Neuroscience, 6*, 1–13.
Taylor, M., Mottweiler, C. M., Aguiar, N. R., Naylor, E. R., & Levernier, J. G. (1985). Paracosms: The imaginary worlds of middle childhood. *Child Development.*
Telkemeyer, S., Rossi, S., Nierhaus, T., Steinbrink, J., Obrig, H., & Wartenburger, I. (2011). Acoustic processing of temporally modulated sounds in infants: Evidence from a combined near-infrared spectroscopy and EEG study. *Frontiers in Psychology, 1*(62).
Temple, E., & Posner, M. I. (1998). Brain mechanisms of quantity are similar in 5-year-olds and adults. *Proceedings of the National Academy of Sciences of the USA, 95*, 7836–7841.
Thibodeau, R. B., Gilpin, A. T., Brown, M. M., & Meyer, B. A. (2016). The effects of fantastical pretend-play on the development of executive functions: An intervention

study. *Journal of Experimental Child Psychology, 145,* 120–138.

Thomson, J. M., Richardson, U., & Goswami, U. (2005). Phonological similarity neighborhoods and children's short-term memory: Typical development and dyslexia. *Memory and Cognition, 33,* 1210–1219.

Thomason, M. E., Grove, L. E., Lozon, T. A., Vila, A. M., Yongquan, Y., Nye, M. J., Manning, J. H., Pappas, A., Hernandez-Andraded, E., Yeo, L., Mody, S., Berman, S., Hassan, S. S., & Romero, R. (2015). Age-related increases in long-range connectivity in fetal functional neural connectivity networks in utero. *Developmental Cognitive Neuroscience, 11,* 96–104.

Toda, S., & Fogel, A. (1993). Infant response to still-face situation at 3- and 6-months. *Developmental Psychology, 29,* 532–538.

Tomasello, M. (1988). The role of joint-attentional processes in early language acquisition. *Language Sciences, 10,* 69–88.

Tomasello, M. (1990). Cultural transmission in the tool use and communicatory signalling of chimpanzees? In S. Parker & K. Gibson (Eds.), *Language and Intelligence in Monkeys and Apes: Comparative Developmental Perspectives,* pp. 274–311. Cambridge: Cambridge University Press.

Tomasello, M. (1995). Joint attention as social cognition. In C. Moore & P. J. Dunham (Eds.), *Joint Attention: Its Origins and Role in Development,* pp. 103–130. Hillsdale, NJ: Lawrence Erlbaum.

Tomasello, M. (1998). *Cognitive and Functional Approaches to Language Structure.* Marwah, NJ: Lawrence Erlbaum Associates.

Tomasello, M. (2000). Do young children have adult syntactic competence? *Cognition, 74,* 209–253.

Tomasello, M. (2008). *First Words: A Case Study of Early Grammatical Development.* Cambridge: Cambridge University Press.

Tomasello, M., Akhtar, N., Dodson, K., & Rekau, L. (1997). Differential productivity in young children's use of nouns and verbs. *Journal of Child Language, 24,* 373–387.

Tomasello, M., & Call, J. (2018). Thirty years of great ape gestures. *Animal Cognition,* 1–9.

Tomasello, M., & Carpenter, M. (2007). Shared intentionality. *Developmental Science, 10*(1), 121–125.

Tomasello, M., Striano, T., & Rochat, P. (1999). Do young children use objects as symbols? *British Journal of Developmental Psychology, 17,* 563–584.

Topal, J., Gergely, G., Miklósi, A., Erdohegyi, A., & Csibra, G. (2008). Infants' perseverative search errors are induced by pragmatic misinterpretation. *Science, 321*(5897), 1831–1834.

Torpey, D. C., Hajcak, G., Kim, J., Kujawa, A., & Klein, D. N. (2012). Electrocortical and behavioral measures of response monitoring in young children during a go/no-go task. *Developmental Psychobiology, 54*(2), 139–150.

Treiman, R., & Baron, J. (1981). Segmental analysis ability: Development and relation to reading ability. In G. E. MacKinnon & T. G. Waller (Eds.), *Reading Research: Advances in Theory and Practice,* Vol. 3. New York: Academic Press.

Treiman, R., & Zukowski, A. (1991). Levels of phonological awareness. In S. Brady & D. Shankweiler (Eds.), *Phonological Processes in Literacy: A Tribute to Isabelle Y. Liberman,* pp. 67–83. Hillsdale, NJ: Erlbaum.

Treiman, R., & Zukowski, A. (1996). Children's sensitivity to syllables, onsets, rimes and phonemes. *Journal of Experimental Child Psychology, 61,* 193–215.

Tremoulet, P., & Feldman, J. (2000). Perception of animacy from the motion of a single object. *Perception, 29,* 943–951.

Trionfi, G., & Reese, E. (2009). A good story: Children with imaginary companions create richer narratives. *Child Development, 80*(4), 1301–1313.

Tulving, E. (2002). Episodic memory: From mind to brain. *Annual Review of Psychology, 53,* 1–25.

Tunmer, W. E., & Nesdale, A. R. (1985). Phonemic segmentation skills and beginning reading. *Journal of Educational Psychology,* 77, 417–527.

Tunteler, R., & Resing, C. (2002). Spontaneous analogical transfer in four-year-olds: A microgenetic study. *Journal of Experimental Child Psychology, 83*(3), 149–166.

Turati, C., Natale, E., Bolognini, N., Senna, I., Picozzi, M., Longhi, E., & Macchi Cassia, V. (2013). The early development of human mirror mechanisms: Evidence from electromyographic recordings at 3 and 6 months. *Developmental Science, 16*(6), 793–800.

Turkeltaub, P. E., Gareau, L., Flowers, D. L., Zeffiro, T. A., & Eden, G. F. (2003). Development of neural mechanisms for reading. *Nature Neuroscience, 6,* 767–773.

Turkewitz, G. (1995). The what and why of infancy and cognitive development. *Cognitive Development, 10,* 459–465.

Tzourio-Mazoyer, N., De Schonen, S., Crivello, F., Reutter, B., Aujard, Y., & Mazoyer, B. (2002). Neural correlates of woman face processing by 2-month-old infants. *NeuroImage, 15,* 454–461.

Uller, C. (2004). Disposition to recognize goals in infant chimpanzees. *Animal Cognition,* 7(3), 154–161.

Ungerer, J. A., Zelazo, P. R., Kearsley, R. B., & O'Leary, K. (1981). Developmental changes in the representation of objects in symbolic play from 18 to 31 months of age. *Child Development, 52,* 186–195.

Ungerleider, L. G., & Mishkin, M. (1982). Two cortical visual systems. In D. J. Ingle, M. A. Goodale, & R. J. W. Mansfield (Eds.), *Analysis of Visual Behaviour.* Cambridge, MA: MIT Press.

Vaish, A., & Striano, T. (2004). Is visual reference necessary? Vocal versus facial cues in social referencing. *Developmental Science,* 7, 261–269.

van den Heuvel, M. I., Turk, E., Manning, J. H., Hect, J., Hernandez-Andrade, E., Hassan, S. S., Romero, R., van den Heuvel, M. P., & Thomason, M. E. (2018). Hubs in the human fetal brain network. *Developmental Cognitive Neuroscience, 30,* 108–115.

vanMarle, K., Chu, F. W., Mou, Y., Seok, J. H., Rouder, J., & Geary, D. C. (2016). Attaching meaning to the number words: Contributions of the object tracking and approximate number systems. *Developmental Science, 21*(1), e12495.

Van Oeffelen, M. O., & Vos, P. G. (1982). A probabilistic model for the discrimination of visual number. *Perception & Psychophysics*, *32*(2), 163–170.

Vargha-Khadem, F., Gadian, D. C., Watkins, K. E., Connelly, A., Van Paesschen, W., & Mishkin, M., (1997). Differential effects of early hippocampal pathology on episodic and semantic memory. *Science*, *277*, 376–380.

Vaughan, W. J., & Greene, S. L. (1984). Pigeon visual memory capacity. *Journal of Experimental Psychology: Animal Behavior Processes*, *10*, 256–271.

Viennot, L. (1979). Spontaneous reasoning in elementary dynamics. *European Journal of Science Education*, *1*, 205–221.

Vihman, M. M., Nakai, S., DePaolis, R. A., & Halle, P. (2004). The role of accentual pattern in early lexical representation. *Journal of Memory and Language*, *50*, 336–353.

Vintner, A. (1986). The role of movement in eliciting early imitations. *Child Development*, *57*, 66–71.

Visé, M., & Schneider, W. (2000). Determinanten der Leistungsvorhersage bei Kindergarten und Grundschulkindern: Zur Bedeutung metakognitiver und motivationalerEinflußfaktoren. *Zeitschrift für Experimentelle Psychologie*, *32*(2), 51–58.

Vogan, V. M., Morgan, B. R., Powell, T. L., Smith, M. L., & Taylor, M. J. (2016). The neurodevelopmental differences of increasing verbal working memory demand in children and adults. *Developmental Cognitive Neuroscience*, *17*, 19–27.

Volterra, V., & Erting, C. (1990). *From Gesture to Language in Hearing and Deaf Children*. Berlin: Springer-Verlag.

Vygotsky, L. S. (1934). *Myshlenie i rech': Psikhologicheskie issledovaniya [Thinking and Speech: Psychological Investigations]*. Moscow and Leningrad: Gosudarstvennoe Sotsial'no-Ekonomicheskoe Izdatel'stvo.

Vygotsky, L. (1978). *Mind in Society*. Cambridge, MA: Harvard University Press.

Walker, C. M., & Gopnik, A. (2014). Toddlers infer higher-order relational principles in causal learning. *Psychological Science*, *25*(1), 161–169.

Walker, C. M., Lombrozo, T., Legare, C. H., & Gopnik, A. (2014). Explaining prompts children to privilege inductively rich properties. *Cognition*, *133*(2), 343–357.

Walker, C. M., Lombrozo, T., Williams, J. J., Rafferty, A. N., & Gopnik, A. (2017). Explaining constrains causal learning in childhood. *Child Development*, *88*(1), 229–246.

Wang, C. H., Tseng, Y. T., Liu, D., & Tsai, C. L. (2017a). Neural oscillation reveals deficits in visuospatial working memory in children with developmental coordination disorder. *Child Development*, *88*(5), 1716–1726.

Wang, S., & Baillargeon, R. (2005). Inducing infants to detect a physical violation in a single trial. *Psychological Science*, *16*, 542–549.

Wang, S., & Baillargeon, R. (2008). Can infants be 'taught' to attend to a new physical variable in an event category? The case of height in covering events. *Cognitive Psychology*, *56*(4), 284–326.

Wang, S., Baillargeon, R., & Paterson, S. (2005). Detecting continuity violations in infancy: A new account and new evidence from covering and tube events. *Cognition*, *95*, 129–173.

Wang, Y., Mauer, M.V., Raney, T., Peysakhovich, B., Becker, B. L. C., Sliva, D. D., & Gaab, N. (2017b). Development of tract-specific white matter pathways during early reading development in at-risk children and typical controls. *Cerebral Cortex*, *27*(4), 2469–2485.

Wason, P. C. (1966). Reasoning. In B. Foss (Ed.), *New Horizons in Psychology*. Harmondsworth: Penguin Books.

Wason, P. C., & Johnson-Laird, P. N. (1972). *Psychology of Reasoning: Structure and Content*. Cambridge, MA: Harvard University Press.

Wass, S.V., Scerif, G., & Johnson, M. J. (2012). Training attentional control and working memory – is younger, better? *Developmental Review*, *32*(4), 360–387.

Watson, J. S. (1994). Detection of self: The perfect algorithm. In S. T. Parker, R. W Mitchell, & M. L. Boccia (Eds.), *Self-Awareness in Animals and Humans: Developmental Perspectives*, pp. 131–148. New York: Cambridge University Press.

Waxman, S. R. (1990). Linguistic biases and the establishment of conceptual hierarchies: Evidence from preschool children. *Cognitive Development*, *5*, 123–150.

Waxman, S. R., & Braun, I. (2005). Consistent (but not variable) names as invitations to form object categories: New evidence from 12-month-old infants. *Cognition*, *95*, B59–B68.

Waxman, S. R., & Gelman, S. A. (2009). Early word-learning entails reference, not merely associations. *Trends in Cognitive Sciences*, *13*, 258–263.

Waxman, S. R., Herrmann, P., Woodring, J. W., & Medin, D. L. (2014). Humans (really) are animals: Picture-book reading influences 5-year-old urban children's construal of the relation between humans and non-human animals. *Frontiers in Psychology*, *5*, 172.

Waxman, S. R., & Markow, D. B. (1995). Words as invitations to form categories: Evidence from 12- to 13-month-old infants. *Cognitive Psychology*, *29*, 257–302.

Waxman, S. R., & Medin, D. (2007). Experience and cultural models matter: Placing firm limits on childhood anthropocentrism. *Human Development*, *50*, 23–30.

Weber, C., Hahne, A., Friedrich, M., & Friederici, A. D. (2004). Discrimination of word stress in early infant perception: Electrophysiological evidence. *Cognitive Brain Research*, *18*, 149–161.

Wellman, H. M. (1978). Knowledge of the interaction of memory variables: A developmental study of metamemory. *Developmental Psychology*, *14*, 24–29.

Wellman, H. M. (1985). The origins of metacognition. In D. L. Forrest-Pressley, G. E. MacKinnon, & T. G. Waller (Eds.), *Metacognition, Cognition and Human Performance*, pp. 1–31. Orlando, FL: Academic Press.

Wellman, H. M. (2002). Understanding the psychological world: Developing a theory of mind. In U. Goswami (Ed.), *Blackwell Handbook of Childhood Cognitive Development*, pp. 167–187. Oxford: Blackwell.

Wellman, H. M., & Gelman, S. A. (1998). Knowledge acquisition in foundational domains. In W. Damon, D. Kuhn, & R. Siegler (Eds.), *Handbook of Child Psychology*, 5th edition, Vol. 2, pp. 523–573. New York: Wiley.

Wellman, H. M., Ritter, K., & Flavell, J. (1975). Deliberate memory development in the delayed reactions of very young children. *Developmental Psychology, 11*, 780–787.

Wellman, H. M., Somerville, S. C., & Haake, R. J. (1979). Development of search procedures in real-life spatial environments. *Developmental Psychology, 15*, 530–542.

Wellman, H. M., & Woolley, J. D. (1990). From simple desires to ordinary beliefs: The early development of everyday psychology. *Cognition, 35*, 245–275.

Wendelken, C., Baym, C. L., Gazzaley, A., & Bunge, S. A. (2011). Neural indices of improved attentional modulation over middle childhood. *Developmental Cognitive Neuroscience, 1*(2), 175–86.

Wendelken, C., Ferrer, E., Ghetti, S., Bailey, S., Cutting, L., & Bunge, S. A. (2017). Fronto-parietal structural connectivity in childhood predicts development of functional connectivity and reasoning ability: A large-scale longitudinal investigation. *Journal of Neuroscience, 3*, 726–716.

Wendelken, C., Munakata, Y., Baym, C., Souza, M., & Bunge, S. A. (2012). Flexible rule use: Common neural substrates in children and adults. *Developmental Cognitive Neuroscience, 2*(3), 329–339.

Werker, J. F., & Hensch, T. K. (2015). Critical periods in speech perception: New directions. *Annual Review of Psychology, 66*, 173–196.

Werker, J. F., & Tees, R. C. (1984). Cross-language speech perception: Evidence for perceptual reorganization during the first year of life. *Infant Behavior and Development, 7*, 49–63.

Wertsch, J. V. (1985). *Vygotsky and the Social Formation of Mind*. Cambridge, MA: Harvard University Press.

Westermann, G., & Mareschal, D. (2014). From perceptual to language-mediated categorization. *Philosophical Transactions of the Royal Society B: Biological Sciences, 369*, 20120391.

Westermann, G., Mareschal, D., Johnson, M., Sirois, S., Spratling, M., & Thomas, M. S. C. (2007). Neuroconstructivism. *Developmental Science, 10*(1), 75–83.

Whitaker, K. J., Vendetti, M. S., Wendelken, C., & Bunge, S. A. (2017). Neuroscientific insights into the development of analogical reasoning. *Developmental Science, 21*(2), e12531.

White, R. E., Prager, E. O., Schaefer, C., Kross, E., Duckworth, A. L., & Carlson, S. M. (2017). The 'Batman Effect': Improving perseverance in young children. *Child Development, 88*(5), 1563–1571.

Whiten, A., & Ham, R. (1992). On the nature and evolution of imitation in the animal kingdom: Reappraisal of a century of research. In P. B. Slater, J. S. Rosenblatt, C. Beer, & M. Milinski (Eds.), *Advances in the Study of Behaviour*, pp. 239–283. San Diego, CA: Academic Press.

Wilcox, T., & Biondi, M. (2015). Object processing in the infant: Lessons from neuroscience. *Trends in Cognitive Sciences, 19*(7), 406–413.

Wilkening, F. (1981). Integrating velocity, time and distance information: A developmental study. *Cognitive Psychology, 13*, 231–247.

Wilkening, F. (1982). Children's knowledge about time, distance and velocity interrelations. In W. J. Friedman (Ed.), *The Developmental Psychology of Time*, pp. 87–112. New York: Academic Press.

Wilkening, F., & Anderson, N. H. (1991). Representation and diagnosis of knowledge structures in developmental psychology. In N. H. Anderson (Ed.), *Contributions to Information Integration Theory: Vol. III. Developmental*, pp. 43–80. Hillsdale, NJ: Erlbaum.

Wilkening, F., & Cacchione, T. (2010). Children's intuitive physics. In U. Goswami (Ed)., *The Wiley-Blackwell Handbook of Childhood Cognitive Development*, 2nd edition. Oxford: Blackwell.

Wilkening, F., & Martin, C. (2004). How to speed up to be in time: Action-judgment dissociations in children and adults. *Swiss Journal of Psychology, 63*, 17–29.

Wimmer, H. (1993). Characteristics of developmental dyslexia in a regular writing system. *Applied Psycholinguistics, 14*, 1–33.

Wimmer, H. (1996). The nonword reading deficit in developmental dyslexia: Evidence from children learning to read German. *Journal of Experimental Child Psychology, 61*, 80–90.

Wimmer, H., & Hummer, P. (1990). How German-speaking first graders read and spell: Doubts on the importance of the logographic stage. *Applied Psycholinguistics, 11*, 349–368.

Wimmer, H., Landerl, K., Linortner, R., & Hummer, P. (1991). The relationship of phonemic awareness to reading acquisition: More consequences than precondition but still important. *Cognition, 40*, 219–249.

Wimmer, H., Landerl, K., & Schneider, W. (1994). The role of rhyme awareness in learning to read a regular orthography. *British Journal of Developmental Psychology, 12*, 469–484.

Wimmer, H., & Perner, J. (1983). Beliefs about beliefs: Representation and constraining function of wrong beliefs in young children's understanding of deception. *Cognition, 13*, 103–128.

Whyte, E. M., Behrmann, M., Minshew, N. J., Garcia, N.V., & Scherf, K. S. (2016). Animal, but not human, faces engage the distributed face network in adolescents with autism. *Developmental Science, 19*(2), 306–317.

Winston, P. H. (1980). Learning and reasoning by analogy. *Communications of the ACM, 23*(12).

Woodward, A. L. (1998). Infants selectively encode the goal object of an actor's reach. *Cognition, 69*, 1–34.

Woodward, A. L. (2003). Infants' developing understanding of the link between looker and object. *Developmental Science, 6*, 297–311.

Woodward, A. L., & Guajardo, J. J. (2002). Infants' understanding of the point gesture as an object-directed action. *Cognitive Development, 17*, 1061–1084.

Woolfe, T., Want, S., & Siegal, M. (2002). Signposts to development: Theory of mind in deaf children. *Child Development, 73*, 768–778.

Wright, K., Poulin-Dubois, D., & Kelley, E. (2015). The animate-inanimate distinction in preschool children. *The British Journal of Developmental Psychology, 33*(1), 73–91.

Wynn, K. (1990). Children's understanding of counting. *Cognition, 36*, 155–193.

Wynn, K. (1992). Addition and subtraction by human infants. *Nature, 358*, 749–750.

Xia, Z., Hancock, R., & Hoeft, F. (2017). Neurobiological bases of reading disorder Part I: Etiological investigations. *Language and Linguistics Compass, 11*, e12239.

Xie, W., Mallin, B. M., & Richards, J. E. (2017). Development of infant sustained attention and its relation to EEG oscillations: An EEG and cortical source analysis study. *Developmental Science, 21*(3), e12562.

Xu, F. (2002). The role of language in acquiring object kind concept in infancy. *Cognition, 85*, 223–250.

Xu, F. (2003). Numerosity discrimination in infants: Evidence for two systems of representations. *Cognition, 89*, B15–B25.

Xu, F., & Carey, S. (1996). Infants' metaphysics: The case of numerical identity. *Cognitive Psychology, 30*, 111–153.

Xu, F., & Spelke, E. S. (2000). Large number discrimination in 6-month-old infants. *Cognition, 74*, B1–B11.

Xu, F., Spelke, E. S., & Goddard, S. (2005). Number sense in human infants. *Developmental Science, 8*, 88–101.

Youngblade, L. M., & Dunn, J. (1995). Individual differences in young children's pretend play with mother and sibling: Links to relationships and understanding of other people's feelings and beliefs. *Child Development, 66*, 1472–1492.

Younger, B. A. (1985). The segregation of items into categories by 10-month-old infants. *Child Development, 56*, 1574–1583.

Younger, B. A. (1990). Infants' detection of correlations among feature categories. *Child Development, 61*, 614–620.

Younger, B. A., & Cohen, L. B. (1983). Infant perception of correlations among attributes. *Child Development, 54*, 858–867.

Younger, B. A., & Cohen, L. B. (1985). How infants form categories. In G. Bower (Ed.), *The Psychology of Learning and Motivation: Advances in Research and Theory*, pp. 211–247. New York: Academic Press.

Yu, C., & Smith, L. B. (2012). Embodied attention and word learning by toddlers. *Cognition, 125*(2), 244–262.

Yuan, S., & Fisher, C. (2009). "Really? She blicked the baby?": Two-year-olds learn combinatorial facts about verbs by listening. *Psychological Science, 20*(5), 619–626.

Yussen, S. R., & Levy, V. M. (1975). Developmental changes in predicting one's own span of short-term memory. *Journal of Experimental Child Psychology, 19*, 502–508.

Zaitchik, D. (1990). When representations conflict with reality: The preschooler's problem with false beliefs and 'false' photographs. *Cognition, 35*, 41–68.

Zelazo, P. D., & Frye, D. (1997). Cognitive complexity and control: A theory of the development of deliberate reasoning and intentional action. In M. Stamenov (Ed.), *Language Structure, Discourse, and the Access to Consciousness*, pp. 113–153. Amsterdam: John Benjamins.

Zelazo, P. D., Frye, D., & Rapus, T. (1996). An age-related dissociation between knowing rules and using them. *Cognitive Development, 11*, 37–63.

Zelazo, P. D., & Muller, U. (2010). Executive function in typical and atypical development. In U. Goswami (Ed.), *The Wiley-Blackwell Handbook of Childhood Cognitive Development*, 2nd edition. Oxford: Blackwell.

Zelazo, P. D., Muller, U., Frye, D., & Marcovitch, S. (2003). The development of executive function in early childhood. *Monograph of the Society for Research in Child Development, 68*(3), 274.

Zeskind, P. S., Sale, J., Maio, M. L., Huntington, L., & Weiseman, J. R. (1985). Adult perceptions of pain and hunger cries: A synchrony of arousal. *Child Development, 56*, 549–554.

Ziegler, J. C., Bertrand, D., Tóth, D., Csépe, V., Reis, A., Faísca, L., Saine, N., Lyytinen, H., Vaessen, A., & Blomert, L. (2010). Orthographic depth and its impact on universal predictors of reading: A cross-language investigation. *Psychological Science, 21*(4), 551–559.

Ziegler, J. C., & Goswami, U. (2005). Reading acquisition, developmental dyslexia, and skilled reading across languages: A psycholinguistic grain size theory. *Psychological Bulletin, 131*, 3–29.

Ziegler, J. C., & Goswami, U. (2006). Becoming literate in different languages: Similar problems, different solutions. *Developmental Science, 9*(5), 429–436.

Ziegler, J. C., Stone, G. O., & Jacobs, A. M. (2010). What's the pronunciation for –OUGH and the spelling for /u/? A database for computing feedforward and feedback inconsistency in English. *Behavior Research Methods, Instruments, & Computers, 29*, 600–618.

Zinober, B., & Martlew, M. (1985). The development of communicative gestures. In M. Barret (Ed.), *Children's Single Word Speech*, pp. 183–215. Chichester: Wiley.

Author index

Subject index

9781138923911